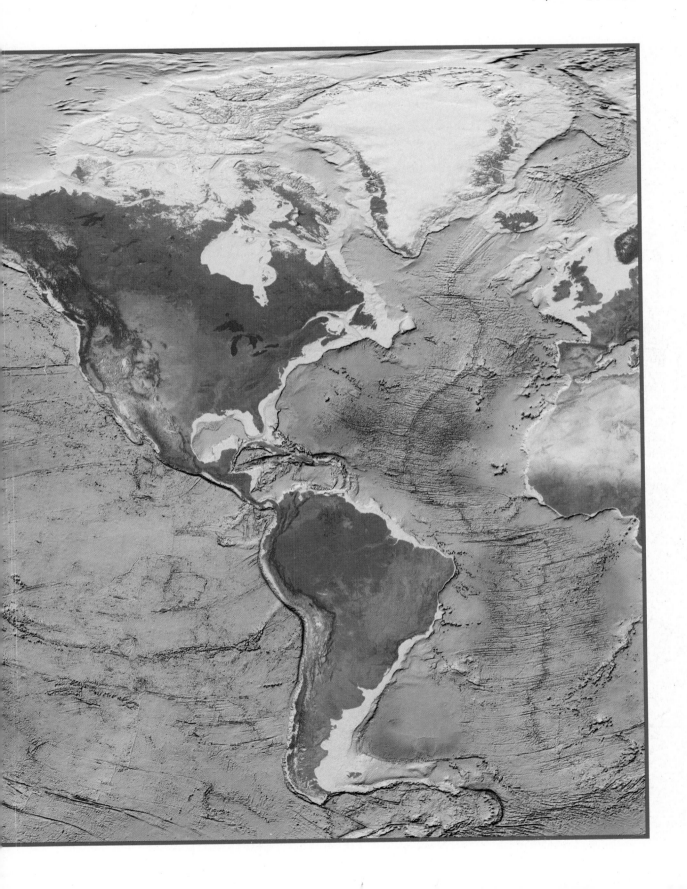

ABOUT THE COVER

This photography by Michael Collier shows tilted, folded, and eroded rock layers along the flank of the Sheep Mountain anticline, Big Horn Basin, Wyoming. The dramatic scenery expresses an intricate geologic history spanning hundreds of millions of years. The oldest rock layer exposed in Sheep Mountain represents accumulations of coral and shells in a tropical sea 340 million years ago when North America was thousands of kilometers farther south, straddling the equator. Overlying younger layers accumulated between 300 and 100 million years ago as loose sand and mud in shallow seas and rivers, and as salt formed by the evaporation of seawater. After the layers were buried and hardened into rock, strong forces compressed the region, folding the layers into an upward-bent fold, or anticline. Erosion of the folded and upturned rock layers removed some rock types more rapidly than others, forming the different-colored ridges and valleys that follow the upturned layers across the terrain. At a smaller scale, erosion of individual layers produced the triangular-shaped rocky protrusions that look like giant bones, teeth, or scales on the back of a dragon. The scene demonstrates the interplay of different aspects of geology—a sequence of rock layers formed in various environments, a geologic structure (the fold), erosion by water and downslope movement, and the immensely long timescale over which most geologic events operate to construct and sculpt landscapes.

exploring GEOLOGY

Second Edition

STEPHEN J. REYNOLDS
Arizona State University

JULIA K. JOHNSON
Arizona State University

MICHAEL M. KELLY
Northern Arizona University

PAUL J. MORIN
University of Minnesota
National Center for Earth-surface Dynamics

CHARLES M. CARTER

McGraw Hill **Higher Education**

Boston   Burr Ridge, IL   Dubuque, IA   New York   San Francisco   St. Louis
Bangkok   Bogotá   Caracas   Kuala Lumpur   Lisbon   London   Madrid   Mexico City
Milan   Montreal   New Delhi   Santiago   Seoul   Singapore   Sydney   Taipei   Toronto

The McGraw·Hill Companies

Higher Education

EXPLORING GEOLOGY, SECOND EDITION

1 2 3 4 5 6 7 8 9 0 DOW/DOW 0 9

ISBN 978–0–07–337668–4

MHID 0–07–337668–X

Publisher: *Ryan Blankenship*
Executive Editor: *Margaret J. Kemp*
Director of Development: *Kristine Tibbetts*
Senior Developmental Editor: *Joan M. Weber*
Senior Marketing Manager: *Lisa Nicks*
Senior Project Manager: *Gloria G. Schiesl*
Senior Production Supervisor: *Sherry L. Kane*
Lead Media Project Manager: *Judi David*
Senior Designer: *David W. Hash*
Cover Designer: *John Joran*
(USE) Cover Image: *front: Sheep Mountain anticline in Big Horn Basin, Wyoming, ©Michael Collier; back: Visualization of Sheep Mountain anticline using data from the USDA National Agricultural imagery program, ©Paul Morin*
Senior Photo Research Coordinator: *Lori Hancock*
Layout: *Stephen Reynolds, Julia Johnson, Cindy Shaw, and Lachina Publishing Services*
Compositor: *Lachina Publishing Services*
Typeface: *9/10.5 Avenir 35 Light*
Printer: *R. R. Donnelley Willard, OH*

Library of Congress Cataloging-in-Publication Data

Exploring geology / Stephen J. Reynolds ... [et al.]. – 2nd ed.
    p. cm.
  Includes index.
  ISBN 978–0–07–337668–4 — ISBN 0-07-337668-X (hard copy : alk. paper) 1. Geology — Textbooks. I. Reynolds, Stephen J.

  QE26.3.E97 2010
  550--dc22

                                                          2009020306

BRIEF CONTENTS

CONTENTS

CHAPTER 1: THE NATURE OF GEOLOGY   2

CHAPTER 2: INVESTIGATING GEOLOGIC QUESTIONS   24

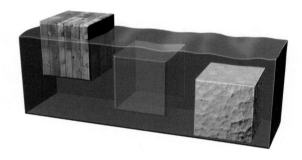

CHAPTER 3: PLATE TECTONICS   50

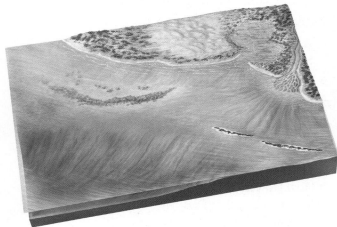

CHAPTER 9:
GEOLOGIC TIME 234

CHAPTER 10:
THE SEAFLOOR AND
CONTINENTAL MARGINS 266

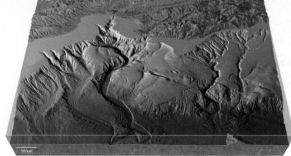

CHAPTER 11:
MOUNTAINS, BASINS, AND CONTINENTS 296

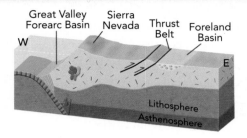

CHAPTER 12:
EARTHQUAKES AND EARTH'S INTERIOR 326

CHAPTER 13:
CLIMATE, WEATHER, AND THEIR INFLUENCES ON GEOLOGY 364

CHAPTER 14:
SHORELINES, GLACIERS, AND CHANGING SEA LEVELS   396

CHAPTER 15:
WEATHERING, SOIL, AND UNSTABLE SLOPES   436

CHAPTER 16:
RIVERS AND STREAMS   466

CHAPTER 17:
WATER RESOURCES   500

CHAPTER 18:
ENERGY AND MINERAL
RESOURCES   526

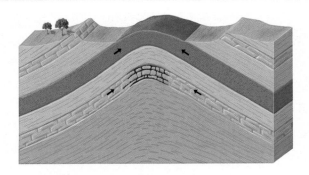

CHAPTER 19:
GEOLOGY OF THE SOLAR SYSTEM   558

PREFACE

TELLING THE STORY . . .

WE WROTE *EXPLORING GEOLOGY* SO THAT STUDENTS could learn from the book on their own, freeing up instructors to teach the class in any way they want. I (Steve Reynolds) first identified the need for this book while I was a National Association of Geoscience Teachers' (NAGT) distinguished speaker. As part of my NAGT activities, I traveled around the country conducting workshops on how to infuse active learning and scientific inquiry into introductory college geology courses, including those with upwards of 200 students. In the first part of the workshop, I asked the faculty participants to list the main goals of an introductory geology college course, especially for nonmajors. At every school I visited, the main goals were similar and are consistent with the conclusions of the National Research Council (see box below):

- to engage students in the process of scientific inquiry so that they learn what science is and how it is conducted,
- to teach students how to observe and interpret landscapes and other aspects of their surroundings,
- to enable students to learn and apply important geologic concepts,
- to help students understand the relevance of geology to their lives, and
- to enable students to use their new knowledge, skills, and ways of thinking to become more informed citizens.

I then asked faculty members to rank these goals and estimate how much time they spent on each goal in class. At this point, many instructors recognized that their activities in class were not consistent with their own goals. Most instructors were spending nearly all of class time teaching content. Although this was one of their main goals, it commonly was not their top goal.

Next, I asked instructors to think about why their activities were not consistent with their goals. Inevitably, the answer was that

Like most geologists, author Steve Reynolds prefers teaching students out in the field, where they can directly observe the geology and reconstruct the sequence of geologic events.

most instructors spend nearly all of class time covering content because (1) textbooks include so much material that students have difficulty distinguishing what is important from what is not; (2) instructors needed to lecture so that students would know what is important; and (3) many students have difficulty learning independently from the textbook.

In most cases, textbooks drive the curriculum, so the author team decided that we should write a textbook that (1) contains only important material, (2) indicates clearly to the student what is important and what they need to know, and (3) is designed and written in such a way that students can learn from the book on their own. This type of book would give instructors freedom to teach in a way that is more consistent with their goals, including using local examples to illustrate geologic concepts and their relevance. Instructors would also be able to spend more class time teaching students to observe and interpret geology, and to participate in the process of scientific inquiry, which represents the top goal for many instructors.

COGNITIVE AND SCIENCE EDUCATION RESEARCH

To design a book that supports instructor goals, we delved into cognitive and science-education research, especially research on how our brains process different types of information, what obstacles limit student learning from textbooks, and how students use visuals versus text while studying. We also conducted our own research on how students interact with textbooks, what students see when they observe photographs showing geologic features, and how they interpret different types of geologic illustrations, including geologic maps and cross sections. *Exploring Geology* is the result of our literature search and of our own science-education research. As you examine *Exploring Geology*, you will notice that it is stylistically different from most other textbooks, which will likely elicit a few questions.

HOW DOES THIS BOOK SUPPORT STUDENT CURIOSITY AND INQUIRY?

Exploring Geology promotes inquiry and science as an active process. It encourages student curiosity and aims to activate existing student knowledge by posing the title of every two-page spread and every subsection as a question. In addition, questions are dispersed throughout the book. Integrated into the book are opportunities for students to observe patterns, features, and examples before the underlying concepts are explained. That is,

we employ a learning-cycle approach where student exploration precedes the introduction of geologic terms and the application of knowledge to a new situation. For example, chapter 15 on slope stability begins with a three-dimensional image of northern Venezuela, and readers are asked to observe where people are living in this area and what processes might have formed these sites.

WHY ARE THE PAGES DOMINATED BY ILLUSTRATIONS?

Geology is an extremely visual science. Typically, geology textbooks contain a variety of photographs, maps, cross sections, block diagrams, and other types of illustrations. These diagrams help portray the distribution and geometry of geologic units on the surface and in the subsurface in a way words could never do. In geology, a picture really is worth a thousand words or more.

Exploring Geology contains a wealth of figures to take advantage of the visual nature of geology and the efficiency of figures in conveying geologic information. This book contains few large blocks of text, and most text is in smaller blocks that are specifically

linked with illustrations. An example of our integrated figure-text approach is shown above and on the next page. In this approach, each short block of text is one or more complete sentences that succinctly describe a geologic feature, geologic process, or both of these. Most of these text blocks are connected to their illustrations with leader lines so that readers know exactly which feature or part of the diagram is being referenced by the text block. A reader does not have to search for the part of the figure that corresponds to a text passage, as occurs when a student reads a traditional textbook with large blocks of text referencing a figure that may appear on a different page.

The approach in *Exploring Geology* is consistent with the findings of cognitive scientists, who conclude that our minds have two different processing systems, one for processing *pictorial* information (images) and one for processing *verbal* information (speech and written words). Cognitive scientists also speak about two types of memory: *working memory*, also called short-term memory, holds information that our minds are actively processing, and *long-term memory* stores information until we need it (Baddeley, 2007). Both the verbal and pictorial processing systems have a limited amount of working memory, and our minds have to use much of our mental processing space to reconcile the two types of information in working memory. For information that has both pictorial and verbal components, the amount of knowledge we retain depends on reconciling these two types of information, on transferring information from working memory to long-term memory, and on linking the new information with our existing mental framework.

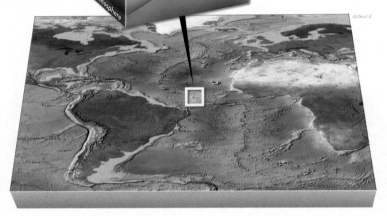

58

3.4 What Happens at Divergent Boundaries?

AT MID-OCEAN RIDGES, Earth's tectonic plates diverge (move apart). Ridges are the sites of many small to moderate-sized earthquakes and much submarine volcanism. On the continents, divergent motion can split a continent into two pieces, forming a new ocean basin as the pieces move apart.

A What Happens at Mid-Ocean Ridges?

Mid-ocean ridges are divergent plate boundaries where new oceanic lithosphere forms as two oceanic plates move apart. These boundaries are also called *spreading centers* because of the way the plates spread apart.

1. A narrow trough, or *rift*, runs along the axis of most mid-ocean ridges. The rift forms because large blocks of crust slip down as spreading occurs. The movement causes faulting, resulting in frequent small to moderate-sized earthquakes.

2. As the plates move apart, solid mantle in the asthenosphere rises toward the surface. It partially melts in response to a decrease in pressure. The molten rock (magma) rises along narrow conduits, accumulates in magma chambers beneath the rift, and eventually becomes part of the oceanic lithosphere.

3. Much of the magma solidifies at depth, but some erupts onto the seafloor, forming submarine lava flows. These eruptions create new ocean crust that is incorporated into the oceanic plates as they move apart.

4. Mid-ocean ridges are elevated above the surrounding seafloor because they consist of hotter, less dense materials, including magma. They also are higher because the underlying lithosphere is thinner beneath ridges than beneath typical seafloor. Lower density materials and thin lithosphere mean that the plate "floats" higher above the underlying asthenosphere. The elevation of the seafloor decreases away from the ridge because the rock cools and contracts, and because the less dense asthenosphere cools enough to become part of the more dense lithosphere.

Ocean
Oceanic Crust
Lithospheric Mantle
Asthenosphere

03.04.a1-2

New experiences from the environment enter the brain via the senses. Images, for example, come in through the eyes, and sounds enter the ears.

Input from the senses is filtered and transferred into two different types of working memory, a *visual* area for images and a *phonetic* area for words. Each type of working memory has a very limited capacity to hold new information.

Information from working memory is processed further and transferred into long-term memory. Ideally, new information is linked to existing knowledge in long-term memory to build a more complete understanding.

When information from long-term memory is needed, it is retrieved into working memory, where it can be processed to make decisions.

WHY ARE THERE SO MANY FIGURES?

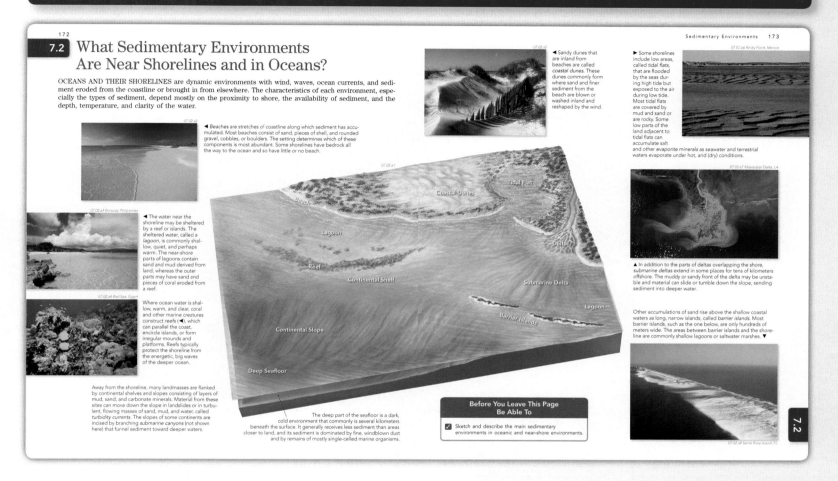

This textbook contains more than 2,600 illustrations, which is two to three times the number in most introductory geology textbooks. One reason for this is that the book is designed to provide a concrete example of each rock type, environment, or geologic feature being illustrated. Research shows that many college students require concrete examples before they can begin to build abstract concepts (Lawson, 1980). Also, many students have limited travel experience, so photographs and other figures allow them to observe places, environments, and processes they have not been able to observe firsthand. The inclusion of an illustration for each text block reinforces the notion that the point being discussed is important. In many cases, as in the example on this page, conceptualized figures are integrated with photographs and text so that students can build a more coherent view of the environment or process.

Exploring Geology focuses on the most important geologic concepts and makes a deliberate attempt to eliminate text that is not essential for student learning of these concepts. Inclusion of information that is not essential tends to distract and confuse students rather than illuminate the concept; thus you will see fewer words. Cognitive and science-education research has identified a redundancy effect, where information that restates and expands upon a more succinct description actually results in a decrease in student learning (Mayer, 2001). Specifically, students learn less if a long figure caption restates information contained elsewhere on the page, such as in a long block of text that is detached from the figure. We avoid the redundancy effect by including only text that is integrated with the figure.

The style of illustrations in *Exploring Geology* was designed to be more inviting to today's visually oriented students who are used to photo-realistic, computer-rendered images in movies, videos, and computer games. For this reason, many of the figures were created by world-class artists who have worked on Hollywood movies, on television shows, for *National Geographic*, and in the computer-graphics industry. In most cases, the figures incorporate real data, such as satellite images and aerial photographs. Our own research shows that many students do not understand geologic cross sections and other subsurface diagrams, so nearly every cross section in this book has a three-dimensional aspect, and many maps are presented in a perspective view with topography. Research findings by us and other researchers (Roth and

Bowen, 1999) indicate that including people and human-related items on photographs and figures attracts undue attention, thereby distracting students from the geologic features being illustrated. As a result, our photographs have nondistracting indicators of scale, like dull coins and plain marking pens. Figures and photographs do not include people or human-related items unless we are trying to illustrate how geoscientists study geologic processes and features.

HOW ARE GEOLOGIC TERMS INTRODUCED IN THIS BOOK?

Wherever possible, we introduce terms after students have an opportunity to observe the feature or concept that is being named. This approach is consistent with several educational philosophies, including a learning cycle and just-in-time teaching. Research on learning cycles shows that students are more likely to retain a term if they already have a mental image of the thing being named (Lawson, 2003). For example, this book presents students with the collection of igneous rocks shown to the right and asks them to think about how they would classify the rocks. Only then does the textbook present a classification of igneous rocks.

Also, the figure-based approach in this book allows terms to be introduced in their context rather than as a definition that is detached from a visual representation of the term. In this book, we introduce new terms in italics rather than in boldface, because boldfaced terms on a textbook page cause students to immediately focus mostly on the terms, rather than build an understanding of the concepts. In this second edition, we have included a glossary for those students who wish to look up the definition of a term to refresh their memory. To expand comprehension of the definition, each entry in the glossary references the page where the term is defined in the context of a figure.

WHY DOES THE BOOK CONSIST OF TWO-PAGE SPREADS?

This book consists of two-page spreads, most of which are further subdivided into sections. Research has shown that because of our limited amount of working memory, much new information is lost if it is not incorporated into long-term memory. Many students keep reading and highlighting their way through a textbook without stopping to integrate the new information into their mental framework. New information simply displaces existing information in working memory before it is learned and retained. This concept of cognitive load (Sweller, 1994) has profound implications for student learning during lectures and while reading textbooks. Two-page spreads and sections help prevent cognitive overload by providing natural breaks that allow students to stop and consolidate the new information before moving on.

Each spread has a unique number, such as 6.9 for the 9th topical two-page spread in chapter 6 (see below). These numbers help instructors and students keep track of where they are and what is being covered. Each two-page spread, except for those that begin and end a chapter, contains a *Before You Leave This Page* checklist that indicates what is important and what is expected of students before they move on. This list contains learning objectives for the spread and provides a clear way for the instructor to indicate to the student what is important. The items on these lists are complied into a master *What-to-Know* list.

SIGNIFICANT ADVANTAGES OFFERED BY *EXPLORING GEOLOGY*

Two-page spreads and integrated *Before You Leave This Page* lists offer the following advantages to the student:

- Information is presented in relatively small and coherent chunks that allow a student to focus on one important aspect or geologic system at a time.
- Students know when they are done with this particular topic and can self-assess their understanding with the *Before You Leave This Page* list.
- Two-page spreads allow busy students to read or study a complete topic in a short interval of study time, like breaks between classes.
- All test questions and assessment materials are tightly articulated with the *Before You Leave This Page* lists so that exams and quizzes cover precisely the same material that was assigned to students via the *What-to-Know* list.

The two-page-spread approach also has huge advantages for the instructor. Before writing this book, the authors wrote the items for the *Before You Leave This Page* lists. We then used this list to decide what figures were needed, what topics would be discussed, and in what order. In other words, *the textbook was written from the learning objectives*. The *Before You Leave This Page* lists provide a straightforward way for an instructor to tell students what information is important. Because we provide the instructor with a master *What-to-Know* list, an instructor can selectively assign or eliminate content by providing students with an edited *What-to-Know* list. Alternatively, an instructor can give students a list of assigned two-page spreads or sections within two-page spreads. In this way, the instructor can tell students what content they are responsible for even if the material is not covered in class.

Before You Leave This Page Be Able To

✓ Describe the characteristics of a volcanic dome.

✓ Explain or sketch the two ways by which a volcanic dome can grow.

✓ Explain or sketch how a volcanic dome can collapse or be destroyed by an explosion.

✓ Describe the types of rocks associated with volcanic domes.

✓ Describe how you might recognize a volcanic dome in the landscape.

Two-page spreads are organized into 19 chapters that are arranged into five major groups: (1) introduction to Earth and the science of geology, (2) earth materials and the processes that form them, (3) geologic time and tectonic systems, (4) climate and surface processes, and (5) capstone chapters on resources and planetary geology. The first three chapters provide an overview of geology, the scientific approach to geology, and plate tectonics—a unifying theme interwoven throughout the rest of the book. The next five chapters cover earth materials, including minerals (chapter 4), different families of rocks and structures (chapters 5-8), and the processes that form or modify rocks. Unlike many geology books, *Exploring Geology* begins the discussion of earth materials with an examination of landscapes—something students can relate to—as a lead-in to rocks, then to minerals, and finally to atoms, the most abstract topic in geology books. The sedimentary environments chapter includes a brief introduction to weathering, setting the stage for the discussion of clastic sediments but saving a more detailed discussion of weathering and soils for the part of the book that deals with surficial processes. Also, this book integrates the closely related topics of metamorphism and deformation into a single chapter.

After earth materials, we cover the principles of geologic time, emphasizing how geologists reconstruct Earth history (chapter 9). We then move on to ocean basins, mountains and basins, and earthquakes (chapters 10-12), all of which integrate and apply information about rocks, structures, geologic time, and plate tectonics. These chapters provide important details about aspects of plate tectonics, after students have gained an understanding of rocks, structures, and geologic time from earlier chapters. We have also incorporated a small component of historical geology, including evolution of the continents and ocean basins.

Next, we briefly discuss weather and climate (chapter 13) to provide a backdrop for subsequent chapters on surface processes and to introduce timely topics, such as hurricanes and climate change. This chapter also discusses deserts, drought, and rain forests. Shorelines, glaciers, and sea-level changes are integrated into a single chapter (chapter 14) to present a system approach to Earth processes and to emphasize the interplay between glaciations, sea level, and the character of the shoreline. Chapter 15 focuses on weathering, soils, and slope stability; chapter 16 presents rivers, streams, and flooding; and chapter 17 covers surface-water and groundwater resources and groundwater-related problems.

We consider the last two chapters to be capstones, integrating and applying previous topics to enable students to understand energy and mineral resources (chapter 18) and planetary geology (chapter 19). These two chapters give students and instructors an opportunity to see how an understanding of rock types, rock-forming processes, geologic structures, geologic time, and the flow of water and other fluids can help us understand important resources and the surfaces of other planetary bodies. The late placement of both chapters allows a more comprehensive treatment of these topics than would be possible if they were incorporated into earlier chapters.

SPECIAL TEXT FEATURES
Concept Sketches

Most items on the *Before You Leave This Page* list are by design suitable for student construction of concept sketches. Concept sketches are sketches that are annotated with complete sentences that identify geologic features, describe how the features form, characterize the main geologic processes, and summarize geologic histories (Johnson and Reynolds, 2005). An example of a concept sketch is shown below.

In our experience, concept sketches are an excellent way to actively engage students in class and to assess their understanding of geologic features, processes, and history. Concept sketches are well suited to the visual nature of geology, especially cross sections, maps, and block diagrams. Geologists are natural sketchers using field notebooks, blackboards, publications, and even napkins, because sketches are an important way to record observations and thoughts, organize knowledge, and try to visualize geometries of rock bodies and sequences of events. A student who can draw, label, and explain a concept sketch generally has a good understanding of that concept.

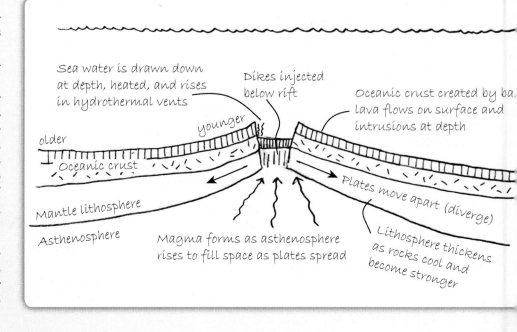

TWO-PAGE SPREADS

Most of the book consists of *two-page spreads*, each of which is about one or more closely related topics. Topical spreads convey the geologic content and help organize knowledge.

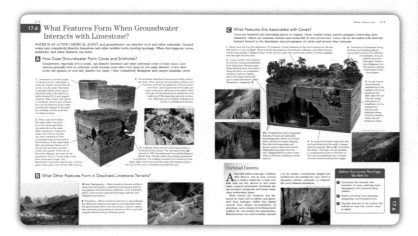

Each chapter has at least one two-page spread illustrating how geology impacts society and another two-page spread that specifically describes how geoscientists study typical problems.

The next-to-last two-page spread in each chapter is a *Connections* spread, which is designed to help students connect and integrate the various concepts from the chapter and to show how these concepts can be applied to an actual location. *Connections* are about real places that illustrate the geologic concepts and features covered in the chapter and explicitly illustrate how a geologic problem is investigated and how geologic problems have relevance to society. The Connections spread also prepares the student for a following Investigation two-page spread.

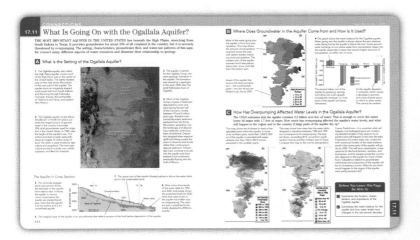

Each chapter ends with an *Investigation* spread that is an exercise in which students apply the knowledge, skills, and approaches learned in the chapter. These exercises mostly involve virtual places that students explore and investigate to make observations and interpretations and to answer a series of geologic questions.

Investigations are modeled after the types of problems geologists investigate, and they use the same kinds of data and illustrations encountered in the chapter. The Investigation includes a list of goals for the exercises and step-by-step instructions, including calculations and methods for constructing maps, graphs, and other figures. These investigations can be completed by students in class, as worksheet-based homework, or as online activities.

NEW IN THE SECOND EDITION

All chapters include additions and improvements. For example:

- A NEW *Glossary* has been added at the end of the book, with each entry linked back to the figure-based text where that term is introduced.

- Six chapters are expanded, condensed, or reorganized, including expansion of sections on glaciers, igneous rocks, and plate-tectonic settings of earthquakes, and shortening of the chapter on deformation.

- Many two-page spreads have been extensively revised with larger photographs and improved layout, illustrations, and text.

- Photographs were upgraded, and new figures were added, including figures showing Earth's four spheres and maps showing moraines of the Midwest and Northeast.

- Spreads titled "Application" in the first edition have been retitled "Connections" to present a case study that connects key principles from the chapter.

The second edition of *Exploring Geology* represents a major revision compared to the first edition. The style, approach, and sequence of chapters are unchanged, but revisions within chapters vary from moderate to major in extent. The second edition contains more than 130 new or revised figures and more than 130 new or replaced photographs. In addition, we eliminated dozens of photographs and figures. We revised most text blocks to improve clarity and conciseness. Most chapters contain the same number and order of two-page spreads, but six chapters were expanded, condensed, or extensively reorganized. We improved the layout of most two-page spreads. Most changes were made in response to comments by reviewers and students or from the authors' recognition that some aspect could be improved. We tried to respond to every reviewer comment unless it was incompatible with the philosophy of the book, caused undue hardships in the organization of chapters and content of figures, or we did not agree with it. We describe the main revisions below.

A complete *Glossary* is included at the end of the book, and defines key terms and concepts. Each definition includes the page number where the term is introduced.

CHAPTER 1 is moderately reorganized, combining information from several spreads into one to allow for a new illustration on Earth's four spheres. We also revised the main figures for the Connections and Investigation spreads.

CHAPTER 2 is totally reorganized with seven spreads appearing in a different order compared to the first edition. There are 22 new or revised figures and photographs.

CHAPTER 3, Plate Tectonics, received moderate revisions but contains a number of new and revised figures.

CHAPTER 4 has six heavily revised spreads, with the addition or elimination of many figures and photographs.

CHAPTER 5 is reorganized with a rearrangement of seven spreads as well as revision of material within these spreads. It contains a new spread to expand igneous rock classification.

CHAPTER 6 mostly has new or replaced photographs and figures and changes to accompanying text blocks.

CHAPTER 7 was extensively revised with the addition of 30 new or replaced photographs. The page layouts, as in most chapters, are heavily revised.

CHAPTER 8 is greatly consolidated compared to the first edition and now features 16 spreads instead of 18. We eliminated some structural content that is probably too detailed for introductory classes.

CHAPTER 9 was partly reorganized so that unconformities followed relative dating.

CHAPTERS 10 and 11 received only minor or moderate revisions, mostly consisting of new or revised figures and text.

CHAPTER 12 was slightly reorganized, and a new spread was added to expand coverage of the plate-tectonic setting of earthquakes.

CHAPTER 13 includes mostly minor or moderate revisions, including 14 new or revised photographs and figures.

CHAPTER 14 represents a significant expansion, gaining four pages to increase the coverage of glaciers, which is almost completely redone. This chapter contains 18 new or revised figures or photographs, mostly within the section on glaciers.

CHAPTER 15 contains minor to moderate changes except in the Connections spread, which is extensively revised.

CHAPTER 16 contains 20 new or revised figures and photographs but otherwise has only moderate revisions.

CHAPTER 17 also contains a significant number of revised figures, including moving water well locations on the Investigation spread to clarify areas of contamination.

CHAPTERS 18 and 19 had moderate revisions, mostly consisting of changes in layout, replacement of photographs and figures, and changes to text blocks. Chapter 19 contains an expanded section on the uniqueness of Earth.

A detailed list of revisions is available from the authors. Users of the first edition will recognize how extensive the changes are within most chapters and will observe significant improvements on every page.

REFERENCES CITED

Baddeley, A.D., 2007. Working memory, thought, and action. Oxford, England: Oxford University Press, 400 p.

Johnson, J.K., and Reynolds, S.J., 2005. Concept sketches—Using student- and instructor-generated annotated sketches for learning, teaching, and assessment in geology courses. Journal of Geoscience Education, v. 53, pp. 85–95.

Lawson, A.E., 1980. Relationships among level of intellectual development, cognitive styles, and grades in a college biology course. Science Education, v. 64, pp. 95–102.

Lawson, A., 2003. The neurological basis of learning, development & discovery: Implications for science & mathematics instruction. Kluwer Academic Publishers, Dordrecht, The Netherlands, 283 p.

Mayer, R.E., 2001. Multimedia learning. Cambridge: Cambridge University Press, 210 p.

Roth, W.M., and Bowen, G. M., 1999. Complexities of graphical representations during lectures: A phenomenological approach. Learning and Instruction, v. 9, pp. 235–255.

Sweller, J., 1994. Cognitive Load Theory, learning difficulty, and instructional design. Learning and Instruction, v. 4, pp. 295–312.

McGraw-Hill offers various tools and technology products to support *Exploring Geology, Second Edition*. Students can order supplemental study materials by contacting their local bookstore or by calling 800-262-4729. Instructors can obtain teaching aids by calling the Customer Service Department at 800-338-3987, visiting the McGraw-Hill website at www.mhhe.com, or contacting their McGraw-Hill sales representative.

INSTRUCTOR RESOURCES

TEXT WEBSITE

http://www.mhhe.com/reynoldsgeology2e

Online resources available on the Text Website include:

- A completely revised and comprehensive Instructor's Manual features Teaching Tips, a general discussion for transitioning into *Exploring Geology*, a chapter-by-chapter discussion on how to approach topics and tie things together, and suggestions for classroom activities that will best promote learning.
- A *What-to-Know* list encompasses the entire book and is editable.
- EZ Test Exam Questions (see detailed description below) were prepared by the authors and correspond to items on the *What-to-Know* list in a multiple-choice format. These questions can also be used for in-class activities, like those using personal response devices (clickers), or online quizzes and self study. Where possible, questions are built around figures and involve problem solving and higher-order thinking.

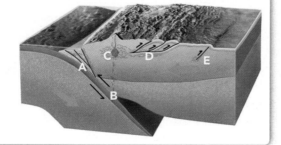

1. **In what site on this figure would you expect high pressure-low temperature metamorphism?**
   a) A and B
   b) B and C
   c) C and D
   d) D and E
   e) none of the above

- Investigation Worksheets with answers are located under Instructor Resources. Blank worksheets are available to students and can be printed as assignable homework, or they can be filled out during class time, preferably in small group activities, or in lab. Alternatively, an instructor can deploy online versions of the worksheets that are automatically graded by Blackboard.
- Presentation Center includes digital files of all the textbook illustrations and photographs suitable for PowerPoint presentations. Digital image files are available by chapter as either PowerPoint or .jpg files. All line drawings, photos, maps, and tables are included. The Presentation Center is a powerful tool that allows you to search and download illustrations from numerous McGraw-Hill science titles.

Instructor's Power Media DVD

This two-DVD set, prepared by the authors, will bring a visual geology experience right into the classroom. It provides lecture PowerPoint files with key points built around the spectacular art from the book. Certain slides in the presentations contain built-in links to launch media files, most of which consist of interactive versions of figures from the book. The media include animations, 3-D maps, terrains, and globes that can be rotated, and images that can be used with a GeoWall.

Computerized Test Bank Online (EZ Test)

Test questions were prepared by the authors to assure that assessment directly correlates to *What-to-Know* items in *Exploring Geology, Second Edition*. The comprehensive bank of test questions is provided as Microsoft Word files and within a computerized test bank powered by McGraw-Hill's flexible electronic testing program EZ Test Online (www.eztestonline.com). EZ Test Online allows you to create paper and online tests or quizzes in this easy-to-use program.

Imagine being able to create and access your test or quiz anywhere, at any time without installing the testing software. Now, with EZ Test Online, instructors can select questions from multiple McGraw-Hill test banks or author their own, and then either print the test for paper distribution or give it online.

TEST CREATION

- Author/edit questions online using the 14 different question type templates.
- Create printed tests or deliver online to get instant scoring and feedback.
- Create question pools to offer multiple versions online— great for practice.
- Export your tests for use in WebCT, Blackboard, PageOut and Apple's iQuiz.
- Compatible with EZ Test Desktop tests you've already created.
- Sharing tests with colleagues, adjuncts, TAs is easy.

ONLINE TEST MANAGEMENT

- Set availability dates and time limits for your quiz or test.
- Control how your test will be presented.
- Assign points by question or question type with drop-down menu.
- Provide immediate feedback to students or delay until all finish the test.

- Create practice tests online to enable student mastery.
- Upload your roster to enable student self-registration.

ONLINE SCORING AND REPORTING

- Automated scoring for most of EZ Test's numerous question types.
- Allows manual scoring for essay and other open response questions.
- Manual re-scoring and feedback is also available.
- EZ Test's grade book is designed to easily export to your grade book.
- View basic statistical reports.

SUPPORT AND HELP

- User's Guide and built-in page-specific help.
- Flash tutorials for getting started on the support site.
- Support Website: www.mhhe.com/eztest
- Product specialist available at 1-800-331-5094.
- Online Training: http://auth.mhhe.com/mpss/workshops/

Course Delivery Systems

With help from our partners WebCT, Blackboard, Top-Class, eCollege, and other course management systems, professors can take complete control of their course content. Course cartridges containing website content, online testing, and powerful student tracking features are readily available for use within these platforms.

TEGRITY

Tegrity Campus is a service that makes class time available all the time by automatically capturing every lecture in a searchable format for students to review when they study and complete assignments. With a simple one-click start and stop process, you capture all computer screens and corresponding audio. Students replay any part of any class with easy-to-use browser-based viewing on a PC or Mac.

Educators know that the more students can see, hear, and experience class resources, the better they learn. With Tegrity Campus, students quickly recall key moments by using Tegrity Campus's unique search feature. This search helps students efficiently find what they need, when they need it across an entire semester of class recordings. Help turn all your students' study time into learning moments immediately supported by your lecture.

To learn more about Tegrity watch a 2-minute Flash demo at http://tegritycampus.mhhe.com

Classroom Performance System and Questions

The Classroom Performance System (CPS) brings interactivity into the classroom or lecture hall. CPS is a wireless response system that gives the instructor and students immediate feedback from the entire class. The wireless response pads are essentially remotes that are easy to use and engage students. CPS allows you to motivate student preparation, interactivity, and active learning so you can receive immediate feedback and know what students understand. A text-specific set of questions is available via download from the Instructor area of *Exploring Geology, Second Edition* website.

Custom Publishing

Did you know that you can design your own text or lab manual, using any McGraw-Hill text and your personal materials to create a custom product that correlates specifically to your syllabus and course goals? Contact your McGraw-Hill sales representative to learn more about this option.

STUDENT RESOURCES

Text Website

http://www.mhhe.com/reynoldsgeology2e

A free website is available to students and provides practice quizzes with immediate feedback. The website includes:

- How to use and study with *Exploring Geology*
- Practice quiz questions
- Investigation worksheets in a printable format
- Instructions for creating concept sketches

McGraw-Hill's Connect Geology

McGraw-Hill's Connect Geology is a web-based assignment and assessment platform that gives students the means to better connect with their coursework, with their instructors, and with the important concepts that they will need to know for success now and in the future.

With Connect Geology, instructors can deliver assignments, quizzes, and tests online. Nearly all the questions from the text are presented in an auto-gradable format and tied to the text's learning objectives. Instructors can edit existing questions and author entirely new problems; track individual student performance—by question, assignment; or in relation to the class overall—with detailed grade reports; integrate grade reports easily with Learning Management Systems (LMS), such as WebCT and Blackboard; and much more.

By choosing Connect Geology, instructors are providing their students with a powerful tool for improving academic performance and truly mastering course material. Connect Geology allows students to practice important skills at their own pace and on their own schedule. Importantly, students' assessment results and instructors' feedback are all saved online – so students can continually review their progress and plot their course to success.

Some instructors may also choose Connect Geology Plus for their students. Like Connect Geology, Connect Geology Plus provides students with online assignments and assessments, plus 24/7 online access to an eBook—an online edition of the text——to aid them in successfully completing their work, wherever and whenever they choose.

LearnSmart

Built around metacognition learning theory, LearnSmart provides your students with a GPS (Guided Path to Success) for your geology course. Using artificial intelligence, LearnSmart intelligently assesses a student's knowledge of course content through a series of adaptive questions. It pinpoints concepts the student does not understand and maps out a personalized study plan for success. Available as an integrated feature of McGraw-Hill's Connect Geology, you can incorporate LearnSmart into your course in a number of ways to

- Gauge student knowledge before a lecture
- Reinforce learning after lecture
- Prepare students for assignments and exams

Discover for yourself how the LearnSmart diagnostic ensures students will connect with the content, learn more effectively, and succeed in your course.

eBook

If you or your students are ready for an alternative version of the traditional textbook, McGraw-Hill has partnered with Course-Smart and VitalSource to bring you innovative and inexpensive electronic textbooks. Students can save up to 50% off the cost of a print book, reduce their impact on the environment, and gain access to powerful web tools for learning, including full text search, notes and highlighting, and email tools for sharing notes between classmates. eBooks from McGraw-Hill are smart, interactive, searchable, and portable.

To review comp copies or to purchase an eBook, go to **www .CourseSmart.com**.

Related Titles of Interest

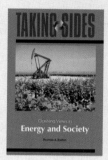

Taking Sides: Clashing Views in Energy and Society

By Thomas A. Easton, Thomas College

ISBN: 978-0-07-812755-7 / MHID: 0-07-812755-6

Taking Sides: Clashing Views in Energy and Society presents current controversial issues in a debate-style format designed to stimulate student interest and develop critical thinking skills. Each issue is thoughtfully framed with an issue summary, an issue introduction, and a postscript. An instructor's manual with testing material is available online for each volume. *Taking Sides: Clashing Views in Energy and Society* is also an excellent instructor resource with practical suggestions on incorporating this effective approach in the classroom. Each Taking Sides reader features an annotated listing of selected World Wide Web sites. Visit **www .mhcls.com** for more information. To purchase an electronic eBook version of this title, visit **www.CourseSmart.com**.

Laboratory Manual for Physical Geology, Seventh Edition
By Charles E. Jones, Norris W. Jones
ISBN: 978-0-07-336939-6 / MHID: 0-07-336939-X

Zumberge's Laboratory Manual for Physical Geology, Fourteenth Edition
By Robert H. Rutford, James L. Carter
ISBN: 978-0-07-305149-9 / MHID: 0-07-305149-7

ACKNOWLEDGMENTS

Writing a totally new type of introductory geology textbook would not be possible without the suggestions and encouragement we received from instructors who reviewed various drafts of this book and its artwork. We are especially grateful to people who contributed entire days either reviewing or attending symposia to openly discuss the vision, challenges, and refinements of this kind of new approach. Many of our colleagues enthusiastically encouraged us onward, including Bruce Herbert, Dexter Perkins, Scott Linneman, Steve Semken, Diane Clemens-Knott, Jeff Knott, Barbara Tewksbury, and Cathy Manduca. The first edition manuscript received special attention from reviewers Scott Linneman, Richard Sedlock, and Grenville Draper. Bill Dupré and Dexter Perkins each gave the published first edition a much-needed and extremely valuable full-book review to guide revisions for the second edition. For all of this we are very grateful.

This book contains over 2,600 figures, two to three times more than a typical introductory geology textbook. This massive art program required great effort and artistic abilities from the artists who turned our vision and sketches into what truly are pieces of art. In addition to our coauthor Chuck Carter, we greatly appreciate the dedication and artistic touches of illustrators Cindy Shaw, Susie Gillatt, Daniel Miller, David Fierstein, Karen Carter, and Ren Olsen. We also benefited from interactions with designers David Hash and Chris Willis, who helped translate our ideas about pedagogy into a workable and aesthetically sound design. Cindy Shaw deserves special praise for acting as Art Director for the second edition, greatly improving the book by standardizing illustrations, nudging and redoing troublesome parts of the layout, and adding arrows and other special touches. She and Susie Gillatt expertly improved the color and fidelity of numerous photographs, many new to the second edition. Terra Chroma, Inc. of Tucson, Arizona, supported many aspects in the development of this book. Shane Reynolds and Courtney Merjil helped prepare illustrations for the PowerPoint presentations. Numerous people went out of their way to provide us with photographs, illustrations, and advice—in some cases going out into the field to take the photographs we needed. These helpful people included Vince Matthews, Ron Blakey, Karen Carr, Michael Collier, Bill Dupré, Tom Sharp, Cheryl Alvarado, Ramón Arrowsmith, Jessica Barone, Doug Bartlett, Don Burt, Phil Christensen, Bill Dickinson, Ed Garnero, Allen Glazner, Jeff Knott, Matthew Larsen, Spencer Lucas, Chris Marone, Tom McGuire, Michael Ort, Peg Owens, Jack Ridge, Nancy Riggs, Steve Semken, James Speer, Barbara Tewksbury, and David Walsh.

We used a number of data sources to create many illustrations. Reto Stöckli of the Department of Environmental Sciences at ETH Zürich and NASA Goddard produced the Blue Marble and Blue Marble Next Generation global satellite composites. We used data from the ZULU server at the NASA Earth Science Enterprise Scientific Data Purchase Program for hundreds of figures in this book. Brian Davis of the USGS EROS Data Center was quick to find elusive data, and Collin Bode of the National Center for Earth-surface Dynamics was indispensable in helping us process GIS data. Debbie Leedy provided mineralogy and chemistry 3D files, and Melanie Busch and Joshua Coyan provided other 3D files and results from their groundbreaking studies with eye-tracking equipment.

We have treasured our interactions with the wonderful Iowans at McGraw-Hill Higher Education, who enthusiastically supported our vision, needs, and progress. We especially thank Marge Kemp, our publishing editor and most steadfast champion, for encouraging our nontraditional approach and for providing timely reality checks. Joan Weber skillfully and cheerfully guided the development of the book during the entire publication process, making it all happen. We also appreciate the support, cooperation, guidance, and enthusiasm from Thomas Timp, Lisa Nicks, Gloria Schiesl, Eric Weber, Judi David, Lori Hancock, Sherry Kane, Mary Jane Lampe and many others who worked hard to make this book a reality. Jeff Lachina and Shawn Vazinski of Lachina Publishing Services were a pleasure to work with while helping prepare the book for internal review and final printing. We thank Dexter Perkins for leading the development of the Instructor's Manual for the second edition.

Finally, a project like this is truly life consuming, especially when the author team is doing the writing, illustrating, photography, near-final page layout, media development, and development of assessments and teaching ancillaries. We are extremely appreciative of the support, patience, and friendship we received from family members, friends, colleagues, and students who shared our sacrifices and successes during the creation of this new vision of a textbook. We thank Susie Gillatt; John, Kay, and Widget Reynolds; Annabelle Louise; Sarah Kelly; Lisa Logan; August Morin; Oliver Morin; and Karen Carter. We thank you all so much!

SECOND EDITION REVIEWERS

Special thanks and appreciation go out to all reviewers. This edition was improved by many beneficial suggestions, new ideas, and invaluable advice provided by these reviewers. We appreciate all the time they devoted to reviewing manuscript chapters, attending focus groups, surveying students, and promoting this text to their colleagues:

Steve Adams   *Guilford Tech Community College*
Jeffrey M. Amato   *New Mexico State University*

Bryan Anderson   *Bethel University*
Jessica Barone   *Monroe Community College*
Margaret H. Benoit   *The College of New Jersey*
Aaron J. Celestian   *Western Kentucky University*
Amanda Colosimo   *Monroe Community College*
Bruce H. Corliss   *Duke University*
George E. Davis   *California State University–Northridge*
Allen Dennis   *University of South Carolina–Aiken*
Kathleen Devaney   *El Paso Community College*
Grenville Draper   *Florida International University*
William Dupré   *University of Houston*
John Encarnacion   *Saint Louis University*

Mark Feigenson   *Rutgers University*
Lydia K. Fox   *University of the Pacific*
Christine A. M. France   *University of Maryland–College Park*
Alan D. Gishlick   *Gustavus Adolphus College*
Frank D. Granshaw   *Portland Community College*
Roy Haggerty   *Oregon State University*
Nelson R. Ham   *St. Norbert College*
Willis Hames   *Auburn University*
Stanley C. Hatfield   *Southwestern Illinois College*
Daniel I. Hembree   *Ohio University*
Michael Westphal Hiett   *Middle Tennessee State University*
Maureen McCurdy Hillard   *Louisiana Tech University*
Ann Holmes   *University of Tennessee–Chattanooga*
Eric Jerde   *Morehead State University*
Steven D. Kadel   *Glendale Community College*
Zoran Kilibarda   *Indiana University Northwest*
David T. King, Jr.   *Auburn University*
Yvette Kuiper   *Boston College*
Jennifer Latimer   *Indiana State University*
Ming-Kuo Lee   *Auburn University*
G. David Mattison   *Sul Ross State University*
James R. Mayer   *University of West Georgia*
Paul McCarthy   *University of Alaska–Fairbanks*
Trent McDowell   *University of North Carolina–Chapel Hill*
Joseph G. Meert   *University of Florida*
Sadredin C. Moosavi   *Tulane University*

Henry T. Mullins   *Syracuse University*
Jacob A. Napieralski   *University of Michigan–Dearborn*
Pamela Nelson   *Glendale Community College*
Philip M. Novack-Gottshall   *University of West Georgia*
Jeffrey A. Nunn   *Louisiana State University*
Richard Orndorff   *Eastern Washington University*
Dexter Perkins   *University of North Dakota*
Mark E. Reinhold   *Northern Essex Community College*
Jeffery G. Richardson   *Columbus State Community College*
Seth Rose   *Georgia State University*
Cassandra J. Runyon   *College of Charleston*
Michael Rygel   *State University of New York, College at Potsdam*
Karen L. Savage   *California State University–Northridge*
Donald J. Sidman   *University of Wisconsin–Eau Claire*
Derek Sjostrom   *University of Alaska–Anchorage*
James H. Speer   *Indiana State University*
David A. Steffy   *Jacksonville State University*
Alycia L. Stigall   *Ohio University*
Eric Straffin   *Edinboro University of Pennsylvania*
Gina Seegers Szablewski   *University of Wisconsin–Milwaukee*
Carol Thompson   *Tarleton State University*
Jennifer Thomson   *Eastern Washington University*
Robert Thunell   *University of South Carolina*
Nicole Wilson   *Mansfield University of Pennsylvania*
Scott White   *University of South Carolina*
Bridget Wyatt   *San Francisco State University*

GEOLOGY FOCUS GROUP ATTENDEES

Elizabeth Heise   *University of Texas–Brownville*
Dexter Perkins   *University of North Dakota*
Shane V. Smith   *Youngstown State University*
Neptune Srimal   *Florida International University*
Gina Seegers Szablewski   *University of Wisconsin–Milwaukee*
Carol Thompson   *Tarleton State University*
Elizabeth Widom   *Miami University of Ohio*

STUDENT FOCUS GROUP PARTICIPANTS

Instructor: Julia Smith Wellner   *University of Houston*
Kristopher Butler
Juan Garcia
Kristin Martinez
Michael Martinez
John Reed Meixner
William Melton
Reyes Ramirez
Carly Sims
Fawad Zafar

FIRST EDITION REVIEWERS

Mead A. Allison   *Tulane University*
Jeffrey Amato   *New Mexico State University*
John Anderson   *Georgia Perimeter College*
Martin Appold   *University of Missouri–Columbia*
Suzanne L. Baldwin   *Syracuse University*
Julie K. Bartley   *University of West Georgia*
Mark Baskaran   *Wayne State University*
Timothy Bralower   *Pennsylvania State University*
Nathalie Nicole Brandes   *Montgomery College*
J. Bret Bennington   *Hofstra University*
Elisa Bergslein   *Buffalo State College*
David M. Best   *Northern Arizona University*
Theodore Bornhorst   *Michigan Technical University*
Steve Boss   *University of Arkansas*
Douglas Britton   *Long Beach City College*
Pamela C. Burnley   *Georgia State University*
John H. Burris   *San Juan College*
T.J. Callahan   *College of Charleston*
James L. Carew   *College of Charleston*
Cinzia Cervato   *Iowa State University*
Sean Chamberlin   *Fullerton College*
Renee M. Clary   *Mississippi State University*
Diane Clemens-Knott   *California State University–Fullerton*
Kevin Cole   *Grand Valley State University*
Chuck Connor   *University of South Florida*
Peter Copeland   *University of Houston*
Ellen A. Cowan   *Appalachian State University*
Randel Tom Cox   *University of Memphis*
Jim Criswell   *Cape Fear Community College*
Margaret E. Crowder   *Western Kentucky University*

Kathleen Devaney   *El Paso Community College*
Craig Dietsch   *University of Cincinnati*
Rebecca L. Dodge   *University of West Georgia*
Grenville Draper   *Florida International University*
William Dupré   *University of Houston*
Stewart S. Farrar   *Eastern Kentucky University*
David E. Fastovsky   *University of Oklahoma*
Mark D. Feigenson   *Rutgers University*
Stan Finney   *California State University–Long Beach*
Mark Fischer   *Northern Illinois University*
Mark Frank   *Northern Illinois University*
Kyle C. Fredrick   *Buffalo State College*
Heather Gallacher   *Cleveland State University*
Yongli Gao   *East Tennessee State University*
Ed Garnero   *Arizona State University*
Dennis Geist   *University of Idaho*
Danny Glenn   *Wharton County Junior College*
Francisco Gomez   *University of Missouri–Columbia*
Andrew M. Goodliffe   *University of Alabama*
G. Michael Grammer   *Western Michigan University*
Ronald Greeley   *Arizona State University*
Nathan L. Green   *University of Alabama*
Roy Haggerty   *Oregon State University*
Willis Hames   *Auburn University*
Duane Hampton   *Western Michigan University*
Thor A. Hansen   *Western Washington University*
Michael J. Harrison   *Tennessee Tech University*
Timothy Heaton   *University of South Dakota*
Bruce Herbert   *Texas A&M University*
David Hirsch   *Western Washington University*
Curtis L. Hollabaugh   *University of West Georgia*
Bernard A. Housen   *Western Washington University*

Mary Hubbard   *Kansas State University*
Paul F. Hudak   *University of North Texas*
Marilyn C. Huff   *New Mexico State University*
John R. Huntsman   *University of North Carolina–Wilmington*
Jason Janke   *Metropolitan State College of Denver*
Steven C. Jaume   *College of Charleston*
Amanda Palmer Julson   *Blinn College*
Steve Kadel   *Glendale Community College*
Jeffrey A. Karson   *Syracuse University*
Michael Katuna   *College of Charleston*
G. Randy Keller   *University of Oklahoma*
David T. King, Jr.   *Auburn University*
Kent C. Kirkby   *University of Minnesota*
Jeffrey Knott   *California State University–Fullerton*
Mark A. Kulp   *University of New Orleans*
Ming-Kuo Lee   *Auburn University*
Robert A. Leighty   *Mesa Community College*
Adrianne A. Leinbach   *Wake Technical Community College*
Stephen D. Lewis   *California State University–Fresno*
Scott R. Linneman   *Western Washington University*
William W. Little   *Brigham Young University*
Brian E. Lock   *University of Louisiana–Lafayette*
Richard Lozinsky   *Fullerton College*
Neil Lundberg   *Florida State University*
James Martin-Hayden   *University of Toledo*
Stephen Mattox   *Grand Valley State University*
Kyle Mayborn   *Western Illinois University*
Joseph Meert   *University of Florida*

Gretchen Miller   *Wake Technical Community College*
Kula C. Misra   *University of Tennessee–Knoxville*
Dan Moore   *Brigham Young University*
Jared R. Morrow   *University of Northern Colorado*
Michael A. Murphy   *University of Houston*
John E. Mylroie   *Mississippi State University*
Thomas Naehr   *Texas A&M University–Corpus Christi*
Ravi Nandigam   *University of Texas–Brownsville*
Pamela Nelson   *Glendale Community College*
Steven R. Newkirk   *University of Memphis*
Peter A. Nielsen   *Keene State College*
Philip M. Novack-Gottshall   *University of West Georgia*
Clair Russell Ossian   *Tarrant County College*
William C. Parker   *Florida State University*
Roy E. Plotnick   *University of Illinois–Chicago*
Kate Pound   *St. Cloud State University*
Steven Ralser   *University of Wisconsin–Madison*
Kenneth Rasmussen   *Northern Virginia Community College*
John Renton   *West Virginia University*
Carl Richter   *University of Louisiana–Lafayette*
Nancy Riggs   *Northern Arizona University*
Bethany D. Rinard   *Tarleton State University*
Delores Robinson   *University of Alabama–Tuscaloosa*
Scott Rowland   *University of Hawaii at Manoa*
Cassandra J. Runyon   *College of Charleston*
Randye L. Rutberg   *Hunter College*
Dewey D. Sanderson   *Marshall University*

Roy Schlische   *Rutgers University*
Richard Sedlock   *San José State University*
Yuch-Ning Shieh   *Purdue University–West Lafayette*
Eric Small   *University of Colorado–Boulder*
Abe Springer   *Northern Arizona University*
Neptune Srimal   *Florida International University*
Paula J. Steinker   *Bowling Green State University*
Mark J. Sutherland   *College of DuPage*
Michael Taber   *University of Northern Colorado*
J. Robert Thompson   *Glendale Community College*
Jan Tullis   *Brown University*
Lensyl Urbano   *University of Memphis*
Stacey Verardo   *George Mason University*
Mari Vice   *University of Wisconsin–Platteville*
Adil M. Wadia   *The University of Akron, Wayne College*
Stephen Wareham   *California State University–Fullerton*
Richard Warner   *Clemson University*
Johnny Waters   *Appalachian State University*
Barry Weaver   *University of Oklahoma*
John Weber   *Grand Valley State University*
David A. Williams   *Arizona State University*
Harry Williams   *University of North Texas*
Wendi J. W. Williams   *University of Arkansas–Little Rock*
Kenneth Windom   *Iowa State University*
Lorraine W. Wolf   *Auburn University*
Aaron Yohsinobu   *Texas Tech University*

ABOUT THE AUTHORS

STEPHEN J. REYNOLDS

Stephen J. Reynolds received an undergraduate geology degree from the University of Texas at El Paso, and M.S. and Ph.D. degrees in structure/tectonics and regional geology from the University of Arizona. He then spent ten years directing the geologic framework and mapping program of the Arizona Geological Survey, where he completed the 1988 *Geologic Map of Arizona*. Steve currently is a professor in the School of Earth and Space Exploration at Arizona State University, where he has taught Physical Geology, Structural Geology, Field Geology, Orogenic Systems, Cordilleran Regional Geology, Teaching Methods in the Geosciences, and others. He helped establish the ASU Center for Research on Education in Science, Mathematics, Engineering, and Technology (CRESMET), and was President of the Arizona Geological Society. He has authored or edited nearly 200 geologic maps, articles, and reports, including the 866-page *Geologic Evolution of Arizona*. He also coauthored *Structural Geology of Rocks and Regions*, a widely used Structural Geology textbook, and *Observing and Interpreting Geology*, a laboratory manual for Physical Geology. His current geologic research focuses on structure, tectonics, and mineral deposits of the Southwest, including northern Mexico. For a dozen years, he has done science-education research on student learning in college geology courses, especially the role of visualization. Steve is known for innovative teaching methods, has received numerous teaching awards, and has an award-winning website. As a National Association of Geoscience Teachers (NAGT) distinguished speaker, he traveled across the country presenting talks and workshops on how to infuse active learning and inquiry into large introductory geology classes. He is commonly an invited speaker to national workshops and symposia on active learning, visualization, and teaching methods in college geology courses. He also has been a long-time industry consultant in mineral, energy, and water resources, and has received outstanding alumni awards from UTEP and the University of Arizona.

JULIA K. JOHNSON

Julia K. Johnson is currently a full-time faculty member in the School of Earth and Space Exploration at Arizona State University. Her M.S. and Ph.D. research involved structural geology and geoscience education research. The main focus of her geoscience education research is on student- and instructor-generated sketches for learning, teaching, and assessment in college geology classes. Prior to coming to ASU, she did groundwater studies of copper deposits and then taught full time in the Maricopa County Community College District, teaching Physical Geology, Environmental Geology, and their labs. At ASU, she teaches Introduction to Geology to nearly 1,000 students per year and supervises the associated introductory geology labs. She also coordinates the introductory geology teaching efforts of the School of Earth and Space Exploration, helping other instructors incorporate active learning and inquiry into large lecture classes. At ASU, Johnson recently coordinated a very successful project focused on redesigning introductory geology classes so that they incorporated more online content and asynchronous learning. Julia is recognized as one of the best science teachers at ASU and has received student-nominated teaching awards and very high teaching evaluations in spite of her challenging classes. In recognition of her teaching, she was a Featured Faculty of the Month on ASU's website in 2005. She has authored publications on geology and science-education research, including an article in the *Journal of Geoscience Education* on concept sketches. She coauthored *Observing and Interpreting Geology* and also developed a number of websites used by geology students around the world, including the *Visualizing Topography* and *Biosphere 3D* websites.

MICHAEL M. KELLY

Michael M. Kelly received an undergraduate geology degree from the University of California, Santa Cruz and an M.S. degree in geology from Northern Arizona University. His graduate research defined ductile structures and strain in Mojave Desert mountain ranges. As a USGS geologist, he mapped in the western U.S., coauthoring several geologic maps and performing paleomagnetic research and laboratory studies on Columbia River Basalts. As Senior Geologist at EMCON Associates, he led environmental investigations into geologically complex groundwater industrial contamination sites across the Pacific Northwest. He returned to the southwest as the director of the Center for Research and Evaluation of Advanced Technologies in Education (CREATE) at NAU and was adjunct faculty in Environmental Sciences. Here, Kelly's research activities centered around the use of virtual reality to enhance undergraduate science education. He was co-investigator on numerous NSF science education projects and is the author or coauthor of numerous publications resulting from the CREATE's efforts. Kelly is lead author on a virtual reality geology laboratory curriculum that has been in use by undergraduates since 1999. Today, Kelly teaches at Northern Arizona University and designs and assesses media-enhanced science curricula, especially displays and other materials for the National Park Service. His recent research focuses on spatial learning, particularly how landscapes and terrain are interpreted and how they can be used as frameworks for understanding connected science domains. Kelly has designed virtual reality exhibits installed in several national parks, and his 3-D terrain software *ROMA* is used nationally for geoscience education in secondary and undergraduate institutions. He published in the *Journal of Geoscience Education* a study on the effectiveness of the GeoWall in undergraduate geology.

PAUL J. MORIN

Paul J. Morin is Director of the Antarctic Geospatial Information Center at the University of Minnesota. He is responsible for supporting Antarctic scientific and research operations through GIS, remote sensing, cartography, visualization, and software development. Morin also has strong interests in the effect of artistic technique and technology on the efficacy of visualizations in the hands of students. Paul co-founded the GeoWall Consortium, a group of over 200 teaching institutions around the world using visualization and stereo projection in the classroom. Over the past five years, Paul has been instrumental in bringing earth-science visualization to science museums around the world. He is a co-investigator and co-developer of earth science museum exhibits that travel the world and have visited the American Museum of Natural History, the Field Museum, and many others. He was a major contributor of interactive visualizations to the NAGT-sponsored laboratory manual for Physical Geology. For several years, he was an NAGT distinguished speaker, visiting universities and colleges to present talks on the role of visualization in geology courses. He is regarded by many people as one of the top visualization developers in the geosciences. Other professional interests include the visualization of data sources that are traditionally viewed as being too complex for students to understand, such as three-dimensional spherical convection, seismic tomography, and paleontology. His visualizations have been published in *Wired, National Geographic*, and *Nature*.

CHUCK M. CARTER

Chuck M. Carter has been working in the artistic end of the science and entertainment industries for more than 20 years. His illustration and animation work has been used extensively by *National Geographic*, and his illustrations and layouts are featured in books published by *National Geographic* to feature the best of their artwork. He was the first freelance artist hired by *National Geographic* to create a three-page digital illustration (dinosaur evolution) on a consumer desktop machine and in 1994 was instrumental in helping launch *National Geographic Online*. His current magazine clients include *Scientific American, Wired*, and many others. He also has worked with Harcourt Education, McGraw-Hill Higher Education, Knight-Ridder News in Motion, and other clients for more than 18 years. He has produced illustrations and animations for museums, the U.S. Navy, U.S. Department of Defense, and various defense contractors. His entertainment projects include being one of two artists on the computer game *Myst*, and he has worked on more than 25 other video games, including the popular *Command and Conquer* series, as a digital artist, animator, writer, art director, and computer-graphics supervisor. While working with Threshold Entertainment, he worked as a digital matte painter for shows like *Babylon 5*. He is lead illustrator and a coauthor of this book. Currently, Chuck manages some 40 artists at Vicarious Visions – an Activision game company in Albany, NY.

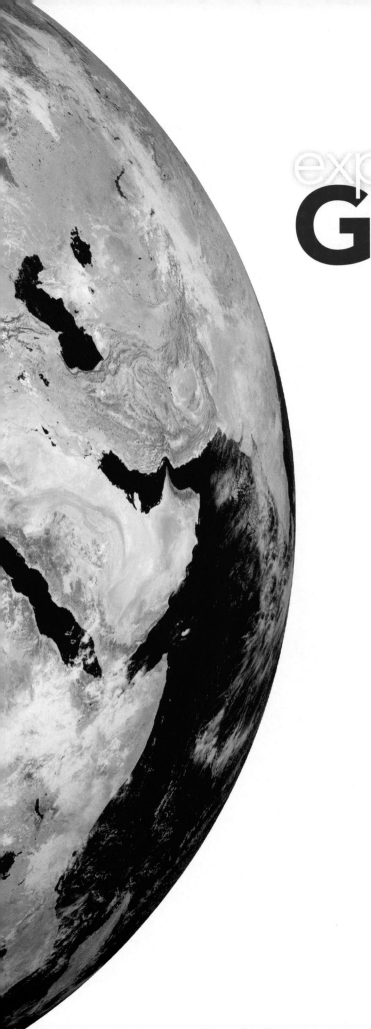

exploring
GEOLOGY

The Nature of Geology

GEOLOGY HAS MANY EXPRESSIONS in our world. Geologic processes reshape Earth's interior and sculpt its surface. They determine the distribution of metals and petroleum and control which places are most susceptible to volcanoes, floods, and other natural disasters. Geology encompasses factors that are critical to ecosystems, such as climate and the availability of water. In this book, we explore geology, *the science of Earth*, and examine why an understanding of geology is important in our modern world.

North America and the surrounding ocean floor have a wealth of interesting features. The large image below (▼) is computer-generated and combines different types of data to show features on the land and seafloor. The shading and colors on land are from space-based satellite images, whereas colors and shading on the seafloor indicate depths below the surface of the sea. Can you find the place where you live? What types of features are there?

01.00.a2 Glacier NP, MT

◄ **The dramatic scenery of Glacier National Park** in Montana features cliffs and rugged mountains that expose a series of intricate gray rock layers. The beautiful valleys preserve evidence of being carved by glaciers during the most recent ice age, approximately 10,000 to 30,000 years ago.

What processes sculpt the land surface and produce such beautiful scenery? What evidence is there for past climate changes, including those that allowed glaciers to cover large parts of North America?

01.00.a1

Glacier
National
• Park

Mount
St. Helens

The 1980 eruption of Mount St. Helens in southwestern Washington (▼) ejected huge amounts of volcanic ash into the air, toppled millions of trees, unleashed floods and mudflows down nearby valleys, and killed 57 people. Geologists study volcanic phenomena to determine how and when volcanoes erupt and what hazards volcanoes pose to humans and other creatures.

How do geologic studies help us determine where it is safe to live?

01.00.a3 Mount St. Helens, WA

TOPICS IN THIS CHAPTER

Rocks of New England and easternmost Canada record a fascinating history, which includes an ancient ocean that was destroyed by the collision of two landmasses. Many of these rocks, such as those in Nova Scotia, have contorted layers (▼), and some rocks provide evidence of having been formed at a depth of 30 km below the surface.

How do layers in rocks get squeezed and deformed, and how do rocks from deep in the earth get to the surface where we now find them?

01.00.a4 Nova Scotia, Canada

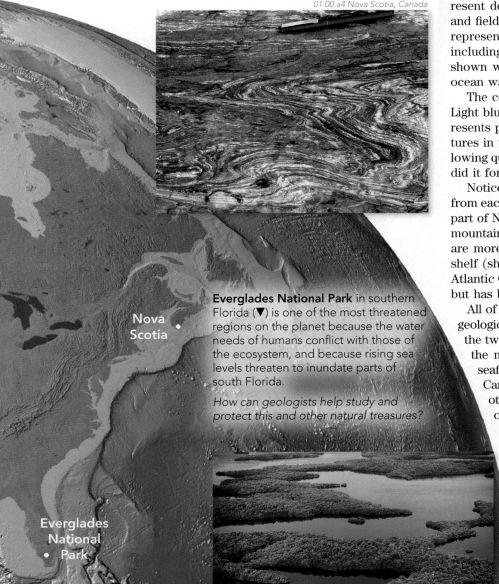

Nova Scotia

Everglades National Park in southern Florida (▼) is one of the most threatened regions on the planet because the water needs of humans conflict with those of the ecosystem, and because rising sea levels threaten to inundate parts of south Florida.

How can geologists help study and protect this and other natural treasures?

Everglades National Park

01.00.a5 Everglades NP, FL

A View of North America

North America is a diverse continent, ranging from the low, tropical rain forests of Mexico to the high Rocky Mountains of northern Canada. In the large image of North America on the left, the colors on land are from satellite images that show the distribution of rock, soil, plants, and lakes. Green colors represent dense vegetation, including forests shown in darker green and fields and grassy plains shown in lighter green. Brown colors represent deserts and other regions that have less vegetation, including regions where rock and sand are present. Lakes are shown with a solid blue color. Note that there are no clouds or ocean waters in this artificial picture.

The color of the ocean floor varies with depth below sea level. Light blue colors represent shallow areas, whereas dark blue represents places where the seafloor is deep. Observe the larger features in this image, both on land and at sea. Ask yourself the following questions: What is this feature? Why is it located here? How did it form? In short, what is its story?

Notice that the two sides of North America are very different from each other and from the middle of the continent. The western part of North America appears more complex because it has many mountains and valleys. The mountains in the eastern United States are more subdued, and the East Coast is surrounded by a broad shelf (shown in a light blue-gray) that continues out beneath the Atlantic Ocean. The center of the continent has no large mountains but has broad plains, hills, river valleys, and large lakes.

All of the features on this image of Earth are part of geology. The geologic history of North America explains why the mountains on the two sides of the continent are so different and when and how the mountains formed. Geology explains how features on the seafloor came to be, and why the central United States and Canada are the agricultural heartland of the continent, whereas other areas are deserts. The relatively high standard of living of people in the United States and Canada is largely due to an abundance of natural resources, especially coal, petroleum, minerals, and fertile soils. Such resources are the result of the geologic history. In short, geology controls the height and shape of the land and seafloor, the types of materials that are present, and the processes that affect the land, sea, and us. As shown throughout this book, geology affects many aspects of our lives.

1.0

1.1 How Does Geology Influence Where and How We Live?

GEOLOGY INFLUENCES OUR LIVES IN MANY WAYS. Geologic features and processes constrain where people can live because they determine whether a site is safe from landslides, floods, or other natural hazards. Some areas are suitable building sites, but other areas are underlain by unstable geologic materials that could cause damage to any structure built there. Geology also controls the distribution of energy resources and the materials required to build houses, cars, and factories. Finally, geologic processes shape the surface of the planet and produce a wonderful diversity of landscapes, including beautiful scenery.

A Where Is It Safe to Live?

The landscape around us contains many clues about whether a place is relatively safe or whether it is a natural disaster waiting to happen. What important clues should guide our choice of a safe place to live?

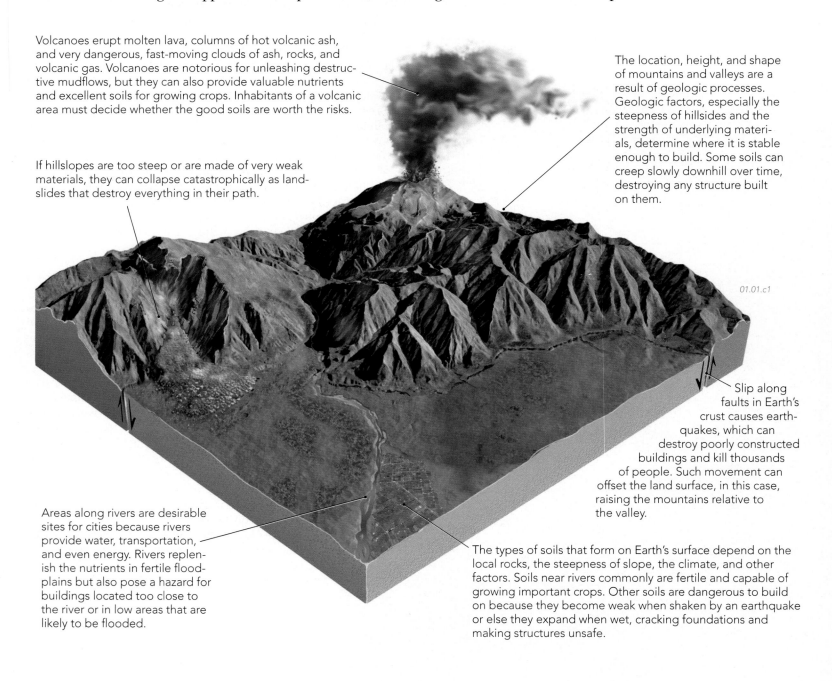

Volcanoes erupt molten lava, columns of hot volcanic ash, and very dangerous, fast-moving clouds of ash, rocks, and volcanic gas. Volcanoes are notorious for unleashing destructive mudflows, but they can also provide valuable nutrients and excellent soils for growing crops. Inhabitants of a volcanic area must decide whether the good soils are worth the risks.

If hillslopes are too steep or are made of very weak materials, they can collapse catastrophically as landslides that destroy everything in their path.

The location, height, and shape of mountains and valleys are a result of geologic processes. Geologic factors, especially the steepness of hillsides and the strength of underlying materials, determine where it is stable enough to build. Some soils can creep slowly downhill over time, destroying any structure built on them.

01.01.c1

Slip along faults in Earth's crust causes earthquakes, which can destroy poorly constructed buildings and kill thousands of people. Such movement can offset the land surface, in this case, raising the mountains relative to the valley.

Areas along rivers are desirable sites for cities because rivers provide water, transportation, and even energy. Rivers replenish the nutrients in fertile floodplains but also pose a hazard for buildings located too close to the river or in low areas that are likely to be flooded.

The types of soils that form on Earth's surface depend on the local rocks, the steepness of slope, the climate, and other factors. Soils near rivers commonly are fertile and capable of growing important crops. Other soils are dangerous to build on because they become weak when shaken by an earthquake or else they expand when wet, cracking foundations and making structures unsafe.

B | How Does Geology Influence Our Lives?

To explore how geology affects our lives, observe this photograph, which shows a number of different features, including clouds, snowy mountains, slopes, and a grassy field with horses and cows (the small, dark spots). For each feature you recognize, think about what is there and what processes might be occurring. Then, think about how geology influences the life of the animals and how it would influence your life if this was your home.

In the distance are snow-covered mountains partially covered with clouds. Snow and clouds both indicate the presence of water, an essential ingredient for life. The mountains have a major influence on water in this scene. As the snow melts, water flows downhill toward the lowlands, to the horses and cows.

The horses and cows roam on a flat, grassy pasture and avoid slopes that are steep or barren of vegetation. The steepness of slopes reflects the strength of the rocks and soils, and the flat pasture resulted from loose sand and other materials that were laid down during flooding along a desert stream. Where is the likely source of the water needed to grow grass in the pasture?

01.01.b1, Henry Mtns., UT

C | What Controls the Distribution of Natural Resources?

This map of North America shows the locations of large currently or recently active copper mines (orange dots) and iron mines (blue dots). What do you notice about the distribution of each type of mine?

Large *copper* mines are restricted to the mountainous western part of the continent (west of the purple line). *Magma* (molten rock) invaded this part of the continent between 160 and 35 million years ago and formed the copper deposits. As described later in this book, these magmas formed only along the western side of the continent, so the copper deposits are here, too.

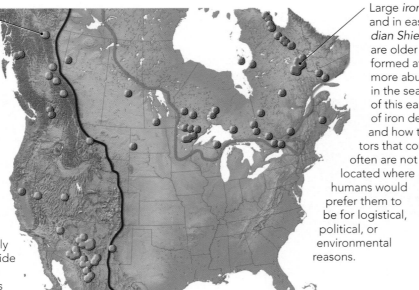

01.01.c1

Large *iron* mines are common in the Great Lakes region and in eastern Canada, within an area called the *Canadian Shield* (inside the red line). Most rocks in this region are older than one billion years, and the iron-rich rocks formed at a time in Earth's history when oxygen became more abundant in the atmosphere, causing iron dissolved in the seas to precipitate into vast iron-rich layers. Rocks of this early age are less common out west, so this type of iron deposit is less common, too. The age of rocks and how the rocks formed are two of many geologic factors that control where mineral resources occur. Resources often are not located where humans would prefer them to be for logistical, political, or environmental reasons.

> ### Before You Leave This Page Be Able To
>
> ✔ Sketch or list some ways that geology controls where it is safe to live.
>
> ✔ Explain how geology influences the distribution of natural resources.

1.1

1.2 How Does Geology Help Explain Our World?

THE WORLD HAS INTERESTING FEATURES at all scales. Views from space show oceans, continents, and mountain belts. Traveling through the countryside, we notice smaller things—a beautiful rock formation or soft, green hills. Upon closer inspection, the rocks may include fossils that provide evidence of ancient life and past climates. Here, we give examples of how geology explains big and small features of our world.

A Why Do Continents Have Different Regions?

Examine this shaded-relief map of part of the northwestern United States (▼). The map shows landscape features around Yellowstone National Park in northwestern Wyoming. Note the different features in different regions.

As you make observations, think about what might control whether a region is mountainous, hilly, or flat, and why mountains are in some areas but not others.

Some regions on this map, such as the *Rocky Mountains*, are high in elevation and have rough landscapes with rugged mountains, steep valleys, and canyons.

Within this region are broad plains, including the *Snake River Plain*, which cuts across the landscape of southern Idaho.

The region south of the Snake River Plain contains alternating mountains and valleys.

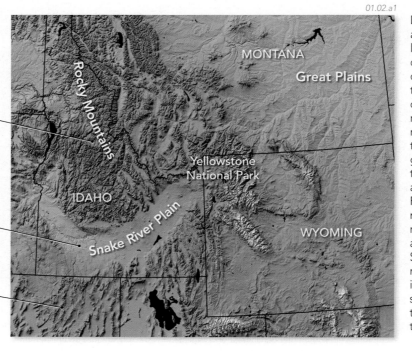

01.02.a1

Different regions exist because each has a different geologic history. Certain geologic events affected some areas but not others. The mountains in the western half of this area formed as a result of activity that occurred along the western edge of the continent, but the effects did not reach inland to the interior of the continent. As a result, the rough land surface to the west reflects its more complex geologic history, but the *Great Plains* to the east has a less complicated appearance and geologic history. The Snake River Plain is relatively smooth because massive outpourings of lava (molten rock) buried older landscapes across this area. Movement on faults south of the Snake River Plain formed the alternating mountains and valleys. Each region has its own geologic history, and the landscape and rocks contain clues as to the types and sequence of geologic events that affected each place.

B What Stories Do Landscapes Tell?

Observe this photograph taken along the edge of a cliff and think of at least two questions about what you see.

The land below the cliff has sunlit *plateaus* (flat-topped areas) and shadowy canyons.

Rocks in the cliff are reddish brown. These include a large, angular block of rock that is perched on the edge of the cliff.

01.02.b1 Muley Point, UT

Several questions about the landscape come to mind. What are the reddish-brown rocks? Why did a cliff form here? How long will it take for the block of rock to tumble off the cliff?

The answer to each question helps explain part of the scene. The first question is about the *present*, the second is about the *past*, and the third is about the *future*. The easiest questions to answer are usually about the present, and the hardest ones are about the past or the future.

The reddish-brown rocks consist of sand grains packed and stuck together to form a rock called *sandstone*. The red color is due to iron-bearing minerals that oxidized and reddened over time, like a rusty piece of metal.

This cliff, like most others, formed because its rocks (the sandstone) are harder to wear away than rocks below.

It is difficult to say when the block will fall, but in geologic terms it will be soon.

C How Has the Global Climate Changed Since the Ice Ages?

These computer-generated images show where glaciers and large ice sheets were during the last ice age and where they are today. Note how the extent of these features changed in this relatively short period of time. What caused this change, and what might happen in the future because of global warming or cooling?

28,000 Years Ago

Twenty-eight thousand years ago, Earth's climate was slightly cooler than it is today. Cool climates permitted continental ice sheets to extend across most of Canada and into the upper Midwest of the United States. Ice sheets also covered parts of northern Asia and Europe.

01.02.c1

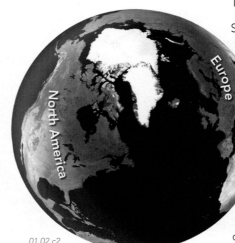

01.02.c2

Today

Since 20,000 years ago, Earth's climate warmed enough to melt back the ice sheets to where they are today. Our knowledge of the past extent of ice sheets comes from geologists who examine the landscape for appropriate clues, including glacial features and deposits that remained after the glaciers retreated.

D What Is the Evidence That Life in the Past Was Different from Life Today?

Museums and action movies contain scenes, like the one below, of dinosaurs lumbering or scampering across a land covered by exotic plants. Where does the evidence for these strange creatures come from?

▶ **1.** This mural, painted by artist Karen Carr, is two stories tall and shows what types of life are interpreted to have been on Earth during the Jurassic Period, approximately 160 million years ago. Dinosaurs roamed the landscape, while the ancestors of birds began to take flight. Flowering plants were not yet abundant and grasses had not yet appeared, so non-flowering trees, bushes, and ground cover dominated the landscape.

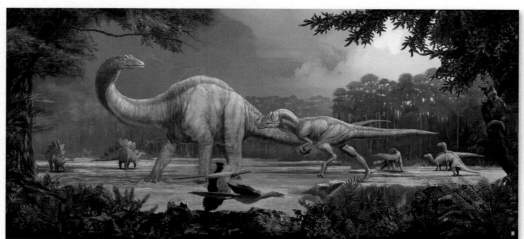

01.02.d1

01.02.d2 Dinosaur NP, UT

◀ **2.** Fossil bones of Jurassic dinosaurs are common in Dinosaur National Park, Utah. From such bones and other information, geologists infer how long ago these creatures roamed the planet, what the creatures looked like, how big they were, how they lived, and why they died. Studying the rocks that enclose the bones provides clues to the local and global environments at the time of the dinosaurs. Rocks and fossils are the record of past geologic events, environments, and prehistoric creatures.

Before You Leave This Page Be Able To

✓ Explain why different regions have different landscape features.

✓ Describe some things we can learn about Earth's past by observing its landscapes, rocks, and fossils.

1.2

1.3 What Is Inside Earth?

HAVE YOU EVER WONDERED WHAT IS INSIDE EARTH? You can directly observe the uppermost parts of Earth, but what else is down there? Earth consists of concentric layers that have different compositions. The outermost layer is the *crust*, which includes *continental crust* and *oceanic crust*. Beneath the crust is the *mantle*, Earth's most voluminous layer. The molten *outer core* and the solid *inner core* are at Earth's center.

A How Does Earth Change with Depth?

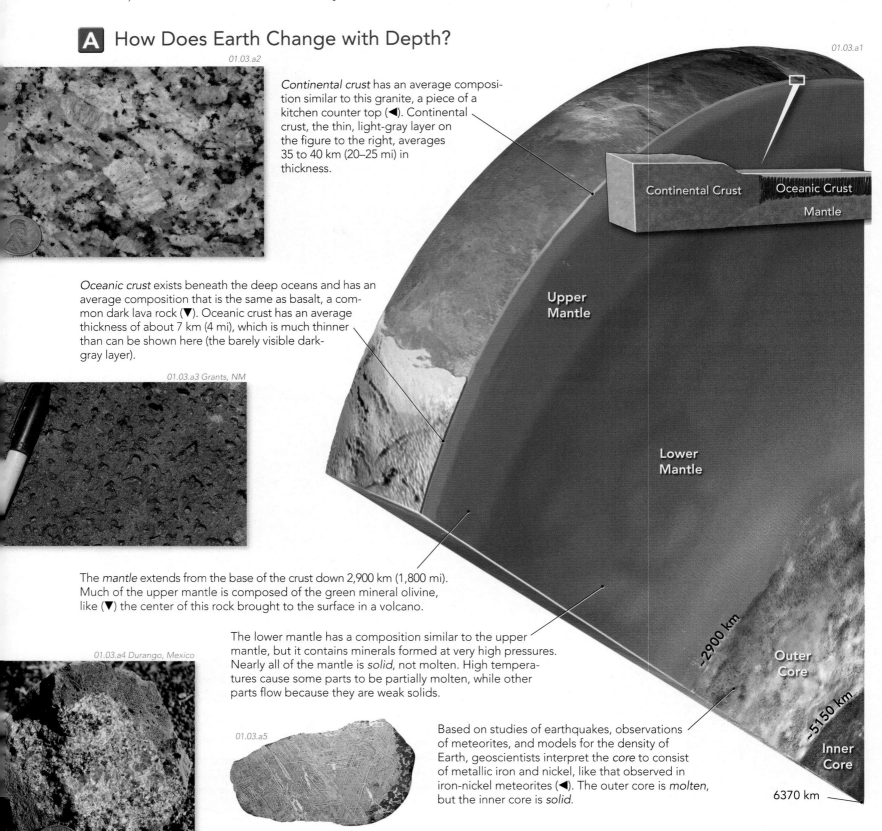

01.03.a2

Continental crust has an average composition similar to this granite, a piece of a kitchen counter top (◄). Continental crust, the thin, light-gray layer on the figure to the right, averages 35 to 40 km (20–25 mi) in thickness.

Oceanic crust exists beneath the deep oceans and has an average composition that is the same as basalt, a common dark lava rock (▼). Oceanic crust has an average thickness of about 7 km (4 mi), which is much thinner than can be shown here (the barely visible dark-gray layer).

01.03.a3 Grants, NM

The *mantle* extends from the base of the crust down 2,900 km (1,800 mi). Much of the upper mantle is composed of the green mineral olivine, like (▼) the center of this rock brought to the surface in a volcano.

01.03.a4 Durango, Mexico

The lower mantle has a composition similar to the upper mantle, but it contains minerals formed at very high pressures. Nearly all of the mantle is *solid*, not molten. High temperatures cause some parts to be partially molten, while other parts flow because they are weak solids.

01.03.a5

Based on studies of earthquakes, observations of meteorites, and models for the density of Earth, geoscientists interpret the *core* to consist of metallic iron and nickel, like that observed in iron-nickel meteorites (◄). The outer core is *molten*, but the inner core is *solid*.

01.03.a1

Continental Crust   Oceanic Crust   Mantle

Upper Mantle

Lower Mantle

~2900 km

~5150 km

Outer Core

Inner Core

6370 km

B Are Some Layers Stronger Than Others?

In addition to layers with different compositions, Earth has layers that are defined by strength and how easily the material in the layers fractures or flows when subjected to forces.

The uppermost part of the mantle is relatively strong and solidly attached to the overlying crust. The crust and uppermost mantle together form an upper, rigid layer called the *lithosphere* (*lithos* means "stone" in Greek). The part of the uppermost mantle that is in the lithosphere is the *lithospheric mantle*.

The mantle directly beneath the lithosphere is mostly solid, but it is hotter than the rock above and can flow under pressure. This part of the mantle, called the *asthenosphere*, functions as a soft, weak zone over which the lithosphere may move. The word *asthenosphere* is from a Greek term for "*not strong.*" The asthenosphere is approximately 80 to 150 km thick, so it can be as deep as about 250 km.

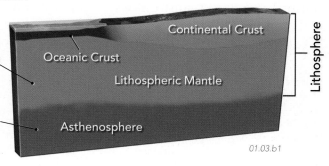

01.03.b1

C Why Do Some Regions Have High Elevations?

Why is the Gulf Coast of Texas near sea level, while the Colorado mountains are 3 to 5 km (2 to 3 mi) above sea level? Why are the continents mostly above sea level, but the ocean floor is below sea level? The primary factor controlling the elevation of a region is the thickness of the underlying crust.

The granitic crust is less dense than the underlying mantle, and so rests, or floats, on top of the mantle. The underlying lithospheric mantle is mostly solid, not liquid.

The thickness of continental crust ranges from less than 25 km (16 mi) to more than 60 km (37 mi). Regions that have high elevation generally have thick crust. The crust beneath the Rocky Mountains of Colorado is commonly more than 45 km (28 mi) thick.

The crust beneath low-elevation regions like Texas is thinner. If the crust is thinner than 30 to 35 km (18 to 20 mi), the area will probably be below sea level, but it can still be part of the continent.

Most islands are volcanic mountains built on oceanic crust, but some are small pieces of continental crust.

Oceanic crust is thinner than continental crust and consists of denser rock than continental crust. As a result, regions underlain only by oceanic crust are well below sea level.

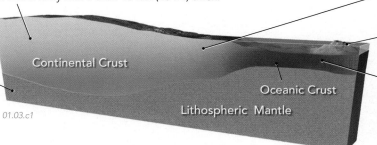

01.03.c1

Density and Isostasy

The relationship between regional elevation and crustal thickness is similar to that of wooden blocks of different thicknesses floating in water (▼). Wood floats on water because it is less dense than water. Ice floats on water because it is less dense than water, although ice and water have the same composition. Thicker blocks of wood, like thicker parts of the crust, rise to higher elevations than do thinner blocks of wood.

For Earth, we envision the crust being supported by mantle that is solid, unlike the liquid used in the wooden-block example. This concept of different thicknesses of crust riding on the mantle is called *isostasy*. Isostasy explains most of the variations in eleva-

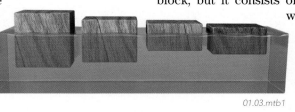

01.03.mtb1

tion from one region to another, and it is commonly paraphrased by saying *mountain belts have thick crustal roots*. As in the case of the floating wooden blocks, most of the change in crustal thickness occurs at depth and less occurs near the surface. Smaller, individual mountains do not necessarily have thick crustal roots. They can be supported by the strength of the crust, like a small lump of clay riding on one of the wooden blocks.

The *density* of the rocks also influences regional elevations. The fourth block shown here has the same thickness as the third block, but it consists of a denser type of wood. It therefore floats lower in the water. Likewise, a region of Earth underlain by especially dense crust or mantle is lower in elevation than a region with less dense crust or mantle, even if the two regions have

similar thicknesses of crust. Temperature also controls the *thickness* of the *lithosphere*, and this affects a region's elevation. If the lithosphere in some region is heated, it expands, becoming less dense, and so the region rises in elevation. Thinner lithosphere also yields higher elevations.

Before You Leave This Page Be Able To

☑ Sketch the major layers of Earth.

☑ Sketch and describe differences in thickness and composition between continental crust and oceanic crust, and contrast lithosphere and asthenosphere.

☑ Sketch and discuss how the principle of isostasy can explain differences in regional elevation.

1.3

1.4 What Processes Affect Our Planet?

EARTH IS SUBJECT TO VARIOUS FORCES. Some forces arise within Earth, and others come from the Sun and Moon. The interactions between these forces and Earth's land, water, air, and inhabitants control most natural processes and influence our lives in many ways.

A How Do Forces and Processes Affect Earth?

1. Earth's *gravity* causes air in the atmosphere to press down on Earth's surface and on its inhabitants. The weight of this air causes *atmospheric pressure*, which generally is greater at sea level than in high elevations — there is less air on top of high elevations than at sea level.

2. *Water*, in either liquid or frozen forms, moves downhill in rivers and glaciers, transporting rocks and other debris and carving downward into the landscape. The downward movement of ice and water is driven by the force of Earth's gravity.

3. The Sun and Moon exert a *gravitational pull* on Earth. Although the Sun is much larger, it exerts less force on the Earth because it is so far away compared to the Moon.

4. *Electromagnetic energy*, including visible light, infrared, ultraviolet, and other forms of energy, radiate from the Sun to Earth. The Sun provides in excess of 99% of Earth's surface-energy budget and so drives surface temperatures, wind, and other processes.

Air Pressure

Earth-Moon Gravity

Sun-Earth Gravity

Electro-magnetic Energy

Wind

Ocean Current

Gravity

Heat

Heat

Magma

Radioactive Decay

Forces

5. Uneven solar heating causes variations in water and air temperatures across the surface of Earth, causing *wind* and *ocean currents*. Blowing wind picks up and moves sand and dust across Earth's surface and makes waves on the surface of oceans and lakes. Rotation of Earth around its axis helps guide the direction of wind and ocean currents as they distribute thermal energy from one part of Earth to another.

6. The mass of Earth causes a downward pull of *gravity*, which attracts objects toward the center of Earth. Earth's gravity is the force that makes water, ice, and rocks move downhill.

9. Temperature increases downward into Earth's interior. *Heat* from deeper in Earth rises upward toward the cooler surface. Some heating is by direct contact between a hotter rock and a cooler rock, whereas other transfer of heat occurs via a moving material, especially rising molten rock (magma).

8. *Radioactive decay* of naturally occurring uranium, potassium, and certain other elements produces heat, especially in the crust where these radioactive elements are concentrated.

01.04.a1

7. Earth's gravity causes the weight of rocks to exert a downward force on underlying rocks. These rocks in turn push against adjacent rocks, causing squeezing of rocks from all directions. This force increases deeper into the interior because more rocks lie above. In many parts of Earth, forces compress the rocks equally from all directions, but additional forces arise by processes deep within Earth, such as from the subsurface movement of rocks and magma. Forces generated in one area can be transferred to an adjacent area, causing sideways pushing or pulling on the rocks.

B How Do Earth's Surface and Atmosphere Interact with Solar Energy?

Critical interactions occur between solar energy and Earth's atmosphere, oceans, and land. These interactions express themselves in wind, clouds, rain, and snow. Our atmosphere shields Earth from cosmic radiation, transfers water from one place to another, and permits life to exist. Like the oceans, the atmosphere is constantly moving, producing winds and storms that impact Earth's surface.

1. The atmosphere includes a low percentage of water vapor, most of which *evaporated* from Earth's oceans. Under certain conditions, the water vapor condenses to produce clouds, which are made of tiny water droplets or ice crystals. Rain, snow, and hail may fall from clouds back to the surface as *precipitation*.

2. The Sun produces vast amounts of energy, including *ultraviolet radiation* and visible light. In the upper levels of the atmosphere, oxygen absorbs most of the Sun's harmful *ultraviolet radiation* and prevents it from reaching Earth's surface, where it would have a detrimental effect on many forms of life. Most of the Sun's energy, including light and other forms of radiation, passes through the atmosphere, eventually reaching Earth, warming the planet and providing light for plants and animals.

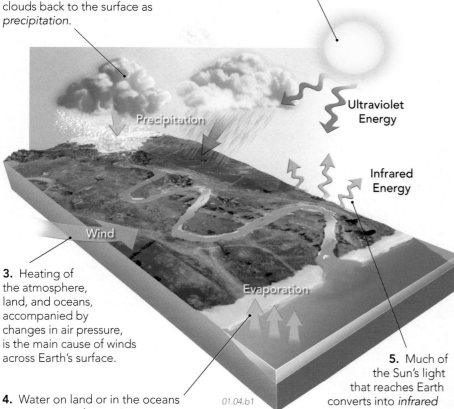

01.04.b1

3. Heating of the atmosphere, land, and oceans, accompanied by changes in air pressure, is the main cause of winds across Earth's surface.

4. Water on land or in the oceans can *evaporate*, becoming water vapor in the atmosphere. Most *water vapor* comes from evaporation in the oceans, but some also comes from evaporation of lakes, rivers, irrigated fields, and other sites of surface water. Plants take moisture from soils, surface waters, or air, and release water vapor into the atmosphere through the process of *evapotranspiration*.

5. Much of the Sun's light that reaches Earth converts into *infrared energy*, a form of energy related to heat. Some of this energy radiates upward and is trapped by the atmosphere, which warms in a process called the *greenhouse effect*. This process regulates global temperatures, which are moderate enough to allow water to exist as liquid water, gaseous water vapor, and solid ice. Water is a key requirement for life.

Energy and Forces

Earth's energy supply originates from internal and external sources. *Internal energy* comes from within Earth and includes heat energy trapped when the planet formed and heat produced by radioactive decay. This heat drives many internally generated processes, including the melting of rocks at depth to produce magma.

The most significant source of *external energy* is the Sun, which bathes Earth in light, thermal energy, and other electromagnetic energy. Thermal energy and light from the Sun are more intense in equatorial areas of Earth than in polar areas, causing temperature differences in the atmosphere and oceans. Temperature differences help drive wind and ocean currents. Sunlight is also the primary energy source for plants, through the process of *photosynthesis*.

Early in Earth's history, meteoroids and other objects left over from the formation of the solar system bombarded the planet. During the impacts, *kinetic energy* (energy due to movement of an object) changed into thermal energy, adding a tremendous amount of heat, some of which remains stored in our planet's hot interior.

Internal forces also affect Earth. All objects that have mass exert a gravitational attraction on other masses. If a mass is large and close, the pull of gravity is relatively strong. Earth's gravity acts to pull objects toward the center of Earth. Gravity is probably the most important agent on Earth for moving material from one place to another. It causes loose rocks, flowing glaciers, and running water to move downhill from higher elevations to lower ones, and drives ocean currents and wind. Moving water, ice, air, and rocks can etch down into Earth's surface, shaping landscapes.

Objects on Earth also feel an *external* pull of gravity from the Sun and Moon. Gravity between the Sun and Earth maintains our planet's orbit around the Sun. The Moon's pull of gravity on Earth is stronger than that of the Sun and causes more observable effects, especially the rise and fall of ocean tides.

Before You Leave This Page Be Able To

✓ Describe the different kinds of energy that impact Earth from the outside, and what effects they have on our planet.

✓ List the different kinds of energy that arise within Earth's interior and explain their origins.

✓ Sketch and explain how Earth's surface and atmosphere interact with solar energy.

1.4

1.5 How Do Rocks Form?

THE VARIOUS PROCESSES THAT OPERATE on and within Earth produce the variety of rocks we observe. Many common rocks form in river bottoms, beaches, or other familiar environments on Earth's surface. Other rocks form in less familiar environments, under high pressure deep within Earth, or at high temperatures beneath a volcano. To understand the different kinds of rock that can form, we explore the types of materials that characterize different modern-day environments.

A What Types of Sediments Form in Familiar Surface Environments?

Much of the surface of Earth is dominated by mountains, rivers, and lakes. Think back to what you have observed on the ground in these types of places—probably mud, sand, and larger rocks. These loose materials are *sediment* and are formed by the breaking and wearing away of other rocks in the landscape. Although more hidden from us, sediment also occurs beneath the sea.

01.05.a2 Switzerland

◄ *Glaciers* incorporate rock debris into their flowing, icy masses. They carry a wide variety of sediment, from large, angular boulders to fine rock powder. They ultimately deposit the sediment along the edges of the melting ice.

Steep mountain fronts exhibit large, angular rocks that broke away from bedrock and moved downhill under the influence of gravity. Steep mountains may produce landslides and unstable, rocky slopes covered with angular blocks (►).

01.05.a3 San Juan Mtns., CO

Sand dunes (►) are mostly sand, which has been shifted along the ground by the wind. They contain sand because wind cannot pick up larger particles, but blows away smaller ones.

01.05.a4 Mojave Desert, CA

Beaches (▼) typically have waves, sand, broken shells, and rounded, well-worn stones. Some beaches are mostly sand, and others are mostly stones.

01.05.a1

River channels contain sand, pebbles, and cobbles, whereas low areas beside the channel accumulate silt and clay. Some rivers flow into *lakes*, which have a muddy bottom with sand around the lake shore.

In deeper water, the *seafloor* consists of mud and the remains of swimming and floating creatures that settle to the bottom. Seafloor closer to the land receives a greater contribution of sand and other sediment derived from the land. Rivers and wind are especially effective in delivering this sediment from the land to the sea.

01.05.a5 Naxos, Greece

B   What Types of Rocks Form in Hot or Deep Environments?

Some rocks form in environments that are foreign to us and hidden from view, deep within Earth. Others form at very high temperatures associated with volcanic eruptions. Distinct families of rocks result from these rock-forming processes, which include solidification of magma, precipitation of minerals from hot water, or high temperatures and pressures that transform one type of rock into another type of rock.

01.05.b2 Philippines

2. In many volcanoes, magma flows onto the surface, creating *lava* that flows downhill or piles up around a vent (▶).

01.05.b3 Hawaii

▲ **1.** Explosive volcanoes erupt *volcanic ash*, which can fall back to Earth and blanket the terrain or can rapidly and dangerously surge down the flanks of a volcano.

5. Distinctive rocks form when *hot waters* cool and minerals precipitate from them. This may occur beneath the surface or on the surface in hot springs (▼).

3. Magma that does not erupt may cool and solidify in a *magma chamber,* forming granite or other rocks at depth. Heat from the magma chamber may bake adjacent rocks, changing them into different kinds of rocks.

Magma

Heating of Rocks

Force

Force

Force

4. Deep within Earth where temperatures and pressures are high, *forces* can squeeze and deform rocks into new arrangements and into new types of rocks. Under such force, solid rocks slowly flow, shear, and bend. Changing a rock by heat, pressure, or deformation is the process of *metamorphism.*

01.05.b1

01.05.b4 Yellowstone NP, WY

Families of Rocks

Diverse environments shown on these pages produce many different types of rocks that, depending on the classification scheme, are grouped into three or four families. To interpret how rocks form, we observe modern environments and note the dominant types of sediment, lava, or other material. We infer that these same types of materials would have been produced in older, prehistoric versions of that environment. By doing this, we use modern examples to interpret ancient rocks and to understand how they formed. In this way, *the present is the key to the past.*

Sedimentary rocks form on Earth's surface, mostly from loose sediment that is deposited by moving water, air, or ice. If loose sediment is buried, it can become consolidated into hard rock over time. Other types of sedimentary rocks form by precipitation of minerals from water or by coral and other organisms that extract material directly from water.

Rocks formed from cooled and solidified magma are *igneous rocks.* These form when volcanoes erupt ash and lava or when molten rock crystallizes in magma chambers at depth.

Rocks changed by temperatures, pressures, or deformation are *metamorphic rocks.* Metamorphism can change sedimentary or igneous rocks, or even preexisting metamorphic rocks. Finally, rocks that precipitate directly from hot water are *hydrothermal rocks.* Some geologists classify these rocks with metamorphic rocks.

Before You Leave This Page Be Able To

☑ Distinguish the four families of rocks by describing how each type forms.

☑ For each family of rocks, describe two settings where such rocks form and the processes that take place in each setting.

☑ Describe what we mean by "the present is the key to the past" and how it is used to interpret the origin of rocks and sediment.

1.5

1.6 What Can Happen to a Rock?

MANY THINGS CAN HAPPEN TO A ROCK after it forms. It can break apart into sediment or be buried deeply and metamorphosed. If temperatures are high enough, a rock can melt and then solidify to form an igneous rock. Uplift can bring metamorphic and igneous rocks to the surface, where they break down into sediment. Examine the large figure below and think of all the things that can happen to a rock.

1. Weathering

A rock on the surface interacts with sunlight, rain, wind, plants, and animals. As a result, it may be mechanically broken apart or altered by chemical reactions via the process of *weathering*. Weathering creates sediment, which ranges from very fine clay to the large boulders shown here (▼).

01.06.a2 San Juan Mtns., CO

2. Erosion and Transport

Rock pieces loosened or dissolved by weathering can be stripped away by *erosion* and moved away from their source. Glaciers, flowing water, wind, and the force of gravity on hillslopes can *transport* eroded material away.

TRANSPORT

01.06.a1

8. Uplift

At any point during its history, a rock may be *uplifted* back to the surface where it is again exposed to weathering. Uplift commonly occurs in mountains, but it can also occur over broad regions that lack mountains.

UPLIFT

6. Melting

A rock exposed to high temperatures may *melt* to produce a magma. Melting usually occurs at great depth, in the lower crust or the mantle.

01.06.a3 Capetown, South Africa

7. Solidification

As magma cools, either at depth or after being erupted onto the surface, it will solidify and harden, a process called *solidification*. If crystals form during solidification, the process is *crystallization*, as illustrated by large, well-formed crystals (◄) that formed by slow cooling of magma at depth. The granite crystallized at depth and much later was uplifted to the surface.

The Life and Times of a Rock — The Rock Cycle

This process, in which a rock may be moved from one place to another or even converted into a new type of rock, is the *rock cycle*. Scottish physician James Hutton first conceived of the rock cycle in the late 1700s as a way to explain the recycling of older rocks into new sediment. Most rocks do not go through the entire cycle, but instead move through only part of the cycle. Importantly, the different steps in the rock cycle can happen in almost any order. Steps are numbered on this page only to guide your reading and follow *possible* sequences of events for a single rock.

Suppose that uplift brings up a rock and exposes it at Earth's surface. Weathering dissolves and breaks up the rock into smaller pieces that can be eroded and transported at least a short distance before being deposited. Under the right conditions, the rock fragments will be buried beneath other sediment or perhaps beneath volcanic rocks that are erupted onto the surface. Many times, however, sediment is not buried, but only weathered, eroded, transported, and deposited again. As an example of this circumstance, imagine a rounded rock in a river. When the river currents are strong enough, they pick up and carry the rock downstream, perhaps depositing it within or near the channel, where it may remain for years, centuries, or even millions of years. Later, a flood that is larger than the last one may pick up the rock and transport it farther downstream.

If the rock is buried, it has two possible paths. It can be buried to some depth and then be uplifted back to the surface to be weathered, eroded, and transported again. Alternatively, it may be buried so deeply that it is metamorphosed under high temperatures and pressures. Uplift can bring the metamorphic rock to the surface.

If the rock remains at depth and is heated to even higher temperatures, it can melt. The magma that forms may remain at depth or may be erupted onto the surface. In either case, the magma eventually will cool and solidify into an igneous rock. Igneous rocks formed at depth may later be uplifted to the surface or they may remain at depth, where the rocks can be metamorphosed or even remelted.

A key point to remember is that the rock cycle illustrates the possible things that can happen to a rock. Most rocks do not complete the cycle because of the many paths, interruptions, backtracking, and shortcuts a rock can take. The path a rock takes through the cycle depends on the specific geologic events that happen and the order in which they occur. There are many possible variations in the path a rock can take.

3. Deposition

When transportation energy decreases sufficiently, water, wind, and ice *deposit* their sediment. Sediment carried by rivers and streams can reside within or next to the channel, or collect near the river's mouth. The river gravels below are at rest for now but could be picked up and moved by a large flood. Some sediment results from the *precipitation of ions* from water or by the actions of organisms.

01.06.a4 Tibet

4. Burial and Lithification

Once deposited, sediment can be buried and compacted by the weight of overlying material. Chemicals in groundwater can coat sedimentary grains with minerals and deposit natural cements that bind adjacent grains. The process of sediment turning into rock is *lithification*.

5. Deformation and Metamorphism

After a rock forms, strong forces can squeeze the rock and fold its layers, a process called *deformation*. If buried deeply enough, a rock can be heated and deformed to produce a *metamorphic* rock. The rock to the left began as some other type of rock but was strongly deformed, metamorphosed, and finally converted into a metamorphic rock.

01.06.a5 Kettle Falls, WA

UPLIFT

BURIAL

Before You Leave This Page Be Able To

☑ Sketch a simple version of the rock cycle, labeling and explaining, in your own words, the key processes.

☑ Describe why a rock might not experience the entire rock cycle.

1.6

1.7 How Do the Atmosphere, Water, and Life Interact with Earth's Surface?

WHAT SHAPES THE SURFACE OF EARTH? The elevation of Earth's surface is a reflection of crustal thickness, mountain building, and other geologic processes, but various processes give the landscape its detailed shape. Three important factors that affect landscapes are water and its movement on the surface, the atmosphere and its movement around Earth, and the impact of diverse life-forms.

A How Does Water Move on Our Planet?

Water on Earth resides in oceans, glaciers, lakes, rivers, and in soil and rock. Much of this water flows continuously under the influence of gravity.

1. Rainfall strikes Earth's surface, leading to weathering and erosion. As raindrops land on the surface, they cover rocks and soil with liquid water that begins the weathering process. With sufficient precipitation, water will accumulate and then flow downhill as *runoff*, causing *erosion*. Rates of weathering, erosion, and runoff are highly variable, sometimes quite rapid and other times very slow.

2. When winter snows don't melt completely, as is common at higher elevations and at polar latitudes, snow accumulates in *glaciers*, which are huge flowing fields or tongues of compressed snow and ice. Glaciers pick up sediment as they move, in part by breaking away and grinding up materials beneath and beside the glacier. As glaciers move, they carve the land beneath and transport sediment that has been incorporated in the ice. They deposit this sediment when the ice melts.

01.07.a2 Mount Cook, New Zealand

3. As moving water and the sediment it carries encounter obstructions, like solid rocks and loose debris, the water's energy helps break apart the obstruction and picks up and transports the pieces (▶). Flowing water is the most important agent sculpting the land surface.

6. Water in oceans, lakes, wetlands, and other places on land can *evaporate* into water vapor in the atmosphere. The water vapor can return to Earth as precipitation. The flow of water from the land and oceans to the atmosphere and back again is the *hydrologic cycle*.

Precipitation

Rivers — Evaporation

Groundwater Flow

Ocean Currents

01.07.a1

4. Water in lakes, rivers, and runoff can sink into the ground and travel through cracks and other empty spaces in rocks and soils. This subsurface *groundwater* can react chemically with rocks through which it flows. It flows toward lower areas, where it may emerge back on Earth's surface as *springs* (▼).

01.07.a3 Grand Canyon, AZ

5. The uppermost part of the oceans is constantly in motion, primarily due to the force of wind blowing across the surface. Winds blowing across the oceans cause *waves* (▶) that erode and shape shorelines.

01.07.a4 Monterey Bay, CA

B What Is Above, At, and Below Earth's Surface?

Earth consists of four overlapping spheres: the lithosphere, hydrosphere, biosphere, and atmosphere. The litho–sphere is the solid Earth, the hydrosphere represents Earth's water, and the atmosphere is its air. The biosphere includes all the places where there is life, from up in the atmosphere to land and beneath the oceans.

01.07.b1

The *atmosphere* is a mix of mostly nitrogen and oxygen gas that surrounds Earth's surface. It includes the air, clouds, and precipitation, and it gradually diminishes in concentration out to a distance of approximately 100 km, the aproximate edge of outer space. It is 78% nitrogen, 21% oxygen, less than 1% argon, and smaller amounts of carbon dioxide and other gases. It has a variable amount of water vapor, averaging about 1%.

The *biosphere* includes life and all of the places it can exist on, below, and above Earth's surface. In addition to the abundant life on Earth's surface, the biosphere extends about 10 km up into the atmosphere, to the bottom of the deepest oceans, and downward into the cracks and tiny spaces in the subsurface. In addition to visible plants and animals, Earth has a large population of diverse micro-organisms.

The *hydrosphere* includes water in all its expressions, including oceans, lakes, rivers, streams, wetlands, glaciers, groundwater, moisture in soil, and water vapor in clouds. Over 96% of water on Earth is salt water in the oceans, and most freshwater is in glaciers and groundwater, not in lakes and rivers.

The *lithosphere* refers generally to the solid upper part of the Earth and specifically to the combination of the crust and strong part of the uppermost mantle. Water, air, and life extend down into the lithosphere, so the boundary between the solid Earth and other spheres is not distinct, and the four spheres partially overlap.

The Role of the Biosphere

How does life (the *biosphere*) interact with the surface of Earth? How does the landscape affect life, and how much does life affect landscapes?

Today, life on Earth is the main source of oxygen and carbon dioxide in the atmosphere. Why does this matter? We need oxygen to survive, and both gases play a critical role in decomposing and recycling material on Earth's surface. Oxygen combines readily with many Earth materials, causing them to decompose so they can be broken apart and carried away by water or wind. Carbon dioxide, when combined with water, makes a weak acid that can attack and weaken rocks. Vegetation changes the way water moves across Earth's surface, slowing it down and allowing it to remain in contact with rocks and soils longer. This increases the rate at which weathering breaks down earth materials.

Humans have affected the surface of Earth by removing vegetation that would compete with crops, villages, and cities. This is shown in the satellite image below, which depicts the extent of clear-cutting of forests near Mount Rainier, Washington. Green areas are forest, whereas tan areas were logged for paper and wood products.

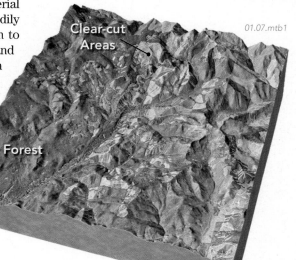

01.07.mtb1

Whenever we clear forests or build cities, we change the balance between Earth and the atmosphere, and we usually increase the power of water to erode into the land.

Before You Leave This Page Be Able To

☑ Draw a sketch that shows the major ways that water moves on, under, and above Earth's surface.

☑ Explain how moving water, ice, and wind can shape the Earth's surface.

☑ Sketch and explain Earth's four spheres, especially what characterizes each sphere and how they interact.

☑ Describe how life, the atmosphere, and landscapes are connected.

1.7

1.8 What Is Earth's Place in the Solar System?

EARTH IS NOT ALONE IN SPACE. It is part of a system of planets and moons associated with the Sun, which together comprise a *solar system*. The Sun is the most important object for Earth because it provides light and heat, without which life would be difficult if not impossible. Earth has a number of neighbors, including the Moon.

A What Are Earth's Nearest Neighbors?

Earth has five nearby neighbors—three other planets, called the inner planets, one moon, and the Sun. The Sun and the Moon have direct effects on Earth. Both exert gravitational pull on our planet, and the Sun is our primary source of energy. The three planets (Mercury, Mars, and Venus), while not affecting us directly, provide glimpses of how Earth might have turned out. Earth and these three other inner planets are rocky and are called *terrestrial planets*.

1. The Sun is the center of our solar system. It is by far the largest object in our solar system, but it is only a medium-sized star compared with other suns in our galaxy. The Sun's gravity is strong enough to keep all the planets orbiting around it. On Earth, a year is defined as the time it takes Earth to complete one orbit around the Sun.

2. The Sun creates light, heat, and other types of energy by fusing together hydrogen atoms in the process of *nuclear fusion*. This process is different than the process of nuclear fission, which causes atoms to break apart and is how Earth generates much of its internal heat. The Sun is the only object in our solar system that generates its own light—all the planets and moons, including our own, are bright because they reflect the Sun's light.

5. Mars is farther from the Sun and is smaller than Earth. Recent exploration of Mars reveals that water once flowed on the planet's surface.

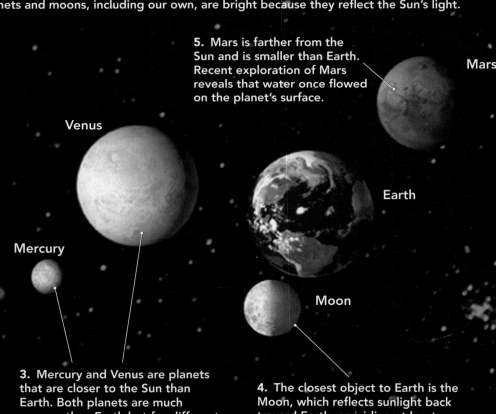

Mars

Venus

SUN

Earth

Mercury

Moon

3. Mercury and Venus are planets that are closer to the Sun than Earth. Both planets are much warmer than Earth but for different reasons. Mercury is close to the Sun and has no atmosphere blocking the Sun's energy. Venus has a thick atmosphere that traps heat like a greenhouse. Venus is shrouded in clouds but is shown here with no clouds.

4. The closest object to Earth is the Moon, which reflects sunlight back toward Earth, providing at least some light on most nights. The Moon's surface is covered with craters produced by meteoroid impacts. Many craters are large enough to be seen from Earth with binoculars. The Moon's gravity (along with that of the Sun) causes tides in the Earth's oceans.

01.08.a1

B What Are Some Characteristics of the Outer Planets?

The five outer planets and many asteroids are farther from the Sun than Mars. The four largest outer planets are called *gas giants* because of their large size and gas-rich character. Pluto is a small, distant object that is no longer considered to be a planet.

1. Hundreds of times larger than Earth, Jupiter is the largest planet in the solar system. Like the Sun, Jupiter is composed mostly of hydrogen and helium. It has a distinctly banded, swirly atmosphere with what appears to be a huge red storm. Jupiter and the other gas giants are much larger than the inner planets.

2. Saturn is a gas giant similar to Jupiter in composition and atmosphere. Saturn has huge, beautiful rings, composed mostly of small chunks of ice and dust (as observed by the Cassini spacecraft).

Jupiter

Uranus

Saturn

3. Uranus and Neptune are smaller gas-giant planets, but these planets are still much larger than Earth. Their atmospheres contain significant methane, which causes their bluish color.

Pluto

Neptune

4. Far from the Sun, Pluto is a small, icy object. An international group of astronomers recently reclassified Pluto as not a planet, devising the name *plutoid* for such objects. This designation leaves our solar system with eight true planets instead of the nine we have traditionally considered. Pluto's size is greatly exaggerated here compared with the rest of the planets.

5. Asteroids are rocky fragments left over from the formation of the solar system. They orbit between Mars and Jupiter and have a composition that is similar to certain meteorites.

C What Is the Shape and Spacing of the Orbits of the Planets?

If we step back from the solar system so we can see the shape of each planet's orbit, this is what the orbits of the inner planets and Jupiter look like. In other words, if you traveled straight up from the Arctic, perpendicular to the orbit of the planets, this is the view you would have.

Observe that all of the planets' orbits, including Earth's orbit, are almost circular. In other words, Earth is at about the same distance from the Sun during all times of the year.

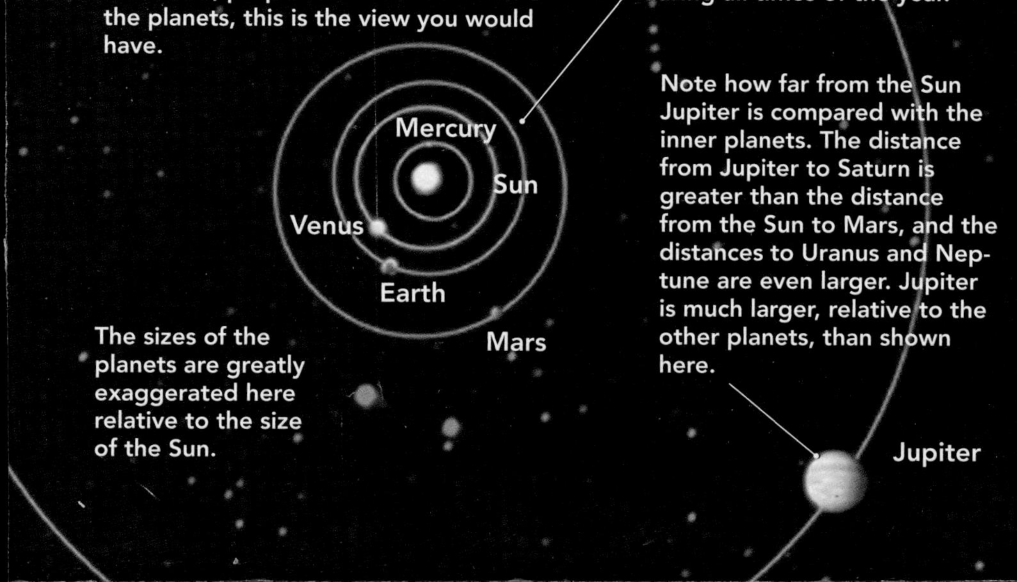

Note how far from the Sun Jupiter is compared with the inner planets. The distance from Jupiter to Saturn is greater than the distance from the Sun to Mars, and the distances to Uranus and Neptune are even larger. Jupiter is much larger, relative to the other planets, than shown here.

Earth's orbit is essentially circular, so Earth receives nearly the same amount of light and heat at all times of the year. Earth's seasons (summer and winter), therefore, are not caused by changes in the distance between Earth and the Sun. The seasons have another explanation, which involves the tilt of Earth's spin axis relative to its orbit, a topic we explore later in this book.

Mercury

Sun

Venus

Earth

Mars

Jupiter

The sizes of the planets are greatly exaggerated here relative to the size of the Sun.

Before You Leave This Page Be Able To

☑ Sketch a view of the solar system, from the Sun outward to Jupiter.

☑ Explain why the Sun and the Moon are the most important objects to Earth.

☑ Summarize how the outer planets are different from the inner planets.

1.9 How Is Geology Expressed in the Black Hills and in Rapid City?

THE BLACK HILLS OF SOUTH DAKOTA AND WYOMING are a geologic wonder. The area is home to three national parks and one national monument. It is famous for its gold and for the presidents' faces carved into granite cliffs at Mount Rushmore. Rapid City, at the foot of the mountains, was devastated by a flash flood in 1972. In this area, the impacts of geology are dramatic and provide an opportunity to examine how geologic concepts presented in this chapter connect together and how they apply to a real place.

A What Is the Setting of the Black Hills?

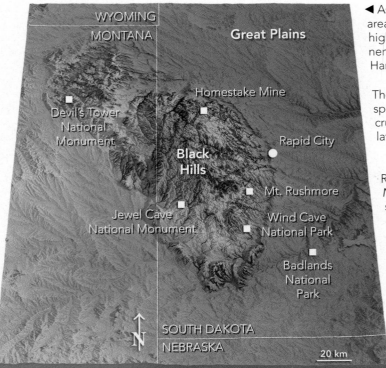

01.09.a1

◀ As seen in this shaded relief map, the *Black Hills* are an isolated mountainous area that rises above the surrounding *Great Plains*. The region has a moderately high elevation, more than 1,000 m (3,000 ft) above sea level, because the continental crust beneath the area is thick (about 45 km, or 28 mi). The highest point, Harney Peak, lies near the center of the Black Hills.

The famous gold deposits of the *Homestake Mine* formed at submarine hot springs nearly 1.8 billion years ago. The rocks were then buried deep within the crust, where they were heated, strongly deformed, and metamorphosed. Much later, uplift of the Black Hills brought the rocks and gold closer to the surface.

Rapid City is on the eastern flank of the Black Hills. To the south, *Badlands National Park*, known for its intricately eroded landscapes, is carved into soft sedimentary rocks. The Black Hills is home to many caves, including those at *Wind Caves National Park* and *Jewel Caves National Park.*

01.09.a3 Mount Rushmore, SD

The presidents' faces at *Mount Rushmore* (▶) were chiseled into a granite that solidified in an underground magma chamber 1.7 billion years ago. The granite and surrounding metamorphic rocks were cooled, uplifted, and overlain by a sequence of sedimentary layers. More recently, they were uplifted to the surface when the Black Hills formed 60 million years ago.

01.09.a2 Devils Tower, WY

Devils Tower (◀) is a well-known landmark that rises out of the Black Hills. The rock formed by solidification of a magma chamber at depth, followed by uplift and erosion to bring the rocks to the surface. The distinctive columns are the result of fracturing as the hot rock cooled.

▼ This figure shows the geometry of rock units beneath the Black Hills. The Black Hills rose when horizontal forces squeezed the area and warped its rock layers. As the mountains were uplifted, erosion stripped off upper layers of rock, exposing an underlying core of ancient igneous and metamorphic rocks (shown in brown). Rapid City is near the boundary between the hard, ancient bedrock in the center of the mountains (shown in brown) and sedimentary rocks of the plains (shown in purples, blues, and greens).

01.09.a4

Sedimentary Rocks

Igneous and Metamorphic Rocks

Sedimentary Rocks

B What Geologic Processes Affect the Rapid City Area?

This view of Rapid City is an aerial photograph superimposed over topography. Where do you think erosion is occurring? Where are sediments deposited? Which places are most susceptible to landslides? Examine this scene and think about the geologic processes that might be occurring in each part of the area.

Rapid City is located along the mountain front, partly in the foothills and partly on the plains. Some parts of the city are on low areas next to Rapid Creek, which begins in the Black Hills and flows eastward through a gap in a ridge and then through the center of the city.

Upturned rock layers form a ridge that divides the city into two halves. Some of the homes are right along the creek, whereas others are on the steep hillslopes.

The plains contain sedimentary rocks, some of which were deposited in a great inland sea and then buried by other rocks. With uplift of the mountains and erosion, the rocks came back to the surface where they are weathered and eroded today.

01.09.b1

This part of the Black Hills consists of hard igneous and metamorphic rocks that form steep mountains and canyons. Farther northwest (not in this view), the world-famous *Homestake Mine* produced 39 million ounces of gold, more than any other mine in the Western Hemisphere. The underground mine is no longer operating, but reached depths of more than 2.5 km (8,000 ft)!

Rapid Creek drains a large area of the Black Hills and flows through the middle of Rapid City. A small dam forms Canyon Lake just above the city.

A flash flood in 1972 destroyed buildings, bridges, and roads along Rapid Creek (▶), leaving the creek littered with shattered houses and other debris, an example of the hazards of living too near flowing water.

01.09.b2 Rapid City, SD

The Rapid City Flash Flood of 1972

In June of 1972, winds pushed moist air westward up the flanks of the Black Hills, forming severe thunderstorms. The huge thunderstorms remained over the mountains, where they dumped as much as 15 inches of rain in one afternoon and evening. This downpour unleashed a *flash flood* down Rapid Creek that was ten times larger than any previously recorded flood on the creek. The swirling floodwaters breached the dam at Canyon Lake, which increased the volume of the flood downstream through Rapid City. The floodwaters raced toward the center of the city. They killed 238 people, destroyed more than 1,300 homes, and caused 160 million dollars in damage. Most of the damage occurred along the creek channel, where many homes had been built too close to the creek, and in areas low enough to be flooded by this large volume of water. Since the flood, the city has removed buildings on many flood-prone sites and developed a wide greenway in what used to be the danger zone.

Before You Leave This Page Be Able To

☑ Briefly sketch the landscape around Rapid City and explain how geology affects this landscape.

☑ Identify and explain ways that geology affects the people of Rapid City.

☑ Describe the events that led to the Rapid City flood and explain why there was so much damage.

How Is Geology Affecting This Place?

GEOLOGY HAS A MAJOR ROLE, from global to local scales, in the well-being of our society. The image below shows an aerial photograph superimposed on topography for an area near St. George, Utah. In this investigation, you will identify some important geologic processes operating in this region and think about how geology affects the people who live here.

Goals of This Exercise:

- Determine where important geologic processes are occurring.
- Interpret how geology is affecting the people who live here.
- Identify a relatively safe place to live that is away from geologic hazards.

Begin by reading the procedures list on the next page. Then examine the figure and read the descriptions flanking the figure.

4. The Virgin River receives water from precipitation in mountains around Zion National Park. It enters the valley through a narrow gorge. Hot springs at the end of the gorge, where the river flows through the cliffs, provide recreation.

1. Most of this region receives only a small amount of rain and is fairly dry. The low areas are part of a desert that has little vegetation and that is hot during the summer. The dry climate, coupled with erosion, provides dramatic exposures of the various rocks. People living here rely on water from wells, reservoirs, and the rivers that flow into the area from distant mountains that receive more rain and snow than this low, dry area.

2. A high, pine-covered mountain range and steep, rocky cliffs flank the valley. The cliffs and mountains receive abundant winter snow and torrential summer rains, which cause flash flooding down canyons that lead into the valley. This photograph (▼) shows the valley, mountains, and cliffs, viewed toward the northwest. The high mountains in the photograph are outside of the area shown in the main figure.

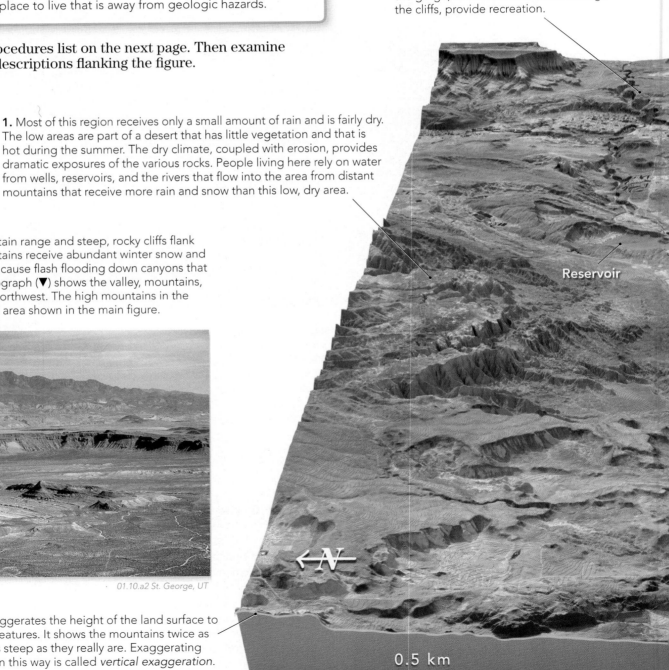

Reservoir

0.5 km

01.10.a2 St. George, UT

3. This figure exaggerates the height of the land surface to better show the features. It shows the mountains twice as high and twice as steep as they really are. Exaggerating the topography in this way is called *vertical exaggeration*.

Procedure:

Use the figure and descriptions to complete the following steps. Record your answers in the worksheet, which will be provided by your instructor in paper form, as a printable file, or as an activity you complete online.

1. Using the image below, explore this landscape. Make observations about the land and the geologic processes implied by the landscape. Next, mark on the provided worksheet at least one location where the following geologic processes would likely occur: weathering, erosion, transport of sediment, deposition, formation of igneous rock, flooding, and landsliding.

2. Using your observations and interpretations, indicate on the worksheet all the ways that geology might influence the lives of the people who live here. Think about each landscape feature and geologic process, and then decide whether it has an important influence on the people. Where would you look for water? Is there a higher potential for a certain type of natural hazard (flooding, earthquakes, etc.) in a particular part of the area?

3. Using all your information, select a location away from geologic hazards that would be a relatively safe place to live compared to more hazardous sites in the area. Mark this location on your worksheet with the word *Here*.

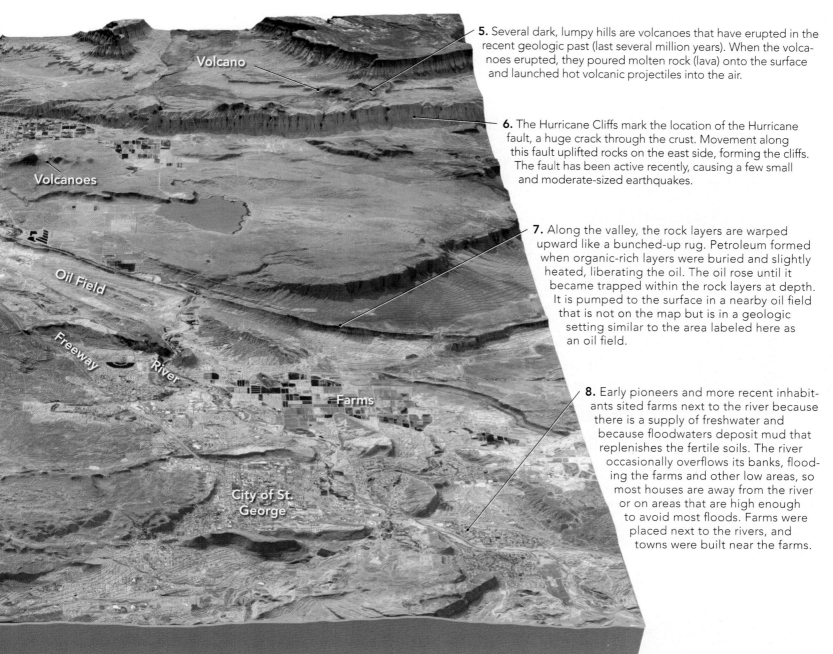

Volcano

Volcanoes

Oil Field

Freeway

River

Farms

City of St. George

5. Several dark, lumpy hills are volcanoes that have erupted in the recent geologic past (last several million years). When the volcanoes erupted, they poured molten rock (lava) onto the surface and launched hot volcanic projectiles into the air.

6. The Hurricane Cliffs mark the location of the Hurricane fault, a huge crack through the crust. Movement along this fault uplifted rocks on the east side, forming the cliffs. The fault has been active recently, causing a few small and moderate-sized earthquakes.

7. Along the valley, the rock layers are warped upward like a bunched-up rug. Petroleum formed when organic-rich layers were buried and slightly heated, liberating the oil. The oil rose until it became trapped within the rock layers at depth. It is pumped to the surface in a nearby oil field that is not on the map but is in a geologic setting similar to the area labeled here as an oil field.

8. Early pioneers and more recent inhabitants sited farms next to the river because there is a supply of freshwater and because floodwaters deposit mud that replenishes the fertile soils. The river occasionally overflows its banks, flooding the farms and other low areas, so most houses are away from the river or on areas that are high enough to avoid most floods. Farms were placed next to the rivers, and towns were built near the farms.

01.10.a1

1.10

Investigating Geologic Questions

OUR WORLD IS FULL OF GEOLOGIC MYSTERIES. Investigating these mysteries requires asking appropriate questions and then knowing what to observe, how to interpret what we observe, and how to analyze the problem from different viewpoints. Also, geologic data and questions require some innovative ways to depict the materials on Earth's surface and in the subsurface. The investigation of geologic questions leads to new ideas and theories about Earth. This chapter explores ways to investigate geologic questions, beginning with a mystery about the Mediterranean Sea.

This image of the Mediterranean region shows the seafloor colored in shades of blue according to depth, with darker blue representing deeper water. On land, satellite data show rock and sand in shades of brown or tan and areas with forests and grasslands in shades of green.

Trace the coast of the Mediterranean Sea. Where does the Mediterranean Sea connect with the Atlantic Ocean or other bodies of water?

Drilling from a research ship encountered thick layers of salt buried beneath the Mediterranean Sea. Such salt layers usually form when large volumes of seawater evaporate, as in hot, dry climates.

How do you get salt deposits at the bottom of a sea?

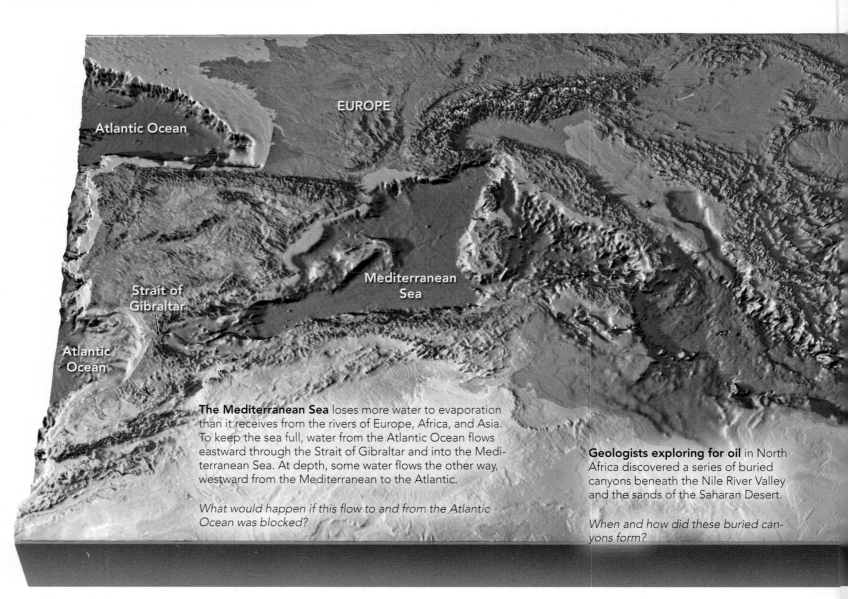

EUROPE

Atlantic Ocean

Mediterranean Sea

Strait of Gibraltar

Atlantic Ocean

The Mediterranean Sea loses more water to evaporation than it receives from the rivers of Europe, Africa, and Asia. To keep the sea full, water from the Atlantic Ocean flows eastward through the Strait of Gibraltar and into the Mediterranean Sea. At depth, some water flows the other way, westward from the Mediterranean to the Atlantic.

What would happen if this flow to and from the Atlantic Ocean was blocked?

Geologists exploring for oil in North Africa discovered a series of buried canyons beneath the Nile River Valley and the sands of the Saharan Desert.

When and how did these buried canyons form?

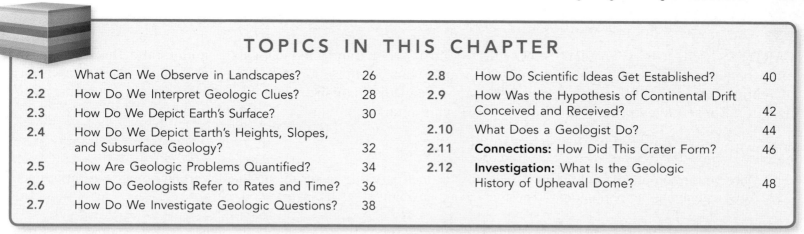

TOPICS IN THIS CHAPTER

About six million years ago, many species of animals around the Mediterranean Sea and Black Sea suddenly became extinct. Marine organisms that lived in the deeper parts of the Mediterranean Sea were most affected. Geologists determined this by collecting and studying fossils.

What dramatic changes in the environment caused these species to perish?

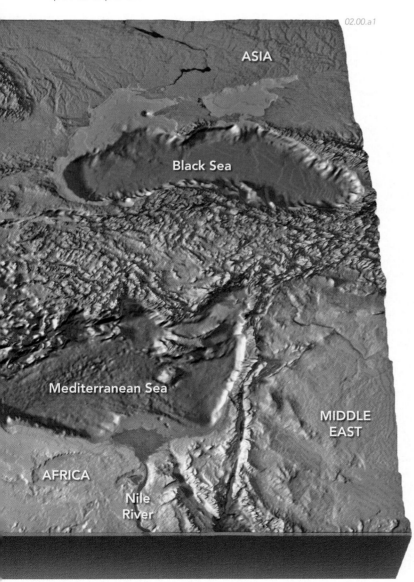

02.00.a1

When the Mediterranean Sea Was a Desert

The Mediterranean Sea is 4,000 km (8,700 mi) long and is surrounded by land, except for its connection to the Atlantic Ocean through the Strait of Gibraltar. The Mediterranean is, on average, 1,500 m (4,900 ft) deep and holds approximately four million cubic kilometers (one million cubic miles) of water.

In the 1960s and early 1970s, geologists made a series of puzzling discoveries in the Mediterranean Sea. Surveys of the seafloor using sound waves revealed some unusual features, similar to those observed for large subsurface layers of salt. To investigate these features, scientists used a research ship, the *Glomar Challenger*, to drill holes in the seafloor and retrieve samples from water depths of several kilometers. This and later drilling documented layers of salt hundreds to thousands of meters thick within sediments on the seafloor. Drilling also revealed sands that showed evidence of having been deposited by wind. From these and other studies arose an amazing idea—in the past the Mediterranean Sea had completely dried up.

According to this idea, the flow of Atlantic water into the Mediterranean Sea was blocked six million years ago by bedrock near the Strait of Gibraltar. The blockage probably occurred because of volcanism or uplift of bedrock by mountain building. As water in the Mediterranean evaporated, it deposited layer upon layer of salt. The great thickness of the salt indicates that seawater spilled into the Mediterranean basin from the Atlantic Ocean many times and then evaporated. After several hundred thousand years, the Mediterranean Sea evaporated completely! It became a dry, hot, salt flat, similar to parts of Death Valley, but 1,500 to 3,000 m (5,000 to 10,000 ft) below sea level. Rivers draining into this new deep basin eroded down through the land, cutting canyons hundreds of meters deep. The drying of the Mediterranean Sea six millions years ago caused profound climate changes in the region, leading to the extinction of many species of animals in and around the Mediterranean and Black Seas.

By 5.3 million years ago, a global rise of sea level caused Atlantic Ocean water to spill over the bedrock and cascade into the Mediterranean Sea. Geologists calculate that it took more than 100 years to refill the Mediterranean basin. As the Mediterranean Sea rose, adjacent rivers deposited sediment that filled and buried the recently cut canyons. The development of this explanation for layers of salt beneath the Mediterranean is a classic example of how science works. Geologists were puzzled by the possible presence of salt and sought an explanation. They gathered additional data and developed an explanation that explained the salt, buried canyons, and extinctions. Geologists continue evaluating the idea today, filling in and refining the story as they complete additional investigations.

2.0

2.1 What Can We Observe in Landscapes?

EARTH'S GEOLOGIC HISTORY IS RECORDED in its rocks and landscapes. To understand this history, we often begin by observing a landscape to determine what is there. Most geologic landscapes display a variety of features, such as different rock layers that we can distinguish by color, texture, and the way the rocks fracture. These pages provide a guide for observing a landscape and reading its story.

A What Features Do Landscapes Display?

Observe the top photograph, trying to identify distinct parts of the scene and then focusing on one part at a time. After examining the photograph, read the accompanying text.

02.01.a1 Canyonlands NP, UT

Color commonly catches our attention. These rocks are various shades of red, tan, and gray. Close examination of these rocks by geologists reveals that the rocks consist of consolidated sand and mud, and therefore are *sedimentary rocks*.

Another thing to notice is that this hill has different parts. A small knob of light-colored rocks sits on the very top.

Below the knob is a reddish and tan slope and a small reddish cliff.

There is a main, light-colored cliff, the upper part of which has a tan color and is fairly smooth and rounded.

Some parts of the cliff have horizontal lines that can be followed around corners of the cliff. These lines are the outward expression of *layers* within the rock. We call such layers in sedimentary rocks *beds* or *bedding*. These beds originally extended across the area prior to more recent erosion.

Lower parts of the cliff have a darker reddish-brown color and display many sharp angles and corners. Some of these corners coincide with vertical cracks, or *fractures*, that extend back into the rock. The red color is a natural stain on the outside of the rocks.

Below the cliff is a slope that has pinkish-red areas locally covered by loose pieces of light-colored rock. A reasonable interpretation is that the loose pieces have fallen off the main cliff.

02.01.a2 Canyonlands NP, UT

In this figure, color overlays accentuate different features and parts of the hill. Compare these features with the photograph above.

The uppermost three rock units (numbered 1, 2, and 3) are shaded tan, yellow, and orange. Rocks of the main cliff (4) are shaded light purple. On the lower slope, the reddish rocks are shaded light orange (5), whereas the covering of loose, light-colored rocks is shaded gray (6).

Brown lines highlight beds in the rock units.

Black lines mark fractures cutting the rocks.

Simplifying this scene into a few types of features makes it easier to observe, describe, and understand the landscape. In this landscape, we observe only a few beds and rock layers but many fractures. Some layers are more resistant to weathering and form cliffs, whereas less resistant ones form slopes. Weathering has rounded off corners on the top of the cliff, removed the reddish stain, and loosened pieces that fell off the cliff, covering a slope of underlying, reddish rocks.

Reexamine the top photograph. Do you look at the scene differently? Try this strategy when observing features close to where you live.

B What Are Some Strategies for Observing Landscapes?

Observe the photograph below and try to recognize the types of features, such as layers and fractures, described on the previous page. After you have made your observations, read the text, which describes aspects to observe and some helpful strategies for looking at any landscape.

1. Most landscapes have a fairly complex appearance when viewed in their entirety, so a useful approach is to focus on one part of the landscape at a time. In the scene below, examine the left side of the image and compare it to the center. What similarities and differences do you observe?

2. Another approach is to let the geology guide your observations from one part to another. In this scene, spend some time looking only at the cliff, and then focus on the reddish slope below the cliff. Next, pay attention to the piles of loose rocks that rest on the reddish slope.

3. Next, try focusing on one *type of geologic feature* at a time. In this photograph, start by concentrating on the fractures in the cliff. Are they steep, and are they evenly spaced? How do they affect the appearance of the cliff? Use the same approach to look at the ledges that cross the reddish slope.

02.01.b1 Monument Valley, AZ

4. Color is one of the first things we notice in any scene. Rocks, sediment, and soils have a range of colors depending on the composition of the materials and the environmental conditions imprinted on those materials. Some colors are integral to the rock, but others are a natural stain on the outside surfaces of the rock. The rocks of Monument Valley are reddish brown to tan inside, but are locally coated by a darker brown stain.

5. Some rock types are more *resistant to erosion* than others and have more dramatic expressions in the landscape. Cliffs and ledges generally represent rock types that are hard to erode, as shown by the cliffs of hard sandstone in this photograph. Slopes or soil-covered areas contain weaker materials, such as the loose wind-blown sand in the foreground of this image.

8. To visualize different components of a landscape, draw a sketch that captures the main features but leaves out less important details. Compare the sketch below with the photograph above. Note how the sketch changes the way you look at the photograph.

7. The *shapes* of eroded rocks depend on the hardness of the rock, thickness of layers, spacing of fractures, and many other factors. Landscapes change over time, so shapes seen today will evolve into different shapes on timescales of years to millions of years.

6. Obvious features in many landscapes are *layers* in the rocks. The cliff represents a thick layer of sandstone, whereas the underlying slopes and ledges are the expression of dozens of layers. In this location, the layers are nearly horizontal, but layers can be tilted or even folded. These differences in orientation have a great impact on the appearance of the resulting landscape. To understand how layers influence the landscape, you first observe the layers and recognize how they are oriented.

02.01.b2

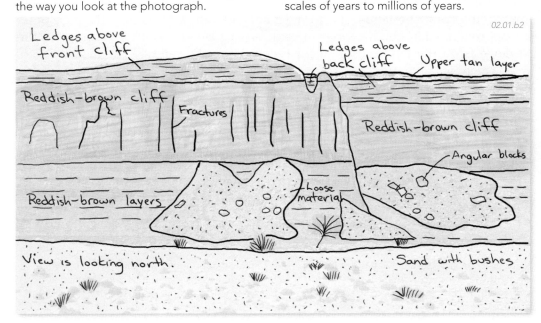

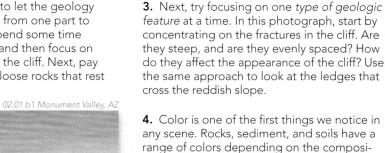

Before You Leave This Page Be Able To

✓ Draw a simple sketch of a landscape photograph, identifying the main components, like those shown on these pages.

✓ Summarize the different aspects or features you can observe in a landscape.

2.1

2.2 How Do We Interpret Geologic Clues?

LANDSCAPES AND ROCKS CONTAIN MANY CLUES about their geologic history. From the characteristics of a rock, we can infer the environment in which the rock formed. We can also apply some simple principles to determine the age of one rock unit or geologic feature relative to another. Changes in landscapes through time provide additional clues about how a place has changed and why it has its present appearance.

A How Can We Infer the Environment in Which a Rock Formed?

To infer how a rock formed, compare its characteristics, such as the size and roundness of pebbles or other material it contains, to those of deposits in modern environments, and decide which environment is the best match. Observe the characteristics of the rock in the large photograph to the left and compare the rock with the photos of sediment deposited in two modern environments. Which of these environments is most similar to the one in which the rock formed?

02.02.a1 Southern AZ

02.02.a3 Salt River, AZ

02.02.a2 Wickenburg, AZ

or

▶ Many river channels contain rounded stones surrounded by a matrix of sand. Note the gray-and-black marking pen for scale.

◀ Steep mountain fronts typically contain angular rocks of many sizes in a matrix of mud, sand, or small rock fragments.

B How Can We Envision the Slow Change of Landscapes Through Time?

Most landscapes evolve so slowly that we rarely notice significant changes in our lifetime. To get around this limitation, geologists use a strategy called *trading location for time*, which uses different parts of a landscape to represent different stages in the evolution of the landscape. In other words, we mentally arrange the different parts into a logical progression of how we interpret the landscape to have changed, or will change, through time. The three models below illustrate this approach; each could represent a different place or a different stage in the evolution of a landscape. The earliest stage is on the left, and the most recent is on the right.

02.02.b1

02.02.b2

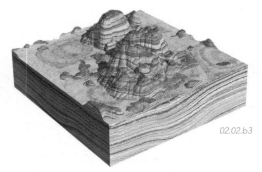

02.02.b3

Erosion cuts into a sequence of rock layers, carving a mountain with steep sides and a broad top. Such a flat-topped mountain is commonly called a *mesa*.

With time, erosion wears away the edges of the mesa, forming a smaller, steep-sided mountain, which geologists commonly call a *butte*.

Erosion continues to wear down the terrain, leaving low, rounded hills and isolated knobs. Given enough time, a mesa or butte might evolve into such hills and knobs.

C How Do We Determine the Sequence of Past Geologic Events?

When exploring the geology of an area, we like to know the sequence of events that formed the rocks and geologic features. We determine the *relative ages* of rocks and geologic features by using commonsense principles, including the four described below.

Principle 1: The Youngest Layer Is on Top, and the Oldest Is on the Bottom

Any rock layer must be younger than any rock unit on which it is deposited. Here, the reddish layer on top is the youngest. The bottom yellowish layer is the oldest rock unit, and overlying layers are successively younger.

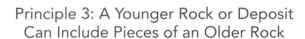

02.02.c1

Principle 2: A Geologic Feature Is Younger Than a Rock Unit or Feature It Crosscuts

02.02.c2

The feature (fault) offsetting the layers must be younger than rock layers it crosses and displaces. The fault does not offset the land surface, so the land surface is younger. There are other types of crosscutting relationships, such as magma that cuts across existing rocks.

Principle 3: A Younger Rock or Deposit Can Include Pieces of an Older Rock

For these gray, pebble-sized pieces to have been incorporated into the tan layer, the pieces had to already exist. The tan layer is younger than both the pieces and the gray bedrock that contributed the pieces.

02.02.c3

Principle 4: A Younger Magma Can Bake or Otherwise Change Older Rocks That Are Nearby

When this dark-gray granite was molten, heat from the magma baked and metamorphosed the adjacent, preexisting rocks, causing a narrow reddish zone next to the granite. Baking can occur underground, as in this case, and beneath molten lava or hot volcanic ash erupted on the surface.

02.02.c4

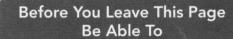
02.02.c5 Death Valley, CA

◀ An exciting part of geology is visiting a new place and figuring out what events happened and in what order. This outcrop shows three rock units. The upper unit contains light-gray, angular pieces. The lowest unit has light-gray layers that are tilted. Between these two units is a thin unit that lacks large clasts and is medium gray. Stop and use the principles described above to interpret the relative ages of the three units. Not all principles may apply.

The upper rock is youngest, because it overlies the other units and contains clasts of the lowest unit. The lower unit is on the bottom and is the oldest.

Before You Leave This Page Be Able To

- ✓ Describe the overall philosophy used to infer the environment in which a rock formed.

- ✓ Describe or sketch what is meant by *trading location for time*.

- ✓ Sketch and summarize four principles used to determine the relative ages of rocks and geologic features.

2.2

2.3 How Do We Depict Earth's Surface?

THE SURFACE OF EARTH displays various features, including mountains, hillslopes, and river valleys. We commonly represent such features on the land surface with *topographic maps* and *shaded-relief maps*. To depict the types of materials on Earth's surface, we use *satellite images* and *geologic maps*. A geologic map is the most important piece of geologic information for an area, because it shows the ages and types of rocks and sediment, as well as geologic features, some of which could pose a hazard.

A How Do Maps and Satellite Images Help Us Study Earth's Surface?

Satellite images and various types of maps are the primary ways we portray the land surface. Some maps depict the shape and elevation of the land surface, whereas others represent the materials on that surface. Views and maps of *SP Crater* in northern Arizona provide a particularly clear example of the relationship between geologic features, the land surface, and different types of maps.

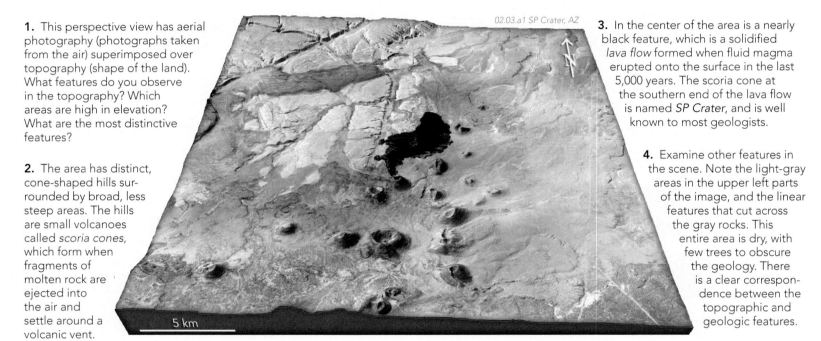

02.03.a1 SP Crater, AZ

1. This perspective view has aerial photography (photographs taken from the air) superimposed over topography (shape of the land). What features do you observe in the topography? Which areas are high in elevation? What are the most distinctive features?

2. The area has distinct, cone-shaped hills surrounded by broad, less steep areas. The hills are small volcanoes called *scoria cones*, which form when fragments of molten rock are ejected into the air and settle around a volcanic vent.

5 km

3. In the center of the area is a nearly black feature, which is a solidified *lava flow* formed when fluid magma erupted onto the surface in the last 5,000 years. The scoria cone at the southern end of the lava flow is named *SP Crater*, and is well known to most geologists.

4. Examine other features in the scene. Note the light-gray areas in the upper left parts of the image, and the linear features that cut across the gray rocks. This entire area is dry, with few trees to obscure the geology. There is a clear correspondence between the topographic and geologic features.

02.03.a2 SP Crater, AZ

◄ This photograph, taken from the air, shows SP Crater and the dark lava flow that erupted from the base of the volcano.

02.03.a3 SP Crater, AZ

▲ This photograph, taken from the large crater south of SP Crater, shows the crater (on the left) and several other scoria cones. The view is toward the north.

Before You Leave These Pages Be Able To

☑ Describe how each of the four types of maps and images depicts Earth's surface.

☑ Describe what contours on a topographic map represent and how contour spacing indicates the steepness of a slope.

☑ Briefly describe what a geologic map shows, using the area around SP Crater as an example.

▼ A *shaded relief map* emphasizes the shape of the land by simulating light and dark shading on the hills and valleys. The individual hills on this map are *scoria cones*. The area is cut by straight and curving stream valleys that appear as gouges in the landscape. Simulated light comes from the upper left corner of the image.
02.03.a4

▼ A *topographic map* shows the elevation of the land surface with a series of lines called *contours*. Each contour line follows a specific elevation on the surface. Standard shaded relief maps and topographic maps depict the shape of the land surface but give no specific information about geology.
02.03.a5

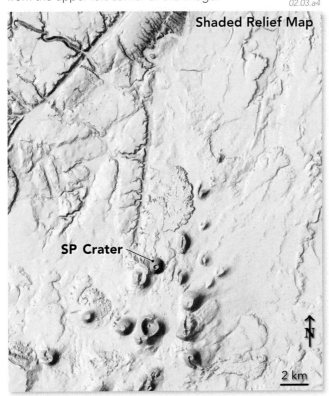

Shaded Relief Map

SP Crater

2 km

Topographic Map

Most topographic maps show every fifth contour with a darker line, to help emphasize the broader patterns and to allow easier following of lines across the map. These dark lines are called *index contours*.

Adjacent contour lines are widely spaced where the land surface is fairly flat (has a gentle slope).

Contour lines are more closely spaced where the land surface is steep, such as on the slopes of the scoria cones. Note how the shapes of the contours reflect the shapes of the different scoria cones.

▼ A *satellite image* commonly uses measurements of different wavelengths of light reflecting from a land surface. This computer-processed image shows the distribution of different types of plants, rocks, and other features. The dark area in the center of the image is the black, solidified lava flow that erupted from the base of SP Crater, and reddish areas are scoria cones.
02.03.a6

▼ A *geologic map* represents the distribution of rock units and geologic features exposed on the surface. This one shows SP Crater and lava flow and older rock units. Compare the colored areas on this geologic map with the different areas visible on the satellite image to the left. Each color on this geologic map represents areas that have a certain type of rock or feature.
02.03.a7

Satellite Image

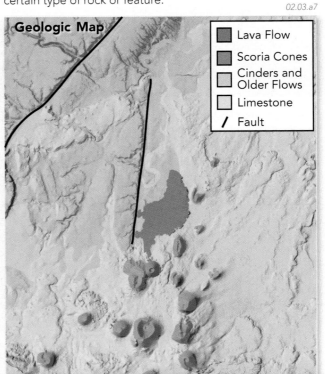

Geologic Map

	Lava Flow
	Scoria Cones
	Cinders and Older Flows
	Limestone
/	Fault

The gray area in the center of the map marks the SP lava flow, and dark pink areas are scoria cones. Light pink represents volcanic cinders and older lava flows. Lavender indicates areas with light-colored rock (limestone) at the surface.

Compare the four maps to match specific features of the area. Which map or image gives you the best information about the shape of the landscape? Which of these gives you the best information about the geology?

2.3

2.4 How Do We Depict Earth's Heights, Slopes, and Subsurface Geology?

DIAGRAMS OF THE LAND SURFACE AND UNDERLYING GEOLOGY are essential tools for visualizing and understanding Earth. We use two-dimensional and three-dimensional diagrams to depict the steepness of slopes, the thickness and subsurface geometry of rock units, and how these units interact with the surface. Some diagrams show interpretations of how present-day landscapes arose via a sequence of geologic events.

A How Do We Refer to Differences in Topography?

Earth's surface is not flat and featureless but instead has high and low parts. Topography is steep in some areas but nearly flat in others. We use common terms to refer to the height of the land and the steepness of slopes.

1. The height of a feature above sea level is its *elevation*. Scientists describe elevation in *meters* or *kilometers* above sea level, but some maps and most signs list elevation in feet.

2. Beneath water, we talk about *depth*, generally expressing it as depth below sea level. We use *meters* for shallow depths and *kilometers* for deep ones.

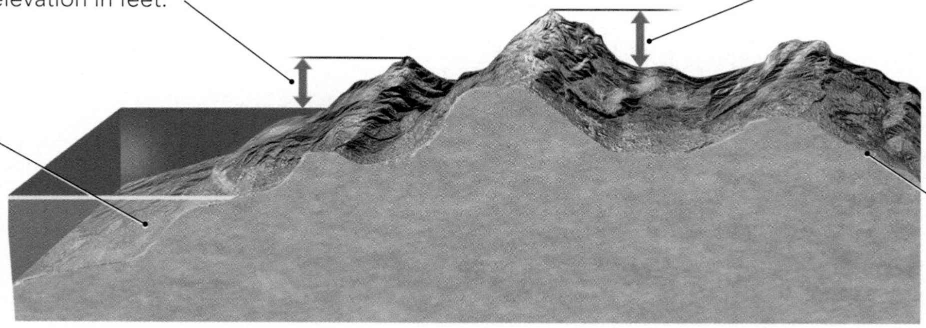

3. We also refer to the height of a feature above an adjacent valley. The difference in elevation of one feature relative to another is *topographic relief*. Like elevation, we measure relief in meters or feet; we refer to rugged areas as having *high relief* and to topographically subdued areas as having *low relief*.

4. Cliffs and slopes that drop sharply in elevation are *steep* slopes, whereas topography that is less steep is referred to as being *gentle*, as in a gentle slope.

02.04.a1

B How Do We Represent Topographic Slopes?

We can depict steepness of the land surface with an imaginary slice through a terrain, like one through SP Crater and its surroundings (▼). This type of portrayal of ups and downs of the land surface is a *topographic profile*.

1. The front of this figure shows the change in elevation across the land surface. It is a type of topo‐ graphic profile.

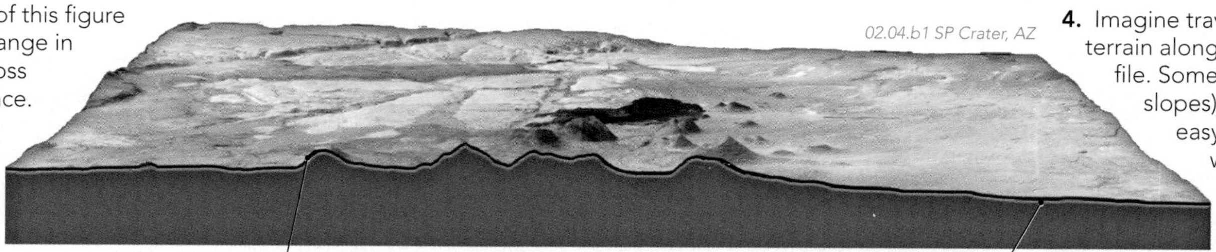

02.04.b1 SP Crater, AZ

4. Imagine traveling across this terrain along the line of the pro‐ file. Some parts (the gentle slopes) are relatively easy to travel across, whereas the steep slopes require more effort.

2. Steeper parts of the profile represent steep slopes on the sides of the small volcanoes. There is moderate relief between the peaks and surrounding plains.

3. Other parts of the profile are less steep, including lower elevation plains surrounding the volcanoes. There is only low relief from one part of the plains to another.

5. Some topographic profiles are simple plots of height of the topog‐ raphy versus distance across the land, like the black line that traces a profile across SP Crater (▶). The profile runs from west (on the left) to east (on the right), so it is an east-west profile. Most topographic profiles have such directions labeled directly on the plot, along with scales for elevation and for horizontal distances.

6. We describe steepness of a slope in *degrees* from hori‐ zontal. The eastern slope of SP Crater has a 26-degree slope (26° slope). We also talk about *gradient* — a 26° slope drops 480 meters over a distance of one kilometer, typically expressed as 0.48, or 480 m/1000 m.

02.04.b2 SP Crater, AZ

C How Do We Represent Geologic Features in the Subsurface?

Most of our planet's geology is beneath Earth's surface, hidden from our view. We are most aware of rock units if they are exposed in *outcrops* on a mountainside, in a deep canyon, or perhaps in a roadcut. However, such units are also present beneath areas of relatively gentle topography. Geologic diagrams help us envision and understand the thicknesses, orientations, and subsurface distributions of rock units. Such diagrams are also important ways in which geologists document and communicate their understanding of an area.

Block Diagram

1. A *block diagram* portrays in three dimensions the shape of the land surface and the subsurface distributions of rock units. It also shows the location and orientation of faults, folds, and other geologic features (if present).

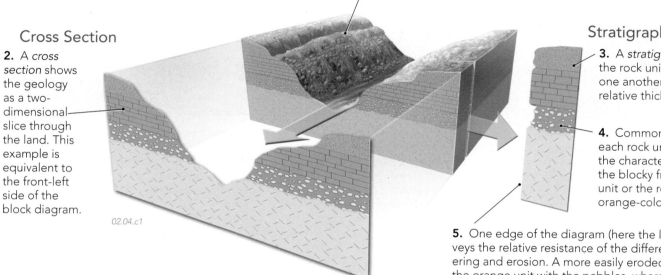

02.04.c1

Cross Section

2. A *cross section* shows the geology as a two-dimensional-slice through the land. This example is equivalent to the front-left side of the block diagram.

Stratigraphic Section

3. A *stratigraphic section* shows the rock units stacked on top of one another (with appropriate relative thicknesses).

4. Commonly, the patterns within each rock unit visually represent the character of the unit, such as the blocky fractures in the gray unit or the rounded pebbles in this orange-colored sedimentary unit.

5. One edge of the diagram (here the left edge) typically conveys the relative resistance of the different rock units to weathering and erosion. A more easily eroded unit is recessed, like the orange unit with the pebbles, whereas more resistant units protrude farther out, like the two gray units.

Evolutionary Diagrams

Evolutionary diagrams (▶) are block diagrams, cross sections, or maps that show the history of an area as a series of steps, proceeding from the earliest stages to the most recent one. Here, an upper tan rock layer is deposited in the sea and later eroded.

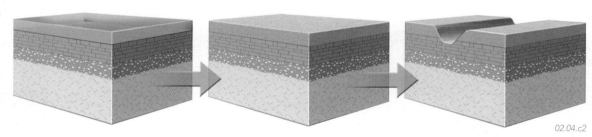

02.04.c2

Earliest Stage:
Arrival of Sea

Intermediate Stage:
Deposition of Layer

Late Stage:
Erosion of Layers

Sketching Geology

A challenge of geology is trying to visualize how geology exposed at the surface continues at depth. Sketches drawn in the field while studying the geology capture one's thoughts while they are still fresh and while ideas can be tested by making additional field observations. The sketch to the right is a simplified geologic cross section drawn to summarize field relationships for some faulted rock layers. A sketch is an excellent way to conceptualize and think about geology, either in the field or from a textbook, because it emphasizes important features.

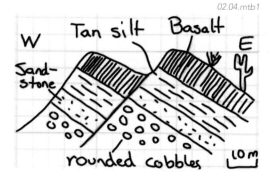

02.04.mtb1

Before You Leave This Page Be Able To

✓ Sketch and describe what we mean by elevation, depth, relief, and slope.

✓ Sketch or describe the types of diagrams geologists use to represent subsurface geology and the sequence of rock units.

✓ Sketch or describe what is shown by a series of evolutionary diagrams.

2.4

2.5 How Are Geologic Problems Quantified?

GEOLOGISTS APPROACH PROBLEMS IN MANY WAYS, asking questions about Earth processes and then collecting data that help answer these questions. Some questions require *quantitative* data, which are numeric and are typically visualized and analyzed using data tables, calculations, equations, and graphs.

A What Is the Difference Between Qualitative and Quantitative Data?

02.05.a1 Augustine Island, AK

02.05.a2 Augustine Island, AK

02.05.a3 Augustine Island, AK

When Augustine volcano in Alaska erupts, geologists make various types of observations and measurements. Some observations are *qualitative*, like simple descriptions, and others are measurements that are *quantitative*. Both types of data are essential for documenting geologic phenomena.

Qualitative data include descriptive words, labels, sketches, or other images. We can describe this picture of Augustine volcano with phrases like "contains large, angular fragments," "releases steam," or "the rocks are mostly gray." Such phrases can convey important information about the site.

Quantitative data involve numbers that represent measurements. Most result from scientific instruments, such as this thermal camera that records temperatures on Augustine volcano, or with simple measuring devices like a compass. Geologists collect quantitative data in the field and in the lab.

B What Quantitative Properties Do We Measure in the Field?

Geologists often describe features qualitatively, but they also collect quantitative data, which consists of numerical values measured with scientific tools or instruments. Some measurements are collected in the field.

02.05.b1 Lake Pleasant, AZ

02.05.b2 Kyrgyzstan

◀ ORIENTATION: Geologists observe and measure the orientation of geologic features, such as layers, fractures, and folds. In this view, a geologist is using a level on a hand-held compass to measure how much the sedimentary layers have been tilted.

▶ SURFACE FEATURES: Most geologists use topographic and other maps to mark locations of data, but some geologists use precise surveying instruments to study geologic phenomena. Such measurements can document the movement of the land surface before or after an earthquake or volcanic eruption.

02.05.b3 Alaska

02.05.b4 Yellowstone NP, WY

◀ GAS COMPOSITION: Volcanoes emit various gases, and the quantity and composition of these gases change over time. Gas measurements provide valuable clues about whether a volcano is preparing to erupt.

▶ WATER FLOW AND CHEMISTRY: We can measure the velocity and volume of flowing water in rivers and groundwater, and chemical analyses, including some performed in the field, document what the water contains.

C   What Quantitative Properties Do We Measure in the Laboratory?

Some data collection requires laboratory environments for preparation of samples and for analysis with sophisticated scientific instruments. Laboratory measurements are the main source of our understanding of the physical properties, chemical composition, and ages of rocks, soils, and other geologic materials.

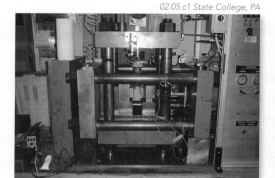
02.05.c1 State College, PA

02.05.c2 Rand Mtns., CA

02.05.c3 Syracuse, NY

PHYSICAL PROPERTIES: Density, strength, and other physical properties of a rock, as measured in the laboratory, form the basis for evaluating how rocks behave when subjected to forces, such as during earthquakes.

COMPOSITION: Chemical analyses and the percentage of different materials in a rock, as measured in the lab, provide information about the composition of the rock and the conditions under which the rock formed.

AGE: Certain rocks can be dated using precise analytical instruments that measure the ratios between different types of radioactive elements. Magnetic measurements also help infer the age of some sequences of rocks.

D   How Do We Calculate Density, and How Does It Differ from Weight?

Density is a very important quantitative property for understanding the interior of Earth. It controls regional elevations and causes forces that result in earthquakes. We determine or estimate density of earth materials by directly measuring a rock in the laboratory, by using instruments to measure the pull of gravity, or by numerically analyzing how fast seismic waves pass through materials between an earthquake and a seismic-recording device.

Density

Density refers to how much mass (substance) is present in a given volume. Below, a wooden block, a "cube" of water, and a stone block all have the same shape and volume but different amounts of mass. The wood is less dense than water and floats, but the stone is more dense and sinks. The cube of water has the same density as the surrounding water and so does not sink to the bottom or float on the surface.

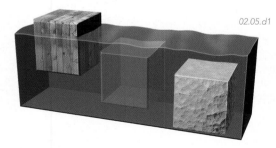
02.05.d1

We calculate density using the following formula:

$$density = mass/volume$$

We measure mass in grams or kilograms, and volume in cubic centimeters, cubic meters, or liters. Density is in units of gm/cm^3 or gm/L. Water has a density of $1 \ gm/cm^3$ at room temperature and pressure, whereas ice, which floats on water, is slightly less dense at $0.92 \ gm/cm^3$. Granite has a higher density of $2.65 \ gm/cm^3$.

Weight

The *weight* of an object is how much downward force it exerts under the pull of gravity. Weight depends on how much mass the object contains and the strength of the gravity field.

02.05.d3

Moon

Earth

02.05.d2

A person has the same mass, whether standing on Earth or on the Moon. If the person weighs 180 pounds on Earth, he or she will weigh only 30 pounds on the Moon (because of the Moon's lower gravity). When addressing scientific issues, scientists rarely talk about weight, instead referring to mass and using metric-system units like gm/cm^3.

Before You Leave This Page Be Able To

✓ Explain how qualitative data differ from quantitative data.

✓ Describe several types of quantitative data that geologists use.

✓ Describe what density is, how it is calculated, and how it differs from weight.

2.5

2.6 How Do Geologists Refer to Rates and Time?

TIME IS ONE OF THE MOST IMPORTANT ASPECTS of geology and is one of the key things that makes geology different from most other sciences. Geologists commonly investigate events that happened thousands, millions, or billions of years ago. Many geologic events and processes occur over long periods of time and at rates that are so slow as to be nearly imperceptible. The magnitude of geologic time is immense, and difficult for even some geologists to comprehend. So, geologic time needs a special language and calendar.

A How Do We Refer to Rates of Geologic Events and Processes?

We calculate the rates of geologic processes in a similar way to how we calculate the speed of a car or a runner. The major difference is that most geologic rates are measured using metric units of millimeters per year or centimeters per year instead of miles per hour or feet per second. Many geologic rates, like uplift of a mountain range, are generally quite slow, but others, like earthquakes, are rapid.

A runner (▶) provides a good reminder of how to calculate rates. A rate is how much something changed divided by the time required for the change to occur.

If this runner sprinted 40 meters in 5 seconds, the runner's average speed is calculated as follows:

$distance/time = 40\ m/5\ s = 8\ m/s$

02.06.a1

Geologic processes faster than the runner include the motion of the energy formed by earthquakes (5 km/sec) or the speed of an explosive volcanic eruption (100s km/hr).

Geologic processes slower than the runner include movement of groundwater (m/day), motion of continents (cm/yr), and uplift and erosion of the land surface (as fast as mm/yr, but typically much slower).

B How Do We Subdivide Geologic Time?

The geologic history of Earth is long, so geologists commonly refer to time spans in millions of years (m.y.) or billions of years (b.y.). If we are referring to times before the present, we use the abbreviation *Ma* (mega-annum) for millions of years before present and *Ga* (giga-annum) for billions of years before present. We also use a special calendar to refer to the four main chapters of Earth's geologic history.

The most recent chapter in Earth history is the *Cenozoic Era*, which began at 65 Ma (65 m.y. ago) and continues to the present.

The next oldest chapter is the *Mesozoic Era*, which refers to the time interval from 251 Ma to 65 Ma. It largely coincides with the time when dinosaurs roamed the planet.

The *Paleozoic Era* started at 542 Ma, a date that marks the widespread appearance of abundant creatures with shells and other hard body parts. It ended at 251 Ma when many Paleozoic organisms became extinct.

The oldest chapter in Earth's history is the *Precambrian*. It includes most of Earth's history and extends from about 4,500 Ma (that is 4.5 billion years before the present, or 4.5 Ga) to 542 Ma, the start of the Paleozoic Era.

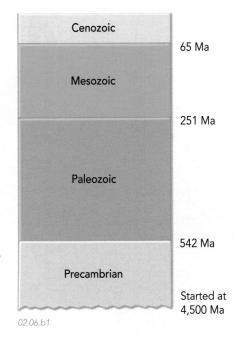

02.06.b1

Each of the four chapters represents a different amount of geologic time.

The figure to the left omits much of the Precambrian because that time interval is so long compared to the other three chapters.

The figure to the right is drawn to scale and shows all of the Precambrian. The Precambrian, often divided into early and late parts, represents approximately 90% of geologic time, whereas the Cenozoic Era represents only 1.4% of geologic time.

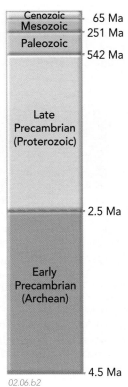

02.06.b2

C What Are Some Important Times in Earth's History?

If the entire 4.5-billion-year-long history of Earth is scaled to a single calendar year, the Precambrian takes up the first 10 months and part of November. On this geologic calendar, Earth formed on January 1.

On this calendar, the oldest dated rocks (about 3.9 to 4.0 b.y. old) would fall in early March. The oldest known fossils are only a little younger (in late March). Both are within *Precambrian* time, a long time interval before the Paleozoic Era. This calendar shows the Precambrian in light brown.

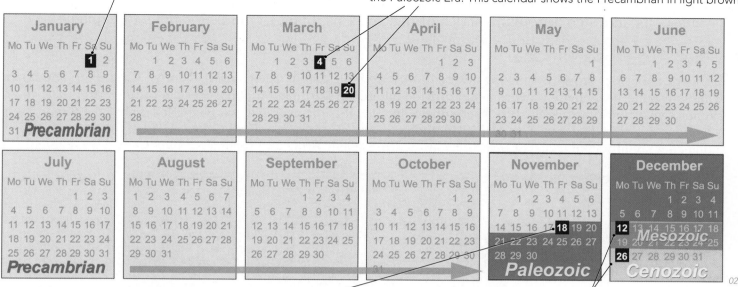

Animals having hard shells became commonplace at about 542 Ma. On our calendar, this is the middle of November. This event is used to define the beginning of the *Paleozoic Era*, the second of Earth's main chapters.

Earth's final two chapters began in December. The Paleozoic Era ended and the *Mesozoic Era* started at 251 Ma, in mid-December. The Mesozoic Era ended and the *Cenozoic Era* began at 65 Ma, equivalent to December 26. The first humans show up in the last 15 minutes of the last day.

Fossils, Numeric Ages, and the Geologic Timescale

Many rocks include *fossils*, which are shells, bones, leaf impressions, and other evidence of prehistoric animals and plants. Geologists discovered that the types of fossils change from one layer to another in a sequence of sedimentary rocks. Younger rocks lie on top of older rocks, so by comparing one location to another, geologists defined certain age periods on the basis of their fossils. Geologists named these age periods for the place where they were first recognized, such as the Pennsylvanian Period for Pennsylvania. When arranged in their proper order, these periods and their subdivisions constitute the *geologic timescale*. ▶

The timescale has the four main divisions discussed above: the *Precambrian*, *Paleozoic*, *Mesozoic*, and *Cenozoic*, from oldest to youngest. These four divisions of geologic time appear throughout this book and are an essential part of the vocabulary of geology.

Geologists calculate the actual, or *numeric age* of rocks and events in thousands, millions, or billions of years before the present. Many rocks contain atoms that change by natural radioactive decay

into a different type of atom, such as atoms of potassium (K) decaying to form argon (Ar). Calculating the numeric age of a rock involves precisely measuring the abundance of both types of atoms (K and Ar). The measurements are done in a laboratory using high-precision instruments.

Such calculations, when combined with a knowledge of fossils, provide ages for the boundaries between the divisions of the geologic timescale shown here. When referring to a geologic event, we use the name of the time period, such as the Mississippian, or the age in millions of years before present (320-359 Ma). The timescale is not set in stone, as evidenced by a recently proposed change (◀) in the subdivisions of the Cenozoic.

Currently Accepted Divisions

Cenozoic	Neogene	23 Ma
	Paleogene	65 Ma
Mesozoic	Cretaceous	
	Jurassic	145
	Triassic	200
		251 Ma
Paleozoic	Permian	
	Pennsylvanian	300
	Mississippian	320
	Devonian	359
	Silurian	416
	Ordovician	444
	Cambrian	488
		542 Ma
Precambrian		

Older Divisions

Quaternary	1.8
Tertiary	
	65 Ma

02.06.mtb1

Before You Leave This Page Be Able To

✓ Calculate a rate and give an example of how a rate is calculated.

✓ List the four main chapters of Earth history, from oldest to youngest, showing which chapter is longest and which one is shortest.

✓ Discuss the geologic timescale and the kinds of data that were used to construct it.

2.7 How Do We Investigate Geologic Questions?

GEOLOGY IS A FIELD OF SCIENCE and aims to solve scientific questions. Every region has a wealth of interesting questions with answers of importance to society. To answer the questions, geologists use their senses and scientific instruments to observe Earth and its processes. They use the resulting observations to answer questions and then, through a series of logical steps, build from observations to explanations.

A What Are Observations?

We learn about our world by making *observations*. We look, listen, smell, and feel so we can record and analyze what is around us. Scientific instruments provide additional information about aspects of the world that we cannot sense, and they allow us to discriminate fine details. For example, we might sense that the temperature outside is near freezing, but if we use a thermometer we can measure a precise value. Every day we make judgments about whether our observations are worth remembering and reliable enough to plan a course of action.

1. Geologists, like other scientists, take special care to make valid observations, such as when examining these layers of volcanic ash. An observation that is judged to be valid becomes a piece of *data* that can be used to develop possible explanations.

3. Evaluating the validity of observations is critical, so geologists commonly repeat measurements to compare values. They may bring other geologists out to the field to check and discuss their observations, measurements, and ideas. ▼

02.07.a1 Gray Mtn., AZ

2. Compasses and other scientific instruments provide quantitative information, provided they are checked and calibrated to ensure that measurements represent valid and trustworthy data. Geologists record data in a field notebook or portable computer and collect samples to permit later reexamination and analysis.

02.07.a2 Gray Mtn., AZ

B How Are Interpretations Different from Data?

Data, by themselves, are not very useful until we analyze them in the context of existing ideas. Perhaps the data will confirm old ideas, or perhaps they will point out a need for a new interpretation. The recent history of volcanic eruptions near Yellowstone National Park illustrates the difference between data and interpretations.

DATA: This map shows a belt of relatively smooth, lower elevation terrain that trends northeast across the mountains of southern Idaho and northern Nevada. It contains mostly volcanic rocks.

INTERPRETATION: Some process related to volcanism formed a belt of low topography and volcanic rocks in the belt outlined in red.

02.07.b1

MONTANA
0.6 m.y.
4.3 m.y.
7 m.y.
10.5 m.y.
15 m.y. NEVADA IDAHO UTAH
WYOMING

DATA: The belt of smooth topography ends near Yellowstone, a recently active volcanic area in the northwestern corner of Wyoming.

INTERPRETATION: Recent volcanism at Yellowstone may be related to the process that smoothed the topography of the belt.

DATA: Samples of volcanic rock analyzed in the laboratory provide ages for when the rock formed. The ages, shown in white as millions of years (m.y.), get younger toward the northeast, from 15 million years in Nevada to less than one million years near Yellowstone.

INTERPRETATION: The very recent volcanism at Yellowstone occurred for the same reasons as the older volcanism to the southwest.

C What Is an Explanation?

When geologists examine a collection of related data, several interpretations may fit together to make a coherent story or *explanation*. The table below summarizes data-interpretation pairs from Part B. The bottom row in the table is a new piece of data obtained from other studies. These data and interpretations combine to form a possible explanation, or *hypothesis*, for how the belt of smooth topography formed.

Data	Interpretation	A Possible Explanation
A belt of smoothed topography, mostly in volcanic rocks, extends in a northeast direction and cuts across the region.	The belt of smoothed topography is related to some process that also produced volcanic eruptions.	For 15 million years, North America and its lithosphere have been moving southwest over a deep thermal disturbance called a *hot spot*. The hot spot involves melting of rocks at depth, resulting in volcanism on the surface. As North America moves southwestward over the hot spot, new volcanoes erupt and then become inactive once that area moves past the hot spot. If North America continues to move southwestward, the hot spot may cause new volcanoes northeast of Yellowstone.
The belt ends at Yellowstone National Park.	Volcanism at Yellowstone is related to the smoothed topography.	
Volcanic rocks along the belt get younger to the northeast.	The smoothed belt did not form all at once but rather sequentially, from southwest to northeast.	
The North American continent is moving slowly to the southwest based on satellite observations.	There is a source of magma beneath Earth's crust. The continent has moved over the magma source, causing volcanic activity to occur in a narrow belt. Volcanism initially occurred in the southwest but migrated to the northeast over time as North America moved to the southwest.	

D What Is Happening at Yellowstone National Park?

Geologists have long recognized that several huge volcanic eruptions occurred in Yellowstone during the past two million years. Geologist Bob Smith studied Yellowstone for decades and in 1973 noticed that lake levels along the southern side of Yellowstone Lake had risen, drowning trees. How did he investigate his observations?

1. To check his observation, Smith and colleagues conducted a new, detailed survey of the area's topography using high-precision surveying equipment.

02.07.d1 Yellowstone NP, WY

2. This view shows Yellowstone Lake, with north to the left. When Smith compared the new survey with the last survey done in 1920, he discovered that the elevation of the area shaded orange had increased (the area had risen) in a remarkably short period of time. What was causing this area along the north side of the lake to rise in elevation?

02.07.d2

3. Smith concluded that the rising area north of the lake caused the lake to spill over its southern shoreline, drowning trees in the areas shown in purple.

Observations, Interpretations, and Hypotheses

Discovery of drowned trees along Yellowstone Lake and the follow-up studies illustrate how we develop and investigate questions. An observation (the drowned trees) led to the question, *What is going on here?* The question led to a possible interpretation that parts of the land around the lake may be actively rising or sinking. An explanation that is developed to explain observations and that allows testing is a *hypothesis*. To test his predictions, Smith used precise surveying equipment to collect new observations of the land surface. He scrutinized and validated the new data and proposed a new hypothesis that rising land beneath the northern part of the lake had dis-placed water that drowned trees along the southern shore.

This example illustrates the strategy of considering different types of data and different scales of observation. Smith interpreted local uplift north of Yellowstone Lake to be the result of a large magma chamber beneath the surface. One interpretation is that the magma originated by deep melting related to a hot spot in the mantle. As the North American continent and lithosphere moved over the hot spot, volcanism and faulting formed the belt of smoothed topography along the *Snake River Plain*. Many geologists accept this explanation but still consider other explanations.

Before You Leave This Page Be Able To

☑ Explain what observations are and how they become valid.

☑ Describe how data differ from an interpretation, and provide one example of each.

☑ Summarize how data and interpretations lead to new explanations.

☑ Describe how a series of observations led to an explanation for regional and local processes at Yellowstone.

2.8 How Do Scientific Ideas Get Established?

HOW DOES A SCIENTIFIC EXPLANATION move through the steps from an initial idea to a testable hypothesis and finally to a widely accepted *theory* supported by a rigorous body of knowledge? Geologists and other scientists begin with observations, propose possible explanations (hypotheses), make predictions based on each hypothesis, and conduct investigations to test each prediction. Science is a way to evaluate which hypotheses are most likely to be correct and which are not. It is a body of knowledge based on supported hypotheses and on accepted theories that have been examined and tested many times.

A How Do We Test Alternative Explanations?

Science proceeds as scientists explore the unknown—making observations and then systematically investigating questions that arise from observations that are puzzling or unexpected. Often, we try to develop several possible explanations and then devise ways to test each one. The normal steps in this process are illustrated below, using an investigation of groundwater contaminated by gasoline.

Steps in the Investigation

Observations

1. Someone makes the *observation* that groundwater from a local well contains gasoline, near an old buried gasoline tank. The first step in any investigation is to make observations, recognize a problem, and state the problem clearly and succinctly. Stating the problem as simply as possible simplifies it into a more manageable form and helps focus our thinking on its most important aspects.

Questions Derived from Observations

2. The observation leads to a *question* — Did the gasoline in the groundwater come from a leak in the buried tank? Questions may be about what is happening currently, what happened in the past or, in this case, who or what caused a problem.

Proposed Explanations and Predictions from Each Explanation

3. Scientists often propose several explanations, referred to as *hypotheses*, to explain what they observe. A hypothesis is a causal explanation that can be tested, either by conducting additional investigations or by examining data that already exist.

4. One explanation is that the buried tank is the source of contamination.

5. Another explanation is that the buried tank is not the source of the contamination. Instead, the source is somewhere else, and contamination flowed into the area.

6. We develop *predictions* for each explanation. For this example, the tank should have some kind of leak and should be surrounded by gasoline. Also, the type of gasoline in the tank should be the same as in the groundwater. Next, we plan some way to *test* the predictions, such as by inspecting the tank or analyzing the gasoline in the tank and groundwater.

Results of Investigation

7. The investigation discovered no holes in the tank or any gasoline in the soil around the tank. Records show that the tank held unleaded gasoline, but gasoline in the groundwater is leaded. We compare the results of any investigation with the predictions to determine which possible explanation is most consistent with the new data.

Conclusions

8. Data collected during the investigation support the conclusion that the buried tank is not the source of contamination. Any explanation that is inconsistent with data is probably incorrect, so we pursue other explanations. In this example, a nearby underground pipeline may be the source of the gasoline. We can devise ways to evaluate this new hypothesis by investigating the pipeline. We also can revisit the previously rejected hypothesis if we discover a new way in which it might explain the data.

02.08.a1–6

B How Does a Hypothesis Become an Established Theory?

A hypothesis that survives scientific scrutiny can be elevated to the higher standard of acceptability of a *theory*. Like a hypothesis, a theory explains existing data and helps predict data not yet collected, but a theory encompasses a more extensive body of knowledge. The scientific process rejects many hypotheses, and few hypotheses survive the intense investigation, experimentation, and testing of predictions to become theories accepted by a majority of scientists. The testing and rejecting of ideas distinguishes science from ways of knowing based on faith.

1. Scientists found fossils of the same land animals in South America and Africa, even though these continents are separated by an ocean. To explain these observations, scientists proposed a hypothesis that long ridges of land, called *land bridges*, once linked the two continents. The hypothetical land bridges would have allowed land animals to walk from one continent to the other. According to the hypothesis, the bridges later collapsed or were submerged beneath the oceans.

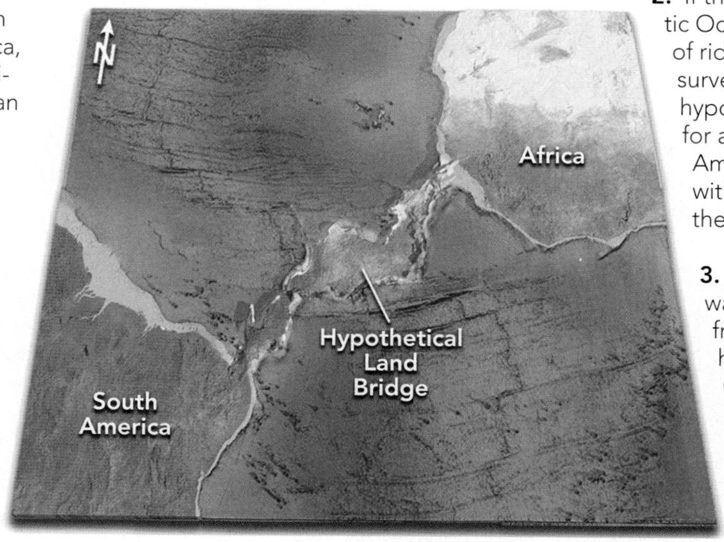

Africa

Hypothetical Land Bridge

South America

02.08.b1

2. If the land bridges once existed, then the South Atlantic Ocean should contain submerged ridges, or remnants of ridges, that once connected the two continents. When surveys of the ocean floor failed to find land bridges, the hypothesis had to be abandoned. So scientists had to look for another way to explain the similarity of fossils in South America and Africa. Compare this hypothetical diagram with the data and an alternative explanation portrayed on the next pages.

3. A land-bridge hypothesis was also proposed as a way to explain the migration of animals and humans from Asia to North America during the Ice Ages. This hypothesis, unlike the hypothesis about the Atlantic Ocean, is supported by a lot of data and by a credible explanation of why a land bridge existed. A submerged ridge does link Alaska and Asia, and it would have been dry land when the growth of glaciers lowered sea level. So the hypothesis that a land bridge existed off Alaska evolved into a theory, while the hypothesis that one existed in the South Atlantic was rejected.

How and Why Scientific Understandings Change Over Time

Geology, like other sciences, is a way of investigating the world around us. It is an evolving framework of knowledge and methods, not a static collection of facts. Explanations and theories accepted by the geologic community can change over time as new data, new scientific instruments, and new ideas become available.

Although many scientific explanations are considered to be "correct" and are supported by many lines of evidence, the history of science warns us not to trust any explanation as "final truth." There is so much evidence supporting some theories that they probably will never be shown to be wrong. On the other hand, scientific scrutiny has caused many proposed hypotheses or theories to be rejected or greatly modified based on new data. Some accepted scientific explanations needed only to be revised slightly to account for new data or other scientific advances. In other cases, the science of the time was not sophisticated enough to produce explanations that could hold up under scrutiny. Scientists operate under the principle that

no explanation in science is ever *proven*, but some are eliminated. There are no final answers, just logical, well-tested explanations based on the best data available.

In the 1700s, for example, the most influential scientists of the time could not accept that stones (*meteorites*), such as the one shown here, fell out of the sky.

02.08.mtb1

For a time, scientists and others believed that meteors and meteorites resulted from lightning that fused dust with other particles in the air. This explanation was rejected when chemists noted that some meteorites consisted of iron-nickel alloys that were not found in any Earth rocks. Also, some meteorites fell in plain view

when there were no lightning storms. Stones really were falling from the sky!

We have gained much understanding using the methods of science, but we still do not know many things about the universe. There are countless interesting questions left to investigate, and many important theories left to imagine. We not only lack reasonable explanations for many scientific phenomena, in many cases we do not yet know the right questions to ask or what data to collect.

Before You Leave This Page Be Able To

✓ Explain the logical steps taken to evaluate an explanation.

✓ Describe how a hypothesis becomes an established theory.

✓ Describe what causes changes in scientific understandings, and discuss why scientific explanations are never proven to be "true."

2.9 How Was the Hypothesis of Continental Drift Conceived and Received?

SOME CONTINENTS HAVE MATCHING SHAPES that appear to fit together like the pieces of a giant jigsaw puzzle. Alfred Wegener (1880–1930) observed the fit of these continents and tried to explain this and other data with a new hypothesis called *continental drift*. Wegener argued that the continents were once joined together but later drifted apart. To illustrate the development of geologic explanations, we review the historically important controversy over continental drift.

A Were the Continents Once Joined Together?

Fairly accurate world maps became available during the 1800s and scientists, including Alfred Wegener, noted that some continents, especially the southern continents, appeared to fit together. After considering many types of data, Wegener arrived at a creative explanation for this pattern.

This figure shows how the southern continents are interpreted to have fit together 150 million years ago. In this figure, we included the *continental shelves* because they are parts of continents that are currently underwater. In this arrangement, the bulge on the eastern side of South America fits nicely into the embayment on the western coast of Africa.

The fit of the continents and other supporting evidence preserved in rocks and fossils inspired Wegener and others to suggest that South America, Africa, Antarctica, Australia, and most of India were once joined but later drifted apart. Even Madagascar can fit into the puzzle.

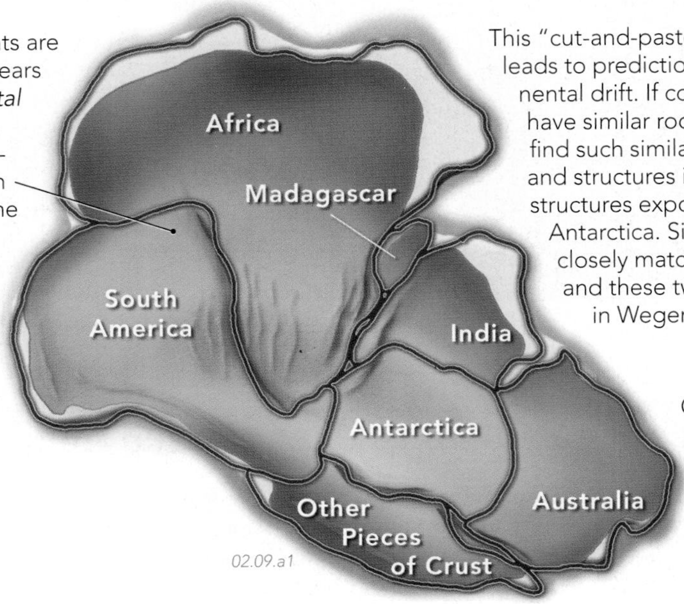

This "cut-and-paste" fit of the continents is intriguing and leads to predictions for testing the hypothesis of continental drift. If continents were once joined, they should have similar rocks and geologic structures. Geologists find such similarities when they compare the rocks and structures in southern Australia with the rocks and structures exposed around the edges of ice sheets on Antarctica. Similarly, the geology of western Africa closely matches that of eastern South America, and these two areas are adjacent to each other in Wegener's reconstruction.

Geologists gave the name *Gondwana* to this hypothetical combination of the southern continents into a single large supercontinent.

02.09.a1

B Is the Distribution of Fossils Consistent with Continental Drift?

Another piece of evidence supporting continental drift is the correspondence of the fossils of plants and land animals on continents now several thousand kilometers apart and separated by wide oceans.

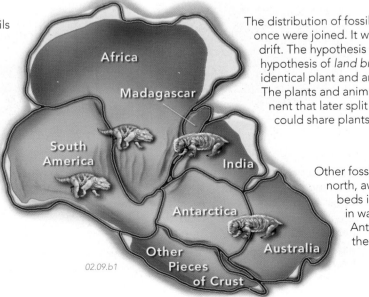

This figure illustrates that fossils of some land animals exist on several continents now separated by wide oceans. The animals lived more than 150 million years ago and are now extinct. These land animals could not swim across the wide oceans that currently separate the continents.

The distribution of fossils is consistent with the idea that the continents once were joined. It was a key piece of evidence in favor of continental drift. The hypothesis of continental drift provided an alternative to the hypothesis of *land bridges*, discussed previously, and it explained why identical plant and animal fossils are found on different continents. The plants and animals were originally on a single huge supercontinent that later split into separate smaller continents. Two continents could share plants and land animals before they split, but not after!

Other fossil data suggest that Antarctica was once farther north, away from the South Pole. Such data include coal beds interpreted to have formed from plants that grew in warm-weather swamps. One explanation is that Antarctica moved to its present polar location after the coal formed more than 150 million years ago.

02.09.b1

C How Did Continental Drift Explain Glacial Deposits in Unusual Places?

Geologists studying continents in the Southern Hemisphere were puzzled by evidence that ancient glaciers had once covered places that today are close to the equator, and much too warm to have major glaciers.

1. This rounded outcrop in South Africa has a polished and scratched surface that is identical to those observed at the bases of modern glaciers. This observation is surprising because South Africa is currently a fairly warm and dry region without any glaciers.

02.09.c1 Kimberly, South Africa

02.09.c2 Kimberly, South Africa

2. Sedimentary rocks above the polished surface contain an unsorted collection of rocks of various sizes. Some of the rocks have scratch marks, like those seen near glaciers.

3. The scratch marks on the polished surface tell geologists the direction that glaciers moved across the land as they gouged the bedrock. Geologists interpret the scratch marks and other observations as evidence that glaciers moved across the area about 280 million years ago.

4. The overall directions of glacial movement inferred from the scratch marks made it seem as if the glaciers had come from the oceans, something that is not seen today. Wegener discovered that these data made more sense when the continents were pieced back together into a larger, ancient continent, as shown in this illustration. According to this model, a polar ice cap was centered over South Africa and Antarctica 280 million years ago, and the directions of glacial ice movement were those shown by the blue arrows.

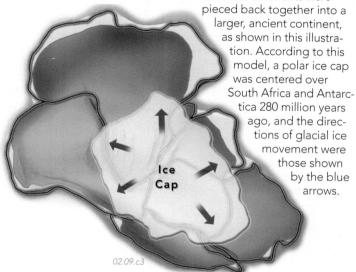

Ice Cap

02.09.c3

Old and New Ideas About Continental Drift

The hypothesis of *continental drift* received mixed reviews at first from geologists and other scientists. Geologists working in the Southern Hemisphere were intrigued by the idea because it explained the observed similarities in rocks, fossils, and geologic structures on opposite sides of the Atlantic and Indian Oceans. Geologists working in the Northern Hemisphere were more skeptical, in part because many had not seen the Southern Hemisphere data for themselves.

We now know that Wegener, with the evidence he knew about, was on the right track. A crucial weakness of his hypothesis was that he could not explain how or why the continents moved. Wegener imagined that continents plowed through or over oceanic crust in the same way that a ship plows through the ocean. Scientists of his day, however, could demonstrate that this mechanism was not feasible. Continental crust is not strong enough to survive the forces needed to move a large mass across such a great distance while pushing aside oceanic crust. Because scientists of Wegener's time could show with experiments and calculations that this mechanism was unlikely, they practically abandoned the hypothesis, in spite of its

other appeals. The hypothesis probably would have been more widely accepted if Wegener or another scientist of that time had proposed a viable mechanism that explained how continents could move.

In the late 1950s, the idea of drifting continents again surfaced with the availability of new information about the topography, age, and magnetism of the seafloor. These data showed, for the first time, that the ocean floor had long submarine mountain belts, such as the Mid-Atlantic Ridge (▼) in the middle of the Atlantic Ocean. Harry Hess and Robert

Dietz, two geologists familiar with Wegener's work, examined the new data on ocean depths, and also new data on magnetism of the seafloor. The magnetic data had largely been acquired in the search for enemy submarines during World War II. Hess and Dietz both proposed that oceanic crust was spreading apart at underwater mountain belts, carrying the continents apart. This process of *seafloor spreading* rekindled interest in Alfred Wegener's idea of continental drift. Wegener's hypothesis morphed into the theory of plate tectonics, the topic of chapter 3.

02.09.mtb1

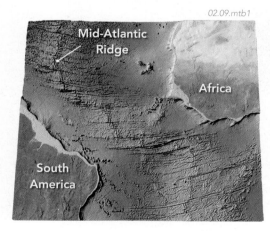

Mid-Atlantic Ridge

Africa

South America

Before You Leave This Page Be Able To

✓ Describe observations Wegener used to support continental drift.

✓ Discuss why the hypothesis was not widely accepted.

✓ List some discoveries about the seafloor that brought a renewed interest in the idea of continental drift.

2.10 What Does a Geologist Do?

GEOLOGISTS ADDRESS A WIDE RANGE OF PROBLEMS important to society. Because the problems are so diverse, there are many types of geologists, each with different questions to answer and interesting problems to explore. Some things can be explored directly in the field, whereas others require sophisticated computers or other technology. Some problems involve studying places that are inaccessible, such as other planets or the interior of Earth. Many questions address active ongoing processes, and others focus on the interpretation of ancient processes based on rocks, structures, fossils, and other geologic data.

A How Do Geologists Investigate Questions in the Field and Laboratory?

The traditional view of a geologist is of a field geologist, a person outdoors with backpack, hiking boots, and rock hammer traversing a scenic mountain ridge. Geologists produce descriptions, maps, and other data needed to reconstruct Earth history and to determine how this history is important to us today. They also investigate questions using laboratories, outdoor experimental facilities, numerical modeling, and other strategies.

02.10.a1 Antarctica

02.10.a2 Hawaii

▶ During field studies, geologists observe various aspects of the natural environment, record these observations, and propose explanations for these observations. A common goal is to understand the area's geologic processes and history, commonly by producing a geologic map and geologic cross sections. Sometimes, just getting to the field site is an outdoor adventure.

◀ Some field geologists study volcanoes, earthquakes, landslides, floods, or other natural hazards. To help us avoid these hazards, geologists study the processes that are operating, determine how often hazardous events occur, and identify areas that are most likely to be affected. This geologist is extracting samples of molten lava on Hawaii.

02.10.a3 South Africa

▲ Geologists find most energy and mineral resources that we use. They conduct field studies to find areas likely to contain a certain resource, and then through drilling or using other techniques determine if the resource is there and how much can be produced. Energy and mineral companies employ many geologists, sending them to various parts of the world. Besides searching Earth's surface, geologists go deep underground, perhaps 3 km (10,000 ft) below the ground in a South African gold mine, or up in helicopters to take photos and collect other kinds of data.

◀ Some geologic systems must be investigated in the laboratory using scientific devices that simulate conditions deep within Earth's crust or mantle. The device shown here subjects geologic materials to very high pressures, equivalent to those in the mantle. The scientist then observes the resulting material with a microscope to document how the material changed and perhaps under what conditions it melted. Such experiments are a main way we investigate the materials and processes that occur within the deep crust and mantle.

02.10.a4 Tempe, AZ

B How Do Geologists Study Inaccessible Places?

Many geologic questions involve places where we cannot do field studies. Geologists study magma chambers beneath volcanoes, rock layers that are kilometers beneath the seafloor, or the geology of Mars and other objects in the solar system. To conduct these sorts of studies, geologists use telescopes, high-resolution imaging devices, various scientific instruments, and remote probes to collect pertinent data.

02.10.b1

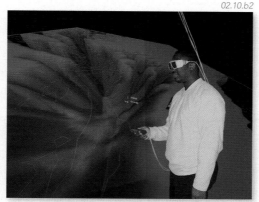
02.10.b2

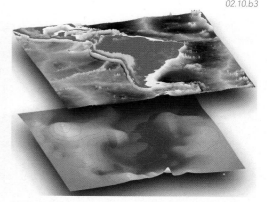
02.10.b3

An exciting field of geology is exploring other planets and their moons, principally by sending spacecraft that orbit a planet or that land on its surface. From such observations, planetary geologists try to understand what processes are reshaping the surface of the moon or planet and whether there is a possibility of water and life. This recent image (▲) shows channels on Mars interpreted to have been carved by water that flowed on the surface a very long time ago.

▲ Most oil, gas, and coal reside within sedimentary rocks, and much lies offshore, beneath the seafloor. Exploration for these resources focuses on understanding the sequence and geometry of rock layers and the conditions that deposited each layer. Petroleum geologists image layers beneath the land and seafloor, commonly using sophisticated computers and virtual-reality environments to visualize data in three dimensions.

▲ Studying Earth's hidden interior utilizes indirect means of observation, such as instruments that measure seismic energy from earthquakes. Seismologists, scientists who study such waves, analyze when and how these seismic waves arrive and then use powerful computers and sophisticated programs to model subsurface parameters, such as temperature. The model above shows the configuration of a boundary in the mantle at a depth of about 400 km beneath South America.

Geologic Studies of Ecosystems

Some geologists study ecosystems to understand the geologic processes that are operating and to determine how different geologic factors affect the health of the system. These studies help ecologists and others to better protect the ecosystems. The types of research that these geologists do are as diverse as the ecosystems, which range from deserts to rain forests to icy glaciers. In such studies, geologists commonly work with biologists, climate specialists, and government agencies that decide how the land will be managed or protected.

The most important factor controlling the viability of many ecosystems is the availability of clean water. In most geologic settings, water moves between the surface and subsurface, and geologic studies are the primary basis for determining how much water is available, how water flows, and the purity of water. Geologists study rock layers and other materials on and beneath the surface to understand how these materials

02.10.mtb1

influence the flow of water. Geologists guide cleanup efforts by sampling water (▲), pinpointing sources of contamination, and predicting which way the contamination is likely to move in the future.

In addition to water, the viability of an ecosystem generally depends on the health of its soils. Geologic studies document how soils develop, how they are changing, and the best ways to prevent erosion or other types of soil deterioration. Such studies also

determine the natural fluxes of organic material and other chemical constituents among the soil, plants, and water. Geologists investigate these factors in order to understand processes that affect the system and how the system responds to environmental changes that are the result of natural or human causes.

Before You Leave This Page Be Able To

☑ Discuss the kinds of questions geologists investigate with field studies, and contrast those with questions studied using techniques other than field studies.

☑ Describe how geologists find energy and mineral resources.

☑ Discuss how geologists study ecosystems.

2.10

2.11 How Did This Crater Form?

A CRATER ON THE PLATEAU OF NORTHERN ARIZONA is similar to those on the Moon. The crater is a huge pit, more than 1,250 m (4,100 ft) across and 170 m (560 ft) deep, with a raised rim of broken rocks. Some dramatic geologic event must have occurred here. What could it be? We explore this mystery to connect different ideas in this chapter and to illustrate how to evaluate competing explanations for this unusual feature.

A What Would You Observe at the Crater?

Observe the photographs below and make a list of what you think are the most important features. From your observations, consider how this crater might have formed and how your observations might support alternative explanations. What other types of information would help you have a better understanding of the crater's origin?

02.11.a1 Northern AZ

▶ The raised rim of the crater consists of large, angular blocks of limestone and sandstone, many of which are fractured and shattered. These blocks are pieces of the rock layers that underlie the region and are exposed in the crater walls, so we interpret them to somehow have been thrown or blasted out of the crater.

02.11.a2 Northern AZ

02.11.a3 Northern AZ

◀ Shattered rocks in the walls of the crater contain unusual microscopic minerals, which laboratory experiments show form only at extremely high pressures. These minerals are not present in rocks away from the crater or elsewhere in the region.

B What Is the Geologic Setting of This Crater?

This geologic block diagram shows the geometry of rock layers near the crater. A geologist constructed this cross section by observing the rocks exposed in the crater and in the surrounding plains, by examining results from drilling in the floor of the crater, and by extrapolating these observations into the subsurface.

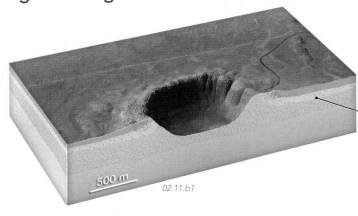

500 m

02.11.b1

Meteorites are scattered on the surface around the crater. Geologists collect small and large pieces to understand what information the meteorites provide about the formation of the crater, as well as the origin of the solar system.

Layers of sandstone and limestone are nearly horizontal away from the crater, but along the rim of the crater they have been bent backward, away from the crater. In places, the layers have been overturned relative to how they started.

C What Are Some Possible Explanations for the Origin of the Crater?

Since the crater was discovered, geologists have proposed and tested several explanations for its origin. Among these explanations are a volcanic explosion, warping by a rising mass of salt, and excavation by a meteoroid impact.

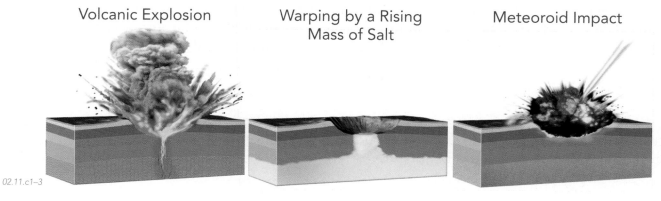

02.11.c1–3

	Volcanic Explosion	Warping by a Rising Mass of Salt	Meteoroid Impact
Possible Explanation	A volcanic explosion blasts open the crater during a violent eruption of gas and magma.	A rising mass of salt warped up the rock layers as it rose toward the surface. Rainwater later dissolves the salt to form the crater.	A large meteoroid streaks through Earth's atmosphere and blasts the crater when it collides with the surface.
Predictions	Volcanic materials, such as ash, should be present around the crater, and solidified magma might underlie the crater floor.	A mass of salt should exist directly beneath the floor of the crater.	Fragments of a meteoroid might remain beneath the crater or on the surrounding plains, but much of the meteoroid could have been vaporized during the blast. Meteoroid impacts cause very high pressures in rocks near the collision.
Results of Testing	No volcanic material was found around the crater or by drilling beneath the crater, so this explanation seems improbable.	No salt was found beneath the crater, despite extensive drilling into the crater floor.	Drilling did not encounter a large mass of meteorite beneath the crater, but small meteorite fragments are scattered across the surrounding plains. Minerals formed by high pressures exist within rocks in the walls of the crater.

An Incident 50,000 Years Ago

Of the three explanations, only the meteoroid-impact explanation seems possible because the crater does not contain the rock types predicted by the other two explanations. Other types of data also support an origin by meteoroid impact. Laboratory experiments show that the unusual minerals in the walls of the crater can form only at very high pressures, such as those caused when a meteoroid impacts Earth. These experiments also show that such impacts shatter rocks near the collision and bend back rock layers away from the point of impact. Computer models of meteoroid impacts predict that large meteoroids, such as the one required to form this crater, are mostly vaporized by the impact.

Therefore, little of the meteoroid would remain in the crater, but scattered pieces might survive around the crater. It is estimated that collectors have gathered 30 tons of meteorite specimens from around the site. The accepted origin for this crater is reflected in its famous name, *Meteor Crater*.

From the currently available information, geologists conclude that the crater formed about 50,000 years ago when a meteoroid 30 to 50 meters in diameter smashed onto the surface at a speed of about 11 km/s (25,000 mi/hr). A recent study suggests that the meteoroid broke apart and slowed before it hit. Humans were not yet living in this area to observe the fireball blazing across the sky, but

animals, such as the woolly mammoth, were around to witness this unexpected catastrophe.

Before You Leave This Page Be Able To

✓ Describe or sketch the three explanations for the origin of the crater and which observations support or do not support each possible explanation.

✓ Describe how geologists interpret Meteor Crater to have formed.

2.11

What Is the Geologic History of Upheaval Dome?

A PERPLEXING GEOLOGIC FEATURE CALLED UPHEAVAL DOME forms a conspicuous feature within Canyonlands National Park of Utah. In most of Canyonlands, colorful sedimentary rocks are nearly horizontal, but around Upheaval Dome they are abruptly warped upward and eroded into a very unusual circular structure. The feature continues to puzzle geologists as to how and when it formed.

Goals of This Exercise:

- Make some observations and develop some questions about Upheaval Dome.
- Determine the sequence of geologic events that formed the rock layers in the dome.
- Suggest some ways to test possible explanations for the origin of the dome.

A What Are Some Observations and Questions About the Dome?

The three-dimensional perspective below shows an unusual circular feature called *Upheaval Dome*. Make some observations about this landscape, and record your observations on a sheet of paper or on the worksheet. As you study this landscape, record any questions or ideas that you have.

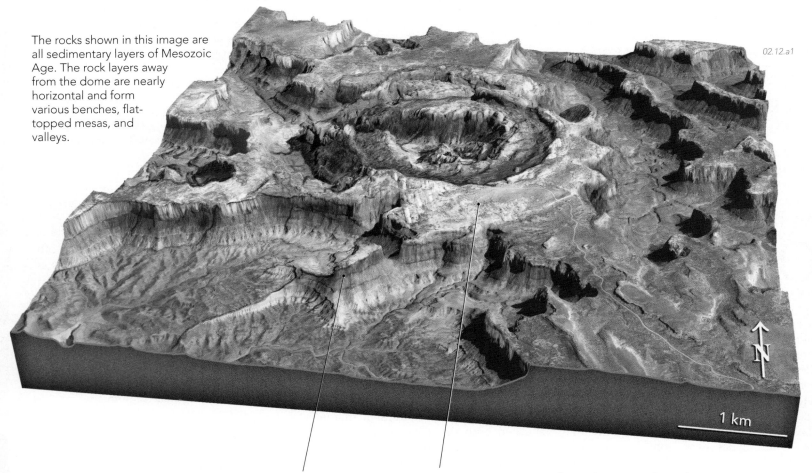

02.12.a1

The rocks shown in this image are all sedimentary layers of Mesozoic Age. The rock layers away from the dome are nearly horizontal and form various benches, flat-topped mesas, and valleys.

1 km

Some layers are resistant to erosion and form cliffs, whereas others are less resistant and erode into slopes. In some areas, erosion of the layers has formed buttes and rounded hills.

The rock layers in the dome form a large ring-shaped feature and are tilted outward in all directions, as shown in the geologic cross section on the next page.

B What Sequence of Geologic Events Formed the Rocks and the Dome?

Shown below are a stratigraphic section with the sequence of rock layers, and a geologic cross section across the dome. Using these two figures and the strategies in section 2.2, determine the order in which the layers were formed, and write your answers on the worksheet.

Stratigraphic Section

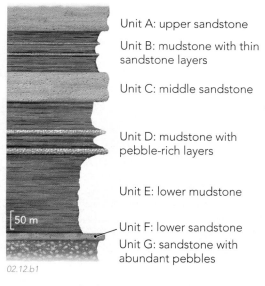

Unit A: upper sandstone

Unit B: mudstone with thin sandstone layers

Unit C: middle sandstone

Unit D: mudstone with pebble-rich layers

Unit E: lower mudstone

50 m

Unit F: lower sandstone

Unit G: sandstone with abundant pebbles

02.12.b1

Units A, B, and C have a Jurassic Age, whereas units D and E are Triassic. Units F and G are Permian (see timescale in section 2.6).

Cross Section

The letters A–G mark the units shown in the stratigraphic section to the left. The letters are assigned in order from top to bottom, not in the order in which the units formed. Some units, such as B and D, contain a series of related sedimentary rock layers rather than just a single type of rock.

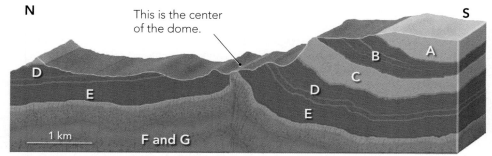

02.12.b2

It is uncertain what types of rocks lie at depth below the dome (i.e., below units F and G).

All of the rock units have been folded into the dome. Any layers deposited after the dome was formed have been eroded away.

Many faults and folds (not shown here) have thickened the rock layers in some places and thinned them in others.

C How Would You Test Possible Explanations for the Origin of the Dome?

The origin of Upheaval Dome is controversial, and geologists currently debate three competing explanations. No single explanation has yet gained widespread acceptance by the geologic community.

1. A rising mass of salt warped the rock layers upward. A thick salt layer is known to be present beneath much of the Canyonlands region, and the salt may have risen upward because it is less dense than the surrounding rocks.

2. The dome formed as a result of rising magma. Igneous rocks formed from magma are common elsewhere in the region, where they have bowed up and baked the surrounding rock layers.

3. The dome is part of a larger, circular crater formed by a meteoroid impact. Many of the larger meteoroid-impact craters on the Moon and elsewhere contain a central peak or dome, which is interpreted to form by converging shock waves. In this case, Upheaval Dome only represents the center of a larger crater.

The age of the dome is poorly constrained. The dome must be younger than all of the rock layers in the vicinity because all of the layers are warped by the dome. When the dome formed, the currently exposed rocks were several kilometers deep, buried beneath overlying rock layers that have since eroded away. This erosion, therefore, removed some key evidence for the origin of the dome.

Procedures for Possible Explanations of the Dome

1. For each of the three explanations, draw a simple sketch on the worksheet illustrating which types of rocks you predict to find at depth.

2. List a prediction that follows from each explanation. Then, explain how that prediction could be tested.

3. List the types of information you would like to know about this location to further constrain the origin of the dome.

2.12

CHAPTER 3 Plate Tectonics

THE SURFACE OF EARTH IS NOTABLE for its dramatic mountains, beautiful valleys, and intricate coastlines. Beneath the sea are unexpected features, such as undersea mountain ranges, deep ocean trenches, and thousands of submarine mountains. In this chapter, we examine the distribution of these features, along with the locations of earthquakes and volcanoes, to explore the *theory of plate tectonics*.

These images of the world show large topographic features on the land, colored using satellite data that show areas of vegetation, rocks, and sand. Colors on the seafloor indicate depths below sea level, ranging from light blue for seafloor that is at relatively shallow depths to dark blue for seafloor that is deep.

03.00.a1

Juan de Fuca Ridge

The seafloor west of North America displays a long, fairly straight fracture that trends east-west and ends abruptly at the coastline. North of this fracture, a ridge called the Juan de Fuca Ridge zigzags across the seafloor.

What are these features on the seafloor and how did they form?

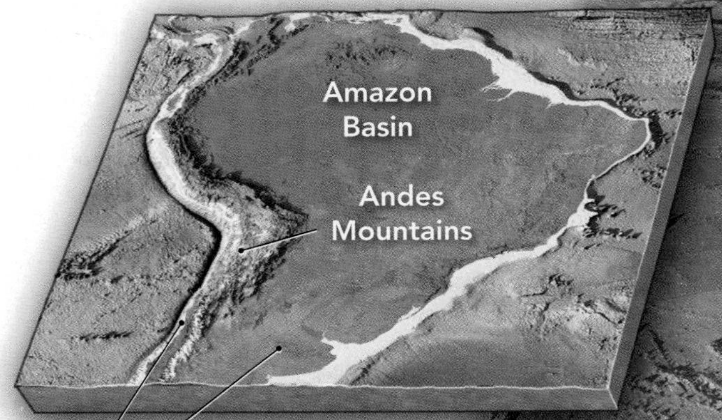

Amazon Basin

Andes Mountains

South America has two very different sides. The mountainous Andes parallel the western coast, but a wide expanse of lowlands, including the Amazon Basin, makes up the rest of the continent. The western edge of the continent drops steeply into the Pacific Ocean and is flanked by a deep trench. The eastern edge of the continent continues well beyond the shoreline and forms a broad bench covered by shallow waters (shown in light blue).

Why are the two sides of the continent so different?

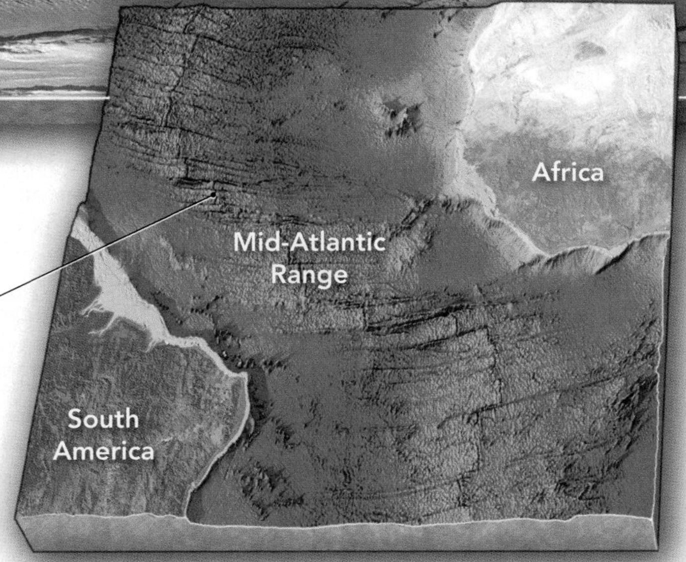

Africa

Mid-Atlantic Range

South America

A huge mountain range, longer than any on land, is hidden beneath the waters of the Atlantic Ocean. The part of the range shown here is halfway between South America and Africa. The ridge zigzags across the seafloor, mimicking the shape of the two continents.

What is this underwater mountain range, and why is it almost exactly in the middle of the ocean?

TOPICS IN THIS CHAPTER

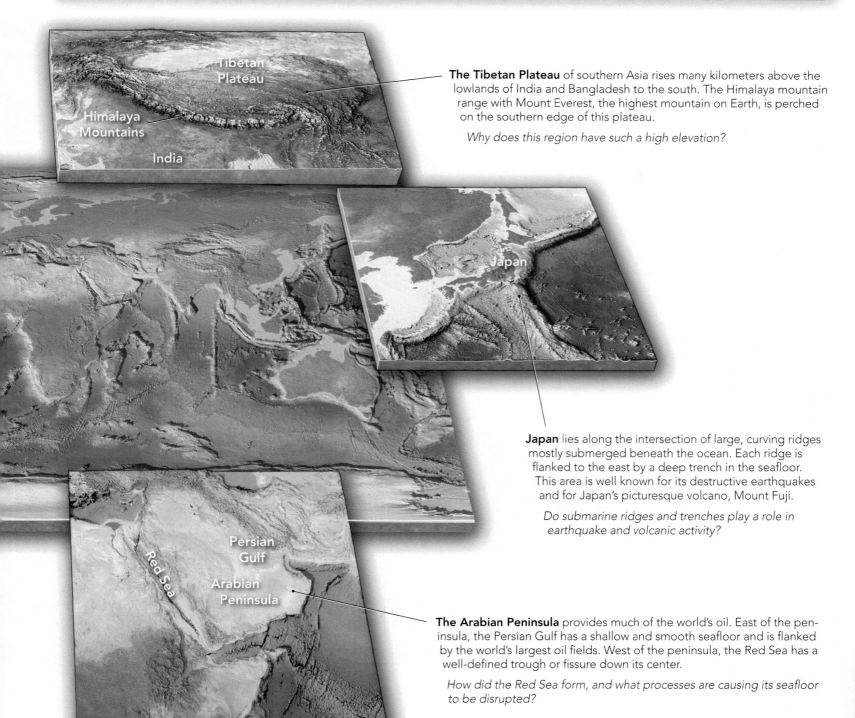

The Tibetan Plateau of southern Asia rises many kilometers above the lowlands of India and Bangladesh to the south. The Himalaya mountain range with Mount Everest, the highest mountain on Earth, is perched on the southern edge of this plateau.

Why does this region have such a high elevation?

Japan lies along the intersection of large, curving ridges mostly submerged beneath the ocean. Each ridge is flanked to the east by a deep trench in the seafloor. This area is well known for its destructive earthquakes and for Japan's picturesque volcano, Mount Fuji.

Do submarine ridges and trenches play a role in earthquake and volcanic activity?

The Arabian Peninsula provides much of the world's oil. East of the peninsula, the Persian Gulf has a shallow and smooth seafloor and is flanked by the world's largest oil fields. West of the peninsula, the Red Sea has a well-defined trough or fissure down its center.

How did the Red Sea form, and what processes are causing its seafloor to be disrupted?

3.0

3.1 What Are the Major Features of Earth?

OCEANS COVER 71% OF EARTH'S SURFACE. Seven major continents make up most of the rest of the surface, and islands account for less than 2%. We are all familiar with the continents and their remarkable diversity of landforms, from broad coastal plains to steep, snow-capped mountains. Features of the ocean floor, not generally seen by people, are just as diverse and include deep trenches and submarine mountain ranges. Islands exhibit great diversity, too. Some are large and isolated, but other islands define arcs, ragged lines, or irregular clusters. What are the characteristics of each type of feature, and how did these features form?

03.01.a1

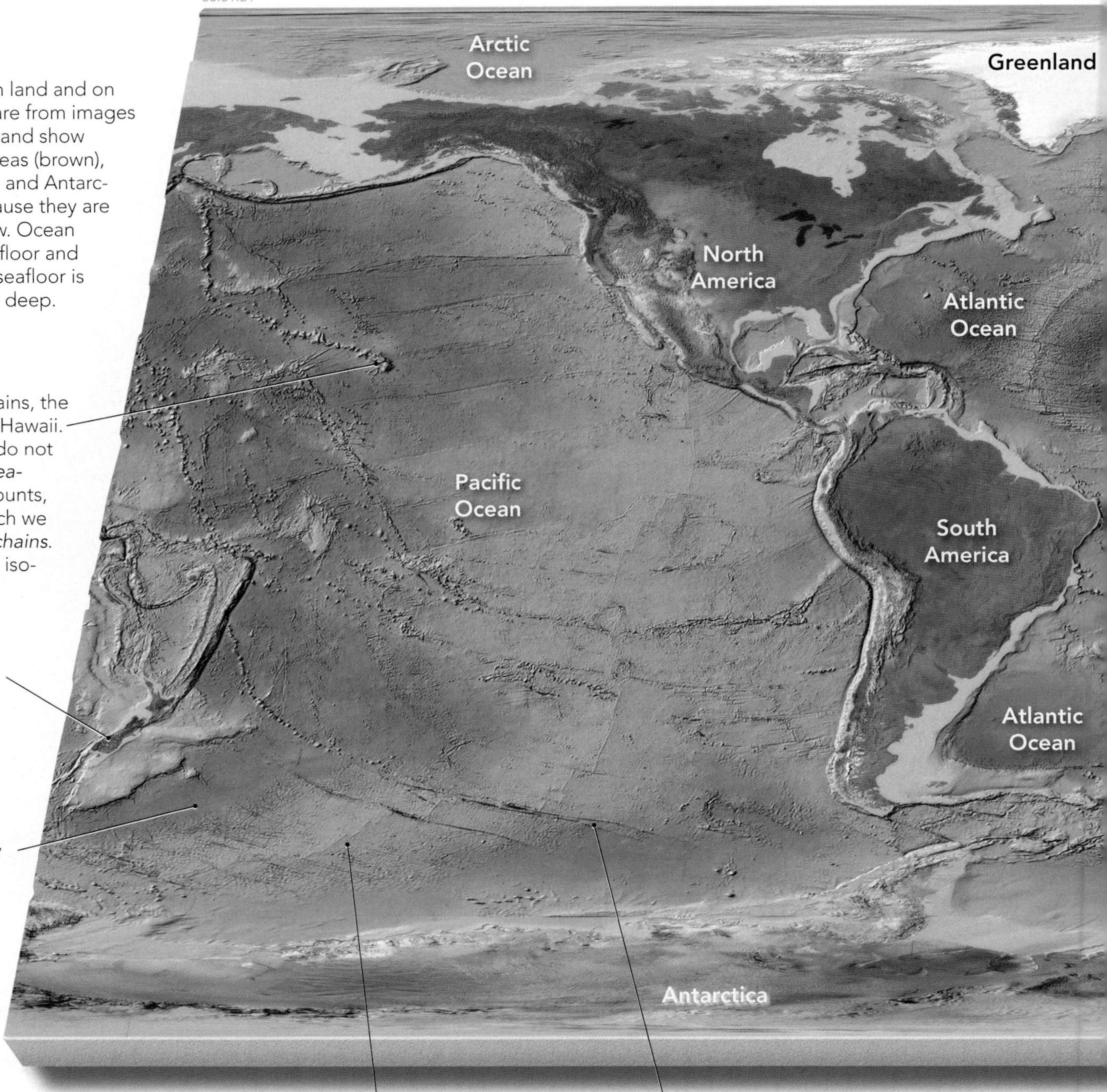

This map shows large features on land and on the seafloor. The colors on land are from images taken by satellites orbiting Earth and show vegetated areas (green), rocky areas (brown), and sandy areas (tan). Greenland and Antarctica are white and light gray because they are mostly covered with ice and snow. Ocean colors show the depth of the seafloor and range from light blue where the seafloor is shallow to darker blue where it is deep.

Parts of the seafloor have mountains, the largest of which form islands like Hawaii. Most mountains on the seafloor do not reach sea level and are termed *seamounts*. Some islands and seamounts, like Hawaii, are in long belts, which we refer to as *island and seamount chains*. Other islands and seamounts are isolated or form irregular clusters.

Some large islands, such as New Zealand, look like a small version of a continent.

Much of the ocean floor is moderately deep—3 to 5 km (9,800–16,000 ft)—and has a fairly smooth surface. These smooth regions are *abyssal plains*.

Mid-ocean ridges are broad, symmetrical ridges that cross the ocean basins. They are 2 to 3 km (6,600–9,800 ft) higher than the average depth of the seafloor. One long ridge, named the East Pacific Rise, crosses the eastern Pacific and heads toward North America. Another occupies the middle of the Atlantic Ocean.

Cracks and steps cross the seafloor mostly at right angles to the mid-ocean ridges. These features are *oceanic fracture zones*.

Arctic Ocean

Greenland

North America

Atlantic Ocean

Pacific Ocean

South America

Atlantic Ocean

Antarctica

Some continents continue outward from the shoreline under shallow seawater (light blue in this image) for hundreds of kilometers, forming submerged benches known as *continental shelves*. Which coastlines have broad continental shelves, like those surrounding Great Britain?

All continents contain large interior regions with gentle topography. Some continents have flat coastal plains, while others have mountains along their edges. Some mountains, like the Ural Mountains, are in the middle of continents.

Most continental areas have elevations of less than 1 to 2 km (3,300 to 6,600 ft). Broad, high regions, called *plateaus*, reach higher elevations, such as the Tibetan Plateau of southern Asia. Continents also contain mountain chains and individual mountains. Mount Everest, the highest point in the world, is almost 9 km (30,000 ft) in elevation.

Before You Leave This Page Be Able To

☑ Identify on a world map the named continents and oceans.

☑ Identify on a world map the main types of features on the continents and in the oceans.

☑ Describe the main characteristics for each type of feature, including whether it occurs in the oceans, on continents, or as islands.

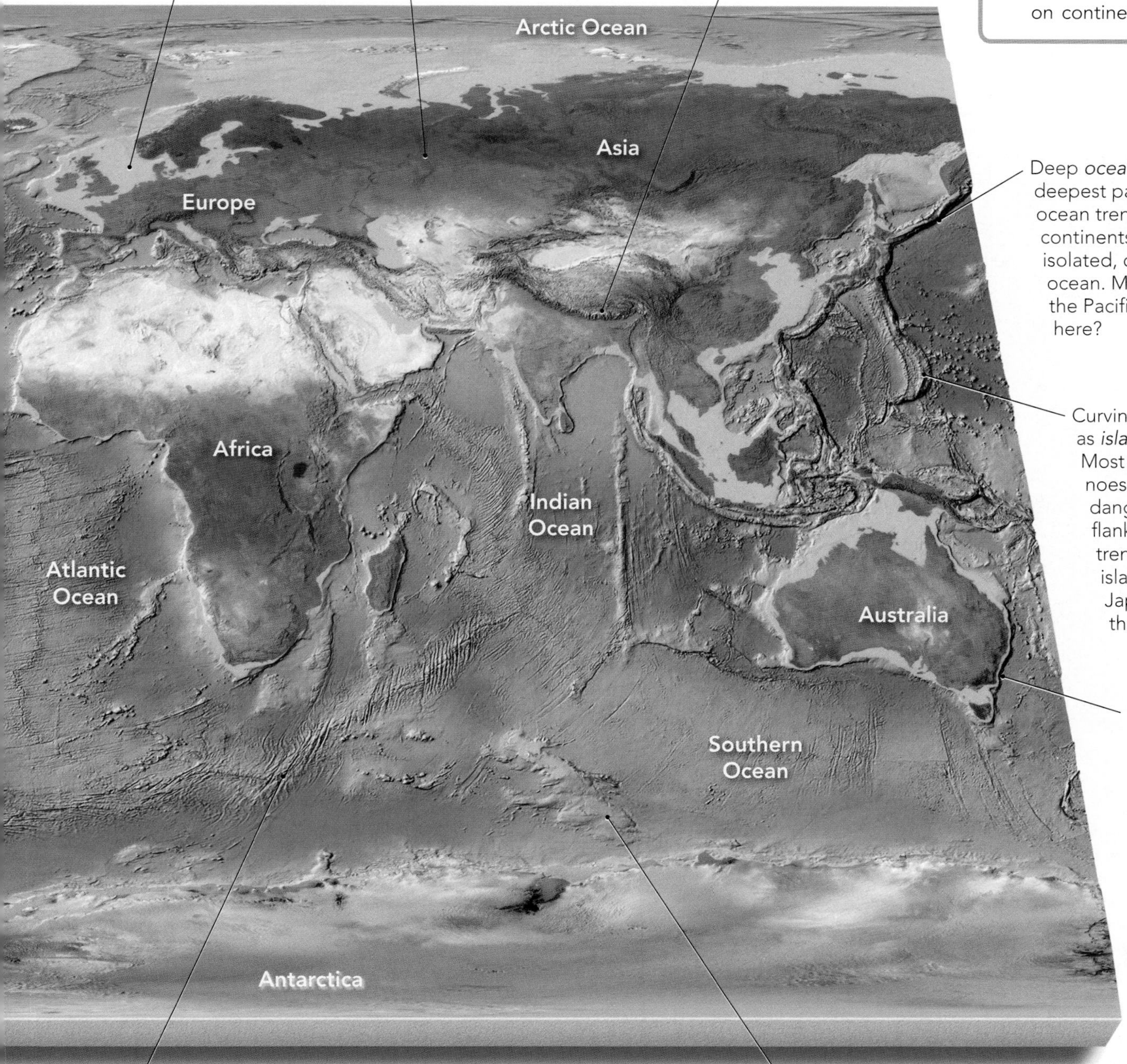

Arctic Ocean

Asia

Europe

Africa

Indian Ocean

Atlantic Ocean

Australia

Southern Ocean

Antarctica

Deep *ocean trenches* make up the deepest parts of the ocean. Some ocean trenches follow the edges of continents, whereas others form isolated, curving troughs out in the ocean. Most ocean trenches are in the Pacific Ocean. Why are they here?

Curving chains of islands, known as *island arcs*, cross the seafloor. Most of the islands are volcanoes, and many are active and dangerous. Most island arcs are flanked on one side by an ocean trench. Offshore of the Mariana island arc, located south of Japan, is the Mariana Trench, the deepest in the world.

Some continents (such as South America) are flanked by an ocean trench, but other continents, such as Australia and Africa, have no nearby trenches.

Mid-ocean ridges and their associated fracture zones encircle much of the globe. In the Atlantic and Southern Oceans, they occupy a position halfway between the adjacent continents.

The oceans contain several broad, elevated regions called *oceanic plateaus*. The Kerguelen Plateau near Antarctica is one example, and another oceanic plateau lies northeast of Australia.

3.1

3.2 Where Do Earthquakes and Volcanoes Occur?

EARTHQUAKES AND VOLCANOES are spectacular manifestations of geology. Many of these are in distant places, but some are close to where we live. The distributions of earthquakes and volcanoes are not random, but instead define clear patterns and show a close association with mountain belts and other regional features. These patterns reflect important, large-scale Earth processes.

A Where Do Most Earthquakes Occur?

On this map, yellow circles show the locations of moderate to strong earthquakes that occurred between 1973 and 2000. Observe the distribution of earthquakes before reading on. What patterns do you notice? Which regions have many earthquakes and which have few? Are earthquakes associated with certain types of features?

Earthquakes are not distributed uniformly across the planet. Most are concentrated in discrete belts, such as one that runs along the western coasts of North and South America.

Most earthquakes in the oceans occur along the winding crests of mid-ocean ridges. Where the ridges curve or zigzag, so do the patterns of earthquakes.

Earthquakes are sparse in some continental interiors but are abundant in others, like the Middle East, China, and Tibet.

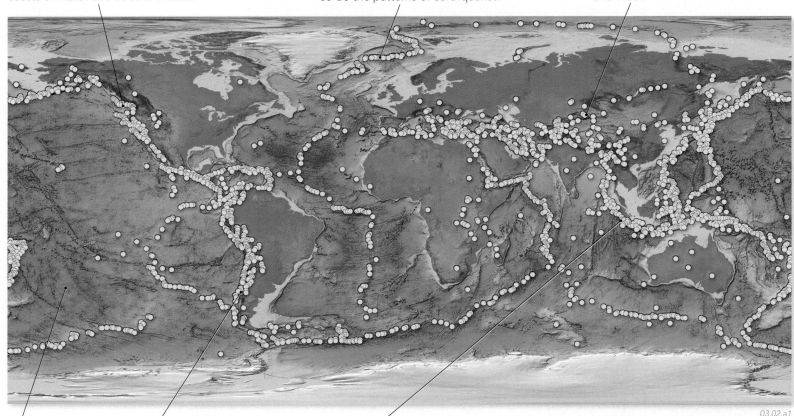

03.02.a1

Large areas of the seafloor, especially the abyssal plains, have few earthquakes. Volcanically active islands, like Hawaii, in the middle of the Pacific Ocean, do have earthquakes.

Some continental edges experience many earthquakes, but other edges have few. Earthquakes are common along the western coasts of South America and North America, and these edges also have narrow continental shelves. There are few earthquakes along the eastern coasts of the Americas, where the continental shelves are wide.

Ocean trenches and associated island arcs have numerous earthquakes. In fact, many of the world's largest and most deadly earthquakes occur near ocean trenches. A recent example was the large earthquake that produced deadly ocean waves in the Indian Ocean in December 2004.

Before You Leave These Pages Be Able To

✓ Show on a world relief map the major belts of earthquakes and volcanoes.

✓ Describe how the distribution of volcanoes corresponds to that of earthquakes.

✓ Compare the distributions of earthquakes, volcanoes, and high elevations.

B Which Areas Have Volcanoes?

On the map below, orange triangles show the locations of volcanoes that have been active in the last several million years. Observe the distribution of volcanoes and note which areas have volcanoes and which have none. How does this distribution compare with the distribution of earthquakes?

Volcanoes, like earthquakes, are widespread, but commonly occur in belts. One belt extends along the western coasts of North and South America.

Some volcanoes occur in the centers of oceans, such as the volcanoes near Iceland. Iceland is a large volcanic island along the mid-ocean ridge in the center of the North Atlantic Ocean.

Volcanoes occur along the western edge of the Pacific Ocean, extending from north of Australia through the Philippines and Japan. Many are part of island arcs, associated with ocean trenches and earthquakes.

03.02.b1

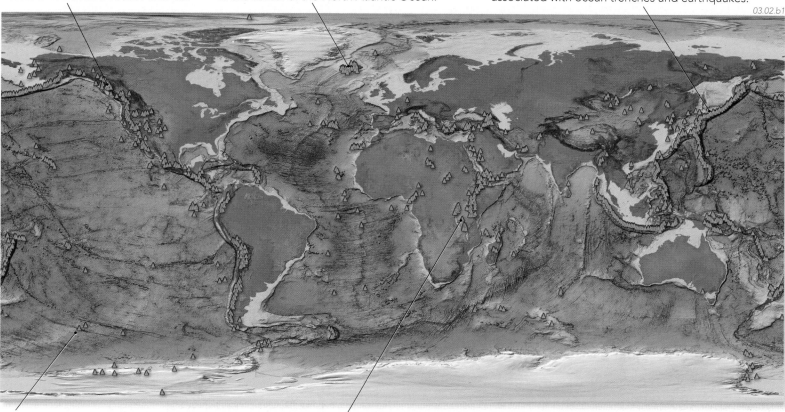

Volcanic eruptions occur beneath the oceans, but this map shows only the largest submarine volcanic mountains. Volcanism is widespread along mid-ocean ridges, but it generally does not form mountains.

Some volcanoes form in the middle of continents, such as in the eastern part of Africa and China.

This map (▼) shows the topography of Earth's surface and seafloor, with high elevations in brown, low land elevations in green, shallow seafloor in light blue, and deep seafloor in dark blue.

Using the three maps shown here, compare the distributions of earthquakes, volcanoes, and high elevations. Identify areas where there are (1) mountains but no earthquakes, (2) mountains but no volcanoes, and (3) earthquakes but no volcanoes. Make a list of these areas, or mark the areas on a map.

03.02.b2

3.2

3.3 What Causes Tectonic Activity to Occur in Belts?

WHY DO EARTHQUAKES AND VOLCANOES occur in belts around Earth's surface? Why are there vast regions that have comparatively little of this activity? What underlying processes cause these observed patterns? These and other questions helped lead to the *theory of plate tectonics*.

A What Do Earthquake and Volcanic Activity Tell Us About Earth's Lithosphere?

1. Examine the map below, which shows the locations of recent earthquakes (yellow circles) and volcanoes (orange triangles). After noting the patterns, compare this map with the lower map and then read the associated text.

2. On the upper map, there are large regions that have few earthquakes and volcanoes. These regions are relatively stable and intact pieces of Earth's outer layers. There are a dozen or so of these regions, each having edges defined by belts of earthquakes and volcanoes.

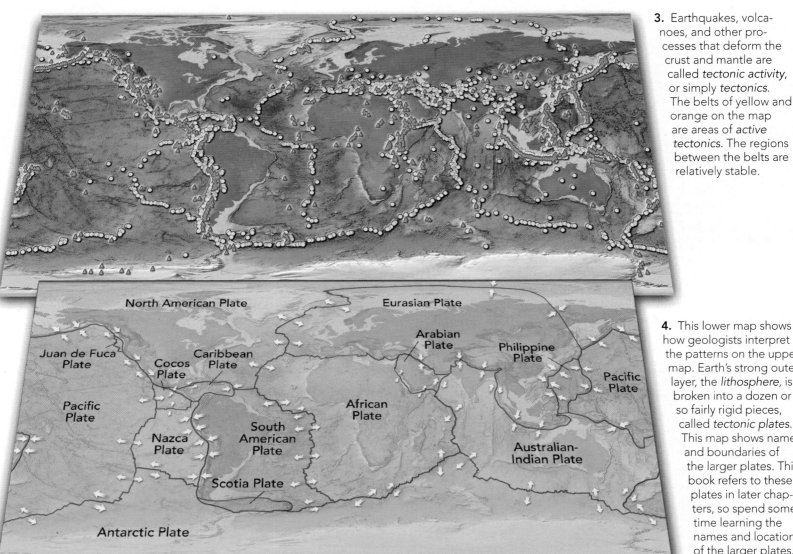

3. Earthquakes, volcanoes, and other processes that deform the crust and mantle are called *tectonic activity*, or simply *tectonics*. The belts of yellow and orange on the map are areas of *active tectonics*. The regions between the belts are relatively stable.

4. This lower map shows how geologists interpret the patterns on the upper map. Earth's strong outer layer, the *lithosphere*, is broken into a dozen or so fairly rigid pieces, called *tectonic plates*. This map shows names and boundaries of the larger plates. This book refers to these plates in later chapters, so spend some time learning the names and locations of the larger plates.

03.03.a1

5. Compare the two maps and note how the distribution of tectonic activity, especially earthquakes, outlines the shapes of the plates. Earthquakes are a better guide to plate boundaries than are volcanoes. Most volcanoes do lie near plate boundaries, but many plate boundaries have no volcanoes. Furthermore, some volcanoes and earthquakes occur in the middle of plates. As a general rule, though, most tectonic activity occurs near plate boundaries.

B How Do Plates Move Relative to One Another?

Plate boundaries have tectonic activity because plates are moving *relative to one another*. For this reason, we talk about the *relative motion* of plates across a plate boundary. Two plates can move away, toward, or sideways relative to one another, resulting in three types of plate boundaries: *divergent, convergent,* and *transform.*

Divergent Boundary

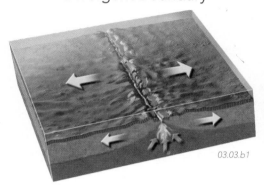

03.03.b1

Convergent Boundary

03.03.b2

Transform Boundary

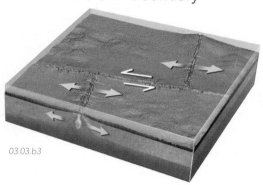

03.03.b3

At a *divergent boundary*, two plates move apart relative to one another. In most cases, magma fills the space between the plates.

At a *convergent boundary*, two plates move toward one another. A typical result is that one plate slides under the other.

At a *transform boundary*, two plates move horizontally past one another, as shown by the white arrows on the top surface.

C Where Are the Three Types of Plate Boundaries?

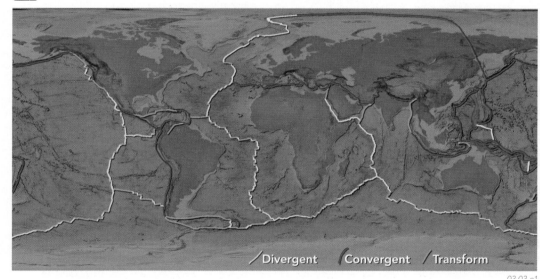

Divergent   Convergent   Transform

03.03.c1

This map shows plate boundaries according to type. Compare this map with the maps in part A and with those shown earlier in the chapter. For each major plate, note the types of boundaries between this plate and other plates it contacts. Then use the various maps to determine whether each type of plate boundary has the following features:

- Earthquakes
- Volcanoes
- Mountain belts
- Mid-ocean ridges
- Ocean trenches

Rigid and Not-So-Rigid Plates

If all tectonic activity were due to purely rigid plates converging, diverging, or moving along transform boundaries, earthquakes and volcanoes would only occur at plate boundaries. The maps of earthquakes and volcanoes show that this is not precisely true. Most tectonic activity occurs at or near inferred plate boundaries, but some tectonic activity happens well away from any boundary. Clearly, Earth is more complicated than a simple plate-tectonic model would imply. This is largely because some parts of the lithosphere are weaker than other parts. Forces can be transmitted through the strong parts of a plate, causing the weaker parts to break and slip. This causes any plate boundaries that are within a continent, like the southern part of Asia, to be very complex and somewhat diffuse. In a few cases, including the seafloor south of India, tectonic activity within a plate indicates that a new plate boundary may be forming.

Before You Leave This Page Be Able To

✓ Describe plate tectonics and how it explains the distribution of tectonic activity.

✓ Sketch and explain the three types of plate boundaries.

✓ Compare the three types of plate boundaries with the distributions of earthquakes, volcanoes, mountain belts, mid-ocean ridges, and ocean trenches.

3.3

3.4 What Happens at Divergent Boundaries?

AT MID-OCEAN RIDGES, Earth's tectonic plates diverge (move apart). Ridges are the sites of many small to moderate-sized earthquakes and much submarine volcanism. On the continents, divergent motion can split a continent into two pieces, forming a new ocean basin as the pieces move apart.

A What Happens at Mid-Ocean Ridges?

Mid-ocean ridges are divergent plate boundaries where new oceanic lithosphere forms as two oceanic plates move apart. These boundaries are also called *spreading centers* because of the way the plates spread apart.

1. A narrow trough, or *rift*, runs along the axis of most mid-ocean ridges. The rift forms because large blocks of crust slip down as spreading occurs. The movement causes faulting, resulting in frequent small to moderate-sized earthquakes.

2. As the plates move apart, solid mantle in the asthenosphere rises toward the surface. It partially melts in response to a decrease in pressure. The molten rock (magma) rises along narrow conduits, accumulates in magma chambers beneath the rift, and eventually becomes part of the oceanic lithosphere.

3. Much of the magma solidifies at depth, but some erupts onto the seafloor, forming submarine lava flows. These eruptions create new ocean crust that is incorporated into the oceanic plates as they move apart.

4. Mid-ocean ridges are elevated above the surrounding seafloor because they consist of hotter, less dense materials, including magma. They also are higher because the underlying lithosphere is thinner beneath ridges than beneath typical seafloor. Lower density materials and thin lithosphere mean that the plate "floats" higher above the underlying asthenosphere. The elevation of the seafloor decreases away from the ridge because the rock cools and contracts, and because the less dense asthenosphere cools enough to become part of the more dense lithosphere.

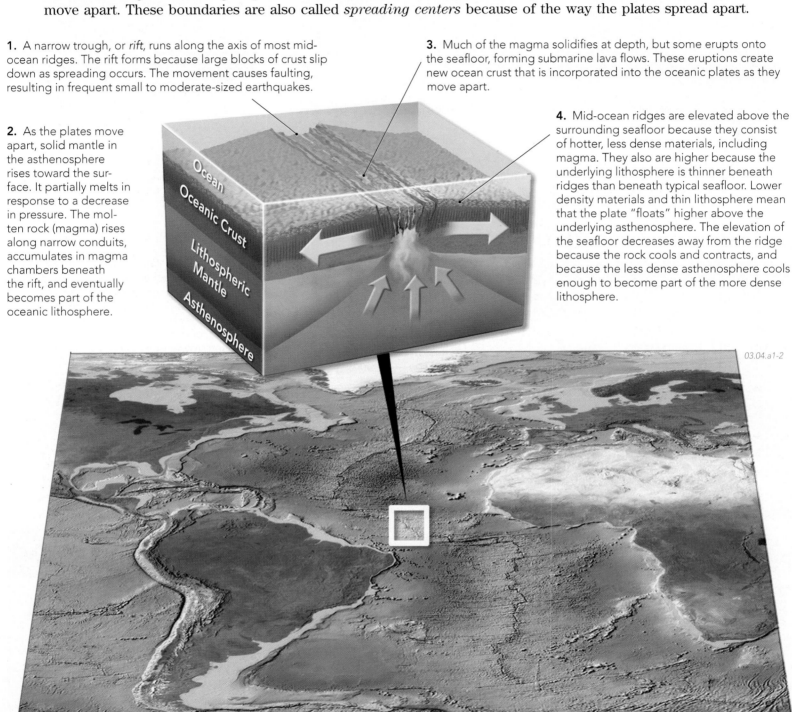

03.04.a1-2

B What Happens When Divergence Splits a Continent Apart?

Most divergent plate boundaries are beneath oceans, but a divergent boundary may also form within a continent. This process, called *continental rifting*, creates a *continental rift*, such as the Great Rift Valley in East Africa. Rifting can lead to seafloor spreading and formation of a new ocean basin, following the progression shown here.

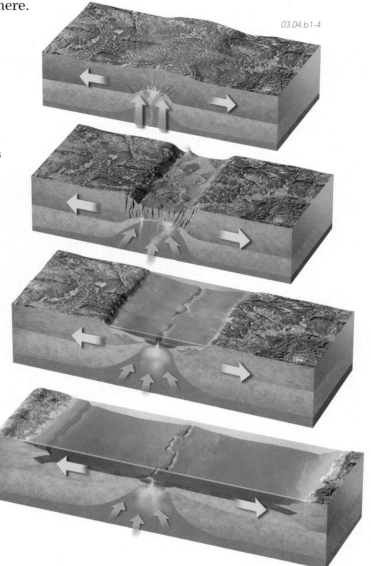

03.04.b1-4

1. The initial stage of continental rifting commonly includes broad uplift of the land surface as mantle-derived magma ascends into and pushes up the crust. The magma heats and can melt parts of the continental crust, producing additional magma. Heating of the crust causes it to expand, which results in further uplift.

2. Stretching of the crust causes large crustal blocks to drop down along faults, forming a *continental rift*, like in the Great Rift Valley. The down-dropped blocks may form *basins* that can trap sediment and water, resulting in lakes. Deep rifting causes solid mantle material in the asthenosphere to flow upward and partially melt. The resulting magma may solidify beneath the surface or may erupt from volcanoes and long fissures on the surface. The entire crust thins as it is pulled apart, so the central rift becomes lower in elevation over time.

3. If rifting continues, the continent splits into two pieces and a narrow ocean basin forms as seafloor spreading takes place. A modern example of this is the narrow Red Sea, which runs between Africa and the Arabian Peninsula. As the edges of the continents move away from the heat associated with active spreading, the thinned crust cools and drops in elevation, eventually dropping below sea level. The continental margin ceases to be a plate boundary. A continental edge that lacks tectonic activity is called a *passive margin*.

4. With continuing seafloor spreading, the ocean basin becomes progressively wider, eventually becoming a broad ocean like the modern-day Atlantic Ocean. The Atlantic Ocean basin formed when North and South America rifted away from Europe and Africa, following the sequence shown here. Continental edges on both sides of the Atlantic are currently passive margins. Seafloor spreading continues today along the ridge in the middle of the Atlantic Ocean, so the Americas continue to move away from Europe and Africa.

5. East Africa and adjacent seas illustrate the different stages of continental rifting. Here, a piece of continent has been rifted away from Africa, and another piece is in the early stages of possibly doing the same (▶).

6. Early stages of rifting occur along the East African Rift, a long continental rift that begins near the Red Sea and extends into central Africa. The rift is within an elevated (uplifted) region and has several different segments, each featuring a down-dropped rift. Some parts of the rift contain large lakes.

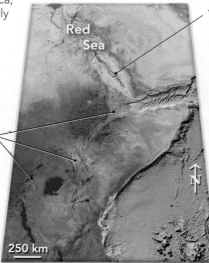

Red Sea

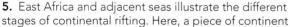
250 km

03.04.b5

7. The Red Sea represents the early stages of seafloor spreading. It began forming about 50 million years ago when the Arabian Peninsula rifted away from Africa. The Red Sea continues to spread and slowly grow wider.

Before You Leave This Page Be Able To

☑ Sketch, label, and explain the features and processes of an oceanic divergent boundary.

☑ Sketch and label the characteristics of a continental rift (i.e., a divergent boundary within a continent).

☑ Sketch, label, and explain the stages of continental rifting, using East Africa and the Red Sea as examples of stages.

3.4

3.5 What Happens at Convergent Boundaries?

CONVERGENT BOUNDARIES FORM when two plates move toward each other. Convergence can involve two oceanic plates, an oceanic plate and a continental plate, or two continental plates. Oceanic trenches, island arcs, and Earth's largest mountain belts form at convergent boundaries. Many of Earth's most dangerous volcanoes and largest earthquakes also occur along these boundaries.

A What Happens When Two Oceanic Plates Converge?

1. Convergence of two oceanic plates forms an *ocean-ocean convergent boundary*. One plate bends and slides beneath the other plate along an inclined zone. The process of one plate sliding beneath another plate is *subduction*, and the zone around the downward-moving plate is a *subduction zone*. Many large earthquakes occur in subduction zones.

2. An *oceanic trench* forms as the subducting plate moves down. Sediment and slices of oceanic crust collect in the trench, forming a wedge called an *accretionary prism*. This name signifies that material is being added (*accreted*) over time to the wedge- or prism-shaped region.

3. As the plate subducts, its temperature increases, releasing water from minerals in the downgoing plate. This water causes melting in the overlying asthenosphere, and the resulting magma is buoyant and rises into the overlying plate.

4. Some magma erupts, initially under the ocean and later as dangerous, explosive volcanoes that rise above the sea. With continued activity, the erupted lava and exploded volcanic fragments construct a curving belt of islands in an *island arc*. An example is the arc-shaped belt of the Aleutian Islands of Alaska. The area between the island arc and the ocean trench accumulates sediment, most of which comes from volcanic eruptions and from the erosion of volcanic materials in the arc.

5. Magma that solidifies at depth adds to the volume of the crust. Over time, the crust gets thicker and becomes transitional in character between oceanic and continental crust. Volcanic islands join to form more continuous strips of land, as occurred to form the island of Java in Indonesia.

Accretionary Prism

Subducting Lithosphere

Asthenosphere

03.05.a1

B What Happens When an Oceanic Plate and a Continental Plate Converge?

1. The convergence of an oceanic and a continental plate forms an *ocean-continent convergent boundary*. Along this boundary, the denser oceanic plate subducts beneath the more buoyant continental plate.

2. An oceanic trench marks the plate boundary and receives sediment from the adjacent continent. This sediment and material scraped off the oceanic plate form an *accretionary prism*.

3. Volcanoes form on the surface of the overriding continental plate in the same way the volcanoes form in an ocean-ocean convergent boundary. These volcanoes erupt, often violently, producing large amounts of volcanic ash, lava, and mudflows, which pose a hazard for people who live nearby. Examples include large volcanoes of the Andes of South America and the Cascade Range of Washington, Oregon, northern California, and southern British Columbia.

4. Compression associated with the convergent boundary squeezes the crust for hundreds of kilometers into the continent. The crust deforms and thickens, resulting in uplift of the region. Uplift and volcanism may produce a high mountain range, such as the Andes.

5. Magma forms by melting of the asthenosphere above the subduction zone. It can solidify at depth, rise into the overlying continental crust before solidifying, or reach the surface and cause a volcanic eruption.

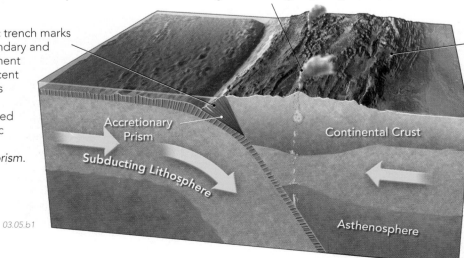

Accretionary Prism

Subducting Lithosphere

Continental Crust

Asthenosphere

03.05.b1

C   What Causes the Pacific Ring of Fire?

Volcanoes surround the Pacific Ocean, forming the *Pacific Ring of Fire*, as shown in the map below. The volcanoes extend from the southwestern Pacific, through the Philippine Islands, Japan, and Alaska, and then down the western coasts of the Americas. The Ring of Fire results from subduction on both sides of the Pacific Ocean.

03.05.c1

1. In the Pacific, new oceanic lithosphere forms along a mid-ocean ridge, the East Pacific Rise. Once formed, new lithosphere moves away from the ridge as seafloor spreading continues.

2. Oceanic lithosphere subducts beneath the Americas, forming oceanic trenches on the seafloor and volcanoes on the overriding, mostly continental, plates.

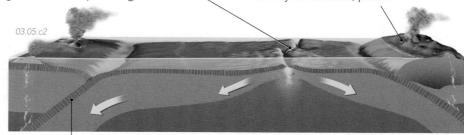

03.05.c2

3. Subduction of oceanic lithosphere also occurs to the west, beneath Japan and island arcs of the western Pacific.

4. More oceanic plate is subducted than is produced along the East Pacific Rise, so the width of the Pacific Ocean is shrinking with time.

D   What Happens When Two Continents Collide?

Two continental masses may converge along a *continent-continent convergent boundary*. This type of boundary is commonly called a *continental collision*, and it produces large mountain ranges.

The large plate in the figure to the right is partly oceanic and partly continental, and the oceanic part is being subducted under another continent at a convergent boundary.

As the oceanic part of the plate continues to subduct, the two continents become closer to each other. Magmatic activity occurs in the overriding plate above the subduction zone. The edge of the approaching continent has no such activity because it is not a plate boundary, yet.

When the converging continent arrives at the subduction zone, it may partially slide under the other continent or simply clog the subduction zone as the two continents collide. Because the two continents are thick and have the same density, neither can be easily subducted beneath the other and into the asthenosphere. Continental collisions form enormous mountain belts and high plateaus, such as the Himalaya and Tibetan Plateau of southern Asia. The Himalaya and Tibetan Plateau are still forming today, as continental crust of India collides with the southern edge of Asia.

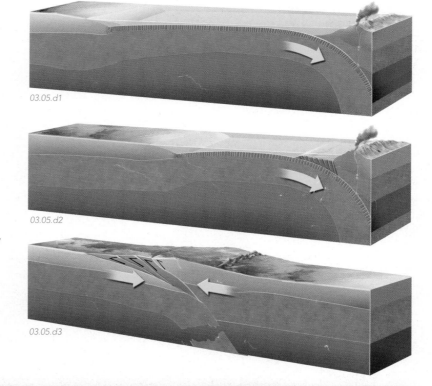

03.05.d1

03.05.d2

03.05.d3

Before You Leave This Page Be Able To

✓ Sketch, label, and explain the features and processes associated with ocean-ocean and ocean-continent convergent boundaries.

✓ Sketch, label, and explain the steps leading to a continental collision (continent-continent convergent boundary).

3.5

3.6 What Happens Along Transform Boundaries?

AT TRANSFORM BOUNDARIES, PLATES SLIP HORIZONTALLY past each other along *transform faults*. In the oceans, transform faults are associated with mid-ocean ridges. Transform faults combine with spreading centers to form a zigzag pattern on the seafloor. A transform fault can link different types of plate boundaries, such as a mid-ocean ridge and an ocean trench. Some transform boundaries occur beside or within a continent, sliding one large crustal block past another, as occurs along the San Andreas fault in California.

A Why Do Mid-Ocean Ridges Have a Zigzag Pattern?

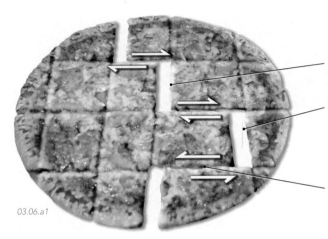

03.06.a1

To understand the zigzag character of mid-ocean ridges, examine how the two parts of this pizza have pulled apart, just like two *diverging* plates.

The break in the pizza did not follow a straight line. It took jogs to the left and the right, following cuts where the pizza was the weakest.

Openings created where the pizza pulled apart represent the segments of a mid-ocean ridge that are spreading apart. However, unlike a pizza, at a mid-ocean ridge, no open gaps exist because new material derived from the underlying mantle fills the space as fast as it opens, forming new oceanic crust.

The openings are linked by breaks, or *faults*, where the two parts of the pizza simply slide by one another. There are no gaps along these breaks, only *horizontal* movement of one plate sliding past the other. Arrows show the direction of relative motion. A fault that accommodates the horizontal movement of one tectonic plate past another is a *transform fault*.

Transform Faults Along the Mid-Ocean Ridge

1. Mid-ocean ridges, such as this one in the South Atlantic Ocean, have a zigzag pattern similar to the broken pizza.

2. In this region, spreading occurs along north-south ridges. The direction of spreading is east-west, perpendicular to the ridges.

3. East-west offsets are *transform faults* along which the two diverging plates simply slide past one another, like the breaks in the pizza. These transform faults link the spreading segments and have the relative motion shown by the white arrows.

4. Transform faults along mid-ocean ridges are generally perpendicular to the axis of the ridge. As in the pizza example, transform faults are parallel to the direction in which the two plates are spreading apart.

03.06.a2

Africa

South America

South Atlantic Ocean

Plate Boundary

6. Continuing outward from most transform faults is an *oceanic fracture zone*, which is a step in the elevation of the seafloor. A fracture zone is a former transform fault that now has no relative motion across it. It no longer separates two plates and instead is within a single plate. Opposite sides of the fracture zone have different elevations because they formed by seafloor spreading at different times in the past and so have had different amounts of time to cool and subside after forming at the spreading center. Younger parts of the plate are warmer and higher than older parts.

5. The zigzag pattern of mid-ocean ridges reflects the alternation of spreading segments with transform faults. In this example, the overall shape of the ridge mimics the edges of Africa and South America and so was largely inherited from the shape of the original rift that split the two continents apart.

B What Are Some Other Types of Transform Boundaries?

The Pacific seafloor and western North America contain several different transform boundaries. The boundary between the Pacific plate and the North American plate is mostly a transform boundary, with the Pacific plate moving northwest relative to the main part of North America.

1. The Queen Charlotte transform fault, shown as a long green line, lies along the edge of the continent, from north of Vancouver Island to southeastern Alaska.

2. The zigzag boundary between the Pacific plate and the small Juan de Fuca plate has two transform faults, shown here as green lines. These transform faults link three ridge segments that are spreading (shown here as yellow lines).

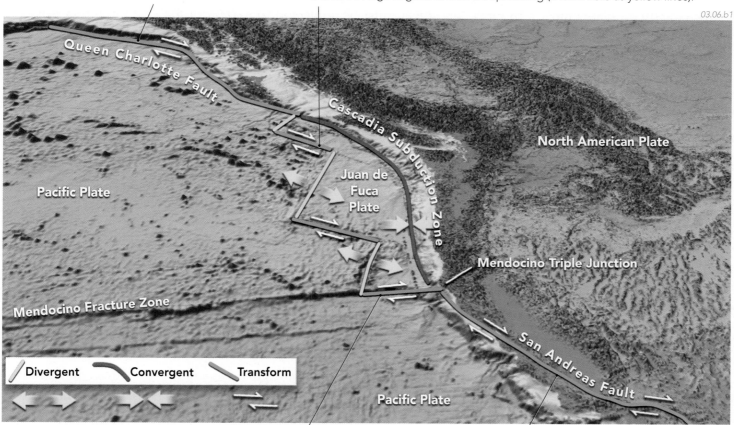

03.06.b1

3. The Mendocino fracture zone originated as a transform fault, but is now entirely within the Pacific plate and is no longer active. Oceanic crust to the north is higher because it is younger than oceanic crust to the south.

4. A transform fault links a spreading center (between the Pacific plate and the Juan de Fuca plate) with the Cascadia subduction zone and the San Andreas fault. The place where the three plate boundaries meet is a *triple junction*. The Mendocino triple junction is the meeting place of two different transform faults and a subduction zone.

5. The San Andreas transform fault extends from north of San Francisco to southeast of Los Angeles. The part of California west of the fault is on the Pacific plate and is moving approximately 5 cm/yr to the northwest relative to the rest of North America. South of this map area, the transform boundary continues across southern California and into the Gulf of California.

03.06.b2 Carrizo Plain, CA

Before You Leave This Page Be Able To

- ✓ Sketch, label, and explain an oceanic transform boundary related to seafloor spreading at a mid-ocean ridge.

- ✓ Sketch, label, and explain the motion of transform faults along the west coast of North America.

6. Californians have a transform fault in their backyard. In central California, the San Andreas fault forms linear valleys, abrupt mountain fronts, and lines of lakes. In the Carrizo Plain (▶), the fault is a linear gash in the topography. Some streams follow the fault and others jog to the right as they cross the fault, recording relative movement of the two sides. In this view, the North American plate is to the left, and the Pacific plate is to the right and is being displaced toward the viewer at several centimeters per year.

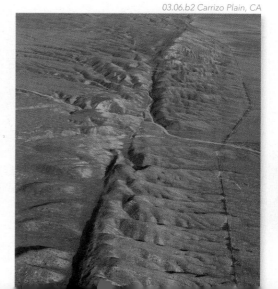

3.6

3.7 How Do Plates Move and Interact?

THE PROCESS OF PLATE TECTONICS circulates material back and forth between the asthenosphere and the lithosphere. Some asthenosphere becomes lithosphere at mid-ocean spreading centers and then takes a slow trip across the ocean floor before going back down into the asthenosphere at a subduction zone. Besides creating and destroying lithosphere, this process is the major way that Earth transports heat to the surface.

A What Moves the Plates?

How exactly do plates move? To move, an object must be subjected to a *driving force* (a force that drives the motion). The driving force must exceed the *resisting forces*—those forces that resist the movement, such as friction and any resistance from other material that is in the way. What forces drive the plates?

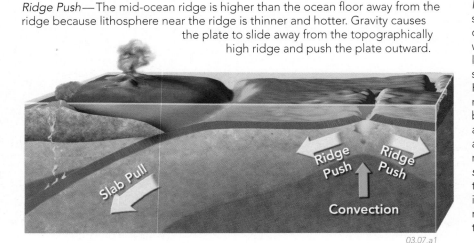

Slab Pull—Subducting oceanic lithosphere is more dense than asthenosphere, so gravity pulls the plate downward into the asthenosphere. Slab pull is a significant force, and a plate being subducted generally moves faster than plates not being subducted. Subduction sets up other forces in the mantle that can work with or against slab pull.

Ridge Push—The mid-ocean ridge is higher than the ocean floor away from the ridge because lithosphere near the ridge is thinner and hotter. Gravity causes the plate to slide away from the topographically high ridge and push the plate outward.

Mantle Convection—The asthenosphere, although a solid, is capable of flow. It experiences convection, where hot material rises due to its lower density, while cold material sinks because it is more dense. Hot material rises at mid-ocean ridges, cools, and eventually sinks back into the asthenosphere at a subduction zone. *Convection* also occurs at centers of upwelling mantle material called *hot spots*, and it can help or hinder the motion of a plate. Another important source of forces is the motion of a plate with respect to the underlying mantle.

03.07.a1

B How Fast and in What Directions Do Plates Move Relative to One Another?

Plates move at 1 to 15 cm/yr, about as fast as your fingernails grow. This map shows velocities and relative motions along major plate boundaries, based on long-term rates. Arrows indicate whether the plate boundary has divergent (outward pointing), convergent (inward pointing), or transform (side by side) motion.

03.07.b1

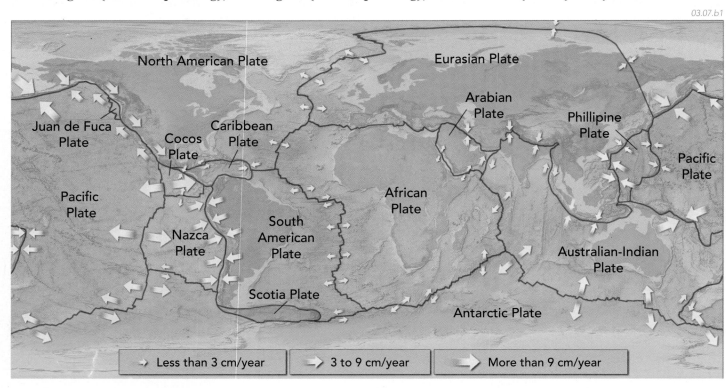

North American Plate
Eurasian Plate
Arabian Plate
Phillipine Plate
Juan de Fuca Plate
Caribbean Plate
Cocos Plate
Pacific Plate
Pacific Plate
African Plate
Nazca Plate
South American Plate
Australian-Indian Plate
Scotia Plate
Antarctic Plate

→ Less than 3 cm/year → 3 to 9 cm/year → More than 9 cm/year

C Is There a Way to Directly Measure Plate Motions?

Modern technology allows direct measurement of plate motions using satellites, lasers, and other tools. The measured directions and rates of plate motions are consistent with our current concept of lithospheric plates and with the theory of plate tectonics.

Global Positioning System (GPS) is an accurate location technique that uses small radio receivers to record signals from several dozen Earth-orbiting satellites. By attaching GPS receivers to sites on land and monitoring changes in position over time, geologists produce maps showing motions for each plate. Arrows point in the direction of motion, and longer arrows indicate faster motion.

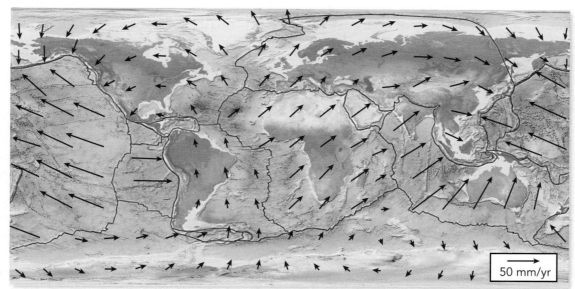

50 mm/yr

03.07.c1

Note the motions of different plates. Africa is moving to the northeast, away from South America. North America is moving westward and rotating counterclockwise in this view. These motions match predictions from the theory of plate tectonics.

D What Happens Where Plate Boundaries Change Their Orientation?

The boundary between two plates may be of a different type in different places. It can change from divergent to transform, for example, depending on its orientation compared to the direction of relative plate movement. Nearly all plate boundaries contain curves or abrupt bends, so most boundaries change type as they cross Earth's surface.

▼ **1.** As these two interlocking blocks pull apart, two gaps (equivalent to spreading centers) form, linked by a transform boundary where the blocks slip horizontally by one another. The boundary changes its type as it changes its orientation.

03.07.d1

3. The boundary between the North American and Pacific plates illustrates how a plate boundary changes its character as it changes orientation. In most of Alaska, the two plates converge, and the Pacific plate subducts beneath North America.

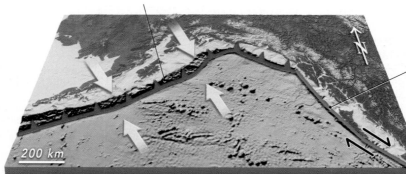

200 km

03.07.d2

4. To the southeast, the plate boundary bends and becomes parallel to the edge of the continent. As it bends, the plate boundary changes from being convergent to largely having a transform motion. The transform fault, named the Queen Charlotte fault, allows the Pacific plate to slide northwestward past North America at a rate of 5 cm/yr.

▶ **2.** A small-scale example of this type of change in motion occurred along a fault in Alaska, where lateral motion on the fault during an earthquake caused local pulling apart of the rock and ice as the fault curved around several bends.

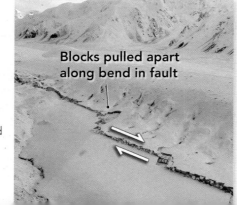

Blocks pulled apart along bend in fault

03.07.d3

Before You Leave This Page Be Able To

☑ Sketch and explain the driving forces of plate tectonics.

☑ Describe the typical rates of relative motion between plates.

☑ Describe one way to directly measure plate motion.

☑ Sketch, label, and explain how a plate boundary can change its type as its orientation changes.

3.7

3.8 What Geologic Features Does Plate Tectonics Help Explain?

MANY OF EARTH'S LARGE FEATURES, including mountains, mid-ocean ridges, and ocean trenches, are the result of plate tectonics. Ridges and trenches are in predictable places on the ocean floor, and most mountain ranges are situated along continental margins. The theory of plate tectonics explains many other features, such as island and seamount chains, continents that look like they would fit together, and variations in the age of the seafloor. The theory built upon earlier ideas about continental drift and seafloor spreading.

A Is the Age of the Seafloor Consistent with Plate Tectonics?

According to plate tectonics, oceanic crust forms from upwelling magma and spreading at a mid-ocean ridge and then moves away from the ridge with further spreading. If so, the crust should be youngest near the ridge, where it was just formed, and should be progressively older away from the ridge. Also, oceanic crust near the ridge will not have had time to accumulate much sediment, but the sediment cover should thicken outward from the ridge.

Since 1968, ocean-drilling ships have drilled hundreds of deep holes into the seafloor. Geologists use drill cores and other drilling results to measure the thickness of sediment and examine the underlying volcanic rocks (basalt). They analyze samples of sediment, rock, and fossils to determine the age, character, and origin of the materials. From this, they interpret the geologic history.

Sediment

Basalt

Drill core samples reveal that sediment is thin or absent on the ridge but becomes thicker away from the ridge. Age determinations from fossils in the sediment and from underlying volcanic rocks show that oceanic crust gets systematically older away from mid-ocean ridges. Drilling results from many parts of the oceans strongly support the theory of plate tectonics.

03.08.a1

B If Continents Have Rifted Apart, Do Their Outlines and Geology Match?

According to the theory of plate tectonics and the earlier hypothesis of continental drift, the South Atlantic Ocean formed when South America rifted away from Africa. If the continents are moved back together, their outlines should match. In this figure, we have moved the continents most of the way back together so their outlines can be compared. Note that where one continent juts out, the other curves in.

In 1973, geologist Edward Bullard showed that continents fit together better if we include the continental shelves. Bullard pointed out that the shelves are parts of each continent, although hidden beneath shallow seas. If included here, the continental shelves would mostly fill the gaps between the continents.

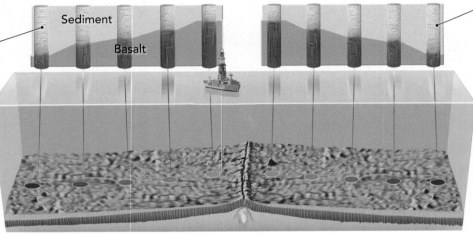

South America

Africa

03.08.b1

If the continents were once joined, they should have common geologic features, including the same ages and types of rocks and fossils. If restored back to their joined positions, the rocks and other features should match. Examine this map, which has brown areas showing large blocks composed of ancient (Precambrian) crust. Green areas and black lines mark younger rocks and geologic structures (faults and folds), respectively. Do the features on the two continents match?

A wealth of geologic data demonstrate a good match between the geology of South America and Africa. Detailed geologic comparisons of the age and character of the Precambrian rocks in South America with those in central Africa confirm the similarities in nature and age of the two areas. These results are consistent with the theory of plate tectonics and provide an excellent test of its validity. The concept of matching geology between the two continents has been very important to the mining industry. When precious minerals are discovered on one continent, geologists understand that they should also look for these commodities on the other continent, if the minerals formed before the continents rifted apart.

C How Does Plate Tectonics Help Explain Island and Seamount Chains?

Fairly straight lines of oceanic islands and submarine mountains (*seamounts*) cross some parts of the ocean floor. These *island and seamount chains* are different in character and origin from curved *island arcs*, which are related to subduction. How do linear chains of islands and seamounts form?

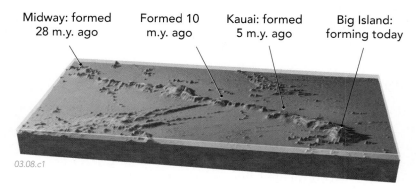

Midway: formed 28 m.y. ago   Formed 10 m.y. ago   Kauai: formed 5 m.y. ago   Big Island: forming today

03.08.c1

1. Most island and seamount chains are in the Pacific Ocean. One begins on Hawaii's Big Island and continues northwest more than 2,000 kilometers, passing through Midway Island, the site of a pivotal air and sea battle during World War II.

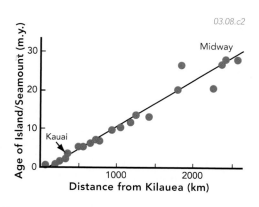

03.08.c2

2. Volcanoes are active on the Big Island today, but not on the other Hawaiian islands. Ages of volcanic rocks in the island and seamount chain increase systematically to the northwest. When we plot the ages of these rocks as a function of distance from Kilauea (the active volcano on the Big Island), there is a clear relationship between age and distance. How can we explain this pattern?

A Model for the Formation of Island and Seamount Chains

Island and seamount chains and most clusters of islands in the oceans have two key things in common: they were formed by volcanism and they are near sites that geologists interpret to be above unusually high-temperature regions in the deep crust and upper mantle. Geologists refer to these anomalously hot regions as *hot spots*.

1. This figure shows how linear island and seamount chains can be related to a plate moving over a hot spot. At a hot spot, hot mantle rises and melts, forming magma that ascends into the overlying plate. If the plate above the hot spot is moving relative to the hot spot, volcanism constructs a chain of volcanoes.

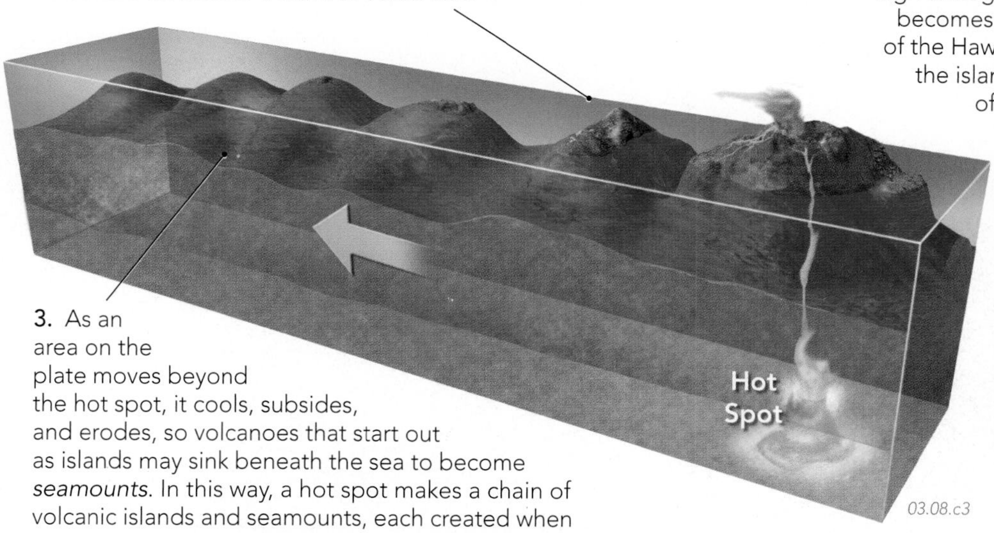

2. Magma generated by a hot spot may solidify at depth or form a volcanic mountain on the ocean floor.
If the submarine volcano grows high enough above the seafloor, it becomes a volcanic island. Each of the Hawaiian Islands, including the island of Maui (▶), consists of volcanoes. Geologists consider the hot spot to be currently below or near the eastern side of the Big Island, near Kilauea volcano.

03.08.c4 Haleakala, Maui, HI

3. As an area on the plate moves beyond the hot spot, it cools, subsides, and erodes, so volcanoes that start out as islands may sink beneath the sea to become *seamounts*. In this way, a hot spot makes a chain of volcanic islands and seamounts, each created when it was over the hot spot. According to this model, volcanoes above the hot spot may be erupting today, those close to but not above the hot spot are relatively young, and those farthest from the hot spot are older. The present volcanic activity and pattern of ages on the graph presented earlier are consistent with the hot-spot model and with the calculated motion of the Pacific plate on which Hawaii rides. The Hawaiian Islands and seamounts provide an excellent test of plate tectonics.

Hot Spot

03.08.c3

4. If a plate is not moving or is moving very slowly, the hot spot forms a cluster of volcanic islands and seamounts instead of a linear chain. The Galápagos, a cluster of volcanic islands in the eastern Pacific, are interpreted to be above a hot spot.

Before You Leave This Page Be Able To

✓ Predict the relative ages of seafloor from place to place using a map of an ocean with a mid-ocean ridge.

✓ Discuss how plate tectonics can explain similar continental outlines and geology on opposite sides of an ocean.

✓ Describe the characteristics of an island and seamount chain, and how it is interpreted to be related to a hot spot.

3.8

3.9 Why Is South America Lopsided?

THE TWO SIDES OF SOUTH AMERICA are very different. The western margin is mountainous while the eastern side and center of the continent have much less relief. The differences are a reflection of the present plate boundaries and of the continent's geologic history during the last 200 million years. South America nicely illustrates many aspects of plate tectonics, including the connections between tectonics of the land and seafloor. It also is an excellent example of how to analyze the major features of a region.

A What Is the Present Setting of South America?

The perspective view below shows South America and the surrounding oceans. Observe the topography of the continent, its margins, and the adjacent seafloor. See if you can find mid-ocean ridges, transform faults, ocean trenches, oceanic fracture zones, and other plate-tectonic features. From these features, infer the locations of plate boundaries in the oceans, and predict what type of motion (divergent, convergent, or transform) is likely along each boundary. Make your observations and predictions before reading the accompanying text.

1. The Galápagos Islands are located in the Pacific Ocean, west of South America. They consist of a cluster of about 20 volcanic islands, flanked by seamounts. Some of the islands are volcanically active and are interpreted to be over a hot spot.

9. The center of the South American continent has low, subdued topography because it is away from any plate boundaries. It is a relatively stable region that has no large volcanoes and few significant earthquakes. It is not tectonically active.

8. In the South Atlantic, the Mid-Atlantic Ridge is a divergent boundary between the South American and African plates. Seafloor spreading creates new oceanic lithosphere and moves the continents farther apart at a rate of 3 cm/yr.

2. The Andes mountain range follows the west coast of the continent and is the site of many dangerous earthquakes and volcanoes. A deep ocean trench along the edge of the continent marks where an oceanic plate subducts eastward beneath South America.

3. The Pacific seafloor contains mid-ocean ridges with the characteristic zigzag pattern of a divergent boundary with offsets along transform faults. The Nazca plate lies north of this ridge, and the Antarctic plate is to the south.

7. The eastern side of South America has a continental shelf that slopes gently toward the adjacent seafloor. There is no trench, no significant tectonic activity, or other evidence for a plate boundary. Instead, the continent and adjacent seafloor to the east are part of the same plate, and this edge of the continent is a passive margin.

6. The curved Scotia Island arc is related to a trench to the east and the westward subduction of the oceanic part of the South American plate.

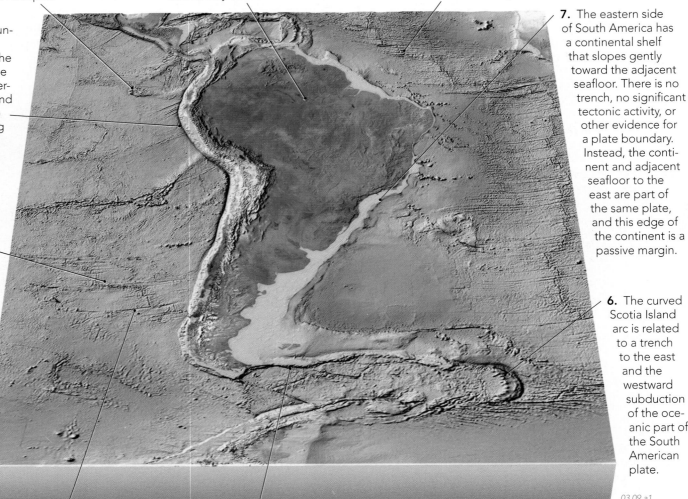

03.09.a1

4. Many oceanic fracture zones cross the seafloor and were formed along transform faults, but they are no longer plate boundaries.

5. The southern edge of the continent is very abrupt and has a curving "tail" extending to the east. This edge of the South American plate is a transform boundary where South America is moving west relative to oceanic plates to the south.

B What Is the Geometry of the South American Plate and Its Neighbors?

This cross section shows how geologists interpret the configuration of plates beneath South America and the adjacent oceans. Compare this cross section with the plate boundaries you inferred in part A.

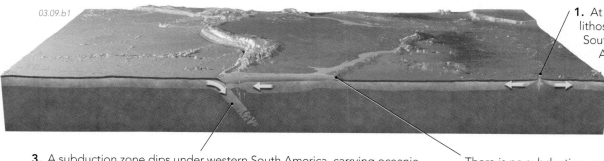

03.09.b1

1. At the Mid-Atlantic Ridge, new oceanic lithosphere is added to the African and South American plates as they move apart. As this occurs, the oceanic part of the South American plate gets wider.

2. Along the eastern edge of South America, continental and oceanic parts of the plate are simply joined together along a passive margin. There is no subduction, no seafloor spreading, or any type of plate boundary. As a result, the eastern continental margin of South America lacks volcanoes, earthquakes, and mountains.

3. A subduction zone dips under western South America, carrying oceanic lithosphere beneath the continent. Subduction causes large earthquakes and produces magma that feeds dangerous volcanoes in the Andes.

C How Did South America Develop Its Present Plate-Tectonic Situation?

If South America is on a moving plate, where was it in the past? When did it become a separate continent, and when did its current plate boundaries develop? Here is one commonly agreed-upon interpretation.

Around 140 million years ago, Africa and South America were part of a single large *supercontinent* called *Gondwana*. At about this time, a continental rift developed, starting to split South America away from the rest of Gondwana and causing it to become a separate continent.

By 100 million years ago, Africa and South America were completely separated by the South Atlantic Ocean. Spreading along the Mid-Atlantic Ridge moved the two continents farther apart with time. While the Atlantic Ocean was opening, oceanic plates in the Pacific were subducting beneath western South America. This subduction thickened the crust by compressing it horizontally and by adding magma, resulting in the formation and rise of the Andes mountain range.

Today, Africa and South America are still moving apart at a rate of several centimeters per year. As spreading along the mid-ocean ridge continues, the Atlantic Ocean gets wider. Earth, however, is not growing through time, and the expanding Atlantic Ocean is balanced by shrinking of the Pacific Ocean, whose oceanic lithosphere disappears into subduction zones along the Pacific Ring of Fire.

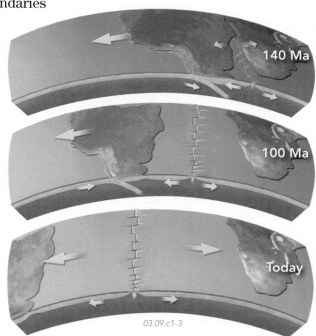

140 Ma

100 Ma

Today

03.09.c1-3

These photographs contrast the rugged Patagonian Andes of western South America with landscapes farther east that have more gentle relief and are not tectonically active.

03.09.c4 Cuernos del Paine, Chile

03.09.c5 Central Argentina

Before You Leave This Page Be Able To

✓ Sketch and describe the present plate-tectonic setting of South America, and explain the main features on the continent and adjacent seafloor.

✓ Discuss the plate-tectonic evolution of South America over the last 140 million years.

3.9

3.10 Where Is the Safest Place to Live?

AN UNDERSTANDING OF PLATE TECTONICS allows us to predict which places are at most risk from earthquakes and volcanoes. The most important things to know in this context are the locations and types of plate boundaries. In this exercise, you will examine an unknown ocean between two continents, make observations of the land and seafloor, and identify plate boundaries and other features. Using this information, you will predict the risk for earthquakes and volcanoes and determine the safest places to live.

Goals of This Exercise:

- Use the features of an ocean and two continental margins to identify possible plate boundaries and their types.
- Use the types of plate boundaries to predict the likelihood of earthquakes and volcanoes.
- Determine the safest sites for two cities, considering the potential for earthquakes and volcanic eruptions.
- Draw a cross section that shows the geometry of the plates at depth.

Procedures for the Map

This perspective view shows two continents, labeled A and B, and an intervening ocean. Use the topography on the land and seafloor to identify possible plate boundaries and then complete the following steps. Mark your answers on the map on the worksheet, which will be provided to you by your instructor in either paper or electronic form. Alternatively, your instructor may have you complete the investigation online.

1. Use the topographic features on land and the depths of the seafloor to identify possible plate boundaries. Draw lines showing the location of each plate boundary on the map in the worksheet. Label the boundaries as either divergent, convergent, or transform. Use colored pencils or different types of lines to better distinguish the different types of boundaries. Provide a legend that explains your colors and lines.

2. Draw circles [O], or use color shading, to show places, on land or in the ocean, where you think earthquakes are likely.

3. Draw triangles [▲] at places, on land or in the ocean, where you think volcanoes are likely. Remember that not all volcanoes form *directly on* the plate boundary; some form off to one side. For different plate-tectonic settings, consider where volcanoes form relative to that type of plate boundary.

4. Determine a relatively safe place to build one city on each continent. Show each location with a large plus sign [+] on the map. On the worksheet, explain your reasons for choosing these as the safest sites.

Continent A

N

Procedures for the Cross Section

The worksheet contains a modified version of this figure for you to use as a starting point for making a cross section. Add lines and colors to the front of the diagram to show the geology in the subsurface. Use other figures in this chapter as guides to the thicknesses of the lithosphere and to the subsurface geometries typical for each type of plate boundary. Your cross section should only show features on the front of the block diagram, not features that do not reach the front edge. Your cross section should clearly:

1. Identify the crust, mantle, lithosphere, and asthenosphere, and show an accurate representation of their relative thicknesses.

2. Show the locations and relationships between lithospheric plates at any spreading center or subduction zone.

3. Include arrows to indicate which way the plates are moving relative to each other.

4. Show where melting is occurring at depth to form volcanoes on the surface.

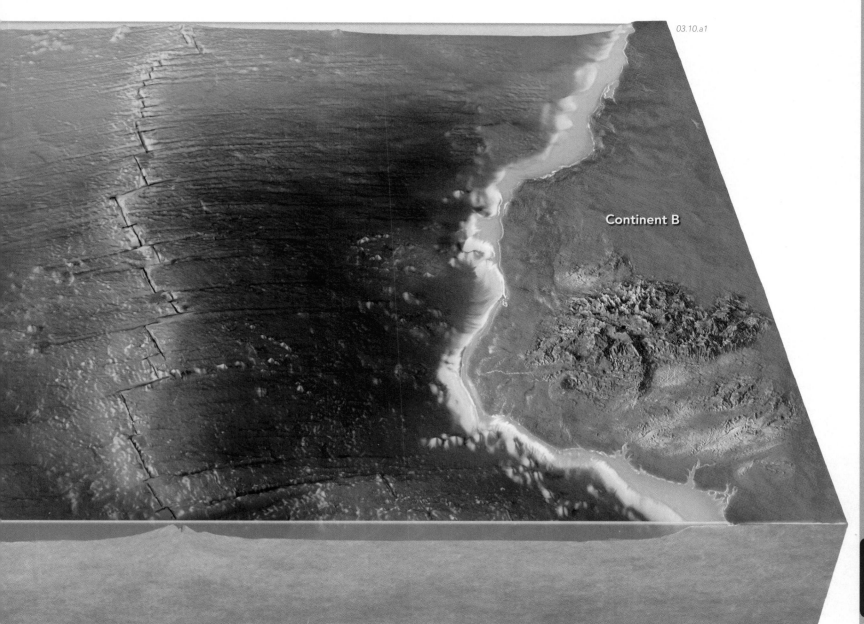

03.10.a1

Continent B

4 Earth Materials

EARTH'S SURFACE IS COMPOSED of many kinds of materials: black lava flows, white sandy beaches, red cliffs, and gray granite hills. Earth provides a treasury of gemstones and other mineral resources, many of which are essential to modern society. What kinds of materials are common on Earth, and how did the less common ones, such as gemstones, form? Here, we explore Earth materials from landscapes to atoms.

This perspective view (▶) shows satellite data superimposed over topography for southernmost California and adjacent Baja California, Mexico. The Peninsular Ranges, a forested mountainous area east of San Diego, are in greens and browns in the center of the image. The white line across the image, added for reference, marks the border between the United States and Mexico.

What are the rocks that make up the hills and mountains of the Peninsular Ranges? ▼

04.00.a2

04.00.a1

Peninsular Ranges

San Diego

UNITED STATES
MEXICO

10 km

The Peninsular Ranges contain many outcrops of grayish-colored rocks, most of which are igneous rocks like granite. When viewed up close (▼), the granite displays four different kinds of crystals: whitish, light pink, transparent gray, and black.

What are rocks made of, and what controls the color and other properties of a rock?

04.00.a4

San Diego County is a famous source of beautiful minerals, including tourmaline crystals (◀), that can be pink, purple, green, or all three colors.

What are crystals, how do they form, and where do we find them?

04.00.a3

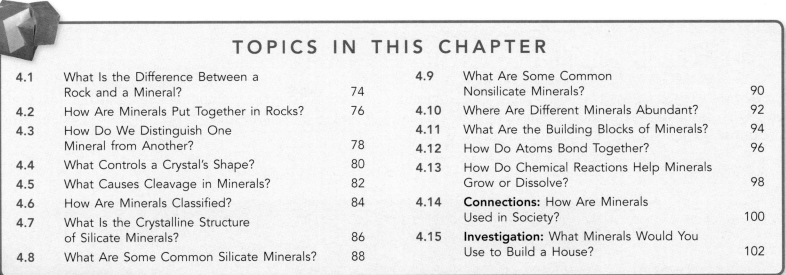

TOPICS IN THIS CHAPTER

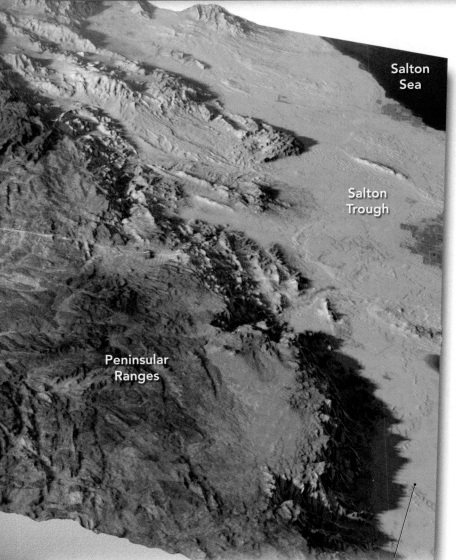

East of the Peninsular Ranges, the land drops down into the lowlands of the Salton Trough, which is characterized by sandy deserts, farmlands of the Imperial Valley, and several large, salty lakes, including the blue Salton Sea. The sand in the Salton Trough was eroded from the adjacent mountains and carried to the area by rivers or strong winds.

What are most sand grains composed of?

The Peninsular Ranges

The area called the Peninsular Ranges is a broad, upland region that stretches 1,500 km across southernmost California and southward onto the Baja Peninsula of Mexico. In this image, the mountains appear green because they are mostly covered by forests and other types of vegetation. The lowlands of the Salton Trough east (right) of the mountains receive much less rain and have a lighter color in this image because vegetation is sparse and sand and rocks cover the surface.

The mountains and lowlands contain a variety of rocks and other Earth materials. Gray granite and darker igneous rocks form most of the mountain range, and these formed at depths of 10 or more kilometers. They crystallized from magma that solidified deep within the crust, mostly between 130 and 80 million years ago. The range also includes similar-aged metamorphic rocks, many of which formed when sedimentary rocks were buried, heated, and deformed. The magma and metamorphism are related to subduction of oceanic plates beneath the western edge of North America (an ocean-continent convergent boundary).

Uplift of the Peninsular Range occurred during the last 10 million years, long after the granite and metamorphic rocks formed. Uplift and tilting of the range brought these deep rocks to the surface, in part at the same time that faulting downdropped large blocks in the Salton Trough. Once exposed at the surface, these rocks were weathered and eroded to produce sand, gravel, and other sediments that were transported into the lowlands to the east and west by streams, gravity, and wind. The sediments are soft and unconsolidated, and therefore easily eroded. Harder rocks, similar to those exposed in the mountains, underlie the sediments. The contrast of landscapes between the flat lowlands and the more rugged, granitic mountains reflects differences in the geologic history, the types of geologic materials exposed at the surface, and the climate.

4.0

4.1 What Is the Difference Between a Rock and a Mineral?

WHAT MATERIALS MAKE UP THE WORLD around us? What do we see if we look closely at a rock outcrop? How does the rock look when viewed with a magnifying glass? We investigate these questions using the beautiful scenery of Yosemite National Park in California.

A What Materials Make Up a Landscape?

1. Observe this photograph of Yosemite Valley, the heart of Yosemite National Park. What do you notice about the landscape?

2. This landscape is dominated by dramatic cliffs and steep slopes of massive gray rock perched above a green, forested valley. The valley is famous for waterfalls and for huge rock faces. The appropriately named Half Dome is in the right side of the photograph. What would we see if we got closer to this landscape?

04.01.a1

04.01.a2

3. From more than a few meters away, the rock making up Yosemite's cliffs looks fairly homogeneous. It all seems to be the same kind of gray rock, a kind of igneous rock called *granite*, cut by fractures.

04.01.a3
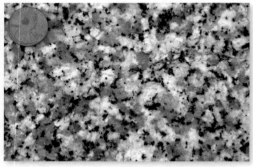

4. Closer examination reveals several different-colored grains in the rock: whitish, clear gray, and black. To better observe a rock at this scale, a geologist may collect a hand-sized piece, called a *hand specimen*.

04.01.a4

5. When examined with a magnifying glass or *hand lens*, the rock contains three or four different minerals with distinct appearances. The clear gray crystals all have similar chemical composition and physical properties, and so represent one kind of *mineral*. The whitish crystals are a different kind of mineral, and the black crystals are a third kind of mineral.

6. To examine the rock in even more detail, a geologist will cut a very thin slice from a hand specimen and glue it to a glass slide to make a *thin section*. The slice is so thin that light can pass through it. Geologists then examine the thin section using a microscope that has polarizing filters.

04.01.a5
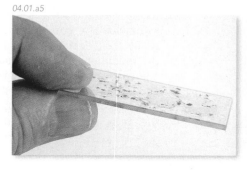

04.01.a6

2 mm

7. When polarized light shines through a thin section and optical filters, the internal structure of crystals interacts with the light in ways that allow us to observe diagnostic characteristics, to identify minerals, and to estimate percentages of minerals.

B What Is and What Is Not a Mineral?

What characteristics define a mineral? To be considered a mineral, a substance must fulfill all of the criteria listed below. A mineral is a naturally occurring, inorganic, crystalline solid with a relatively consistent composition.

Natural

04.01.b1 Fluorite

04.01.b2

A mineral must be *natural*. Crystals on the left grew naturally from hot water flowing through a rock, but synthetic crystals on the right grew in a laboratory. Natural diamonds are minerals, but synthetic diamonds grown in the lab are not.

Inorganic

04.01.b3 K-Feldspar

04.01.b4

The crystal on the left is *inorganic* and a mineral. The shells on the right have the same composition as the crystal, but they were made by clams and other creatures; they are not considered to be a mineral by most geologists.

Solid

04.01.b5

04.01.b6

All minerals are *solid*, not liquid or gaseous. Ice, a solid, is a mineral, but liquid water is not, even though it has the same composition. Liquid mercury, although natural and found in rocks, is not considered a mineral.

Ordered Internal Structure

04.01.b7 Calcite

04.01.b8

A mineral has an *ordered internal structure,* which means that atoms are arranged in a regular, repeating way. Such substances are considered to be *crystalline,* and they can form nice geometric crystals. The mineral on the left is crystalline, and the shape of the crystals reflects the internal arrangement of its atoms. The volcanic glass (obsidian) on the right is not crystalline. Its atoms are arranged in a random way, so volcanic glass is not a mineral.

Specific Chemical Composition

04.01.b9

04.01.b10

Minerals are homogeneous and so have specific chemical compositions that do not depend on the size of the sample that is analyzed. Table salt, which is the mineral *halite,* contains atoms of the chemical elements *sodium* (Na) and *chlorine* (Cl) in equal proportions, no matter how big or small the specimen. The rock on the right is not a mineral because different parts of the rock have very different compositions. Most minerals have a specific chemical formula, like NaCl for halite.

Rocks, Minerals, and "Minerals"

When we hear the word *mineral* used in the context of *vitamins and minerals*, are these minerals the same as the minerals described above? The answer is *no*. In a kitchen or pharmacy, the term *mineral* refers to a chemical *element*, such as potassium (K). This type of (nutritional) mineral is different from the crystalline mineral of geologists.

In geology, most minerals consist of at least two different chemical elements, i.e., naturally occurring chemical compounds, such as the *sodium* (Na) and *chlorine* (Cl) that make up the mineral *halite* (salt). Many minerals have three, four, or even more chemical elements. A few minerals, however, include only one chemical element,

and these are called native elements. The mineral *diamond*, for example, consists entirely of the element *carbon*.

Some rocks contain only a single mineral. Limestone may be 100 percent of the mineral calcite. Sandstone may be 100 percent quartz. Most rocks, like the granites from Yosemite National Park, include several different minerals, and each mineral is made of one or more elements. So, rocks are made of minerals, and minerals are made of elements. Although the vitamin pill you take with breakfast may not contain any geologic minerals, most of the nutritional elements in the pill were extracted from geologic minerals.

Before You Leave This Page Be Able To

☑ Explain the relationship between rocks, minerals, and chemical elements.

☑ Explain each characteristic that a material must have to be a mineral, listing an example that is a mineral and an example that is not.

☑ Explain the difference between a mineral in a vitamin pill and a geologic mineral.

4.1

4.2 How Are Minerals Put Together in Rocks?

THERE ARE MANY KINDS OF ROCKS, varying greatly in texture, color, and the minerals they contain. Geologists use the term *texture* to refer not only to the roughness or smoothness of a rock, but also to the way its grains and minerals are arranged. What controls the texture of a rock? How are minerals in a rock connected to one another? What can we determine about a rock from its texture and the types of minerals it contains?

A How Are Minerals Put Together in Rocks?

The beautiful and geologically interesting Engineer Mountain in the San Juan Mountains of southwestern Colorado contains two different types of rocks, providing examples of the two main ways that minerals occur in rocks.

04.02.a1

◄ The main mountain has an upper gray part and a lower reddish-brown part. Loose pieces of the upper gray part tumble down the hillside, forming gray slopes that cover some of the red rocks.

04.02.a2

▲ A closer view of the mountainside reveals differences between the gray and reddish parts. Both parts contain vertical fractures, but the reddish-brown part also has well-defined, nearly horizontal layers, whereas the gray part does not.

Two Types of Rocks

04.02.a3 San Juan Mtns., CO

04.02.a4 San Juan Mtns., CO

Crystalline—The gray rock displays light-colored crystals surrounded by a gray, fine-grained material (called a *matrix*) with crystals too small to see in this photograph. A rock composed of interlocking minerals that grew together is a *crystalline rock*. Crystalline rocks typically form in high-temperature environments by crystallization of magma, by metamorphism, or by precipitation from hot water. Some crystalline rocks consist of crystals formed from the precipitation of minerals in cooler waters, like when a lake evaporates.

Clastic—A close look at the reddish-brown layer reveals that it includes distinct pieces of rock derived from older weathered and eroded rocks. These pieces, called *clasts*, range from small sand grains to larger pebbles. A rock consisting of pieces derived from other rocks is a *clastic rock*. Most clastic rocks form on Earth's surface in low-temperature environments, such as sand dunes, rivers, and beaches—any place sediment is deposited. Clasts also compose some volcanic rocks, including those formed by explosions.

B What Different Attributes Do Rocks Display?

Crystalline rocks and clastic rocks can have various attributes, depending on the sizes and shapes of minerals and clasts in the rocks and on how the minerals and clasts are arranged. The photographs below, mostly of cut and polished rock slabs, show some common rock textures. The slabs are 5 to 30 cm (2 to 12 in.) across.

Crystalline Rocks

Clastic Rocks

Types of Minerals

Rocks can consist of one mineral or many minerals, but most contain more than one mineral. This crystalline rock (◄) is an igneous rock with several types of minerals, each having a distinctly different color. This clastic sedimentary rock (►) also includes different types of minerals, including pebbles of quartz in various shades of gray.

Sizes of Crystals or Clasts

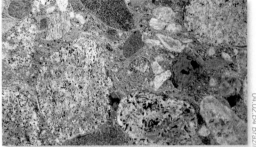

Rocks can contain various sizes of crystals and clasts. This granitic crystalline rock (◄) has coarse crystals, including some whitish ones that are more than 5 to 10 cm (2 to 4 in.) in diameter. This clastic sedimentary rock (►) includes various sizes of larger clasts in a matrix of smaller pebbles and sand.

Shapes of Crystals or Clasts

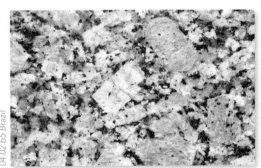

Crystals and clasts in rocks can have various shapes. This crystalline igneous rock (◄) includes some large, nearly rectangular crystals surrounded by smaller, more irregularly shaped crystals. This clastic sedimentary rock (►) includes mostly rounded pebbles (clasts) in a matrix of sand and smaller clasts. Some clastic rocks contain clasts that are sharp and angular.

Layers or No Layers

Crystalline rocks and clastic rocks may or may not have distinct layers. This crystalline metamorphic rock (◄) has dark- and light-colored, folded layers or bands. This clastic sedimentary rock (►) also has distinct layers, distinguished by differences in the size and composition of the clasts.

Before You Leave This Page Be Able To

☑ Explain the difference between a clastic rock and a crystalline rock and the differences between the general environments in which clastic and crystalline rocks form.

☑ Describe or sketch four general characteristics to observe in crystalline and clastic rocks.

4.2

4.3 How Do We Distinguish One Mineral from Another?

MINERALS HAVE MANY PROPERTIES that allow us to distinguish them from each other. Some properties are reflected by the shape of the mineral or the way the mineral breaks. Others are physical properties, like hardness and magnetism, that we can evaluate with simple tests.

A What Clues Does the Appearance of a Mineral Provide?

The first thing that we notice about a mineral is usually its outward appearance. We may note its size, shape, color, or how light reflects from its surface. These properties provide clues about the identity of a mineral. The figures below illustrate some physical properties that are relevant for identifying a mineral.

Prismatic

Cubic

Hexagonal

04.03.a1–3

04.03.a4-6 Quartz

Crystal Shape—A mineral that grows unobstructed by its surroundings can have a distinctive geometric shape. The shape of the crystal reflects the arrangement of atoms within the mineral and therefore provides a clue about the mineral's identity. Common crystal shapes include, but are not limited to, cubes, rectangular prisms, and hexagons (six-sided shapes).

Color—The color of a mineral is a useful, but not always reliable, property for mineral identification. Bright or unique colors are easily noticed, but a mineral can occur in several color varieties, such as the different-colored versions of quartz shown here. Other minerals always have the same color. It is the color and the crystal shape that make some minerals so beautiful and so highly valued as gemstones or mineral specimens.

Cleavage—Some minerals break in specific ways because of their internal arrangement of atoms. If a mineral breaks preferentially along a specific set of planes, the mineral has *cleavage*. Some minerals, like the ones shown here, break along one set of cleavage planes and cleave into thin sheets, but other minerals break along several sets of cleavage planes having different orientations.

04.03.a7 Biotite

Luster—The way that light bounces off a mineral is a property called *luster*. A mineral can be highly reflective, dull, or somewhere in between. It can be partially transparent or opaque. It can look like metal, a pearly shell, a silky material, or a simple piece of earth. The names of different types of luster reflect the quality and intensity of the reflection. Luster terms include metallic, nonmetallic, glassy, pearly, silky, resinous, and earthy.

04.03.a8-10

No Cleavage—Some minerals have an internal atomic arrangement that does not contain planes along which the mineral breaks. These minerals do not have cleavage but instead break along *fractures*. Fractures in minerals tend to have rough, irregular surfaces, like one shown here cutting a quartz crystal. In contrast, cleavage planes tend to be more regular.

04.03.a11 Quartz

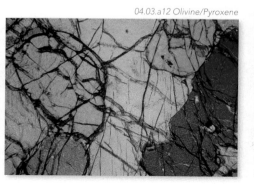
04.03.a12 Olivine/Pyroxene

Microscopic Observations—To identify minerals in rocks, especially fine-grained rocks, we often examine a *thin section* (◄) using a microscope. When a thin section is viewed between two polarizing filters, light shining through the thin section causes the different minerals to exhibit distinctive and diagnostic colors.

B What Tests Can We Perform to Help Us Identify a Mineral?

We determine some mineral properties by conducting tests. We may touch a mineral with a magnet to check its magnetism, or we may try to scratch it to determine its hardness.

1. Hardness—Some minerals are very hard, and some are quite soft. A mineral can be scratched by a material that is harder than the mineral but not by one that is softer. To estimate mineral hardness, we often conduct scratch tests using common objects, such as a fingernail, penny, or knife blade. We may also use other minerals of known hardness for comparison.

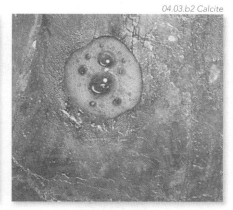
04.03.b1 Gypsum

04.03.b2 Calcite

3. Effervescence—If a drop of dilute hydrochloric acid (HCl) is placed on a mineral, a reaction may cause vigorous bubbling, or *effervescing*. The mineral *calcite*, which is the main mineral in limestones, effervesces strongly with HCl, but no other common minerals do. Any sample that effervesces when tested is most likely to be calcite.

2. Streak—If a mineral is rubbed against an unglazed porcelain plate (called a streak plate), it may leave a trail of powdered material called a *streak*. Some minerals have a diagnostic streak color. The iron-oxide mineral *hematite*, for example, has a reddish streak.

04.03.b3 Hematite

04.03.b4 Magnetite

4. Magnetism—A few iron-bearing minerals are naturally magnetic. The mineral *magnetite* is the strongest natural magnet. It is attracted to other magnets, and its magnetism can be strong enough to deflect a compass needle. Other magnetic minerals are less magnetic than magnetite, but magnetism may still help identify them.

5. Density—Some minerals are more dense than others. This property can often be detected by simply holding a mineral and noting how heavy it feels. We call this approach a *heft test*. In the lab, scientists precisely determine the ratio of the density of a substance to the density of freshwater, a property called *specific gravity*.

04.03.b5

In this example (▶), crushed crystals are placed into a glass beaker on one side of a balance scale and weigh as much as two beakers of water. This sample of dry, crushed crystals is therefore twice as dense as water and has a specific gravity of 2. It would be more dense if it were a solid crystal without air between the crushed pieces. A typical specific gravity of rock is 2.7 (i.e., 2.7 times more dense than freshwater).

Mohs Hardness Scale

Mohs Hardness Scale (▶) consists of 10 common minerals ranked in order of hardness, from 1 to 10. The softest mineral (talc) is 1, and the hardest mineral (diamond) is 10. These numbers describe the *relative* hardnesses of the minerals, but the numbers do not provide a real comparison of their actual hardnesses. Quartz (hardness of 7) is twice as hard as apatite (hardness of 5), and diamond (hardness of 10) is about five times as hard as corundum (hardness of 9).

Hardness and Mineral		Common Objects
1	Talc	
2	Gypsum	
3	Calcite	Fingernail (2.5)
4	Fluorite	Copper wire (3.5)
5	Apatite	
6	K-feldspar	Window glass or knife blade (5.5)
7	Quartz	
8	Topaz	
9	Corundum	
10	Diamond	

Before You Leave This Page Be Able To

✓ Explain the properties of a mineral that can be observed without using a test.

✓ Describe how to test for hardness, streak, effervescence, and magnetism.

✓ Explain the meaning of a mineral's specific gravity.

✓ Explain Mohs Hardness Scale.

4.3

4.4 What Controls a Crystal's Shape?

CRYSTALS CAN HAVE BEAUTIFUL SHAPES. The outward shape of a crystal reflects a combination of factors, including the arrangement of atoms in the crystal and how the crystal's growth was affected by the material around it. What, at an atomic scale, controls the shape of a crystal?

A How Is the Shape of a Mineral Related to Its Internal Structure?

If the growth of a mineral is unconstrained by surrounding materials, the outward shape of the crystal mimics the mineral's internal structure of atoms. The relatively simple outward shape and internal structure of halite nicely illustrate this relationship between the interior and the exterior of a mineral.

04.04.a1 Halite

The photograph above shows natural crystals of table salt, which is the mineral *halite* (NaCl). These crystals grew together and look like a number of cubes connected together.

Mineralogists (geologists and other scientists who study minerals) have documented that halite consists of equal proportions of sodium (Na) and chlorine (Cl) atoms. It has the chemical formula NaCl. Mineralogists have investigated the atomic arrangement of atoms within halite and find that sodium and chlorine atoms have a geometric arrangement that is like a cube. In this figure, sodium atoms are yellow, chlorine atoms are green, and chemical bonds that link adjacent atoms are represented by stick-like connectors. Note that the green chlorine atom in the center of the structure is surrounded by and bonded with six sodium atoms. Other minerals have more complicated shapes or chemical formulas, but we use halite here because it is so simple.

04.04.a3

04.04.a2

In a crystal, one part of the atomic arrangement repeats indefinitely to make the entire crystal. In halite, the smallest part is one pair of sodium (Na) and chlorine (Cl) atoms. Sodium and chlorine atoms alternate in three perpendicular directions. Note that in this figure, whether you go up-down, left-to-right, or front-to-back, Na and Cl alternate in the crystalline structure.

04.04.a4

A different way to represent crystals is to show atoms as spheres that fit together and touch (◄). This type of model more accurately represents the relationship between adjacent atoms and their electrons, but it is more difficult to see the internal structure. Note that for halite, the relative sizes of sodium (Na) and chlorine (Cl) atoms allow them to pack together tightly in a cube-shaped arrangement. Atoms are so tiny that a one-inch cube of halite contains more than 100,000,000,000,000,000,000,000 [10^{23}] pairs of Na and Cl atoms.

In addition to growing as cubic crystals, halite will also break into cube-shaped or shoebox-shaped fragments. If you examine table salt with a magnifying glass, you will observe that most salt grains are tiny cubes or slightly elongated boxes, like the larger, broken halite crystals shown here.

04.04.a5 Halite

B How Are Atoms Arranged in a Mineral?

Atoms fit together in a limited number of ways. How closely atoms can be packed together depends on their electrical charge (positive versus negative) and the relative sizes of different kinds of atoms (e.g., smaller Na atoms fit between larger Cl atoms). A single atom typically bonds to 3, 4, 6, 8, or even more atoms. Atoms of similar charge repel each other, whereas atoms of opposite charge attract, and so atoms are generally arranged in geometric patterns. Three common arrangements of atoms are shown below, but other arrangements are common.

Atoms can be arranged in the shape of a cube. This type of structure is referred to as *cubic*.

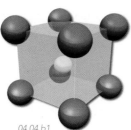

04.04.b1

One atom can be surrounded by four other atoms, arranged as a pyramid with three sides and a base. This arrangement and four-sided shape is called a *tetrahedron* (tetra = four).

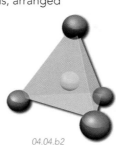

04.04.b2

Atoms can be arranged in a shape that is like two oppositely pointing, four-sided pyramids joined at their bases. This shape is an *octahedron* (octa = eight).

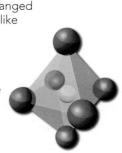

04.04.b3

C How Is the Shape of a Crystal Affected by the Environment in Which It Grows?

For a crystal to attain a perfect shape, it must grow unimpeded by surrounding material. Most nicely shaped crystals grew in an open space, in water or in magma, and so did not grow into other crystals. When crystals do grow within solid rock or around preexisting crystals, they generally do not have such well-formed shapes.

These crystals have well-defined shapes, flat crystal faces, and sharp ends called *terminations*. Most such crystals grew into a space filled with hot or cold water rather than solid rock.

04.04.c1 Calcite

In this rock, partially transparent gray quartz is in irregularly shaped masses that fill the spaces between and around the white and red minerals. The quartz grew after the other minerals were already there, so it had to conform to their shapes.

04.04.c2

The Depths to Which Open Spaces Can Exist in the Crust

How deep can we go into the earth and still find open spaces where crystals can grow unimpeded? Some movies depict huge open spaces in Earth's mantle or even its core, but these movies are not an accurate portrayal of the subsurface of Earth.

For an open space to exist, as between the crystals shown here, rocks that form the walls of the cavity must be strong enough to support the pressure exerted by the weight of the overlying rocks. If the pressure is too great, the cavity will collapse under the load. The existence of caves confirms that rocks are strong enough to hold open spaces to a depth of at least several kilometers. At greater depth, such as in deep mines, open spaces have to be held open by steel or wooden beams or else the walls and roof will collapse. At depths greater than several kilometers, fluid pressure from pockets of water or magma can prop open enough space for well-formed crystals to grow, but such fluid-filled openings disappear quickly if the fluid pressure drops, allowing adjacent rocks to collapse inward into the unsupported opening.

04.04.mtb1 Apophyllite/Stillbite

Before You Leave This Page Be Able To

☑ Explain or sketch how the internal arrangement of atoms is reflected in the crystal form of halite.

☑ Explain what it means to say that crystals have an ordered atomic arrangement.

☑ Sketch or describe three common ways in which atoms are arranged in a mineral.

☑ Sketch or explain how the shape of a crystal is affected by the environment in which the crystal grows.

4.4

4.5 What Causes Cleavage in Minerals?

CLEAVAGE IS THE TENDENCY OF MINERALS TO BREAK along parallel planes. Some minerals cleave into cubes, and others cleave into thin sheets. Still other minerals break along irregular fractures instead of cleavage planes. Cleavage is controlled by the arrangement of atoms in a mineral and the strengths of the bonds between atoms.

A What Happens at an Atomic Scale When a Mineral Cleaves?

The same orderly arrangement of atoms that causes crystals to form with specific shapes can also affect the way crystals break. Breaking a mineral requires applying enough force to break bonds between atoms. In many minerals, different bonds have different strengths, so the mineral breaks preferentially (cleaves) along the easiest directions and through the weakest links.

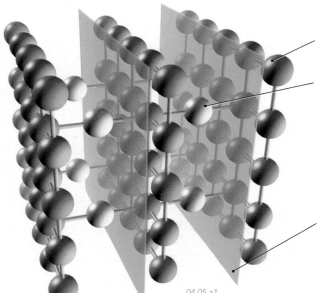

04.05.a1

1. This mineral consists of three kinds of atoms shown here in brown, blue, and gray.

2. The brown atoms are linked with (bonded to) the blue atoms, forming flat sheets.

3. Adjacent sheets are joined together by long bonds between the gray atoms and the brown atoms in the sheets on either side.

4. Bonds between the brown and blue atoms (within the sheets) are stronger than bonds between the brown and gray atoms (linking the sheets). If the mineral is subjected to sufficient force, the force will break the weakest bonds (those between the brown and gray atoms). The breaks will occur along the cleavage planes shown in yellow.

5. With this type of arrangement of atoms and bonds, the mineral will cleave along one set of planes, splitting into thin sheets, like the mica mineral shown to the right. ▶

04.05.a2 Biotite

B What Happens if All of the Bonds Have the Same Strength?

In the example above, one set of bonds is relatively weak and so forms a natural place for breaking across the mineral. How does the mineral break if all the bonds have similar strengths or if the arrangement of atoms and bonds does not allow the crystal to break along any planes?

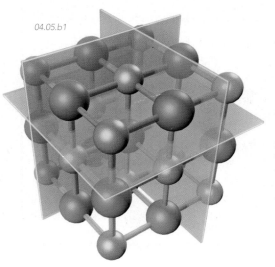

04.05.b1

The bonds in this mineral (◀) all have a similar strength but are arranged in such a way that the mineral can break along three sets of planes without passing through an atom. In this example, the three planes are mutually perpendicular (at 90° to each other).

04.05.b2 Quartz

The bonds in this mineral are not arranged into a configuration that allows any cleavage planes to form. Instead, the left end of the crystal has broken like glass, along an irregular curved *fracture* instead of along cleavage planes. The well-defined planes bounding most of the rest of the crystal were formed during the growth of the crystal and illustrate that the way in which a crystal grows can be different than the way it breaks.

C What Are Some Common Types of Cleavage?

If a mineral has cleavage, it can cleave along one or more sets of parallel planes. Two sets of planes might be perpendicular (90°) to one another or might intersect at some other angle. In the diagrams below, colored planes show the orientation of possible cleavage planes.

One Direction of Cleavage

04.05.c1

If a mineral has a single direction of cleavage, it cleaves along one set of parallel planes, forming thin sheets. Examples of a single direction of cleavage are minerals of the mica family.

Two Perpendicular Directions of Cleavage

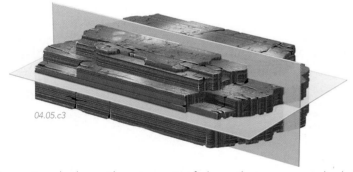

04.05.c3

Many minerals cleave along two sets of planes that are perpendicular to one another. This type of cleavage results in right-angle (90°) steps along broken crystal faces. The pyroxene mineral group has right-angle cleavage.

Two Non-Perpendicular Directions of Cleavage

04.05.c5

Two planes of cleavage can intersect at angles other than 90°. Minerals with this type of cleavage can break into pieces having corners that do not form right angles. The amphibole group of minerals has this type of cleavage.

Three Perpendicular Directions of Cleavage

04.05.c2

If a mineral cleaves along three perpendicular sets of planes, broken faces have a stair-step geometry and the mineral commonly breaks into cubes, as is typical of halite.

Three Non-Perpendicular Directions of Cleavage

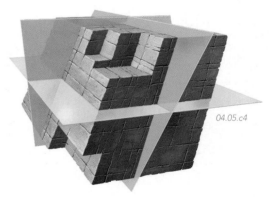

04.05.c4

Minerals that cleave along three directions of planes that are not mutually perpendicular break into pieces that are shaped like a *rhomb*, or a sheared box. Calcite is the most common mineral that cleaves into rhombs.

Before You Leave This Page
Be Able To

☑ Explain or sketch the relationship between cleavage and the arrangement and strengths of bonds.

☑ Explain what happens if a mineral lacks planes along which it may cleave.

☑ Sketch and describe five types of cleavage.

4.5

4.6 How Are Minerals Classified?

WITH NEARLY 100 NATURALLY OCCURRING ELEMENTS, it should not be a surprise that there are thousands of different minerals. Some minerals are so rare that they occur only in unusual environments, but others are so common they are almost everywhere on Earth's surface. Here, we concentrate on minerals that are very common and are critical to our understanding of Earth.

A How Are Similar Chemical Elements Grouped in the Periodic Table?

Chemical elements are the fundamental building blocks of minerals, so geologists classify minerals into several *mineral groups* based on the main chemical components within those minerals. Before discussing these mineral groups, we take a tour of the chemical elements via the *Periodic Table*, a useful way to organize the elements.

1. Each element in the Periodic Table has an *atomic symbol*, one or two letters representing the name of the element (commonly the name in *Latin*) and an *atomic number* (shown to the upper left of the symbol). Elements that share a background color on the table share some similar chemical properties.

2. The table begins with hydrogen (H), the lightest element, and advances to higher atomic numbers and heavier elements from left to right and from top to bottom.

3. Elements shaded orange are the *alkali* and *alkali earth metals* and include sodium (Na), potassium (K), calcium (Ca), and magnesium (Mg) on the left side of the table and aluminum (Al) in the right half of the table.

4. Elements colored yellow are called *transition metals*. They include many familiar metals, such as chromium (Cr), iron (Fe), nickel (Ni), copper (Cu), zinc (Zn), silver (Ag), and gold (Au).

5. The elements colored green are *nonmetals* and include carbon (C), silicon (Si), and oxygen (O). The nonmetals typically bond with both types of metallic elements to form minerals.

6. The last column includes elements called *noble gases* because they are gases that do not readily combine with other elements.

04.06.a1

7. The elements colored purple and blue include some familiar elements, such as uranium (U), and many that are less familiar. Elements with atomic numbers higher than 92 are not known in natural settings (these are produced only in the laboratory), except for plutonium (Pu), which is produced in unusual circumstances by natural nuclear reactions.

8. The lightest and simplest elements, hydrogen (H) and helium (He), are the most abundant elements in the universe. The elements oxygen (O) and silicon (Si) make up 74% of Earth's crust, with the rest being mostly aluminum (Al), iron (Fe), calcium (Ca), sodium (Na), potassium (K), and magnesium (Mg). Consequently, the most common minerals that we see are made of oxygen and silicon, with lesser amounts of these other common elements.

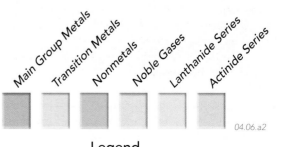

Legend: Main Group Metals | Transition Metals | Nonmetals | Noble Gases | Lanthanide Series | Actinide Series

04.06.a2

B   What Are the Major Classes of Rock-Forming Minerals?

The most important rock-forming minerals can be divided into several classes based on their chemistry. The Periodic Table provides a useful framework.

04.06.b2

1. *Silicates*, including the mineral *quartz* (◄) are the most important mineral group on Earth. They contain silicon and oxygen, the two most abundant elements in the crust, and so are very common. In silicates, each silicon atom is bonded only to oxygen. In most silicate minerals, the silicon-oxygen units are linked by bonds to metals, such as Fe, Mg, Na, K, Ca, and Al.

2. *Carbonates* contain carbon and oxygen bonded together in a triangular arrangement. The triangles are linked by other elements, most commonly the metal calcium (Ca). An example is the mineral *calcite* ($CaCO_3$). ►

04.06.b3

3. *Oxides* consist of oxygen bonded with a metal, such as iron in the mineral *hematite* (Fe_2O_3). ►

04.06.b4

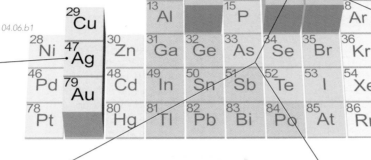
04.06.b1

7. *Native minerals* are minerals that contain only a single element. The metals copper (Cu), silver (Ag), and gold (Au) can occur as native minerals or in combinations, called *alloys*, with other metals. Nonmetallic elements that occur as native minerals include sulfur (S) as *native sulfur* and carbon (C) as *graphite* and *diamond*.

4. *Halides* contain chlorine (Cl) or fluorine (F), both of which are nonmetals that typically bond with a metal from the left side of the table. *Halite* (NaCl) is a halide mineral. ►

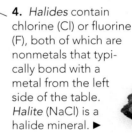
04.06.b5

6. *Sulfides* contain sulfur (S) bonded with a metal, such as iron (Fe) or copper (Cu). The mineral *pyrite* (FeS_2), also called "fool's gold," is a common sulfide. ►

04.06.b6

5. *Sulfates* contain sulfur (S) that is only bonded to oxygen. The sulfur-oxygen units are bonded to a metal, such as calcium (Ca) or iron (Fe). *Gypsum* ($CaSO_4 \cdot 2H_2O$) is a common sulfate. ►

04.06.b7

Asbestos: The Importance of Mineral Classification

Asbestos has long had the reputation of being a dangerous material. Asbestos was used on pipes, ceilings, and ships because of its excellent fire-retarding and insulating properties. Some health studies have shown that asbestos dust, when inhaled, can cause cancer and other serious health problems. But what is asbestos, and is the health story simple?

The term *asbestos* does not refer to a single mineral, but instead refers to a number of silicate minerals whose common characteristic is that they tend to form fibers. Five types of asbestos belong to a group of silicate minerals called *amphiboles*, including a variety known as "blue" asbestos, which forms individual, straight fibers. Medical studies have shown conclusively that "blue" asbestos poses a severe health risk.

In contrast, the type of asbestos most commonly used in the United States is *chrysotile*, a fibrous form of the silicate mineral *serpentine*. Chrysotile has a totally different mineral structure than the other types of asbestos and occurs in fibers that are curved and form interlocking bundles. Studies of this "white" asbestos show that it is much less hazardous than blue asbestos, mostly causing problems if breathed in large amounts for a long time, as occurs with chrysotile miners. The medical community and Environmental Protection Agency (EPA) are at odds with many scientists because the medical studies and the EPA have lumped all asbestos into the same category rather than considering the health risks of each kind separately. As a result, hundreds of billions of dollars may be spent, perhaps need-

lessly, removing *chrysotile* asbestos from schools and businesses.

Before You Leave This Page Be Able To

☑ Describe the Periodic Table, including the locations of the main groups of chemical elements (metals, transition metals, nonmetals, and noble gases).

☑ List the major classes of minerals and discuss the main chemical characteristic of each class.

☑ Describe two types of asbestos and the health controversy over asbestos.

4.6

4.7 What Is the Crystalline Structure of Silicate Minerals?

SILICATE MINERALS ARE THE MOST IMPORTANT rock-forming minerals because they comprise most of Earth's crust and mantle. There are different groups of silicate minerals, and the different groups have distinctive cleavage and other mineral characteristics. Here, we explore the types of silicate minerals, from atoms to crystals.

A What Do Silicate Minerals Contain?

In most silicates, one silicon atom is bonded with four oxygen atoms to form the negatively charged SiO_4^{-4} complex. This SiO_4^{-4} complex has a very important shape, called a *tetrahedron*, that controls many aspects of silicate minerals. The silicon-oxygen tetrahedron forms a building block for the vast majority of the minerals on Earth.

Silicon-Oxygen Tetrahedron

1. The four oxygen atoms and one silicon atom combine in an SiO_4^{-4} complex, which can be represented by a four-pointed pyramid called a *tetrahedron*.

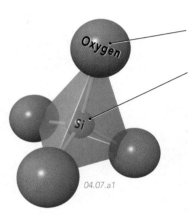

2. An oxygen atom is at each corner of the pyramid.

3. A much smaller silicon atom is in the center of the pyramid.

4. The SiO_4^{-4} complex takes the shape of a tetrahedron because the four oxygen atoms have similar atomic charges and so repel each other. The oxygen atoms move as far as possible from each other, taking positions that define the shape of a tetrahedron.

04.07.a1

Linked Silicon-Oxygen Tetrahedra

5. A silicon-oxygen tetrahedron has a negative electric charge that allows it to bond with other tetrahedra. Each oxygen atom in the tetrahedron is a naturally protruding site ready to bond to other elements and chemical complexes.

6. An oxygen atom can be shared by two adjacent tetrahedra. In this manner, silicon-oxygen tetrahedra can link together to form different types of silicate minerals.

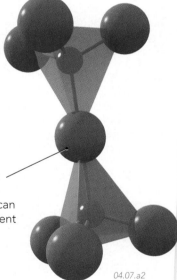

04.07.a2

Silicon-Oxygen Tetrahedra and Metallic Elements

7. In addition to bonding to one another, silicon-oxygen tetrahedra bond with other elements, such as the green atoms shown in this figure. The bonds to the green atoms are not shown.

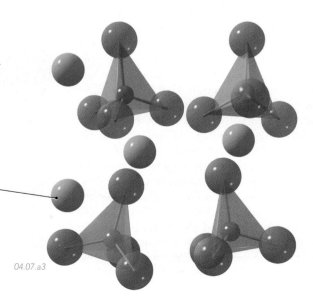

8. Silicon-oxygen tetrahedra have a negative electrical charge and so attract positively charged atoms, called *cations* (the green atoms).

9. A huge variety of minerals results from the ability of silicon-oxygen tetrahedra to bond with other silicon-oxygen tetrahedra, with various cations, and with other chemical substances. There are more than 4,000 known minerals, but most are uncommon to rare. Several dozen minerals, many of which are silicate minerals, compose most rocks we encounter at the surface. A typical rock contains one to five main minerals, with a small number of less abundant minerals, many of which are visible only under a microscope. So learning a few minerals helps you to identify most rocks.

04.07.a3

B What Are the Different Types of Silicate Minerals?

Silicon-oxygen tetrahedra can be connected in five main ways, each producing a major group of silicate minerals that share common characteristics. Bonds that link one tetrahedron to another are strong, but bonds to other elements between tetrahedra provide planes of weaker bonds, allowing silicate minerals to cleave. A silicate mineral's cleavage, or lack of cleavage, reflects how the tetrahedra are arranged. The silicon-oxygen tetrahedra are shown below with the oxygen atoms on each corner.

Independent Tetrahedra

04.07.b1

1. Some minerals contain silicon-oxygen tetrahedra that are bonded to other elements (not shown), but not to other tetrahedra. Minerals in this group, including *olivine*, do not break along clearly defined planes because bonds are more or less equally strong in all directions.

Single Chains

04.07.b2

2. Tetrahedra may form *single chains* by sharing two oxygen atoms. The chains are strongly bonded and difficult to break, so cleavage cuts parallel to, rather than across, the chains. This results in two planes of cleavage that are nearly perpendicular (90° angle) to each other. Minerals that belong to this group are called *pyroxenes*.

Double Chains

04.07.b3

3. Tetrahedra can also form *double chains* if half the tetrahedra share two oxygen atoms and half share three, as shown here. In this case, the mineral cleaves parallel to the double chains and along two planes of cleavage separated by angles of 60° and 120°. Minerals that belong to this group are called *amphiboles*.

Sheets

04.07.b4

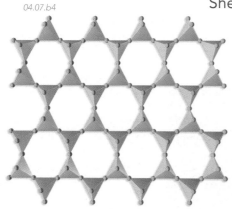

4. In *sheet silicates*, tetrahedra share three oxygen atoms to form continuous sheets. Other elements and water molecules can fit between the sheets, forming minerals with layered structures. Bonds between sheets are weak, so these minerals have one main direction of cleavage parallel to the sheet structure. The most common sheet-silicate minerals are *micas* and *clay minerals*.

Frameworks

04.07.b5

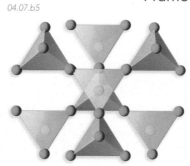

5. Tetrahedra in *framework silicates* share all four oxygen atoms, forming a structure bonded well in three dimensions. *Quartz*, and a few rarer framework silicates, contain only silicon-oxygen tetrahedra bonded to each other. Quartz is hard and has no cleavage, so it fractures instead. Some framework silicates have other elements in the structure between the silicon-oxygen tetrahedra, providing several planes of cleavage. Minerals belonging to the *feldspar group* are good examples.

Silicon, Silica, and Silicone

These three similar words can be confusing, so let's explore what each one means.

Silicon is the fourteenth element of the Periodic Table, having atomic symbol Si. The name silicon also is used for a *synthetic* material—a material produced by humans that does not occur naturally. Synthetic silicon is a semiconductor used to make computer chips.

Silica refers to a compound containing only silicon and oxygen in a ratio of 1:2, so it has the formula SiO_2. Quartz is 100% silica. Although each silicon atom in quartz is bonded to four oxygen, each oxygen is shared between two silicon, so the ratio Si:O is 1:2. Geologists speak about silica more than about silicon because silicon is nearly always bonded with oxygen in rocks and minerals.

Silicone is a *synthetic* material in which carbon is bonded to silicon atoms to keep the material in long chains. These chains make silicone a material that can be used as a type of grease or as caulk for sealing around windows and doors.

Before You Leave This Page Be Able To

☑ Sketch or explain a silicon-oxygen tetrahedron and how one can join with another tetrahedron or a cation.

☑ Explain or sketch how silicon-oxygen tetrahedra link in different geometries to produce five silicate mineral groups.

☑ Explain the differences between silicon, silica, and silicone.

4.7

4.8 What Are Some Common Silicate Minerals?

SILICATE MINERALS ACCOUNT FOR OVER 90% of the minerals in Earth's crust. Most silicate minerals also contain other elements, commonly aluminum (Al), calcium (Ca), sodium (Na), potassium (K), iron (Fe), and magnesium (Mg). The presence and amounts of these elements influence the crystalline structure, which in turn determines mineral properties, such as color and cleavage.

A What Are Some Light-Colored Silicate Minerals?

The most common silicate minerals in the upper part of the continental crust have light colors and typically are white, light gray, and light pink. Some of these minerals are almost transparent, and some have a reflective, silvery color. Light-colored silicate minerals predominate in the upper continental crust and are present in smaller amounts in rocks of the oceanic crust and the mantle.

04.08.a1

Quartz—This very common mineral, with a formula of SiO_2, is generally transparent to nearly white, but it can be pink, brown, or purple. Its silicon (Si) and oxygen (O) atoms are strongly bonded in a tight, three-dimensional *framework*, so quartz is hard (Mohs hardness of 7) and does not cleave. Instead, it breaks along fractures that have smoothly curving surfaces that are described as being *conchoidal*, as on the left end of this broken crystal.

04.08.a2

Potassium Feldspar—Potassium feldspar, often just called *K-feldspar*, contains potassium (K), aluminum (Al), silicon, and oxygen, with lesser amounts of sodium (Na) and calcium (Ca). It generally is a pink to cream-colored mineral, but in volcanic rocks it can be nearly transparent. Many K-feldspar crystals display two directions of cleavage, and some show wavy, light-colored lines on crystal surfaces, as shown here. K-feldspar is abundant in all granites, and it is common in many other igneous, sedimentary, and metamorphic rocks.

04.08.a3

Plagioclase—Plagioclase is one of the two most common feldspar minerals. Feldspars are a group of framework silicates that contain varying amounts of potassium (K), sodium (Na), calcium (Ca), and aluminum (Al), in addition to silicon and oxygen. In plagioclase, the potassium content is close to zero percent. Plagioclase exhibits a complete gradation from Na-rich varieties to Ca-rich varieties, accompanied by a color gradation from nearly white to dark gray or brown. However, most plagioclase has a whitish to light-gray color. Some crystal faces display straight lines called *striations*, as shown here.

04.08.a4

Muscovite—This sheet-silicate mineral is part of the *mica* family, whose members all have one direction of cleavage and so break into flakes and sheets. It typically is partially transparent, a clear to silvery-gray color, and somewhat shiny because the flat surfaces of the sheets reflect light. Muscovite contains potassium (K) and aluminum (Al), in addition to silicon and oxygen. Its atomic structure contains a component of water, expressed in its chemical formula as $(OH)^-$. The bonds holding the sheets together are stronger than the bonds between sheets, so sheets can be peeled apart with your fingers.

Before You Leave These Pages Be Able To

☑ Describe the main light- and dark-colored silicate minerals, including their general characteristics, such as cleavage and main elements.

☑ Discuss the characteristics of clay minerals and how they form.

B What Are Some Dark-Colored Silicate Minerals?

Dark silicate minerals predominate in dark igneous and metamorphic rocks and also are many of the dark crystals scattered within otherwise light-colored rocks. They form most of the oceanic crust and the mantle and are present in variable amounts in continental crust, especially the lower crust. Dark-colored silicate minerals are also called *mafic minerals* to acknowledge their high magnesium (Mg) and iron (Fe) content.

Amphibole—The term *amphibole* refers to a group of related silicate minerals. Amphibole minerals contain magnesium, iron, calcium, sodium, and aluminum, in addition to silicon and oxygen. They can be black, dark green, pale green, or nearly white. They commonly form crystals that are long

04.08.b1

compared to their width, like the long green crystals shown here that radiate outward from several centers. Amphiboles are double-chain silicates and so cleave along planes that meet to form angles of 60° and 120°.

04.08.b2

Pyroxene—The term *pyroxene* refers to a group of single-chain silicate minerals that share a similar crystal structure. Pyroxene minerals can include various amounts of calcium, sodium, aluminum, iron, and magnesium in addition to silicon and oxygen. Their color can be black, dark brown, green, or nearly white. Most pyroxenes tend to form crystals that are roughly *equant*, meaning that all dimensions are about the same. Pyroxenes have two nearly perpendicular directions of cleavage (90° angles), which helps distinguish them from amphiboles.

04.08.b3

Olivine—Olivine is the most common mineral in the upper mantle and usually has a distinctive olive-green color. It has independent tetrahedra linked by iron or magnesium, and no cleavage. Its composition varies between iron-rich and magnesium-rich end members, but samples from the mantle are magnesium rich.

04.08.b4

Garnet—Garnets are silicates that can be just about any color, but a deep red color is very common. The crystals are distinctive, having 12 diamond-shaped faces when perfectly formed. The reason color is so variable is because chemistry is variable. Garnets contain silica with variable amounts of calcium, iron, magnesium, manganese, and aluminum.

04.08.b5

Biotite—Biotite is a dark-colored mica (sheet silicate) that is typically black or brown. All biotite contains potassium, aluminum, silicon, and oxygen, with variable amounts of iron and magnesium. Brown biotite, commonly having a tint of bronze, is rich in magnesium and contains little iron. Like all micas, biotite has one dominant direction of cleavage.

Clay Minerals

The term *clay* is used in two ways in geology. It refers either to a family of minerals or to any very fine sedimentary particles that are less than 0.002 millimeters in diameter. Clay minerals have a sheet-silicate structure similar to that of mica, but the bonds holding the atoms together are much weaker. The sheets in clays are weakly held together, so they easily slip past one another, giving

04.08.mtb1

clays their slippery feel. When some clay minerals get wet, water pushes apart the weakly bonded sheets, causing the clay to expand.

Most clay minerals have light colors but may appear dark if mixed with other material (dark minerals or organic debris). Most clay minerals form by weathering of rocks at Earth's surface or from chemical reactions that occur when hot water

interacts with rocks containing feldspar, volcanic ash, and other reactive materials. Fine grain size and low density mean that clay particles are easily transported. Once formed, fine particles of clay can be picked up by wind and water and then transported long distances. Some clay makes it to the open ocean, where it finally settles to the ocean floor.

4.9 What Are Some Common Nonsilicate Minerals?

MANY MINERALS DO NOT INCLUDE SILICON and so are classified as *nonsilicates*. Some of the most common nonsilicate minerals are *carbonates* and *halides*, which typically form by precipitation from water. *Oxides* and *sulfides* form when metal atoms bond with oxygen or sulfur, respectively. Nonsilicate minerals are an important resource for our society and are used widely in industry, highways, and homes.

Carbonates

Carbonate minerals contain a metallic element, such as calcium (Ca) or magnesium (Mg), linked with a carbon-oxygen combination called *carbonate* $(CO_3)^{-2}$. The most common carbonate minerals are *calcite* and *dolomite*. Others include *malachite* and *azurite*, striking green and blue copper carbonates. *Trona*, a sodium carbonate, is an important mineral used to manufacture many products. Carbonates typically precipitate from water or have an organic origin (e.g., corals).

04.09.a1

04.09.a2

04.09.a3 Ios, Greece

Calcite—This mineral is the most common calcium-carbonate mineral $(CaCO_3)$ and occurs in a variety of water-related environments. It may be almost clear but commonly has a cream to light gray color. It is the only common mineral that effervesces with dilute hydrochloric acid (HCl) because HCl breaks bonds in calcite and releases carbon dioxide (CO_2) gas.

Dolomite—This mineral is similar to calcite, but magnesium (Mg) substitutes in the structure for some calcium (Ca). It has the formula $CaMg(CO_3)_2$. The mineral is cream-colored, light gray, tan, or brown and may not effervesce with HCl unless pulverized into a fine powder. A rock composed mostly of the mineral dolomite is a *dolostone*. Rocks that contain dolomite also commonly contain calcite.

Most limestones are nearly 100% calcite, and some carbonate rocks contain a mix of calcite and dolomite. Carbonate minerals also occur in coral and shells, including the mineral *aragonite*, which has the same composition as calcite but a different atomic arrangement. When limestone is heated and metamorphosed, calcite grows into larger crystals and the limestone becomes *marble*.

Oxides

Oxide minerals consist of oxygen bonded with iron (Fe), titanium (Ti), aluminum (Al), or other metals. Iron-oxide minerals are the most common oxides, except for ice, which is a hydrogen-oxide mineral (the solid phase of H_2O).

04.09.a4

04.09.a5

04.09.a6 Sherman Mine, Ontario, Canada

Hematite—This iron oxide (Fe_2O_3) can be black, brown, silvery gray, or earthy red, but it consistently has a red streak. Hematite is the red color in rust, provides color in some paints, and is responsible for many red-rock landscapes. It commonly forms when other iron-bearing minerals oxidize.

Magnetite—This iron oxide (Fe_3O_4) is typically black and is strongly magnetic, here attracting a paper clip. It is present as small black grains in many kinds of igneous, sedimentary, and metamorphic rocks, as well as in beach sands and other sediments.

Magnetite and hematite occur together in beautifully layered sedimentary rocks called *banded iron formations*. Some Precambrian iron formations are mined for iron in the Great Lakes region of the United States and Canada, and are mined elsewhere.

Sulfides

Sulfide minerals contain sulfide ions $(S)^{-2}$ bonded with iron (Fe), lead (Pb), zinc (Zn), or copper (Cu). Sulfide minerals, including the copper-iron sulfide mineral *chalcopyrite*, are the principal metal ores in many large mines.

Pyrite—Pyrite is a common iron-sulfide mineral (FeS_2). It has a pale bronze to brass-yellow color for which it earns the name "fool's gold." It commonly forms cube-shaped crystals with faces showing straight lines (*striations*).

04.09.a7

04.09.a8, Nova Scotia, Canada

Small crystals of pyrite, such as the brass-colored ones shown here, are common within and along hydro-thermal veins. Weath-ering of pyrite can cause adjacent rocks to become coated with yellow and orange, sulfur-rich material, like the stained quartz vein shown here.

Galena—This mineral is made of lead sulfide (PbS). It forms distinc-tive metallic-gray cubes with a cubic cleavage. It has a high density (specific gravity), which can be felt easily by picking up a sample. In the United States, many galena crystals are from lead mines near the Mississippi Valley.

04.09.a9

04.09.a10

There are many other important sulfide min-erals, including copper sulfides (shown here) and zinc sulfides. We mine sulfides because of their high metal contents. Most sulfide-rich mineral deposits formed when hydro-thermal fluids passed through rock.

Salt and Related Minerals (Halides and Sulfates)

Halide minerals (salts) consist of a metallic element, such as sodium (Na) or potassium (K), and a halide element, usually chlorine (Cl). *Sulfate minerals*, especially *gypsum*, commonly occur with salt. They consist of an element such as calcium (Ca) and a sulfur-oxygen complex ion called sulfate $(SO_4)^{-2}$. Many halides and sulfates form when water evaporates in a lake or from precipitation in a shallow sea with limited connection to the ocean.

04.09.a11

04.09.a12

Gypsum—This hydrated calcium-sulfate mineral ($CaSO_4 \cdot 2H_2O$) is typically gray, white, or clear and can be scratched with a fingernail. Most gypsum forms in environ-ments similar to those in which halite forms, and the two minerals commonly occur together. Gypsum also precipitates from hot or warm water that circulated underground through fractures in rocks.

Halite—Halite (NaCl) has cubic cleavage and a salty taste. It generally forms from the evap-oration of salty water, such as a drying lake or a part of a sea that becomes cut off from the rest of the oceans. When concentrated in thick beds to make a rock, it is called *rock salt*.

Before You Leave This Page Be Able To

☑ Discuss the key chemical constituents for each of the five nonsilicate mineral groups.

☑ Describe the major nonsilicate minerals, including their general characteristics such as color, cleavage, and any diagnostic attributes.

4.9

4.10 Where Are Different Minerals Abundant?

MINERALS ARE NOT UNIFORMLY DISTRIBUTED within Earth's interior or on its surface. Some are common in many different geologic settings on the surface and at various depths within Earth, whereas others are restricted to a relatively small number of places. Knowing where different kinds of minerals occur provides a useful framework for understanding the composition of Earth's interior and surface.

A What Types of Minerals Compose Different Parts of Earth?

Most of Earth consists of minerals. Some small parts of the crust and mantle, and all of the outer core, are liquid and so are not minerals. Small amounts of water, volcanic glass, and organic matter are also present in places, but overall, Earth is a complex mix of various crystals of minerals.

1. Earth's solid surface displays a wide variety of minerals and other materials. The surface of Earth is mineralogically diverse because it is compositionally diverse and exposes rocks and minerals formed under many different conditions, some on Earth's surface and some from deep within Earth. Once on the surface, these materials are exposed to water, oxygen, carbon dioxide, and different climates. This diversity in ingredients and environments produces a huge suite of minerals for us to examine.

2. In most places, the uppermost part of the continental crust consists of sedimentary rocks. These rocks typically contain some combination of quartz, clays, feldspars, and carbonate minerals, with small amounts of oxides and other minerals. In other places, the exposed bedrock is igneous and metamorphic rocks. Much of the continental crust is granite or rock close to *granite*, which is composed of the light-colored silicate minerals quartz and feldspar (◄). The sample shown here also contains black biotite and oxide minerals.

04.10.a2

3. The oceanic crust generally has a thin veneer of soft, weakly consolidated sediment overlying rocks. Oceanic sediments can contain carbonate, oxide, and sulfide minerals, and organic materials that are not minerals. Beneath the soft sediments, the oceanic crust is *basalt*, an igneous rock dominated by plagioclase and mafic silicate minerals, including black pyroxene and green olivine. Oceanic crust also contains the igneous rock *gabbro* (►), which contains silicate minerals, including dark-colored *pyroxene* and *amphibole* and lesser amounts of whitish *plagioclase feldspar*.

04.10.a3

4. The upper mantle, in both the lithosphere and asthenosphere, is mostly solid and consists of silicate minerals similar to those found in oceanic crust: plagioclase, pyroxene, and olivine. It also contains oxides and related minerals. With increasing depth, the amount of plagioclase decreases until it eventually disappears, and olivine and pyroxenes make up most of the upper mantle. This piece of upper mantle (►), brought up in a volcano, consists almost entirely of the green silicate mineral *olivine*. The abundance of this mineral in the upper mantle is the reason why the illustrations in this book show the mantle with a greenish-brown color.

04.10.a4

5. The lower mantle consists of minerals stable only at very high pressures. We have no real samples, but synthetic equivalents are produced in special laboratories where lower mantle conditions can be recreated. These experiments show that the most common minerals in the lower mantle are Fe-Mg silicates and oxides.

6. The outer core is molten and is interpreted to be composed mostly of molten iron, with lesser amounts of nickel and other elements. It likely contains scattered crystals, but they are not abundant. The inner core is interpreted to be composed of a crystalline iron and iron-nickel alloy, similar to what makes up some iron-rich meteorites. ►

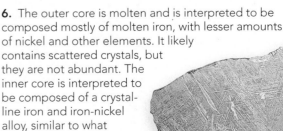

04.10.a5

04.10.a1

B What Elements Are Common in the Crust, Whole Earth, and Universe?

What chemical elements are common in Earth's crust, Earth as a whole, and the universe? Examine the figures below, and see if some of the elemental abundances surprise you.

Average Abundances in Earth's Crust

The crust consists almost entirely of the dozen elements highlighted below. Those that stick up highest on this figure are most abundant. The other elements, although locally very important and sometimes essential to life, add up to less than 1% of the crust.

The most abundant element in the crust is oxygen (O). Although a gas in Earth's atmosphere, in the crust it is almost all tied up in solid minerals.

The second most abundant element in the crust is silicon (Si), which combines with oxygen (O) to form silicon-oxygen tetrahedra in minerals. As a result, silicate minerals containing oxygen and silicon are widespread in the crust, especially quartz and feldspar.

The most abundant alkali and alkali-earth metals are calcium (Ca), potassium (K), sodium (Na), and magnesium (Mg), all of which combine with silicon (Si), oxygen (O), and aluminum (Al) in silicate minerals, such as feldspar and mica.

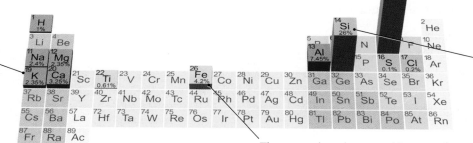

04.10.b1

The most abundant transition metal is iron (Fe), a key element in a number of silicate and iron-oxide minerals, like hematite and magnetite.

Average Abundances for the Entire Earth

If we consider the composition of the whole Earth, we get a different picture. Geologists and chemists estimate that Earth consists mostly of four elements with only two others that are abundant enough to show here.

Silicon is the third most abundant element on Earth and is nearly all bonded with oxygen in silicate minerals.

Oxygen is one of the two most abundant elements in the whole Earth, just as it is in the crust. Almost all of it is in silicate and oxide minerals.

Magnesium (Mg) is one of the most abundant metals. In the crust and upper mantle, it occurs mostly in silicate minerals, such as *olivine* and *pyroxene*. In the lower mantle, it is also present in oxide minerals.

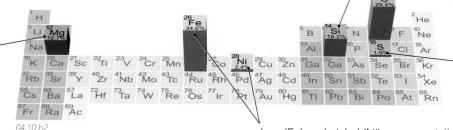

04.10.b2

Sulfur is common in alloys and magmas in the core. It is also present in sulfide minerals in both the crust and mantle, and in some sulfate minerals at or near Earth's surface.

Iron (Fe) and nickel (Ni) are especially abundant in Earth's core, where they occur in magma, as native metals, or in alloys, some including sulfur (S) and oxygen (O). Iron and very small amounts of nickel also occur in silicate and oxide minerals in the mantle and crust.

Average Abundances for the Universe

The overall composition of the universe is dominated by hydrogen (H) and helium (He), because these two lightest elements are the main component in suns, nebulae, and many other large astronomical objects. The universe contains much smaller amounts of carbon (C), nitrogen (N), oxygen (O), and neon (Ne).

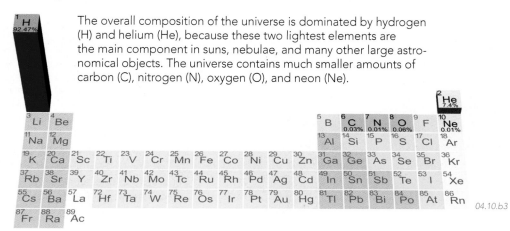

04.10.b3

Before You Leave This Page Be Able To

✓ Identify the most common class of minerals in the crust, mantle, and inner core.

✓ List the three most abundant elements in the crust and in Earth as a whole, and explain why silicate minerals are so abundant in the crust and mantle.

4.10

4.11 What Are the Building Blocks of Minerals?

MINERALS ARE COMPOSED OF CHEMICALLY BONDED ELEMENTS. An element is a type of atom that has a specific number of protons (e.g., all hydrogen atoms have one proton, whereas oxygen atoms have eight protons). The mineral halite can be broken into smaller pieces of halite, but if separated into its chlorine and sodium atoms, it is no longer halite.

A How Are Minerals Related to Elements and Atoms?

An atom is the smallest unit of an element that retains the characteristics of the element. Atoms are made of even smaller particles (including electrons, protons, and neutrons), but if, for example, a single atom of gold were broken apart, its pieces would no longer be gold.

04.11.a1

◄ The mineral halite consists of atoms of two chemical elements— chlorine and sodium. If halite is dissolved in water, it dissolves (►) to produce salt water containing individual atoms of chlorine and sodium.

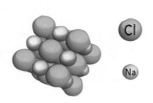

04.11.a2

Chlorine (Cl) and sodium (Na) atoms (►) each have a central nucleus surrounded by electrons at various distances from the nucleus.

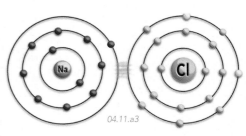

04.11.a3

B What Is a Model for the Structure of an Atom?

Atoms are too small to observe, so we use conceptual models to visualize them. The simple two-dimensional model of atoms shown in the previous figure does not fully represent atoms, which are three-dimensional spheres.

1. Atoms have a tiny central core called the *nucleus*. The nucleus is much smaller than the entire atom but is shown enlarged here.

2. The nucleus has two kinds of particles— *protons* and *neutrons*. Protons, shown in blue, have a positive (+) electrical charge, and neutrons, shown in red, do not have a charge.

3. The number of protons is called the *atomic number* of an element. The number of neutrons and protons is the atom's *atomic mass*. The number of neutrons can vary for an element, but the number of protons is consistent.

4. Negatively charged (−) electrons, shown in red, surround and orbit the nucleus. To be electrically neutral, an atom must have the same number of electrons (−) and protons (+). The proton's positive charges attract the atom's electrons, binding them to the nucleus. The area where the electrons orbit is called the *electron cloud*, but it really is not a cloud. It is simply a way of showing the area in which the electrons can reside. The outer edge of the cloud defines the size of an atom, but nearly all of the atom is empty space.

04.11.b1

Electron Shells

5. Groups of electrons orbit the nucleus at different distances, called *electron shells*. Each shell has a different level of energy, increasing away from the nucleus. The atom below has three shells, numbered 1, 2, and 3.

6. The inner shell (1), closest to the nucleus, can hold two electrons. Moving outward, successive shells can hold 8, 18, and 32 electrons. Electrons fill inner shells before they fill outer shells, so the inner shells are full but the outermost shell may only be partially full.

7. Atoms are most stable when their outermost shell is full, so atoms with only a few electrons in an unfilled outer shell may donate electrons to another atom in order to become more stable. Alternatively, atoms with a nearly full outer shell may borrow electrons from another atom to get a full shell and become stable.

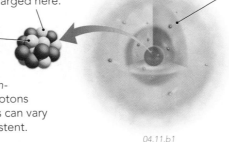

04.11.b2

8. This chlorine atom has seven electrons in its outer shell and so can accommodate one more. It would try to gain an electron to fill its outer shell (3).

9. This atom (sodium) has only one electron in its outer shell and so has a tendency to lose this electron, perhaps donating it to an atom such as the one on the left.

10. If an atom *gains* an electron, it acquires an overall *negative* charge. If it *loses* an electron, it acquires a *positive* charge. Charged atoms are called *ions*.

04.11.b3

How Does the Periodic Table Organize the Characteristics of Elements?

Chemists use the Periodic Table to organize the elements according to the elements' *atomic number* and *electron orbitals*. The table begins with the lightest element (hydrogen) and advances to the heaviest elements. Below we consider the two left-hand columns and the six right-hand columns because these are the most straightforward.

1. The columns are numbered from I to VII with Roman numerals to indicate the *number of electrons* in the outermost shell.

2. The rows correspond to the *number of electron shells*. Elements in the top row have one shell, those in the second row have two shells, and so forth.

3. Elements in the first column have only one electron in their outer shell. Hydrogen (H) only has one shell (it is in row one), whereas sodium (Na) is in the third row and so has three shells. Recall that the number of outer electrons influences whether an atom loses or gains electrons.

4. The first shell can hold only two electrons, which is why a large gap exists between the right and left sides of the first row.

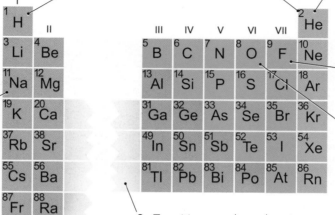

04.11.c1

5. The last column is reserved for noble gases, which do not easily gain or lose an electron because they have complete outer shells. This column could be numbered with both II and VIII—helium (He) only has two electrons, enough to fill its outer shell, whereas other noble gases in the column have eight or more electrons filling their outer shell.

6. Fluorine (F), in the second row, has two shells. It has seven electrons in its outer shell and so is in column VII. If it could borrow another electron, its outer shell would be full.

7. Oxygen (O) has two shells and six electrons in its outer shell; it needs two more electrons to fill this shell. Two oxygen atoms can fill their outer shells by bonding with a silicon atom, which has four electrons in its outer shell.

8. Transition metals, such as iron, occupy columns in the central part of the table (not shown). They lose and gain electrons from several shells, not just the outermost shell.

Some Practice with the Periodic Table

We can use the Periodic Table to predict how many shells each element has and how many electrons are in its outer shell. Try this for the elements listed in the table to the right before you check your answers. We try this out with chlorine, to show how it works.

Chlorine (Cl): Chlorine is in the third row, so it has three shells. It is in the seventh column (VII), so has seven electrons in its outer shell. It seeks one more electron to complete its outer shell, and this can be accomplished by borrowing an electron from sodium (Na), as in halite (NaCl).

Lithium (Li)—2 shells, 1 electron in outer shell

Magnesium (Mg)—3 shells, 2 electrons in outer shell

Nitrogen (N)—2 shells, 5 electrons in outer shell

Potassium (K)—4 shells, 1 electron in outer shell

Portraying the Atom

Atoms are tiny, but they can be detected by high-powered electronic microscopes. We cannot *look* down an optical microscope and see an atom with its nucleus and electrons. Our view of an atom is a *model*—a human-generated representation, or approximation, of what we think is there. Many tests have confirmed that the basic model is valid, but there are limitations. In particular, drawing an atom presents unavoidable problems.

The first problem is one of scale. The nucleus is so tiny compared to the size of the atom that you cannot show an accurately scaled nucleus and still fit the atom on a page. A hydrogen atom, for example, is nearly 150,000 times larger than its nucleus. The electrons, too, are extremely small when compared to an entire atom and so cannot be plotted to scale.

A second problem is how to show the electrons. They are in motion but do not travel around the nucleus in a regular manner. It is tempting to draw electron orbitals the same way we draw planets orbiting our Sun. In reality, an electron can be nearly anywhere within the cloud of electrons, although at any instant it is most likely to be somewhere near the center of its shell. Electron shells represent different *energy levels* more than they represent specific distances away from the nucleus.

Finally, atoms are not hard spheres with well-defined edges. Atoms are more empty space than matter, and their edges are defined by how far out the outermost electrons travel away from the nucleus. We often show atoms as hard-edged spheres because it is easier to see relationships between solid objects than between fuzzy,

partially overlapping clouds. Such depictions, however, are incorrect in detail.

Before You Leave This Page Be Able To

✓ Describe the relationship between a mineral and the elements of which it is composed.

✓ Explain or sketch the structure of an atom, including its main particles.

✓ Sketch the general shape of the Periodic Table and explain the significance of its rows and columns.

4.11

4.12 How Do Atoms Bond Together?

ELEMENTS COMBINE TO FORM MINERALS. The kind of bond that develops to hold two atoms together depends on the way in which the two atoms borrow, donate, or share electrons. The Periodic Table helps explain whether an element will gain, lose, or share electrons, and therefore how the element will bond with other atoms to form a mineral.

A How Do Atoms Bond Together?

Two atoms bond together by sharing, donating, or borrowing electrons from their outermost orbital shells. This process, called *chemical bonding*, can be illustrated by the ways people in a room might share or transfer money.

04.12.a1-4

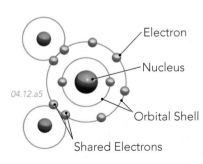	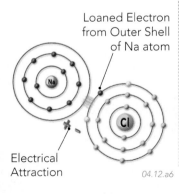	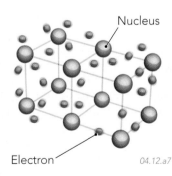	
Two people in the same room could both hold, or *share*, a single bill. The people have to stay close together to hold the bill, so sharing a bill forms a strong bond.	One person could loan one (or more) dollars to another person. A bond formed by *loaning* money is not as strong as one formed by sharing money, but it is still a bond.	If we had many dollar bills, we could keep passing (transferring) them around so that each person had a few. People and bills could move around freely while staying in the room.	If we stack the bills they may cling together a little, but we could easily pull the bills apart. This type of bond is very weak.

Covalent Bond

When two atoms *share* an electron, the bond is a *covalent bond*. The figure below shows a covalent bond between hydrogen and oxygen. Together the two hydrogen and one oxygen atom make the compound H_2O, or water.

04.12.a5

Electron

Nucleus

Orbital Shell

Shared Electrons

Ionic Bond

An *ionic bond* forms because of the attraction of two oppositely charged ions, such as when one atom *loans* one or more electrons to another atom.

Loaned Electron from Outer Shell of Na atom

Electrical Attraction

04.12.a6

Metallic Bond

Electrons in a *metallic bond* are *shared widely* by many atoms. This holds the material together in a nonrigid way, which is why many metals are pliable.

Nucleus

Electron

04.12.a7

Intermolecular Force

Several types of weak bonds can attract a molecule (a combination of atoms) to another molecule. Such bonds are relatively weak, such as those that connect sheets in micas and clays.

Strongly Bonded Sheets

04.12.a8

Weak Bonds

Bond Type	Strength	Mechanism	Mineral Examples
Covalent Bond	Strongest	*Sharing* of electrons between atoms	Diamond, bonds within sheets of graphite
Ionic Bond	Moderate	*Transfer (loaning)* of electrons from one atom to another, resulting in attraction between opposite charges	Halite, fluorite
Metallic Bond	Low	Widespread *sharing* of electrons among many atoms	Gold, copper
Intermolecular Force	Lowest	Attraction due to polarity of molecules, which are bonded combinations of atoms	Water ice, bonds between sheets of mica, clay, and graphite

B How Does the Periodic Table Reflect These Bond Types?

The Periodic Table provides general guidance about which type of bond two elements are likely to form. If one element is much better at attracting and holding onto electrons than the other element, the bond will be *ionic*. When two elements have nearly equal ability to attract and hold electrons, they *share* the electrons and the bond is *covalent*. A measure of this ability to attract electrons is called *electronegativity*. Electronegativity changes predictably across the Periodic Table, as shown by the relative height of each element in the figure below. A greater height indicates a greater ability to pull electrons (higher electronegativity).

1. Elements on the left side have only one or two electrons in their outer shell. So, except for hydrogen, they easily give up those electrons and do not have a strong ability to attract other electrons. When they combine with elements on the right side (group VII), which do exert a strong attraction, an *ionic* bond forms. Halite (NaCl) forms when sodium (Na) transfers an electron to chlorine (Cl), resulting in two oppositely charged ions that attract each other. Halite is one of the minerals that is nearly 100% ionically bonded.

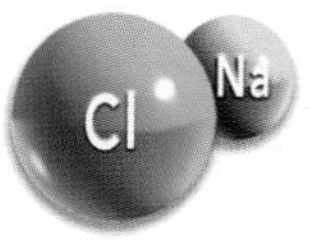

04.12.b2

2. A water molecule (H_2O) forms when two hydrogen (H) atoms share electrons with one oxygen (O) atom. The electronegativities of hydrogen and oxygen are about the same, so the bond is covalent. When frozen, water molecules combine in an orderly structure to form the mineral *ice*.

04.12.b3

3. Elements on the right side, except for the very last column, have a strong ability to attract electrons because their outer shells have one or two vacancies available for electrons. The mineral fluorite forms when a calcium atom transfers electrons to two fluorine atoms, and the bonds are ionic. Elements in the last column to the right have completely filled shells and so do not attract electrons.

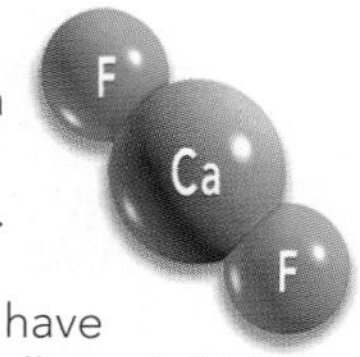

04.12.b4

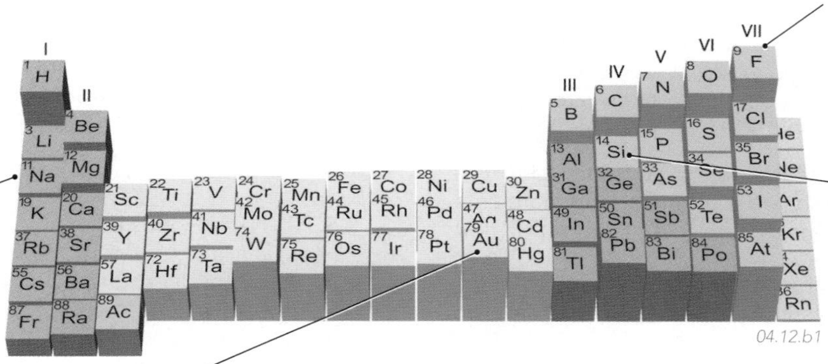

04.12.b1

4. Some elements have similar abilities to attract electrons (similar heights in this table) and can join via covalent bonds. A silicon atom can share electrons with four oxygen atoms, forming strong, covalent bonds in an SiO_4 tetrahedron.

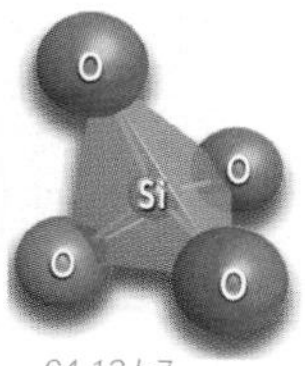

04.12.b7

6. Metals in the center of the table, like gold (Au) and copper (Cu), share electrons freely among their atoms to form metallic bonds. This makes many of these elements, especially copper, good conductors of electricity.

04.12.b5

5. In calcite ($CaCO_3$), bonds between the carbon atom and three surrounding oxygens are covalent, but bonds between the calcium atom and the carbonate group (CO_3) are ionic because the calcium atom transfers two electrons to the carbonate group.

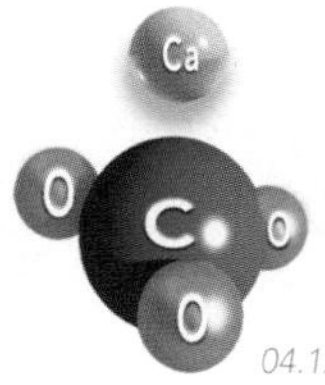

04.12.b6

C How Do Bonds Explain the Difference Between Diamond and Graphite?

The type and arrangement of bonds that hold a mineral together control many of the mineral's physical properties, like hardness. The minerals *diamond* and *graphite* both consist solely of carbon. Diamond is a very hard crystal, whereas graphite is soft and greasy feeling. Why are these two minerals so different?

Diamond contains strong covalent carbon-to-carbon bonds that form a strong interconnected framework. Diamond is the hardest naturally occurring mineral, with a Mohs hardness of 10, and is suitable for cutting and polishing into sparkling gemstones.

04.12.c1

Graphite has several kinds of bonds. It contains sheets that are covalently bonded and therefore strong, but weak intermolecular bonds hold adjacent sheets together. These weak bonds allow sheets to slide apart easily, making graphite soft enough to use as the "lead" in pencils and as an industrial lubricant.

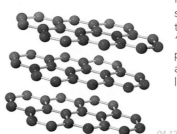
04.12.c2

Before You Leave This Page Be Able To

☑ Explain the different types of bonds and how electrons cause each type.

☑ Explain how the Periodic Table helps predict which kind of bond will form, and provide a mineral example for each kind of bond.

☑ Explain how differences in bonds cause diamond and graphite to have very different properties.

4.12

4.13 How Do Chemical Reactions Help Minerals Grow or Dissolve?

ELEMENTS COMBINE TO FORM MINERALS under various temperatures and pressures, but minerals also can be destroyed under a wide variety of conditions. The types of chemical bonds in a mineral determine whether the mineral grows, dissolves, or is unaffected by its physical and chemical environment. Water is often a key factor, so here we discuss how minerals crystallize or dissolve in water.

A What Are the Properties of Water, and How Does It Dissolve Some Solid Materials?

H_2O can be a gas, a liquid, or a solid. It has some unusual properties, setting it apart from many other chemical compounds. A water molecule consists of two hydrogen atoms covalently bonded to one oxygen atom.

The Water Molecule and Its Polarity

1. In a water molecule, the two hydrogen atoms are on one side of the oxygen. The water molecule has no overall charge (because the 10 protons are balanced by 10 electrons), but the electrons are not evenly distributed.

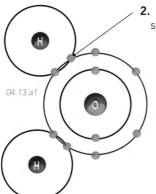
04.13.a1

2. Oxygen more strongly attracts electrons, so the shared electrons spend more time around the oxygen.

3. The molecule therefore has a polarity, with a negative side and two positive ends on the other side (▼). This polarity causes water molecules to be attracted to charged atoms (*ions*).

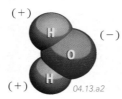
04.13.a2

Hydrogen Bonding

4. Water molecules are attracted to each other as well as to ions. In water, a weak bond called a *hydrogen bond* forms between one molecule's hydrogen atom and another molecule's oxygen atom. The hydrogen bond is responsible for some of water's unique properties (e.g., viscosity, surface tension, etc.).

5. This bond forms as hydrogen is attracted in two different directions. The covalent bond *inside* the water molecule pulls the hydrogen inward, keeping the molecule together.

04.13.a3

6. Hydrogen bonds from another water molecule pull the atoms outward, causing a weak bond to form between the molecules.

The Ability of Water to Dissolve Minerals

7. We all know that water can dissolve some solid materials, like salt, but how does water do this? When we add salt crystals, like the one in the center of this block, to a pot of water, each crystal is surrounded by water molecules. In halite, sodium atoms have loaned an electron and so have a positive charge (Na^+). Such positive ions are called *cations*. Chlorine has gained an electron and so has a negative charge (Cl^-). Such negative ions are called *anions*.

04.13.a4

8. The negatively charged chlorine anion is attracted to the positive (H) end of the water molecule. If this attraction is strong enough, it can pull the chlorine away from the halite crystal and into the water.

9. In a similar manner, sodium in halite is a positively charged cation (Na^+) and so is attracted to the negative side of any adjacent water molecule. This attraction can pull the sodium ion away from the halite crystal and into the water.

10. Once dissolved in water, the positively charged sodium cation (Na^+) will be surrounded by the negative sides of water molecules. The encircling water molecule may prevent sodium from rejoining the halite crystal.

11. The negatively charged chlorine anion (Cl^-) is likewise surrounded by the positive side of the encircling water molecules. It is water's polarity that enables it to dissolve some solid materials and makes it a good solvent and cleaning agent.

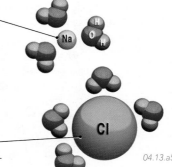
04.13.a5

B How Do Minerals Precipitate from Water?

To understand how a mineral can grow, we need to think about the environment where it is growing. Minerals on Earth are always surrounded by other materials, perhaps other minerals, magma, water, or air. Minerals contain specific atoms, and a crystal needs a nearby source of these atoms in order to grow. A growing crystal of halite (NaCl) needs additional sodium and chlorine ions. So the environment in which crystallization occurs is important because it places physical constraints and chemical constraints on crystal growth.

What happens if salty water evaporates? Over time, salt crystals precipitate (grow) on the bottom and sides of the container. How did the salt crystals form, and where did the material come from?

If there is lots of water, the few sodium (Na⁺) ions and chloride (Cl⁻) ions are kept apart by the water molecules and by the constant movement of the ions. As a result, the ions rarely come into direct contact with each other.

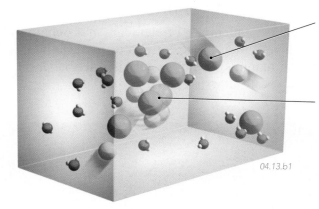
04.13.b1

As the water evaporates into the air, sodium and chlorine are left behind and so become more concentrated in the water. As a result, the two ions begin to find each other and begin to bond.

The resulting NaCl pairs combine and, to keep local charges balanced, begin to organize into an ordered structure with alternating cations and anions, forming a salt crystal.

Writing a Chemical Reaction for Salt in Water

Dissolving halite in water, or precipitating halite crystals from water, like any chemical reaction, can be expressed with an equation that describes what is happening. For this reaction, we put halite (NaCl) on one side of the equation and the sodium (Na⁺) and chlorine (Cl⁻) ions on the other side.

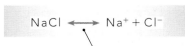

$$NaCl \longleftrightarrow Na^+ + Cl^-$$

Because this reaction can go either way (halite may dissolve or precipitate), we link the two sides of the equation with a two-headed arrow.

Whether halite precipitates or dissolves depends on how much halite is in the water. If we add more NaCl, the reaction will eventually move to the right, and halite crystals will grow because there is too much dissolved Na⁺ and Cl⁻. If, however the water solution is not already saturated with the two ions, we may be able to add quite a bit of salt before crystallization begins. Temperature is also important. Most minerals dissolve more easily in hot water than in cold water.

The Properties of Ice

Depending on pressure and temperature, water molecules can exist as a gas, liquid, or solid. The solid form of H_2O is ice, which is a mineral with an orderly crystalline structure. Ice is clear when pure but generally is cloudy or blue because of trapped air bubbles. It is a soft mineral, but it can erode into landscapes if it carries harder rocks, as it does in glaciers.

When water freezes, the weak hydrogen bonds form as the molecules rearrange to form a crystal. This rearrangement of bonds results in water molecules that are farther away from each other in ice than they are in liquid water. As a result, ice is less dense than water and so floats in water, whether it is ice cubes in our glass or icebergs adrift in the sea (▶). Water is one of the few compounds that has a solid form that is less dense than the liquid form. Most other materials, including most rocks, become less dense as they melt, so the liquid form is less dense than the solid form.

The lower density of ice has many important implications for landscapes and life. Water that freezes and expands can pry apart rocks and soil, loosening pieces that can be transported away. If ice was more dense than water, ice that formed on the surface of a lake would sink to the bottom, allowing new ice to form on the surface.

This process would repeat until the lake was frozen solid, from top to bottom. Few creatures could survive freezing of all the water in a lake. Our world would be very different if ice was more dense than water and sank rather than floated.

04.13.mtb1 Antarctica

Before You Leave This Page Be Able To

☑ Sketch a water molecule and illustrate why it has polarity.

☑ Describe the properties of water that are attributable to polarity and those that are attributable to hydrogen bonding.

☑ Describe how halite dissolves and crystallizes in water.

☑ Describe why ice is less dense than water and why this is important.

4.13

4.14 How Are Minerals Used in Society?

PEOPLE HAVE ALWAYS USED MINERALS, and they have become essential to our modern society. We need minerals to build our houses, cars, roadways, and buildings. Sometimes we use minerals in their natural form. Other times, we extract key elements from the minerals and use them for manufacturing. On average, each American uses, either directly or indirectly, 21 metric tons of minerals and rock per year.

A How Are Minerals Used for Their Chemical Components?

One of the major uses of minerals is as a source of elements and compounds that we then use to manufacture other products. We mine minerals and then process them to extract the required elements or compounds. The resulting materials are used in the manufacture of materials, such as glass, metals, and computers, or are then combined with other elements to create new useful compounds.

Iron—This element is mostly mined from iron formations (▶) containing *hematite* and *magnetite*—both iron oxides. Iron is the main ingredient in steel, which is used in many products, from kitchen utensils and appliances to automobiles, construction equipment, skyscrapers, railroads, and ships.

04.14.a1 Sherman Mine, Ontario, Canada

04.14.a2, Kidd Creek Mine, Ontario, Canada

Copper—Copper conducts electricity and so is used for electrical wires in telephones, computers, automobiles, and nearly everything that is electric. It is also used in brass and bronze. Most copper comes from copper-sulfide minerals (◀) and from various blue-green copper minerals.

Sodium and Halite—Sodium has some uses as a pure element, and it is extracted from *halite* (NaCl) and from the mineral *trona*, a sodium carbonate. Halite is used in human and animal diets, as a highway deicer, and in water softeners. It and trona are used to make soaps, metals, and many household items. Trona is important for manufacturing glass.

04.14.a3 Trona, CA

04.14.a4

Phosphorus—We use phosphorus in fertilizers, soft drinks, and consumer devices, including some televisions. Many large phosphate mineral deposits form by accumulations of marine sediments. The main phosphorous ore mineral is *apatite*, a calcium phosphate mineral similar in composition to human teeth.

Silicon—The element silicon, used to create computer chips (▶) and solar panels, is mostly derived from *quartz*, a very common silicon-oxide mineral. Although quartz is present in granite and other rocks, it is most concentrated in certain sandstones, loose sands, and quartz veins, so that is where most silicon-rich materials are mined.

04.14.a5

04.14.a6

Calcium—Calcite, the most common carbonate mineral, is the chief source of calcium and is used to help construct many parts of our infrastructure. Calcite is processed into the main ingredient in cement, which is used in roadways, sidewalks, bridges, airports, large buildings, and the foundations of homes.

Minerals in Your Medicine Cabinet

Minerals are used to make many items you find in a house, and you may be surprised at some of the unexpected places minerals show up, such as in your medicine cabinet. Most toothpastes contain calcite as the "scouring agent," and they also contain fluoride, derived from the mineral *fluorite*, and various sodium compounds derived from *trona*. The major ingredient in many antacids is calcium carbonate, in some cases derived from ground-up calcite. The abrasive material on nonmetallic nail files, or emery boards, is finely ground *garnet* or the mineral *corundum*, which also occurs as rubies and sapphires. Most makeup consists of *clay minerals* but may also include small flakes of mica as a glitter. Foot and body powders, also called talcum powder, may contain *kaolinite* (a clay mineral) and *talc*, the softest mineral in Mohs Hardness Scale. Finally, the medicine cabinet is composed of steel, derived from *hematite* and *magnetite*, or aluminum, derived from fine-grained, clay-like materials. The mirror of the cabinet consists mostly of glass derived from *quartz* and is coated with a silver compound or some other reflective substance.

B How Are Minerals Used for Their Physical Properties?

In addition to being sources of chemicals, we use many minerals intact because of some special property they possess. Minerals are used because of their color, density, resistance to heat or abrasion, shininess, or the ease with which they can be shaped. We use minerals for ceramics and as fillers to thicken and extend the volume of materials like paint, and we use huge volumes of crushed stone from rocks containing quartz, feldspar, or calcite.

Quartz—Large quantities of quartz are melted and mixed with other materials to make glass windows and glass block (▶). Quartz is used as filler materials in paint, paper, and in some food and vitamin products (listed as silicon dioxide). Synthetically grown quartz crystals are used in halogen bulbs and timing devices.

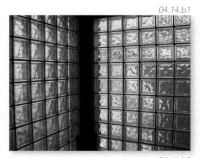
04.14.b1

04.14.b2

Clay Minerals— We use many clay minerals, all of which are sheet silicates. Clay is used to produce brick, cement, and ceramics (such as bathtubs), as well as tile for roofs (◀), floors, and walls. Large quantities are also used for cat litter and as fillers in paper, paint, and food products.

Feldspars—These very common silicate minerals are used in ceramics, including tile (▶) and china, and in glass-fiber insulation. They are also used in glass making to improve hardness and durability. In the United States, most feldspar is mined from granite and other igneous rocks.

04.14.b3

04.14.b4

Gypsum—Gypsum is a sulfate mineral that mostly forms from the evaporation of water in salty lakes or some inland sea. It is mostly used for wallboard (sheetrock) and plaster products. Additionally, it is used for cement production, agricultural applications, glass making, and other industrial processes.

C In What Geologic Environments Do Gem-Quality Minerals Form?

Most gems are minerals—very beautiful ones! Some gems are not minerals because they are not natural (e.g., cubic zirconium) or they do not have an ordered crystalline structure. Some organic materials, including pearls and amber, are sometimes considered gems. What environments enable beautiful gem minerals to grow?

Diamonds form deep in the mantle under conditions of high temperature and extremely high pressures. They are brought to the surface through volcanic conduits called *kimberlite pipes.* Diamond is mined in pipes or in sediment eroded from diamond pipes.

Opal shimmers as it moves in light, showing various shades of blue, green, red, and other colors. Opal is not a mineral because it does not have an orderly crystalline structure. It consists of microscopic spheres of silica that include trapped water, giving opals their distinctive spectrum of color. Opal commonly forms in volcanic rocks and some sedimentary rocks that have fractures and other natural openings (called *voids*). Silicon-rich water fills these fractures and voids and deposits the opal.

Ruby and *sapphire* are both varieties of the mineral *corundum* (aluminum oxide). *Emerald* and *aquamarine* are varieties of the mineral *beryl* (an aluminum silicate). All four gemstones mostly form in *pegmatites,* which are coarse-grained igneous rocks that crystallized from magma containing relatively high amounts of water. Extra water promotes the growth of very large crystals. Ruby, sapphire, emerald, and aquamarine may also form during metamorphism of some sedimentary rocks.

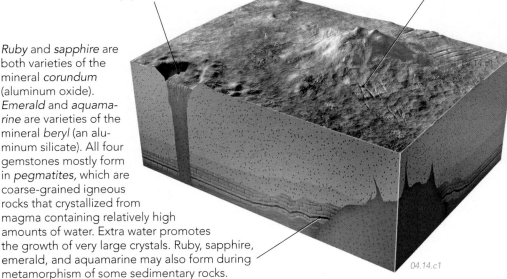
04.14.c1

Before You Leave This Page Be Able To

✓ Distinguish the two main ways that minerals are used in society.

✓ Describe some chemical uses of common minerals.

✓ Describe how minerals are used in some of the products that are in your medicine cabinet.

✓ Describe some ways we use the physical properties of minerals.

✓ Describe the geologic environments in which some gemstones form.

4.14

4.15 What Minerals Would You Use to Build a House?

MINERALS AND ROCKS ARE USED to make many objects around us. Minerals, rocks, and products derived from them compose our homes, cars, streets, buildings, electrical grid, and water-supply system. If something is not grown, it comes from rocks, minerals, or petroleum. In this exercise, you will decide what minerals are used in materials to build the important parts of a house.

Goals of This Exercise:

- Make some observations about minerals based on their appearance in a photograph or from samples provided by your instructor.
- Identify minerals based on their appearance and diagnostic properties.
- Determine, based on each mineral's characteristics and how it is commonly used, which mineral(s) to use for each part of a house.

A Describe and Identify These Minerals

Examine each mineral in the photographs below or from samples provided by your instructor. For each mineral, make observations, such as crystal form, luster, color, and cleavage. Write these observations on the accompanying worksheet or a sheet of paper. Then, read the accompanying text blocks that provide additional information about each mineral. If you have access to mineral samples, perform tests, such as determining hardness, on each mineral. For each mineral, the worksheet contains additional important information that will help in identification.

04.15.a1

04.15.a2

04.15.a3

This six-sided mineral has a hardness of 7 and a conchoidal fracture instead of cleavage. It does not effervesce.

This mineral is partially transparent, has a hardness of 3, cleaves into rhombs, and effervesces with dilute HCl.

This mineral is very soft, feels sticky when wet, and does not effervesce. It contains very fine material. It is not talc or graphite.

Each of these spherical masses consists of a number of intergrown crystals of a cream-colored to partially transparent mineral. The mineral can be scratched with a fingernail and does not effervesce. ▶

 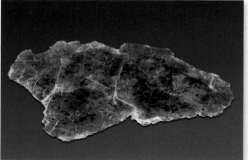

◀ This mineral has one direction of cleavage and flakes into thin sheets. It is nonmagnetic and does not effervesce. When held up to the light, thin sheets are partially transparent and have a silvery-gray color.

04.15.a4

04.15.a5

04.15.a6

Copper Carbonate and Other Copper Minerals

Native Copper

◀ These blue-green and copper-colored minerals contain copper. They include copper-carbonate minerals, such as malachite (green) and azurite (blue). The metallic material is native copper. These minerals were not discussed in detail.

This mineral has a metallic luster and a distinctive red streak. It is nonmagnetic and in some samples has a reddish tint. ▶

04.15.a7

B Devise Ways to Build a House Using Minerals and Mineral Products

This illustration shows parts of a house for which you need to find a mineral or mineral-derived product. Using the minerals you identified in part A along with information about the uses of minerals in the chapter and in the worksheet, consider options for which mineral or mineral product you will use to construct different parts of the house. Identify the mineral by name and list the properties this mineral had that were useful for the house.

Roof—A roof is a barrier to rain and snow. Some type of mineral product is used to cover the plywood sheets on the roof.

Mineral Name and Useful Mineral Properties:

_____

_____

Insulation—To keep the house at a comfortable temperature, a material that conducts heat slowly is placed outside, inside, or within the exterior walls. Commonly, this material is fiberglass, which is produced by melting a common and inexpensive silicate rock and turning the melt into glass fibers.

Mineral Name and Useful Mineral Properties:

_____

_____

Exterior Walls—The outside walls act as a barrier to rain and snow and support the roof and the rest of the structure.

Mineral Name and Useful Mineral Properties:

_____

_____

04.15.b1

Windows—These let in visible light and other solar energy and provide visibility to the outside.

Mineral Name and Useful Mineral Properties:

_____

_____

Electrical Wiring—A material that conducts electricity is used for electrical wiring. Most wire is made from a metal because metals are conductive and ductile (can be shaped easily into wire).

Mineral Name and Useful Mineral Properties:

_____

_____

Cement Slab—Cement is used to make a fairly smooth, stable base for floor tile, wood, or carpet. It is also used as a foundation to support the walls.

Mineral Name and Useful Mineral Properties:

_____

_____

Inside of Walls—Interior walls separate the house into rooms but commonly do not support the structure. They typically have vertical beams (called studs) of a strong material that supports sheets of wallboard that form the actual wall. The covering sheets should be soft enough so that holes can be cut for electrical outlets and switches.

Mineral Name and Useful Mineral Properties:

_____

_____

Plumbing—Metal pipes are commonly used to carry fresh water into the house and from one part of the house to another.

Mineral Name and Useful Mineral Properties:

_____

_____

4.15

Igneous Environments

MOLTEN ROCK MAY REACH EARTH'S SURFACE and erupt in a volcano, or it may solidify underground, later to be uplifted and exposed by erosion. Igneous rocks form some very distinctive landscapes, including huge gray mountains, cone-shaped volcanoes, and precipitous volcanic buttes. How does molten rock form, move, and solidify, and what is the relationship, if any, between the formation of magma and plate tectonics? Finally, what types of landscape features do igneous rocks and processes produce?

An unusual circular depression crowns the top of the Jemez Mountains near Los Alamos, New Mexico. This feature, called the *Valles Caldera*, is outlined by a dashed line below. The rocks within and near the caldera are of volcanic origin, and most are less than two million years old. Examine this feature and the other features shown on this satellite image.

How did this caldera form, and what is its relationship to the nearby, relatively recent volcanic rocks?

In this satellite image, green colors show areas covered by trees and bushes, mostly in the mountains. Yellow and orange colors in the Cerros del Rio area represent volcanic rocks that have less plant cover. Tan, brown, and some orange colors are areas of other rock types, loosely consolidated sediment, and soil.

The Valles Caldera formed when a huge magma chamber erupted, covering the region with hot, suffocating volcanic ash. As the ash erupted, the roof of the underground magma chamber collapsed, forming the circular caldera. After the main collapse, slow-moving lava flows built up dome-shaped hills within and next to the caldera. Note how these lava-formed hills form a nearly circular ring within the caldera.

How do igneous features such as calderas form, and how do we recognize them in the landscape?

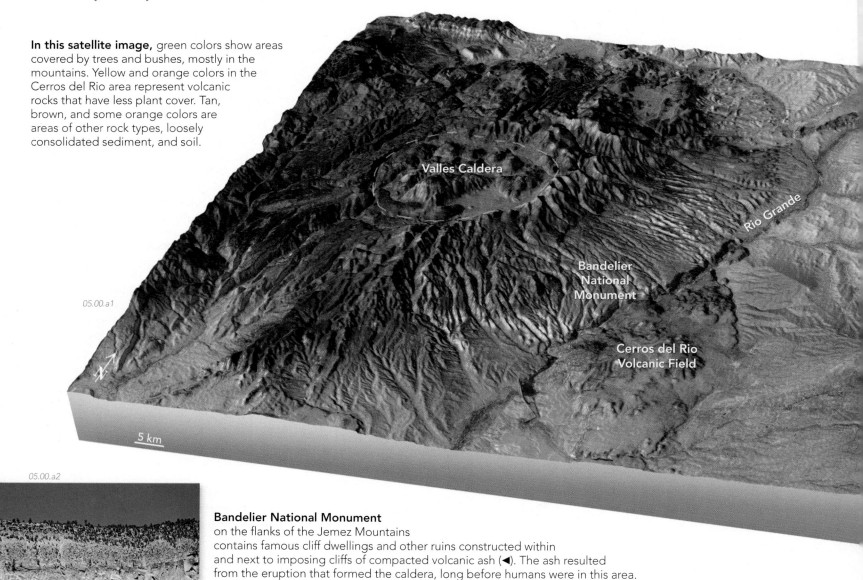

05.00.a1

Valles Caldera

Rio Grande

Bandelier National Monument

Cerros del Rio Volcanic Field

5 km

05.00.a2

Bandelier National Monument
on the flanks of the Jemez Mountains
contains famous cliff dwellings and other ruins constructed within
and next to imposing cliffs of compacted volcanic ash (◄). The ash resulted
from the eruption that formed the caldera, long before humans were in this area.

How do we recognize volcanic ash, and what makes some ash strong enough to form a cliff?

TOPICS IN THIS CHAPTER

The Harding Pegmatite Mine, east of the Rio Grande, has igneous rocks (▼) with large crystals, some as long as two meters (about six feet). Compare the size of the long crystals with the rock hammer in the photograph below. These unusual rocks must have formed in a very different igneous environment than the volcanic rocks to the west.

What factors control whether crystals in igneous rocks are microscopic or are meters long?

Harding Pegmatite

05.00.a3

Many small volcanoes and dark lava flows form the small Cerros del Rio volcanic field across the Rio Grande from Bandelier National Monument. Unlike the explosive eruptions of the Valles Caldera, magma from the smaller volcanoes flowed onto the surface in a less violent manner and constructed dark volcanic layers, like these exposed along the Rio Grande in the photo below. ▼

Where and how does magma form, and what factors determine whether magma erupts as an explosion of hot ash or an outpouring of less explosive lava?

05.00.a4

Valles Caldera and Bandelier National Monument

The Valles Caldera of the Jemez Mountains is one of the most studied volcanic features in the world. It was here that geologists first figured out how the collapse of a caldera is related to explosive eruptions of volcanic ash. The caldera has been explored using deep drill holes to study its subsurface geometry, to investigate the potential for geothermal energy, and to better understand these large volcanic features. Geologists use the volcanic eruptions and collapse of the Valles Caldera as a model of what could occur in future eruptions in Yellowstone National Park of Wyoming.

About 1.2 million years ago, a huge volume of magma rose from deep in the crust and accumulated in a *magma chamber* several kilometers below the surface. Subsequently, some of the magma reached the surface and erupted explosively, forming a turbulent cloud of pumice, volcanic ash, rock fragments, and hot, toxic gases that raced outward at speeds of hundreds of kilometers per hour. As magma escaped from the underground chamber, the roof of the chamber collapsed, forming the roughly circular depression visible today. After the main explosive eruption, smaller volumes of magma reached the surface, producing slow-moving lava that piled up into dome-shaped mounds within the caldera.

Volcanic ash that was erupted from the caldera blanketed most of the area. Some ash layers became compacted by the weight of additional ash that accumulated on top. Streams later eroded steep canyons, within which the ancient puebloan peoples of the Southwest built the cliff dwellings and other structures preserved within Bandelier National Monument.

5.0

5.1 What Textures Do Igneous Rocks Display?

IGNEOUS ROCKS FORM BY SOLIDIFICATION OF MAGMA. Most igneous rocks have millimeter- to centimeter-sized crystals, but some have meter-long crystals and others are noncrystalline glass. Igneous rocks vary from nearly white to nearly black, or they can have mixed colors. They may contain holes, fragments, or ash that has been compacted. What do the different textures tell us about how the magma solidified?

A What Textures Are Common in Igneous Rocks?

The *texture* of a rock refers to the sizes, shapes, and arrangement of different components. The texture of an igneous rock depends mostly on overall crystal size, the variation in crystal size within that rock, and the presence of other features, such as holes and rock fragments.

The most obvious textural distinction among igneous rocks is whether or not a rock has crystals that are visible to the unaided eye (i.e., without using a hand lens or microscope). The crystals in the rock shown here are large and easily observed without a hand lens or microscope. Igneous rocks with crystals that are visible to the unaided eye are said to be *phaneritic*.

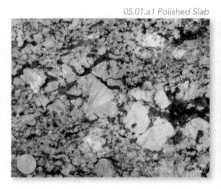
05.01.a1 Polished Slab

05.01.a2 Wickenburg, AZ

Some igneous rocks, like the one here, do not contain crystals that are visible to the unaided eye. Instead, these rocks consist of microscopic crystals, fine-grained volcanic ash, volcanic glass without any crystals, or a combination of these. Such rocks are *aphanitic* and result from magma that solidifies too rapidly to grow crystals that are visible in a hand specimen.

05.01.a3 Northern AZ

Some igneous rocks contain very large crystals, which may be centimeters to meters long. We call very coarse igneous rocks, like the one shown above, *pegmatite*.

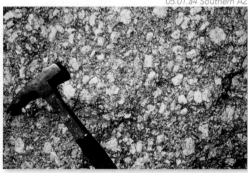
05.01.a4 Southern AZ

This rock is *coarsely crystalline* (also described as being *coarse grained*). Most crystals are larger than several millimeters, and many are several centimeters across.

05.01.a5 South Africa

Medium-grained rocks have crystals that are easily visible to the unaided eye. Crystals in such rocks are typically millimeters across, but not centimeters across.

05.01.a6 Southern AZ

Crystals in *fine-grained* igneous rocks can be too small to see without a hand lens. In some fine-grained rocks, the crystals are visible only with a microscope.

05.01.a7 Greece

Some igneous rocks consist of glass rather than crystals of minerals. A rock may be 100% *volcanic glass* or may be mostly *glassy* with some crystals or rock fragments.

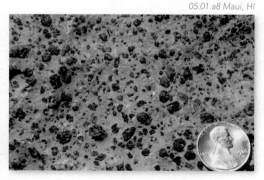
05.01.a8 Maui, HI

Igneous rocks that include larger crystals in a finer grained *matrix* are *porphyritic*. The crystals in a porphyritic rock are termed *phenocrysts*.

05.01.a9 Northern NM

Many volcanic rocks contain small holes known as *vesicles*, and we describe such rocks with the adjective *vesicular*.

05.01.a10 Southern, CA

Volcanic ash and pumice, when still hot, can be compacted by overlying materials, becoming a hard rock with a *welded* texture.

05.01.a11 Southern CA

Some volcanic rocks contain angular fragments in a finer matrix and are called a *volcanic breccia*.

B In What Settings Do the Different Igneous Textures Form?

The different textures of igneous rocks reflect the environment in which the magma solidified. Perhaps the magma solidified at depth, or maybe it erupted onto the surface as molten *lava*. Examine the figure below and think about where each texture in the photographs on these two pages might form.

Vesicles form when gases dissolved in magma accumulate as bubbles. They can form only under low pressures on the surface or very near the surface. Many lavas are vesicular, and much of the material in volcanic ash forms when the thin walls between vesicles burst, shattering partially solidified magma into sharp particles. Most volcanic ash is broken vesicles.

Volcanic breccia can form in many ways, including from explosive eruptions of ash and rock fragments, from a lava flow that breaks apart as it partially solidifies while flowing, or from volcano-triggered mudflows and landslides on the steep and unstable slopes of the volcano.

Volcanic glass forms when magma erupts on the surface and cools so quickly that crystals do not have time to form. This can happen in a lava flow or in volcanic ash.

For a *porphyritic* texture to form, magma needs sufficient time in a subsurface magma chamber to grow visible crystals. Later, the magma rises closer to the surface, where the remaining magma solidifies rapidly into the fine-grained matrix around the larger crystals (phenocrysts).

Pegmatite may form if magma is water rich. The dissolved water allows atoms to migrate farther and faster and so helps large crystals to grow. This generally occurs near the sides and top of a magma chamber and in local pockets within the magma. Most pegmatite forms at moderate to deep levels within Earth's crust.

05.01.b1

Some *volcanic ash* erupts vertically in a column and settles back to Earth. This ash cools significantly before accumulating on the surface. Because it is relatively cool and strong, the ash may not become welded; thus it is said to be *nonwelded*.

Other volcanic ash erupts in thick clouds of hot gas, ash, and rock fragments, called *pyroclastic flows*, that flow rapidly downhill under the influence of gravity. The ash deposited by pyroclastic flows is very hot, and so most parts are *welded* to some extent.

Fine-grained igneous rocks form if the magma only has enough time to grow small crystals. This commonly occurs when magma solidifies on the surface in a thick lava flow or at shallow depths beneath the surface, because cooling in these settings is fairly rapid. Medium-grained rocks form deeper, where cooling occurs more slowly.

Coarse-grained igneous rocks form at greater depths, where magma cools at a rate that is slow enough to allow large crystals to grow.

Before You Leave This Page Be Able To

☑ Sketch or describe the various textures displayed by igneous rocks.

☑ Sketch an igneous system and show where the main igneous textures form.

5.2 How Are Igneous Rocks Classified?

IGNEOUS ROCKS VARY IN chemical composition and therefore in mineral content. Some are composed entirely of dark minerals, whereas others contain only light-colored minerals. We classify igneous rocks so that we can use a single name to identify rocks that form in a similar way and have a similar composition.

A How Do the Characteristics of Igneous Rocks Vary?

Compare the colors and sizes of crystals in these rock samples, each of which is 5 to 10 cm (2 to 4 in.) across. What criteria would you use to sort these rocks if you wanted to classify them or give them names?

05.02.a1

B How Do We Examine and Identify Different Minerals?

To better identify the minerals in a rock and to estimate their percentages, geologists observe coarse-grained rocks by cutting a slab and by using a hand lens. Fine-grained rocks require a microscope.

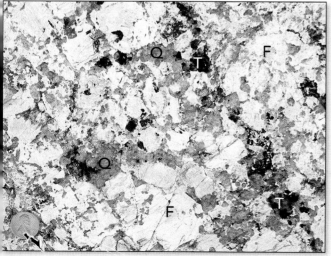

05.02.b1

Rocks we call *granite* are all dominated by feldspar and quartz, but they can differ in overall crystal size, in the other minerals they contain, and in their proportion of quartz, feldspar and these other minerals.

This photograph shows part of a slab of very coarse-grained granite. The rock has several distinct minerals as well as less distinct areas stained yellowish brown by iron oxide minerals. How many different kinds of minerals do you see in the rock, and what percentage of each are present? Examine the photograph and then read the list of minerals below.

F: Feldspar (cream colored)

Q: Quartz (gray, transparent)

T: Tourmaline (black crystals)

Before You Leave These Pages Be Able To

☑ Sketch and describe how igneous rocks are classified.

☑ List some common igneous rocks and a few characteristics of each.

☑ Describe the main differences between felsic and mafic rocks.

C How Does the Composition of Igneous Rocks Vary?

Geologists organize igneous rocks according to the *size of crystals* and the *kind of minerals* in a rock. Below, images in the left column feature rocks with coarse crystals. Each rock in the right column has a composition similar to the rock on the left but a smaller grain size. From top to bottom, the rocks contain less light-colored minerals. Rocks with a light color and abundant quartz and feldspar are *felsic* rocks, whereas rocks that are dark and contain minerals rich in magnesium and iron are *mafic* or *ultramafic* rocks. *Intermediate* rocks are in between.

Coarsely Crystalline   Finely Crystalline or Glassy

Felsic

Granite is a coarsely crystalline, light-colored igneous rock. The light color is due to an abundance of the light-colored, felsic minerals feldspar and quartz. Most granites also contain some biotite (black mica), and some contain white mica (muscovite), and garnet.

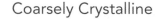
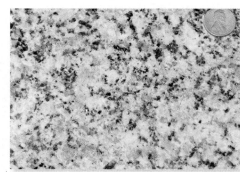

05.02.c1 Polished Slab

05.02.c2 Southern CA

Rhyolite is the fine-grained equivalent of granite. It is mostly a finely crystalline rock, but it can contain glass, volcanic ash, pieces of pumice, and variable amounts of visible crystals (phenocrysts) of quartz, K-feldspar, or biotite.

Intermediate

Diorite contains more mafic minerals than does granite. It is *intermediate* between felsic and mafic compositions. It generally contains abundant plagioclase feldspar and amphibole, and it can contain variable amounts of either biotite or pyroxene.

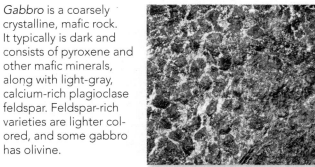

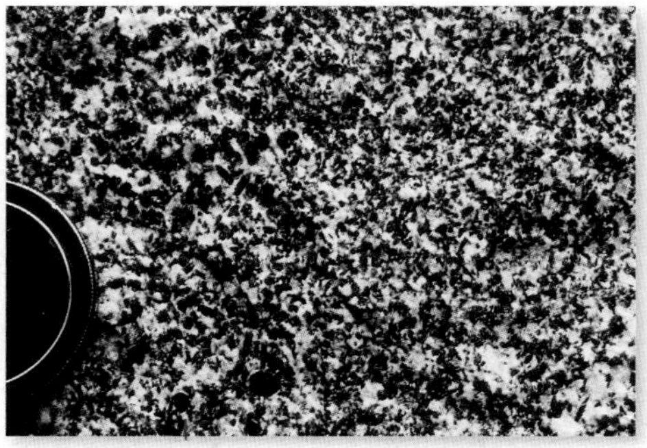

05.02.c3 Northern CA

05.02.c4 Northern AZ

Andesite is the fine-grained equivalent of diorite. It is commonly gray or greenish, but it can also have a slight maroon or purplish tint. Andesite commonly has phenocrysts of cream-colored feldspar or dark amphibole.

Mafic

Gabbro is a coarsely crystalline, mafic rock. It typically is dark and consists of pyroxene and other mafic minerals, along with light-gray, calcium-rich plagioclase feldspar. Feldspar-rich varieties are lighter colored, and some gabbro has olivine.

05.02.c5 Southern AZ

05.02.c6 Northern NM

Basalt is a dark lava rock. Most basalt is dark gray to nearly black, and many outcrops have vesicles, as shown here. Basalts can contain phenocrysts of dark pyroxene, green olivine, or cream-colored plagioclase feldspar.

Ultramafic

Peridotite is the main coarsely crystalline *ultramafic* rock. Compared to mafic rocks, it contains more magnesium-rich and iron-rich minerals, especially green olivine and dark pyroxene. Some varieties are mostly olivine. The upper mantle is composed of peridotite.

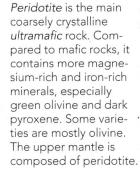

05.02.c7 South Africa

05.02.c8 South Africa

Ultramafic lavas erupted early in Earth's history, and so such rocks are preserved only in the oldest parts of some continents. The magma was very hot and commonly grew olivine or pyroxene crystals that are unusually long for a lava flow.

5.2

5.3 What Are Some Other Igneous Rocks?

SOME IGNEOUS ROCKS HAVE DISTINCTIVE TEXTURES that allow them to be identified with a unique name and mode of origin. Such characteristics include holes, fragments, or extremely large crystals. These textures are more common in some igneous compositions than in others.

A What Are Some Other Common Igneous Rocks?

Some common igneous rocks fit into the classification system presented on the previous two pages, appropriately being called granite or basalt, but they possess some attribute that caused geologists to assign the rock a special name to convey the texture and, by inference, the specific way in which the rock formed.

Obsidian is a shiny volcanic glass that is normally a medium gray to black color. Most obsidian has a composition equivalent to that of rhyolite. It forms when a lava flow cools too rapidly to form crystals. Some obsidian contains phenocrysts or fragments.

05.03.a1 Iceland

05.03.a2 Northern NM

Tuff is a volcanic rock composed of a mix of volcanic ash, pumice, crystals, and rock fragments. If the particles of ash and pumice cool before being buried by overlying materials, the rock remains only weakly consolidated and is *nonwelded tuff.*

Volcanic glass is unstable, eventually changing from noncrystalline glass into rhyolite consisting of very small crystals. The conversion to rhyolite can produce blobby or layered patches of distinct crystals, in this case forming a rock called *snowflake obsidian.*

05.03.a3 Inyo Mountains

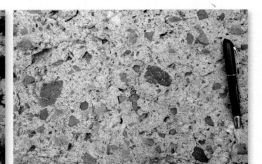

05.03.a4 Hieroglyphic Mtns., AZ

If tuff gets buried while still hot, as within a thick pyroclastic flow, the weight of overlying materials compacts ash and pumice into lenses, forming *welded tuff.* Tuff commonly contains angular fragments of older rocks, which do not compact, as shown here.

Pumice is a volcanic rock containing many vesicles (holes). The holes are so numerous that most pumice floats on water. The solid material in pumice begins as volcanic glass, but over time it can convert into microscopic crystals.

05.03.a5 Katmai, AK

05.03.a6 Wickenburg, AZ

Volcanic rocks with fragments form in other ways, such as the breaking apart of lava that solidifies during flow. Fragmental rocks also form from mixtures of volcanic rock, ash, and mud. In either case, the resulting fragmental rock is a *volcanic breccia.*

Scoria is a dark gray, black, or reddish volcanic rock that contains many vesicles. It usually has the composition of basalt or andesite. In outcrops, scoria consists of a jumbled mass of rock fragments as large as several meters across.

05.03.a7

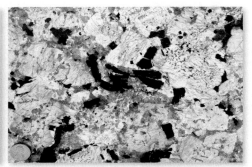

05.03.a8 Polished Slab

When magma crystallizes deep within the crust, a crystallizing magma may contain enough dissolved water that it grows exceptionally large crystals. If the crystals are larger than several centimeters (they can be meters across), the rock is *pegmatite.*

B How Do Mineral Abundances and Igneous Rocks Relate to Our Classification?

Our classification of igneous rocks mostly considers the composition of the rocks (e.g., percentage of felsic versus mafic minerals) and the *size of crystals* in the rock. How do rocks such as scoria and tuff fit into this classification system? The table below places scoria, obsidian, and other rocks into this classification system and also illustrates how the abundance of common minerals varies for different igneous rocks.

1. *Pegmatite* can be any composition, but most is felsic (granitic), containing large crystals of feldspar and quartz. Granitic pegmatite typically also contains one or more mica minerals (muscovite and biotite). Some includes less common minerals, which can form gem-stones, like garnet, tourmaline, and beryl (emerald and aquamarine).

2. Mafic and ultramafic magmas do form pegmatite, but such pegmatite is less common than granitic ones. Mafic and ultramafic pegmatite are important in some parts of the world because they contain chromium, platinum, and other important mineral resources.

3. *Obsidian* is mostly felsic (rhyolite), but some is barely into the intermediate field on this graph. Intermediate and mafic volcanic rocks can be glassy, but most geologists do not call such rocks obsidian.

4. *Pumice* is light-colored and felsic to intermediate. The silica-rich magmas trap gas, forming abundant vesicles. Pumice is usually present as millimeter- to centimeter-sized pieces within tuff, but can also form volcanic units that are nearly all pumice. Highly vesicular basalt is *scoria*.

5. Like pumice, most *tuff* is felsic to intermediate, especially tuff formed by pyroclastic flows or huge eruption columns. Tuff can also be mafic (basaltic), but this type of tuff mostly forms from ash particles that settle out of the air from smaller eruption columns.

05.03.b1

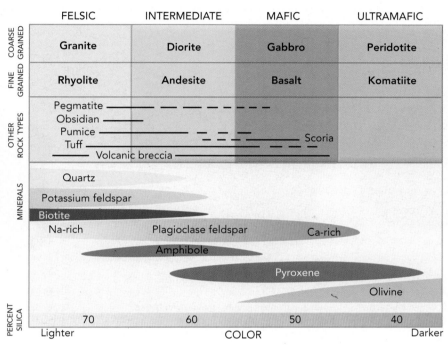

6. *Volcanic breccia* can be almost any composition, since forming one only requires that a volcanic material break into angular fragments during or soon after an eruption. Rhyolite and andesite produce breccia because their silicate-rich magmas do not flow easily and so break apart during flow or are fragmented by explosive eruptions from trapped gas. Mafic magma can flow more easily, but mafic lava flows can cool, partly solidify, and break apart during flow.

7. Minerals present in igneous rocks vary as a function of composition, as does the silica (SiO_2) content. Light-colored (felsic) minerals, like quartz, K-feldspar, and Na-rich plagioclase feldspar, reside in silica-rich felsic and some intermediate rocks, but they are uncommon in mafic rocks, which contain less silica. Biotite is present in many felsic rocks, but it and amphibole are more abundant in intermediate rocks. Mafic minerals pyroxene and olivine, along with Ca-rich plagioclase, are the dominant minerals in mafic and ultramafic rocks.

The Chemical Composition of Igneous Rocks

The chemical composition of a rock largely determines the percentages of different minerals in the rock. Silica is the main ingredient in most igneous rocks, which are typically between 44% and 77% SiO_2 (see figure above). Granite and related *felsic* igneous rocks contain high amounts of silica (SiO_2), commonly 70% to 77%, and they consist mostly of quartz and feldspar. Rhyolite and some obsidian have the same chemical composition as granite but contain mostly microscopic mineral grains or glass, respectively.

Silica is also the dominant chemical constituent of *mafic* rocks, such as basalt and gabbro, but at lower concentrations (44% to 50% SiO_2). Compared to felsic rocks, mafic rocks contain more magnesium, iron, and calcium, and these elements cause darker, mafic minerals, such as pyroxene and olivine, to be more abundant.

Intermediate rocks, including andesite and diorite, contain intermediate amounts of silica (about 60% SiO_2) compared to felsic and mafic rocks. They also contain intermediate amounts of magnesium, iron, calcium, potassium, and other elements. They contain abundant feldspar, with variable amounts of quartz and mafic minerals, especially amphibole and biotite.

Before You Leave This Page Be Able To

✓ List the main characteristics of obsidian, pumice, scoria, tuff, breccia, and pegmatite, and indicate where each of these rock types fits into an igneous classification system based on composition.

✓ Summarize the main minerals that are present in felsic, intermediate, mafic, and ultramafic rocks.

5.3

5.4 How Do Temperature and Pressure Vary Inside Earth?

THE INTERIOR OF EARTH IS HOT. Heat, which is a transfer of thermal energy, flows from deeper, hotter regions of Earth to the cooler surface. This heat drives plate tectonics and provides the energy required to melt rocks. All igneous rocks are a direct result of Earth's internal heat. Where does this heat come from?

A How Did Earth Get So Hot, and Why Hasn't It Cooled Off More?

Earth is 4.5 billion years old, so why is Earth's interior still hot? Although the early history of Earth included several events that increased its temperature, these early events only account for 20% of Earth's overall heat today. Most of Earth's high internal temperature is instead due to energy released by radioactive decay of unstable atoms.

Why the Early Earth Was Hot

05.04.a1

05.04.a2

05.04.a3

Earth is still hot partly as a result of its formation about 4.5 billion years ago. The materials that formed the early planet became hot as they collided, coalesced, and were compressed under the force of gravity.

As Earth grew in size it became a larger target for asteroids and meteoroids that were attracted by its gravity. As these objects collided with Earth, the energy of the impacts heated the surface even more.

Early in its history, Earth was mostly molten. Masses of iron and nickel, which are very dense materials, were pulled by gravity toward the planet's core, adding heat as they sank.

Heating of Earth by Radioactive Decay

Radioactive decay of elements heats Earth's interior. This process began early in Earth's history and accounts for most of Earth's total thermal energy today. Three types of radioactive decay can occur: *alpha*, *beta*, and *gamma decay*. Alpha decay is the most important heating process in Earth today, so it is discussed in more detail below.

1. The nucleus of an atom consists of protons and neutrons. Some nuclei are unstable and undergo radioactive decay, becoming more stable by losing a proton or neutron, and in the process releasing energy in the form of particles or electromagnetic radiation.

2. During *alpha decay*, an unstable atom releases a speedy particle that impacts and heats surrounding materials. The ejected particle is equivalent to a helium nucleus (two protons and two neutrons, but no electrons).

05.04.a4

3. The original atom lost two protons and two neutrons and so becomes a different element (different atomic number). For example, uranium-238 (uranium with an atomic weight of 238) decays to form a completely different element, thorium-234, by alpha decay.

4. The new element may be stable or may itself undergo radioactive decay by alpha, beta, or gamma decay. Each step in the decay process generates additional thermal energy. *Beta decay* involves an atom losing a beta particle, which is an electron. In *gamma decay*, an atom emits an energetic particle called a photon. Alpha decay, however, is the main source of heat energy for Earth's hot interior.

B   How Do Pressure and Temperature Change with Depth?

Earth's layers vary in composition, pressure, and temperature. Overall, pressure and temperature increase with depth. The temperature increase with depth is called the *geothermal gradient*, and is somewhat variable. Temperature generally increases by 25-40°C per kilometer of depth into the crust.

Rocks on the surface are typically 10°C to 35°C (air temperature). Temperature gradually increases with depth, from these low temperatures near the surface to 900°C in the lower crust. Hotter areas exist in some places, especially those associated with magma.

Mantle temperatures are typically 1,400°C to 1,500°C, but melting is uncommon because high pressures keep most of the mantle solid. Radioactive decay occurs in the crust and core, but radioactive decay in the mantle accounts for most of Earth's thermal energy.

The hottest part of Earth is its metallic core, which is 3,000°C to 5,000°C. The outer part is molten, but the inner core is solid.

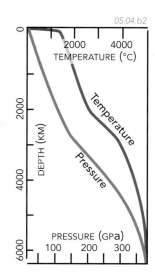
05.04.b1

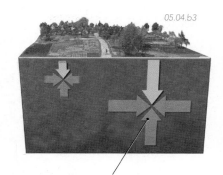
05.04.b2

◄ This plot shows how temperature and pressure increase downward from the surface to the core. Pressure is in units of gigapascals (GPa).

05.04.b3

The term *pressure* refers to the force that compresses a rock. The pressure on a rock at some depth in Earth is the force exerted by the weight of rock above it. This force increases with depth and is mostly balanced by forces pushing in from the sides.

Transferring Thermal Energy from One Place to Another

Heat is the transfer of thermal energy from high-temperature to low-temperature objects. In Earth, heat moves from the hotter interior to the cooler surface. Heat transfer, or *heat flow*, results when two adjacent masses have different temperatures. The three mechanisms of heat transfer are *conduction, radiant heat transfer,* and *convection*.

Conduction—A water-filled pan placed directly on the burner gets hot as thermal energy is transferred by direct contact between the burner and pan, and the pan and water. Heat transfer by direct contact is *conduction*.

Radiant Heat Transfer—A hot burner on a stove can warm your hands a short distance away. Such warming occurs because heat from the burner radiates through the air, a process called *radiant heat transfer* or *thermal radiation*.

Convection—Water near the bottom of the pan gets warmer, expands, and rises because it is less dense than the cooler water around it. When the rising water reaches the surface, it cools and flows back down the sides. This type of heat transfer by flow of a liquid or by a solid but weak material is *convection*. If the material flows around a circular path, as in the pan, we use the term *convection cell*.

Moving a pan full of hot water away from the stove also transfers heat from one place to another. In this case, the movement of the heated material does not follow a circular path and so is not part of a convection cell.

05.04.b4

Transferring Thermal Energy Via Plate Tectonics

1. As solid asthenosphere rises beneath a mid-ocean ridge, it brings hot rocks upward by *convection*, adding material to the oceanic lithosphere.

2. Seawater is drawn into the hot crust of the mid-ocean ridge, where it gets hotter and rises, forming a *convection cell* of seawater (too small to show here), which helps cool the oceanic crust.

05.04.b5

3. The hot, newly created lithosphere begins to cool by *conduction* of heat to adjacent cooler rocks and to seawater. Some of the underlying asthenosphere cools, hardens, and becomes part of the lithosphere.

4. The cooled oceanic lithosphere subducts back into the asthenosphere. This downward motion, coupled with upward motion of material beneath mid-ocean ridges, completes a kind of convection cell.

Before You Leave This Page Be Able To

✓ Describe the events that made the early Earth hot.

✓ Describe alpha decay and how radioactivity heats Earth.

✓ Describe three ways that heat is transferred from a warmer mass to a cooler one and an example of conduction and convection by plate tectonics.

5.4

5.5 How Do Rocks Melt?

IGNEOUS ROCKS FORM when magma (molten rock) solidifies. Where does such magma come from? How do rocks melt, under what conditions do they melt, and where in Earth does melting occur?

A What Happens When a Substance Changes from a Solid to a Liquid?

What is the difference between a solid and a liquid, and what happens at a molecular level when a solid melts?

05.05.a1

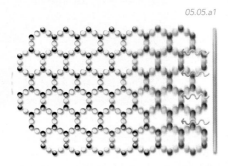

05.05.a2

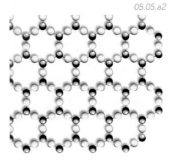

05.05.a3

In solids, atoms and bonds are always vibrating. A temperature increase causes the vibrations to increase, eventually to the point where bonds break and melting begins. An increase in pressure has the opposite effect, compressing the solid and making it more resistant to melting.

A mineral or rock will remain solid if the pressure and bond strength are sufficient to overcome vibrations due to temperature. Beneath Earth's surface, pressure arises mostly from the weight of overlying rocks, gradually increasing with depth of burial.

When bond strength and pressure are inadequate to hold a solid material together, melting will begin. Different bonds break at different temperatures, so magma generally contains some partially bonded, or weakly bonded, molecules and material within the melt.

B Under What Conditions Is a Material Solid or Liquid?

If pressure tends to keep a rock solid and temperature causes it to melt, which one prevails? The graph below shows temperatures and pressures under which a material can exist either as a solid or as a liquid.

1. Temperature is plotted on the horizontal axis, and pressure is plotted on the vertical axis. The conditions for any place within Earth can be shown as a point, such as point A, that represents a specific temperature and pressure.

2. Pressure increases downward within the earth, so pressure is plotted on this graph as increasing from top to bottom. In this manner, the graph mimics the earth, with pressure increasing with depth, but it is a graph, not a cross section.

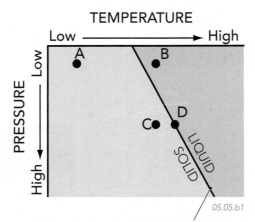

05.05.b1

4. A rock at the low temperature and low pressure represented by point A is solid.

5. A rock at point B is at the same low pressure as the rock at point A, but it has a higher temperature. It plots in the liquid field and so under these conditions is liquid (magma).

6. A rock at point C is at the same high temperature as the magma at point B, but it is solid because the higher pressure helps hold the atoms together and prevents melting.

7. If pressure-temperature conditions plot directly on the melting curve, like at point D, the rock is in the process of melting or solidifying. Under these conditions, some solid rock and some magma are present.

8. Take a moment to think about the following questions, considering that rocks plotting at points A and C are solid whereas point B represents conditions under which the rock would melt:

• What changes in temperature or pressure are required to move from point A to point B?

• What changes are needed to change the conditions from point C to point B?

• Both of these changes, from A to B, and from C to B, will cause melting. Think about what processes within the earth could cause a rock to follow one of these two paths.

3. A line, called the *melting curve* or *solidus*, divides the graph into two areas, called *fields*. If a rock is at a pressure and temperature that plot to the left of the line, the rock remains solid. If the pressure-temperature conditions plot to the right of the line, the rock will be completely melted (magma). The melting curve slopes down to the right because higher temperatures are needed to melt a rock under higher pressure. The position of the melting curve depends on the composition of the rock, shifting to the left (lower melting temperature) for more felsic compositions. Also, different minerals have different melting curves, so not all minerals will melt at once.

C What Causes a Rock to Melt?

When we think of melting, we normally think of heating something, an ice cube for example, until it turns into a liquid. Heating does cause rocks to melt, but there are complicating factors. Rock melting is influenced by three main factors: temperature, pressure, and water content.

Melting by Heating

1. When a rock is heated, some or all of its minerals can melt. On this graph, melting would occur if a rock were heated so that its temperature increased from point A to point B. Therefore, a temperature increase caused by heating can melt a rock. Most rocks contain different minerals with different melting temperatures, so an increase in temperature causes only partial melting, unless temperature becomes very high.

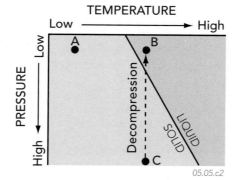

05.05.c1

2. If an increase in temperature is accompanied by an increase in pressure, as from point A to point C, the higher pressure may be enough to keep the rock from melting. The path from point A to point C is similar to the change in conditions that occurs when a rock is simply buried—temperature increases, but the rock does not melt because pressure also increases.

Melting by Decompression

3. Pressure decreases if a rock moves up from depth, getting closer to the surface. So a rock that is uplifted will experience a decrease in pressure, as from point C to point B. If the rock is already hot (point C), it may melt as the pressure decreases (to point B), a process called *decompression melting*.

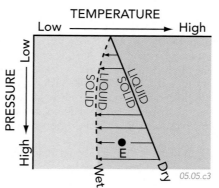

05.05.c2

4. For decompression melting to occur, the rock has to be fairly hot and must be uplifted fast enough so that it cannot cool significantly during uplift. If a rock is uplifted slowly, it can cool enough to stay solid. A hot, deeply buried rock following a path from point C to point A (cooling during uplift) would stay within the solid field on the diagram. In other words, the rock would not melt.

Melting by Adding Water

5. Adding water can significantly lower, by as much as 500°C, the temperature at which a rock will melt. On the graph, the dashed lines show the position of the melting curve if the rock contains water. Adding water moves the melting curve to lower temperatures. So, adding water to a dry rock at point E puts it on the liquid side of the melting curve, and the rock will melt.

05.05.c3

6. A hot rock can melt, therefore, if water moves into the system, even with no change in pressure or temperature. A rock at point E will be solid under normal, dry conditions (it is to the left of the non-dashed melting curve). If a small amount of water is added to the small spaces within and between crystals, the dashed line becomes the boundary between solid and liquid, so the rock at point E now begins to melt.

Determining the Conditions Under Which Rocks Melt

Laboratory experiments let scientists determine the conditions under which a particular rock melts. Scientists place a small sample of the rock in a special oven and raise the pressure to an amount equal to some pressure, like P1 (rock is at condition A on the graph).

The rock is then heated to a specific temperature (T1) and held at this temperature and pressure (position B) for days or weeks. At the end of the time, scientists inspect the sample for signs that it melted. If the rock did not melt, the scientists repeat the experiment at higher temperatures until melting is detected. The procedure is repeated for different temperatures (T2, T3, and T4) and pressures (P1, P2, and P3) until enough data points define the liquid-solid boundary line for the rock being investigated.

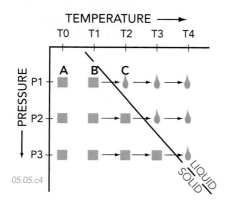

05.05.c4

5.5

5.6 How Do Igneous Rocks Form?

DIFFERENT IGNEOUS ROCK COMPOSITIONS AND TEXTURES reflect the type of material that was melted, the way the magma solidified, and whether the magma solidified at depth or was erupted onto the surface. How do geologic processes create so many different types of igneous rocks, and how can we use the composition and texture of an igneous rock to infer something about the rock's origin?

A What Processes Are Involved in the Formation of Igneous Rocks?

The igneous process begins when magma forms by melting at depth, followed by movement of the magma toward the surface and then solidification of the magma into solid rock. Given this history, igneous systems are best described from the bottom up, so begin with number 1 at the bottom of this page.

05.06.a1

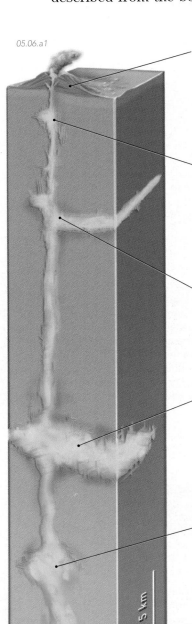

05.06.a2 Laki, Iceland

05.06.a3 Wickenburg, AZ

05.06.a4 Polished Slab

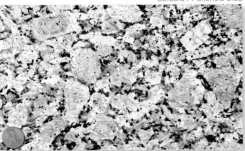

6. Magma that reaches the surface erupts as lava (molten rock that flows on the surface) or as volcanic ash. Volcanic ash forms when dissolved gases in the magma expand and blow the magma apart into small fragments of volcanic glass. Any igneous rock that forms on the surface is called an *extrusive* rock because it forms from magma *extruded* onto the surface (▶). More commonly, we simply call it a *volcanic* rock.

5. Many magma chambers are only several kilometers below the surface, as beneath a volcano. Magma may be added a little at a time to the chamber, and some magma may solidify before the next batch arrives. Some of the magma may crystallize in the chamber, while some rises to the surface. In this case, the rising magma may carry some of the early-formed crystals all the way to the surface, forming a porphyritic volcanic rock. ▶

4. As magma rises through the crust, it may stop in, or pass through, a series of magma chambers. A body of molten rock in the subsurface is referred to as an *intrusion* because of the way the magma intrudes into (invades) the surrounding rocks. Any igneous rock that solidifies *below the surface* is called an *intrusive* rock. Although there is a subtle difference, most geologists use the terms *plutonic* and *intrusive* synonymously.

3. Magma can accumulate to form a *magma chamber*. The magma may solidify in this chamber and never reach the surface, or it may reside in the chamber temporarily before continuing its journey upward. An igneous rock that solidified at a considerable depth (more than several kilometers) is referred to as a *plutonic* rock, and the body of rock that forms is called a *pluton*. Granite (▶) is a very common plutonic rock and forms granitic plutons.

2. Once magma begins to form, separate pockets of magma may accumulate to make a larger volume of magma. The magma rises because it is less dense than rocks around it.

1. The first stage in the formation of an igneous rock is melting, typically 40–150 km beneath the surface, in the deeper parts of the crust or in the mantle. The place where melting occurs is called the *source area*. Complete melting is rare, and most magmas result from *partial melting*, leaving most of the source area unmelted.

5 km

Before You Leave These Pages Be Able To

✓ Sketch and describe the processes involved in forming igneous rocks.

✓ Sketch or describe how melting can influence magma composition.

✓ Sketch or describe how partial crystallization, assimilation, and magma mixing can change a magma.

B What Processes Influence the Composition of a Magma?

The initial composition of a magma depends on the kind of rock that was melted in the source area and whether rocks in the source area were completely melted or only partially melted. Once a magma forms, its composition can be changed by several processes, including the formation of crystals, melting of rocks adjacent to the magma chamber, and mixing of two different types of magma.

Partial and Nearly Complete Melting

1. If a magma was generated by *complete melting* of the source region, it would have a composition identical to that of the source. For a number of reasons, complete melting is not common.

2. Most rocks melt by partial melting as some minerals melt before others. Felsic minerals melt at lower temperatures than mafic minerals, so partial melting produces a magma that is *more felsic* than the source. For example, partial melting of a mafic source can yield an intermediate magma.

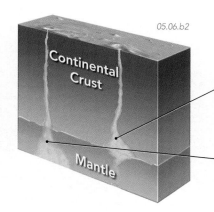

05.06.b1

Type of Source Area

4. If a more felsic source area, such as continental crust, is melted, the magma will be felsic. If an intermediate source is almost completely melted, the magma will have an intermediate composition, but partial melting more commonly produces a felsic magma.

3. The overall composition of the mantle is ultramafic but, due to partial melting, magmas generated in the mantle are mostly mafic. Most mafic magma is derived by partial melting of the mantle.

05.06.b2

Crystallization

1. As magma cools, mafic minerals crystallize first, which makes the composition of the remaining magma less mafic (more felsic). Consequently, partial crystallization of a mafic magma typically produces a magma of more intermediate composition.

2. Once formed, heavy mafic minerals may settle (sink) through the magma and collect in layers at the bottom of the magma chamber. This process, called *crystal settling*, will make lower parts of the magma chamber more mafic, leaving the remaining magma more felsic.

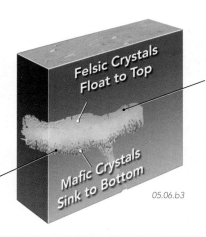

05.06.b3

3. Felsic crystals may be less dense than magma and so may float upward. This makes the top of the magma chamber more felsic.

▶ **4.** Light-colored, less dense feldspar crystals floated to the top of the magma that formed these rocks, while heavier, dark-colored minerals *settled* to the bottom.

05.06.b4 South Africa

Assimilation and Magma Mixing

1. If two different magmas come into contact, they may mix, a process called *magma mixing*. Magma mixing produces a magma that has a composition intermediate between the two magmas that mixed. In this photograph (◀), felsic magma, dominating most of this image, engulfed round pockets of a coexisting mafic magma (the dark patches). Light-colored crystals from the felsic magma mixed into the edges of the mafic magma.

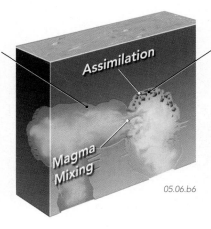

05.06.b5 Southern AZ

05.06.b7 South Africa

2. Mafic magma is hotter than the melting temperature of felsic rocks, so mafic magma can melt felsic wall rocks. If wall rocks around a magma melt, they may be incorporated into the magma, a process called *assimilation*. The gray metamorphic rocks below have been incorporated and partly *assimilated* into a lighter colored intrusion.

5.6

5.7 How Does Magma Move?

MAGMA RISES, MOVING CLOSER TO EARTH'S SURFACE, after it forms. What processes cause magma to begin moving, and what processes allow it to rise through the solid mantle or crust? How does a rising magma make space for itself, and how does it accumulate in underground magma chambers?

A How Does Magma Rise Through the Crust?

When a rock melts, the volume of melt produced is about 10% greater than the volume of the original rock. The magma therefore is *less dense* and will rise if the overlying rock will let it through. Pressures from the magma and tectonic forces, which pull and push crustal rocks, help produce fractures and other weaknesses through which magma can move.

05.07.a1 Big Maria Mtns., CA

05.07.a2, Krafla, Iceland

05.07.a3 Harcuvar Mtns., AZ

Magma can travel through and fill a fracture, forming a feature called a *dike*, like the light-colored ones shown above. The cracks needed for dikes to form are created by tectonic or magmatic forces that pull the rocks apart as the magma forces its way in.

Most magma forms in regions where tectonic forces exist. These forces can fracture solid rock, opening space for the magma. The open fissure in this site in Iceland resulted from tectonic pulling along a rift or divergent plate boundary.

Solid rocks above a magma chamber can break off and drop down into the magma, providing space and letting the magma move up. These dark fragments broke off the walls of a magma chamber and were incorporated into the light-colored magma.

B What Determines How Far a Magma Can Rise Toward Earth's Surface?

Most magmas solidify at depth. Magma gets trapped and crystallizes at depth because of the difficulty of rising through solid rock. There are four main constraints on how magma ascends.

Magma Pressure

Pressure from the weight of the overlying rocks is directed in toward the magma from all sides. The pressure pushes the magma into any available openings and drives it toward the surface. The confining pressure exerted on the magma decreases as the magma rises higher into the crust.

Density

Differences in density drive the flow of magma. Mafic magma generated from partial melting of the mantle is *less dense* than the surrounding solid rocks and so rises. When the rising mafic magma reaches the base of the crust, its density may be *greater* than that of the crustal rocks. The mafic magma may then stop and form a magma chamber within the crust.

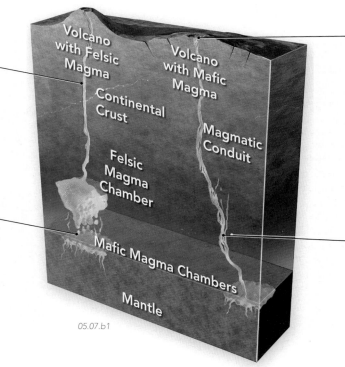
05.07.b1

Gas Pressure

Magma contains dissolved gases, such as carbon dioxide, sulfur dioxide, and water vapor (steam). As magma rises to shallower levels, decreasing pressure allows the gases to form bubbles in the magma. If this occurs, the density of the magma decreases and the magma rises faster. If magma initially has a low content of dissolved gases, bubbles may not form or assist the magma upward.

Stress

Tectonic stress can help the magma to open steep fractures that provide a pathway to the surface. Alternatively, stress can trap a magma at depth by keeping fractures closed or by creating horizontal fractures that direct a magma sideways.

What Controls How Easily Magma Moves?

Viscosity is a measure of a material's resistance to flow. A *viscous* magma does not flow easily, whereas a *fluid* (less viscous) magma flows more easily. Magmas are considerably more viscous than most other hot liquids with which you are familiar. A magma's viscosity is controlled by its temperature, composition, and crystal content.

1. *Viscous magma* strongly resists flowing. When viscous magma erupts on the surface, it does not spread out but piles up, forming mounds or domes of lava.

2. *Fluid* (less viscous) *magma* flows more easily and may spread out in thin layers on the surface. This magma can travel longer distances from its source and cover large areas with lava.

Temperature

3. *Low Temperature*— The temperature of a magma is the most important control of viscosity. Magma at relatively low temperature, such as one barely hot enough to be molten, is *viscous* and flows only with difficulty—it is very viscous.

4. *High Temperature*— Magma that is very hot has low viscosity and so flows very easily. Mafic magma is hotter than felsic magma, and it is less viscous than felsic magma if the two are at the same temperature.

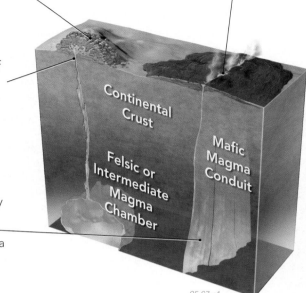

Continental Crust

Felsic or Intermediate Magma Chamber

Mafic Magma Conduit

05.07.c1

Composition

5. *Abundant Silicate Chains*—Silicon and oxygen tetrahedra in magma can link into long silicate chains that do not bend or move easily out of the way of one another. Felsic and most intermediate magmas have a high silicon and oxygen content, and the resulting long silicate chains make the magmas very viscous.

6. *Few Silicate Chains*—Mafic magma contains less silicon and oxygen than intermediate or felsic magma. Consequently, silicon-oxygen tetrahedra are less connected or are in short chains. This allows the magma to flow more easily—it is *less* viscous.

7. *Volatiles*—Silicon content and the abundance of silicate chains may be the most significant compositional variable controlling viscosity, but water dissolved in magma disrupts long chains, decreasing the viscosity. Water and other volatiles decrease viscosity in other ways not discussed here.

Percentage of Crystals

8. *Abundant Crystals*—As a magma cools, crystals begin to form. The crystals in the flowing magma get in each other's way and cause the magma to flow more slowly. A magma with abundant crystals is more viscous than a magma with fewer crystals.

9. *Few Crystals*—A magma that has few crystals has few internal obstructions and flows more easily (is *less* viscous). Such magma flows more smoothly and thus can flow faster and farther. Recall that the amount of crystals in a magma is partly controlled by how much time the magma spends cooling in a magma chamber, so this part of a magma's history influences how the magma flows later, like when it reaches the surface.

05.07.c2   05.07.c3

Abundant Crystals   Few Crystals

How Viscous Is Your Breakfast?

One way to think about viscosity is to examine a typical Sunday breakfast that might be eaten by a student while visiting a relative's well-stocked home. The fluids you encounter at breakfast are much less viscous than magma, but they illustrate important aspects of viscosity.

The orange juice that begins the feast has *low viscosity* and so pours easily, like a very hot basalt without many crystals. Next on the menu is oatmeal, which is *more viscous*, like a crystal-rich, felsic magma with long silicate chains. When thick, it piles up in a dome-shaped mound that slowly spreads out over time. Adding milk separates the oatmeal flakes, allowing them to move past one another, like adding water to scattered crystals in a magma.

Pancakes, next on the agenda, will be topped with butter and maple syrup. If a stick of butter is out on the counter too long, it softens, becoming *less viscous*, and starts to flow. It was firmer and *more viscous* when cold. Likewise, maple syrup flows very slowly when cold but is more fluid when heated. Temperature clearly has an effect on viscosity. Now, what to put on the hash brown potatoes—low-viscosity catsup, low-viscosity salsa, or high-viscosity chunky salsa? Who knew that thinking about viscosity could be such an important part of breakfast?

Before You Leave This Page Be Able To

☑ Describe three ways in which magmas rise through the crust.

☑ Discuss factors that influence how far a magma rises toward the surface.

☑ Explain the factors that control the viscosity of a magma.

☑ Describe what factors might be combined to form very high-viscosity magma or very low-viscosity magma.

5.7

5.8 How Does Magma Solidify?

MAGMAS EVENTUALLY COOL AND SOLIDIFY. The general term *solidify* is used here instead of the more specific term *crystallize* because a magma can cool so rapidly that crystals do not have time to form. Such rapidly cooling magma instead solidifies to a volcanic glass. The rate of cooling affects the size and shape of any crystals that form.

A Under What Conditions Does Magma Solidify?

For a magma to solidify, it must lose enough thermal energy to its surroundings so that it can cool and change from a liquid to a solid. This generally happens when a hot magma has risen to a place that is cooler, whether on the land surface, on the seafloor, or still underground.

When magma reaches the land surface, it transfers thermal energy, through *conduction* and *radiation*, to the atmosphere, to materials below the volcanic unit, and to any water that is present on the surface.

At depth, magma loses thermal energy to surrounding rocks by *conduction*. As the wall rocks are heated, their temperature increases, possibly causing them to be metamorphosed or even melted.

05.08.a1

Magma also loses heat when it releases gases, including water vapor, into wall rocks or at the surface.

Water in rocks near the magma receives heat by conduction from the magma or from hot wall rocks near the magma. As the water gets hotter, its density decreases and the water rises. The upward-flowing water is replaced by an inflow of cooler water, causing *convection*. Such convection of water may be the primary way some magma cools.

A magma solidifies when minerals crystallize or glass forms. The size of any crystals largely reflects the rate at which the magma cools. Magma cools slowly when it is in hot surroundings or is insulated by wall rocks. The photographs below show four rocks, all of which are felsic and contain the same minerals (mostly feldspar and quartz). The four rocks have different-sized crystals because the magmas had different cooling histories.

Slow Cooling	Medium Cooling	Very Fast Cooling	Slow Then Fast Cooling

Coarse Granite Pegmatite	Fine-Grained Granite	Very Fine-Grained Rhyolite	Porphyritic Intrusive Rock

05.08.a2 Southern AZ · 05.08.a3 Tibet · 05.08.a4 Northern AZ · 05.08.a5 South Africa

Cooling History of a Magma

This graph shows conditions of temperature versus depth. Pressure is proportional to depth, increasing with greater depths. We can track the history of a magma by plotting temperature-depth points that show the path it follows as it cools and reaches the surface. Follow the numbered changes by starting at the bottom.

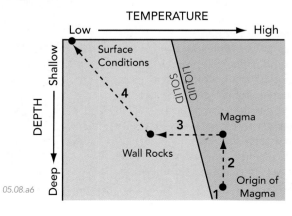

05.08.a6

4. At some later time, the now-solidified magma and its wall rocks are uplifted to the surface, where they cool to low temperature and where we can inspect them for clues about their history.

3. The magma cools by losing thermal energy to the surrounding wall rocks. It crosses the melting curve along path 3 on the graph, and so it crystallizes.

2. Some of the magma rises some distance in the mantle or crust, along path 2 on the graph, and so it is in a place where it is surrounded by cooler rocks.

1. A magma forms at depth, where temperature is high enough to overcome pressure and to cause melting.

B In What Order Do Minerals Crystallize?

Minerals melt at different temperatures—felsic minerals melt before mafic ones. Minerals crystallize in the opposite order from which they melt—mafic minerals crystallize before felsic ones. One way to think about how minerals crystallize is through an idealized sequence of mineral crystallization called *Bowen's Reaction Series*.

1. Mafic minerals, like olivine and pyroxene (◄), are the first to crystallize from a mafic magma. They typically do not crystallize from a felsic magma because felsic magmas lack sufficient magnesium to form these minerals.

05.08.b2

2. Amphibole (▼) and biotite are most common in rocks of intermediate composition but are also present in mafic rocks or felsic rocks. They crystallize at temperatures lower than olivine and pyroxene, but before most of the felsic minerals.

05.08.b5

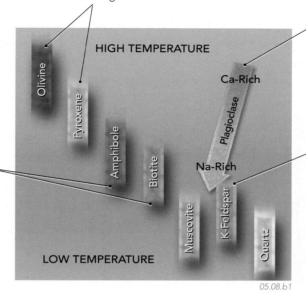

05.08.b1

3. Plagioclase feldspar (►) may be calcium-rich (Ca) or sodium-rich (Na), or somewhere between. Calcium-rich plagioclase crystallizes at high temperature, sometimes with olivine and pyroxene. Plagioclase with less calcium crystallizes at lower temperatures.

05.08.b3

4. The light-colored felsic minerals quartz (►), K-feldspar(▼), and muscovite crystallize at the lowest temperatures. These minerals, along with Na-rich plagioclase, may be the only minerals formed from *felsic* magmas, which lack the chemical components required to grow abundant *mafic* minerals. Felsic minerals rarely grow from mafic magmas, which lack sufficient silicon.

05.08.b4

05.08.b6

05.08.b7 South Africa

5. Minerals that crystallize early in the crystallization sequence can grow unimpeded in the magma and so commonly have well-defined crystal shapes, like these well-formed crystals of light-colored plagioclase. ◄

6. Minerals that crystallize late in the sequence must grow around preexisting crystals, so they may grow in irregular, poorly defined crystal shapes. The white plagioclase crystals (►) in this rock grew late and so had to fill around dark mafic crystals that formed early.

05.08.b8 South Africa

How Crystallization Changes the Composition of a Magma

As minerals crystallize from a cooling magma, they remove the chemical constituents that are incorporated into the crystals. Therefore, the chemical composition of the remaining magma changes as minerals crystallize.

When *mafic* minerals crystallize from a magma, they extract the mafic components, such as magnesium, iron, and calcium. As these crystals are removed from the magma, the remaining magma contains less of the elements that were used to grow the mafic minerals. That is, the magma becomes less mafic (more intermediate or felsic).

This graph illustrates the effects of crystallization on a magma. It shows magnesium content (expressed as magnesium oxide, MgO) and silicon content (expressed as

SiO_2) for a series of rocks produced by a crystallizing magma. This example is typical: the magma starts out crystallizing the Mg-rich minerals olivine and pyroxene, causing the Mg content of the remaining magma to decrease over time. In other words, crystallization of mafic minerals is making the remaining magma less mafic.

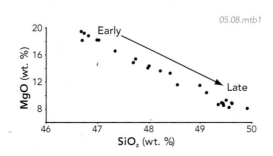

05.08.mtb1

Graph: MgO (wt. %) on vertical axis (8, 12, 16, 20) versus SiO_2 (wt. %) on horizontal axis (46, 47, 48, 49, 50), with "Early" labeled at upper left and "Late" at lower right.

Before You Leave This Page Be Able To

✓ Explain or sketch the processes by which a magma cools.

✓ Describe or sketch the cooling history of a magma as it rises in the crust.

✓ Explain the order in which minerals crystallize from a magma (Bowen's Reaction Series), and compare it to the order in which they melt.

✓ Describe how the rate of magma cooling affects the size and shape of crystals.

✓ Explain how the crystallization of minerals can change the composition of remaining magma.

5.8

5.9 How Does Magma Form Along Divergent Plate Boundaries?

ABOUT 60% OF EARTH'S MAGMA forms at mid-ocean ridges, where two oceanic plates spread apart. Magma also forms during the rifting of continents. What causes melting in these two settings, and what types of rocks result?

A What Causes Melting Along Mid-Ocean Ridges?

Two plates move away from one another (diverge) along mid-ocean ridges. To understand how melting occurs here, examine the magmatic system, beginning with processes at the bottom, within the mantle.

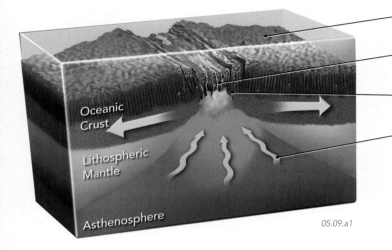

05.09.a1

5. Older oceanic crust moves away from the ridge in a conveyor-belt manner as new oceanic crust forms along the axis of the ridge.

4. Magma rises upward through magma-filled fractures that form as the plates *pull apart. S*ome magma erupts as lava within the rift.

3. The buoyant magma rises away from the unmelted residue in the mantle and accumulates in magma chambers in the crust and upper mantle.

2. As the plates separate, solid asthenosphere rises to fill the area between the plates. As the asthenosphere rises, pressure decreases and the rock partially melts (decompression melting). A plot of decompression melting is on the next page under the heading *Melting in the Mantle.*

1. Mantle rocks, including those in the asthenosphere, are mostly solid and crystalline, not molten. The mantle's high pressures and temperatures allow these rocks to flow as a weak solid while maintaining a crystalline structure. Parts of the asthenosphere are close to their melting temperature.

B What Types of Igneous Rocks Form Along Mid-Ocean Ridges?

New oceanic crust formed at mid-ocean ridges consists of several different kinds of rocks. The rocks are all mafic, but they have different textures and features depending on how and where the magma solidified. The mafic magma forms by partial melting of the ultramafic mantle. Begin at the top, and work your way down through the oceanic crust.

05.09.b2 San Juan Islands, WA

The upper part of oceanic crust consists of basaltic lava flows. When these lavas erupted into water, they formed a series of overlapping mounds called *pillows*. These distinctive rocks, called *pillow basalts,* have in some cases become uplifted above sea level, where we can now observe them (▶). The pillows shown are about 0.5 to 1 meter across.

Countless thin, vertical intrusions of finely crystalline basalt cut across the pillow basalts from below. These thin intrusions, called *dikes*, are so closely spaced that they are called *sheeted dikes*. Each dike represents a thin, tabular conduit through which magma passed. Most dikes are oriented parallel to the oceanic rift and perpendicular to the direction of spreading.

05.09.b3 Smartville, CA

Sheeted dikes merge downward into *gabbro*, the coarsely crystalline equivalent of basalt. The gabbro represents magma chambers beneath the rift and locally displays layers formed by settling of light and dark crystals. ▶

The base of the gabbro is the base of the oceanic crust, below which are *ultramafic* rocks of the mantle. The mantle rocks show evidence of having been partially melted to form all of the overlying mafic rocks in the crust (pillow basalt, sheeted dikes, and gabbro).

05.09.b1

How Are Magmas Generated in Continental Rifts?

Continental rifts form where tectonic forces attempt, perhaps successfully, to split a continent apart. Such rifts have a central trough where faults drop down huge crustal blocks. Rifts are characterized by a diverse suite of igneous rocks because melting takes place in both the mantle and the crust. The sequence of events begins in the mantle.

Continental Rift

4. Some felsic and intermediate magmas solidify underground as granite and related igneous rocks, while others erupt on the surface in potentially explosive volcanoes.

3. Heat from the hot mafic magma melts the adjacent continental crust, producing felsic magma. Intermediate magma forms from mixing of felsic and mafic magmas or from the assimilation of continental crust by a mafic magma.

2. The mantle-derived mafic magma rises into the upper mantle and lower continental crust and accumulates in large magma chambers. Some of the mafic magma reaches the surface and erupts as mafic (basaltic) lava flows.

1. Solid asthenosphere rises beneath the rift and undergoes *decompression melting* (see graph below for melting in the mantle). Partial melting of the ultramafic mantle source rock yields mafic magma.

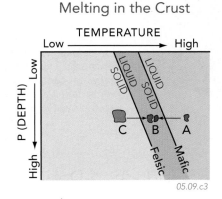

05.09.c1

▶ Melting of *mantle* beneath rifts is caused by *decompression*. The asthenosphere rises into shallower, lower pressure regions, and a decrease in pressure allows the rocks to melt. This produces mafic magma that can erupt onto the surface, forming basalt.

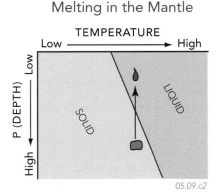

05.09.c2   05.09.c3

◀ This graph shows a melting curve for mafic rock (basalt) and a lower temperature melting curve for felsic rock (granite). A hot, mantle-derived mafic magma rises into continental crust and is hotter (at point A) than adjacent crust (at point C). The hot mafic magma heats the continental crust (from C to B). As the temperature of the crust crosses the felsic line, the granitic crust melts to produce felsic magma. The mafic magma loses heat to the crust (from A to B) and solidifies.

Ophiolites — Slices of Oceanic Crust on Land

How do we know what is in oceanic crust? After all, it is hidden deep beneath the sea. The sequence of rocks in oceanic crust has been reconstructed by dredging samples from the seafloor, by drilling into oceanic crust, and by studying ancient examples on land. Perhaps of most significance, geologists have gained much data by using research ships that have completed more than 1,700 drill holes, some more than 1,400 meters (nearly a mile) deep. Drill cores retrieved from these sites are important because they provide samples of oceanic crust from the surface to moderate depths.

If we know the right places, we can examine oceanic crust on a hike across dry land. Tectonic movements have sliced off pieces of oceanic crust and thrust them onto the edges of continents and onto islands. These slices contain a consistent sequence, from top to bottom, of oceanic sediment, pillow basalt, sheeted dikes, and gabbro. This distinctive sequence is called an *ophiolite complex* and is identical to the sequence of newly formed oceanic crust shown on the previous page, except it contains an additional layer of oceanic sediment on top. Such sediment accumulates on top of the pillow basalts, and the sedimentary cover gets thicker with time. Many ophiolites are probably sections of oceanic crust created at long-vanished mid-ocean ridges.

Before You Leave This Page Be Able To

✓ Sketch or describe why melting occurs along mid-ocean ridges and why the resulting magmas are basaltic (mafic).

✓ Describe the types of igneous rocks that form along mid-ocean ridges.

✓ Describe how melting occurs in continental rifts and how it results in diverse igneous rocks.

✓ Discuss how an ophiolite compares to a section through oceanic crust.

5.10 How Does Magma Form Along Convergent Plate Boundaries?

MANY MAGMAS ARE GENERATED ALONG CONVERGENT BOUNDARIES, especially at subduction zones, which produce dangerous, explosive volcanoes. What type of melting produces this magma? Are there differences in the causes of melting and the resulting magmas at the three types of convergent boundaries: ocean-ocean, ocean-continent, and continent-continent?

A How Is Magma Generated Along Subduction Zones?

About a fifth of Earth's magma forms where an oceanic plate subducts into the mantle, at an ocean-ocean or ocean-continent convergent boundary.

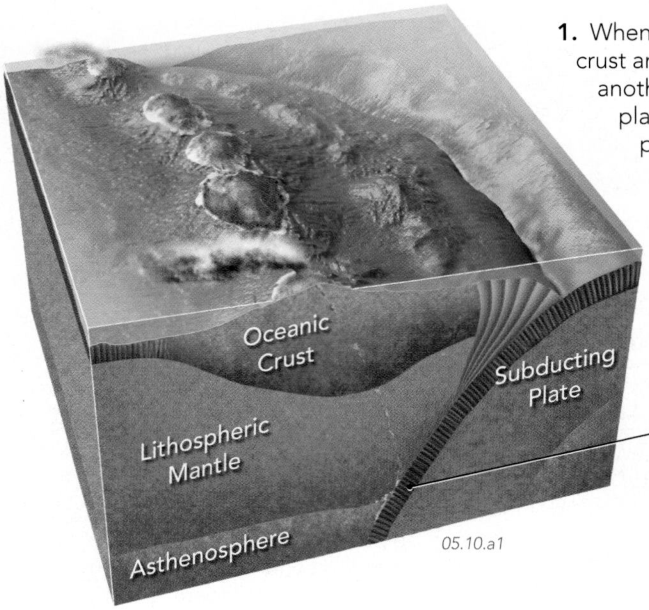

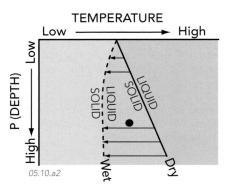

05.10.a2

1. When an oceanic plate, composed of oceanic crust and lithospheric mantle, converges with another oceanic plate, or with a continental plate, subduction occurs. As the subducted plate descends, both pressure and temperature gradually increase.

2. In response to the changes in pressure and temperature, existing minerals in the subducting plate convert into new ones through the process of *metamorphism*. Water-bearing minerals, such as mica, break down, which forces water out of the crystalline structures. This water, along with any water carried down by wet sediments during subduction, then rises into the overlying asthenosphere.

3. The added water lowers the melting temperature of the mantle above the subducting plate (▲). If the temperature is high enough, melting occurs, and mantle-derived magmas rise into the overriding plate. The magma then may crystallize at depth or eventually erupt at the surface.

Oceanic Crust
Subducting Plate
Lithospheric Mantle
Asthenosphere
05.10.a1

B What Happens When Subduction-Derived Magmas Encounter the Crust?

Subduction-derived magmas rise into the overriding plate, which may be an oceanic plate or a continental plate. The magmas interact with and modify the crust they encounter and may themselves be modified by that interaction. Begin with step 1 at the bottom.

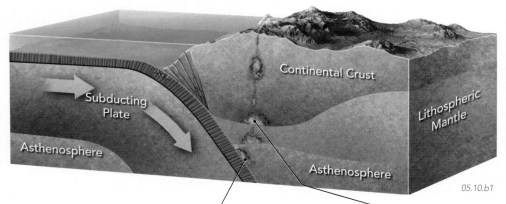

Subducting Plate
Asthenosphere
Continental Crust
Lithospheric Mantle
Asthenosphere
05.10.b1

3. Most subduction-related magma probably never reaches the surface, but some erupts, forming clusters or belts of volcanoes. If the overlying crust is continental, the volcanoes are usually part of a mountain belt. If the overlying crust is oceanic (as shown above in *part A*), subduction-generated magma creates individual volcanoes along an *island arc*. In both settings, the subduction-generated magma mostly has an intermediate composition (andesite). Island arcs can erupt mafic magma, and continental magma can be felsic. In both cases, magmas added at depth and on the land surface thicken the crust.

1. Most magma generated above the subducting slab begins with a *mafic* composition because it forms by partial melting of the ultramafic mantle. Less commonly, partial melting may generate magma of *intermediate* composition.

2. If the overriding plate is a continental plate, the rising magma encounters thick continental crust that slows its upward journey. The magma heats the surrounding rocks, commonly causing localized *partial melting* that produces *felsic* or *intermediate* magma.

How Does Water Get into a Subduction Zone?

During subduction and collision, water is released by sediments and minerals in the descending crust. Where does the water come from?

An oceanic plate being subducted formed originally along a mid-ocean ridge, where seawater flows into the hot crust and forms water-bearing metamorphic minerals. These water-bearing minerals, shown with blue spots in the oceanic crust, travel with the plate as it moves away from the ridge. Once formed, oceanic crust is slowly covered by sediment derived from continents, islands, and creatures living in the sea. As time passes, the sedimentary layer thickens. This sediment contains trapped seawater and minerals, including clay, that have water in their mineral structure.

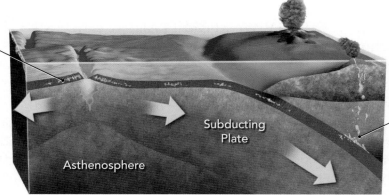

Only the water trapped in minerals gets deep enough to cause melting. Other water probably is driven off further up the zone.

05.10.c1

D What Magmatism Accompanies Continental Collisions?

When two continents, such as Asia and India, converge, the encounter is best described as a *collision* because continental crust is buoyant and difficult to subduct. One plate may partially slide beneath the other but, because the descending plate is continental rather than oceanic, continental collisions result in different types of magmas than those in a typical subduction zone.

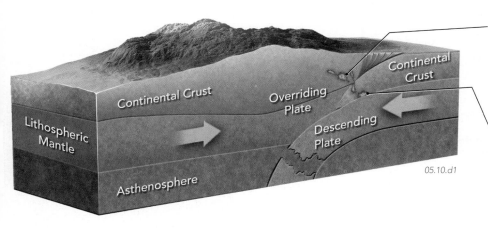

05.10.d1

3. Magmas produced by continental collisions typically do not reach the surface, partly because they have to pass through thick continental crust. Also, some magmas produced have a relatively high water content compared to mantle-derived magmas and so pass through the wet solidus (and therefore solidify) as they rise. So, continental collisions, unlike other convergent boundaries, do not have many volcanoes.

2. Water may be released by metamorphism of water-bearing minerals and, if the descending continental crust gets hot enough, it undergoes partial melting, producing *felsic* magmas.

1. During a continental collision, one continental plate may slide beneath another continental plate. The descending continental crust gets hotter and experiences increased pressure.

Partial Melting of Continental Crust

Most rocks include more than one mineral, each of which melts at a different temperature. Rocks within continental crust typically include plagioclase feldspar, K-feldspar, quartz, mica, and amphibole. Of these minerals, quartz and K-feldspar melt at the lowest temperatures and so melt first as a rock is heated. If only the low-melting-temperature minerals melt, the rock undergoes *partial melting*.

As quartz and K-feldspar melt, the first small amount of magma will contain relatively high amounts of silicon, potassium, and other elements common in these minerals. Such magma is felsic and will form igneous rocks rich in quartz and feldspar, like *granite*. If some mafic minerals also melt, the resulting magma has an *intermediate* composition. Therefore, different compositions of magma form, depending on the amount of partial melting (how much of the rock melts).

As felsic magma forms and rises, it carries felsic material upward into the *upper* crust, which becomes more *felsic*. Partial melting leaves behind mafic minerals and so makes the *lower* crust more *mafic*. Over time, partial melting has helped remake the continental crust into an upper, felsic part and a lower, intermediate to mafic part.

Before You Leave This Page Be Able To

✓ Describe and sketch how magma is generated in a subduction zone.

✓ Describe what happens when subduction-derived magma encounters overlying crust.

✓ Describe how water gets into a subduction zone.

✓ Explain and sketch how magma forms during continental collisions.

✓ Explain partial melting of continental crust, the kinds of magma formed, and the effect on the crust.

5.10

5.11 How Is Magma Generated at Hot Spots and Other Sites Away from Plate Boundaries?

SOME MAGMATISM OCCURS AWAY FROM PLATE BOUNDARIES. For example, magma is associated with hot spots and with places where a continent is beginning to rift apart. What causes rock to melt at these sites, and what type of magma results from these settings?

A What Type of Magmatic Activity Occurs at Hot Spots?

A hot spot is a site of intense magmatic activity that has certain types of igneous activity that cannot be explained easily by its plate-tectonic setting (e.g., not near a subduction zone or mid-ocean ridge). Many hot spots, like Hawaii, do not coincide with a plate boundary, but some, like the Galápagos Islands, are close to a plate boundary. Hot spots have different igneous manifestations depending on whether the hot spot is within a continental plate or an oceanic plate.

Hot Spots and Mantle Plumes

05.11.a1

◄ Most hot spots are considered to be the crustal expression of a rising *plume* of hot mantle material. There is current debate about how deep in the mantle such plumes originate, but some may begin at the core-mantle boundary and ascend all the way through the lower mantle and into the asthenosphere.

▶ The movement of plumes through the mantle is commonly compared to the rising blobs within a lava lamp. The teardrop shape of the blobs nicely matches experimental models of mantle plumes, but, unlike the liquid substance in lava lamps, the plume and surrounding mantle are both solid. Like the blobs in the lamp, a mantle plume rises because it is hotter and less dense than material around it.

05.11.a2

Hot Spots in Oceans

05.11.a3

◄ When magma generated by a mantle plume encounters the lithosphere, it spreads out along the boundary. There, it causes melting of the overlying lithosphere, and additional melting occurs by decompression. Magma from the lithosphere and plume can reach the surface, creating large volcanoes on the seafloor. This is occurring on the Big Island in Hawaii and on the seafloor farther to the southeast.

▶ A lithospheric plate may be moving above a plume that is anchored in the deep mantle. An active volcano overlies the hot spot, but volcanic activity will cease once the volcano has moved off the hot spot. The hot spot creates a succession of volcanoes along a linear chain of islands and seamounts as the overlying plate moves across the hot spot.

05.11.a4

Hot Spots in Continents

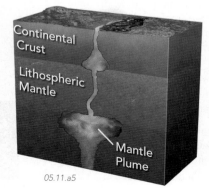

05.11.a5

When a rising mantle plume encounters continental lithosphere, its high temperatures cause melting. If the melting occurs in the lower part of the lithosphere (in the mantle), it produces mostly mafic magma (◄). If the mantle-derived magma causes melting in the upper part of the lithosphere (in the crust), it can generate felsic magma.

Mafic and felsic magmas may mix in the crust to produce intermediate magmas. Consequently, when crustally derived magmas make it to the surface, the resulting eruptions may be of many types. If the magma is felsic, there is a tendency for the eruptions to be explosive and to form a large volcanic depression called a *caldera* (▶). If the magma is more mafic, the eruptions will be less explosive.

05.11.a6

B Where Does Magmatism Occur Away from Plate Boundaries?

Magmatism occurs in a variety of settings not associated with plate boundaries. Tectonic forces may cause faults, fractures, and other structural features totally *within* a plate. Faulting and fracturing create rifts in the crust and provide a pathway for the easy ascent of mantle-derived or crustal magmas. Such rifts are commonly associated with magmatism that is not directly related to a plate boundary and may not be related to a hot spot.

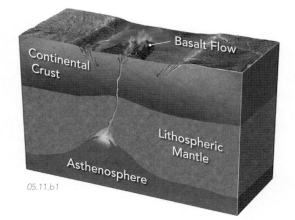
05.11.b1

◄ Rifts in continental interiors commonly produce basaltic lava flows interpreted to be derived from melting of the lithosphere or asthenosphere. The magmatism could be caused by *decompression melting* as the asthenosphere rises in response to stretching and thinning of the overlying plate. Rifting can eventually split the continent in two, as the rift evolves to become a divergent plate boundary.

► In the southwestern United States, some volcanism appears to be caused by asthenosphere encroaching on and melting the lithosphere. This process is expressed on the surface by basalt derived from melting of the lower lithosphere and by intermediate to felsic rocks derived from partial melting of the continental crust. The crust and mantle part of the lithosphere have been stretched and thinned in part of the region, breaking the landscape into a series of basins (valleys) and mountain ranges.

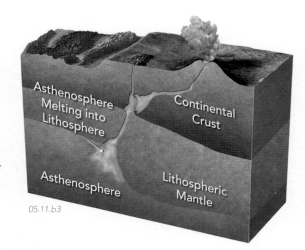
05.11.b3

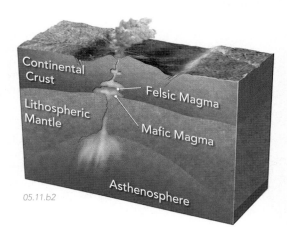
05.11.b2

◄ If mantle-derived magma accumulates in magma chambers in the continental crust, it may heat the crust and melt parts of it. Such melting yields felsic magmas that may or may not make it to the surface. Much mafic magma gets trapped in the lower part of the continental crust because it loses heat and solidifies as it melts the crust.

Investigating the Sources of Magma

How can we infer the tectonic setting in which ancient volcanic rocks formed? One approach is to analyze the chemistry of the volcanic rocks and compare these results with analyses of modern volcanic rocks for which the tectonic setting is known.

The graph on the right is based on vanadium (V) and titanium (Ti) analyses of many mid-ocean ridge basalts and many oceanic island basalts. The two different kinds of basalts have different chemistries. Samples of mid-ocean ridge basalt typically plot above the line because they have relatively high amounts of vanadium compared to oceanic island basalts. Likewise, samples of oceanic island basalts generally plot below the line. The many data points used to orig-

inally define the line are not shown. The blue boxes are analyses of basalt samples collected by drilling in the Philippine Sea. The samples are similar to basalts from mid-ocean ridges, so we interpret the samples to have probably formed in this setting.

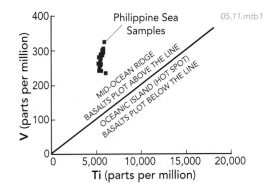
05.11.mtb1

Philippine Sea Samples

MID-OCEAN RIDGE BASALTS PLOT ABOVE THE LINE
OCEANIC ISLAND (HOT SPOT) BASALTS PLOT BELOW THE LINE

V (parts per million) — 0, 100, 200, 300, 400
Ti (parts per million) — 0, 5,000, 10,000, 15,000, 20,000

Before You Leave This Page Be Able To

☑ Sketch or explain a mantle plume and its magmatic expression in both oceanic and continental plates.

☑ Sketch or describe how a hot spot can form a sequence of volcanic islands on a moving oceanic plate.

☑ Sketch or describe how magmatism occurs away from plate boundaries.

☑ Sketch or explain how we infer from chemical analyses the tectonic setting of an ancient basalt.

5.12 How Do Large Magma Chambers Form and How Are They Expressed in Landscapes?

MAGMA OFTEN ACCUMULATES UNDERGROUND IN CHAMBERS containing thousands of cubic kilometers of molten rock. How do these chambers form, what are their shapes, and what processes occur within them? What do they look like after they have solidified and are uplifted to the surface?

A What Is a Magma Chamber and What Processes Occur in Large Chambers?

A *magma chamber* is an underground body of molten rock. Think of it as an always-full reservoir or holding tank that allows magma to enter from below and perhaps exit out the top. Magma chambers are very dynamic, with magmas evolving, crystallizing, and being replenished by additions of new magma. The figure below, which represents a cubic kilometer beneath the earth's surface, illustrates some of the main processes.

Large magma chambers can consist of a single magma type but generally involve more than one influx of magma.

During crystallization, early-formed minerals that remove chemical components from the magma may rise or sink (*crystal settling*) within the chamber.

As a new pulse of magma rises into the chamber, it may mix with existing magma or may remain distinct, but it adds a new pulse of thermal energy.

A partially crystallized magma could be heated by a new, hotter pulse of magma. The additional heat may cause minerals to melt back into the magma.

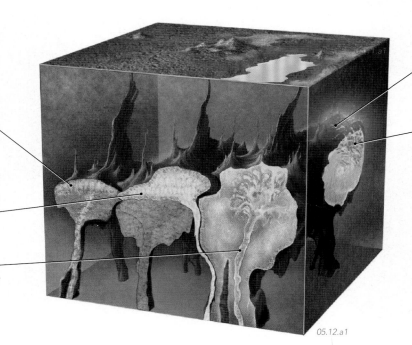

05.12.a1

The magma can heat and partially melt the wall rocks, forming a new magma with a different composition. This melting is aided by heat brought into the chamber by new batches of magma from below.

Magma produced by partial melting of the wall rocks can be assimilated into the existing magma or can rise out of the chamber without interacting chemically with the original magma.

How two magmas mix depends on their relative densities, crystal content, viscosities, and the temperatures at which they crystallize. Magmas of similar density may form well-mixed magma, whereas magmas of different densities may form a patchwork of magma types.

B In What Settings Do Large Magma Chambers Form?

A large influx of magma is required to form a large magma chamber. This, in turn, requires melting on a large scale that is possible only in certain tectonic settings.

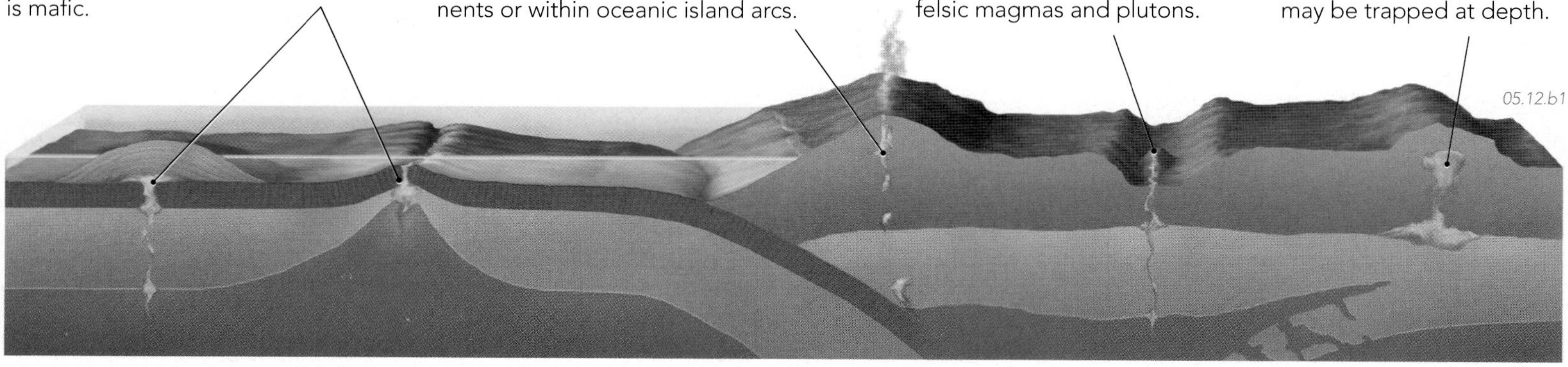

In oceanic lithosphere, large magma chambers form above hot spots and within mid-ocean ridges. In both cases, the mantle-derived magma is mafic.

Large magma chambers, including thick dikes, of intermediate and felsic composition form above subduction zones, either within magmatic arcs on continents or within oceanic island arcs.

Hot spots and rifts within continents produce large amounts of mantle-derived magma that can melt continental crust to form large felsic magmas and plutons.

Continental collisions cause crustal thickening, which can lead to melting of continental crust. Large amounts of felsic magma may be trapped at depth.

05.12.b1

How Are Large, Solidified Magma Chambers Exposed at the Surface?

A solidified magma chamber is called a *pluton*. A pluton can be cylindrical, sheetlike, or very irregular in shape. Several generations of magma may intrude into the same region, forming a complex mass of plutons with various compositions, textures, and shapes. Plutons are classified according to their size and geometry.

Irregular Plutons

Many plutons have irregular shapes, somewhat like vertical cylinders. A pluton with an exposed area of less than 100 km² is a *stock*.

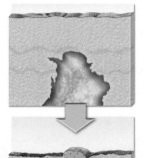

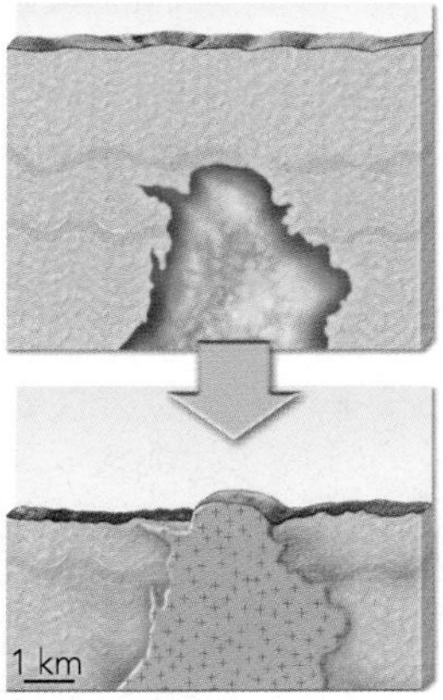

Most stocks are irregularly shaped. Many have a shape like a steeply oriented cylinder or downward-widening, bumpy mass.

On the surface, most stocks have steep boundaries and may resist erosion more than surrounding rocks.

05.12.c1–2

05.12.c7 Toyabe Range, NV

A stock of bold, gray rocks represents a magma that solidified at depth and later was uplifted.

Sheetlike Plutons

Some plutons have a tabular shape, like a thin or thick sheet. The sheet can be vertical, horizontal, or at some other angle.

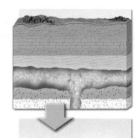

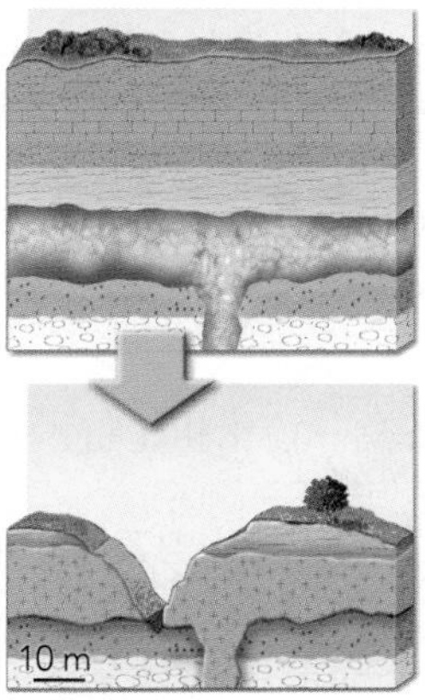

Plutonic sheets can be horizontal, vertical, or inclined, and may be parallel to or cutting across layers in the wall rocks.

When horizontal sheets are exposed at the surface, their tops and bottoms may be visible.

05.12.c3–4

05.12.c8 Cuernos del Paine, Chile

The gray granitic rocks were a horizontal sheet of magma that squeezed between dark metamorphic rocks that are above and below.

Batholiths

A *batholith* is one or more contiguous plutons that cover more than 100 km². Most batholiths include a number of rock types.

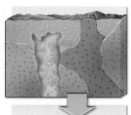

Most batholiths form from multiple magmas emplaced into the same part of the crust over a long time.

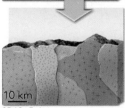

Exposed batholiths are characterized by plutonic rocks that cover a huge region.

05.12.c5–6

05.12.c9

A huge expanse of gray granite characterizes the Sierra Nevada batholith of California, seen here from the east.

The White Mountain Batholith of New England

The White Mountain batholith is centered in the middle of New Hampshire. Granitic rocks of the batholith form high peaks of the White Mountains and many of the area's scenic landmarks.

The batholith consists of several dozen individual plutons (shown in red and yellow) that were emplaced between 200 and 155 million years ago. The plutons represent separate injections of magma, some emplaced at somewhat different times. Some plutons are cylindrical; others are like curved dikes.

05.12.mtb1

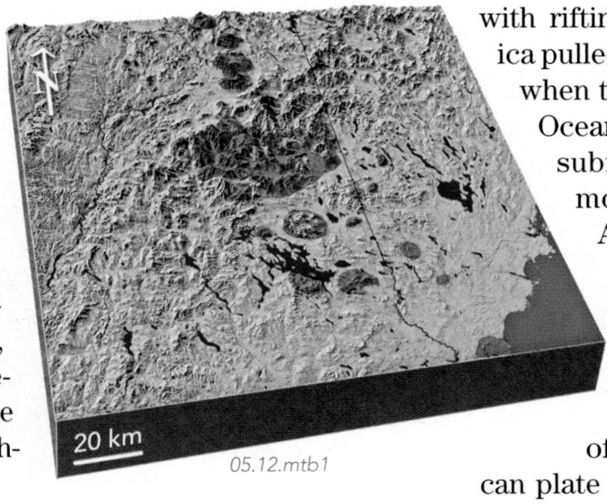

Geologists interpret the White Mountain batholith as being related to a hot spot that melted its way into continental crust. The age of the batholith coincides with rifting as North America pulled away from Africa when the central Atlantic Ocean formed. A line of submerged volcanic mountains in the Atlantic Ocean, called the *New England Seamount Chain*, is interpreted to mark the path of the North American plate over the hot spot.

Before You Leave This Page Be Able To

☑ Describe what a magma chamber is and the processes that occur in one.

☑ Sketch or summarize the tectonic settings in which large magma chambers form.

☑ Sketch the different geometries of large magma chambers and summarize how these are expressed in the landscape.

☑ Describe the character of the White Mountain batholith and how it is interpreted to have formed.

5.12

5.13 How Are Small Intrusions Formed and Expressed in Landscapes?

MANY INTRUSIONS ARE RELATIVELY SMALL OR THIN FEATURES, small enough to be exposed on a single small hill or in a roadcut. Small intrusions can have a sheetlike, pipelike, or even lumpy geometry. Where exposed at the surface, small intrusions can form distinctive landscape features, like a volcanic neck.

A What Features Form When Magma Is Injected as Sheets?

Many small intrusions have the shape of thin or thick sheets, typically ranging in thickness from several centimeters to several tens of meters. These form when underground forces allow magma to generate new fractures or to open up and inject into existing fractures. In some cases, magma squeezes between preexisting layers in the wall rocks, commonly between the horizontal layers of sedimentary rocks.

Dike

05.13.a1

◄ A *dike* is a sheetlike intrusion that cuts across any layers present in the host rocks. Most dikes are steep because the magma pushes apart the rocks in a horizontal direction as it rises vertically and fills the resulting crack to form a dike. Dikes are also common within larger plutons.

► The Greek island of Santorini erupted catastrophically around 1650 BCE. Steep dikes in the walls of the volcanic crater cut across the volcanic layers. Some of the dikes are along faults related to collapse of the volcano during the eruption.

05.1.a2

Sill

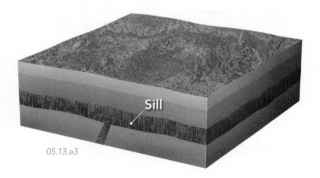

05.13.a3

◄ An intrusion that is parallel to layers in the host rocks is called a *sill*. Most sills are subhorizontal and form by pushing adjacent rocks upward rather than sideways.

► These dark-colored mafic sills intruded parallel to layers of light-colored, sedimentary wall rocks. Like most sills, these contain steep fractures formed by cooling of the sills after they solidified.

05.13.a4 Salt River Canyon, AZ

Laccolith

In some areas, ascending magma encounters gently inclined layers and begins squeezing parallel to them, forming a *sill*. The magma then begins inflating a lump- or bulge-shaped magma body called a *laccolith*. As the magma chamber grows, the layers over the laccolith tilt outward and eventually define a dome. ▼

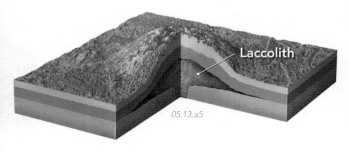

05.13.a5

► The Four Corners region of the American Southwest contains some of the world's most famous stocks and laccoliths, including these in the Henry Mountains of southern Utah. The laccoliths formed 25 million years ago at a depth of several kilometers and were later uncovered by erosion. Igneous rocks of the laccolith are medium grained and porphyritic and have an intermediate composition.

05.13.a6

B What Kind of Magma Chambers Form Within and Beneath Volcanoes?

Magma that erupts from volcanoes is fed through conduits that may be circular, dike shaped, or both. After the volcano erodes away, the solidified conduit can form a steep topographic feature called a *volcanic neck*.

05.13.b1 Mount Taylor, NM

05.13.b3

05.13.b5

▶ **1.** A small volcano has been partially eroded, revealing a cross section through the volcano. A resistant and jointed volcanic conduit marks the center of the volcano.

◀ **2.** Many volcanic necks, like the one to the right, form as erosion wears down a volcano, exposing the harder, more resistant rocks that solidified inside the magmatic conduit of the volcano.

3. Shiprock is a famous volcanic neck that rises above the landscape of New Mexico (▼). It consists of fragmented mafic rocks (breccia), and it connects to dikes (not shown) that radiate out from the conduit.

05.13.b4

05.13.b6

◀ **4.** Some volcanic necks, including Shiprock, were not originally *inside* a volcano but instead were magmatic conduits that formed well *beneath* the surface. The volcano above Shiprock was not a mountain, but a crater (pit) excavated by a violent explosion. The explosion occurred when magma ascending up a conduit encountered groundwater and generated huge amounts of steam, which expanded violently, causing an explosion. After the volcanic eruption, erosion removed the crater and hundreds of meters of rock that once overlay the area around the conduit.

05.13.b2

Columnar Joints

Many igneous rock bodies display distinctive fracture-bounded columns of rocks, like the ones in Devil's Postpile National Monument in California (▶). These fractures, known as *columnar joints*, form when a hot but solid igneous rock contracts as it cools. The fractures carve out columns that commonly have five or six sides. Columnar joints are common in basaltic lava flows, felsic ash flows, sills, dikes, and some laccoliths. In a tabular unit, like a flow, sill, or dike, columnar joints tend to be perpendicular to the tabular unit—they are vertical in a horizontal lava flow, ash flow, or sill, but horizontal in a vertical dike.

05.13.mtb1

Before You Leave This Page Be Able To

☑ Sketch the difference between a dike and a sill, and explain why each has the orientation that it does.

☑ Sketch or discuss the geometry of a laccolith.

☑ Sketch and explain two ways that a volcanic neck can form.

☑ Describe how columnar joints form.

5.13

5.14 How Did the Sierra Nevada Form?

ONE OF THE WORLD'S MOST STUDIED BATHOLITHS makes up the scenic granite peaks of the Sierra Nevada of central California. The batholith contains a diverse suite of plutonic rocks that cover an area of 40,000 km² (16,000 mi²). It nicely illustrates the connections between tectonic setting, cause of melting, processes within magma chambers, and the resulting rock types and scenery.

A What Is the Nature of the Sierra Nevada Batholith?

The Sierra Nevada batholith includes hundreds of individual plutons, some of which cover more than 1,000 km² (380 mi²). The batholith includes small stocks that are only hundreds of meters across, as well as countless dikes and sills of various compositions. The batholith was constructed by separate pulses of magma that invaded the crust, mostly between 140 and 80 million years ago. Rocks within and around the batholith tell its geologic story.

05.14.a2 Yosemite National Park, CA

◄ 1. The scenery of the Sierra Nevada is dominated by peaks, cliffs, and rounded domes of massive gray granite. It also contains intermediate and mafic rocks in stocks, dikes, and sills.

2. The figure below shows the landscape of the region colored according to rock type. The Sierra Nevada is the broad, high mountain range and is mostly granitic rocks (colored gray) with smaller areas of metamorphic rocks (colored green). Patches of volcanic rocks much younger than the batholith are shown in red and pink. The valley east of the Sierra Nevada is the Owens Valley, which is underlain by recent sediments (colored yellow).

05.14.a3

▲ 3. The most common rocks in the batholith are light- to medium-gray granite and other plutonic rocks. The plutons solidified slowly, so they have medium-grained to coarse-grained crystals.

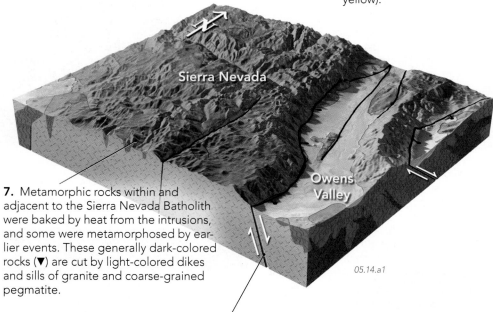

05.14.a1

7. Metamorphic rocks within and adjacent to the Sierra Nevada Batholith were baked by heat from the intrusions, and some were metamorphosed by earlier events. These generally dark-colored rocks (▼) are cut by light-colored dikes and sills of granite and coarse-grained pegmatite.

05.14.a4

▲ 4. Some outcrops, including this one, show great diversity in composition, including light-gray granite, dark intermediate and mafic rocks, and thin cream-colored felsic dikes. The different rock types represent distinct magmas, mostly emplaced at different times.

05.14.a5

6. The steep east side of the Sierra Nevada is a fault that downdropped Owens Valley relative to the mountains. During faulting, the entire Sierra was tilted, raising the eastern side of the range so that it is now higher and steeper than the western side.

5. Some plutons display compositional variations that record crystallization and settling of early-formed crystals. In this photograph a lighter colored, intermediate part is to the left and a darker mafic one is to the right. ▶

05.14.a6

B What Is the Tectonic History of the Batholith and Surrounding Areas?

The Sierra Nevada batholith is a product of plate tectonics—it formed by subduction-related partial melting of mantle and lower continental crust. Its origin illustrates how different magmas are generated.

Plate-Tectonic History of the Batholith

1. This figure shows the interpreted setting of the batholith 100 million years ago, when North America was converging with oceanic plates in the Pacific Ocean. Most of the batholith formed between 140 and 80 million years ago when oceanic plates in the Pacific Ocean were being subducted eastward beneath North America.

2. Water driven from minerals in the subducting slab rose into the overlying mantle, causing partial melting because the water lowered the melting temperature. Melting of ultramafic mantle generated mafic magma that rose through the lower lithosphere and toward the overlying crust.

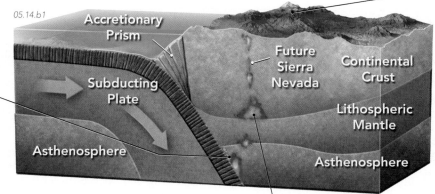

05.14.b1

5. While the batholith formed underground, large volumes of magma reached the surface and erupted in explosive volcanoes. The volcanoes were mostly eroded away, but their record is preserved in sedimentary rocks deposited between the batholith and the offshore trench, in what is now the Great Valley of central California. The Great Valley and Coast Ranges also contain rocks from the accretionary prism that formed near the trench.

4. Magma rose in the crust and large volumes solidified at depth as plutons and dikes in the future site of the Sierra Nevada. Sixty million years of sustained magmatism, with many discrete magmas, constructed the regional batholith.

3. Heat transfer from mantle-derived mafic magma caused partial melting of continental crust, which generated felsic magma. Mixing of felsic and mafic magmas, along with partial crystallization, produced a wide range of igneous compositions.

6. About 80 million years ago, a change in the plate-tectonic setting shut off magmatism in the Sierra Nevada and shifted magmatism eastward into Nevada and Arizona.

7. Between 80 and 30 million years ago, the Sierra Nevada batholith (shown here in gray) was slowly uplifted until the plutonic rocks, which formed at depths of 10 to 20 km, and their metamorphic wall rocks (shown here in green) became exposed at the surface. The rocks originally on top of the batholith were eroded away during the uplift.

05.14.b2

8. During the last 5 million years, faulting along the eastern side of the batholith uplifted the Sierra Nevada to the majestic mountain range it is today. This faulting, part of an episode of continental rifting, was accompanied by the eruption of felsic domes and other volcanic units shown in red and pink.

How Do Geologists Study the Sierra Nevada?

Many geologists study the Sierra Nevada Range to reconstruct the geologic history of this special place and to study the processes of magma chambers.

To study the batholith and its magmatic processes, geologists first do geologic field work by hiking up and down the ridges and valleys, examining the rocks, identifying boundaries between different plutons, and collecting samples for later analysis. From the field studies, geologists construct geologic maps and geologic cross sections that represent the distribution of the different plutons. Geologists cut thin sections from the rock samples to determine what minerals are present and in what order the

minerals crystallized from the magma. Chemical analyses of the samples for potassium, silicon, and other elements document how the magma evolved over time. Analysis of isotopes helps determine the age of the rocks and the types of source rocks that were melted to form the magmas. The Sierra Nevada Batholith is an excellent and scenic place to study igneous processes.

05.14.mtb1

Before You Leave This Page Be Able To

✓ Describe the Sierra Nevada batholith and what rocks it contains.

✓ Sketch the plate-tectonic setting that formed the Sierra Nevada batholith.

✓ Sketch or describe how magmas of the batholith formed.

✓ Describe how the deep batholithic rocks ended up on Earth's surface.

✓ Briefly summarize the kinds of data geologists collect in studying the batholith.

5.14

5.15 What Types of Igneous Processes Are Occurring Here?

IGNEOUS ACTIVITY IS NOT DISTRIBUTED UNIFORMLY ON EARTH. As a result, some regions are more likely to experience volcanic eruptions and other igneous activity. In this exercise, you will investigate five sites to interpret the types of igneous rocks likely to be present, the style of eruption, and the probable causes of melting.

Goals of This Exercise:

- Use the regional features of an ocean and two continents to infer the tectonic setting and cause of melting at five sites.
- Observe and identify nine rock types and infer the cooling history of each rock based on its texture.
- For the volcanic rocks, predict the viscosity of the magma and what type of eruption probably formed the rock.

A Tectonic Settings of Igneous Activity

The perspective view below shows two continents and an intervening ocean basin. The area has five sites, labeled A, B, C, D, and E, where igneous activity has been observed. For each site, investigate the igneous processes responsible for the activity and enter your results in the worksheet or online using the steps listed below.

1. Use the features on this map to infer whether the tectonic setting of each site is associated with a plate boundary and, if so, which type of plate boundary is present. The possible tectonic settings for this region are: (1) an oceanic or continental divergent boundary, (2) one of the three types of convergent boundaries, or (3) a hot spot in a continent, ocean, or both. However, not all of these settings are present in this area.

2. For each site, determine the likely cause of melting. The options are (1) decompression melting, (2) melting by adding water, and (3) melting of continental crust caused by an influx of mantle-derived magmas. More than one of these causes might apply to each site. Think about the kinds of igneous rocks you would expect to find at each site, including those that solidify at depth (plutonic) and those erupted onto the surface (volcanic). Your instructor may ask you to list the predicted rock types.

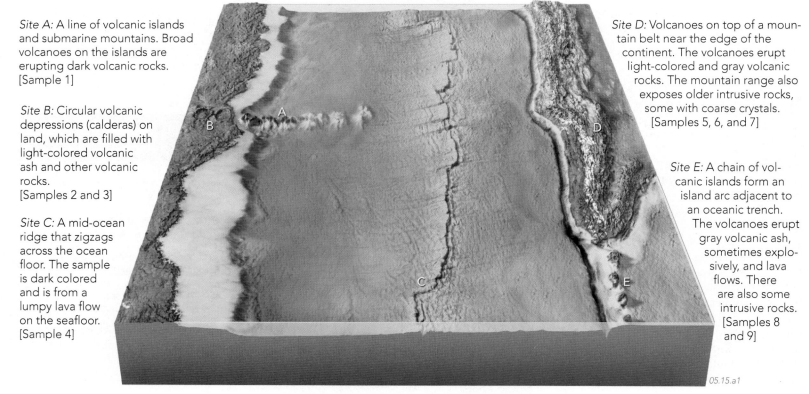

Site A: A line of volcanic islands and submarine mountains. Broad volcanoes on the islands are erupting dark volcanic rocks. [Sample 1]

Site B: Circular volcanic depressions (calderas) on land, which are filled with light-colored volcanic ash and other volcanic rocks. [Samples 2 and 3]

Site C: A mid-ocean ridge that zigzags across the ocean floor. The sample is dark colored and is from a lumpy lava flow on the seafloor. [Sample 4]

Site D: Volcanoes on top of a mountain belt near the edge of the continent. The volcanoes erupt light-colored and gray volcanic rocks. The mountain range also exposes older intrusive rocks, some with coarse crystals. [Samples 5, 6, and 7]

Site E: A chain of volcanic islands form an island arc adjacent to an oceanic trench. The volcanoes erupt gray volcanic ash, sometimes explosively, and lava flows. There are also some intrusive rocks. [Samples 8 and 9]

05.15.a1

B Predicting the Types of Igneous Rocks and Eruptions at Each Site

The photographs below show nine different rocks. Your instructor may provide you with samples of each rock or may substitute a different suite of rocks. Observe each of these rocks and complete the steps below.

1. Your instructor may have you write a short description of each photograph or of actual samples.

2. On the worksheet, indicate (1) whether each rock shown is coarsely crystalline, finely crystalline, or has other distinctive igneous textures, (2) whether it is probably mafic, intermediate, or felsic, and (3) the name you would apply to such a rock. For certain samples, some of this information is provided.

3. Predict the cooling and solidification history (slow, moderate, fast, slow then fast, or slow cooling in the presence of water) for each rock sample based on its texture.

4. For each site, use the rock samples that you interpret to be volcanic to predict whether the magma for that site has a high or low viscosity, and what type of volcanic eruption probably formed the rock sample.

5. Your instructor may have you use the various types of information to explain how the samples are consistent with the tectonic setting of each site. Alternatively, your instructor may have you infer the entire sequence of events including (1) what caused the initial melting event, (2) what processes might have occurred in the magma chamber, (3) where and how the rock cooled and solidified, and (4) whether uplift and erosion are required to expose the rock.

Descriptions of Rocks

Some descriptions are provided, because it is more difficult to describe and identify rocks from a photograph than if you could pick up and closely examine each rock. Make your own observations first, and then read the hints here.

Rock 1. This rock is dark colored, has no visible crystals, and is not glassy. It does have many vesicles (holes) and is a volcanic rock. A chemical analysis revealed a relatively low silica content.

Rock 4. This rock is finely crystalline and lacks visible crystals. It has a dark-gray color. A chemical analysis of the sample indicates that it is a mafic rock.

Rock 5. This porous rock in the center of the photograph has numerous holes and is extremely lightweight. It is light colored and is composed of volcanic glass. There are no visible crystals, but the rock is likely to be felsic in composition.

Rock 2. This light-colored rock is finely crystalline and not glassy. It contains some fine-grained gray crystals and a few small holes. A chemical analysis documents that it is a felsic rock. It was collected on a volcano.

Rock 3. This brown-colored rock contains large, light-colored lenses of flattened pumice in a finely crystalline matrix. Under the microscope, the matrix contains compacted volcanic ash, along with small crystals and fragments of other rocks. The crystals are mostly quartz and K-feldspar.

Rock 6. This rock has very large crystals, some of which are 5 to 10 cm long. Most of the rock is K-feldspar and quartz, and the dark crystals are a type of amphibole that is most common in felsic rocks.

Rock 9. This rock contains large crystals of K-feldspar in a brown-colored matrix of small to medium-sized crystals. There is no glass, ash, or vesicles, so the rock is probably not a volcanic rock.

05.15.b1

Rock 7. This medium-grained rock is a plutonic rock. It has a salt-and-pepper appearance, caused by the presence of felsic minerals (feldspar and quartz) and mafic minerals (mostly biotite mica). It is intermediate in composition.

Rock 8. This rock has scattered visible crystals of amphibole and biotite in a medium-gray, finely crystalline matrix. It reportedly was collected either on a volcano or from dikes exposed near a volcano.

5.15

Volcanoes and Volcanic Hazards

A VOLCANIC ERUPTION is one of nature's most spectacular events. Volcanoes blast scalding volcanic ash into the air, as orange streams of molten rock pour down the volcano's flank. Volcanoes represent an obvious geologic hazard, and eruptions claim the lives of tens of thousands of people at a time. In this chapter, we explore volcanoes and their associated landforms and hazards.

Mount St. Helens in southwestern Washington was once one of the most beautiful and symmetrical high peaks (▼) in the Cascade Range of the Pacific Northwest. Its shape changed forever in May 1980 when the sleeping volcano erupted violently. The eruption blew apart the volcano's north flank and excavated a huge crater where the mountain peak used to be. Within the newly formed crater, continuing eruptions built the steaming lava dome shown in the larger photograph below.

What is a volcano, and how do we recognize one?

06.00.a2

Mount St. Helens

06.00.a1

06.00.a3

Pre-1980 View from West

2005 View from North

TOPICS IN THIS CHAPTER

06.00.a4

The May 1980 eruption started with a northward-directed blast that knocked over millions of trees and unleashed a *pyroclastic flow,* a swirling, hot cloud of dangerous gases, volcanic ash, and angular rock fragments. The pyroclastic flow swept downhill and across the landscape, burying and killing almost all living things in its path. This was followed immediately by a huge column of volcanic ash that rose 25 km (15 mi) into the atmosphere (◄). The ash was carried eastward by the wind and blocked sunlight as it settled back to Earth across a large area of Washington, Idaho, and Montana.

What are the different ways that volcanoes erupt, and what hazards are associated with each type of eruption?

06.00.a5

Since the main eruption, magma rising through the throat of the volcano has collected on top of the vent, forming a *lava dome* (▲). Periodic collapse of part of the unstable dome unleashes explosions or avalanches of hot volcanic ash and rocky fragments.

What factors determine whether magma erupts as an explosion of hot ash or a slow outpouring of lava?

The May 1980 Eruption

With eruptions continuing into 2008, Mount St. Helens is the most active of the 15 large volcanoes that crown the Cascade Range of the Pacific Northwest. The mountain is the youngest volcano in the range, being entirely constructed during the last 40,000 years. Before 1980, a team of geologists from the U.S. Geological Survey studied the geology of the mountain and recognized that past eruptions had unleashed vast amounts of volcanic ash, lava, and volcanic mudflows. Prior to 1980, the volcano last erupted in the mid-1800s.

After more than 100 years of quiescence, the volcano reawakened in March 1980 when it vented steam, shook the area with many earthquakes, and pushed out an ominous bulge of rock on its north flank. At 8:32 a.m., on May 18, 1980, an earthquake caused the oversteepened north flank to collapse downhill in a huge landslide that carried rock pieces as large as buildings. This catastrophic removal of rock released pressure on the magma inside the volcano, which exploded northward in a cloud of scalding and suffocating volcanic ash. The pyroclastic flow raced across the landscape at speeds of up to 1,000 km (600 mi) per hour. The eruption blasted away most of the north flank of the mountain and forever changed the peak's appearance. It turned the surrounding countryside into a barren wasteland smothered by a thick blanket of volcanic ash. Early evacuations helped limit the loss of life, but 57 people perished and damage estimates for the eruption exceeded one billion dollars, making it the most expensive and deadly volcanic eruption in U.S. history.

Although the level of activity at Mount St. Helens greatly diminished in 2007 and 2008, geologists continue to monitor the volcano by keeping track of any ongoing volcanic and seismic activity, and carefully measuring changes in the mountain. Remotely operated cameras keep watch over the crater and the domes. Geologists also use instruments to monitor temperatures, gas emissions, and tilting of the land surface.

6.1 What Is and Is Not a Volcano?

AN ERUPTING VOLCANO IS UNMISTAKABLE—glowing orange lava cascading down a hillside, molten fragments blasting into the air, or an ominous, billowing plume of gray volcanic ash rising into the atmosphere. But what if a volcano is not erupting? How do we tell whether a mountain is a volcano?

A What Are the Characteristics of a Volcano?

How would you describe a volcano to someone who had never seen one? Examine the two volcano photographs below and look for common characteristics.

06.01.a1

A volcano is a *vent* where magma and other volcanic products erupt onto the surface. The volcano on the left (the Pu`u `Ō`ō volcano in Hawaii Volcanoes National Park) produces large volumes of molten lava, whereas the one on the right (Kanaga volcano, Alaska) is erupting volcanic ash. Many geologists reserve the term *volcano* for hills or mountains that have been *constructed* by volcanic eruptions. Some eruptions do not produce hills or mountains, and we consider them to be volcanoes, too.

Most volcanoes have a *crater*, a roughly circular depression usually located near the top of the volcano. Other volcanoes have no obvious crater or are nothing but a crater.

Volcanoes consist of *volcanic rocks*, which form from lava, pumice, volcanic ash, and other products of volcanism.

Besides erupting from volcanoes that have the classic shape of a cone, magma erupts from fairly linear cracks called *fissures* and from huge circular depressions called *calderas*.

06.01.a2

Many volcanoes display evidence of having been active during the last several hundred to several million years, or even during the last several days. Such evidence can include a layer of volcanic ash on hillslopes (left side of volcano shown above) or lava and ash flows that are relatively unweathered and that lack a well-developed soil. Over time, erosion degrades and disguises volcanoes and volcanic craters, making them less obvious.

B Is Every Hill Composed of Volcanic Rocks a Volcano?

If a landscape lacks most of the diagnostic features described above, it is probably not a volcano. Many mountains and hills are not volcanoes, and some volcanoes do not make mountains or hills.

▼ The flat-topped hill below, called a *mesa*, has a cap of volcanic rocks, but it is not a volcano. It did not form over a volcanic vent. Instead, it is an eroded remnant of a once extensive lava flow that covered the region, as shown in the figure to the right.

06.01.b1 Hopi Buttes, AZ

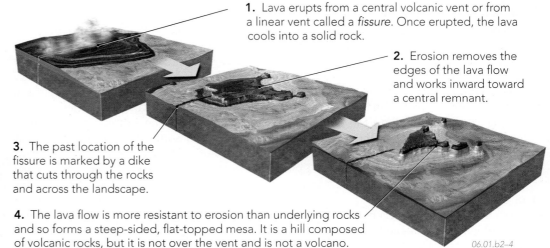

1. Lava erupts from a central volcanic vent or from a linear vent called a *fissure*. Once erupted, the lava cools into a solid rock.

2. Erosion removes the edges of the lava flow and works inward toward a central remnant.

3. The past location of the fissure is marked by a dike that cuts through the rocks and across the landscape.

4. The lava flow is more resistant to erosion than underlying rocks and so forms a steep-sided, flat-topped mesa. It is a hill composed of volcanic rocks, but it is not over the vent and is not a volcano.

06.01.b2–4

C What Are Some Different Types of Volcanoes?

Volcanoes have different sizes and shapes and contain different types of rocks. These variations reflect differences in the composition of the magmas and the style of the eruptions. There are four common types of volcanoes that are shaped like hills and mountains: *scoria cones*, *shield volcanoes*, *composite volcanoes*, and *volcanic domes*. Later in this chapter, we describe other types of volcanoes that are not hills or mountains.

Scoria Cone

06.01.c1

◄ *Scoria cones* are cone-shaped hills several hundred meters high, or higher, usually with a small crater at their summit. They also are called *cinder cones* because they contain loose black or red, pebble-sized volcanic *cinders* (scoria), along with larger volcanic *bombs*. The scoria is basaltic or, less commonly, andesitic in composition. Some scoria cones form next to, or on the flanks of, composite and shield volcanoes.

Shield Volcano

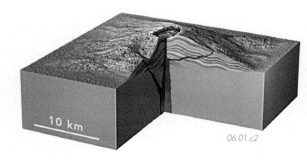

06.01.c2

► *Shield volcanoes* have broad, gently curved slopes and can be relatively small (less than a kilometer across) or can form huge mountains tens of kilometers wide and thousands of meters high. They commonly contain a crater or line of craters and have *fissures* along their summit. Shield volcanoes consist mostly of basaltic lava flows with smaller amounts of scoria and volcanic ash.

Composite Volcano

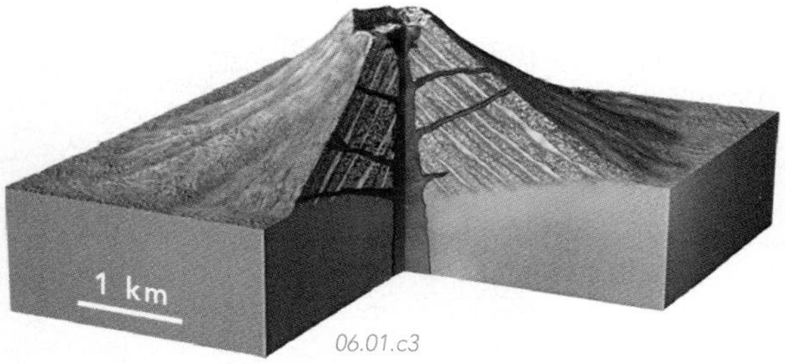

06.01.c3

◄ *Composite volcanoes* are typically fairly symmetrical mountains thousands of meters high, with moderately steep slopes and a crater at the top. They may be large, but are, on average, much smaller than shield volcanoes. Their name derives from the interlayering of lava flows, pyroclastic deposits, and volcanic mudflows. They consist mostly of intermediate-composition rocks, such as andesite, but can also contain felsic and mafic rocks.

Volcanic Dome

06.01.c4

► *Volcanic domes* are dome-shaped features that may be hundreds of meters high. They consist of solidified lava, which can be highly fractured or mostly intact. Domes include some volcanic ash intermixed with rock fragments derived from solidified lava in the dome. They form where felsic or intermediate magma erupts and is so viscous that it piles up around a vent. Many domes are within craters of composite volcanoes.

The Relative Sizes of Different Types of Volcanoes

Volcanoes vary from small hills less than a hundred meters high to broad mountains tens of kilometers across. Although sizes vary quite a bit, we can make some generalizations about the relative sizes of the different volcano types.

The figure below illustrates that some types of volcanoes are larger than others. The volcanoes on this figure cannot be drawn to their true scale relative to one another because the largest shield volcanoes are so large that we cannot show them on the same drawing with small scoria cones. The figure does accurately show which volcanoes are the largest and which ones are the smallest.

Scoria cones and domes, which typically form during a single eruptive episode, are the smallest volcanoes. Shield volcanoes and composite volcanoes are much larger because they are constructed, layer by layer, by multiple eruptions. Shield volcanoes have more gentle slopes than scoria cones, domes, or composite volcanoes.

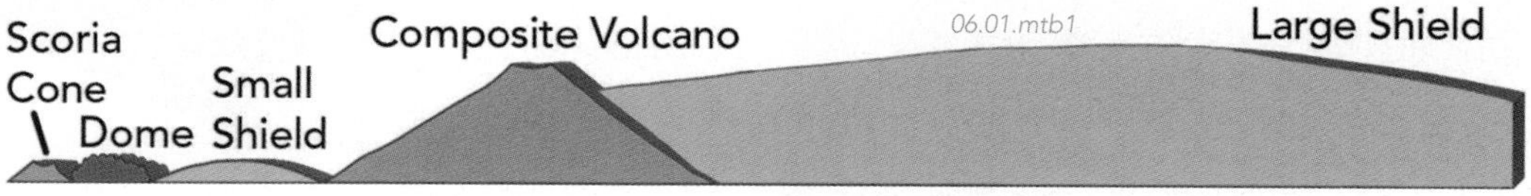

Before You Leave This Page Be Able To

✓ Sketch or describe the diagnostic characteristics of a volcano.

✓ Describe or sketch why every hill composed of volcanic rocks is not a volcano.

✓ Sketch and describe the four main types of volcanoes that construct hills and mountains.

✓ Sketch or describe the relative sizes of different types of volcanoes.

6.1

Scoria
Cone   Small   Composite Volcano   06.01.mtb1   Large Shield
   Dome Shield

6.2 What Controls the Style of Eruption?

THE DIFFERENT SHAPES OF VOLCANOES reflect differences in the style of eruption. Some eruptions are explosive, whereas others are comparatively calm. What causes these differences? The answer involves both magma chemistry and gas content, both of which control how magma behaves near the surface.

A What Are Ways That Magma Erupts?

Magma may behave in several different ways once it reaches Earth's surface. Explosive *pyroclastic eruptions* throw bits of lava, volcanic ash, and other particles into the atmosphere. During nonexplosive eruptions, lava issues from a vent and flows onto the surface. Both types of eruptions can occur from the same volcano.

Lava Flows and Domes

▶ When magma erupts onto the surface and flows away from a vent, it creates a *lava flow*. Erupted lava can be fairly fluid, flowing downhill like a fast river of molten rock. Some lava flows are not so fluid and travel only a short distance before solidifying.

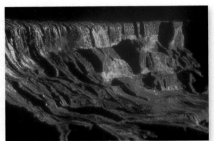
06.02.a1 Kilauea, HI

▶ A *lava dome* forms from the eruption of highly viscous lava. The high viscosity of the lava is generally due to a high silica content and causes the lava to pile up around the vent, instead of flowing away. Domes are often accompanied by several types of explosive eruptions.

06.02.a3 Mount St. Helens, WA

Pyroclastic Eruptions

▶ Some explosive eruptions send molten lava into the air. A *lava fountain*, such as shown here, can accompany basaltic volcanism and results from a high initial gas content in a less viscous lava. The gas propels the lava and separates it into discrete pieces.

06.02.a2 Kilauea, HI

▶ Other explosive eruptions eject a mixture of volcanic ash, pumice, and rock fragments into the air. Such airborne material is called *tephra*, and tephra particles that are sand sized or smaller are *volcanic ash*. Ash mostly forms when bubbles blow apart bits of magma. Tephra is derived from pumice, fragmented volcanic glass, and shattered preexisting rocks.

06.02.a4 Augustine volcano, AK

Two Different Eruptions of Tephra from the Same Volcano

The Augustine volcano in Alaska produces tephra in two eruptive styles—an *eruption column* and a *pyroclastic flow*.

06.02.a5

▲ *Eruption Column*—Tephra, which forms when magma is blown apart by volcanic gases, erupts high into the atmosphere, forming an *eruption column*. The tephra falls back to Earth as solidified and cooled pieces of rock. Finer particles of ash drift many kilometers away from the volcano and slowly settle down to the ground.

▼ *Pyroclastic Flow*—Some ash does not jet straight up but collapses down the side of the volcano as a dense, hot cloud of ash particles and gas. This eruption style is a *pyroclastic flow* or simply an *ash flow*. A pyroclastic flow can be devastating because of its high speed (more than 100 km/hr) and high temperature (exceeding 500°C).

06.02.a6

The two kinds of eruptions differ primarily because of gas content of the magma. An *eruption column* forms when large volumes of volcanic gas come out of the magma and overcome gravity to carry the cloud of tephra up into the atmosphere.

A *pyroclastic flow* forms when the amount of gas is less and cannot support the eruption column, so the column rapidly collapses and flows downhill under the force of gravity.

06.02.a7

B How Do Gases Affect Magma?

1. To envision dissolved gas in magma, think what happens when you open a bottle or can of soda. The liquid may have no bubbles until it is opened, at which time bubbles appear in the liquid, rise to the top, and perhaps cause the soda to spill out. The dissolved gas was always in the liquid, but it only became visible when you opened the top and released the pressure that held the gas in solution.

06.02.b1

2. Magma, like the soda, contains some dissolved gases, including H_2O (water vapor), CO_2 (carbon dioxide) and SO_2 (sulfur dioxide). These gases have a critical effect on eruption style and help the magma rise toward the surface.

3. As shown in this enlargement of the magma, confining pressure at depth keeps most of the gases in solution and keeps bubbles from forming.

4. As the magma approaches the surface, pressure decreases and the gases cannot remain in solution. Bubbles of gas form in the magma. If enough bubbles form quickly, the expanding bubbles cause the magma to be more buoyant and help it rise toward the surface and erupt out of the volcano.

06.02.b2

C How Does Viscosity Affect Gases in Magma?

Viscosity, the resistance to flow, dictates how fast a magma can flow and how fast crystals and gas can move through the magma. When gas in a magma comes out of solution, movement of the resulting bubbles is resisted by the magma's viscosity. If the bubbles cannot escape, the magma is potentially more explosive.

06.02.c1 Mount St. Helens, WA

More Viscous

◄ Felsic magmas contain a lot of silica, and so they are relatively viscous. The high viscosity prevents gas from escaping easily. Gas builds up in the magma and, when it expands, greatly increases the pressure on the surrounding rock. This can cause explosive eruptions.

Less Viscous

► Less viscous magma, such as one with a basaltic composition, allows gas bubbles to escape relatively easily. This can lead to a fairly non-explosive eruption, such as this basaltic lava flow that flows smoothly downhill from the vent.

06.02.c2 Kilauea, HI

Composition, Viscosity, and Eruptive Style

Composition of magma is the main control on a volcano's eruptive style, shape, associated rock types, and potential hazards. This is because composition, especially the amount and length of silicate chains in the melt, controls viscosity and whether gas builds up in the magma.

Mafic (basaltic) magma has fewer and shorter silicate chains than felsic or intermediate magma, and so is relatively less viscous. The lower viscosity allows mafic magma to flow from the volcano in a relatively fluid lava flow. The fluidity of mafic lavas accounts for the relatively gentle slopes of shield volcanoes, which largely consist of basaltic (mafic) lava flows. Explosive gases can build up in mafic magma, as demonstrated by lava fountains, but the resulting explosive eruptions are relatively small and localized, scattering basaltic scoria and ash close to the scoria cone.

Felsic and intermediate magma have more silicate chains, and the chains are longer, restricting the flow of the magma and making it more viscous. The high viscosity of felsic and intermediate lavas produces steep volcanic domes and steep composite volcanoes. Magma in domes, composite volcanoes, and large volcanic calderas can trap gas and erupt explosively, producing gas-propelled pyroclastic eruptions of volcanic ash, tephra, and rock fragments. As a result, these volcanoes produce a mix of pyroclastic rocks and lava flows, mostly of felsic and intermediate composition. Composition controls viscosity, eruptive style, the shape of the volcano, and the rock types that compose that volcano.

Before You Leave This Page Be Able To

✓ Describe four ways that magma erupts.

✓ Describe the difference between an eruption column and a pyroclastic flow, and the role that gas plays in eruptive style.

✓ Explain how gas behaves at different depths in a magma and how it influences eruptive style.

✓ Describe how viscosity influences how explosive an eruption is.

6.2

6.3 What Features Characterize Scoria Cones and Basaltic Lava Flows?

ERUPTIONS OF BASALTIC MAGMA can form a variety of rock types and landforms. This variety is largely controlled by the gas in the magma, because gas affects the style of eruption and the solidification of lava. A single eruption of basaltic magma can produce a wide range of volcanic features and rock textures.

A What Are Scoria Cones and Basalt Flows?

Basaltic magma has a relatively low viscosity compared to other magmas, and it erupts in characteristic ways. A basaltic eruption can from a fluid lava flow and throw pieces of molten rock into the air.

Basaltic Eruptions—At the beginning of many basaltic eruptions, gases carry bits of lava into the air, forming a *lava fountain*. The airborne bits of lava cool and then fall around the vent as loose pieces of scoria. The lava fountain may be followed by or accompanied by eruption of a basaltic lava flow.

06.03.a1 Hawaii

Scoria Cones—Pieces of scoria from the lava fountain gradually create a cone-shaped hill called a *scoria cone* (also called a *cinder cone*). Ejected fragments can be as small as sand grains or as large as huge boulders. Scoria cones typically form in a short amount of time, from a few months to a few years, and generally are no more than 300 m (~1,000 ft) high.

Basaltic Lava Flows—Fluid basaltic lava pours from the vent and flows downhill. Sometimes, as shown here, the lava fills up and overtops the crater in the scoria cone. At other times, a lava flow issues from cracks near the base of the scoria cone after most of the cone has been constructed.

06.03.a2 Iceland

Rock Types

◄ *Vesicular basalt* contains abundant gas pockets (*vesicles*). The vesicles were gas bubbles that expanded in the magma (as pressure decreased) and were trapped when the lava solidified. Vesicles occur in lava flows and in ejected material, such as highly vesicular scoria, which represents frothy, gas-rich magma.

06.03.a4 Flagstaff, AZ

◄ Basaltic magma may not contain enough gas to form bubbles, so the lava solidifies into *nonvesicular basalt*. A magma that forms nonvesicular basalt may have a low content of gas because it started out with a low content of dissolved gas or because it lost gas somewhere along the way.

06.03.a6 Northern AZ

Scoria cones contain scoria and other fragments explosively ejected from the volcano during a lava fountain. The fragments may have been liquid or solid when ejected. Small blobs cool and solidify in the air to form scoria (cinders). Large blobs of magma and solid angular blocks are ejected as *volcanic bombs*. ◄

Features of Lava Flows

► *Lava tubes* form when the surface of a lava flow solidifies to form an insulating roof over the hot, still-moving interior of the flow. Lava flows insulated by lava tubes can flow farther than lava flows on the surface because the lava stays hotter longer. If the tube drains, it becomes a curving, tube-shaped cave.

AA lava (pronounced "ah-ah") is a type of rough-surfaced lava flow, formed when the lava breaks apart into a mass of jumbled rocks as it flows. AA flows occur in open channels or as irregularly shaped flows. Angular blocks of hardened lava tumble down the front of the flow as it moves. An aa flow has a very rough surface covered with dark, jagged rocks. ►

Pahoehoe is a type of lava flow that has an upper surface with small billowing folds that form a "ropy" texture. A pahoehoe lava flow is usually fed by a lava tube and grows as a series of tongues. As the front of the flow solidifies, the lava breaks out and forms a new tongue, as shown here (►). Pahoehoe lava flows relatively smoothly and easily compared to aa.

06.03.a3 Hawaii

06.03.a5 Hawaii

06.03.a7 Hawaii

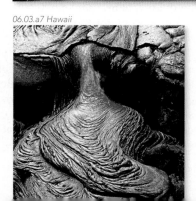

B How Do Scoria Cones and Basalt Flows Form Around the Same Vent?

Early Formation of a Scoria Cone—If basaltic magma contains enough dissolved gas, the gas comes out of solution as the magma approaches the surface. The gas expands dramatically and propels clots of frothy lava out of the conduit, piling up around the vent and forming a scoria cone. This generally occurs early in a basaltic eruption because the magma has not had time to degas in the magma chamber.

The Switch to Lava Flows—After most of the scoria cone is built, magma that contains less gas reaches the surface and erupts nonexplosively as a lava flow. Taking the easiest way out of the vent, the magma can squeeze out near the base of the scoria cone rather than rising to erupt from the summit crater. Some scoria cones are not accompanied by a lava flow, and vice versa.

06.03.b1

C What Do Scoria Cones and Basalt Flows Look Like?

Scoria cones are loose piles of material that erode rapidly. Basalt flows, on the other hand, are generally solid and considerably more resistant to weathering and erosion. A number of scoria cones and lava flows can occur together, forming a basaltic *volcanic field*.

Scoria Cones

▶ Most scoria cones begin with a conical shape and a central crater at the top of the cone. Young scoria cones have little soil or vegetation on them, and commonly are associated with dark, fresh-looking lava flows.

▶ Over time, erosion wears away the summit of a scoria cone, making the cone into a rounded hill without a central crater. Erosion cuts into the slopes, and the slopes gradually build up a veneer of soil and plants.

06.03.c1 Galápagos Islands

06.03.c3 Northern AZ

Lava Flows

▶ Young lava flows have steep flow fronts commonly with discrete, protruding lobes and embayments. The top of the flow is typically rough and displays flow features characteristic of aa and pahoehoe.

▶ Older lava flows are more subdued because they are eroded and sediment has accumulated in low spots on the surface. Flow tops lose their small features, such as pahoehoe, and become covered with soil and plants, as with this overgrown lava flow.

06.03.c2 Northern AZ

06.03.c4 Northern AZ

Battling Lava Flows with Seawater in Iceland

In 1973, the volcano Eldfell ("Fire Mountain" in Icelandic) erupted next to a fishing village on the island of Heimaey in Iceland. Scoria from the basaltic eruption accumulated on roofs and caused houses to collapse and burn. Blocky lava flows (aa) issued from the base of the crater, buried buildings, and encroached on the harbor, threatening to destroy the fishing economy of the island. Local fishermen

06.03.mtb1

and others began pumping cold seawater on the advancing flow, trying to solidify it and save their harbor. By the end of the eruption, 1.5 billion gallons of seawater was pumped onto the flow. There is debate about how effective the pumped water was in slowing down the flow, but the lava flow did stop before it totally closed the harbor. The town is somewhat back to normal and a tourist destination today.

Before You Leave This Page Be Able To

✓ Explain the characteristics of scoria cones and basalt flows, including the associated rock types and features.

✓ Sketch and describe how basaltic magma may form a scoria cone, lava flow, or both.

✓ Describe how you might distinguish between young and old examples of scoria cones and lava flows.

6.4 How Do Shield Volcanoes Form?

SHIELD VOLCANOES ARE THE LARGEST VOLCANOES on Earth. Many of the world's volcanic islands, including the Hawaiian and Galápagos Islands, are shield volcanoes. How do we recognize a shield volcano, how is one formed, and from where does all the magma come?

A What Is a Shield Volcano?

Shield volcanoes have a broad, *shield-shaped* form and fairly gentle slopes when compared to other volcanoes. They are built up by a succession of basaltic lava flows and lesser amounts of scoria and ash. Shield volcanoes can form in any tectonic setting that produces basaltic magma, but the largest ones, such as those in Hawaii, form on oceanic plates in association with hot spots.

1. This image shows satellite data superimposed on topography of the Big Island of Hawaii. The island consists mainly of three large volcanoes. Green areas are heavily vegetated, and recent lava flows are brown or dark gray.

2. Mauna Loa, the central mountain, is the world's largest volcano. It rises 9,000 m (29,520 ft) above the seafloor and is 4,170 m (13,680 ft) above sea level. From seafloor to peak, Mauna Loa is Earth's tallest mountain. Nearby Mauna Kea is an inactive shield volcano and the site of astronomical observatories.

3. Kilauea volcano, probably the most active volcano in the world, is on the southeastern side of the island. Recent lava flows (shown in dark grayish brown) flowed eastward, destroying roads and housing subdivisions.

06.04.a2

4. At Kilauea's summit is a roughly circular depression (▲), called a *caldera*, which in this case contains a smaller circular crater (both shown above). A caldera forms when magma is removed from an underground chamber, which causes the land above to collapse downward.

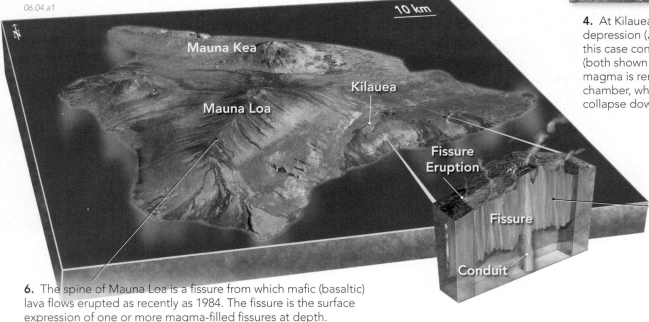
06.04.a1

5. Eruptions from Kilauea issue from linear fissures that are interpreted to form over the top of vertical, sheetlike magma chambers (fissures). Magma in the fissures is fed from below by larger and deeper magma chambers.

6. The spine of Mauna Loa is a fissure from which mafic (basaltic) lava flows erupted as recently as 1984. The fissure is the surface expression of one or more magma-filled fissures at depth.

◄ 7. At the start of the 1983 Kilauea eruption, a new volcanic vent formed, and early, gas-rich magma shot into the air in a lava fountain. The partly cooled and hardened pieces fell back around the vent to form a scoria cone. Such scoria cones are common companions to shield volcanoes.

8. Kilauea's basaltic lava flows have low viscosity and flow fluidly downhill, either on the surface or in lava tubes. Some lava flows reach the ocean, where the molten rock causes seawater to boil in rising clouds of steam (▶). These flows add new land to the island, extending the island outward.

06.04.a3 Hawaii

06.04.a4 Hawaii

B How Do Shield Volcanoes Erupt?

Shield volcanoes erupt mostly low-viscosity basaltic lava and so are dominated by relatively nonexplosive outpourings of lava from fissures and vents. Early phases of eruptions are commonly marked by spectacular fountains in which molten rock is ejected hundreds of meters into the air from fissures or central vents.

06.04.b1 Mauna Loa, HI

06.04.b2 Kilauea, HI

Fissure Eruption—A fissure eruption occurs when magma rises through a fracture and erupts onto the surface from a long fissure. Large volumes of lava can flow out of the fissure, and escaping gas throws smaller amounts of molten rock into the air as a fiery curtain.

Lava Flow—Fluid basaltic lava typically flows downhill as a river of molten rock. Flows can divide, rejoin, spread out, or constrict as they encounter variations in topography. They can even have waves and rapids, like those in a river of water.

06.04.b3

06.04.b4 San Juan Islands, WA

Shield volcanoes, like Mauna Loa, have broad, gentle slopes because the basaltic magma had a relatively low viscosity that allowed it to flow downhill and spread out. Shield volcanoes contain mostly dark-colored, basaltic magma with local hills and layers or reddish and black scoria.

Pillow Basalt— When fluid lava erupts into water, the lava grows forward as small, individual tongues that form rounded shapes called *pillows*. Pillows are reliable evidence that lava erupted into water.

Mauna Loa and Kilauea—Two Hawaiian Volcanoes

Two volcanoes on the Big Island are extremely active, building and reshaping the island before our eyes. Mauna Loa, which in Hawaiian means "long mountain," is the larger of the two and has erupted 33 times since 1843. The most recent eruption was in 1984, and the U.S. Geological Survey reports that the volcano is "certain to erupt again." USGS geologists closely monitor the volcano to anticipate, or perhaps predict, the next eruption.

Kilauea volcano, a short distance east of Mauna Loa, is regarded as the home of *Pele*, the Hawaiian volcano goddess. Kilauea has been even more active than Mauna Loa, erupting nearly continuously since the 1800s, with 61 historic eruptions. It has erupted 34 times since 1952 and nearly nonstop since 1983. The 1983 eruption began with the construction of a new scoria cone during an initial lava-fountaining event. Since that time, lava from the volcano has flowed down toward the sea. When the molten lava enters the ocean, it cools and solidifies, and new land is added to the island. All of the Hawaiian Islands formed by this process—the eruption of basaltic lava flows that make new land where there once was sea.

Before You Leave This Page Be Able To

✔ Describe the type of magma and other general characteristics of a shield volcano.

✔ Explain how shield volcanoes erupt.

✔ Sketch or summarize how you would recognize a shield volcano in the landscape.

✔ Discuss volcanic activity of the two main volcanoes on Hawaii, Mauna Loa and Kilauea.

6.4

6.5 What Causes Flood Basalts?

FLOOD BASALTS INVOLVE HUGE VOLUMES OF MAGMA and represent the largest igneous eruptions on Earth. They cover tens of thousands of square kilometers in Siberia, South Africa, Brazil, India, and the Pacific Northwest. Flood basalts also form large oceanic plateaus at or below sea level. What causes such huge eruptions to occur, and what effects do such large eruptions have on climate and life?

A What Are Flood Basalts?

Flood basalts are basaltic lava flows covering vast areas and commonly being several kilometers thick. They generally involve multiple eruption events, but individual lava flows can cover thousands of square kilometers and contain more than 1,000 cubic kilometers of magma—equal to emptying a cube-shaped magma chamber that is 10 km (over 6 mi) on a side. Flood basalts are fed by a series of long fissures.

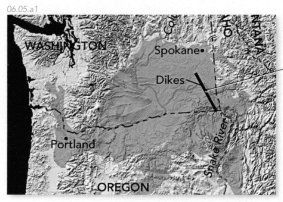
06.05.a1

2. Some of the lava flows were fed by long fissures, now exposed in canyons as dikes, shown here as a single black line.

3. Many of the Columbia Plateau canyons have walls made of multiple basalt flows, each forming a ledge, resulting in a series of steps in the canyon walls. Each flow represents a single eruption, separated by thousands of years in which not much happened.

06.05.a2

4. This fissure eruption, at Mauna Loa in Hawaii, is similar to those that formed basalts of the Columbia Plateau. The volume of lava that poured from the fissures on the Columbia Plateau was much greater.

1. The Columbia Plateau in Washington, Oregon, and western Idaho consists of a thick sequence of over 300 basalt flows that erupted between 17 and 5 million years ago, with about 85% of the total basalt volume being formed between 16.5 and 14.5 million years ago. Some of these flows cover over 50,000 square miles; one lava flood is interpreted to have flowed from Spokane to the Pacific Ocean in a little over a month (i.e., about one-half mile per hour).

06.05.a3

B How Do Flood Basalts Erupt onto the Surface?

Instead of erupting from a single, central volcanic vent, flood basalts erupt from one or more long, nearly continuous fissures or from a discontinuous string of vents. There are typically multiple eruption centers, allowing large volumes of magma to erupt onto the surface in a geologically short period of time. Examine the figure below, starting on the lower left.

2. The combination of a wide fissure and low-viscosity magma allows large volumes of magma to reach the surface and erupt along a linear fissure. A large volume of rapidly erupting magma results in individual lava flows that remain hot and molten for a longer time, flow long distances, and cover large areas. One basalt flow on the Columbia Plateau covers more than 1,000 cubic kilometers. Isotopic dating studies reveal that it probably erupted very quickly, perhaps in only several decades.

1. Mantle-derived basaltic magma rises through the crust along vertical fissures. Pressure from the magma pushes outward against the wall rocks, holding them apart and allowing magma to pass through.

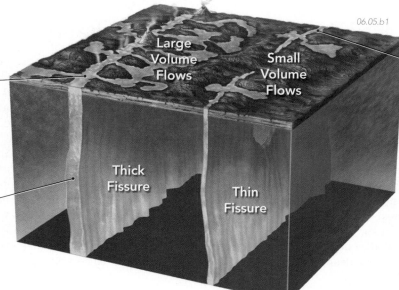
06.05.b1

Large Volume Flows

Small Volume Flows

Thick Fissure

Thin Fissure

3. If a fracture is narrow, less magma can flow through the conduit, and resulting lava flows have smaller volumes and travel shorter distances away from the vent.

C Where Does the Magma for Flood Basalts Originate?

Most flood basalts probably begin at hot spots as rising mantle plumes first encounter the lithosphere. It is uncertain whether the magma comes from melted lithosphere, melted asthenosphere, or directly from the plume.

06.05.c1

A mantle plume rising through the mantle is mostly solid and acquires an inverted teardrop shape as it flows upward.

06.05.c2

When the rising plume encounters the base of the lithosphere, it meets increased resistance and spreads laterally.

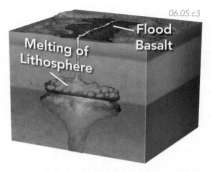
06.05.c3

The rising plume can melt because of decompression, or the plume can melt the adjacent asthenosphere or lithosphere.

Some Important Flood Basalts of the World

Flood basalts on the Columbia Plateau were probably related to the hot spot now beneath Yellowstone.

Large volumes of basalt erupted about 133 million years ago near the southeastern coast of South America. Some of the flows are on land and others are offshore. The magma probably formed when the continent was over a hot spot that is currently in the South Atlantic.

Thick sequences of basalt along the margins of Greenland and Scotland formed above a hot spot that is presently below Iceland.

Flood basalts in Siberia erupted about 250 million years ago, the same time as a major extinction event that affected life in the seas and on land. Were these events related?

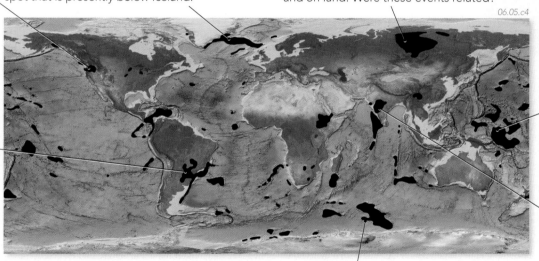
06.05.c4

Vast oceanic plateaus of basalt form the ocean floor in the western Pacific Ocean but are mostly below sea level. They are interpreted to have formed over oceanic hot spots.

Flood basalts cover large areas of India, where they are called the *Deccan Traps*. The basalts erupted 65 million years ago, at the same time as the dinosaurs became extinct.

Basaltic eruptions related to a hot spot constructed the huge *Kerguelen Plateau* near Antarctica.

Flood Basalts, Fissure Eruptions, Climate, and Life

Eruptions of large volumes of flood basalt can potentially change Earth's climate and negatively impact life. Eruptions release large amounts of sulfur dioxide (SO_2) gas that causes acid rain and reflects sunlight, leading to atmospheric cooling. Eruption of flood basalt also releases carbon dioxide (CO_2) gas, which acts as a greenhouse gas, causing global warming. It is unclear whether cooling or warming would dominate, but either would impact life.

Some of Earth's largest flood-basalt eruptions coincide with times when many marine and land animals went extinct. Eruption of the Deccan Traps of India coincides with the extinction of the dinosaurs at the Mesozoic-Cenozoic boundary (65 Ma). Vast quantities of flood basalts in Siberia erupted 250 million years ago, a time when more than 90% of marine species became extinct. This massive extinction, called the *Great Dying*, defines the boundary between the Paleozoic and Mesozoic Eras. Geologists are investigating how flood basalts change regional or global climate, and whether flood basalts had a role in the extinctions.

Before You Leave This Page Be Able To

✓ Describe the characteristics of flood basalts and how they erupt.

✓ Sketch and describe the interpreted relationship between flood basalts and mantle plumes.

✓ Identify at least three areas on Earth that contain flood basalts.

✓ Discuss how flood basalts could affect climate and life on Earth.

6.5

6.6 What Are the Hazards of Basaltic Eruptions?

BASALTIC ERUPTIONS usually only directly affect areas very close to the vent or more distant areas that are in the path of lava flows. Basaltic scoria is very localized, but basaltic flows are relatively fluid and can destroy buildings and crops tens of kilometers away from the volcanic vent. Surprisingly, they can also cause huge floods, if they occur in areas with ice sheets.

A What Is Meant by a Hazard and a Risk?

The terms *hazard* and *risk* may seem more appropriate for a lesson about insurance, but geologists frequently apply these terms when discussing the effects geologic events can have on humans and society. What is the difference between a hazard and a risk?

06.06.a1 Kilauea, HI

06.06.a2 Goma, Congo

◄ A *hazard* is the existence of a potentially dangerous situation or event, such as a potential landslide of a steep slope or a lava flow erupting from a volcano. The hazard in this photograph was a basaltic lava flow.

Risk is an assessment of whether the hazard might have some *societal impact*, such as loss of life, damage to property, loss of employment, destruction of fields and forests, or implications for local or global climates. Remnants of destroyed houses, cars, and roads demonstrate that this area had a high risk for volcanic hazards.

The risk was extreme for people living on the flanks of an active volcano in central Africa (▲). In 1977, a fast-moving (50 km/hr) lava flow killed as many as 300 people living in villages near the volcano. If no people were living near this volcano, a hazard would still exist but there would be essentially no risk.

B What Are the Hazards Associated with Scoria Cones?

Scoria cones can be deadly and destructive, especially to nearby areas. They hurl lava and solid rock into the air and spew out dangerous gases. Fine ash ejected high into the air can cause damage that is more widespread.

Falling Objects

Most scoria falls back to Earth near the vent and piles up on the scoria cone. Hazards that exist nearby include being struck and burned by cinders and being struck by blobs of magma and other projectiles. Larger ejected pieces, called *volcanic bombs* (▼), pose a severe hazard for anyone close to the erupting cone.

06.06.b1 Hawaii

06.06.b2 Flagstaff, AZ

Gases

Volcanic gases are a significant hazard associated with many types of volcanoes, including those with basaltic eruptions. Gases such as carbon dioxide (CO_2) cause asphyxiation if concentrated. Other gases, including hydrogen sulfide (H_2S), cause death by paralysis. Gaseous sulfur dioxide (SO_2), hydrochloric acid (HCl), sulfuric acid (H_2SO_4), and fluorine compounds expelled during eruptions can destroy crops, kill livestock, and poison drinking water for people and animals.

06.06.b3 Iceland

Volcanic Ash

Sand-sized cinders and finer particles of ash can bury nearby structures, and may cause breathing problems for people and livestock. This photograph of the Parícutin volcano in Mexico shows scoria and ash settling out of the column erupted from the volcano. Eruption columns from scoria cones typically reach heights of several kilometers and may be carried away by the wind, impacting areas many kilometers from the volcano.

06.06.b4 Mexico

C What Are the Hazards Associated with Lava Flows?

Lava flows usually move slowly enough that people can get out of the way, but such flows can completely destroy any structures in their path. Destruction may be caused by burning or by burial beneath the encroaching lava.

When Lava Comes to Town

◀ A lava flow will cause wooden structures and vegetation to catch fire. This house in Hawaii burst into flames when touched by a basaltic lava flow from Kilauea. Any structure in the path of a moving lava flow will likely be engulfed by fire and then crushed or bulldozed by the weight of the advancing lava.

▶ Lava flows, such as these encroaching into a subdivision in Hawaii in 1983, cover the land, vegetation, roads, and other human structures. The rough, lava-covered terrain becomes an uninhabitable environment, and communities rarely can be rebuilt after such an event.

06.06.c1 Kalapana, HI

06.06.c2 Kalapana, HI

D How Do Basaltic Eruptions Cause Floods of Water, Ice, and Debris?

A special type of hazard is associated with erupting volcanoes beneath ice sheets. The heat from such eruptions can melt large quantities of ice and produce huge floods.

06.06.d1 Trolladyngja, Iceland

◀ Iceland is a land of ice and fire, due to its location near the Arctic Circle and its position on top of a mid-ocean ridge and a hot spot. It has many glaciers (the Icelandic term for glacier is *jökull*), including a large ice sheet that covers 25% of the country. Beneath the ice are a half dozen basaltic volcanoes, similar to this snow- and ice-covered one.

▶ In 1996, a volcanic eruption beneath an ice sheet melted the ice, releasing a catastrophic flood of meltwater (jökulhlaup in Icelandic). The huge flood carried blocks of ice, rock, and other debris, causing widespread destruction and covering vast areas with sediment. Note the red plane for scale.

06.06.d2 Iceland

Laki and the Summer That Wasn't

In 1783, an Icelandic fissure at a place called Laki unleashed Earth's largest known recorded eruption (16 cubic kilometers of magma). The eruption caused the climate to cool in most of Europe because it released a large amount of ash and sulfur dioxide (SO_2) gas. Sulfur dioxide gas combines with water in the atmosphere to form sulfuric acid (H_2SO_4) in very small drops called *aerosols*. These drops and the volcanic ash drifted over northern Europe for eight months and were thick enough to dim the sunlight. The summer of the eruption was dismal, and the winter was unusually cold. The following summer was marked by crop failure and famine, which continued for the next three years because the climate remained cooler than normal. The cooling effects of sulfur dioxide in the upper atmosphere are interpreted to last only a few years after a big eruption, but can cause a regional, short-term crisis.

06.06.mtb1 Laki, Iceland

Before You Leave This Page Be Able To

☑ Explain how risk is different than hazard, and provide an example of each.

☑ Describe the difference between hazards associated with scoria cones and hazards associated with basaltic flows.

☑ Explain how a volcanic eruption can cause a flood.

☑ Discuss the effects of the Laki eruption of 1783.

6.7 What Are Composite Volcanoes?

COMPOSITE VOLCANOES FORM STEEP, CONICAL MOUNTAINS that are hard to mistake for anything other than a volcano. They are common above subduction zones and are especially numerous along the Pacific Ring of Fire. They contain diverse volcanic rock types that reflect different compositions of magma and several styles of eruption. They are an extremely dangerous type of volcano.

A What Are Some Characteristics of a Composite Volcano?

Composite volcanoes are constructed of interlayered lava flows, pyroclastic flows, tephra falls, and volcano-related mudflows and other debris. They also contain dikes, sills, and other intrusions. Composite volcanoes, also called *stratovolcanoes*, erupt over long time periods, which explains their large size and complex internal structure.

06.07.a1 Mount St. Helens, WA

◄ **1.** *Eruption Column*—Composite volcanoes produce a distinctive column of tephra and gas that rises upward many tens of kilometers into the atmosphere. Coarser pieces of tephra settle around the volcano, but finer particles (volcanic ash) can drift hundreds of kilometers in the prevailing winds.

06.07.a3 Mount Mayon, Philippines

▶ **2.** *Pyroclastic Flows*—These are the most violent eruptions from the volcano. They form when the eruption column collapses downward as a dense, swirling cloud of hot gases, volcanic ash, and angular rocks. Pyroclastic flows are one of the main mechanisms by which these volcanoes are constructed.

6. *Shape*—Composite volcanoes display the classic volcano shape because most material erupts out of a central vent and then settles nearby. They have steep slopes because they form from small eruptions of viscous lava flows that pile up on the flanks of the volcano and help protect pyroclastic material from erosion. The shape represents one snapshot in a series of stacked volcanic mountains that have been built over time.

5. *Lava Flows and Domes*— Lava flows and domes can erupt from any level of a composite volcano. Lava may erupt from the summit crater or escape through vents on the volcano's sides or base. Most lavas associated with composite volcanoes are felsic or intermediate in composition, so the lava is moderately to highly viscous, and moves slowly and with difficulty. The lava may break into blocks (▼) that fall, slide, or roll downhill.

3. *Landslides and Mudflows*—Composite volcanoes can be large mountains that collect rain or snow. Rain and snowmelt mix with loose ash and rocks on the volcano's flanks, causing a volcano-related mudflow called a *lahar*. Landslides and debris flows (▼) are especially hazardous because of the steep slopes, loose rocks, and abundant clay minerals produced when hot water interacts with the volcanic rocks.

06.07.a2

06.07.a4 Augustine volcano, AK

4. *Rocks*—Composite volcanoes consist of alternating layers of pyroclastic flows, lava flows, and deposits from landslides and mudflows. The volcanoes we see today, formed during eruptions from long-lived vents, are built on and around earlier versions of the volcanoes. The present peaks hide a complex interior that was constructed by multiple eruptions over a long time.

06.07.a5 Augustine volcano, AK

B What Types of Rocks and Deposits Form on Composite Volcanoes?

Composite volcanoes consist mostly of intermediate composition lava and ash, but they can also include felsic and mafic materials. The combination of diverse magma compositions and different eruptive styles produce a variety of rock types and volcanic features. Both modern and ancient examples reflect these complexities.

◄ This exposure contains a dark, cliff-forming lava flow (at the top) above a sequence with different kinds of volcanic rock fragments. The fragments vary in size, and they range in composition from basalt to andesite to rhyolite.

▼ The most common rock produced by composite volcanoes is andesite, a mostly gray or greenish-gray volcanic rock that may contain dark- or light-colored phenocrysts, like these cream-colored feldspar phenocrysts.

06.07.b1 Mount Shasta, CA

▶ An eruption column deposits *tephra*, containing fragments of pumice, crystals, and rock in a matrix of volcanic ash. Consolidation of tephra produces a rock called *tuff*, with distinct layers that record variations in the eruption over time.

▶ A pyroclastic flow forms *ash-flow tuff*, which contains an unsorted collection of pumice, rock fragments, crystals, and ash. The tuff accumulates while still hot and can be compacted by overlying ash, becoming harder and more dense (welded).

▶ Volcano-derived mudflows (lahars) and landslide deposits consist of angular rock fragments, usually in a matrix of finer materials, such as mud. Some mud in these settings is derived from volcanic ash that has been weathered or hydrothermally altered.

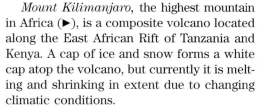

06.07.b2 Tucson Mtns., AZ

06.07.b3 Northern AZ

06.07.b4 Mount Redoubt, AK

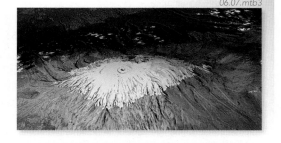

06.07.b5 Mount Redoubt, AK

Some Famous Composite Volcanoes of the World

Composite volcanoes are not distributed uniformly on Earth. Most are above subduction zones at ocean-ocean or ocean-continent convergent boundaries. Many composite volcanoes have names and appearances that show up on newscasts, nature shows, or history and geography courses. Here, we describe a few of the more famous or interesting ones.

The beautiful and symmetrical *Mount Fuji* (▼) is the landmark composite volcano of Japan, which is part of the Pacific Ring of Fire. Mount Fuji last erupted in 1708 and is above a subduction zone where the Pacific plate subducts westward beneath Japan.

Mount Kilimanjaro, the highest mountain in Africa (▶), is a composite volcano located along the East African Rift of Tanzania and Kenya. A cap of ice and snow forms a white cap atop the volcano, but currently it is melting and shrinking in extent due to changing climatic conditions.

Mount Etna, on the island of Sicily in the Mediterranean Sea south of the Italian mainland, is often shown on newscasts as erupting rivers of lava or ejecting glowing volcanic bombs from the crater. Visitors flock to the volcano to witness the spectacle from a distance. Eruptions are visible in this photograph from the *International Space Station* (▼).

06.07.mtb3

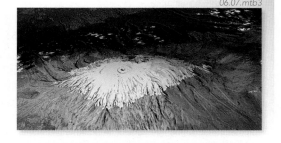

Before You Leave This Page Be Able To

☑ Describe or sketch the characteristics of a composite volcano, including its internal structure.

☑ Describe the processes on composite volcanoes and the rocks they form.

☑ Describe the tectonic setting of most composite volcanoes.

☑ Identify some examples of composite volcanoes from around the world.

06.07.mtb1

06.07.mtb2

6.8 What Disasters Were Caused by Composite Volcanoes?

COMPOSITE VOLCANOES ARE DANGEROUS because they can be very explosive and can unleash pyroclastic flows, toxic gases, and other deadly materials. They are responsible for horrific human disasters, including the destruction of Pompeii in Italy, St. Pierre in Martinique, and the area around Mount St. Helens.

A How Did Vesuvius Destroy Pompeii?

Vesuvius is an active composite volcano near the city of Naples in southwestern Italy. In A.D. 79, a series of pyroclastic flows moved down the flank of the volcano, destroyed the coastal towns of Pompeii and Herculaneum, and killed the cities' inhabitants, estimated at 20,000 people and 5,000 people, respectively.

1. This image is an artist's conception of an explosive Vesuvius eruption striking the city of Naples, which covers most of the region shown. Naples and the surrounding area currently are home to over three million people.

2. Archeologic and geologic evidence from Pompeii indicate that the catastrophe began with earthquakes and the formation of an eruption column that deposited a layer of loose tephra over Pompeii, killing some inhabitants

3. The tephra fall was immediately followed by six pyroclastic flows that raced down the mountainside. Three of these flows hit Pompeii. The first probably burned most of the remaining survivors, and the last was strong enough to complete the destruction of standing buildings. People smothered, suffocated, died from thermal shock, or were crushed by collapsing buildings. The bodies of victims in the ash decomposed, leaving mostly hollow molds, which archeologists filled with plaster to make models of the victims' last moments. ▶

4. The dashed red line marks the outward limit of pyroclastic flows from Vesuvius, but tephra from the eruption column covered a wider area. Note how much of the present city of Naples is within the area devastated by the eruption of 79 A.D.

Herculaneum
Limit of Pyroclastic Flows
Pompeii
10 km
06.08.a1

06.08.a2

B What Happened at St. Pierre, Martinique?

Mount Pelée, a composite volcano on the Caribbean island of Martinique, is part of an island arc over a subduction zone. On May 8, 1902, the volcano erupted and sent a pyroclastic flow into the town of St. Pierre.

1. This view shows the island of Martinique, which consists of several distinct volcanoes, including Mount Pelée, the northernmost peak. Mount Pelée is a composite volcano.

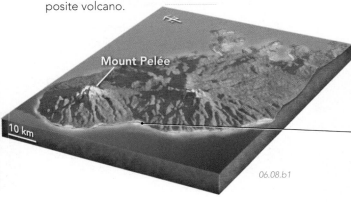
Mount Pelée
10 km
06.08.b1

2. The coastal town of St. Pierre is in a bay, at the foot of Mount Pelée. Before the main eruption, the volcano gave obvious warning signs, including noisy explosions, earthquakes, sulfurous gases, and small eruptions that dusted nearby areas with ash. People from the surrounding countryside sought shelter in the town of St. Pierre, where they witnessed minor eruptions of ash, the formation of a lava dome in the crater, and some small pyroclastic flows.

3. During the main eruption, a massive pyroclastic flow, estimated to have traveled at 500 km/hr, entered the town. Every building was mostly or completely destroyed (▼). Almost all of the 30,000 residents died within minutes. Most deaths were probably caused by asphyxiation as people breathed hot gas and ash. After the main eruption, additional eruptions formed an eruption column and more pyroclastic flows. The lessons learned from Mount Pelée and other eruptions saved lives in 1995, when volcanic eruptions started on Montserrat, a similar volcanic island to the south.

06.08.b2

C What Events Preceded and Accompanied the Mount St. Helens Eruption?

The Cascade Range of the Pacific Northwest has produced some large and notable eruptions, such as the one at Mount Lassen in 1915. Geologists consider these volcanoes to be dangerous, and native people remember, through oral traditions, other cataclysmic eruptions at places like Crater Lake in the Cascades. The eruption of Mount St. Helens in Washington was the first major composite volcano eruption to occur in the age of television, and the world watched the event.

Geologic Studies Before the Eruption

1. Geologists studied Mount St. Helens before the eruption. They mapped the volcano and its surroundings and constructed a geologic cross section through the mountain. This cross section shows that, before the eruption, the volcano consisted of interlayered lava flows, pyroclastic rocks, and mudflows, and had a domelike central conduit. In other words, it was a typical composite volcano. ▶

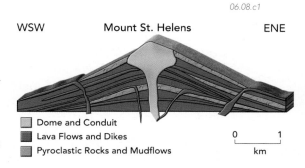

06.08.c1

WSW   Mount St. Helens   ENE

☐ Dome and Conduit
■ Lava Flows and Dikes
■ Pyroclastic Rocks and Mudflows

0   1
km

2. The geologists mapping and studying the area determined that the volcano's eruptive history during the last 40,000 years included pyroclastic flows, tephra falls, mudflows, lava flows, and dome building. The geologists also recognized evidence for horizontal blasts of ash, one of the first places where this threat was recognized.

A Volcano Awakens–Precursors to the Eruption

06.08.c2

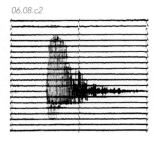

3. In March 1980, Mount St. Helens began to shake from earthquakes that geologists interpreted to be caused by magma moving beneath the mountain. These moderate quakes, including the one recorded on this seismogram (◀), were the signal to geologists that something was going to happen at Mount St. Helens.

▶ **4.** In April 1980, a bulge formed on the north side of the mountain and then continued to grow. Geologists inferred that the bulge was caused by upward-moving magma, and they recognized that the bulge was dangerous and unstable. They monitored its growth carefully.

06.08.c3

"Vancouver! Vancouver! This is it!"

06.08.c4

06.08.c5

5. David Johnston, the 30-year-old USGS geologist who spoke these words, was monitoring Mount St. Helens from a nearby ridge at the moment of the eruption. His observation post was considered to have low risk provided that the volcano erupted out the top as expected. At the time, available data suggested that an eruption was not imminent. His last recorded scientific observation, spoken at the catastrophic start of the eruption, was undeniably correct. This was it!

6. On May 18, 1980, at 8:32 a.m., an earthquake triggered a massive avalanche, as the bulge slid off the north side of the mountain. As this sequence of images shows (◀), the lowering of pressure on the magma caused a lateral blast and an upward growth of an eruption column that spread ash over several states. Pyroclastic flows moved horizontally and ravaged the landscape near the volcano. In addition to David Johnston, 56 other people died in the eruption, mainly from asphyxiation by hot gas and ash.

06.08.c7

7. Pyroclastic flows and other eruptions continued on Mount St. Helens until 1986. After a lull of several years, the volcano built, destroyed, and rebuilt several new domes before activity subsided in 2008.

Before You Leave This Page Be Able To

- ☑ Describe the type of deadly eruption that occurred at Vesuvius.

- ☑ Discuss the eruption at Mount Pelée and events that preceded it.

- ☑ Discuss the eruption of Mount St. Helens, and the data that warned of its dangers.

6.8

06.08.c6

6.9 How Do Volcanic Domes Form?

MANY VOLCANIC AREAS CONTAIN DOME-SHAPED HILLS called *volcanic domes*. The domes form when viscous lava mounds up above and around a vent. When domes collapse, they sometimes release deadly *pyroclastic flows* that rush downhill at hundreds of kilometers an hour. Volcanic domes form distinctive rocks and features in the landscape.

A What Are Some Characteristics of a Volcanic Dome?

Some volcanic domes have a nearly symmetrical dome shape, but most have a more irregular shape because some parts of the dome have grown more than other parts or because one side of the dome has collapsed downhill. Domes may be hundreds of meters high and one or several kilometers across, but they can be much smaller, too.

06.09.a1

◀ This rubble-covered dome formed near the end of the 1912 eruption in the Valley of Ten Thousand Smokes in Alaska. Volcanic domes commonly have this type of rubbly appearance because their outer surface consists of angular blocks as small as several centimeters across to as large as houses. The blocks form when solidified lava fractures as it is pushed from below, and when pieces and blocks slide down steep slopes on the side of a dome.

▶ Most domes do not form in isolation but occur in clusters or in association with another type of volcano. Domes can form within the craters of composite volcanoes, like these within the crater in Mount St. Helens, or within large calderas. In composite volcanoes and calderas, domes commonly are minor eruptions of viscous magmas that remain after a major eruptive event (e.g., the explosion of Mount St. Helens).

06.09.a2

B How Are Volcanic Domes Formed and Destroyed?

Domes form as viscous lava reaches the surface, flows a short distance, and solidifies near the vent. Domes can grow in two different ways—from the inside or from the outside. Domes can also be destroyed in two different ways—collapse or explosion.

Growth of a Dome

Domes mostly grow from the inside as magma injects into the interior of the dome. This new material causes the dome to expand upward and outward, fracturing the partially solidified outer crust of the dome. This process creates the blocks of rubbly, solidified lava that coat the outside of the dome.

Domes can also grow as magma breaks through to the surface and flows outward as thick, slow-moving lava. As the magma advances, the front of the flow cools, solidifies, and can collapse into angular blocks and ash.

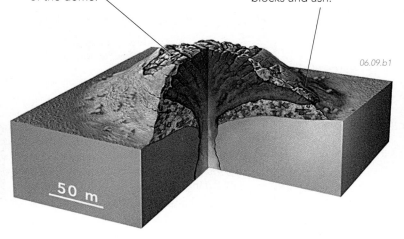

06.09.b1

50 m

Collapse or Destruction of a Dome

Domes can be partially destroyed when steep flanks of the dome collapse and break into a jumble of blocks and ash that flow downhill as small-scale pyroclastic flows. ▶

06.09.b2

06.09.b3

◀ Domes can also be destroyed by explosions originating within the dome. These typically occur when magma solidifies in the conduit and traps gases that build up until the pressure can no longer be held.

C What Types of Rocks and Landscapes Characterize Domes?

Most volcanic domes consist of andesite, rhyolite, or rocks with a composition between andesite and rhyolite. They are distinctive features when they form and harden, and even after they have been partially eroded. They consist of solidified lava that has several different textures, and typically they are associated with pyroclastic rocks and other debris that formed when the dome partially collapsed or was blown apart.

06.09.c1 Torfa, Iceland

Rock Types

◀ Some parts of domes cool rapidly into volcanic glass (*obsidian*) which, although dark, has a felsic composition. Obsidian can be almost entirely glass or can contain vesicles, crystals, and pumice. This example has layers, called *flow bands*, formed by shearing and other processes during flow.

06.09.c2 Wickenburg, AZ

▶ The outer parts of domes cool, solidify, and fracture into angular blocks that can become incorporated into the magma to produce *volcanic breccia*. Such breccias vary from containing mostly blocks to being mostly intact lava, with some blocks.

◀ Obsidian and other volcanic glasses are unstable and over time begin to change from unordered glass into rhyolite composed of very small crystals. The conversion, when not complete, creates a mottled rock with lighter colored rhyolite and darker areas that are still partially glass.

▶ When a volcanic dome collapses, avalanches of rock and other debris can rush downhill in a pyroclastic flow of blocks and ash. The resulting deposits are *tuff* or *volcanic breccia* consisting of pieces of the dome in an ash-rich matrix.

06.09.c3 Wickenburg, AZ

06.09.c4 Flagstaff, AZ

Expression in the Landscape

◀ Some domes are intermediate between a simple dome shape and a lava flow with lobes that spread out from the magmatic conduit. This flow-dome formed 1,300 years ago and so has a relatively uneroded shape and contains unaltered obsidian.

▶ This peak is the remnant of a dome that has been extensively eroded. Over time, the glass has converted to finely crystalline rhyolite.

06.09.c5 Newberry Mtns., OR

06.09.c6 Castle Mtns. NV

Deadly Collapse of a Dome at Mount Unzen, Japan

Mount Unzen towers above a small city in southern Japan. The top of the mountain contains a steep volcanic dome that formed and collapsed repeatedly between 1990 and 1995. The collapsing domes unleashed more than 10,000 small pyroclastic flows (top photograph) toward the city below. In 1991, the opportunity to observe and film these small pyroclastic flows attracted volcanologists and other onlookers to the mountain. Unfortunately, partial collapse of the dome caused a pyroclastic flow larger than had occurred previously. This larger flow killed 43 journalists and volcanologists and left a path of destruction through the valley (lower photograph). Note that damage was concentrated along valleys that drain the mountain.

06.09.mtb1

06.09.mtb2

Before You Leave This Page Be Able To

☑ Describe the characteristics of a volcanic dome.

☑ Explain or sketch the two ways by which a volcanic dome can grow.

☑ Explain or sketch how a volcanic dome can collapse or be destroyed by an explosion.

☑ Describe the types of rocks associated with volcanic domes.

☑ Describe how you might recognize a volcanic dome in the landscape.

6.9

6.10 Why Does a Caldera Form?

CALDERA ERUPTIONS ARE AMONG NATURE'S most violent phenomena. They can spread volcanic ash over huge areas, and the largest erupt more than one thousand cubic kilometers of magma. As the magma withdraws from the magma chamber, the roof of the chamber collapses to form a depression tens of kilometers across. The depression may then fill with ash, lava flows, and sediment.

A What Is a Caldera?

A *caldera* is a large, basin-shaped volcanic depression, which typically has a low central part surrounded by a topographic escarpment, referred to as the *wall* of the caldera. The Valles Caldera of New Mexico nicely illustrates the important features of a caldera. It is relatively young and uneroded because it formed only two million years ago, and its subsurface has been explored by drilling and other geologic studies.

1. This image shows satellite data superimposed on topography. The circular Valles Caldera contains a central depression, about 22 km (14 mi) across, surrounded by steep walls.

2. The caldera formed when a huge volume of magma erupted from a shallow magma chamber, producing a large eruption column and pyroclastic flows.

3. As shown in this cross section, the caldera contains a series of faulted blocks that have been downdropped relative to rocks outside the caldera. Faulting and ground subsidence occurred at the same time as the main eruption of tephra. As a result, thicker amounts of tephra (now consolidated into welded tuff and shown in light maroon) were trapped within the caldera than accumulated outside the caldera.

4. Small, rounded *rhyolite domes* formed within the caldera after the main pyroclastic eruption. The domes were fed by fissures that tapped leftover magma. Some of this magma has solidified at depth, forming granite, but some may still be molten.

06.10.a1

Valles Caldera

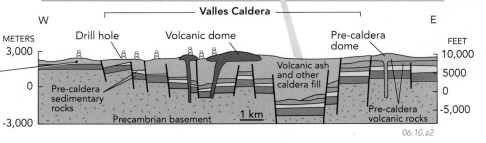

06.10.a2

Crater Lake and the Eruption of Mount Mazama

06.10.mtb1

Crater Lake National Park contains a spectacular caldera, now filled by a beautiful deep-blue lake. The caldera formed about 7,700 years ago when a huge eruption and associated collapse destroyed the top of a large composite volcano called *Mount Mazama*. The eruption was more than 50 times larger than the 1980 eruption of Mount St. Helens. As the magma erupted in a huge eruption column, the roof of the magma chamber subsided, forming the main crater (caldera). Ash from this eruption can be traced all the way to southern Canada, 1,200 km (750 mi) from its source. A small scoria cone, named Wizard Island, grew on the floor of the caldera after the main explosive eruption. The beauty and serenity of Crater Lake today seem incompatible with its fiery, cataclysmic origin 7,700 years ago.

Before You Leave These Pages Be Able To

☑ Describe or sketch the characteristics of a caldera, including its geometry in the subsurface.

☑ Sketch and explain the stages in the formation of a caldera.

☑ Describe how you could recognize a recent caldera and an ancient caldera.

☑ List the kinds of rocks you might find in a caldera from top to bottom.

☑ Explain the formation of Crater Lake.

B How Does a Caldera Form?

The formation of a caldera and the associated eruption occur simultaneously—the caldera subsides in response to rapid removal of magma from the underlying chamber. The largest caldera eruptions produced volcanic ash layers more than 1,000 m thick.

Formation of a Caldera

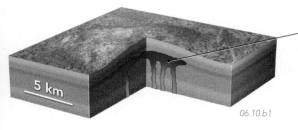

5 km

06.10.b1

1. The first stage is the generation, mostly by crustal melting, of felsic magma. The magma rises and accumulates in one or more chambers that can be kilometers thick and tens of kilometers across. The chamber or chambers may be within several kilometers of the surface.

06.10.b2

2. Next, magma reaches the surface and eruptions begin. As the magma chamber loses material, the roof of the chamber subsides to occupy the space that is being vacated. Curved fractures allow crustal blocks to drop, outlining the edges of the caldera and providing many conduits for magma to reach the surface.

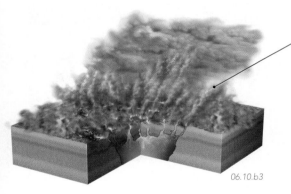

06.10.b3

3. The erupting felsic magma forms eruption columns and pyroclastic flows. Much of the tephra falls back into the caldera, creating a thick pile of ash, which later becomes tuff. Landslides from steep caldera walls produce large blocks and clasts that become part of the caldera deposit. Some ash and other tephra escapes the caldera and covers surrounding areas in thinner layers of tuff.

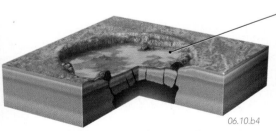

06.10.b4

4. As the eruption subsides, magma rises through fissures along the edges or in the interior of the caldera, erupting on the surface to make volcanic domes. If the caldera remains a closed depression, it may become a lake or may become filled with later sedimentary and volcanic deposits.

06.10.b8

5. Calderas that formed recently still have a clear expression in the landscape, but ancient examples are more difficult to recognize because they may have been eroded, covered by other rocks, or disrupted by faulting or other events. Long Valley caldera (◄), near the eastern border of California, formed 760,000 years ago and has a central depression containing hills that formed as domes. Part of the caldera wall forms the mountain front on the left.

A Section Through a Caldera

The rocks below are arranged from top to bottom in the order they occur in a caldera.

06.10.b5 Long Valley, CA

The upper parts of a caldera typically contain tuff formed when pyroclastic eruptions emptied the underlying magma chamber. Pyroclastic deposits that came form the Long Valley caldera (▲), called the *Bishop Tuff*, spread over much of the western United States.

06.10.b6 Tucson Mtns., AZ

Lower down in the caldera, ash and pumice may become strongly welded into a hard, compact rock, commonly with fragments of older rocks from the caldera walls.

06.10.b7 Silverbell, AZ

At the lowest levels in the caldera, finely crystalline granite and porphyry represents the crystallized magma chamber.

6.11 What Disasters Were Related to Calderas?

CALDERA ERUPTIONS ARE AMONG THE MOST LETHAL natural disasters. Evidence of their past destruction is recorded by geology and by historical accounts in many parts of the world. These pages explore two such ancient disasters, Thera and Krakatau, and one possible future disaster.

A Did the Eruption of a Caldera Destroy a Civilization Near Greece?

Santorini, east of the Greek mainland, is a group of volcanic islands with geology that records a major caldera collapse 3,500 years ago, about the time of the collapse of the Minoan civilization on the island of Crete.

06.11.a1

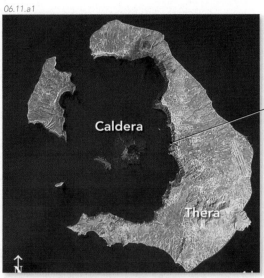

◄ **1.** This satellite view shows the islands of Santorini, including *Thera*, the largest island. Thera and the other islands encircle a submerged caldera that formed when the center of a larger volcanic island collapsed, leaving the modern islands as remnants. The steep cliffs around the caldera are eroded segments of the original wall of the caldera. The curving cliffs (►) expose volcanic layers, products of explosive volcanism that began one to two million years ago. Islands in the middle of the caldera were constructed by more recent eruptions.

06.11.a2

2. The main eruption produced an ash column perhaps 40 km (25 mi) high, followed by pyroclastic flows. The erupted ash buried towns, now excavated (►), with up to 50 m of pumice and ash. Caldera collapse occurred as the eruption of ash emptied a large magma chamber.

3. The collapse of the caldera evidently unleashed a large destructive wave that traveled southward across the sea, probably helping lead to the downfall of the Minoan civilization on the island of Crete. This destruction of the civilization on Santorini and collapse of the volcanic island into the sea may have started legends about the sinking of a land mass and city (Atlantis) into the sea.

06.11.a3

B What Happened During the Deadly Eruption of Krakatau?

One of the largest historic volcanic eruptions struck the Indonesian Island of Krakatau in 1883. A large eruption formed a caldera, destroying several islands. The eruption and caldera collapse unleashed pyroclastic flows and huge waves that spread out across the sea, resulting in the deaths of almost 40,000 people. One explosion was heard thousands of kilometers away and is the loudest sound in recorded history!

06.11.b1　　　　　06.11.b2　　　　　06.11.b3

1. Before 1883, the area contained three islands, the largest of which, *Krakatau*, was made of three volcanoes. The region was densely populated, and many people lived along the coast of these and neighboring islands. Both factors contributed to the heavy death toll from the 1883 eruption.

2. When the eruption began, massive amounts of magma erupted from a magma chamber beneath the islands, forming a high eruption column and pyroclastic flows. The eruption was accompanied by huge explosions, landslides, caldera collapse, and destruction of two of the volcanoes and nearly half of another. Large waves struck ships and adjacent coasts.

3. After the eruption, only part of Krakatau remained. In 1927, a small volcano began to grow within the caldera, forming a new island called *Anak Krakatau* (child of Krakatau). Today, Indonesia is densely populated, with more than 150 volcanoes in a curved line across Sumatra, Java, and smaller islands. It has very high risk of future deadly eruptions.

C Could the Yellowstone Caldera Cause a Future Disaster?

Yellowstone is one of the world's largest active volcanic areas. Abundant geysers, hot springs, and other hydrothermal activity are leftovers from its recent volcanic history. During the last two million years, the Yellowstone region experienced three huge, caldera-forming eruptions. What is the possibility that Yellowstone could erupt again and rain destructive ash over the Rocky Mountains and onto the Great Plains?

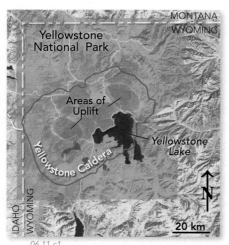

06.11.c1

◄ **1.** This image shows the outline of the youngest Yellowstone caldera, which formed 640,000 years ago. The boundaries of the caldera have been partially obscured by erosion, deposition of sediment, and lava flows that erupted after the caldera formed. Several areas within the caldera have been experiencing uplift.

▶ **2.** Ash from the three Yellowstone eruptions was carried by the wind and deposited over a huge area that extends from northern Mexico to southern Canada and as far east as the Mississippi River. A repeat of such an eruption could devastate the region around Yellowstone and cause extensive crop loss in the farmlands of the Great Plains and Midwest.

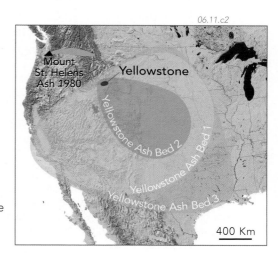

06.11.c2

▼ **3.** A cross section of the youngest caldera in Yellowstone shows associated geysers, earthquakes, and magma. In this model, upwelling mafic magmas melted overlying crust, yielding felsic magma responsible for the large caldera eruptions.

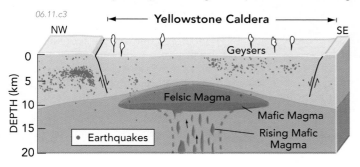

06.11.c3

▶ **4.** To study hazards posed by the caldera, geologists use satellite-based radar to precisely measure ground movements over time. Colors on this computer-generated image show a bull's-eye of very recent uplift within the caldera, perhaps indicating that magma is rising beneath the surface and could erupt in the future.

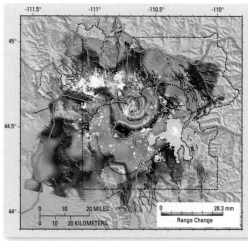

06.11.c4

The Yellowstone Hot Spot and Related Volcanic Features

Large caldera-forming eruptions in the Yellowstone region occurred three times during the last 2.1 million years: 2.1, 1.3, and 0.64 million years ago. The average time between eruptions is about 700,000 years. Because 640,000 years have passed since the last eruption, Yellowstone could perhaps erupt again in the next 100,000 years, maybe much sooner.

Where is all the magma coming from, and is melting still occurring beneath the region? The underlying cause of volcanism is interpreted to be a *hot spot* currently under Yellowstone. According to this model, as North America moved southwestward over the hot spot, the hot spot burned a path across southern Idaho, forming the mostly basaltic Snake River Plain. Ages of basaltic and felsic volcanic rocks on the Snake River Plain are youngest near Yellowstone and become older to the southwest, consistent with the movement of North America over the hot spot. The Columbia Plateau flood basalts of Washington and Oregon could have formed when the same mantle plume reached the lithosphere and spread to the north.

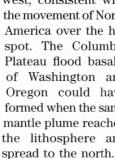

06.11.mtb1

Before You Leave This Page Be Able To

✓ Explain how a volcanic eruption destroyed Santorini.

✓ Describe what happened during the eruption of Krakatau.

✓ Describe the volcanic history of Yellowstone, including the distribution of volcanic ash.

✓ Describe how volcanism at Yellowstone could be related to a hot spot.

6.12 What Areas Have the Highest Potential for Volcanic Hazards?

IN SOME PLACES, THE RISK POSED by volcanic hazards is great. In others, it is inconsequential. Volcanic eruptions are more likely in Indonesia than in Nebraska. Additionally, different types of volcanoes have different eruptive styles, so some volcanoes are more dangerous than others. What factors should we consider when determining which areas are the most dangerous and which are the safest?

A How Do We Assess the Danger Posed by a Volcano?

Potential hazards of a volcano depend on the type of volcano, which we infer from its shape and rock types, and on its history. Examine the volcano below for clues about what type of volcano is present and how it might erupt.

06.12.a1 Augustine, AK

Shape—The shape of a volcano provides important clues about how dangerous the volcano might be. Volcanoes with steep slopes, such as composite volcanoes, are more dangerous because they form from potentially explosive, viscous magma and also are prone to landslides. Volcanoes that have relatively gentle slopes, like most shield volcanoes, result from less explosive basaltic eruptions.

Rock Type—The types of rocks on a volcano reflect the magma composition and style of eruption. If a volcano contains welded tuffs, it has erupted felsic pyroclastic flows. If it consists of rhyolite or andesite, it is more dangerous than a volcano composed of basalt. We can use chemical analyses to help classify the rocks and thereby assess a volcano's potential danger. The volcano shown above is a steep, composite volcano composed of andesitic lava and pyroclastic rocks.

Age and History—The age of a volcano is essential information. If the volcano has not erupted for a long time, maybe it is dormant. The shape of a volcano, especially whether it still has a fresh-looking volcano shape or has been eroded, is one indicator of a volcano's age. Important clues are also provided by a volcano's history, if recorded in historical records, including oral histories from nearby people. Isotopic measurements on volcanic units can provide an accurate indication of a volcano's age. Geologic studies of the sequence of volcanic layers, combined with isotopic ages, provide insight into how often eruptions recur. The volcano above clearly has recent activity, as expressed by the recent dark deposits and the steam and other gases escaping from the summit dome.

B What Areas Around a Volcano Have the Highest Risk?

Once we have determined the type of volcano that is present, we consider other factors that help identify which areas near the volcano have the highest potential risk.

06.12.b1

Village 1
Village 2
Village 3

1. *Proximity*—The biggest factor determining potential risk is proximity, closeness to the volcano. The most hazardous place is inside an active crater. The potential risk decreases with increasing distance away from the volcano.

2. *Valleys*—Lava flows, small pyroclastic flows, and mudflows are channeled into valleys carved into the volcano and surrounding areas. Such valleys are more dangerous than nearby ridges.

3. *Wind Direction*—Volcanic ash and pumice that are thrown from the volcano are carried farthest in the direction that the wind is blowing at the time of the eruption. Most regions have a prevailing wind direction, so a greater hazard of falling material exists in this direction from a volcano.

4. *Particulars*—Each volcano has its own peculiarities, and these influence which part of the volcano is most dangerous. Steeper parts of a volcano pose special risks, and one side of a volcano may contain a dome that could collapse and form pyroclastic flows. This image shows three small villages around a volcano. Is one village at greater risk than the others? Which one is in the least hazardous place, and what ideas led you to this conclusion?

C What Regions Have the Highest Risk for Volcanic Eruptions?

We can think on a broader regional scale about which regions are most dangerous. In North America, volcanoes are relatively common along the west coast and virtually absent east of the Rocky Mountains. *Tectonic setting*, especially proximity to certain types of plate boundaries, is the major factor making some places more prone to volcanic hazards than others. The map below shows locations of recently active volcanoes (orange triangles).

The largest concentration of composite volcanoes is along the Pacific Ring of Fire. The volcanoes form above subduction zones, in island arcs and in mountain ranges along continental margins. Some subduction-zone volcanoes erupt so vigorously that they form calderas.

Much fluid basaltic lava erupts on the seafloor at mid-ocean ridges. Such eruptions pose little risk to humans because almost all of these occur at the bottom of the ocean. The island of Iceland, where a mid-ocean ridge coincides with a hot spot, is an exception.

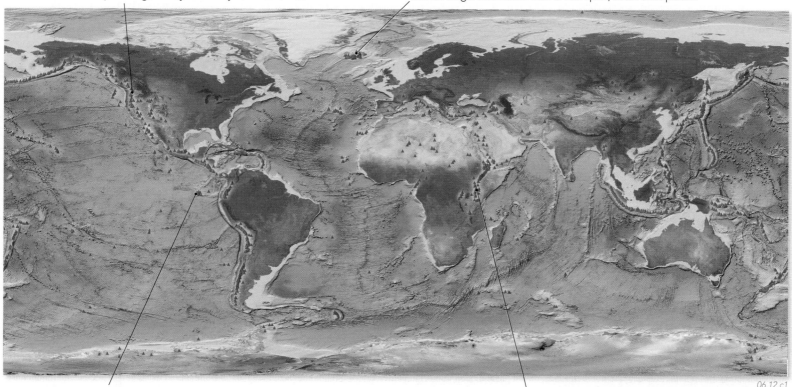

06.12.c1

Many shield volcanoes occur along lines of islands and submarine mountains in the Pacific and other oceans. Most of these linear island chains, and a few other clusters of islands, formed above hot spots. Hawaii and the Galápagos Islands are good examples. Shield volcanoes also occur in other settings, including on continents.

Some volcanic features, including basalt flows, scoria cones, and composite volcanoes, are in the middle of continents. Most of these form over hot spots or in continental rifts, like the East African Rift.

Forecasts, Policy, and Publicity

Predicting volcanic eruptions is currently an imprecise science. There have been some fabulous successes and some disappointing failures. Volcanologists have successfully predicted some eruptions by studying clusters of small earthquakes generated as magma rises through the crust, by measuring changes in the amount of gas released by volcanoes, and through other types of investigations. Some predictions (e.g., Mount St. Helens and Mount Pinatubo) have saved lives because government officials and citizens acted on the scientific evidence. Some predictions have been unsuccessful because an eruption that was considered pos-

sible or even likely did not occur. In other cases, predictions, policy, and publicity interacted in a bad way, with deadly results. In 1985, geologists working on Nevado del Ruiz, a composite volcano in the Colombian Andes, warned of an impending eruption. The city of Armero, with an estimated 29,000 inhabitants, lay in a valley that drained the steep volcano. Local government officials downplayed the risk and assured the citizens that there was no danger. A pyroclastic eruption occurred at night, melting snow and ice on the volcano and unleashing a mudflow that moved at hundreds of miles per hour, engulfing most of Armero and killing more than 20,000 people.

Before You Leave This Page Be Able To

☑ Summarize ways to assess the potential danger of a volcano based on its characteristics.

☑ Describe ways to identify which areas around a volcano have the highest potential hazard.

☑ Describe how the plate tectonic setting of a region influences its potential for volcanic hazards.

6.12

6.13 How Do We Monitor Volcanoes?

GEOLOGISTS MONITOR VOLCANOES using instruments that measure changes in topography, ground shaking, heat flow, gas output, and water chemistry. Any of these changes may indicate that an eruption is imminent. Some monitoring is by geologists in the field area, but much can be done remotely using computer-operated instruments that transmit data from a volcano to an office or laboratory.

A Can Seismic Activity Signal an Eruption?

As magma moves, expands, or contracts, it exerts force on the surrounding rocks and causes them to break, crack, or bend. This deformation causes *ground shaking*, which can be recorded with seismic instruments.

Seismometers are instruments that measure ground shaking, or *seismic activity*. Such activity accompanies movement of magma. Seismometers allow scientists to monitor changes and look for increases in seismic activity. At Mount Pinatubo, Philippines, seismic activity increased before a major eruption on June 15, 1991.

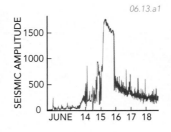
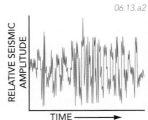

Magma flowing through conduits produces several distinctive types of seismic activity, including a fairly rhythmic, repeating pattern on seismic plots, like the one shown here. Such patterns often precede and accompany a volcanic eruption and so have been used to predict that an eruption is imminent.

B How Does Gas Output Change Before an Eruption?

Gases dissolved in a rising magma may come out of solution, expand, and provide the driving force for an eruption. An increased flow of gases from a volcano may indicate that magma is rising and losing its gas.

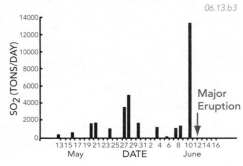

Sulfur dioxide (SO₂) gas emissions have increased just before some volcanic eruptions. Such increases, when integrated with seismic data and other information, may indicate that a volcanic eruption is likely.

Various ground and airplane-mounted instruments can measure the amount of SO₂ coming from a volcano. These instruments allow measurements of the gas flow from a volcano to be made from a safe distance.

This graph shows measured SO₂ emissions just before the 1991 eruption of Mount Pinatubo in the Philippines. The amount of gas coming from the volcano increased dramatically before the eruption.

C How Are Changes in Heat Flow Measured?

Movement of magma can warm parts of a volcano, indicating a potential eruption. Geologists can measure changes in temperature, and the amount of heat being emitted by a volcano, using special instruments. They can also gather this kind of information by satellites, which are especially useful for measuring temperatures and monitoring volcanoes in remote locations.

Steam eruptions on volcanoes happen when water or ice come in contact with hot magma. Heat from the steam and magma can be detected by instruments that measure thermal energy.

Geologists processed this satellite image to emphasize hot areas. Yellow and red mark magma and hot tephra. Bright blue marks cooler ash deposits and volcanic rocks.

D How Are Changes in a Volcano's Topography Monitored?

Before an eruption, the surface of a volcano may change shape by centimeters to hundreds of meters as magma inflates the mountain. Such changes in Earth's surface, called *ground deformation*, alert geologists to volcanic activity and can be measured by several methods.

06.13.d1 Hawaii

◄ **1.** *Global Positioning Satellite (GPS)* units are relatively small devices that use satellites to determine precise locations and elevations. If a volcano's surface deforms, even slightly, a GPS station can track its changing position.

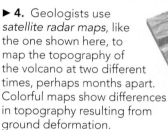
06.13.d2

◄ **3.** Special *surveying instruments* can accurately and precisely measure the distance between the instrument and a distant target.

▶ **2.** *Tiltmeters* are instruments used to determine whether a measuring station is being tilted in one way or another. This is one of the oldest methods for monitoring the inflation or deflation of a volcano in response to movement of magma.

06.13.d3 Mount St. Helens, WA

▶ **4.** Geologists use *satellite radar maps*, like the one shown here, to map the topography of the volcano at two different times, perhaps months apart. Colorful maps show differences in topography resulting from ground deformation.

06.13.d4

North Sister
South Sister
Broken Top
Mt. Bachelor
5 km
0   28 mm
RANGE CHANGE

E Can Mudflows Caused By Volcanoes Be Detected Remotely?

Many composite volcanoes are covered with snow and ice that will melt during an eruption and mix with loose volcanic material on the slopes of the volcano. This thick, muddy mixture, called a *mudflow* or a *lahar*, travels down river channels, destroys houses and bridges, and poses a threat to people downstream.

06.13.e1 Mount St. Helens, WA

◄ This mudflow, caused by the 1980 eruption of Mount St. Helens, destroyed a highway bridge. To detect such mudflows, geologists use a network of remote monitoring stations (▶) to detect the characteristic rumble made by the mudflows as they rush down valleys. This monitoring allows for rapid warning and evacuation of downstream communities. The 1985 mudlfow disaster near Nevado del Ruiz, Columbia, highlighted the need for such remotely activated monitoring systems.

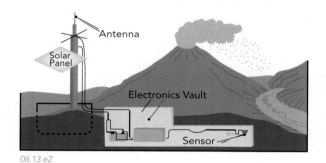

Antenna
Solar Panel
Electronics Vault
Sensor
06.13.e2

Monitoring a Volcano Using Various Approaches

Geologists use as many methods as possible to monitor a volcano and better understand its behavior and potential dangers. Such monitoring, therefore, requires a team of geologists, each with a different field of expertise (seismic records, volcanic gases, and others). Team members compare and discuss various data and observations to develop an interpretation that is consistent with all of the data.

Gathering data using satellites or other remote techniques is efficient and allows geologists to monitor multiple volcanoes at the same time. Collecting some data requires visits to the volcano, including characterization of the type of material produced by an eruption. When an eruption occurs, a geologist knows what to observe, what data to collect, and which area is least hazardous.

06.13.mtb1 Mount St. Helens, WA

Before You Leave This Page Be Able To

- ✓ Discuss why seismic measurements are helpful for predicting an eruption.
- ✓ Explain why and how volcanic gases and thermal energy are measured.
- ✓ Discuss how topographic changes on a volcano could precede an eruption and how they are measured.
- ✓ Briefly discuss how a volcano-related mudflow can be detected remotely.

6.13

6.14 What Volcanic Hazards Are Posed by Mount Rainier?

MOUNT RAINIER IS PART OF A CHAIN OF VOLCANOES above the Cascadia subduction zone of the Pacific Northwest. What kind of volcano is Mount Rainier, how did it form, when and how did it last erupt, and what risks does it pose to people living in the valleys below? Mount Rainier provides an opportunity to examine some important aspects of volcanoes, connecting eruption styles, tectonic setting, and potential hazards.

A What Kind of Volcano Is Mount Rainier?

Mount Rainier rises ominously above the city of Tacoma (▶). The steep, symmetrical shape of the mountain identifies it as a dangerous composite volcano. A composite volcano plus a city equals high risk.

06.14.a2

06.14.a1

▲ This image shows the position of Mount Rainier and the suburbs of Tacoma. The top of the volcano is covered by glacial ice and snow. River valleys, only one of which is labeled, begin on the flanks of the volcano and continue into the suburbs. These provide a pathway from the volcano to the people.

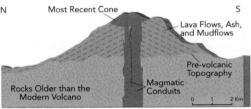
06.14.a3

▲ A geologic cross section of Mount Rainier shows that the andesitic composite volcano was built on an eroded surface of granitic rocks and was fed by a pipelike magmatic conduit. The top of the mountain, largely covered by ice, is a younger volcanic cone that was constructed within an older crater.

B What Is the Plate-Tectonic Setting of Mount Rainier?

Mount Rainier is one of the volcanoes that cap the Cascade Range. The Cascade volcanoes exist because of melting associated with the Cascadia subduction zone, which is an ocean-continent convergent boundary.

The large composite volcanoes of the Cascades are related to a plate boundary between the North American plate to the east and the small Juan de Fuca plate to the west. The Juan de Fuca plate is moving eastward with respect to North America and subducting into the mantle, beneath the edge of the continent.

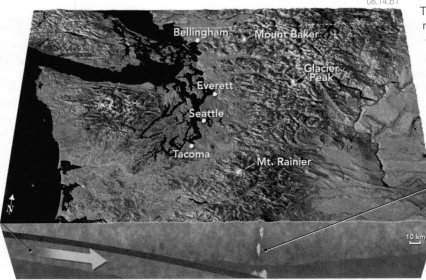
06.14.b1

The Cascade Range is a north-south belt of mountains and is capped by snow-covered composite volcanoes, including the three labeled here. Additional Cascade volcanoes are in Canada to the north and Oregon and California to the south. Mount Rainier and the other Cascade volcanoes have historically erupted mostly viscous, intermediate (andesitic) magmas that form thick, slow-moving lava flows, domes, and explosive pyroclastic materials.

Magma forms by partial melting of mantle above the subduction zone and rises to interact with overlying continental crust. The result is intermediate-composition magma and dangerous composite volcanoes.

C   What Hazards Does Mount Rainier Pose to the Surrounding Area?

Mount Rainier is considered to be a very dangerous volcano. It has had at least eleven significant pyroclastic eruptions in the last 10,000 years. The most recent occurred in 1820. According to the U.S. Geological Survey, mudflows from Mount Rainier constitute the greatest volcanic hazard in the Cascade Range.

1. This large figure shows hazard zones for lava flows, pyroclastic flows, and mudflows. The green zone around Mount Rainier has the highest hazard for lavas and pyroclastic flows. The yellow and orange colors show the potential for mudflows of different sizes and different recurrence intervals (how often, on average, they occur). Yellow is used for least frequent but large mudflows, whereas reddish orange is used for more frequent but small mudflows, especially near the volcano.

2. The hazards from lava flows are mostly near the volcano. Small explosive eruptions also have the most impact close to the summit, but pyroclastic flows can travel tens of kilometers away from the summit, in part following valleys. During a major eruption, a large eruption column could spread ash and pumice across the region. Prevailing winds would probably spread the ash to the east but could blow in any direction depending on the weather conditions.

06.14.c2   Explanation

- Small mudflow event, not necessarily associated with volcanism
- Moderate-size mudflow event
- Large mudflow event
- Lava flows and pyroclastic flows

4. Mudflows could flow northwest all the way into Tacoma and its suburbs. Houses have been built directly in potential mudflow paths and even on top of a huge mudflow that occurred only 600 years ago. Besides eruptions, some of these mudflows were caused by avalanches of rock and ice unrelated to volcanic activity. The risk posed by mudflows is very great! Would you buy one of these houses? How would you know if a house was at risk? Hazard maps like the one above are a good place to start.

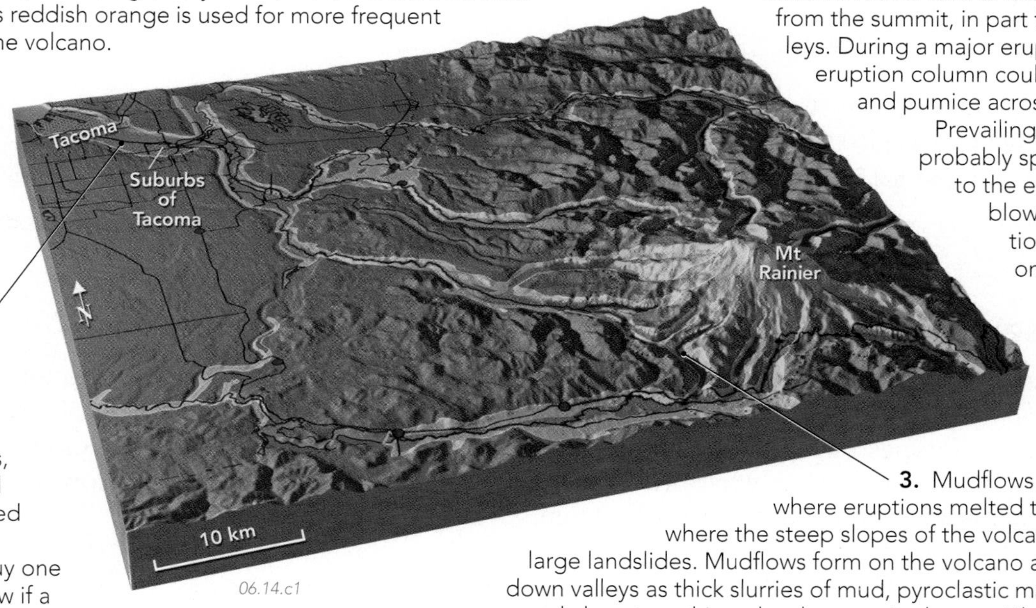

06.14.c1

3. Mudflows have formed where eruptions melted the ice cap or where the steep slopes of the volcano produced large landslides. Mudflows form on the volcano and then flow down valleys as thick slurries of mud, pyroclastic material, rocks, and almost anything else that gets in the way. The hazards are greatest close to the volcano and in valleys that drain the volcano.

Recent Eruptions of Volcanoes in the Cascade Range

Mount Rainier and Mount St. Helens are only two of the dangerous volcanoes in the Cascades. Ten other volcanoes in the U.S. part of the Cascades erupted during the last 4,000 years. Cascade volcanoes also continue a short distance into Canada.

This figure (▶) shows the locations of the large Cascade volcanoes and when they erupted during the last 4,000 years. Although Mount St. Helens is the most active of the Cascade volcanoes, Glacier Peak, Medicine Lake, and Mount Shasta have each erupted six or more times during the last 4,000 years. Seven of the volcanoes, including Mount Rainier, have

erupted during the last 200 years. Nearly all of these volcanoes are dangerous composite volcanoes, so living on the flanks or in the valleys below one of these volcanoes carries

the risk of mudflows, ash falls, and even pyroclastic flows. All of the volcanoes and the associated hazards exist because the Juan de Fuca plate is subducting beneath the continental crust of North America.

06.14.mtb1

Cascade Eruptions During the Past 4,000 Years

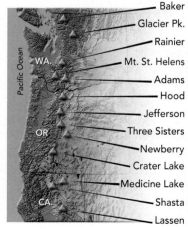
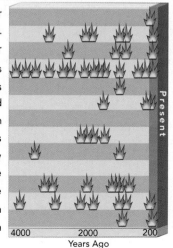

Baker
Glacier Pk.
Rainier
Mt. St. Helens
Adams
Hood
Jefferson
Three Sisters
Newberry
Crater Lake
Medicine Lake
Shasta
Lassen

WA.
OR
CA.
Pacific Ocean
Present

4000   2000   200
Years Ago

Before You Leave This Page Be Able To

- ✓ Describe the type of volcano that Mount Rainier represents.
- ✓ Explain the plate-tectonic setting of Mount Rainier.
- ✓ Discuss the volcanic hazards near Mount Rainier.
- ✓ Briefly summarize how active the Cascade volcanoes have been during the last 4,000 years.

6.14

6.15 How Would You Assess Hazards on This Volcano?

DECIDING WHERE TO LIVE requires careful consideration. An overriding factor is whether a place is safe. In this exercise, you will investigate a volcanic island to determine what types of eruptions have occurred in the past, to assess the volcanic hazards, and to find the least dangerous place on the island to live.

Goals of This Exercise:

- Observe the physical characteristics of the volcano and the rock types it contains.
- Use your observations to determine what type of volcano is present and how it would likely erupt.
- Assess the potential for volcanic hazards in different parts of the island and determine the least dangerous place to live.

A Observing the Characteristics of the Volcano

The study of a volcano begins by observing its physical characteristics, such as its size, shape, steepness of slopes, locations of ridges and valleys, and any unusual topographic features. The next step is to observe the types of rocks that are present, determine the aerial distribution of each rock type, and interpret how and in what order each rock formed. Follow the steps below, and record your answers from each step in the accompanying worksheet or online.

1. Observe the image to the right. Record any important characteristics of the volcano on the worksheet or in your notes.

2. Observe the photographs of rock samples and describe each rock's key attributes, such as whether it contains fragments. Use these attributes to identify the rock types (basalt, rhyolite, tuff, for example) by comparing the photographs and descriptions to those in chapters 5 and 6. Alternatively, your instructor may provide hand specimens of the rocks for you to observe and identify.

3. Use your rock identifications to infer the style of eruption by which each rock formed.

B Assessing the Volcanic Hazards of the Island

Assess the general volcanic hazards of the volcano, and then assess the relative hazards of each part of the island compared to the others. Using your hazard assessments, determine the most dangerous places and the relatively least dangerous place to live.

1. Consider your rock identifications in the context of the topography and geologic features in the areas where the samples were collected. From these combined data, interpret what types of volcanic features, such as craters and domes, are present in different parts of the island. A newspaper account of previous eruptions (bottom of next page) provides some useful clues.

2. Assess how each volcanic feature contributes to the hazard potential in different parts of the island. On the map in the worksheet, draw boundaries around and label those areas that have high, medium, or low hazard potential compared to the rest of the island. The differences between the three hazard zones will be fairly subjective. Use your best judgement and be consistent.

3. From your investigations, identify the areas you interpret to be the most dangerous and the least dangerous places to live. When choosing between two sites that are equally safe, you may consider other factors, such as the scenery, whether the sites are level enough to build on safely, and whether they are subject to storms, landslides, floods, and other natural hazards.

▲ *Rock 1:* Dark gray, glassy igneous rock with small vesicles and some bands produced by flow when the rock was molten. A chemical analysis indicates a felsic composition.

▲ *Rock 2:* Hard, igneous rock that contains flattened pieces of pumice and small crystals of quartz and feldspar. It does not seem to be a recently formed rock.

▲ *Rock 3:* Unit consisting of angular pieces of a light- to dark-gray igneous rock in a matrix of powdery volcanic ash and smaller rock pieces. Many of the dark pieces are glassy and banded and contain scattered vesicles. The volcanic deposit has baked (heated up) the underlying soil.

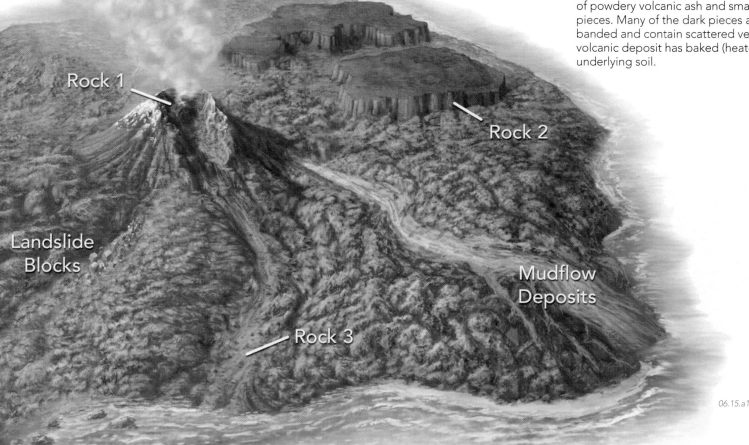

Rock 1

Rock 2

Landslide Blocks

Mudflow Deposits

Rock 3

The following is a newspaper account:

Volcano Erupts!

The *Juanannita volcano* began erupting in early September of 1952, and dozens of small eruptions have occurred since that time. For 10 years before the 1952 eruption, residents of the area observed plumes of white steam rising from the summit of the crater. In the summer of 1952, local inhabitants reported an increase in the output of steam, an increased smell of sulfur, and a series of small earthquakes.

The first eruption was a single explosive burst that lasted about three hours and that was accompanied by clouds of ash that rose kilometers into the air. Heavy ash fell around the volcano, and a light dusting of ash was reported on adjacent islands up to 20 kilometers away. The eruption melted snow and ice high on the crater, forming a mudflow that moved along stream channels and inundated many areas in valleys downstream from the volcano. After the main eruption, a lava dome started growing in the crater.

All subsequent eruptions have been smaller and of a different style. They have been similar to one another. In each eruption, a cloud of ash and rocks moves rapidly downhill and is mostly restricted to stream channels. After each eruption, geologists noted that one side of the dome in the crater had collapsed into a pile of ash and rocks.

CHAPTER 7

Sedimentary Environments

MOST OF EARTH'S EXPOSED SURFACE IS COVERED by sediment and sedimentary rocks. Sediment produced by weathering and erosion is transported by moving water, ice, and wind, and deposited in a variety of environments, ranging from dry deserts to the bottoms of deep oceans. The loose sediment hardens, or *lithifies*, into sedimentary rock. Sediment and sedimentary rock dominate many of Earth's landscapes and are important hosts of energy, mineral, and water resources.

This image shows satellite data and topography for the Coast Range and the Fraser River Valley of southwestern British Columbia. The river enters the Strait of Georgia, an inlet of the Pacific Ocean. The large lavender area along the river includes the cities of Vancouver, British Columbia and neighboring Bellingham, Washington.

Most valleys in these mountains were originally carved by glaciers but now contain lakes and steep mountain streams. The streams are eroding into the mountains and transporting sediment toward the Fraser River.

Where does sediment come from and how is it transported?

Large lakes occupy many of the valleys. The lake waters come mostly from streams and rivers draining the mountains next to the lakes.

What happens to sediment being carried by a river when the river enters a lake?

07.00.a1

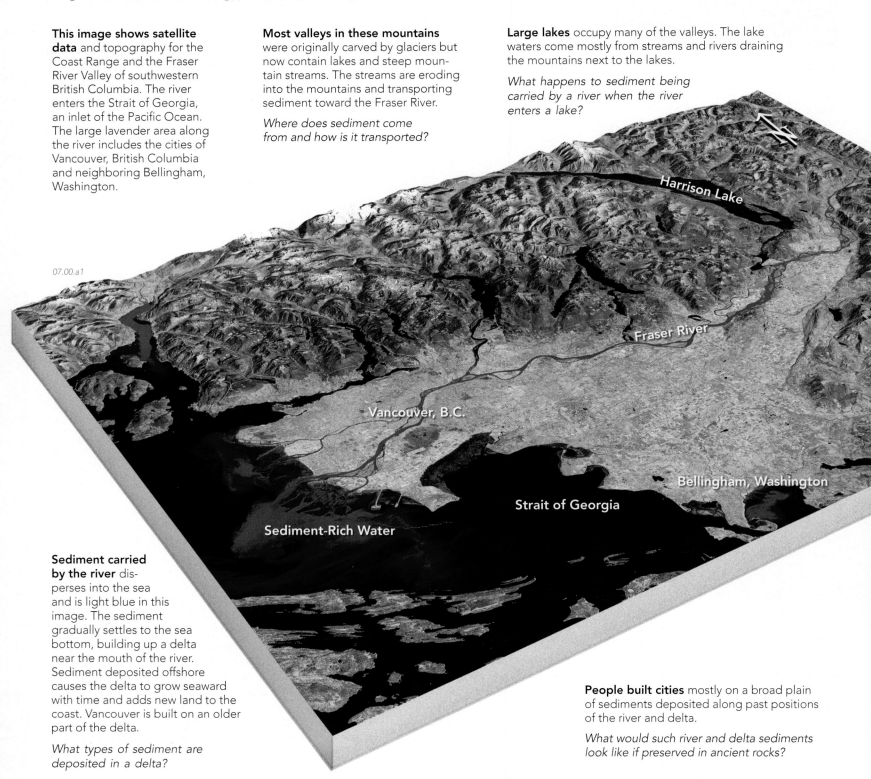

Harrison Lake

Fraser River

Vancouver, B.C.

Bellingham, Washington

Strait of Georgia

Sediment-Rich Water

Sediment carried by the river disperses into the sea and is light blue in this image. The sediment gradually settles to the sea bottom, building up a delta near the mouth of the river. Sediment deposited offshore causes the delta to grow seaward with time and adds new land to the coast. Vancouver is built on an older part of the delta.

What types of sediment are deposited in a delta?

People built cities mostly on a broad plain of sediments deposited along past positions of the river and delta.

What would such river and delta sediments look like if preserved in ancient rocks?

TOPICS IN THIS CHAPTER

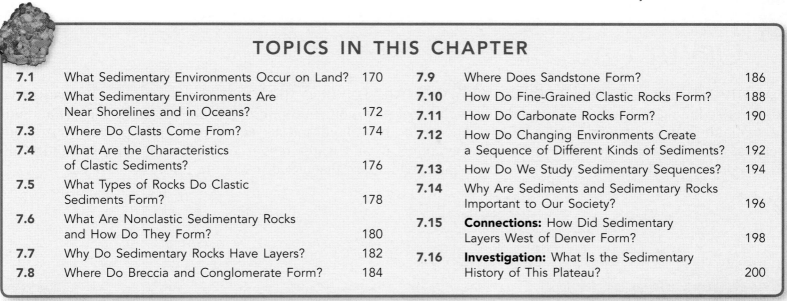

07.00.a2

Mount Baker

20 Kilometers

▲ **Mount Baker, a snow-capped volcanic peak,** rises above the surrounding landscapes of northern Washington.

What types of sediment are deposited near steep mountains, including those covered by ice and snow?

07.00.a3

In this closer view looking upstream of Vancouver (▶), the Fraser River has multiple channels that branch apart and rejoin, producing a distinctly *braided* appearance.

What types of sediment do rivers deposit, and do all rivers deposit the same types of sediment?

The Fraser River

The Fraser River, one of Canada's largest rivers, winds through mountains and a broad valley on its journey to the Pacific Ocean. Tributaries of the river are fed by lakes, seasonal rains, springs, and melting snow and glaciers. Rain and snowfall are heavy because of the region's coastal location and prevailing winds that bring in moisture-laden ocean air from the west.

The physical and chemical processes of *weathering* act on near-surface materials and loosen pieces of rock to produce *sediment*. Loose pieces of rock and soil can be eroded and transported downhill by streams, rivers, glaciers, and the force of gravity. The Fraser River deposits this sediment on river banks, in a broad plain near the river's mouth, and in the delta formed where the river meets the sea.

How does this sediment affect people and businesses in Vancouver? The location of the city largely coincides with the broad, sediment-covered valley and delta. The sediment forms fertile farmlands and provides a relatively flat surface on which to build. In the subsurface, the sediment contains groundwater that supplements the large supply of water provided by the Fraser River and surrounding lakes. This chapter explores how sediment forms, how it is transported and deposited, and the types of rock that sediment becomes. It also examines sedimentary rocks in the landscape, how we study these rocks, and why sediments and sedimentary rocks are important to society.

7.0

7.1 What Sedimentary Environments Occur on Land?

EARTH'S LAND INCLUDES DIFFERENT SEDIMENTARY ENVIRONMENTS, each characterized by distinctive kinds of sediment. The environments differ because of variations in topography, local geology, and the amount of available water. Examining these environments, both on land and at sea, helps us interpret modern landscapes, ancient sedimentary rocks, and energy, mineral, and water resources. These two pages identify the most common sedimentary environments on land, which are discussed in detail on following pages.

07.01.a2 South Fork, CO

Mountain environments are characterized by steep slopes developed on bedrock. Many mountains, but not all, also have high elevation. Erosion is vigorous on such steep slopes and provides abundant sediment, such as the large angular blocks in this photograph (◀). Once it is produced, the sediment can be transported out of the mountains and into other settings.

07.01.a1

Glaciers

Mountains

Braided River

Sand Dunes

07.01.a3 Denali NP, AK

Streams and rivers in mountains typically have steep gradients and are confined by bedrock canyons. As streams and rivers leave the mountains, they can develop a *braided* appearance defined by channels that split apart and rejoin (▲). We use the adjective *fluvial* to refer to the processes and sediment of streams and rivers.

In dry climates, wind picks up and moves sand grains and finer particles. The moving grains form fields of *sand dunes* (▶), which in some places stretch for hundreds of kilometers.

07.01.a4 Namibia, Africa

07.01.a5 French Alps

In high mountains or at high latitudes (close to the North or South Pole), snow can accumulate faster than it is removed by melting or other processes. Over time, the snow becomes compacted into ice, which may flow downhill as a *glacier* (◄). As glaciers move, they erode underlying materials and carry sediment away. The sediment and water are released upon melting of the ice, mostly at the *terminus* (end) of the glacier.

▶ Rivers that flow over gentle terrain commonly meander gracefully. Most rivers are flanked by relatively flat land that may be covered when the river floods (a flood-plain). *Floodplains* of meandering rivers are built, layer upon layer, by mud and sand carried by floodwaters.

07.01.a6

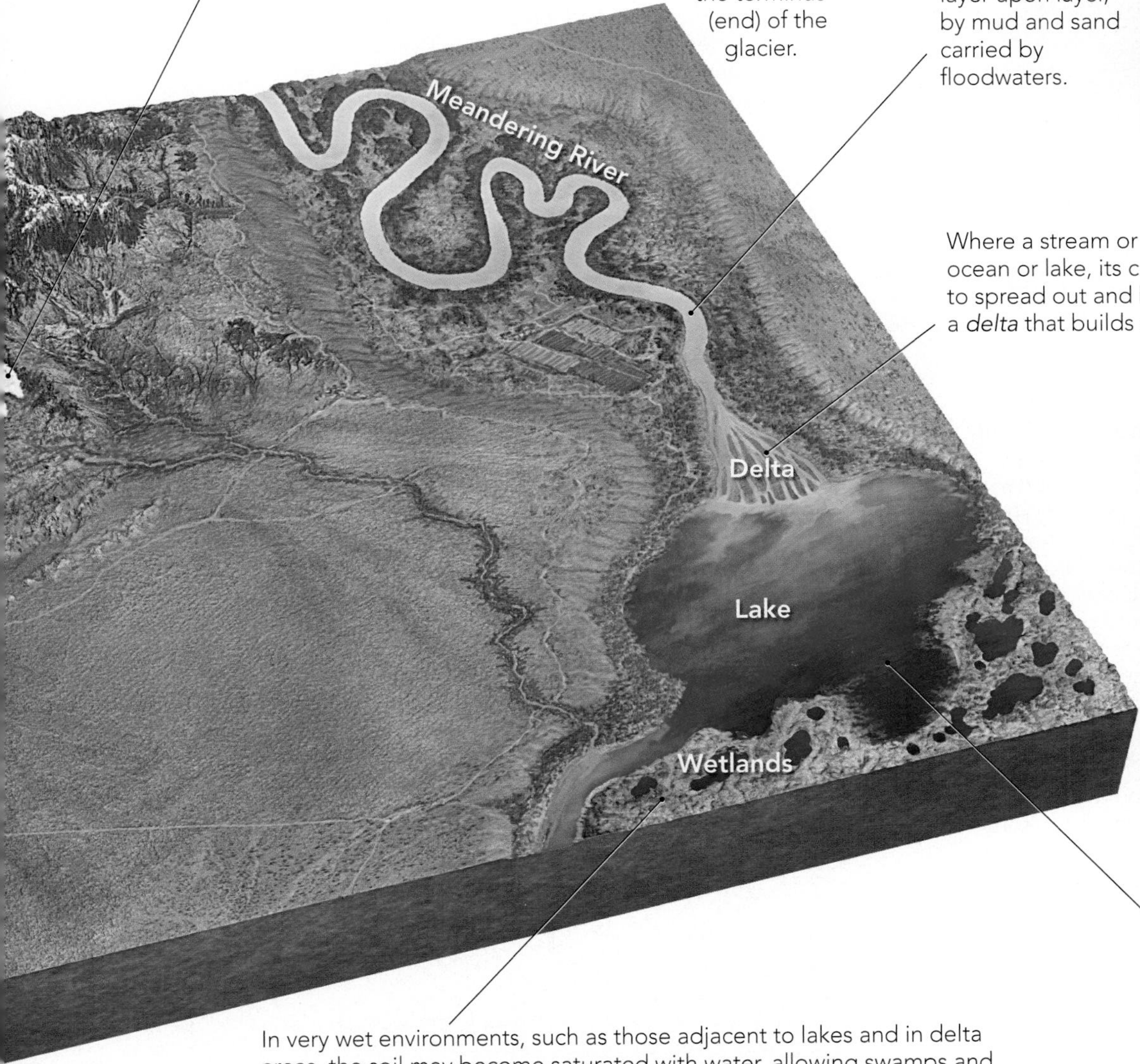

Where a stream or river enters a standing body of water, such as an ocean or lake, its current slows, which causes most of its sediment to spread out and be deposited. The sediment piles up and forms a *delta* that builds out into the ocean or lake.

07.01.a7 Lake Superior, MI

Meandering River

Delta

Lake

Wetlands

▲ Lakes contain a range of environments, from quiet, deep water in the center, to more active water with wind-driven waves along the shoreline. Some lakes are always filled with water, but others dry completely when the water evaporates or when it seeps into underlying materials.

In very wet environments, such as those adjacent to lakes and in delta areas, the soil may become saturated with water, allowing swamps and ponds to form (▼). Such wetlands typically have abundant vegetation, which may become an important component of the sediment.

07.01.a8 Florida

Before You Leave This Page Be Able To

✓ Sketch or describe the main sedimentary environments on land, and describe some characteristics of each.

7.1

7.2 What Sedimentary Environments Are Near Shorelines and in Oceans?

OCEANS AND THEIR SHORELINES are dynamic environments with wind, waves, ocean currents, and sediment eroded from the coastline or brought in from elsewhere. The characteristics of each environment, especially the types of sediment, depend mostly on the proximity to shore, the availability of sediment, and the depth, temperature, and clarity of the water.

07.02.a2

◄ *Beaches* are stretches of coastline along which sediment has accumulated. Most beaches consist of sand, pieces of shell, and rounded gravel, cobbles, or boulders. The setting determines which of these components is most abundant. Some shorelines have bedrock all the way to the ocean and so have little or no beach.

07.02.a1

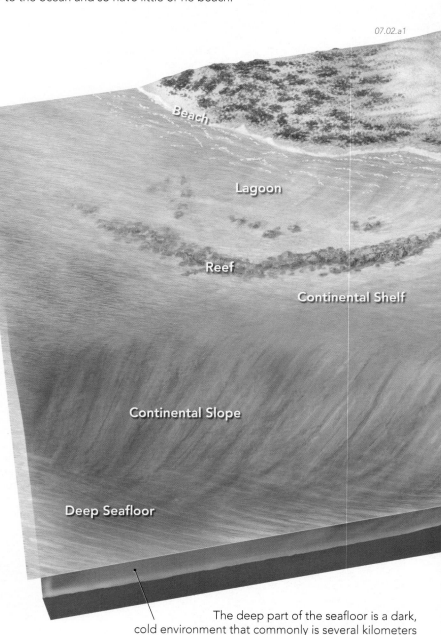

Beach

Lagoon

Reef

Continental Shelf

Continental Slope

Deep Seafloor

07.02.a3 Boracay, Philippines

◄ The water near the shoreline may be sheltered by a reef or islands. The sheltered water, called a *lagoon*, is commonly shallow, quiet, and perhaps warm. The near-shore parts of lagoons contain sand and mud derived from land, whereas the outer parts may have sand and pieces of coral eroded from a reef.

07.02.a4 Red Sea, Egypt

Where ocean water is shallow, warm, and clear, coral and other marine creatures construct *reefs* (◄), which can parallel the coast, encircle islands, or form irregular mounds and platforms. Reefs typically protect the shoreline from the energetic, big waves of the deeper ocean.

Away from the shoreline, many landmasses are flanked by continental shelves and slopes consisting of layers of mud, sand, and carbonate minerals. Material from these sites can move down the slope in landslides or in turbulent, flowing masses of sand, mud, and water, called *turbidity currents*. The slopes of some continents are incised by branching *submarine canyons* (not shown here) that funnel sediment toward deeper waters.

The deep part of the seafloor is a dark, cold environment that commonly is several kilometers beneath the surface. It generally receives less sediment than areas closer to land, and its sediment is dominated by fine, windblown dust and by remains of mostly single-celled marine organisms.

07.02.a5

◄ Sandy dunes that are inland from beaches are called *coastal dunes*. These dunes commonly form where sand and finer sediment from the beach are blown or washed inland and reshaped by the wind.

► Some shorelines include low areas, called *tidal flats*, that are flooded by the seas during high tide but exposed to the air during low tide. Most tidal flats are covered by mud and sand or are rocky. Some low parts of the land adjacent to tidal flats can accumulate salt and other *evaporite minerals* as seawater and terrestrial waters evaporate under hot, arid (dry) conditions.

07.02.a6 Rocky Point, Mexico

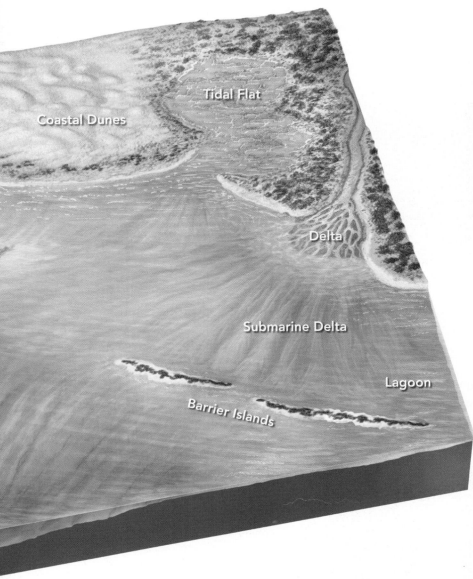

Coastal Dunes

Tidal Flat

Delta

Submarine Delta

Barrier Islands

Lagoon

07.02.a7 Mississippi Delta, LA

▲ In addition to the parts of deltas overlapping the shore, *submarine deltas* extend in some places for tens of kilometers offshore. The muddy or sandy front of the delta may be unstable and material can slide or tumble down the slope, sending sediment into deeper water.

Other accumulations of sand rise above the shallow coastal waters as long, narrow islands, called *barrier islands*. Most barrier islands, such as the one below, are only hundreds of meters wide. The areas between barrier islands and the shoreline are commonly shallow lagoons or saltwater marshes. ▼

07.02.a8 Santa Rosa Island, FL

Before You Leave This Page Be Able To

☑ Sketch and describe the main sedimentary environments in oceanic and near-shore environments.

7.2

7.3 Where Do Clasts Come From?

SEDIMENTARY ROCKS CONSIST OF MATERIALS that came mostly from other locations. Most sediment is pieces of other rocks, or *clasts*, formed by weathering and transport. Other sediment is extracted from water when dissolved material is *precipitated* by chemical reactions or by coral and other aquatic creatures.

A How Do Physical and Chemical Weathering Produce Sediment?

Most sediment forms by *weathering,* which involves physical and chemical processes that attack rocks on or near Earth's surface, loosening pieces and dissolving some material. Different sizes, shapes, and types of sediment form depending on the material that is weathered and the conditions during weathering. The processes of physical and chemical weathering are summarized below and are discussed in detail in a later chapter.

Physical Weathering

Physical weathering is the physical breaking apart of rocks that are exposed to the environment. There are four major causes of physical weathering.

Near-Surface Fracturing—Many processes on or near the surface break rock into smaller pieces. These include fracturing caused by rocks pulling away from a steep cliff. Fractures also result when rocks expand as they are uplifted toward the surface and are progressively exposed to less pressure.

Frost and Mineral Wedging—Rocks can be broken as water freezes and expands in fractures. When the ice melts, the fractured pieces may become dislodged from the bedrock. Crystals of salt and other minerals that grow in thin fractures can also cause rocks to break apart.

Thermal Expansion—Rocks are heated by wildfires and by the sun during the day. As rocks heat up, they expand, often irregularly, and may crack. This process probably plays a relatively minor role in weathering, and geologists currently are debating its importance.

Biological Activity—Roots can grow downward into fractures and pry rocks apart as the root diameter increases. Burrowing animals can transport rock and soil from depth and move it to the surface where it is exposed to the elements, weathered, and eroded.

07.03.a1

Chemical Weathering

Chemical weathering includes several types of chemical reactions that affect a rock by breaking down minerals, causing new minerals to form, or by removing soluble material from the rock. Chemical weathering attacks both solid rock and loose rock fragments, and produces ions in solution, loose grains and other pieces, and a covering of soil.

Dissolution—Some minerals are soluble in water, especially the weakly acidic waters that are common in nature. These minerals, along with the rocks, sediment, and soil that contain them, can dissolve. The dissolved material may be carried away in rivers, streams, or groundwater, or be used locally by plants.

Oxidation—Some minerals, especially those containing iron, are unstable when exposed to Earth's atmosphere. These minerals can combine with oxygen to form oxide minerals, such as iron oxides, which compose the reddish and yellowish material that forms when metal rusts.

Hydrolysis—When silicate minerals are exposed to water, especially water that is somewhat acidic, the water reacts chemically with the minerals. This process commonly converts the original materials to clay minerals, and produces leftover dissolved material that is carried away by the water. Hydrolysis is responsible for the formation of many clay-rich soils.

Biological Reactions—Decaying plants produce acids that can attack rocks, and some bacteria consume certain parts of rocks. These biological processes cause minerals to break down into their constituent elements.

07.03.a2

B Why Are There So Many Different Products of Weathering?

Physical and chemical weathering affect various starting materials with different compositions, grain size, and solubility. The two kinds of weathering work with each other, but one or the other may dominate, depending on the climate and other conditions. The interplay results in various sizes and compositions of sediment.

Type of Material	Different Parts of Material	Importance of Fracturing

07.03.b1 San Juan River, UT

07.03.b2 South Park, CO

07.03.b3 Echo Cliffs, AZ

Rocks, sediment, and soils exposed on Earth's surface have various compositions, which affects how they respond to physical and chemical processes. Materials that are soluble in water and weak acids commonly weather to a pitted or grooved appearance, as displayed by the weathered limestone above. The water dissolves the soluble material and carries it away as dissolved ions, which can later be precipitated in minerals. Other rocks, especially quartz-rich ones, are much less soluble during weathering.

Many rocks, sediment, and soils contain more than one mineral, and each mineral reacts differently to weathering. Pink and cream-colored feldspar crystals in the weathered granite shown here chemically weather to clay minerals or physically weather to sand grains. In general, quartz crystals physically weather to sand grains, whereas dark, mafic minerals chemically weather into clay and iron oxides. As shown in this photograph, the grain size, texture, and other aspects of a rock can affect how it weathers.

Rocks that are fractured are weaker and more easily weathered than rocks that are intact. Fracturing increases the amount of surface area that is exposed to the environment. Fractures permit water, air, and organisms to invade the rock, which causes more chemical and physical weathering. Parts of a rock unit that are more fractured will weather faster than less fractured parts. As a result, highly fractured areas tend to form low parts of the topography, or may form linear notches across ridges and hills.

C How Do Transportation and Erosion Affect Sediment?

Once weathering has loosened pieces of rock, the pieces can be transported by rivers, glaciers, waves, wind, and other forces. During transportation, larger clasts are broken and abraded to produce smaller ones.

07.03.c1 San Juan River, UT

07.03.c2 Adirondack Mtns., NY

◄ Silt, sand, and larger clasts carried in water can cause abrasion of other clasts in a channel or of bedrock along the channel. This process is akin to sandpapering, and so it smooths rough edges, scours pits and recesses into bedrock, and removes small pieces, which become additional sediment.

▲ As large clasts, such as boulders and cobbles, are transported, they can break when they collide with, or grind against, other large clasts. Through this process, boulders can become rounded cobbles, and cobbles can break down into smaller pebbles. Some clasts end up as sand or even smaller grains.

Before You Leave This Page Be Able To

☑ Describe the main processes of physical and chemical weathering.

☑ Describe how the type of material and degree of fracturing influence the type of sediment that results.

☑ Describe how rocks can be broken during transport.

7.3

7.4 What Are the Characteristics of Clastic Sediments?

SEDIMENT CONSISTS OF LOOSE FRAGMENTS of rocks and minerals, or *clasts*. When *clastic* sediment becomes sedimentary rock, the name assigned to the rock depends on the size and shape of the clasts. Other characteristics of the clasts, such as sorting of clasts, can be used to further describe the resulting rock.

A How Are Clastic Sediments Classified?

We primarily use the sizes of clasts to classify loose particles of sediment and the resulting sedimentary rocks into which they are lithified. In addition to clast size, we use the shapes of clasts to further characterize sediment.

Sediment Name	Size Range (millimeters)	Particle Name
Gravel	larger than 256 mm	Boulder
	64 to 256 mm	Cobble
	4 to 64 mm	Pebble
	2 to 4 mm	Granule
Sand	1/16 to 2 mm	Sand
Mud	1/256 to 1/16 mm	Silt
	less than 1/256 mm	Clay

Sizes of Clasts

1. The largest clasts are *boulders*, which are more than 0.25 m long. Note the rock hammer near the center. ▶

2. *Cobbles* are smaller than boulders, being about the size of a softball. The next smaller sediment size is a *pebble*, which has a size that can be held comfortably in one hand. A pebble can be a little longer than 6 cm. This pen rests on a cobble, and the smaller nearby stones are pebbles. ▶

3. *Sand* is smaller than 2 mm and can have a coarse-, medium-, or fine-grain size. *Medium sand* has grains between 1/4 and 1/2 mm in diameter, and *fine sand* has diameters smaller than 1/10 mm. In these photographic enlargements, the left image is coarse sand, the middle one is medium-grained sand, and the right one is fine-grained sand. ▶

07.04.a1 Baja California, Mexico

07.04.a3 Salt River, AZ

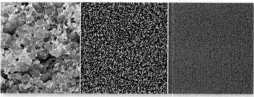
07.04.a4-6

07.04.a2 Salt River, AZ

4. *Clay* and *silt* are particles finer than sand; together they compose *mud* (◀). The term clay refers both to grain size and to a group of minerals. Besides being very small, clay grains are usually made of clay minerals. Clay minerals commonly feel slippery between fingertips. In contrast, silt, which is generally made of quartz and is coarser than clay, feels gritty.

Shapes of Clasts

07.04.a7 Wolf Creek Pass, CO

◀ *Angular clasts* have sharp corners and edges. They can be blocky, triangular, or shaped like chips or plates with angular edges.

07.04.a8 Fairplay, CO

▶ Many clasts have an angular shape with edges and corners that have been partially rounded.

07.04.a9 San Juan Islands, WA

◀ *Rounded clasts* have smooth, curved surfaces and shapes like eggs, flattened balloons, or objects that are nearly spheres.

Before You Leave These Pages Be Able To

✓ Sketch and describe how sediments are classified according to size, sorting, and shape of clasts.

✓ Describe how clast transport affects the size, shape, and sorting of clasts.

✓ Explain four factors that influence the type of sediment transported.

Amount of Sorting

07.04.a10 Wickenburg, AZ

07.04.a11 Ios, Greece

07.04.a12

Sorting describes the size range of clasts in sediment. *Poorly sorted* sediment, like this example, contains a wide range of clast sizes.

Many sediments have moderate sorting of clasts, perhaps containing sand and small pebbles, or perhaps silt and clay.

Well-sorted sediment consists of clasts that all have the same size. Sand on dunes (shown here) is typically well-sorted sand.

B   What Controls the Size, Shape, and Sorting of Clasts?

How do clasts of such different sizes and shapes form? Once they are eroded from a bedrock source, clasts are transported by water, wind, or ice. During transport, the tumbling, collisions, and abrasion may reduce clast size, increase roundness, and sort the clasts by size. The distance that a clast has been transported is therefore a key factor.

Bedrock exposed in mountains or along cliffs breaks off to form blocks the size of boulders and cobbles. Clasts near their source are usually large, angular, and poorly sorted.

As boulders and cobbles are transported by streams, their sharp corners break off because they are the most exposed and weakest parts of a clast. The clasts become more rounded and smaller as pieces break off.

Far from their bedrock source, the clasts are worn into well-rounded pebbles and sand grains. River currents, beach waves, and wind separate clasts by size, eventually producing better-sorted sediments. Coarse materials are slowly left behind, and only the smaller clasts are carried far from the sediment source.

07.04.b1

07.04.b2 San Juan Mtns., CO

Steepness of Slope— Steep slopes, such as the one shown here, commonly have clasts that are larger, more angular, and more poorly sorted than the clasts observed on gentle (less steep) slopes.

07.04.b4

Sediment Supply—A river, beach, or other agent of transport can only move the sediment that is available. Bedrock on this beach provides large boulders that must be worn down into smaller clasts.

07.04.b3 Cascade Creek, CO

Strength of Current—Strong, turbulent river currents and ocean waves can move large clasts, such as these meter-wide boulders, but slow, less turbulent currents move only fine-grained sediment.

07.04.b5 Morocco

Agents of Transport— Wind can carry only sand and finer particles, but rivers, glaciers, mudflows, and other agents of transport can pick up and carry large clasts. These dunes consist of nothing but well-sorted sand because wind cannot bring larger clasts to the area, and smaller material is blown away.

7.4

7.5 What Types of Rocks Do Clastic Sediments Form?

MANY SEDIMENTARY ROCKS HAVE FAMILIAR NAMES, for example, *sandstone*. Sediment is converted into sedimentary rock through a process called *lithification* that involves *compaction* by overlying sediment and *cementation* by calcium carbonate or other materials, ultimately producing clastic sedimentary rock.

A What Are Some Common Clastic Sedimentary Rocks?

We describe and classify clastic sedimentary rocks based primarily on their clast size. Other features, including grain composition and roundness, are also important for making distinctions. Common clastic sedimentary rocks are shown below and described using these criteria of classification.

07.05.a1 Roosevelt Dam, AZ

07.05.a2 Wickenburg, AZ

Gravel-Sized Clasts

◄ *Conglomerate* has rounded pebbles, cobbles, or boulders with sand and other fine sediment between the large clasts. This conglomerate has well-rounded cobbles in a matrix of mostly quartz sand.

▶ *Breccia* is similar to conglomerate except that the clasts are angular. Breccia usually has a jumbled appearance because most is poorly sorted. This example has angular boulders and smaller clasts in a mud-rich matrix.

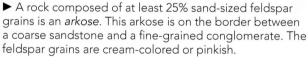
07.05.a3 Durango, CO

07.05.a4 Engineer Mtn., CO

Sand-Sized Clasts

◄ *Sandstone*, consisting of sand-sized grains, generally has better-defined layers than conglomerate or breccia. There are different types of sandstone, as shown in the following photographs.

▶ A rock composed of at least 25% sand-sized feldspar grains is an *arkose*. This arkose is on the border between a coarse sandstone and a fine-grained conglomerate. The feldspar grains are cream-colored or pinkish.

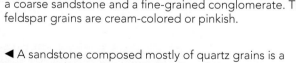
07.05.a5 Holbrook, AZ

07.05.a6 Petrified Forest, AZ

◄ A sandstone composed mostly of quartz grains is a *quartz sandstone*. To accumulate mostly quartz grains, either the source was already quartz rich, like another sandstone, or weathering and transport destroyed grains of weaker minerals.

▶ Some sandstones, called *graywacke*, contain grains of several different compositions. They typically include quartz, feldspar, iron-oxide minerals, mica minerals, and small fragments of other rocks.

07.05.a7 Holbrook, AZ

07.05.a8 Durango, CO

Mud-Sized Clasts

◄ *Siltstone* consists of silt-sized particles, generally quartz, and is commonly in thin to medium-thick layers.

▶ *Shale* consists mostly of very fine-grained clay minerals. The minerals are aligned, so the rock breaks in sheetlike pieces or chips, as displayed here.

B How Do Clastic Sediments Become Clastic Sedimentary Rock?

Compaction

As sediment is buried beneath more sediment or other materials, increasing pressure pushes clasts together, the process of *compaction*. Compaction forces out excess water and causes sediments to lose up to 40% of their volume or more. Originally loose sediment becomes more dense and more compact.

Sand Grains—Sediments near Earth's surface, such as these sand grains, are a loose collection of clasts. The grains rest on one another but do not fit together tightly, so spaces, called *pore spaces*, exist between the grains. Pore spaces are generally filled with air and water.

As sand grains are buried, the weight of overlying sediment forces the grains closer together. The amount of pore space decreases as air or water is expelled, so the layer of sand loses thickness. The amount of compaction is slightly exaggerated in this image.

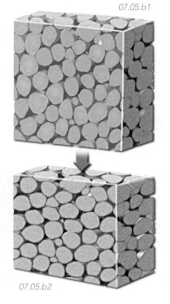
07.05.b1

07.05.b2

Clay Particles—Clay particles, much smaller than sand grains and mostly made of platy clay minerals, may have random orientations when first deposited. The particles prop each other up, preserving abundant open pore space. In this figure, the open space is shown filled with water (blue). When clay-rich sediment is compressed, as for example when we step in mud, the amount of open space decreases, expelling some water.

As clay particles compact from above, they rotate into similar, near-horizontal orientations. Such compaction decreases the size and number of pores, and decreases the thickness of the clay layer. Clay can compact to half its original thickness.

07.05.b3

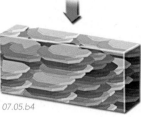
07.05.b4

Cementation

Even after sediment is compacted, adjacent clasts do not fit together perfectly and some openings remain. These pore spaces are commonly filled with water containing dissolved materials. The dissolved materials can precipitate to form minerals that act as a natural *cement* that holds the pieces of sediment together.

When sand grains and other sediment are deposited, and even after they are compacted, abundant pore spaces exist between the grains. These spaces are typically interconnected, which allows water to flow slowly through the sediment, carrying chemical components into or out of the sediment.

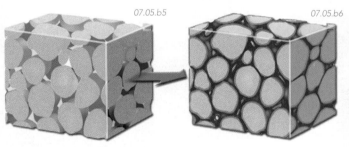
07.05.b5   07.05.b6

As the sediment is buried, minerals can precipitate from water moving through the pore spaces, coating the surfaces of the grains, sticking them together. Minerals that form in pore spaces, *cements*, decrease the amount of pore space, bind the grains together, and turn the sediment into hard sedimentary rock.

Types of Natural Cement

There are four main types of natural cements that hold grains together: calcite, silica, clay minerals, and iron oxides. Other materials, like gypsum, can function as cement but are less common.

Calcite ($CaCO_3$) is a common cement in sandstone and other sedimentary rocks. It holds grains together moderately well, but it is easily redissolved, so a calcite-cemented sandstone may become friable (crumbly).

Silica (SiO_2) acts as a cement in some sandstone and other sedimentary rocks. It

forms a strong cement that can tightly bind grains and form a tough, resistant rock.

Clay minerals can cement together larger grains, including sand. They may have been deposited with the sediment or formed from the alteration of feldspar or volcanic ash.

Iron oxide minerals, like hematite, precipitate from water as a natural cement between the grains. Iron oxide minerals commonly give sediment deposited on land a reddish color, as displayed in the spectacular red-rock landscapes of the Southwest.

Before You Leave This Page Be Able To

☑ Describe or diagram the classification of common clastic sedimentary rocks.

☑ Describe what happens to clastic sediment as it becomes buried and converted into rock.

☑ Describe the natural cements that are common in sedimentary rocks.

7.5

7.6 What Are Nonclastic Sedimentary Rocks and How Do They Form?

SOME SEDIMENTARY ROCKS ARE NOT COMPOSED OF CLASTS and are therefore *nonclastic rocks*. Some *nonclastic rocks* form when dissolved chemicals precipitate from water, as for example when water evaporates. Other nonclastic rocks are extracted directly from the water by coral and other organisms as they make shells and other hard parts, and these are called *biochemical rocks*, to signify the importance of both biological and chemical processes.

A What Types of Sediment Can Precipitate by Water or Be Deposited by Organisms?

Processes capable of producing nonclastic sedimentary rocks may be purely chemical, biological, or have both chemical and biological aspects. The photographs below pair up a modern environment with a sedimentary rock that forms in this type of environment.

Water in oceans and lakes contains dissolved ions, such as sodium, calcium, and chloride. These substances can be left behind as salts or as carbonate minerals if the water *evaporates*.

07.06.a1 Death Valley, CA

When lake water or seawater evaporates, salts and other minerals precipitate, perhaps as a crust along the bottom or edges of the water body. The precipitated materials are called *evaporite minerals* and include the mineral halite (rock salt), pictured here.

07.06.a2

Various types of chemical reactions deposit minerals, including when chemical-rich water is heated or cooled or when two different types of water, such as groundwater and seawater, mix.

07.06.a3 Havasu, AZ

Springs and creeks may contain high amounts of dissolved calcium and can deposit layered calcite in the form of travertine, as shown in this photograph.

07.06.a4 Havasu, AZ

07.06.a5 Red Sea, Egypt

Coral and other creatures, including abundant microscopic organisms, remove chemicals, such as calcium and silica, from water to produce their shells and other structures.

07.06.a6 San Juan River, UT

When cemented together, shells and coral skeletons form a variety of rock called *limestone*, made mostly of calcium carbonate. This limestone contains many easily seen fossils.

07.06.a7

Plant debris can accumulate in great thicknesses in wetlands and bogs, where it partially or wholly decays and is compacted into sedimentary rock.

07.06.a8 Morgantown, WV

Coal, a carbon-rich rock, forms when trees, vines, and other plants die, are buried, and become lithified. The black layer in the photograph is coal.

B What Are Some Common Nonclastic Sedimentary Rocks?

07.06.b1 Death Valley, CA

07.06.b2

07.06.b3 Payson, AZ

1. *Rock salt* refers to halite (NaCl) or to a rock mostly composed of halite. Halite commonly precipitates from evaporation of water.

2. *Gypsum* refers to a mineral and a rock. Like halite, it mostly forms when seawater evaporates in tidal flats and narrow seas.

3. Some *limestone* forms from the calcium carbonate remains of animals, such as coral, clams, and stemmed animals called *crinoids*.

07.06.b4 Havasu, AZ

07.06.b5

07.06.b6 Durango, CO

4. Limestone also forms inorganically, by precipitation from water or through other chemical reactions. Some limestone forms through a combination of biological and inorganic processes.

5. *Chalk* is a soft, very fine-grained limestone that forms from the accumulation of the calcium carbonate remains of microscopic organisms that float in the sea. Chalk forms the famous White Cliffs of Dover, England.

6. If groundwater with dissolved magnesium encounters limestone, an exchange of magnesium in the water for calcium in the rock can change calcite into the mineral dolomite, forming a rock called *dolostone*.

07.06.b7 Battle Mtn., NV

07.06.b8 San Juan River, UT

07.06.b9 Sherman Mine, Canada

7. *Chert* is a silica-rich rock that forms in several ways. One way chert forms is in layers from the accumulation and compaction of tiny, silica-rich plankton shells that fall to the ocean bottom.

8. Chert also forms when seawater and groundwater mix, causing chert to precipitate. Such chert can form as irregular masses called *nodules*, as shown above, or as layers in limestone and dolostone.

9. *Iron formation* is a rock composed of centimeter-thick layers of iron oxide, iron carbonate, and iron silicate minerals, commonly with quartz. Most iron formations precipitated from seawater early in Earth's history.

07.06.b10

07.06.b11 Witbank, South Africa

10. *Peat* forms through the accumulation of plant material, usually in swampy environments. Peat is porous and retains much of the textural character of the original plant material.

11. *Coal* is peat that has been buried, compacted, and heated, losing most of its water and oxygen in the process. Depending on the amount of heat and pressure, coal can be soft and dull or hard and shiny.

Before You Leave This Page Be Able To

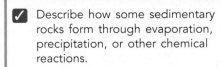

✓ Describe how some sedimentary rocks form through evaporation, precipitation, or other chemical reactions.

✓ Describe how limestone forms.

✓ Describe how some sedimentary rocks form from the accumulation of plant materials.

7.6

7.7 Why Do Sedimentary Rocks Have Layers?

MOST SEDIMENT IS DEPOSITED IN LAYERS. A sedimentary unit, such as a limestone or a sandstone, may be a single thick layer or may include many smaller layers called *beds*. What defines such layers, how do the layers form, and what do they tell us about the conditions that existed during deposition?

A What Types of Layers Do Sedimentary Rocks Contain?

Almost all sedimentary rocks contain layers. The layers vary greatly in thickness, lateral continuity, and characteristics that define the layers. Boundaries between layers also vary; some are quite sharp and others are more gradational.

Thickness of Layers

07.07.a1 Round Rock, AZ

◀ Many sedimentary terrains contain thick layers that have consistent colors and expression wherever they appear in the landscape. A rock unit that is distinct and laterally traceable, like this cliff-forming sandstone, can be called a *formation*.

▶ Sedimentary rocks commonly have layers that may be from millimeters to meters thick. These layers are referred to as *beds* or *bedding*.

07.07.a2 Green River, UT

Definition of Layers

07.07.a3 Durango, CO

◀ Layers, whether they are thick enough to be considered a formation or are very thin, may be distinct because of different grain sizes than adjacent layers. This rock has beds of tan sandstone (coarser grains) and dark-gray shale (finer grains).

▶ Adjacent beds may contrast different compositions, such as from a tan or cream-colored quartz-rich sandstone layer to a layer having more clay and red iron oxide. Layers in this photograph vary in composition from one bed to the next.

07.07.a4 Holbrook, AZ

Boundaries of Layers

07.07.a5 Ruby Canyon, CO

◀ The boundary between two layers can be very sharp. In this photograph, a slope-forming, thin-bedded sandstone partially covered with vegetation in the upper part of the photograph sharply overlies a thick layer of reddish-brown sandstone that forms a cliff. The cliff-forming sandstone changes downward into beds of sandstone and reddish siltstone.

▶ The boundaries between some layers are gradational, involving a gradual change of grain size, composition, thickness of individual beds, or all of these. In this photograph, a gray shale grades progressively upward into a tan sandstone, and then a brownish sandstone. Near the top of the shale, sandstone layers become progressively thicker and more numerous. The change records where a marine environment was replaced by a delta.

07.07.a6 Shiprock, NM

07.07.a7 Punchbowl, CA

◀ Some sharp contacts show that the underlying unit was cut into (*scoured*) by erosion before the overlying layer was deposited. Such scouring is common when rivers erode one layer before depositing another layer on top. This photograph shows a gray conglomerate whose base scours into an underlying brown sandstone.

B How Do Layers in Sedimentary Rocks Form?

07.07.b1 Central CA

07.07.b2 Punchbowl, AZ

07.07.b3 Southern NM

Discrete Event—Some individual layers, or boundaries between layers, mark a discrete event, like a flood. Each thin, light-colored layer in this photograph represents a single, rapid influx of sand onto a muddy seafloor.

Change in Current—The change from layer to layer may reflect a change in the strength or direction of the current that deposited the sediment. Coarse-grained sediment is deposited by strong, turbulent currents.

Sediment Supply—Some layers record a change in the type or amount of sediment being supplied. The tan, quartz-rich sandstone in this photograph is between layers of conglomerate that contains limestone clasts.

07.07.b4 Goosenecks, UT

07.07.b5 Vermilion Cliffs, AZ

Sea-Level Change—A global rise or fall in sea level can cause sedimentary environments to change. Here, sea level rose and fell many times, causing ledge-forming marine limestone to alternate with slope-forming layers of marine and nonmarine shale and siltstone.

Climate Change—Some boundaries between layers reflect regional or global changes of climate. The lower gray layers of rock here formed from clay deposited during a wet period, and the top layers formed from sands deposited in a desert.

Graded Beds

07.07.b6

Fill a jar with water and sediment of mixed size, and then shake the jar. The coarser material will settle first, followed by successively finer sediment, as shown here. Such variation, called a *graded bed*, forms if a strong current loses velocity and drops progressively finer sediment. The graded bed below contains pebbles in its lower part and grades upward into sand. The current slowed over time and deposited finer sediment over coarser sediment.

Cross Beds

07.07.b7

When sand and silt move over a dune or underwater ripple, grains accumulate in thin beds on the down-current side of the dune or ripple, as shown above. Such beds are at an angle to other beds in the same rock and so are called *cross beds*. Cross beds that are centimeters to more than a meter high form within rivers, deltas, and shorelines. Larger cross beds (▼) typically form in large sand dunes. Cross beds preserve the curved profile of the dune and the direction the wind was blowing when the sediment was deposited.

Parallel Beds

07.07.b8 Labyrinth Canyon, UT

Most beds are parallel. Parallel beds form under a variety of conditions and in many cases simply reflect the piling of one layer on another.

Before You Leave This Page Be Able To

☑ Sketch or describe the types of layers that sedimentary rocks contain, including how their thickness varies, what defines the layers, and whether their boundaries are sharp or gradational.

☑ Describe how layers, including graded beds and cross beds, form.

07.07.b9 Casitas Sespe, CA

07.07.b10 Sedona, AZ

7.7

7.8 Where Do Breccia and Conglomerate Form?

BRECCIA AND CONGLOMERATE are the coarsest kinds of clastic sedimentary rocks. Both include large clasts in a matrix of finer material, which may be sand, silt, and clay, or commonly a mixture of all three. *Breccia* contains angular clasts, whereas *conglomerate* has rounded clasts. What are the characteristics of these coarse-grained rocks, and what do their clast size and shape tell us about where these rocks form?

A What Are the Characteristics of Breccia, and Where Does Breccia Form?

The large clast sizes of breccia and conglomerate reflect deposition in an energetic environment, where large, heavy clasts could be picked up and moved. Clasts in breccia are angular, which suggests minimal transport. Longer transport would have rounded the corners and edges of the clasts, making conglomerate instead of breccia.

07.08.a1 Hieroglyphic Mtns., AZ

Characteristics

◀ Breccia is typically a jumble of large, angular clasts in a matrix of sand, mud, and pieces of rock. Breccia contains clasts having a wide range of sizes and so is poorly sorted.

▶ Most breccia forms massive layers, lenses, or wedges that are meters to tens of meters thick, or locally much thicker. There is typically little visible bedding within a single breccia unit, because many breccia units are deposited in sudden, chaotic events.

07.08.a2 Hopi Buttes, AZ

07.08.a3 Hieroglyphic Mtns., AZ

◀ Some breccia, like this one, has a muddy or sandy matrix. Other breccia has a matrix that has smaller angular clasts of the same rock type as the large clasts. In volcanically active areas, the matrix can contain volcanic ash and other rock fragments.

▶ Some breccia, usually breccia formed from landslide deposits, consists of thoroughly shattered rock. Some clasts may be shattered but still loosely held together depending on how strongly the rock was lithified.

07.08.a4 Artillery Mtns., AZ

07.08.a5 Grand Canyon, AZ

Environments of Formation

◀ Some breccia forms from thick slurries of mud and larger clasts that originate in mountains and flow down steep slopes and through canyons. The slurries are called *mudflows* or *debris flows*.

▶ Some breccias, especially those that consist of shattered, angular fragments of rock, represent some type of landslide. The rocks shatter as they travel downhill, collide with one another, and shake apart. The angular fragments shown here are part of a *rockslide*.

07.08.a6 Gros Ventre, WY

07.08.a7 Death Valley, CA

◀ Breccia also forms if steep mountain streams deposit sediment along the mountain front. These deposits can form fan-shaped piles of sediment in an *alluvial fan*.

▶ Glaciers pick up loose clasts or grind them into fine dust. They commonly deposit the combination of large clasts and fine sediment as an unsorted breccia with only minor rounding of clasts.

07.08.a8 Southeastern AK

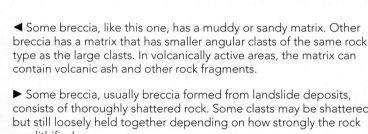

B What Are the Characteristics of Conglomerate, and Where Does Conglomerate Form?

Conglomerate differs from breccia by having rounded, instead of angular, clasts. The greater rounding generally reflects longer distances of transport by a stream or river, or perhaps pounding by waves along a shoreline.

07.08.b1 Baja California, Mexico

Clast Size

◄ Conglomerate contains large clasts, characteristically pebbles, cobbles, or even boulders. The large clasts rest in a matrix of sand and mud; they are well rounded in some conglomerates and only partially rounded in others, like the one shown here.

► Some conglomerate is finer grained, having few clasts larger than a centimeter or two. Such conglomerate represents less turbulent conditions, a source region that lacked large clasts, extreme abrasion of clasts during long transport, or some combination of these factors.

07.08.b3 Petrified Forest, AZ

07.08.b2 Great Smoky Mtns, TN

Sorting

◄ Some conglomerate is mostly clasts with relatively little matrix. In the conglomerate in this photograph, many clasts rest directly on another clast, instead of being completely separated by the sandy matrix.

► Other conglomerate is less well sorted, containing scattered large clasts in a fine-grained matrix. The well-rounded pebbles shown here are surrounded by a matrix of sand and silt.

07.08.b4 Petrified Forest, AZ

07.08.b5 San Juan River, UT

Environments of Formation

◄ Conglomerate can form from sediment deposited in or near a river or stream channel. The clasts can be rounded by even a moderate amount of transport in a river, and clasts that are transported tens or hundreds of kilometers can become well rounded.

► *Braided* rivers and streams migrate back and forth across *alluvial plains*, depositing coarse sediment that can later harden into conglomerate. Most meandering rivers do not carry such coarse sediment, and so generally do not result in conglomerate.

07.08.b6 Tibet

07.08.b7

◄ Waves pound and churn stones on many beaches, rounding their corners and edges. Smaller pebbles and sand quickly abrade and round, but even larger cobbles and boulders may be moved and abraded during extremely high tides or during storms.

Conglomerate forms in other environments, including deltas and their offshore equivalents. It also forms, as did this example (◄), in underwater turbidity currents near continental slopes. Additionally, some conglomerate forms when large storm waves impact a reef, breaking off and partially rounding vulnerable pieces. High energy is the common characteristic of all the different environments where conglomerate forms.

07.08.b8 Carmelo, CA

Before You Leave This Page Be Able To

☑ Sketch or describe the characteristics of a breccia, and identify some environments in which this rock forms.

☑ Sketch or describe the characteristics of a conglomerate, and identify some environments in which this rock forms.

☑ Contrast breccia and conglomerate, and explain reasons why one rock type might form instead of the other.

7.8

7.9 Where Does Sandstone Form?

SANDSTONE IS A COMMON SEDIMENTARY ROCK because sand occurs in many environments, including sand dunes on land, river channels, and submarine canyons beneath the oceans. Different environments produce different varieties of sandstone, each having distinctive characteristics. In what environments does sandstone form, and how do the resulting sandstones differ?

A What Are the Characteristics of Sandstone?

All sandstones, even if formed in different environments, share some common characteristics. By definition, sandstone is mostly or wholly composed of sand-sized grains, and commonly is moderately to well sorted. Quartz is the dominant mineral in most sandstone, but feldspar, iron oxides, calcite, and other minerals are commonly present.

07.09.a1 Durango, CO

Sandstone contains mostly sand grains. Some sandstone, such as the one shown here, consists entirely of sand-sized grains. The individual sand grains can be well rounded or angular, and variation in the proportions of minerals can define the beds.

07.09.a2 Ruby Canyon, CO

Most sandstone has layers that differ in color, grain size, or composition of the grains. Such layers can be parallel beds, as shown here, or cross beds centimeters to tens of meters high. Many layers mean many changes in conditions during deposition.

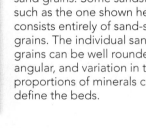
07.09.a3 Petrified Forest, AZ

Sandstone can contain other clast sizes, such as silt (smaller than sand) or scattered pebbles and granules (larger than sand). Such rock is still called sandstone as long as sand is the dominant clast size.

07.09.a4 Green River, UT

Some sandstone layers appear massive from a distance because they have little variation of grain size, as in the massive sandstone cliff in the center of this photograph. Thick layers of sandstone generally were deposited by wind as sand dunes.

B In What Land Environments Does Sandstone Form?

The sand needed to produce sandstone is deposited in rivers, migrating sand dunes, and in other land environments. Terrestrial sandstone (sandstone formed from sand deposited on land), commonly has a pinkish or red color because of oxidation of iron-bearing minerals during exposure to air and groundwater.

07.09.b1 Morocco

Sand dunes form sandstone that is very well sorted. Wind blows away finer grains and cannot pick up large clasts, so sand dunes usually do not include grains larger than sand.

07.09.b2 Green River, UT

Rivers deposit sand in channels and on the adjacent floodplain. In this photograph, the floodplain is the relatively flat area with grass and trees. It is only covered with water, sand, and silt during large floods.

07.09.b3 Snow Canyon, UT

This sandstone, deposited by desert dunes, is made entirely of sand-sized grains and has a reddish color because of iron oxide minerals that cement the sand grains together. The unit has large cross beds that reflect the shapes of original dunes. ◄

07.09.b4 Grand Canyon, AZ

Sandstone deposited by a river usually has discrete layers reflecting floods and shifting positions of the channel. It can be interlayered with siltstone and other rock types. These sandstone ledges alternate with more easily eroded siltstone. ◄

C How Does Sandstone Form Along Shorelines?

Most beaches are dominated by sand, so many beach deposits become sandstone. Sand also dominates many parts of deltas, especially the channels and shallow-water parts of the delta.

07.09.c1 Crete

A typical ocean beach includes sand with shells, pebbles, and locally some larger blocks derived from nearby rock exposures. The resulting sandstone may include these same things, although the high energy of a beach environment may destroy shells before they can be preserved.

07.09.c2 Durango, CO

This outcrop changes from a lower marine shale with thin beds of sandstone, to upper, thicker sandstone beds with less shale. The shale formed from finer grained clays and muds deposited in a near-shore marine environment, which was overrun by a delta that deposited the sandstone.

07.09.c3 Green River, UT

The sandstone in the right side of this photograph formed from beach deposits. The beach sands banked up against an island formed from the darker rocks in the lower left of the image.

07.09.c4 Tonto Creek, AZ

Sandstones can form in other nearshore environments. The cross-bedded sandstone beds shown here formed near a barrier island, kilometers or tens of kilometers offshore of the mainland.

D How Does Sandstone Form in Offshore Environments?

Sand also accumulates at sites farther from shore, in deeper water. The sand is derived from the erosion of continents, islands, deltas, reefs, and barrier islands but may be carried to deeper water by ocean currents.

07.09.d1 Northern CA

Continental shelves and other offshore areas can accumulate sand derived from erosion on land. The sands are moved out onto the shelf by waves and currents and are buried by later sediment.

The composition of sands deposited near volcanic islands reflects the kinds of rocks found on the islands. It usually contains less quartz than do most sands, and may contain pieces of volcanic rock. Such settings can produce black or even green sandstone.

07.09.d2 Big Island, HI

07.09.d3

If loose sand on the continental shelf or slope becomes unstable, it flows down the sloping seafloor as a thick slurry of sediment and water, creating a turbidity current. The beds deposited by these currents contain sand, mud, and larger clasts and commonly display graded bedding that reflects a settling of coarse material before fine material. This example of a turbidity current was created in a laboratory tank.

07.09.d4 Salt Point State Park, CA

These tan sandstone beds are interpreted to represent a turbidity current because they exhibit graded bedding and are interbedded with dark, deep-water marine shales.

Before You Leave This Page Be Able To

✓ Describe the characteristics of sandstones, including their expression in landscapes.

✓ Describe the land environments in which sandstone forms, and how you might distinguish sandstone formed by sand dunes from those formed by rivers.

✓ Describe how sandstone forms along beaches, in deltas, and in offshore environments.

7.10 How Do Fine-Grained Clastic Rocks Form?

THE MOST ABUNDANT CLASTIC SEDIMENTARY ROCKS are fine grained, consisting of grains that are smaller than sand. Compared to coarser sediment, fine sediment is easily transported by water or wind, even by slow-moving water or wind. Fine sediment can remain in transit until it reaches fairly quiet conditions in a lake, sea, or floodplain. The resulting sedimentary rock may be siltstone or shale, one of the most common rocks exposed on land.

A What Are the Characteristics of Fine-Grained Clastic Rocks?

Fine-grained clastic rocks consist mostly of clasts of fine grains of *silt* and *clay*. Silt is slightly finer than sand, and a rock dominated by silt is called a *siltstone*. Clay particles, generally finer grained than silt, may become *shale* when lithified. *Mud* includes both silt and clay, and the resulting rock is called *mudstone* or *mudrock*.

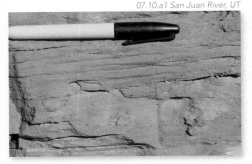
07.10.a1 San Juan River, UT

Siltstone consists of small grains that are not visible with the unaided eye. Siltstone can contain some fine sand grains, as in this example. It commonly occurs in centimeter-thick beds, but some wind-deposited siltstone beds are meters thick.

07.10.a2 Holbrook, AZ

Mudstone is a fine-grained rock that is similar in many ways to shale, so some people do not distinguish the two rock types. Mudstone breaks into pieces that tend to be more rounded than the thin chips into which shale weathers.

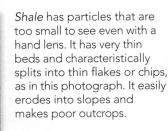

07.10.a3 Durango, CO

Shale has particles that are too small to see even with a hand lens. It has very thin beds and characteristically splits into thin flakes or chips, as in this photograph. It easily erodes into slopes and makes poor outcrops.

07.10.a4 Northern NM

Shale and mudstone can form thick sequences that are almost entirely fine-grained materials but sometimes are interbedded with sandstone or limestone. Depending on the environment, shale and mudstone can have various colors.

B In What Land Environments Do Siltstone and Shale Form?

Silt and clay accumulate in several terrestrial (on-land) settings. In most of these settings, deposition occurs in slow-moving or even stagnant water. Wind can also deposit silt and clay.

07.10.b1 Southern UT

Floodplains of meandering rivers are dominated by silt and fine-grained sand. During floods, silt is carried farther from the channel than are coarser sediments like sand and gravel. As the floodwaters slow, they deposit silt.

07.10.b2 Tibet

Wind transports and deposits silt over large areas. Windblown silt was especially abundant during periods of glaciation, when moving ice sheets ground rock into powdery silt-sized particles. Such wind-blown silt layers can be tens of meters thick.

07.10.b3 Northern NM

The bottoms of lakes are covered with mud carried to the lakes by streams, rivers, wind, and erosion of adjacent hillslopes. Lakes produce soft, thin-bedded rocks that can be dark-colored due to organic material. ◄

07.10.b4 Nepal

Chemical weathering converts many minerals into clay, which then accumulates as a layer of soil on the surface. If such soils are lithified, they usually form fine-grained rocks. ◄

C Where Along Shorelines and Farther Offshore Do Silt and Clay Accumulate?

Silt and clay form in several shoreline and ocean environments. The ocean basins are vast, and clay covers more of the seafloor than any other type of sediment, so shale is a common sedimentary rock.

07.10.c1 Drakes Estero, CA

Some shorelines have *mud flats*, which are flooded by high tides and during storms. When dry, these muddy flats may expose salt, gypsum, and other evaporite minerals that form when seawater evaporates.

07.10.c2 Southwestern CO

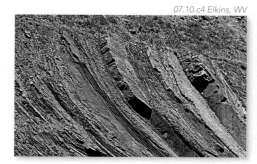

Mud can also accumulate in shallow continental seas, on continental shelves, and on adjacent continental slopes. The multicolored shales shown here formed in a shallow sea within the North American continent.

07.10.c3 Grand Canyon, AZ

The relatively calm water of a lagoon is an efficient trap for mud and clay carried from land by streams, rivers, and wind. The green and gray shales shown here formed from sediment that accumulated slightly offshore.

07.10.c4 Elkins, WV

Many shales form from sediment deposited in seas and ocean basins, where fine particles are carried by wind and ocean currents. Deep-water shales are dark gray, due to a high organic content and relatively low oxidation on the deep seafloor.

D How Are Fine-Grained Clastic Rocks Expressed in the Landscape?

07.10.d1 Grand Junction, CO

Shale, mudstone, and siltstone are relatively easily eroded rocks. Where exposed, these rocks typically form soft slopes covered by small, loose chips derived from weathering of the thinly bedded rocks.

Fine-grained rocks commonly are partially or entirely covered by soil or loose debris (*talus*) from overlying, more-resistant rocks. Here, outcrops of red mudstone project through a surficial cover of light-colored talus.

07.10.d2 Wilson Cliffs, NV

07.10.d3 Comb Ridge, UT

During erosion of landscapes, rivers preferentially carve their channels into shale and siltstone because these rocks are so easily eroded. Many rivers follow shale-rich units across the land surface, and may even follow the shales around folds. This desert wash follows a layer of fine-grained rocks around the bend of a broad fold.

07.10.d4 Petrified Forest, AZ

Shale and associated fine-grained rocks form another distinctive type of landscape— badlands. Badlands have a soft, rounded appearance that reflects the softness of the rocks. Badlands also have an intricate network of small drainages and eroded ridges because erosion is not restrained by strong beds in the rocks.

Before You Leave This Page Be Able To

- ✓ Describe the main characteristics of shale and siltstone, including which rock has the finest particles.

- ✓ Describe the land environments in which shale, mudstone, and siltstone form.

- ✓ Describe the environments near shorelines and farther offshore in which shale and siltstone form.

- ✓ Describe how some shale and siltstone are expressed in the landscape, including some of the landscape features they form.

7.11 How Do Carbonate Rocks Form?

LIMESTONE AND RELATED SEDIMENTARY ROCKS are called *carbonate rocks* because they consist of a carbonate ion combined with calcium, magnesium, or other elements. Most carbonate rocks form directly from water through chemical or biological processes, but some are clastic rocks consisting of pieces derived from shells, coral, or the erosion of carbonate bedrock.

A What Are the Characteristics of Carbonate Rocks?

Limestone is a common rock and exists in many varieties, all of which consist mostly of the mineral *calcite* ($CaCO_3$). Calcite can convert to the mineral *dolomite* by the addition of magnesium (Mg), which produces the carbonate rock *dolostone*. Limestone and dolostone commonly occur together.

07.11.a1 San Juan River, UT

Limestone typically is a gray rock. Its color ranges from almost white to dark gray, but it can also have shades of yellow, tan, or brown. It is soluble, so it frequently has a "dissolved" appearance.

07.11.a2 Durango, CO

Dolostone, made mostly of the mineral dolomite, resembles limestone, but it is more resistant to weathering and erosion because it is less soluble. Dolostone can be gray, but it commonly is tan, light brown, pinkish, or even slightly orange.

07.11.a3 San Juan River, UT

Limestone frequently includes fossils of shells, corals, and other marine organisms, as shown here. Limestones that form in lakes may have fossils of nonmarine organisms, such as freshwater fish.

07.11.a4 San Juan River, UT

Some sedimentary rocks contain a mixture of carbonate with clay, quartz, or some other noncarbonate clastic material. Intermixed clay may give the rock a greenish, tan, dark gray, or pinkish tint. It can also give the rock a mottled appearance.

B In What Nonmarine Environments Do Carbonate Rocks Form?

Most limestones form in marine environments, but limestone can also be deposited around springs, in lakes, and as coatings and other features on the floor, roof, and walls of caves. Carbonate layers also accumulate during soil development in dry climates.

Limestone forms in lakes in hot, dry climates that experience large amounts of evaporation. These limestones usually have creamy tan or brown colors, like the limestones and darker shales shown here.

07.11.b1 Green River, UT

Travertine, a variety of limestone, is usually cream-colored and porous (has open spaces). It can precipitate in cold springs, hot springs, lakes, and caves. Most travertine is layered because different layers precipitate at different times, coating preexisting materials.

07.11.b2 Tonto Natural Bridge, AZ

Limestone in some lakes occurs as coatings or irregular masses of white carbonate material. In some exceptional cases, this carbonate forms pillars, such as these at Mono Lake in eastern California.

07.11.b3 Mono Lake, CA

C How Do Carbonate Rocks Form in Marine and Nearshore Environments?

Most carbonate rocks form in marine settings, including reefs and other shallow-water environments on the continental shelf. Carbonate can also accumulate in deeper water environments, but not the deepest seafloor. Carbonates also form on low-lying mud flats and along shorelines dominated by carbonate sand.

07.11.c1 Goosenecks, UT

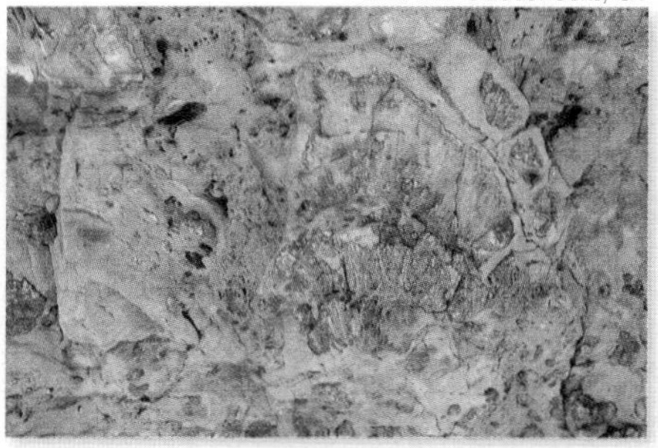

Reefs are important carbonate environments. Coral and other reef organisms extract calcium carbonate from the water to build their skeletons, shells, and stems, which then become incorporated in the reef. Reef-formed limestones normally contain these fossils.

07.11.c2 Goosenecks, UT

Storms and waves break off pieces of reef and grind the pieces of coral and shells into calcite sand. Such calcite-rich debris forms white, sandy beaches. If buried and lithified, the calcite beach sand becomes a type of clastic limestone.

07.11.c3 Provo, UT

Lime muds, formed from the remains of carbonate-secreting organisms or through chemical processes, accumulate on continental shelves. Such deposits can form thick sequences of gray limestone, whose layers record changing conditions on the shelf, especially changes in sea level.

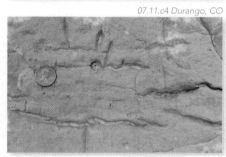
07.11.c4 Durango, CO

Dolostone forms when magnesium-carrying fluids interact with limestone. The fluids cause the calcite to be replaced by dolomite. This replacement is usually not complete, so the rocks are part limestone and part dolostone, each weathering in its distinctive way.

D How Are Carbonate Rocks Expressed in the Landscape?

Limestone is a very common and distinctive sedimentary rock that covers large areas of North America and other continents. It is generally recognized from a distance by its gray color, well-bedded character, and common pock-marked or irregular erosion texture.

07.11.d1 Chamonix, French Alps

In some relatively dry climates, limestone and dolostone are fairly erosion-resistant rocks. The rocks can form gray cliffs and steep slopes composed of beds that may vary slightly in thickness and color.

In wet climates, weathering, erosion, and dissolution of limestone may not affect all areas equally but instead may leave behind pillars of gray limestone. Many pillars represent areas that are less fractured or were otherwise more difficult to dissolve.

07.11.d2 China

07.11.d3 Austrian Alps

Limestone, and to a lesser extent dolostone, are very soluble and so in very wet climates they weather and erode quickly. They commonly contain caves and small openings formed where the carbonate rocks dissolved away. Dissolving of limestone is often most pronounced along fractures, as is shown here.

07.11.d4 Florida

Many caves form when groundwater dissolves limestone in the subsurface. The groundwater carries away the soluble carbonate and leaves an opening behind. If the roof of such a cave collapses, a closed depression, called a *sinkhole*, forms on the surface. Sinkholes can damage buildings, roads, and utilities.

Before You Leave This Page Be Able To

✓ Describe the characteristics of limestone and dolostone.

✓ Describe the environments in which limestone and other deposits of calcium carbonate form.

✓ Describe how dolostone forms.

✓ Describe how carbonate rocks are expressed in the landscape, including sinkholes and limestone pillars.

7.11

7.12 How Do Changing Environments Create a Sequence of Different Kinds of Sediments?

MOST SEDIMENTARY ROCKS ARE IN A SEQUENCE, with multiple layers formed one on top of another. The layers in a sequence may all be the same rock type, perhaps limestone, but more commonly include a variety of rocks. What changes in environment occur that result in a sequence of different rock types?

A What Happens When Environments Shift Through Time?

Environments that move across Earth's surface over time can result in a sequence of different sedimentary rocks. One common change is for seas to advance across a region, covering more land with time, due either to a rise in sea level or a lowering of the land. Such an advance is called a *transgression* and is illustrated below.

1. At the earliest time, a shoreline separates marine environments to the left from beach and land environments to the right.

2. Sand and pieces of shell are deposited along the beach and outward into nearby shallow water, forming a layer of sand.

Time 1

3. Sediment is being deposited on the beach and offshore. The land, however, is being eroded and there are no sediments deposited on land at the time when the beach sand is forming. The time is instead represented on land only by an erosion surface.

07.12.a1

4. By some later time, sea level has risen, and so the sea and beach have moved in across the land.

5. The area that used to be covered by beach sand is now far enough from the shoreline to have clear water in which a coral reef flourishes.

6. A lagoon is between the reef and shoreline. The relatively tranquil water of the lagoon traps mud that accumulates on the seafloor.

Time 2

7. As the sea advances, beach sand is deposited over areas that used to be land. The base of the sand layer is a buried erosion surface from the previous time period.

8. The lagoon is located where a sandy bottom used to exist, so lagoon mud is deposited over the older layer of beach sand.

07.12.a2

9. As the sea moves farther inland, the center of the area becomes a reef. Comparing this figure to previous ones reveals that the area of the reef was originally near a beach and then was a lagoon before becoming a reef.

10. The offshore progression results in a sequence of beach sand overlain by lagoon mud, which is overlain by limestone.

11. Therefore, when the sea advances across the land, shoreline deposits are progressively overlain by sediments that represent areas farther and farther offshore. This progression of sediment types is characteristic of a transgression.

Time 3

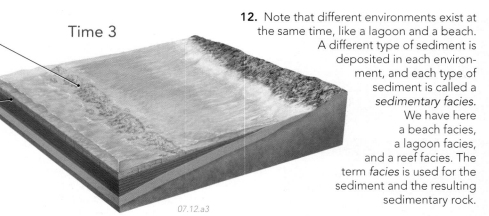

12. Note that different environments exist at the same time, like a lagoon and a beach. A different type of sediment is deposited in each environment, and each type of sediment is called a *sedimentary facies*. We have here a beach facies, a lagoon facies, and a reef facies. The term *facies* is used for the sediment and the resulting sedimentary rock.

07.12.a3

B What Happens When the Sea Moves Out?

The opposite of a transgression occurs when the sea retreats and more land is exposed. A retreat of the sea is called a *regression* and occurs when sea level becomes lower or the land is uplifted. A regression is illustrated below, starting with the sequence formed by the transgression (previous page).

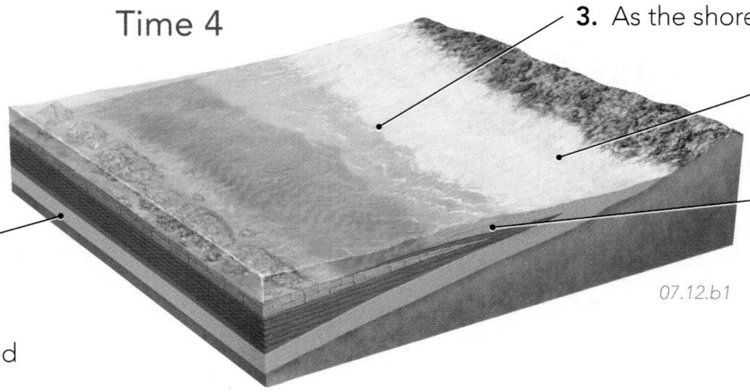

1. As the sea moves out during a regression, the sedimentary facies shift toward the sea (to the left in this series of figures).

2. The sedimentary sequence deposited during the regression will be built on the previous three layers that formed during the transgression.

3. As the shoreline retreats toward the sea, the beach and lagoon follow.

4. Sand that previously was close to the beach is now being eroded and is available for reworking by the wind, rain, and slumping due to gravity.

5. Farther from shore, beach sand is deposited over lagoon mud, and lagoon mud is deposited over reef limestone.

Time 4

Time 5

07.12.b1

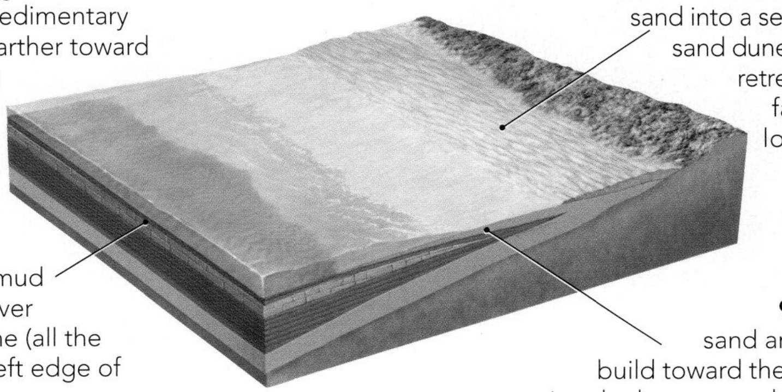

6. As the regression continues, the sedimentary facies shift farther toward the sea. The reef is now out of view.

7. Lagoon mud builds out over the limestone (all the way to the left edge of the model).

8. Wind remobilizes beach sand into a series of coastal sand dunes. As the sea retreats, the dune facies can follow the shoreline toward the sea.

9. The beach sand and dune sand build toward the sea, partially covering the lagoon mud, which in turn overlies the earlier formed limestone.

07.12.b2

Time 6

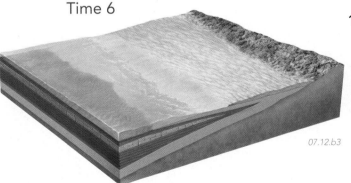

07.12.b3

10. During a regression, the sea retreats, and deeper marine sediment is successively overlain by shallower marine sediment, shoreline deposits, and, if the sea retreats far enough, land facies.

Why Sedimentary Layers End

All sedimentary layers eventually end. That is, they covered only a limited part of Earth's surface. One reason layers end is because a facies ends, as shown by the thinning out and disappearance of the lagoon facies toward the right side of the previous illustrations.

A layer also can end because it is deposited only within a channel, such as the river channel shown below. Coarse gravel accumulates inside the channel (a channel facies) but does not extend outside the channel.

07.12.mtb1

Before You Leave This Page Be Able To

☑ Sketch or describe what happens during a transgression and during a regression, including which way sedimentary facies shift.

☑ Sketch an example of a sequence of rocks formed during a transgression and contrast it with a sequence formed during a regression.

☑ Sketch or describe two reasons why sedimentary layers end.

Stratigraphic Sections

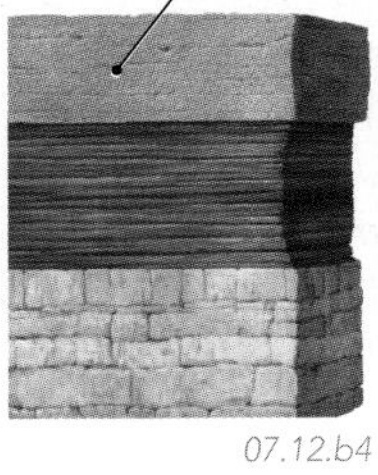

11. This stratigraphic section shows the sequence deposited during the regression (sea moving out). Limestone is overlain by mud, which is overlain by beach sandstone.

12. This section shows the sequence of sediments deposited during the transgression (sea moving in). An erosion surface is successively overlain by beach sand, mud, and limestone.

07.12.b5

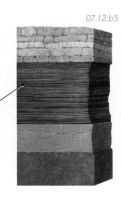

07.12.b4

7.12

7.13 How Do We Study Sedimentary Sequences?

An important goal of geology is to reconstruct past events and environments. We do this by observing a sequence of rocks and noting characteristics that provide clues to the environment in which each rock unit formed. We then interpret the past environments by comparing these characteristics to sediment in modern environments. Studying a sequence of rocks allows us to infer how conditions changed over time.

A What Attributes of Sedimentary Rocks Are Indicators of Environment?

Sedimentary rocks contain many clues about the environment in which they formed. Nearly every attribute, such as the sizes or shapes of clasts, provides some information we can use to infer the environment of deposition.

Color of Rocks

Red sedimentary rocks generally form on land where they can be oxidized (rusted) by the atmosphere, whereas dark gray sedimentary rocks usually form under water and in low-oxygen conditions.

07.13.a1 San Juan River, UT

Clast Size, Shape, and Sorting

Large, angular, poorly sorted clasts indicate strong currents and limited transport, whereas small clasts indicate weak currents. Rounded, well-sorted clasts reflect more transport or reworking by waves.

07.13.a2 Rawhide Mtns, AZ

Thickness of Bedding

Thick bedding implies bigger events, faster rates of deposition, or longer times between environmental changes. Thin bedding implies smaller events or more rapidly changing conditions.

07.13.a3 Monument Valley, AZ

Types of Bedding

Certain types of bedding reflect specific conditions of formation. These graded beds indicate that the strength of the current decreased through time or that suspended material settled out of water during floods.

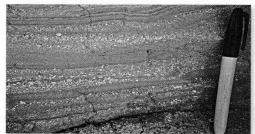
07.13.a4 New River, AZ

07.13.a7 Cedar Mesa, AZ

Large-scale cross beds in a well-sorted sandstone generally indicate deposition by wind as a series of large sand dunes.

Mudcracks

When wet mud dries, the sediment contracts and produces polygon-shaped pieces surrounded by cracks that can fill with sand, as in the modern example (with animal tracks) below. These are called *mudcracks*.

07.13.a5 Northern AZ

07.13.a8 Grand Canyon, AZ

Sedimentary rocks having mudcracks must have been deposited on land and in environments where wet sediment could dry out.

Fossils

Fossils of land plants, like these ferns, indicate that sediment was deposited on land. They also provide information about temperature, elevation, amount of rainfall, and other environmental conditions.

07.13.a6

07.13.a9 Apollo Bay, Australia

Marine fossils are diagnostic of deposition in seawater and can provide information about temperature, salinity, and clarity of the water.

Indicators of the Direction that Water or Wind Currents Flowed

07.13.a10 St. George, UT

07.13.a11 Salt River, AZ

Cross beds, whether they form in rivers (as these did), in sand dunes, or beneath the sea, slope down in the direction in which the current flowed, in this case toward the left.

Large clasts in a river normally are inclined opposite to the direction in which the current flows (current is to the left). This property, called *imbrication*, can be used to infer past flow direction from some conglomerate.

07.13.a12

Ripple marks are small ridges and troughs formed by moving currents. Some ripples have a steeper side toward the direction of current flow. In other examples, back-and-forth waves create symmetrical ripples.

B What Can We Observe and Interpret in a Sequence of Sedimentary Rocks?

We often can infer past events, environments, and changes in the environment by carefully observing a sequence of rock layers. We examine the kinds of rocks, what order they are in, whether there are fossils, and many other things. These observations are the basis for interpreting the environment that each rock layer represents. Try this reasoning by reading each observation and interpretation below, starting with the oldest rocks on the bottom.

Sequence of
Sedimentary Rocks

5. A reddish-gray breccia is at the top of the sequence. It is poorly sorted with angular boulders of granite in a mud-rich matrix. Interpretation—the large, angular clasts indicate only minor transport of the clasts, perhaps in a debris flow from a nearby steep terrain, such as a mountain. The mountainous terrain consisted of granite.

4. Near the top is a tan sandstone. Lower parts of the unit include broken shells, but the upper part has fossils of land plants and coal. Interpretation—the change of the type of fossils in the sandstone is consistent with the unit having formed along a shoreline, perhaps in a beach or delta as the sea was retreating.

3. The middle of the sequence is thick, gray shale with an intervening layer of limestone. The shale contains shallow-water marine fossils, including clams, and the limestone contains fossil coral. Interpretation—the shale and limestone accumulated in the shallow part of a sea. The shale may represent offshore muds or a lagoon; the limestone represents a reef.

2. A layer of tan sandstone overlies the red sandstone and includes marine shell fragments. Interpretation—this tan sandstone is interpreted as a beach sand or a sand that formed in shallow ocean water.

1. A red sandstone at the base of the sequence is well sorted and contains large cross beds. Interpretation—the well-sorted sand and large cross beds are consistent with the sandstone having been deposited on land as a series of large sand dunes.

07.13.b1

Interpretation of the Change in Sediment Type over Time

In addition to specific interpretations about each unit, we can infer *changes in the environment* by comparing each unit with the unit above it. Begin at the bottom with the oldest unit, and work upward toward younger units.

Units 1 to 2. The change from the sand dunes to overlying beach sands and marine shales with limestone is evidence of a transgression, an advance of the sea toward the land.

Unit 3. The sea probably reached its maximum advance during deposition of the limestone. The shoreline at this time was far enough away to allow coral to grow in clear water.

Units 3 to 4. Sometime after the limestone formed, the sea retreated during a regression. Delta sands with land plants were deposited over the marine shales.

Unit 5. Finally, steep, granite mountains formed during some tectonic event. The mountains shed large granite clasts onto nearby areas, perhaps in a series of debris flows.

Before You Leave This Page Be Able To

✓ Describe the attributes that we observe in sedimentary rocks and how each indicates something about the rock's origin and environment.

✓ Given a sequence of rocks and a list of key attributes, interpret the environment of each rock and how the environment changed.

7.13

7.14 Why Are Sediments and Sedimentary Rocks Important to Our Society?

SEDIMENTARY ROCKS AND THE RESOURCES we get from them are essential to our modern society. Sediments and sedimentary rocks are our main sources of groundwater, oil and natural gas, coal, salt, and material for making cement and construction aggregate. Besides providing resources, sedimentary rocks are important because they help us understand the geologic history of Earth, including climate change and how life originated and changed through time.

 How Do Sedimentary Rocks Control the Distribution of Resources?

Sedimentary rocks host many of our most important resources. Some resources, such as coal and salt, originated as sedimentary deposits, whereas other resources are most common in sedimentary rocks because these rocks permit the flow and entrapment of fluids, including water and oil.

Groundwater

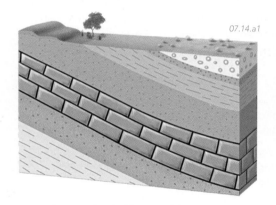

Groundwater, shaded here in blue, occurs predominately in sediment and sedimentary rocks. Most groundwater resides in the pore spaces between sedimentary grains and in fractures, and most of the liquid freshwater on Earth is groundwater.

Petroleum

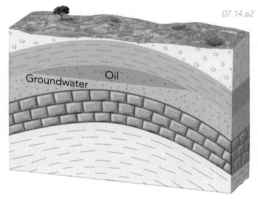

Almost all *oil and natural gas* comes from sedimentary rocks. Oil and gas form in organic-rich sedimentary rocks and then migrate upward until they reach the surface or are trapped at depth. Petroleum is vital to our society.

Coal

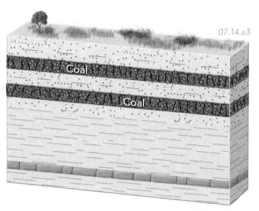

Coal, shown here as dark layers, forms a sedimentary rock through the consolidation of plant remains that accumulate in swamps, deltas, and other wetland environments. Most coal is used to generate electrical energy.

Cement from Limestone

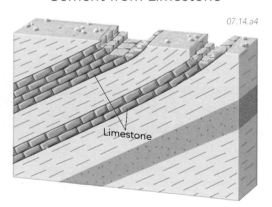

Cement is produced from limestone that is relatively free of sand, silt, chert, and other impurities. We use cement to make concrete for highways, bridges, building foundations, and other construction projects.

Salt

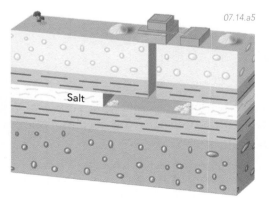

Salt is either mined from ancient sedimentary salt layers or is harvested by evaporating salty water. It is used in the preparation of food, medicine, and various industrial products.

Uranium

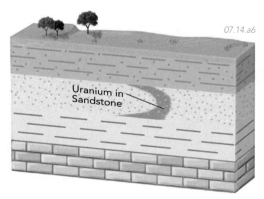

Uranium deposits commonly occur in sandstone and other sedimentary rocks, but the uranium commonly was brought into the area by migrating groundwater. It is used to generate electricity.

B How Do Sedimentary Rocks Help Us Understand Earth's Geologic History?

Sedimentary rocks are the primary source of information about ancient environments, climate change, and past events. Fossils in sedimentary rocks are the main record of how life originated and evolved.

07.14.b1 Indonesia

Geologists study *modern environments* to understand the processes that are occurring, the types of sediment produced, and how these environments may impact where we live and what we do. The scientist above is studying the effects of a large destructive wave associated with a deadly earthquake.

07.14.b2 Grand Canyon, AZ

By studying ancient sedimentary rocks, we observe the record of *past environments.* From these observations, we can interpret the character and distribution of different facies, how environments changed through time, and how resources formed in the environments.

07.14.b3 Calico, CA

Sedimentary rocks provide important data for investigating *climate change.* By understanding the severity and possible causes of past climate changes, we can better understand possible consequences of future climate changes, such as global warming or cooling.

07.14.b4 Buckskin Mtns., AZ

Sedimentary deposits and rocks allow us to *examine the record of past events,* including landslides (shown here), storms, and earthquakes. Studying these deposits and rocks enables us to infer the processes that occurred and consider how the events affect the landscape and life around them.

07.14.b5 Green River, UT

Sedimentary rocks are the main way we study the *sequence of past geological events,* such as advances of the seas, migration of ancient deserts, and erosion of mountains. The succession of rock layers above, from beach sandstone at the base to upper limestone cliffs, records a transgression.

07.14.b6 Durango, CO

Fossils allow us to study *ancient life,* including the types of organisms that lived at different times and the environments in which they lived. By studying the succession of fossils from one layer to the next, we observe how life on Earth evolved and we may infer the causes of the observed changes.

Sand and Gravel—The Most Used Sedimentary Resource

Resources such as gold, oil, and diamonds easily capture our interest because they are so precious or because we depend so highly on them. However, we use sand and gravel in much greater quantities.

The phrase *sand and gravel,* defined as a resource, refers to sediment that commonly is excavated from pits and used in various types of construction. It includes clasts of various sizes, from clay, silt, and sand, to pebbles, cobbles, and boulders. The material sometimes is used as it is, but it more commonly is poured through large screens to sort the clasts by size. In some cases, clasts are crushed to achieve a smaller desired size.

After the material is sorted into the correct sizes, it can be added to cement to make concrete and concrete blocks, added to clay to make tile, or used as fill beneath buildings and roads. In 2006, more than 900 billion tons of sand and gravel were used in the United States, approximately 3 tons per person.

Before You Leave This Page Be Able To

☑ Describe or sketch some of the main resources that occur in sedimentary rocks.

☑ Describe how sedimentary rocks help us understand modern and ancient environments, events, and life.

☑ Describe sand and gravel as a resource and how we use these important materials.

7.14

How Did Sedimentary Layers West of Denver Form?

THE FOOTHILLS OF THE FRONT RANGE west of Denver, Colorado, contain spectacular exposures of sedimentary rocks. The layers have been folded and tilted along the mountain front and form dramatic landscapes. The area provides an example of how to integrate various aspects of sedimentary rocks to interpret the geologic history of a region.

A How Are the Sedimentary Layers Exposed?

The figure below shows a geologic map superimposed on topography of an area west of Denver. The colors show the distribution of different sedimentary layers and other rock types. The front of the figure is a geologic cross section that shows the interpreted geometry of rock layers at depth. Begin on the lower left.

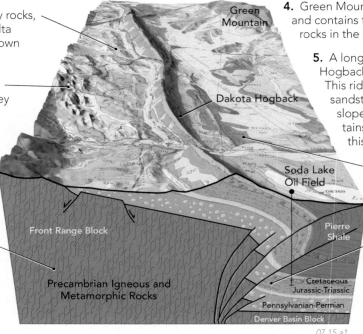

3. A sequence of diverse sedimentary rocks, including marine shale, beach and delta sandstone, river deposits, and windblown sandstone, overlies the red rocks.

2. Red sandstone and conglomerate (colored blue on this map because they are Paleozoic in age) were deposited on top of the metamorphic and igneous basement. The layers were later tilted and now dip eastward off the mountain front. These rocks form the dramatic exposures at Red Rocks Amphitheater, a famous venue for music concerts.

1. The Front Range is part of the Rocky Mountains and consists mostly of Precambrian metamorphic and granitic rocks. These are the oldest rocks in the area and underlie all other rock units, and so they are called *basement rocks*.

4. Green Mountain, a round hill, is east of the mountains and contains the highest and youngest sedimentary rocks in the map area.

5. A long, gently curving ridge called the Dakota Hogback is a dominant feature of the landscape. This ridge is formed by relatively resistant sandstone of the Dakota Formation, which slopes down eastward away from the mountains. This sandstone is colored green on this figure.

6. Low areas on either side of the hogback are underlain by more easily eroded, fine-grained sedimentary layers, mostly shale.

7. The rock sequence continues into the subsurface where it is folded and cut by a series of faults. An oil field was found by studying the sedimentary layers and predicting where to drill to find oil several kilometers beneath the surface.

Labels on figure: Green Mountain; Dakota Hogback; Soda Lake Oil Field; Pierre Shale; Front Range Block; Precambrian Igneous and Metamorphic Rocks; Cretaceous; Jurassic-Triassic; Pennsylvanian-Permian; Denver Basin Block

07.15.a1

B How Are the Sedimentary Layers Expressed in the Landscape?

07.15.b1

07.15.b2

07.15.b3

Red Rocks Amphitheater nestles within the lowest sedimentary unit, a series of reddish conglomerate and sandstone layers that dip away from the mountain. Precambrian basement rocks compose the mountains to the left.

This view looks along the Dakota Hogback, which follows a tilted, resistant sandstone layer of the Dakota Formation. The sandstone is tilted to the right (east), and Red Rocks is to the left (west).

A spectacular roadcut along I-70 exposes tilted Jurassic and Cretaceous rocks. A trail through the sequence is accompanied by descriptions of each rock formation and its ancient depositional environments.

C What Is the History of the Sedimentary Rocks?

The history is depicted below in a stratigraphic section and a summary of key characteristics, photographs, and maps illustrating the interpreted environments. Begin at the bottom of the section with the oldest layer.

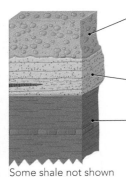

Some shale not shown

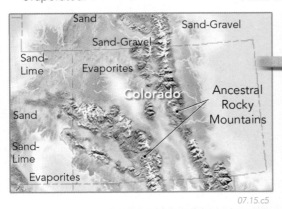

07.15.c1

7. Cenozoic conglomerate and sandstone, exposed at Green Mountain, at the top of the section include clasts derived from the Precambrian crystalline rocks of the Front Range and from some local volcanic terrains. These rocks resulted from the formation and erosion of the mountains 70 m.y. to 40 m.y. ago.

6. The shale is overlain by late Mesozoic (Cretaceous) sandstone, shale, and coal with plant fossils. These layers were deposited on land and along shorelines.

5. A thick sequence of gray shale is next. It contains marine shells, shark teeth, and fish bones. The unit includes beds of limestone and sandstone deposited during minor transgressions and regressions, respectively.

4. The *Dakota Formation*, lying above the Morrison Formation, contains plant fossils, dinosaur tracks, and sedimentary features, such as ripple marks, that indicate it represents shoreline and beach facies. The upper two photographs show ripple marks and dinosaur tracks on the tilted beds.

3. The *Morrison Formation*, famous for its dinosaur fossils and tracks, is middle Mesozoic in age (Jurassic). It includes shale, sandstone, and freshwater limestone, and is interpreted to have been deposited in wetlands, rivers, and lakes.

2. A sequence of mostly red late Paleozoic and early Mesozoic rocks overlies the red conglomerate and breccia. The upper part contains mudstone and marine limestone and is interpreted to have been deposited in a coastal mud flat. It also contains well-sorted, cross-bedded sandstone, interpreted to represent sand dunes.

1. The lowest sedimentary unit is a reddish sequence of sandstone and poorly sorted, coarse conglomerate and breccia (▶). This unit is interpreted to have been deposited by rivers and debris flows that drained an ancient mountain range called the *Ancestral Rockies*. The unit is late Paleozoic in age and rests on Precambrian basement rocks.

07.15.c2 Southwest CO

07.15.c3 Southwest CO

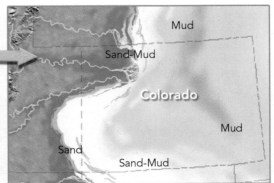

07.15.c4 Red Rocks, CO

Paleogeographic Maps for Three Chapters in the Geologic History of Colorado

1. *Late Paleozoic*—Sandstone and coarse, reddish conglomerate formed from sediment shed off the Ancestral Rockies. Salt and other evaporites formed in inlets of seawater that evaporated. ▼

2. *Early and Middle Mesozoic*—The region is a continental environment dominated by mud flats, sand dunes, lakes, and river systems. These environments change their distribution over time and so deposit a sequence of different sedimentary layers, including mudstone, sandstone, and conglomerate. Dinosaurs roamed the landscape, leaving tracks and bones in the Morrison Formation of Jurassic age. ▼

3. *Late Mesozoic* — A shallow sea stretches from the Arctic to the Gulf of Mexico and is later overrun by deltas from the west. The Dakota Formation accumulated during the transgression, and marine shales accumulated in the shallow sea. ▼

07.15.c5    07.15.c6   07.15.c7

Before You Leave This Page Be Able To

☑ Describe how the characteristics and sequence of sedimentary rocks can be used to reconstruct the geologic history of an area. Use examples from sedimentary rocks west of Denver.

7.16 What Is the Sedimentary History of This Plateau?

A plateau in northern Arizona exposes a sequence of different kinds of sedimentary rocks. Some sedimentary units were deposited on land, and others were deposited by shallow seas. Using key observations about each rock unit, you will reconstruct the history of these sedimentary rocks.

Goals of This Exercise:

- Use photographs or samples to make observations about the sedimentary layers.
- Interpret a possible environment for each sedimentary layer.
- Use a stratigraphic section to infer how the environment changed through time as layers were deposited.

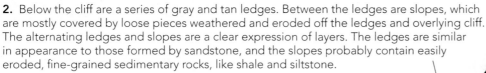

A Observe the Sequence and Characteristics of Sedimentary Layers

Observe this photograph, which shows a sequence of layers, and try to identify boundaries between different sedimentary units. Next, compare your observations with the observations and interpretations next to the photograph and with the information on the next page. Record your observations and ideas on a sheet of paper, perhaps built around a simple sketch of the scene. Your instructor may also provide you with rock samples.

1. The highest rock unit exposed in this area forms an uppermost tan and brown cliff and steep slope. There are some color variations within the cliff, which probably reflect slightly different colors of adjacent layers. The colors and other aspects of the cliff are common in many sandstones. Although not visible here, the cliff-forming unit and nearby underlying layers contain beds of coal, some of which are large enough to mine nearby for the production of electricity.

07.16.a1 Black Mesa, AZ

2. Below the cliff are a series of gray and tan ledges. Between the ledges are slopes, which are mostly covered by loose pieces weathered and eroded off the ledges and overlying cliff. The alternating ledges and slopes are a clear expression of layers. The ledges are similar in appearance to those formed by sandstone, and the slopes probably contain easily eroded, fine-grained sedimentary rocks, like shale and siltstone.

3. Below the ledges is a gray slope, which nearby locally has badlands topography. The rock forming the gray slope has some faint layers, but all of the layers look similar. It looks fairly soft and nonresistant to erosion, as is common for fine-grained rocks like shale. The gray color of the rock implies that the unit was deposited in conditions that were not rich in oxygen.

4. The soft gray rocks directly overlie a tan and cream-colored cliff. The cliff-forming unit is resistant to erosion and weathers like many sandstones do. Elsewhere in the region, the sandstone is more tilted and forms a series of resistant ridges, or hogbacks.

5. The cliff overlies a series of soft, thinly layered rocks that are maroon, reddish brown, gray, and cream colored. The rocks are poorly exposed and composed of fine-grained, easily eroded sedimentary rocks. Most rocks that have this reddish color were deposited on land.

B | Interpret the Sedimentary History of the Sequence of Layers

The stratigraphic section below shows the relative thicknesses of the units. The oldest unit is on the bottom and the youngest is at the top. Photographs and brief descriptions of each rock unit accompany the stratigraphic section. Your instructor may provide you with samples of similar rocks. Follow the steps below to propose a plausible interpretation for the environment of deposition for each sedimentary unit and for how the environment changed from one rock unit to the next. Write your answers to the following questions on the worksheet, on a sheet of paper, or in an online form.

1. What is your interpretation of the environment for each of the four rock units? List two key attributes of each unit that support your interpretation.

2. What is the oldest environment represented by this rock sequence?

3. Does the change of environment from the base of the section up to the thick gray shale indicate an advance (transgression) or retreat (regression) of the sea? Explain the reasons for your answer.

4. Does the change from the thick gray shale to the overlying sandstone indicate a transgression or a regression? Explain the observations that support your answer.

5. Which of the following phrases summarizes the history of the entire sequence: (a) a transgression, (b) a regression, (c) a transgression followed by a regression, or (d) a regression followed by a transgression?

6. Compare this sedimentary sequence to the one exposed west of Denver (in the previous two-page spread). What name from the Denver area would you apply to the lower, yellowish-tan sandstone in this plateau?

Stratigraphic Section

07.16.b1

This unit includes sandstone, mudstone, and layers of coal (shown in black). The upper part of the unit contains sandstone beds with small cross beds. The mudstone has mudcracks and plant fossils. The lowest part of the unit contains tan sandstone with broken marine shells. This photograph (▶) shows thin layers of black coal in this unit.

This shale is medium to dark gray because it has a high amount of organic matter. It contains fossils of clams and other marine organisms. Thin limestone beds are locally present in the middle of the unit but are not shown in the section. The shale and limestone contain abundant marine fossils. The photograph (▶) shows a close-up of the transition from the shale to the overlying sandstone.

This unit is mostly a yellowish-tan sandstone containing quartz sand with small pieces of marine shells. As shown in the photograph (▶), the very base of the unit is a thin conglomerate that overlies a scoured erosion surface. This lower part locally contains fossils of wood and leaves.

The lowest unit includes a conglomerate with moderately rounded pebbles and coarse sand containing scattered rounded pebbles and pieces of fossilized wood. The conglomerate is overlain by reddish, maroon, and gray shale and mudstone with plant fossils. This photograph (▶) shows a nearly circular dinosaur track where a large, plant-eating dinosaur with huge round feet stepped into and pressed down the then-soft sediment.

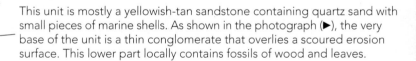

07.16.b2 Black Mesa, AZ

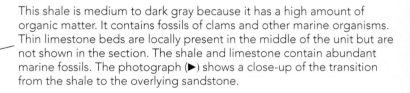

07.16.b3 Shiprock, NM

07.16.b4 Farmington, NM

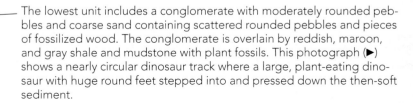

07.16.b5 Church Rock, AZ

7.16

Deformation and Metamorphism

ROCKS ARE HEATED AND SUBJECTED TO HIGHER PRESSURE as they are buried. The new conditions can cause *deformation*, where rocks are squeezed, stretched, sheared, or broken. The conditions can also cause *metamorphism*, expressed by the growth of existing mineral grains, the formation of new minerals, the development of different rock textures and fabrics, or typically all of these. Deformation and metamorphism can occur independently of each other, or can affect a rock at the same time, working together to remake the rock into a different rock type, a *metamorphic rock*.

The Appalachian Mountains and adjacent parts of the eastern United States display a wide variety of landscapes. This image shows satellite data superimposed on topography for part of southeastern Pennsylvania. The image includes curving mountains and ridges (green) alternating with lowlands (pinkish brown). The large river is the Susquehanna River, which flows south and cuts across the ridges and lowlands.

Choose a ridge and follow it across the region. What does it do? What other features do you observe as you examine this image?

This distinctive region has alternating ridges and valleys. Some of the ridges and valleys are straight, but others curve back and forth across the landscape. This region is named the *Valley and Ridge Province*.

How did these unusual landscapes form, and what do they tell us about the architecture of the underlying rocks?

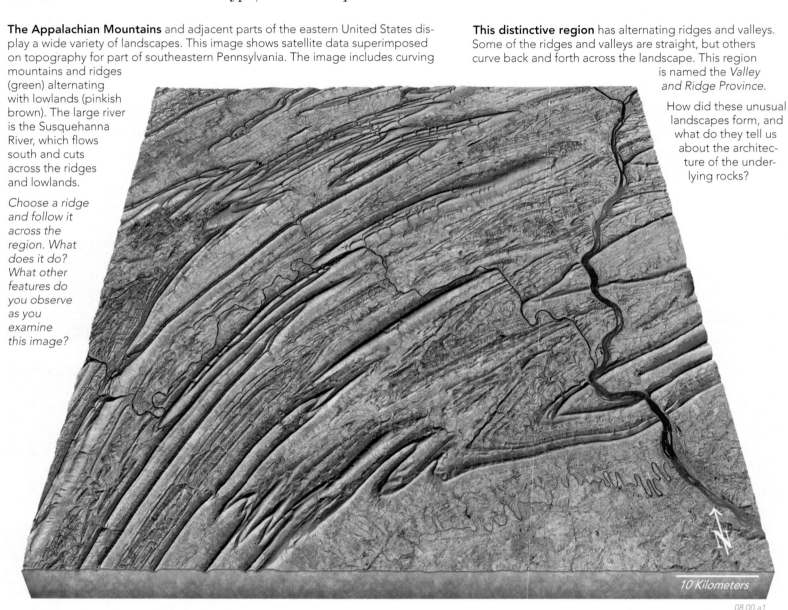

10 Kilometers

08.00.a1

The landscapes of the Valley and Ridge Province, as shown in this cross section for an area south of the map, reflect large folds in the Paleozoic sedimentary rocks. Some of the folds are tens of kilometers across and more than 100 km (62 mi) long. ▶

How do we determine that large folds are present in an area?

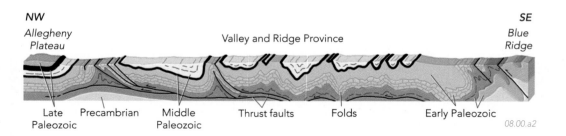

NW
Allegheny
Plateau

Valley and Ridge Province

SE
Blue
Ridge

Late Paleozoic   Precambrian   Middle Paleozoic   Thrust faults   Folds   Early Paleozoic

08.00.a2

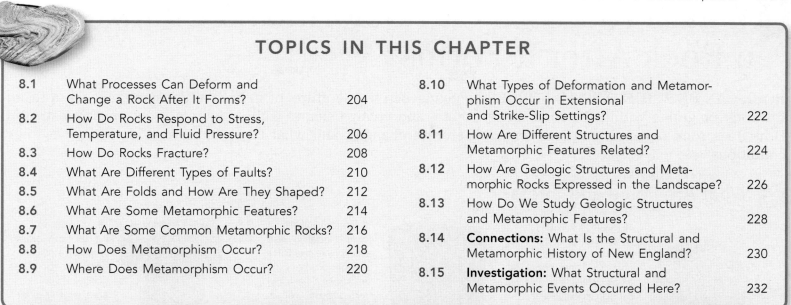

TOPICS IN THIS CHAPTER

08.00.a3 Tennessee

The cross section on the facing page has dark lines (with arrows) that represent faults offsetting the rock layers. These faults commonly stack one layer of rock on top of another, as does the fault in this photograph; the fault is highlighted with a dashed line and arrow.

Which type of fault are these, and what caused the faults to move? Are the faults and folds somehow related?

08.00.a4 Sideling Hill, MD

Large folds warp rock layers, in this view folding the layers into a U-shape. During folding, imposed forces and slightly increased temperatures caused metamorphism, where mineral grains rotated, changed shape, and partially recrystallized or dissolved. These changes produced breaks, called *rock cleavage*, which here cut steeply across the folded layers.

How and under what conditions do folds and cleavage form?

Origin of the Valley and Ridge Province

The Appalachian Mountains have a complex geologic history that includes four main periods of tectonism that caused rocks to deform. One of these mountain-building episodes resulted in the folds, faults, and cleavage of the Valley and Ridge Province.

A broad region of the eastern United States, including the Valley and Ridge Province, is covered by Paleozoic sedimentary rocks, which formed between 540 and 250 million years ago. These rocks were deposited in shallow seas, shorelines, rivers, and other environments. The sedimentary rocks of the Appalachian Mountains were folded, faulted, and heated several times during the Paleozoic. The deformation culminated with a *continental collision* between Africa and eastern North America approximately 300 million years ago. The collision uplifted the central part of the Appalachian range and forced huge slices of rock up and over sedimentary rocks west of the mountains. The rock layers within and below the slices, such as those in the Valley and Ridge Province, responded to the forces by folding, faulting, and squeezing out of the way. As rocks were squeezed, they were buried and heated. Mineral grains grew together or underwent chemical reactions to produce new minerals, a key process of *metamorphism*. Subsequently, when the region was uplifted, the folds and faults guided erosion, which carved away some rock layers faster than others to produce the region's distinctive valleys and ridges that reflect the shape of the folds.

8.0

8.1 What Processes Can Deform and Change a Rock After It Forms?

ROCK CAN BE SUBJECTED TO FORCES, to changes in temperature, or to both at the same time. As a result, the rock or grains within the rock move, rotate, and change shape, while minerals grow or are destroyed. How does rock respond to forces and temperature changes, and what types of features form in these new conditions?

A What Are Force and Stress?

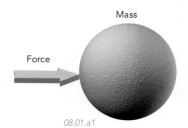

Mass

Force

◄ *Force* is a push or a pull that causes, or tends to cause, change in the motion of a body. It is commonly expressed as the amount of acceleration experienced by a mass.

08.01.a1

◄ The amount of force divided by the area where the force is applied is called the *stress*. The force from a metal weight is distributed evenly across the top of a broad, wooden pillar.

08.01.a2

◄ If the same amount of weight is on a much thinner pillar, the stress (force per unit area) on the pillar is greater. It might cause the pillar to splinter or break.

08.01.a3

B How Do Rocks Respond to Force and Stress?

Rocks within Earth are subjected to forces from the weight of overlying rocks, from tectonic forces pushing or pulling on the rocks, from cooling and heating, and from pressurized fluids, such as water and magma. Just like the wooden pillar above, if a force is concentrated (i.e., high stress), a rock can break or otherwise deform. So, geologists normally talk about stress instead of force. These figures show stress with a blue arrow.

A volume of rock may remain unchanged if subjected to only a small amount of stress (►). If the imposed stresses are greater, three things can happen. The rock may be *displaced* from one place to another, it may be *rotated*, or it may have its shape *modified, or strained*. All three responses may occur at the same time.

08.01.b1

Displacement

In response to stress, a volume of rock may be moved, or displaced, from one place to another. During *displacement*, a rock can behave as a rigid object or can change shape as it moves. In the photograph below, a thin, light-colored granite has been displaced by movement along fractures. ▼

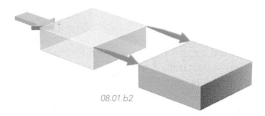

08.01.b2

Rotation

A volume of rock may be rotated in response to stresses. Rotation can tilt the rock or spin it horizontally. The rock layers in the photograph below were deposited as horizontal layers, but the layers have since been rotated. Rotation can be expressed by tilting, folding, or a partial spin of the rock. ▼

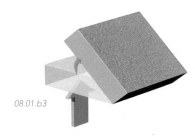

08.01.b3

Strain

A rock can respond to stress by deforming internally—changing size or shape without layers breaking or rotating. A change of size or shape is called *strain*. Stress is the cause, and strain is the effect. Below, originally rounded pebbles in a conglomerate were strained in response to stress squeezing the rock. ▼

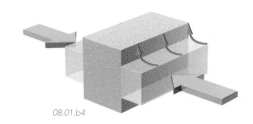

08.01.b4

08.01.b5 Tortolita Mtns., AZ

08.01.b6 Tibet

08.01.b7 Granite Wash Mtns., AZ

C What Processes Affect Rocks at Different Depths?

The structural behaviors of rocks change as temperature and stress increase with depth. The changing conditions also influence the stability of minerals and the way rocks react to water and other fluids.

Rock Behavior	Mineral Response	Effect of Fluids
Rocks can respond to stress by brittle fracture or by flowing like toothpaste squeezed from a tube.	A mineral may become unstable or it may not be affected as temperature and pressure change.	Fluids, such as water, can help minerals grow or dissolve and can affect the strength of a rock.

Shallow

08.01.c1 Whipple Wash, CA

▲ At cool, shallow levels of the crust, rocks usually exhibit *brittle* behavior and fracture.

08.01.c2 Chino Valley, AZ

▲ At low temperature, many minerals, like the quartz in these cobbles, are stable, barely responding to the conditions.

08.01.c3 Lovell Wash, NV

▲ Low-temperature fluids have little effect on many rocks but may form mineral-filled fractures, called *veins*.

Deep

08.01.c4 Barneys Junction, WA

▲ At deeper levels, where temperature and pressure are higher, rocks usually respond to stress by flowing as a weak solid, which is called *ductile* behavior.

08.01.c5 Clearwater, ID

▲ At high temperature and pressure, minerals commonly *recrystallize* into larger or smaller crystals, and new minerals may grow at the expense of existing minerals.

08.01.c6 Northern Cascades, WA

▲ At depth, hot water and other fluids can mobilize chemical constituents, form high-temperature veins, and promote recrystallization of minerals.

Determining the Conditions at Which a Mineral Is Stable

To study the interior of Earth, geologists investigate conditions where different minerals are stable. One approach is to observe rock samples from drill holes, but such observations are usually limited to the upper several kilometers of the crust. To investigate deeper environments, geologists study rocks that were at depth but were later uplifted by tectonics and exposed by erosion.

Geologists also investigate deep environments by doing laboratory experiments. Special laboratory devices, like this one,

08.01.mtb1

permit geologists to subject rocks and minerals to high temperatures and pressures, simulating conditions within the deep Earth. Geologists place a small sample inside the device and then raise the temperature and pressure. After a specific time, the sample is cooled, depressurized, removed from the device, and then examined with a microscope and other analytical instruments to determine which minerals were stable under the high temperatures and pressures and which ones were not.

Before You Leave This Page Be Able To

☑ Describe or illustrate the concept of stress.

☑ Sketch or describe the three ways that a mass of rock can respond to stress.

☑ Describe the differences in structural behavior, mineral response, and effect of fluids between shallow and deep environments.

☑ Briefly describe how we study the conditions under which a specific mineral is stable.

8.1

8.2 How Do Rocks Respond to Stress, Temperature, and Fluid Pressure?

HOW ROCKS RESPOND TO STRESS depends on three main factors: the type and magnitude of stress, the pressure and temperature conditions, and the amount of fluid present in the rock. The interaction of these factors leads to many different geologic structures, including fractures, folds, and veins.

A What Kinds of Stress Affect Rocks?

Rocks within Earth are subject to stress applied by the surrounding rocks. Any point within the earth is affected by stresses from all directions, and the entire array of stresses is called the *stress field*. We simplify the stress field by showing only the stresses applied from three mutually perpendicular directions. The size of the blue arrows in the figures below corresponds to the amount of stress—larger arrows signify more stress.

Confining Pressure

Any point within Earth is pushed downward by the weight of overlying rocks. Adjacent rocks also experience this weight and so push outward in all directions against other rocks. The rock experiences the same amount of force from each direction. We use the term *confining pressure,* rather than stress, when the force imposed on the rock is the same amount from all directions. Water in the pore spaces of a rock exerts a *fluid pressure* that pushes outward in all directions and opposes the inward-directed confining pressure on the rock. High fluid pressure acts to decrease the confining pressure.

08.02.a1

Differential Stress

If stress from tectonics or another source affects the rock, the imposed stress may add to or subtract from the confining pressure. As a result, the amount of combined stress will be greater in some directions than in others, and the rock is subjected to *differential stress*. Differential stress is what deforms rocks.

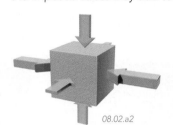

08.02.a2

B What Is the Strength of a Rock and How Does It Vary with Depth?

Strength of Rock

08.02.b1

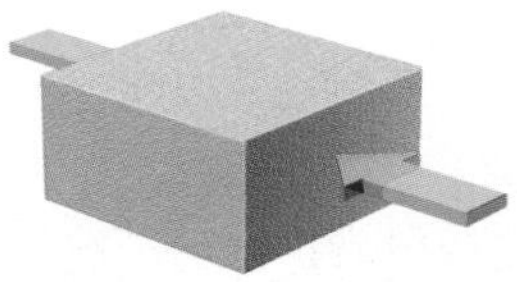

When a small amount of stress is applied to a rock, the rock may contract slightly like an elastic material but otherwise is strong enough to be undeformed.

08.02.b2

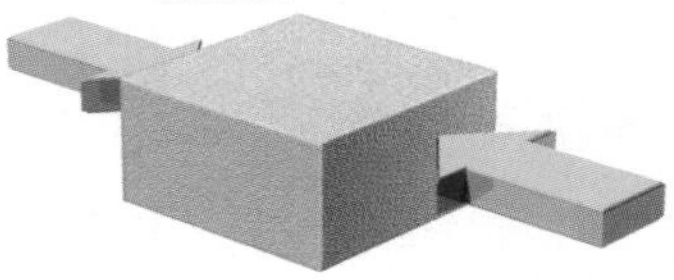

As stress increases, the rock remains essentially undeformed as long as the *strength* of the rock is greater than the amount of differential stress.

08.02.b3

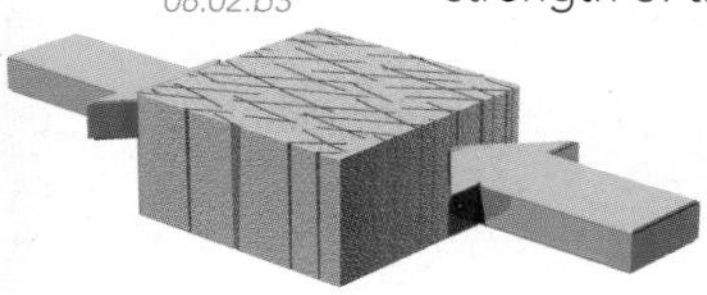

If the imposed stress exceeds the strength of the rock, the rock fails structurally, either by fracturing, folding, or flowing as a ductile solid.

Strength of Continental Crust at Depth

The strength of continental crust varies as a function of depth because temperature and pressure both increase downward. ▼

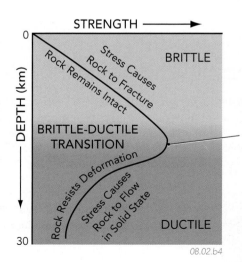

08.02.b4

At shallow levels of the crust, rocks deform by fracturing and other types of *brittle deformation*. Rocks in the upper crust become stronger with depth because increasing confining pressure acts to hold rocks together and makes slip along any fractures more difficult.

Deeper, where pressure and temperature are greater, rocks may deform (flow) by *ductile deformation*. There is a gradational boundary, or transition, between the *upper brittle* and *lower ductile* parts of the crust. This typically occurs at a depth of approximately 15 km and temperatures of more than 300°C. The depth is shallower in anomalously hot regions, including continental rifts.

At greater depths, the effects of temperature dominate over the effects of pressure. Rocks become progressively weaker as they become hotter and can flow more easily in the solid state. The strength of the crust decreases rapidly downward and responds by ductile flow.

C How Do Rocks Respond to Differential Stress?

Rocks may be subject to three types of differential stress: *compression*, *tension*, or *shear*. The way in which rocks respond to these stresses varies as a function of depth because rock strength changes with depth.

	Compression	Tension	Shear

Type of Stress

08.02.c1

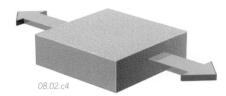
08.02.c4

08.02.c7

1. When a stress pushes in on rock, the stress is called *compression*, shown by the inward-directed arrows above.

4. When stress is directed outward, pulling the rock, the stress is called *tension*. Tension is shown with stress arrows pointing away from the rock.

7. A third type of stress acts to *shear* the rock as if stresses on the edges of a block were applied in opposite directions.

Shallow Levels of Crust

08.02.c2

08.02.c5

08.02.c8

2. Compression in shallow levels of the crust can cause rocks to deform by brittle processes, perhaps causing the rock to fracture and slip.

5. Tension can form fractures that help the rock stretch as it is pulled apart. Fluids, if present, can deposit minerals in the fracture, forming a vein. Tension may also cause slip along fractures (not shown).

8. Shearing in shallow parts of the crust usually forms a *fault*, which is a fracture along which two rock masses have slipped past one another.

Deeper Levels of Crust

08.02.c3

08.02.c6

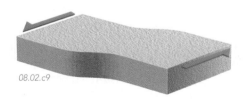
08.02.c9

3. In deep parts of the crust, where rocks are hot enough to flow, compression can squeeze the rocks and form tightly squashed folds and new metamorphic structures.

6. Tension is difficult to maintain deep in the crust because the high confining pressure pushes inward and generally is greater than any forces pulling the rock apart, but differential stress can stretch the rock. If accompanied by high fluid pressure, stress can fracture the rock and form veins.

9. In deep, ductile environments, shearing commonly is distributed across a wide zone. Rocks within the zone of shearing deform and flow as weak solids. A zone of shearing is called a *shear zone*.

How We Determine the Strength of Rocks

The strength of any material, including a rock, is normally determined in the laboratory by gradually increasing the amount of stress on a sample until it deforms. The samples in this photograph were compressed end-on until they bulged or fractured and slipped. Some rocks are stronger than others, so deformation experiments must be performed on many different kinds of rocks. These experiments are conducted under temperatures and pressures appropriate for different depths in the crust and mantle. The strength of rocks can also be investigated by examining how rocks, such as those in deep mines and drill holes, respond to natural stresses.

08.02.mtb1

Before You Leave This Page Be Able To

☑ Sketch and describe the difference between confining pressure and differential stress.

☑ Sketch and summarize how the strength of rocks varies with depth.

☑ Sketch and describe the three types of stress, and provide examples of the structures that each type forms at shallow and deep levels of the crust.

8.2

8.3 How Do Rocks Fracture?

FRACTURES ARE THE MOST COMMON geologic structures. They range from countless small cracks visible in an outcrop to huge faults hundreds or thousands of kilometers long. What are the different types of fractures, how do the different types of fractures form, and how are fractures expressed in landscapes?

A In What Different Ways Do Rocks Fracture?

There are two main types of fractures: *joints* and *faults*. Joints and faults both result from stress but have different kinds and amounts of movement across the fracture.

Joints

08.03.a1

◀ **1.** Most fractures form as simple cracks representing places where the rock has pulled apart by a small amount. These cracks are called *joints* and are the most common type of fracture.

2. These sandstone ledges (▶) are cut by a series of near-vertical joints. The layers are not offset by the joints but are simply pulled apart by a very small amount.

08.03.a3 San Juan River, UT

Faults

▶ **3.** A *fault* is a fracture where rocks have slipped past one another. Rocks across a fault can slip up and down, as shown here, or they can slip sideways or at some other angle. A fault displaces the rocks on one side relative to the other side.

08.03.a4 Shoshone, CA

08.03.a2

◀ **4.** The long fracture in the center of this photograph cuts across and offsets the rock layers. That is, the layers across the fault have been displaced relative to each other.

B How Do Joints Form?

Joints form when stress pulls a rock apart. The orientation of joints is controlled by the amounts of stress imposed from different directions. In the diagrams below, the size of each arrow reflects the magnitude of stress in that direction. Larger arrows indicate that greater stress is being applied in the direction shown.

Stress Environments in Which Joints Form

08.03.b1

◀ The simplest way that a joint can form is by tension, where stresses pull on the rock. The joint forms as a plane that is perpendicular to the direction of tension. Tension joints only form in shallow levels of the crust because deeper levels have too much confining pressure from the weight of overlying rocks.

▶ At most crustal depths, joints form because fluid pressure opposes the inward push of confining pressure. The block shown here is subjected to differential stress, with the least amount of compression being in the direction shown by the smallest arrows. Joints form perpendicular to this direction of least stress.

08.03.b2

Stress Orientations Control the Orientations of Joints

08.03.b3

◀ *Vertical joints* form when the stress field allows the rock to be pulled apart in a horizontal direction. The vertical joints can form in any compass direction, depending on the orientation of the stresses. A rock pulled in a north-south direction, for example, will have joints oriented east-west.

▶ *Horizontal joints* form if a rock is pulled apart in a vertical direction. This can occur, as shown here, when tectonic stresses push on the sides of the rock, which causes the vertical stress to be the smallest stress.

08.03.b4

C What Other Stresses Form Joints?

08.03.c1

The stresses that form joints arise from many sources, but they mostly are due to burial and to tectonic forces. Tectonic forces may push, pull, or shear the rock. These volcanic rocks (▼) are cut by vertical joints formed by tectonic stresses.

08.03.c2

Stresses build up as rocks get warmer or cooler. As some igneous rocks cool, they contract into polygon-shaped columns bounded by joints that commonly meet at 120° angles. The photograph below shows an example of such *columnar joints*. ▼

08.03.c3

Stresses also arise during uplift of buried rocks, causing rocks to fracture due to reduced pressure. These joints, called *unloading joints*, form parallel to the surface and slice off thin sheets of rock. ▼

08.03.c4 South Fork, CO

08.03.c5 Iceland

08.03.c6 Yosemite, CA

D How Does a Fault Form?

Most faults form when horizontal or vertical compressive stress exceeds the rock's strength. One way to study fault formation is by subjecting a cylinder of rock to increasing stress until it breaks.

◄ A cylinder of unfractured rock is used to investigate how rocks deform. We commonly apply compressive stress parallel to the axis of the cylinder.
08.03.d1

◄ As the stress increases, the rock sample experiences some internal strain, bulging slightly as it is shortened (decreased in length).
08.03.d2

◄ When the applied stress exceeds the strength of the rock, a fracture forms at an angle to the compressive stress.
08.03.d3

◄ Once a fracture forms, continued application of stress causes rock on one side of the fracture to slip relative to rock on the other side. This fracture is a fault.
08.03.d4

To apply these laboratory experiments to faults in the crust, envision the cylinder as part of a block of rock being subjected to stress.

◄ 1. Vertical compression forms a fault having an orientation like that of the fault in this cylinder.
08.03.d5

◄ 2. If we put the block on its side and compress it horizontally, we can form a vertical fault that shears the rocks sideways (laterally).
08.03.d6

◄ 3. If we place the block on its other side, the horizontal compression can form a gently inclined fault along which the top of the cylinder moves upward and over the lower part. Whether horizontal compression forms this type of fault or a vertical fault like the one above depends on stresses from the other directions.
08.03.d7

Before You Leave This Page Be Able To

✓ Sketch and describe the two main types of fractures and how each forms.

✓ Sketch and summarize how the orientation of joints reflects the orientation of stresses.

✓ Summarize how different joints form and are expressed in the landscape.

✓ Summarize how a fault forms relative to compression and why faults form with different orientations.

8.3

8.4 What Are Different Types of Faults?

FAULTS ARE FRACTURES along which rocks have slipped. Stresses on rocks are highly variable, and rock properties also vary, so we should anticipate that there can be different types of faults.

A How Do We Describe Faults?

Fault surfaces can have any orientation, from vertical to horizontal, and slip along a fault can be up-and-down, side-to-side, or somewhere in between. We describe the orientation and movement directions of faults with some special geologic terms.

Dip

The right side of this block is a fault surface that is inclined to the right. If we pour water on this surface, it will flow, or drip, directly down the fault. The water is flowing down what geologists call the *dip* of the fault surface. We say that this fault dips to the right. We describe the amount of dip in terms of degrees from horizontal or use terms such as steep, moderate, and gentle. This fault has a moderate dip.

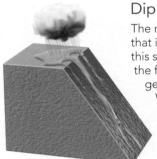

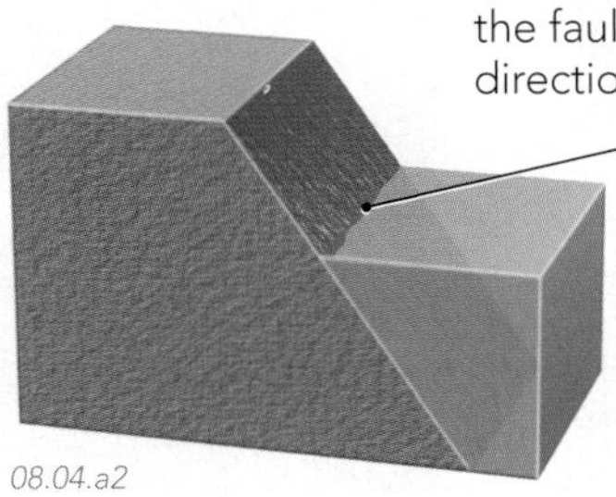

08.04.a1

Strike

If water floods the area, it will intersect, or strike, the fault surface along a horizontal line. The direction of a horizontal line on an inclined surface is the *strike*. We describe the orientation of a surface's strike relative to true north. We might say, for example, that a fault has a "northeast strike." Strike and dip are used to describe layers, joints, and other planar features, in addition to faults.

08.04.a2

Dip-Slip Fault

Slip along a fault can be parallel to the dip—one block moves up or down relative to the other block. This type of fault is called a *dip-slip fault*. In this example, the right block moved down the dip. ▼

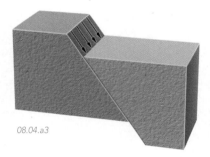

08.04.a3

Strike-Slip Fault

Slip along a fault can occur in a horizontal direction, parallel to the strike. In this example, the right block slipped horizontally back relative to the left block. Slip was parallel to strike, and this fault is a *strike-slip fault*. ▼

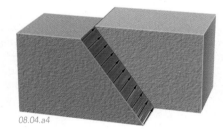

08.04.a4

Oblique-Slip Fault

Slip along a fault can also occur in a direction that is *oblique*, being neither parallel to the dip nor the strike. Here, the right block moved up and back relative to the block on the left. This is an *oblique-slip fault*. ▼

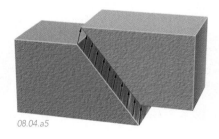

08.04.a5

B What Do We Call the Rocks Above and Below a Fault?

A fault separates the rocks into two *fault blocks*, shown as different colors in the illustration below. The names we apply to these two blocks come from miners exploring for minerals along faults and veins.

The *hanging wall* is the block above the fault. It hung over the miners' heads and was a place to hang lanterns.

Miners could walk on the block below the fault, and this block is the *footwall*.

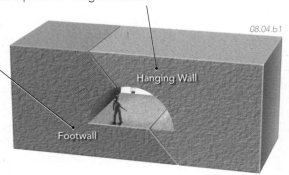

Hanging Wall

Footwall

08.04.b1

Forming an overhang, the hanging wall of this fault is on the left. The geologist is standing on the footwall block.

The footwall of this fault consists of folded and steeply tilted red sedimentary rocks.

08.04.b2 Lincoln Ranch Fault, AZ

Hanging Wall

Footwall

C   What Are the Main Types of Faults?

We classify faults based on the motion of one block relative to the other. There are three main types of faults: *normal faults*, *reverse faults*, and *strike-slip faults*. Black arrows show relative movement, and blue arrows show stress.

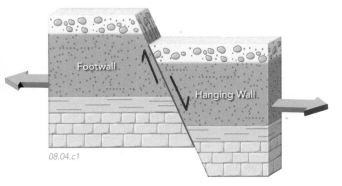

08.04.c1

Normal Fault

If the hanging wall moves down relative to the footwall, the fault is called a *normal fault*. A normal fault forms when the rock units are pulled apart and lengthened, as for example by tension. ◄

A normal fault can have a gentle dip, either because it formed that way initially or because it formed with a steep dip but rotated during or after faulting. ►

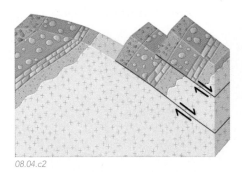

08.04.c2

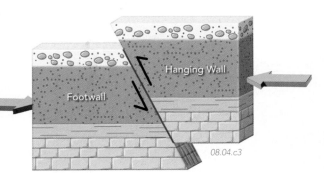

08.04.c3

Reverse Fault

If the hanging wall moves *up* relative to the footwall, the fault is a *reverse fault*. A reverse fault forms as a result of horizontal compression and shortens the rock units in a horizontal direction. ◄

A reverse fault that has a gentle dip is a *thrust fault*. The sheet of rock above the fault is called a *thrust sheet* and is pushed up and over footwall rocks. We depict thrust faults with teeth on the thrust sheet, as shown here. ►

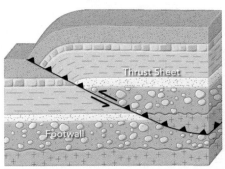

08.04.c4

Strike-Slip Fault

When rocks along a fault move with a side-to-side motion, parallel to the strike of the fault surface, the fault is a *strike-slip fault*. Relative motion is horizontal, offsetting the blocks laterally in one direction or the other.

To refer to the direction in which the two sides moved, imagine standing on one side of the fault and observing which way the other side moved relative to you. In this figure (◄), the opposite side is displaced to the left across the fault, and we call this kind of strike-slip fault a *left-lateral fault*.

If the opposite side is offset to the right, the strike-slip fault is a *right-lateral fault*. ►

08.04.c5

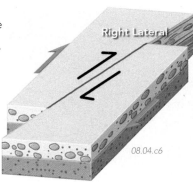

08.04.c6

Relative Displacement of Fault Blocks

Faults commonly break the crust into large, fault-bounded blocks. If one fault block is dropped down along normal faults relative to blocks on either side, the resulting feature is called a *graben*. This term graben can refer to the downdropped block or to the valley above it.

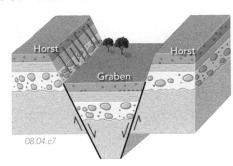

08.04.c7

A block that is uplifted relative to blocks on either side is called a *horst*. Large regions of the American Southwest have horsts (mountains) and grabens (valleys).

8.4

8.5 What Are Folds and How Are They Shaped?

DEFORMATION CAN FOLD ROCK LAYERS, both at depth and at the surface. The folded layers may be bent in gentle arcs or squeezed tightly into sharp angles. We classify folds based on their shape and orientation. Knowing the names of different types of folds gives us a convenient way to describe what we observe in landscapes and outcrops.

A What Is a Fold and What Are the Main Types of Folds?

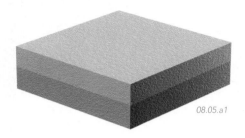

08.05.a1

◄ **1.** Before folding, most rock layers are horizontal because most sedimentary and volcanic layers form with a more or less horizontal orientation.

2. Compressive stress causes shortening, often accommodated by folding of the layers. When you scrunch up a rug, the folds (creases) are perpendicular to the direction of shortening, as shown here for rocks (►). Compression can form folds, faults, or both.

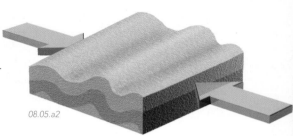

08.05.a2

Anticlines and Synclines

08.05.a3 Tibet

◄ If the rock layers warp up, in the shape of an *A*, the fold is generally called an *anticline*. In an anticline, the *oldest* rocks are in the center of the fold. This fold is an anticline.

08.05.a4 Tibet

If rocks fold down in the shape of a *V* or *U*, the fold is generally called a *syncline*. In a syncline, the *youngest* rocks are in the center of the fold. This fold is a syncline. ◄

08.05.a5

0.25 km

Synclines and anticlines occur together, usually as part of a series of folds. The upward fold of the layers on the left of the diagram is an anticline.

The downward fold of layers on the right side of the diagram is a syncline. The beds that dip to the right in the center of the diagram are part of both folds.

Dome

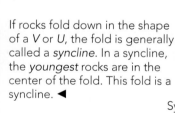

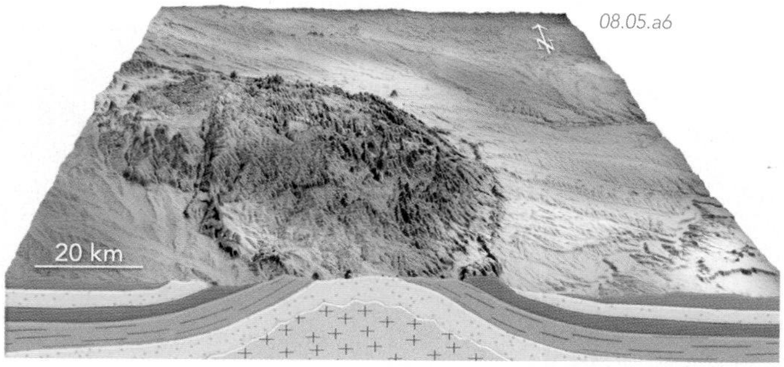

08.05.a6

20 km

Layers that are uplifted in a circular or elliptical area and dip away in all directions form a *dome*. Erosion exposes deeper and older rocks in the center of this dome.

Basin

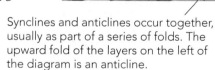

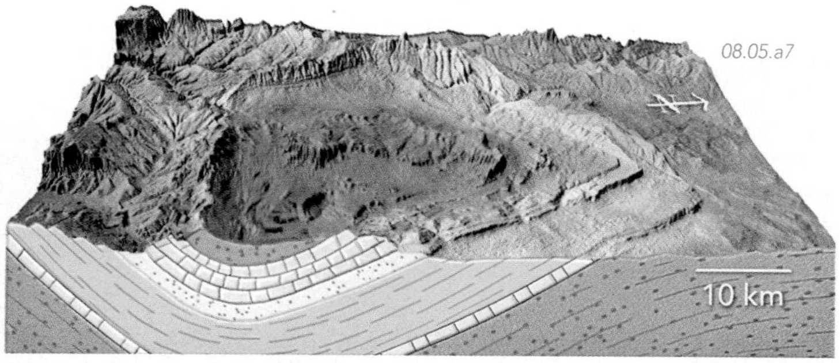

08.05.a7

10 km

A *basin*, formed by folding, is the opposite of a dome. Layers dip toward the center of the basin from all directions. The center of a basin usually preserves younger layers than surrounding areas.

Monocline

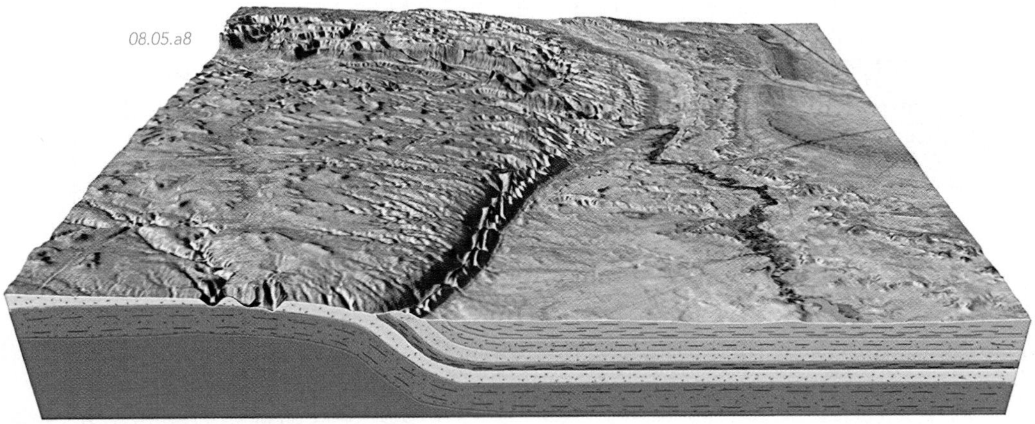

08.05.a8

08.05.a9 Grand Junction, CO

In some folds, nearly flat layers bend down (dip) in one direction and then flatten out again. This type of fold is a *monocline,* a name that indicates that the fold only has one dipping segment. Monoclines can be tens of kilometers long or exposed in an outcrop. Some monoclines have great names, like the Coxcomb, Waterpocket fold, and Comb Ridge in Utah and Arizona. The monocline shown here is part of the San Rafael Swell in central Utah.

This photograph shows a medium-sized monocline in sandstone layers. Horizontal layers on the left bend down in the center of the image (dip to the right) and then fold back to horizontal on the right side of the image.

B. What Is the Geometry of Folds?

A fold has different parts, including the place where it is the most curved and places where the layers are hardly bent at all. These parts and their orientations are used to describe folds.

Hinge, Limbs, and Plunge

The part of a fold that is most sharply curved is the *hinge.* Some geologists also refer to a hinge as a *fold axis.*

The planar or less curved parts of a fold are the *limbs* of the fold. They can be relatively planar, as shown here, or somewhat curved.

08.05.b1

08.05.b2

The hinge of a fold can be horizontal or be inclined from horizontal. If the fold hinge is not horizontal, as in the fold shown here (◀), the fold is said to be *plunging.* The *plunge* of a fold refers to the direction and the amount that the fold plunges (points down). Based on the orientation of its hinge, a fold can be *horizontal (non-plunging),* or *gently, moderately,* or *steeply plunging.*

Axial Surface

An imaginary plane or curved surface, called the *axial surface,* can be fitted through the hinges of each folded layer within a fold. If the axial surface is vertical, a fold is said to be *upright.* Some axial surfaces can be traced across the landscape, as shown on this photograph (▶). Is this fold an anticline, syncline, or monocline?

08.05.b3

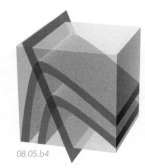

08.05.b4

◀ This axial surface is *inclined* and so the fold does not look symmetrical with respect to Earth's surface. Such a fold is an *asymmetric fold.* The two limbs dip by different amounts.

▶ In this fold, the left limb has been rotated more than 90°, so the limb is said to be *overturned.* The fold is an *overturned fold.*

08.05.b5

08.05.b6 Provo Canyon, UT

Location of Axial Surface

Before You Leave This Page Be Able To

☑ Sketch a cross section of an anticline, syncline, and monocline.

☑ Describe a dome and a basin.

☑ Sketch a fold showing its hinge, limb, and axial surface, and summarize how the orientation of these features can vary.

8.5

8.6 What Are Some Metamorphic Features?

DEFORMATION CAN BE ACCOMPANIED BY METAMORPHISM, during which temperature and pressure can cause the rearrangement of existing materials and the formation of new minerals and new structural features. What characteristics of metamorphic structures can we observe in exposures of metamorphic rock?

A What Is Rock Cleavage?

When rocks are deformed under conditions of low to moderate temperature (less than about 300°C), they may develop a planar fabric along which they break, or *cleave*. This fabric, called *rock cleavage* or simply *cleavage*, is not a fracture but a type of structural discontinuity in the rock. It also is not related to *mineral cleavage*.

08.06.a3 Great Smoky Mtns., TN

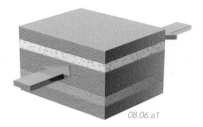

When stress is applied to a rock, the rock can become strained and change shape. The rock typically begins to squash or shorten in the direction of the maximum applied stress. ◄

08.06.a1

▶ Some cleavage is expressed as closely spaced planes that cause the rock to cleave into thin slivers. The marking pen is aligned parallel to cleavage. The folded layers are bedding.

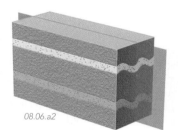

As the rock shortens, it can develop new planar features (marked by the purple plane) that are oriented perpendicular to the direction in which the rock is shortened the most. One type of a planar feature formed in this way is rock cleavage, which is a weakness along which the rock breaks. ◄

08.06.a2

▶ Cleavage typically cuts across bedding, as in this photograph and the previous one. Cleavage is parallel to the red knife and cuts across the gray and brown layers, forcing a sliver of hard, brown rock out of the way.

08.06.a4 Inyo Mtns., CA

B What Is Foliation and How Is It Expressed in Rocks?

Foliation is used as a general term for any planar metamorphic fabric, including cleavage, but it commonly is reserved for more strongly metamorphosed rocks. Foliation forms because of differential stress and is expressed in a variety of ways.

One type of foliation occurs in metamorphic rocks that are rich in mica minerals. This type of foliation, called *schistosity*, is defined by a parallel orientation of mica and other platy minerals. Schistosity makes most rocks shiny, and the rock is *schist*.

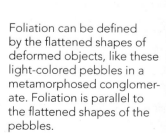
08.06.b1 Ios, Greece

08.06.b2 Swiss Alps

Where metamorphic rocks have alternating lighter and darker-colored *bands*, with different proportions of light and dark minerals, the foliation is a *gneissic* (pronounced "nice-ick") foliation. A metamorphic rock with this type of foliation is a *gneiss*.

Foliation can be defined by the flattened shapes of deformed objects, like these light-colored pebbles in a metamorphosed conglomerate. Foliation is parallel to the flattened shapes of the pebbles.

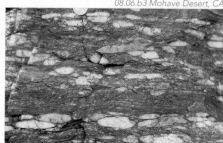
08.06.b3 Mohave Desert, CA

08.06.b4 San Gabriel Mtns., CA

Ductile shearing can form a foliation by flattening and smearing out mineral grains. This rock has a foliation largely defined by lens-shaped, deformed crystals and smeared out light and dark (gneissic) bands.

C What Is Lineation and How Is It Expressed in Metamorphic Rocks?

In addition to planar fabrics (foliation), metamorphism and deformation can form *linear features*, such as aligned minerals or long, deformed pebbles. A linear feature in metamorphic rocks is called *lineation*.

08.06.c1 Naxos, Greece

08.06.c2 Harquahala Mtns., AZ

08.06.c3 Chemehuevi Mtns., CA

Lineation in this metamorphic rock is defined by elongated, blue crystals that grew in a preferred, linear orientation during metamorphism. The mineral is kyanite.

This lineation is defined by the long axes of light-colored feldspar crystals that were stretched out in a horizontal direction. The rock is a metamorphosed granite.

These linear streaks formed as minerals were smeared out during metamorphism and ductile shearing. This type of lineation is parallel to the direction of shearing.

D What Are Some Other Features in Metamorphic Rocks?

08.06.d1 Vredefort, South Africa

08.06.d2 Santa Catalina Mtns., AZ

08.06.d3 Harcuvar Mtns., AZ

If they are not too strongly deformed, metamorphic rocks can preserve features that existed in the rock before it was metamorphosed. This metamorphic rock has tan layers that were sandstone beds with curved tops that represent sedimentary ripples.

This rock began as conglomerate with pebbles. The light-colored pebbles look a bit like stretched pebbles, but the tan and gray pebbles became flattened and folded during deformation and metamorphism, and are not easily recognized as stretched pebbles.

Metamorphic rocks can contain zones that show evidence of intense shearing. Such features are *shear zones*, and are partly a structural feature and partly a metamorphic one. In this photograph, a thin shear zone cuts across metamorphic layers.

08.06.d4 Old Woman Mtns., CA

◄ Most metamorphic rocks have folds, some small and some large and spectacular. This spectacular fold is in banded, gneissic rocks. For scale, note the geologist examining the bottom of the outcrop.

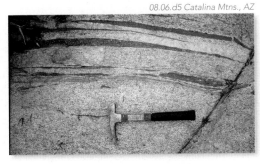
08.06.d5 Catalina Mtns., AZ

Metamorphic rocks, especially those with gneissic foliation, are commonly intimately associated with layers, lenses, pods, and dikes of igneous material. Here, light-colored granite occurs with dark metamorphic layers. ▲

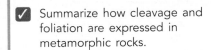

Before You Leave This Page Be Able To

✓ Summarize how cleavage and foliation are expressed in metamorphic rocks.

✓ Summarize the types of features that define lineation and how each type of lineation forms.

✓ Describe some other features that may be present in metamorphic rocks.

8.6

8.7 What Are Some Common Metamorphic Rocks?

ROCKS THAT HAVE BEEN CHANGED (or metamorphosed) by increased temperature and pressure are *metamorphic rocks*. There are many types of metamorphic rocks, and these rocks reflect diverse starting rock types and various conditions under which rocks can be metamorphosed and deformed.

A What Rocks Form When Sedimentary Rocks Are Metamorphosed?

Some metamorphic rocks form when sedimentary rocks are subject to a change in temperature, pressure, or both. The changed conditions may change the rock's grain size, types of minerals that are present, and the rock's texture. Metamorphism is *low grade* if the temperature of metamorphism is low, and *high grade* if temperature is high. Different kinds of sedimentary rocks will produce different kinds of metamorphic rocks, even if metamorphosed at the same grade. Photographs below show what happens to three common sedimentary rocks—shale, sandstone, and limestone.

(left margin, vertical) Increasing Metamorphic Grade

Shale

08.07.a1 Larder Lake, Ontario, Canada
Slate

When a shale is metamorphosed at low to moderate temperature, it can develop cleavage and become *slate* (◄). Slates are dull (not shiny) and commonly dark.

08.07.a2 Southern AZ
Phyllite

At slightly higher temperature, small (microscopic) mica crystals give the rock a shiny aspect or *sheen* (◄). Such a rock is *phyllite*.

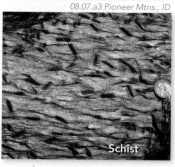
08.07.a3 Pioneer Mtns., ID
Schist

At higher grades, crystals of mica and other minerals become large enough to see. The resulting rock has a *schistosity* and is a *schist*. ◄

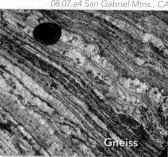
08.07.a4 San Gabriel Mtns., CA
Gneiss

At even higher grades, chemical constituents are mobilized and light- and dark-colored minerals separate, forming a *gneissic* foliation and banded rock called *gneiss*. ◄

Sandstone

08.07.a5 Phoenix Mtns., AZ
Quartzite

Most sandstones are predominantly quartz, a mineral that is stable over a wide range of temperature and pressure conditions. During metamorphism, quartz grains grow together and become so tightly bonded that fractures break across the grains rather than around them. This type of rock is a *quartzite*. Quartzite is made of quartz, just like the original sandstone, and can preserve original sedimentary features, such as these cross beds. ▲

08.07.a6 Joshua Tree, CA
Coarse Quartzite

▲ With higher temperatures of metamorphism, the quartz in the rock begins to merge into larger crystals and can become a coarser grained quartzite, in some cases with no individual grains left. Quartz is soluble and mobile in metamorphic fluids and so at high grades can be redistributed into quartz veins, which are common in metamorphic rocks.

Limestone

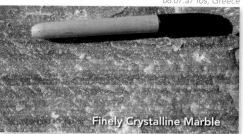
08.07.a7 Ios, Greece
Finely Crystalline Marble

Limestone consists mostly of calcite, a chemically reactive mineral. Low-grade metamorphism of limestone causes calcite to slightly recrystallize, forming a finely crystalline *marble*, but few new minerals form. ▲

08.07.a8 Mojave Desert, CA
Impure Marble

At moderate temperatures, marble becomes medium grained (▲). Impurities, such as clay and chert within the original limestone, may chemically react with calcite to form new metamorphic minerals.

08.07.a9 Naxos, Greece
Coarse Marble

▲ At higher grades, calcite crystals grow coarser to produce a coarsely crystalline *marble*. The one shown here is nearly 100% calcite, but coarse marbles commonly also contain other minerals.

B What Rocks Form When Igneous Rocks Are Metamorphosed?

Metamorphic rocks also can be formed from igneous rocks. Igneous rocks of different compositions and initial grain sizes become different types of metamorphic rocks. Photographs below show several examples.

Increasing Metamorphic Grade

Basalt and Andesite

08.07.b1 Noranda, Quebec, Canada

Greenstone

Most mafic and intermediate volcanic rocks (e.g., basalt and andesite), are gray or black, but they can become fine-grained, greenish rocks, called *greenstone*, when metamorphosed.

08.07.b4 Zermatt, Switzerland

Greenschist

At moderate grades, crystals of green or black mica and amphibole grow in greenstone to produce a schistose greenish rock, *greenschist*. At higher metamorphic grades, the green minerals recrystallize into black amphibole and the rock becomes a type of gneiss.

Rhyolite

08.07.b2 Malartic, Quebec, Canada

Metarhyolite

Slightly metamorphosed, felsic volcanic rocks, such as rhyolite, become light-colored and shiny, cleaved or foliated rocks, called *metarhyolite*. Such rocks may retain original volcanic crystals and fragments.

08.07.b5 Pontresina, Switzerland

Schist

At moderate metamorphic grades, crystals of light-colored muscovite mica become larger and the rock becomes light-colored *schist*. At even higher metamorphic grades, metamorphic processes begin to form coarse-grained layers and pods, and the schist can become gneiss.

Plutonic Rocks

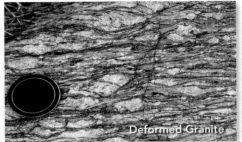
08.07.b3 Harquahala Mtns., AZ

Deformed Granite

After low-grade metamorphism and deformation, plutonic rocks, like granite, may show little change other than developing a weakly developed foliation. Geologists simply call such rocks *deformed granite*.

08.07.b6 Halloran Hills, CA

Gneiss

Plutonic rocks can be metamorphosed and deformed into schists if they are sheared with abundant water. More commonly, they become gneiss at higher temperatures, like the example above. In some cases, the lighter colored layers in these rocks formed by partial melting under high temperatures.

Some Other Distinctive Metamorphic Rocks

A special family of metamorphic rocks forms when preexisting rocks are baked by an igneous intrusion. Metamorphism is dominated by heat from the magma, not by pressure or differential stress, and so the rock contains metamorphic minerals but does not have foliation or lineation. A nonfoliated metamorphic rock is *hornfels* and can form from any kind of starting rock, including limestone, mudstone, shale, or igneous rocks. This hornfels, which includes red garnet, was a shale that was baked by hot mafic magma.

08.07.mtb1 Vredefort, South Africa

Hornfels with Garnet

Before You Leave This Page Be Able To

☑ Describe the changes different sedimentary rocks undergo (as they are metamorphosed) and the metamorphic rocks they become.

☑ Describe the changes different igneous rocks undergo (as they are metamorphosed) and the metamorphic rocks they become.

☑ Describe the origin of hornfels.

8.7

8.8 How Does Metamorphism Occur?

METAMORPHISM OCCURS WHEN A ROCK BECOMES UNSTABLE. A rock may be unstable because the minerals it contains are unstable and as a result it develops new (metamorphic) minerals. It may be unstable because of the way the grains or layers are arranged, so it develops a new (metamorphic) texture. Depending on the conditions, a rock may develop new metamorphic minerals and a new metamorphic texture.

A What Causes Metamorphism?

For a rock to be metamorphosed, it must be subjected to conditions of temperature, pressure, and fluid chemistry that make it unstable. This generally occurs when temperature (T) and pressure (P) increase, either with or without associated tectonic stress, shearing, and abundant fluids. There are two main kinds of metamorphism: *contact metamorphism* is caused by local heating by magma, typically without deformation, whereas *regional metamorphism* involves deformation along with heating over a broader region.

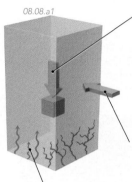

08.08.a1

Pressure increases with depth in Earth because rocks are more deeply buried. Higher pressures compress the rocks and, in combination with high temperatures, may cause some minerals to be unstable. *Tectonic stresses* can cause rocks to move and deform.

Temperature increases with depth and near magma. An increase in temperature usually causes new minerals to grow or existing minerals to grow larger. It can also weaken the rocks, allowing them to deform. The crust contains abundant water and other *fluids*, shown here as blue in water-filled fractures. Such fluids interact with minerals and carry material into, through, and out of rocks.

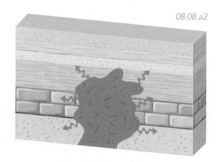

08.08.a2

Contact Metamorphism—Rising magma efficiently brings thermal energy higher into the crust, heating the wall rocks. This is called contact metamorphism because it occurs near contacts (boundaries) of magma. Heating causes new minerals to grow or existing minerals to increase in size. Heating may not be accompanied by deformation, so contact metamorphism commonly forms metamorphic rocks that lack foliation and lineation.

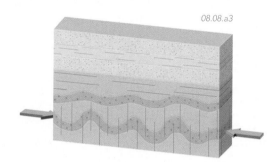

08.08.a3

Regional Metamorphism—In regional metamorphism, heating is accompanied by enough stress to cause deformation. The increase in stress can result from tectonics or burial. Regional metamorphism causes new minerals to grow and existing minerals to increase in size, but deformation during metamorphism generally results in foliation, lineation, and other metamorphic fabrics.

B What Causes Regional Metamorphism?

Metamorphism occurs when rocks are subjected to conditions different than those in which they formed. In the case of regional metamorphism, rocks typically are buried and heated, and they deform in response to stresses.

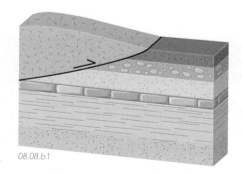

08.08.b1

Burial—Rocks move deeper in the crust when sedimentary and volcanic rocks are deposited on top, or when they are overridden by a thrust sheet, as shown here. As they go deeper, the rocks experience higher pressure-temperature (P-T) conditions.

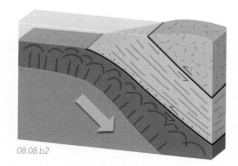

08.08.b2

Subduction—An extreme case of downward-moving rock occurs in subduction zones where two plates converge. Subduction carries the oceanic plate, and slices of other rocks, to great depth and greater pressure and temperature.

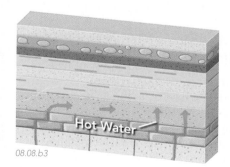

08.08.b3

Heat and Hot Fluids—A flux of heat from depth, such as from deeper magma, can raise the temperature of a region and cause metamorphism. Rising hot waters can introduce dissolved chemicals, changing a rock's composition and growing new minerals.

C What Chemical Processes Occur During Metamorphism and Deformation?

Many processes that operate during metamorphism change the minerals in a rock or change the arrangement of those minerals. Some metamorphic processes are related to heating or are chemical in character. These processes cause grains to grow, recrystallize, redistribute themselves, or even dissolve in response to temperature, pressure, and any imposed stress. In the figures below, red arrows link the *before* and *after* conditions.

Recrystallization

08.08.c1

Formation of cleavage and foliation is aided by the recrystallization of existing minerals and the growth of new minerals. Adjacent minerals can grow with a similar planar or linear orientation (i.e., become aligned), defining a foliation, lineation, or both.

Remobilization

08.08.c2

During metamorphism, chemical constituents in a rock can be remobilized, meaning they diffuse, dissolve, or partially melt in one place and then form crystals in another place. Such processes help form light- and dark-colored bands in gneiss.

Pressure Solution

08.08.c3

Formation of cleavage commonly involves a process called *pressure solution*. Material dissolves from highly stressed edges of grains and precipitates elsewhere in the rock or is carried away by fluids. A rock can lose a significant volume during metamorphism.

D What Physical Processes Can Occur During Metamorphism?

Some processes are physical and may deform or rotate individual crystals, grains, and layers, producing foliation or lineation. In contact metamorphism and some other settings, metamorphism occurs without any of these physical processes, and so the rock lacks foliation or lineation. Blue arrows indicate the type of stress (compression or shearing), and red arrows link the *before* and *after* conditions.

Deformed Grains and Clasts

08.08.d1

Deformation can flatten grains and clasts that were initially somewhat spherical into shapes like pancakes or the thin, long top of a skateboard. If a rock is flattened in one direction during metamorphism, deformed objects will become shaped like pancakes.

Rotation

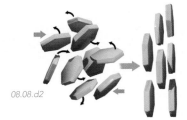

08.08.d2

During compression of some rocks, elongated and platy minerals can rotate so that they become nearly parallel to one another. The rotated minerals produce a foliation or cleavage, and they can also form a lineation if they become aligned during deformation.

Shearing

08.08.d3

Shearing under metamorphic conditions can align or stretch out unoriented crystals. It can also form thin shear zones in which minerals are smeared out, broken, and recrystallized.

Metamorphism Without Deformation

If a rock is only subjected to heating and burial pressures during metamorphism, it essentially will not deform. The rock will grow new minerals or become coarser grained due to the new, hotter conditions. Without the controlling influence of differential stress, new minerals usually grow in random directions, so the rock will lack a foliation or lineation. Also, initially spherical objects, such as pebbles, will remain spherical. ▶

08.08.d4

Before You Leave This Page Be Able To

✓ Summarize causes of metamorphism.

✓ Describe or sketch the physical processes that can accompany metamorphism.

✓ Describe or sketch the chemical processes that can accompany metamorphism.

8.8

8.9 Where Does Metamorphism Occur?

METAMORPHISM OCCURS IN VARIOUS SETTINGS, but especially within mountain belts along *convergent* plate boundaries. Different types of structures and metamorphic features form in different parts of a convergent system and reflect differences in the types of rocks involved, the metamorphic temperatures and pressures, the way the rocks deform, and the role of magma, if any.

A How Do Metamorphic Conditions Vary with Depth and Tectonic Setting?

Temperature and pressure in Earth vary greatly, but both conditions generally increase with depth. Pressure increases with depth because the weight of overlying rocks increases. Temperature also increases with depth, but it varies more because some parts of Earth are hotter than others, even at the same depth.

1. As rocks are subjected to increasing temperature (T) and pressure (P), they are progressively metamorphosed. The colors on this graph illustrate where different metamorphic equivalents of shale form.

2. Where temperature and pressure are very low, as they are near the surface of Earth, most rocks remain unmetamorphosed (no growth of metamorphic minerals).

3. In some tectonically active regions, rocks are taken to depth faster than they can heat up. This results in a *high-P/low-T environment* of metamorphism (on the left part of the diagram). Some minerals only form under high-P/low-T conditions, including those that form a tectonically important subduction-zone rock called *blueschist*. High-P/low-T metamorphism occurs in accretionary prisms and deeper in subduction zones where cold material is rapidly subducted down into the mantle. The rock shown here (◄), called *eclogite*, was taken tens of kilometers down a subduction zone and then uplifted back to the surface.

08.09.a2 Ios, Greece

4. Some shallow areas of the crust, where pressures are low, are abnormally hot and represent a *low-P/high-T environment* (along the top of the diagram). To heat such shallow levels to high temperatures requires the input of thermal energy from magma. This is the setting of contact metamorphism. The crystals weathering out of the rock in the photograph below grew in random orientations during contact metamorphism. ▼

08.09.a3 Vredefort, South Africa

TEMPERATURE ⟶

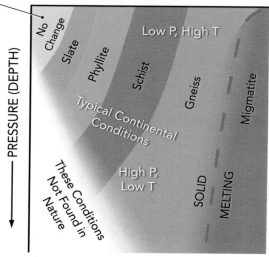

PRESSURE (DEPTH)

No Change

Slate

Phyllite

Schist

Gneiss

Migmatite

Low P, High T

Typical Continental Conditions

These Conditions Not Found in Nature

High P, Low T

SOLID

MELTING

08.09.a1

6. The center of the graph represents temperature and pressure conditions inferred for many common metamorphic rocks. Metamorphism under these conditions can occur over large regions, that is, regional metamorphism. Most metamorphic rocks form under these conditions.

5. If temperature increases enough, rocks may begin to melt. The resulting magma can mingle with the metamorphic rocks and form a rock that is part metamorphic and part igneous. Such a rock is a *migmatite*. ▼

08.09.a4 Inyo Mtns., CA

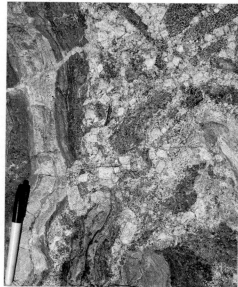

Diagnostic Regional Metamorphic Minerals

7. As pressure and temperature increase, causing increased metamorphism, a succession of minerals form. This diagram shows the stability range of minerals in a regionally metamorphosed shale. Colors used for each mineral in this figure (▶) correspond to temperature zones shown in the figure above. Some minerals, like garnet, are diagnostic of moderate- to high-grade metamorphic conditions. Note that some minerals overlap in the temperatures at which they form, which allows geologists to better infer metamorphic temperatures if a rock has both minerals.

INCREASING METAMORPHISM ⟶

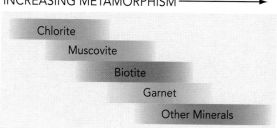

Chlorite

Muscovite

Biotite

Garnet

Other Minerals

08.09.a5

B What Activity Occurs Along Ocean-Continent Convergent Boundaries?

In ocean-continent convergent settings, an oceanic plate subducts beneath a continental plate, forming magma that invades the overriding continental plate. The type of metamorphism varies greatly between the hot environs near magma and the less hot conditions in shallow parts of the subduction zone.

1. A subducted slab is underthrust beneath the overridding plate to form a huge thrust fault, which many geologists call a *megathrust*. Thrusts also form in the overlying, intensely deformed accretionary prism, which consists largely of material scraped off the oceanic plate or sediment deposited in the oceanic trench.

2. Subduction is a relatively fast tectonic process, so the upper parts of the downgoing oceanic plate can be taken to great depth faster than they can be heated by surrounding rocks. As a result, metamorphism in this setting occurs under *high-P/low-T conditions*. The rocks are strongly cleaved, folded, sheared, and sliced by thrust faults. Water that is driven out of the heated rocks forms numerous veins.

3. High-P/low-T conditions continue downward within the *subducted slab*, which moves down into the mantle faster than it can be heated up. Metamorphism of minerals in the oceanic crust releases water that rises into the overlying mantle and causes melting. The mantle-derived magma rises into the continental crust.

4. The crust is heated near magma rising from the underlying subduction zone. The magmatic heat causes contact metamorphism, which is characterized by high-T/low-P conditions. Such contact metamorphism can occur with or without deformation. The contact metamorphism grades into regional metamorphism, especially at deeper levels in the crust.

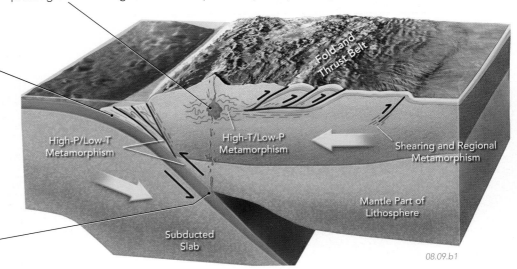

08.09.b1

5. Folds and thrust faults commonly form on the landward side of the main belt of magma. Thrust sheets move into the continent and override and bury underlying rocks, causing regional metamorphism.

C What Deformation and Metamorphism Occur Along Continental Collisions?

Thrust faults form during continental collisions when one continent under-thrusts beneath another. Thrust sheets typically form in a broad zone between the two plates, with sheets of rock sliced off both the underthrusting and over-riding plates. Thrusting causes shearing as well as burial and heating of rocks below the thrust sheets. On this figure, the most intense metamorphism is shaded purple.

Rocks within the thrust sheets are strongly folded and sheared especially near major thrust faults and shear zones. Large zones of regional metamorphism develop below and near the bases of the thrust sheets. The metamorphic conditions are typical regional metamorphism, and they occur at typical crustal temperatures and pressures.

▶ These schists in the Swiss Alps began as sediments and volcanic rocks that were deposited in an ancient ocean that was south of Europe. Later, this oceanic material was buried, heated, and sheared by subduction and a continental collision between Europe and a slice of continental crust.

08.09.c2 Zermatt, Swiss Alps

08.09.c1

Before You Leave This Page Be Able To

✓ Summarize and graph the difference between low-P/high-T, high-P/low-T, and normal P-T metamorphism and their tectonic settings.

✓ Sketch and describe metamorphism along ocean-continent convergent boundaries and continental collisions.

8.9

8.10 What Types of Deformation and Metamorphism Occur in Extensional and Strike-Slip Settings?

CRUSTAL EXTENSION ON LAND OR BENEATH THE OCEANS produces normal faults and other characteristic features. Metamorphism occurs in extensional settings because of shearing, heating near magmas, and the circulation of hot water near mid-ocean spreading centers. Deformation, with or without metamorphism, also occurs along strike-slip faults and shear zones, forming some distinctive features on the land surface.

A What Type of Deformation and Metamorphism Accompanies Divergence?

Continental Rifting

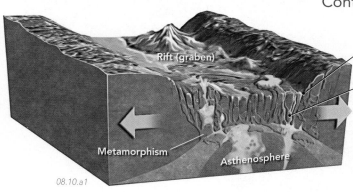

08.10.a1

◄ Normal faults form where two continental plates begin to break apart, forming a continental rift. The faults downdrop a central block, or graben, and help to horizontally stretch and vertically thin the crust.

Tension allows magma to rise along faults and to pry open joints on its way to the surface. The orientation of faults and volcanic fissures on the surface is controlled by the orientation of the stresses.

Metamorphism, shown in purple, occurs due to heat from nearby magma (contact metamorphism), and in some cases heating from mantle-derived magma causes crustal melting. Metamorphism and deformation also occur due to shearing along deep faults and due to the circulation of hot water.

Onset of Seafloor Spreading

► If rifting within a continent continues, it can lead to the onset of seafloor spreading and the formation of a narrow ocean basin. Normal faults formed during the rifting stage are exposed on land and covered by sediments deposited on the coastal plain and continental shelf. The edge of the continent drops below sea level because the crust is thinned by rifting.

As the two parts of the continent begin to separate and seafloor spreading begins, rocks along the continental margin experience metamorphism because they are close to upwelling mantle and to magma along the new mid-ocean ridge. With time, these metamorphic effects diminish as the ridge forms new oceanic crust and moves farther offshore, away from the edge of the continent.

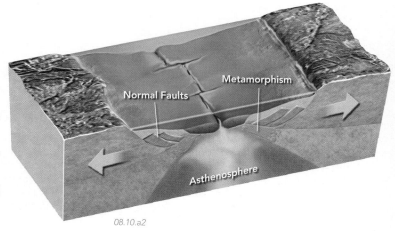

08.10.a2

Mid-Ocean Rifting

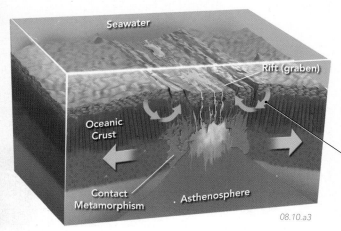

08.10.a3

◄ Extension occurs along oceanic divergent plate boundaries where two oceanic plates spread apart along a mid-ocean ridge. As the plates move apart, normal faults that dip inward from both sides downdrop the floor of the central rift to form a graben. Outward-directed tension (pulling) allows magma to open up joints and other fractures, forming dikes and fissures parallel to the axis of the rift.

Heat from magma causes contact metamorphism (shown in purple) of adjacent rocks. Also, the entire oceanic crust remains warm for some time after it is formed at the ridge, so it experiences a type of low-temperature, regional metamorphism related to these slightly elevated temperatures.

Hot rocks beneath the rift cause heating of seawater, which circulates through and alters or metamorphoses the volcanic rocks and any overlying sediments. Alteration and metamorphism typically change volcanic glass and crystals into various greenish minerals, such as *chlorite* (a green mica). The volcanic rock and sediment can also be replaced by fine-grained quartz, carbonate minerals, or other hydrothermal minerals.

B Where Do Strike-Slip Faults and Shear Zones Form?

During strike-slip movement, one block of rock is sheared sideways past another block of rock. This can occur in various settings, including transform plate boundaries and within the interiors of plates.

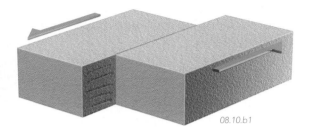

08.10.b1

Stresses can form a strike-slip zone that functions as a plate boundary or that is totally within a tectonic plate (▶). A strike-slip zone may offset the rocks hundreds of kilometers or less than a meter. A strike-slip fault with relatively small amounts of displacement is typically a single fault or several adjacent faults, but zones with larger displacements are thick zones of shear (shear zones).

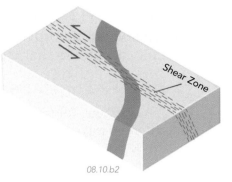

08.10.b2

▲ Rocks can be subjected to horizontal shear stresses, which act to shear the two sides of a block in opposite directions. As a result of the stresses, shearing moves rocks horizontally past one another. Shearing in the upper parts of the crust occurs along a fault, as shown here, and is accompanied by fracturing of adjacent rocks. Shearing at depth will occur along a zone of ductile deformation and will be accompanied by metamorphism and the formation of foliation and lineation.

08.10.b3

◀ All transform boundaries are strike-slip faults that accommodate the lateral displacement of one plate past another. Most are a boundary between two oceanic plates, as are the ones marked here by small white arrows, but a transform fault can also separate two continental plates or can separate an oceanic plate from a continental one.

C What Features Form Along Strike-Slip Faults?

Strike-slip faults result in a number of distinctive features, including offset streams. They also can have folds formed where one block of rock shears past another or where rocks are forced around a bend in the fault.

Strike-slip faults displace rocks on either side horizontally relative to one another, so in a simple case would not uplift or downdrop either side. However, many strike-slip faults have bends, where the fault changes its trace across the land surface from one orientation to another. Right-lateral motion on the fault shown here causes compression along the bend, forming ridges and troughs that are the surface expression of folds and thrust faults.

Horizontal displacement can offset surface features, including roads, agricultural fields, and streambeds. Over time, offset streams develop a characteristic pattern, where they jog parallel to the fault, before continuing along their prefaulting course. The direction of the jog reflects the direction of relative movement across the fault.

08.10.c1

08.10.c2 Carrizo Plain, CA

◀ Faults that are currently active can offset streams, ridges, and other topographic features. The San Andreas fault in central California is the linear feature cutting across drainages in the center of the photograph. The large offset stream takes a jog as it crosses the fault. Is this fault a left-lateral or right-lateral strike-slip fault? Hint: imagine you are standing in the streambed on the near side of the fault, and then observe which way the streambed on the opposite side has been displaced relative to you.

Before You Leave This Page Be Able To

✓ Describe or sketch how deformation and metamorphism occur in continental rifts, rifted continental margins, and mid-ocean ridges.

✓ Describe strike-slip faults, some settings where they occur, and some features formed on the land surface.

8.10

How Are Different Structures and Metamorphic Features Related?

DIFFERENT TYPES OF GEOLOGIC STRUCTURES can be related to one another and may form during the same episode of deformation. Metamorphism may accompany deformation and form cleavage and other metamorphic structures that can be closely related to folds and other large-scale features.

A How Are Fractures, Folds, Tilting, and Shear Zones Related?

Joints, faults, folds, and shear zones can be associated with one another, helping to accommodate deformation in different ways. Displacement on faults and shear zones can cause folding and tilting of layers that are displaced.

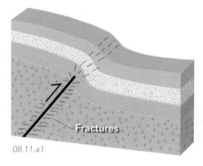

08.11.a1

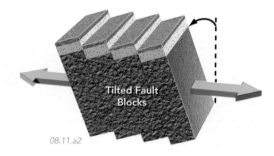

08.11.a2

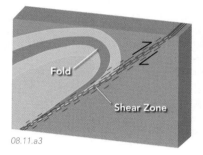

08.11.a3

Joints and folds are common along faults because of the stresses associated with shearing and bending of the rocks. Joints and small folds may imply a nearby fault.

Normal faults can break the crust into a series of *fault blocks*, each bounded by faults. As the faults slip, the blocks can rotate like books on a shelf, tilting any layers.

During displacement on a shear zone, rocks deform by ductile flow, producing metamorphic fabrics. Shearing can also ductilely fold and stretch layers within and near the zone.

Folds Related to Fault Growth

Some folds form in front of a fault advancing through the crust.

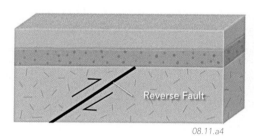

08.11.a4

Movement on most faults begins within crystalline rocks at depth, and the fault grows or propagates upward through the crust.

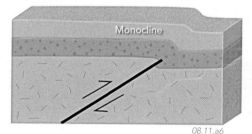

08.11.a6

Sedimentary layers on top of the crystalline rocks can fold in advance of the fault and resemble a rug draped over two boxes of unequal height. Such folding can form a monocline, like the one below.

Folds Related to Changes in Fault Dip

Folds form where a thrust or other type of fault changes dip.

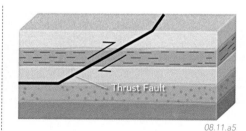

08.11.a5

Faults slicing through layered rocks may cut along a weak rock type, like shale, or along a boundary between two rock layers, before continuing upward.

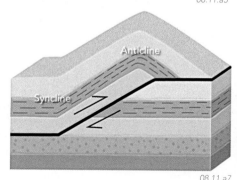

08.11.a7

As the layered rocks are pushed up the fault, they fold as they are forced to conform to the bend in the underlying fault. This forms an anticline near the front of the thrust sheet (as shown below) and a syncline farther back.

08.11.a8 Mexican Hat, UT

08.11.a9 Northeastern WA

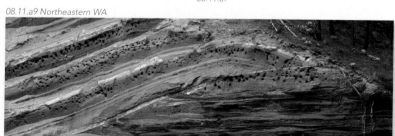

B What Structural and Metamorphic Features Form Near Folds and Thrust Faults?

Thrust faults and folds commonly occur together in regional belts, called *fold and thrust belts.* These form where thrust faults cut through a thick sequence of layered rocks.

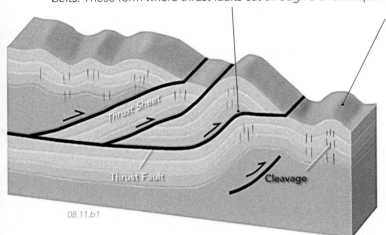

08.11.b1

Other folds develop from overall shortening of rocks in a thrust belt or above thrust faults propagating to the surface from below.

Most thrust belts contain variably developed cleavage (shown with thin dashed lines) related to the folds and to shear along the thrusts. The deformed rocks also contain many joints.

Thrust faults shuffle rock layers by displacing older rocks over younger ones (green and blue over tan in this image). Large folds form where the layers are forced up and over bends in the thrusts.

▶ These anticlines and synclines, each of which are several meters high, formed by shortening of layers of slightly metamorphosed black shale (now slate) and tan sandstone (now quartzite) in a Precambrian fold and thrust belt.

08.11.b2 Barnhardt Canyon, AZ

▶ Larger folds, formed by thrusting and overall shortening, deform shale and sandstone layers during regional thrusting in the foothills of the Patagonian Andes in Argentina, South America. The Rocky Mountains and Appalachian Mountains of North America contain similarly large folds.

08.11.b3 Patagonia, Argentina

C How Is Cleavage Related to Folds?

Cleavage has a close and consistent geometric relationship to folds formed during the same event.

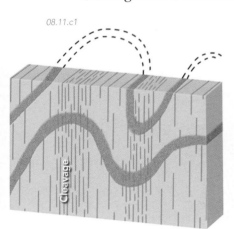

08.11.c1

Cleavage typically is *parallel to the axial surface* of folds that formed during the same episode of deformation. Such cleavage generally cuts across bedding and other layering in a systematic way. Cleavage can form vertically, as shown here, horizontally, or at some orientation in between.

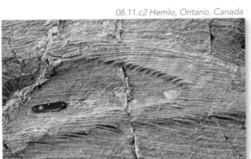

08.11.c2 Hemlo, Ontario, Canada   08.11.c3 Grand Canyon, AZ

These two photographs show cleavage that is parallel to the axial surface of folds. The one on the left is looking down on a single fold with cleavage. The one on the right shows dark cleavage planes parallel to the axial surfaces of a series of small folds.

Using the Relationship Between Bedding and Cleavage to Locate Folds

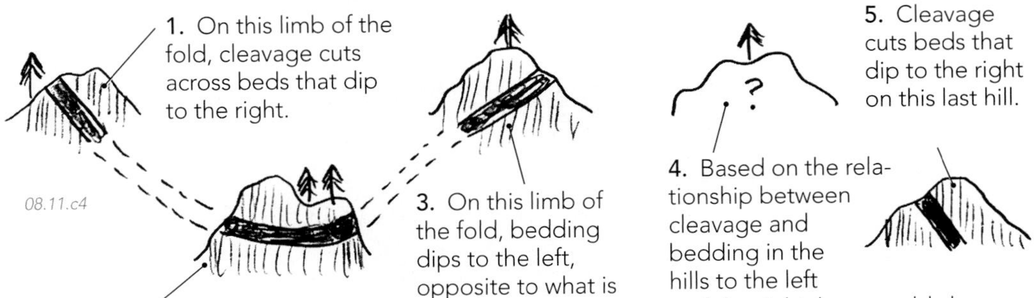

08.11.c4

1. On this limb of the fold, cleavage cuts across beds that dip to the right.

2. In the hinge of a fold, cleavage and bedding are perpendicular to one another.

3. On this limb of the fold, bedding dips to the left, opposite to what is seen on the first hill (the other limb of a syncline).

4. Based on the relationship between cleavage and bedding in the hills to the left and the right, how would cleavage and bedding be oriented on this hill, and what type of fold is present?

5. Cleavage cuts beds that dip to the right on this last hill.

Before You Leave This Page Be Able To

☑ Sketch or describe how joints, faults, folds, tilting, and shear zones can be associated with one another.

☑ Sketch and describe a fold and thrust belt and its associated structures.

☑ Describe and sketch how cleavage relates to bedding in folds.

8.11

8.12 How Are Geologic Structures and Metamorphic Rocks Expressed in the Landscape?

ROCKS AND GEOLOGIC STRUCTURES on Earth's surface are exposed to weathering and erosion, which remove some rocks faster than others. Joints, faults, and cleavage provide easy access for water into the rocks and so may be the focus of weathering and erosion. Folding, faulting, and metamorphism can tilt and otherwise deform rocks, which are then eroded into distinctive landforms that provide clues about what type of structure is present and what events might have happened.

A How Are Joints Expressed in Landscapes?

08.12.a1 Tasmania

08.12.a2 Zion NP, UT

08.12.a3 Grand Canyon, AZ

Most joints occur in *joint sets* that contain many joints with a similar orientation. Most rocks contain several joint sets, which can cut rock layers into a series of rectangular blocks. In this photograph, horizontal beds are cut by a set of vertical joints.

Joints are largely responsible for the appearance of many cliffs, ledges, and other outcrops of rock. In this cliff, a near-vertical joint set cuts across a thick horizontal layer of sandstone. Without joints, the cliff would probably be relatively smooth.

Columnar joints form by the cooling and contraction of solidified igneous rocks and are distinctive in outcrop. The size and orientation of the columns reflect how the rock cooled but most columns, like these, are steep and tens of centimeters in diameter.

B How Are Faults Expressed in Landscapes?

08.12.b1 Borah Peak, ID

◄ When fault movement offsets Earth's surface, it can cause a step in the landscape, called a *fault scarp*. This dirt-colored fault scarp formed during the 1983 Borah Peak earthquake in southeastern Idaho.

08.12.b2 Echo Cliffs, AZ

◄ We commonly recognize faults because of *offsets* or abrupt *terminations* of layers. Also, rocks along faults are highly fractured and easily eroded, so they erode into linear topographic notches. This fault truncates layers and forms a linear notch.

08.12.b3 Crete

◄ Most faults are accompanied by intense fracturing and shattering of rocks in a *fault zone*. Fractured and crushed rocks within the zone are *fault breccia* and indicate that a fault is near.

08.12.b4 Hieroglyphic Mtns., AZ

◄ Most fault surfaces are smooth and polished, and many have linear scratch marks. The scratch marks form parallel to slip along a fault and can be used to infer which way a prehistoric fault moved.

C How Do Tilted and Folded Layers Erode?

08.12.c1 Lime Ridge, UT

08.12.c2 Split Mtn., UT

08.12.c3 The Hogback, NM

Erosion can strip off easily eroded layers, but it slows upon encountering an underlying hard layer. Erosion of soft and hard layers can carve a *dip slope* parallel to planar, dipping layers or to gently curving layers.

As layers change dip, the landscape expression changes too. These layers form a dip slope near the top of the mountain but form steep fins of rock where the layers are nearly vertical near the base of the mountain.

Erosion of dipping layers in a tilted fault block or on the limb of a fold can create a landscape with linear or curved ridges formed from more resistant rock layers. If a ridge has a dip slope on one side, it is a *hogback*.

D How Are Metamorphic Rocks and Shear Zones Expressed in Landscapes?

08.12.d1 Pamour, Ontario, Canada

Metamorphic rocks can be shiny even from a distance, if their mica minerals share a similar orientation and reflect light. This shiny phyllite and quartzite preserves wavy ripple marks and a set of veins that form small ridges. ◄

08.12.d2 Patagonia, Argentina

Metamorphic rocks have many different expressions in landscapes because they can involve different rock types, metamorphic histories, and structural orientations. This folded rock is a schist formed by metamorphism of a dark shale. ◄

Many metamorphic rocks have cleavage, foliation, and layers that form platy, jagged outcrops and tabular slabs of rock, as in the left photograph below. They can include numerous dikes, sills, and pods of granite and other igneous rocks, some of which can help us observe how folded and deformed these metamorphic rocks really are.

Below, two thin shear zones cut across and displace metamorphic layering in a metamorphosed volcanic rock (▼). Rocks on the left side of each shear zone are displaced down relative to rocks on the right.

08.12.d3 Aurland Trail, Norway

08.12.d4 Grand Canyon, AZ

08.12d5 Jerome, AZ

Before You Leave This Page Be Able To

☑ Identify joints in a photograph of a landscape and describe how joint sets weather and erode.

☑ Summarize or sketch how you might identify a fault in the landscape.

☑ Summarize or sketch the features that form when tilted or folded layers are eroded.

☑ Describe some characteristics displayed by metamorphic rocks and shear zones.

8.12

8.13 How Do We Study Geologic Structures and Metamorphic Features?

UNDERSTANDING STRUCTURAL AND METAMORPHIC FEATURES is a key step when reconstructing the history of Earth. Geologists usually begin the process by mapping, making field observations, and by collecting other data and samples. Field studies can be followed by laboratory studies to better understand the timing and conditions of the different deformational and metamorphic events represented in the field.

A What Can We Learn from Field Studies?

The primary way in which geologists collect structural data is by doing field studies. Such field studies involve hiking around an area while observing, describing, and measuring the various geologic and metamorphic features. Also, field studies are the main way geologists collect samples of rocks, which can be taken back to the laboratory where they can be examined with a microscope and analyzed for their chemical content.

Reddish brown sandstone w/ well-rounded grains of quartz. No pebbles or finer grains.

N S

2m

Gray shale that underlies sandstone and is thin bedded with a few fossils

08.13.a2

One of the first steps in the field is to carefully observe the exposed geology. Geologists hike across the area and describe the different rock units, geologic structures, and other features. They then record the observations and descriptions in a notebook, which typically includes sketches (◄) to document in pictures what is difficult to describe in words. Sketching is an important way to explore ideas and possible alternative explanations in the field.

Geologists pay special attention to aspects that are diagnostic of a certain type of geologic structure, such as highly fractured rocks along a fault zone (►). Such exposures also allow us to observe the character and orientation of a feature.

08.13.a3 Grand Canyon, AZ

08.13.a1

08.13.a4

Soil cover

sandstone dips to northeast 25° degrees

red sandstone 25 *Soil cover*

gray shale

Soil cover

tan sandstone

fault cuts Soil

red ss 28

gray shale

Soil cover *faults*

brown cong.

Soil cover

Soil

tan Sandstone

Soil cover

brown conglomerate

Soil

Geologists use a base map, or sometimes an aerial photograph, to plot locations of observations, descriptions, measurements, and samples. The resulting geologic map (◄) shows the distribution of each rock unit, the orientations of beds, faults, and other structures, and possibly zones of key metamorphic minerals.

Some geologic features can be measured in the field as well as described. A geologist can use a compass to measure orientations of bedding, fractures, folds, and other features, including these scratches on a polished fault surface. ►

08.13.a5 Lake Pleasant, AZ

B How Can We Determine When a Fault Was Active?

To illustrate how a structural problem is approached, the figure below shows how we might determine the timing of movement on two faults. Ideally, we would like to know when a fault formed, through what time period it was active, and the age of the most recent movements. It is rarely possible to know all these ages.

Overlap—A lava flow can be erupted or a sedimentary unit can be deposited across a fault, overlapping it without showing any offset. Such a relationship demonstrates that the fault has not moved since the overlapping unit was emplaced.

Faulted Units—A fault must be younger than any units it cuts across, such as these sedimentary and volcanic layers. If we can determine the ages of such layers, then these ages help us infer the age of the fault.

Units Deposited During Faulting—If faulting displaces the land surface and forms a fault scarp, sedimentary and volcanic units may accumulate in downdropped fault blocks. These might include coarse sediment derived from the fault scarp.

Intrusion—Dikes and plutons that cut across a fault and are themselves unfaulted indicate that the fault is older than the intrusion. From the relations shown here, the left fault is older than the gray intrusion at the bottom of the figure and the dark lava flow at the top.

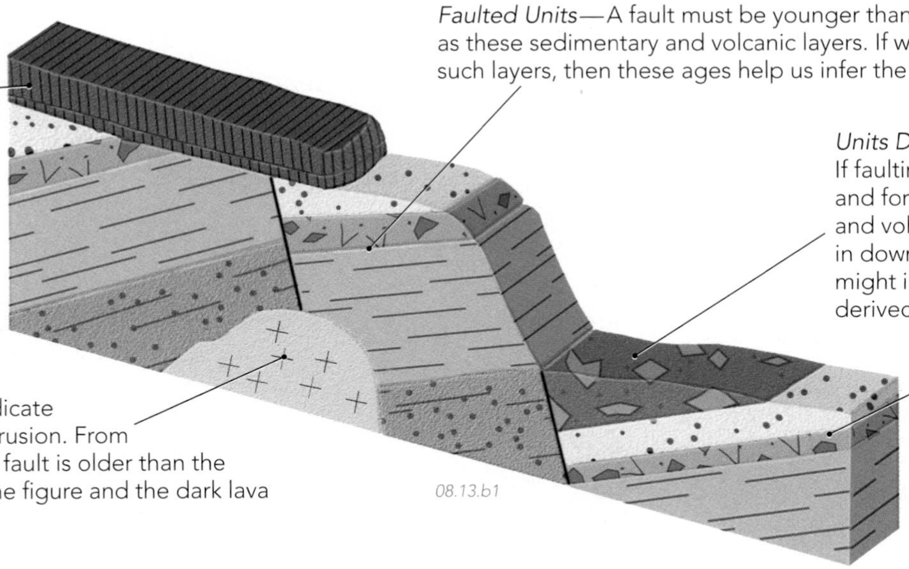

Tilting—Some faulting is accompanied by tilting of the rock units, and so we can infer the age of a fault from the age of tilting.

08.13.b1

C How Do We Investigate Metamorphic Rocks?

We use a similar approach to study metamorphic rocks. Many areas have experienced more than one episode of deformation and metamorphism, so geologists look for key localities that demonstrate the relationships between deformation and metamorphism. Additionally, geologists collect rock specimens that they take back to the laboratory in order to make thin sections, analyze minerals with special microscopes, and make interpretations based on (1) minerals that are present, (2) the observed sequence of minerals, (3) and mineral chemistry.

08.13.c1 Harquahala Mtns. AZ

Metamorphic minerals can grow before, during, or after a deformation event. These white crystals are oriented in all directions, show no foliation or lineation, and therefore grew after deformation had ceased. ◄

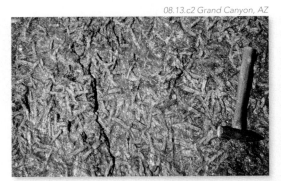
08.13.c2 Grand Canyon, AZ

Certain metamorphic minerals provide constraints on temperatures and pressures reached during metamorphism. These unusually large crystals of the mineral andalusite indicate a specific range of P-T conditions. ◄

Minerals and other features may be large enough to be easily observed, but some require observing the rock with a hand lens. This metamorphic rock has crystals of garnet, which can be chemically analyzed to determine the temperature of metamorphism. ▼

Many studies focus on trying to reconstruct the sequence of events. This meter-wide marble slab has dark bands of horizontal cleavage cutting older metamorphic layering and folded white veins. Each event occurred under different metamorphic conditions. ▼

08.13.c3

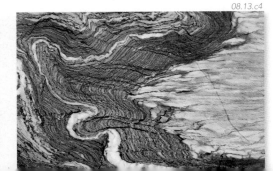
08.13.c4

Before You Leave This Page Be Able To

☑ Summarize how we observe and measure geologic features and the ways we record this information.

☑ Sketch or describe aspects we would observe to infer the age of a fault.

☑ Summarize some aspects we might observe in a metamorphic rock to learn something about its history of metamorphism and deformation.

8.13

8.14

What Is the Structural and Metamorphic History of New England?

NEW ENGLAND CONTAINS A WEALTH of geologic structures and metamorphic rocks, ranging from domes of high-grade gneiss to slightly cleaved sedimentary and volcanic rocks. This area provides an opportunity to examine the regional context of structures and metamorphic features.

A What Rocks and Structures Are Exposed on the Surface?

This figure is a moderately detailed geologic map of New England superimposed over topography. The different colors on the map show the distribution of different kinds and ages of rocks. The color patterns illustrate that the region contains belts of different rocks that trend north-south across the region.

Thin lines are contacts (boundaries) between rock units, whereas thick lines are faults.

Brown and olive-green colors show Precambrian rocks, such as those in the Adirondack and Green Mountains.

Purple, pink, and blue colors represent Paleozoic rocks, which are the most widespread rocks in the area. In the western part of New England, as near the Catskill Mountains, rocks are relatively unmetamorphosed sedimentary rocks that are locally folded and cleaved. ▼

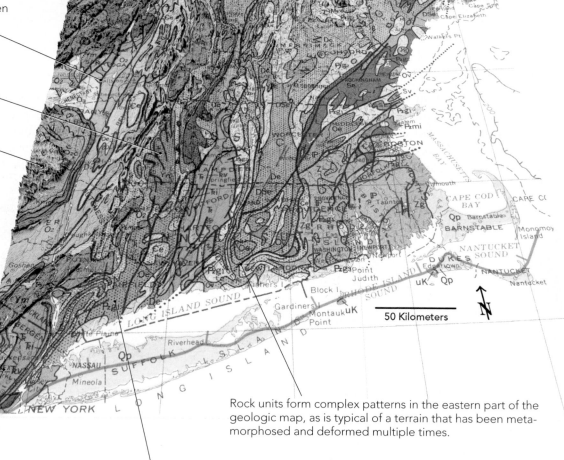

08.14.a1

08.14.a2 Catskill Mtns., NY

Blue-green and orange colors in the lower half of the map show Mesozoic sedimentary and basaltic rocks that are preserved within tilted and downdropped fault blocks. The faults formed during continental rifting of North America from Africa. Yellow colors, such as those on Long Island and Cape Cod, show recent glacial and shoreline sediments.

Metamorphism is shown on this map with a stipple (dotted) pattern over the different colors. Paleozoic and older rocks in the center of the region were metamorphosed at high temperatures and pressures. They include garnet-bearing schist, banded gneiss, and migmatite. ▶

Rock units form complex patterns in the eastern part of the geologic map, as is typical of a terrain that has been metamorphosed and deformed multiple times.

08.14.a3 Glendale Falls, Middlefield, MA

B What Is Below the Surface?

This cross section shows the generalized geometries of rock units and structures across central New England. Geologists measure the orientations of rock layers, metamorphic fabrics, faults, and folds on the surface and then project the layers and structures downward using strikes, dips, and other observations.

1. This cross section is drawn from west to east across the central part of New England, mostly in Vermont.

2. The Taconic Mountains and nearby areas expose folds and thrusts within Paleozoic sedimentary rocks. A main Taconic thrust sheet (shown in tan) contains rocks that were thrust into the area from the east.

3. In the Green Mountains, deformed and metamorphosed Precambrian rocks have been uplifted to the surface. The interpreted projections of these rocks into the air are shown with dashed lines.

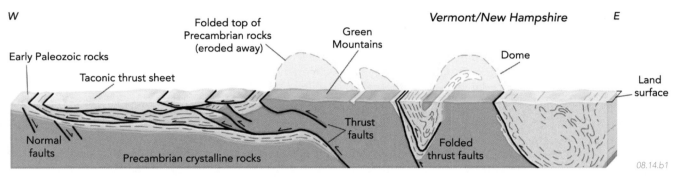

08.14.b1

4. The degree to which these rocks have been metamorphosed varies across the region. The most highly metamorphosed rocks (i.e., rocks heated to the highest temperatures) are on the east.

5. The eastern part of the area contains thrust faults that have been folded and large domes that contain high-grade metamorphic rocks, including banded gneiss. Some areas are called *gneiss domes.*

C How Did Metamorphic Conditions Vary Across the Region?

The metamorphic rocks of New England are famous and have been studied by generations of geologists with the goal of determining the conditions of deformation and metamorphism that affected the rocks.

1. This map below shows different grades of metamorphism across New England. Metamorphic conditions were the most intense in the center of the area (reddish colors) and decreased toward the west and east. Compare this map with the stipple pattern on the geologic map.

2. The diagram below shows the temperature and pressure history inferred from numerous metamorphic studies in two regions of central New England. The curved arrows show how geologists interpret the conditions in each region to have changed over time. The arrows were reconstructed by examining and chemically analyzing minerals under the microscope. Gray lines are boundaries separating metamorphic conditions where three important metamorphic minerals (labeled simply as K, A, and S) are stable.

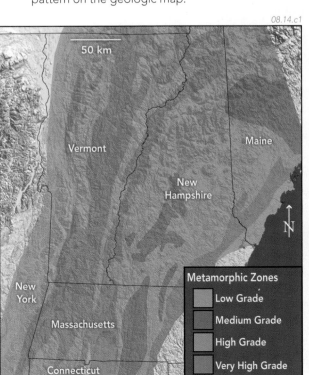

08.14.c1

3. The graph shows that rocks in eastern Vermont were buried (they moved down on the diagram), probably by thrust sheets. They were later heated. Some rocks were buried to depths of 15–30 km (9–19 mi) before being returned to the surface.

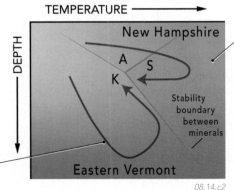

08.14.c2

4. Rocks in New Hampshire were heated (they moved right on the diagram) and then were buried. Initial metamorphism was the high-T/low-P type associated with magma. The rocks locally reached temperatures of 700°C.

Before You Leave This Page Be Able To

✓ Briefly summarize the types of rocks exposed in New England.

✓ Summarize the types of geologic and metamorphic structures in the region.

✓ Briefly summarize how the grade of metamorphism varies from east to west across the region.

What Structural and Metamorphic Events Occurred Here?

VARIOUS GEOLOGIC STRUCTURES AND METAMORPHIC ROCKS are exposed in the area shown below. Much of the area is covered by vegetation, but bedrock exposures in isolated, blocky hills contains clues to the area's structural and metamorphic history. You will use these clues to interpret the history of events.

Goals of This Exercise:

- Use observations of outcrops and a subsurface cross section to locate the position of a major fold.
- Observe several metamorphic rocks to determine their relative metamorphic grade and the type of starting rock from which each metamorphic rock was probably derived.
- Use the orientations of structures to infer the direction in which the rocks were shortened and to consider which structures could have been formed by the same event.
- Use differences in metamorphic grade across the area, along with small-scale structural observations, to infer the structural history.

08.15.a1

This perspective to the north shows the locations of hills of bedrock that are surrounded by grassy areas in which no rocks are exposed.

Each blocky hill has a shape that reflects the orientation of bedding. Each has a *dip slope* facing in toward the center of the area.

Bedding is defined by tan and gray layers.

Cleavage is expressed by near-vertical notches and lines down the front of each outcrop.

The field sketch below shows a cross section of an outcrop at the very top of the large southern hill, as viewed to the north.

A fairly straight stream cuts through the eastern part of the field area and flows mostly along a topographically low notch.

East of this stream are well-exposed banded rocks with steep metamorphic layering and large folds.

Four samples (1, 2, 3, and 4) were collected from outcrops in the area. Photographs of all four samples, numbered 1 through 4 to correspond to the sample numbers, are presented on the facing page.

Observations were made at a fifth site (5), but no sample was collected.

08.15.a2

Looking North

bedding

Cleavage   rock hammer

10 m

Some Field Observations

Field observations for locations 1 through 5 are listed below:

1. The eastern terrain exposes a black-and-white, banded rock (sample 1), which contains coarse crystals of biotite, quartz, and garnet. The rock has abundant folds and a foliation that dips steeply.

2. Small rock exposures in the grassy area expose a gray rock (sample 2). The rock cleaves apart and is shiny, but in hand specimen the grains are too finely crystalline to distinguish. When observed under a microscope, the rock contains abundant fine-grained biotite and muscovite mica aligned parallel to the platy fabric in the rock.

3. On the flanks of the southern hill and forming the tan layers in the blocky hills is a tan and gray rock consisting mostly of visible, sand-sized quartz grains. The rock (sample 3) is very hard and its fractures cut across, not around, the sand grains.

4. The top of the southern hill is a cream-colored to light-gray rock that consists of finely crystalline calcite. The rock locally contains some deformed objects that were probably fossils. The hill contains a large fold with near-vertical cleavage, as shown in the field sketch. This rock is sample 4.

5. Along the western edge of the banded rocks is a stream that flows nearly in a straight line. Erosion along this stream evidently was aided by the fact that the rocks are intensely fractured and shattered into angular pieces all along the stream.

Procedures

Use your observations from this area to complete the following steps, and enter your answers in the appropriate places on the worksheet. Your instructor may provide you with specimens of these rocks.

1. Make observations of the four samples shown in the photographs below. Use your observations, along with the information provided above in the field descriptions, to identify each metamorphic rock type (i.e., provide an appropriate name for the rock, such as schist). For samples 2–4, interpret what type of sedimentary rock each sample was before it was metamorphosed; this task is too difficult to do for sample 1.

2. What type of fold is exposed on the front cross section, an anticline, a syncline, or a monocline?

3. On the map on the worksheet, mark where the fold goes through the grassy areas and hills based on the orientations of bedding as reflected by the dip slopes and bedding-cleavage relationships exposed on the fronts of the blocks.

4. Based on your observations of the samples and from the field observations presented above, determine which rocks are higher in metamorphic grade: the banded rocks to the east or the rocks in the grassy area and southern hill.

5. In the worksheet, present any evidence you have for what type of feature is located near the main stream. Use any observations from the map, cross section, samples, and field descriptions to interpret what type of deformation or displacement has occurred along this feature.

6. From the orientation of the fold and its relationship to cleavage, in what direction were the rocks shortened to form the fold and the cleavage?

▼ Sample 1 is a coarse metamorphic rock with swirled, black-and-white bands that define a foliation. It has coarse crystals.

08.15.a3

08.15.a4

▲ Sample 2 is a shiny metamorphic rock with a strongly developed cleavage. It has no visible grains, but a thin section of the rock contains abundant mica, including muscovite. It does not contain garnet or any high-temperature minerals.

▼ Sample 3 is a metamorphic rock with visible, sand-sized grains of quartz. Cleavage (parallel to the marking pen) is at an angle to bedding, which is represented by the folded layers.

08.15.a5

08.15.a6

▲ Sample 4 consists of fine crystals of calcite. In this photograph, taken from the very top of the southern hill, steep cleavage is perpendicular to nearly horizontal bedding.

8.15

Geologic Time

EARTH HAS A LONG HISTORY. Rocks around the world contain evidence that seas advanced and retreated across the land many times, that vast mountains were uplifted and eroded away, and that various types of creatures arose, left their remains preserved in the rock record, and became extinct. This chapter explains the story of geologic time and how we use geologic principles to reconstruct Earth history.

09.00.a2 Siccar Point, Scotland

◀ **Siccar Point, east of Edinburgh, Scotland** is one of the most important geologic sites in the world. Scottish geologist James Hutton realized that rock layers exposed at Siccar Point require Earth to have a long and protracted history.

What types of evidence indicate that our planet's geologic history is long, some 4.5 billion years old?

Observe the photograph of Siccar Point below, and make observations about the rocks, geometry of layers, and other features.

09.00.a1 Siccar Point, Scotland

TOPICS IN THIS CHAPTER

◄ **The geologic feature** for which Siccar Point is famous is a boundary, or *contact*, that separates two chapters in Earth's history. This contact crosses this photograph from upper right to lower left. Observe this photograph, focusing on the nature of this contact and the features in the rock types on either side of the contact. Below the contact are gray sandstone and shale, whose beds are nearly vertical. Above the contact are reddish sandstone and conglomerate, which contain gently dipping beds and angular pieces of the underlying gray sandstone. The contact between these two rock sequences with very different geologic histories is what inspired Hutton's profound insight.

How does a contact like the one exposed at Siccar Point form, and what does it imply about the length of Earth history?

09.00.a3 Siccar Point, Scotland

Ruins of an Earlier World

As James Hutton explored the rocky coasts of Scotland in the late 1700s, he encountered the remarkable geologic exposures at Siccar Point. The insight he gained on that day in 1788 changed the world. James Hutton's profound realizations provided a new way to think about Earth.

At Siccar Point, James Hutton's attention was drawn to the enigmatic contact, which even from a distance is striking, with vertical gray beds below and gently inclined red beds above. Hutton pondered what had happened to produce such an arrangement of rock types. He wondered if the ancient contact represented the same processes currently occurring on the beach next to the outcrop—modern beach sand was being deposited in horizontal layers over the vertical beds of gray sandstone. In other words, Hutton's insight was that you might be able to use modern processes to

interpret events that had occurred in Earth's past. This principle, today called *uniformitarianism*, was the key step in the development of geology as a science. Uniformitarianism is an important tenet of the modern science of geology, being based on the logical idea that processes operating today are the same or are similar to processes that operated in the past. Uniformitarianism is often stated as "*the present is the key to the past.*"

Following this new logic, Hutton realized that to explain the relationships at Siccar Point, the gray sandstones below the contact must have been tilted and eroded before the red sandstone was deposited across the upturned layers. In essence, Hutton realized that this contact represented an *ancient erosion surface*, which we now call an *unconformity*. Hutton concluded that the gray rocks below the unconformity repre-

sented a mountainous landscape that had been eroded away, and he called these rocks "the ruins of an earlier world."

Hutton noted that erosion and many other geologic processes could be observed to occur relatively slowly compared to the life span of a human, so he realized that the contact at Siccar Point required Earth to have a very long history, much longer than was perceived at the time. Hutton concluded that the history of Earth was very long and partially shrouded, with "no vestige of a beginning, no prospect of an end." The ideas of Hutton were elaborated in books and other writings by Scottish Professor John Playfair, a contemporary of James Hutton, and later by Sir Charles Lyell, a Scottish geologist, who published very influential books in the 1830s and 1840s. Hutton and Lyell are among those people cited as the fathers of geology.

9.0

9.1 How Do We Infer the Relative Ages of Events?

TO DECIPHER THE GEOLOGIC HISTORY OF AN AREA, geologists use several strategies to determine the ages of geologic units, features, and events. The first strategy is to determine the age of one rock relative to another, using a series of commonsense approaches collectively called *relative dating*. Geologists then try to assign actual numbers, in thousands to billions of years, to this relative chronology, using other analytical dating methods, or *numeric dating*. Also, *fossils* allow us to compare ages of different rock layers and to construct the *geologic timescale*. We start here with five main principles of relative dating.

Principle 1: Most Sediments Are Deposited in Horizontal Layers

Most sediments and many volcanic units are deposited in layers that originally are more or less horizontal, a principle called *original horizontality*. If layers are no longer horizontal, some event affected the layers after they formed. The few exceptions to the principle are small in scale and in special environments, such as the face of a sand dune or the undersea slopes of a delta.

These canyon walls expose horizontal gray and reddish layers. These layers were deposited horizontally in Paleozoic time and have remained nearly so for 300 million years.

09.01.a1 Goosenecks of the San Juan, UT

09.01.a2 San Juan River, UT

Just to the east, the same gray and reddish layers are folded. They are no longer horizontal, so something (deformation) must have happened.

Principle 2: A Younger Sedimentary or Volcanic Unit Is Deposited on Top of Older Units

When a layer of sediment is deposited, any rock unit on which it rests must be older, a concept called the *principle of superposition*. This principle is illustrated below.

09.01.a3

1. A layer of tan sediment is deposited over older rocks.

09.01.a4

2. A series of horizontal red layers are then deposited over the first layer.

09.01.a5

3. A third series of layers is deposited last and is on top. In this sequence, the oldest layer is on the bottom and the youngest layer is on the top.

4. Observe all the different layers in this rock sequence. The sediments were deposited and they lithified to form sedimentary rock long before the river eroded the canyon. Which exposed layer is oldest, and where would you look to find the youngest rock layer? ▶

09.01.a6 Dead Horse Point, UT

5. Where is the oldest layer in this tilted sequence? It is most likely on the left, in the lowest part of the section. However, tectonic forces in some exceptional places have actually overturned layering, placing the oldest rock on top instead of on the bottom, but not here. ▶

09.01.a7 San Juan River, UT

Principle 3: A Younger Sediment or Rock Can Contain Pieces of an Older Rock

When a rock or deposit forms, it may incorporate pieces, or clasts, of older rock. A cobble eroded from bedrock and carried by a river cannot exist unless the bedrock already was there. The presence of clasts of an older rock in a younger rock clearly indicates the relative ages, even if you cannot see the two rock units in contact with one another.

The dark, lower basalt contributed clasts into an overlying layer of tan conglomerate. The conglomerate contains *clasts* of—and is therefore younger than—the basalt. The conglomerate also filled fractures in the basalt. ▶

09.01.a8 Lake Pleasant, AZ

A light-colored granite contains dark pieces, called *inclusions*, of older metamorphic rocks that fell into the magma. The metamorphic rocks, and their metamorphic layering, are contained within, and are older than, the granite. ▶

09.01.a9 Harcuvar Mtns., AZ

Principle 4: A Younger Rock or Feature Can Cut Across Any Older Rock or Feature

Many rocks are cross cut by fractures (joints and faults), so the rocks were there before the fractures formed. Dikes, sills, and veins can also intrude into or across preexisting rock units, also showing *cross-cutting relations*.

Several fractures cut across the limestone layers, so they formed after the rock already existed. The fractures are said to be *cross cutting*. ▶

09.01.a10 Little Colorado River, AZ

Light-colored dikes of granite cross cut through darker igneous rocks. The cross-cutting dikes are younger than the dark igneous rocks that host the dikes. ▶

09.01.a11 Santa Catalina Mtns., AZ

Principle 5: Younger Rocks and Features Can Cause Changes Along Their Contacts with Older Rocks

Magma comes into contact with preexisting rocks when it erupts onto the surface or solidifies at depth. In either setting, the magma may locally bake adjacent rock, or fluids from the magma may chemically alter nearby rocks. These changes, called *contact effects*, indicate that the magma is younger than the rocks that were altered.

A dike of basalt intrudes across a grayish sedimentary rock. Heat and fluids from the magma affected the older sedimentary rock, causing a reddish baked zone next to the dike (▼). If an intrusion is a sill (not shown) injected between existing layers, it bakes rocks above and below the sill.

09.01.a12 Bloody Basin, AZ

A lava flow or hot pyroclastic flow can bake and redden older underlying rocks, as shown here (▼). Sediments deposited on top of the volcanic unit after the eruption will not be baked. This contrasts with a sill, which bakes rocks above and below.

09.01.a13 Flagstaff, AZ

Before You Leave This Page Be Able To

✓ Sketch and explain each of the five principles of relative dating, providing an example of each principle.

✓ Apply the principles of relative dating to a photograph or sketch showing geologic relations among several rock units, or among rock units and structures.

9.1

9.2 How Do We Study Ages of Landscapes?

KNOWING THE RELATIVE AGES OF ROCKS AND STRUCTURES provides only one piece of the geologic story. We also need to understand when and how landscape features, such as mountains and valleys, formed.

A How Does a Typical Landscape Form?

Most landscapes have a similar history—rocks form and then are eroded. The histories of many regions typically include the deposition of a sequence of sedimentary layers, lithification into rocks, and later erosion of the rocks.

▼ 1. The sequence begins with deposition of a new sedimentary unit on top of preexisting metamorphic and igneous rocks. Most sediments, such as the layer of sand shown, are deposited as nearly horizontal layers.

09.02.a1

09.02.a2

2. Through time, the depositional environment changes and a series of different sedimentary layers accumulate, with each younger layer being deposited on top. ◄

7. The canyon exposes five or six main sedimentary units and a number of smaller layers. There is a dominant light-colored cliff of sandstone. Layers below the sandstone are the oldest in this area, and red cliffs in the distance expose higher and younger layers. The far mountains are igneous intrusions that baked, and are younger than, the layers. All the layers were deposited and lithified, and the intrusions were emplaced, before erosion began carving the canyon.

3. Over time, the layers are lithified. At some point, deposition stops, and all the layers that will be deposited are there (►). Weathering and erosion can begin.

09.02.a3

09.02.a6 Grand View Point, Canyonlands NP, UT

6. Observe this photograph and think about how the landscape formed before reading the text above the image. How many main rock layers do you observe? Use relative dating principles, like superposition, to infer the relative ages of the different rock layers you noted. Then, guided by the sequence of figures to the right, visualize how the area probably evolved over time.

4. If the region is uplifted or the seas withdraw, the area can begin to be eroded by rivers, streams, glaciers, and the wind. Erosion can more or less uniformly strip the entire land surface, removing the top layers. More likely, erosion will be faster in some areas, like along a river cutting downward in a small canyon. ►

09.02.a4

5. Erosion by a river cuts downward, carving a deeper canyon. The canyon widens as small drainages erode outward from the main river and as the steep canyon walls move downhill in landslides and slower movements. The combination of downcutting, widening, and development of subsidiary drainages, called *tributaries*, sculpts a deeper, wider, and more intricate canyon. ►

09.02.a5

B How Do We Infer the Age of a Landscape Surface?

To investigate when a landscape surface formed, we commonly try to find a rock unit or other geologic feature that was there before the surface formed or one that came after the surface already existed.

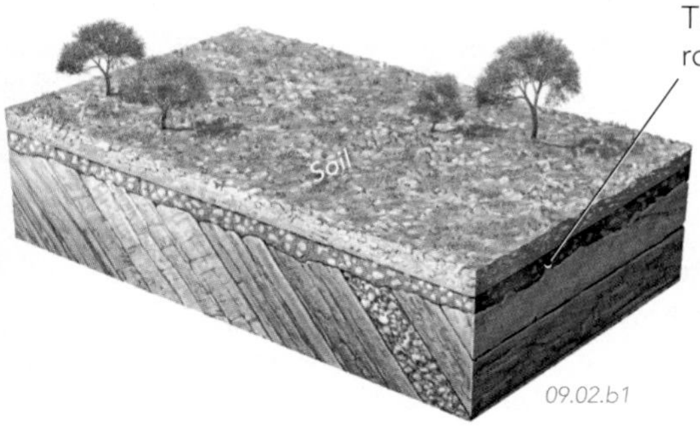

The age of a landscape surface must be younger than any rocks on which it is carved. In this example (◄), erosion beveled across an older series of tilted layers, which were then covered by a thin veneer of sediment and soil.

A landscape surface is older than any rock that is deposited on top of the surface. A lava flow (►) is ideal for dating a surface because it formed during a short time and its age can usually be determined by analytic dating methods.

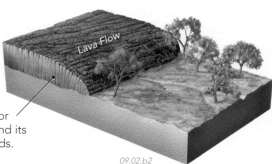

09.02.b1

09.02.b2

Sometimes the age of a landscape surface cannot be dated directly, but we can infer its age relative to other features. Many rivers are flanked by raised, gentle surfaces called *terraces*. A terrace was formed sometime in the past, before the river eroded down to its present level (◄). The terrace is older than the modern channel.

A landscape surface progressively develops more soil if it remains undisturbed by erosion and deposition. A surface with well-developed soil, such as the uplands shown with thick red clay and white carbonate accumulations (►), must be thousands of years old. Recent sediment along the stream has no soil.

09.02.b3

09.02.b4

09.02.b5 Picture Rocks State Park, AZ

In many climates, especially deserts, rock surfaces develop a dark coating if left undisturbed for hundreds to thousands of years. This coating, called *rock varnish* or *desert varnish*, consists of iron-oxide and manganese-oxide materials, which are mostly derived from windblown dust. Rock varnish becomes darker the longer a rock is exposed at the surface, with very dark varnish requiring thousands of years. The darkness of varnish is therefore an indicator of how long that rock surface has been exposed. Rock varnish can be weathered or worn away, resetting the process. The dark varnished boulders shown here (▲) sat undisturbed on the surface for thousands of years, before parts of the varnish were scraped off by Native Americans to form artistic petroglyphs.

09.02.b6 Little Harquahala Mtns., AZ

In some settings, stones become concentrated on the surface through time, forming a feature called *desert pavement*. Over time, finer materials wash away, blow away, or move down into the soil, while pebbles and larger clasts remain on the surface or move up from just below the surface. If left undisturbed, the pavement becomes better developed over time, and exposed stones get coated with desert varnish, like the ones shown here. It takes more than ten thousand years to form a well-developed pavement with darkly varnished stones. Desert pavement can therefore be used as an indication of age.

Cosmic Rays

Stones on the surface progressively accumulate telltale amounts of certain chemical elements produced when cosmic rays strike the stones. A form of isotopic dating is used to determine how long the stones have been on the surface. Geologists collect samples of the stones and analyze them in the laboratory. ◄

09.02.b7

Before You Leave This Page Be Able To

☑ Describe the sequence of events represented in a typical landscape of flat-lying sedimentary rocks.

☑ Describe or sketch how you could assess the age of a landscape surface.

9.3 What Is the Significance of an Unconformity?

EROSION SURFACES CAN BE BURIED AND PRESERVED beneath later deposits. These buried erosion surfaces, called *unconformities*, can represent large intervals of time missing from a rock sequence. They provide a glimpse of the shape and longevity of ancient landscapes. There are three types of unconformities: *angular unconformities*, *nonconformities*, and *disconformities*. What do these features look like, how do they form, and what do they tell us about past geologic events?

A What Does an Angular Unconformity Represent?

Erosion surfaces, formed in the past, can be buried and preserved within a sequence of rocks. If underlying rocks are tilted before formation of the erosion surface, the unconformity is an *angular unconformity*.

1. A gray limestone is deposited under the sea in nearly horizontal layers. The blue in the figure below represents water.

2. Later, the sea withdraws and the limestone beds are folded. As the folded beds are uplifted, they are beveled by erosion. This results in tilted beds being exposed on the surface.

3. A conglomerate is deposited over the eroded beds, forming an unconformity. If the underlying layers have been tilted, as in this example, it is an angular unconformity.

09.03.a1

09.03.a2

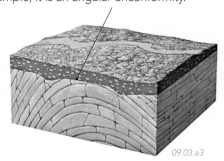
09.03.a3

09.03.a4 Grand Canyon, AZ

Rocks Above Unconformity

Unconformity

Rocks Below Unconformity

◄ 4. Examine this photograph of the eastern Grand Canyon. There is an angular unconformity between tilted layers below and nearly flat-lying layers (beds) above. The rocks below the unconformity are approximately 1,100 million years (1.1 billion years) old, whereas those above are 540 million years old. There is a long time span represented by the unconformity, for which there is no record, except that there was tilting followed by erosion, before deposition of the upper layers. Other events could have—and did—occur, but we have no record of them at this site.

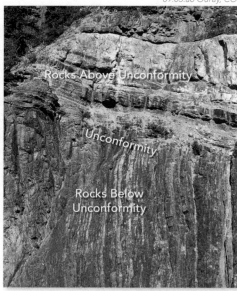
09.03.a6 Ouray, CO

Rocks Above Unconformity

Unconformity

Rocks Below Unconformity

09.03.a5 Grand Canyon, AZ

Rocks Above Unconformity

Unconformity

Rocks Below Unconformity

◄ 5. In this view of the same angular unconformity, the unconformity slopes down from right to left as it cuts across layers in the underlying reddish rocks. The ledge of brownish-red, cliff-forming sandstone directly above the unconformity is thicker to the left and thins toward the right. This thinning indicates that the rocks below the unconformity formed a small hill, against which the sandstone was deposited. Such preserved ancient topography is common along unconformities.

6. The same unconformity (separating rocks that are more than 1 billion years old from those that are less than 550 million years old) is exposed across many parts of the United States and is called the *Great Unconformity*. In this view (▲), gently tilted sedimentary layers unconformably overlie vertical layers within the underlying metamorphic rocks. The time represented by the unconformity is 1.3 billion years (from 1.7 billion years below to 400 million years above).

B How Does a Nonconformity Form?

Some erosion surfaces form on top of rocks that are not layered, especially igneous rocks like granite. This type of unconformity is called a *nonconformity*.

09.03.b3 Grand Canyon, AZ

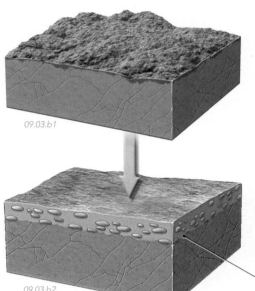

09.03.b1

◀ The formation of a nonconformity begins when a granite, or other nonlayered rock, is uplifted to the surface and eroded. The surface is weathered into a reddish zone of sand, clay, and iron oxides.

◀ Subsequently, conditions change, and the erosion surface is buried by sand and cobbles, perhaps derived in part from weathering of the granite. Ultimately, the sediment lithifies to produce sandstone and conglomerate. The contact between the granite and overlying sedimentary rock is a nonconformity.

09.03.b2

▲ This nonconformity has dark sandstone over tan granite. It is the same Great Unconformity as shown on the previous page, but here it overlies granite instead of layered and tilted metamorphic or sedimentary rocks.

C What Is a Disconformity and What Does It Indicate About an Area's History?

If rock layers are not tilted before they are overlapped by younger layers, but the boundary still represents millions of years of time, the contact is a *disconformity*. A disconformity can involve erosion or just a long time period with little or no deposition. Disconformities may be overlooked because they are parallel to rock layers.

09.03.c1   *09.03.c2*   *09.03.c3*

1. The first step in the development of a disconformity is deposition of horizontal layers producing sedimentary rock. In the figure above, the layers are limestone but could be any type of sedimentary rock.

2. Next, the rock is exposed at the surface because the region is uplifted or because sea level drops. Sedimentation stops, and weathering and erosion affect the now-exposed land surface.

3. After some time, sedimentation resumes, and the surface of older rock is eventually buried by a younger layer of sediment, forming a disconformity. This new layer can be deposited by water or can be deposited on land, perhaps as sand in a desert.

09.03.c4 Buckfarm Canyon, Grand Canyon, AZ

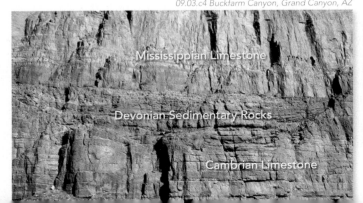

4. This photograph, colored lightly for emphasis, shows three series of rock layers separated by two disconformities. One disconformity is above the reddish lens of Devonian sedimentary rocks, and an older disconformity is below the reddish lens.

Before You Leave This Page Be Able To

☑ Sketch an angular unconformity, a nonconformity, and a disconformity, and describe what sequence of events is implied by each.

9.3

9.4 How Are Ages Assigned to Rocks and Events?

DETERMINING THE RELATIVE AGES OF ROCKS, and the order of events, is only one part of deciphering the geologic history of a location. We also want to know how long ago the rocks formed and when the events occurred, and to assign *ages* in hundreds, thousands, millions, or billions of years. This is done by using *analytical dating methods*, most of which involve chemically analyzing a rock for the products of natural radioactive decay. Determining the ages of rocks using analytical measurements is called *numeric dating*.

A How Does Radioactive Decay Occur?

All atoms of any given element must have the same number of protons, but some differ in the number of neutrons they contain. Thus, different varieties of the same element may have different *atomic weights;* these varieties of the same element are called *isotopes*. Some isotopes are unstable through time, changing into a new element or isotope by the process of *radioactive decay*.

09.04.a1–3

1. This schematic figure shows atoms before any radioactive decay. These starting atoms are called the *parent atom* or *parent isotope*. Over time, some of the parent isotope will decay into a different element called the *daughter product* or *daughter atom*.

2. At a later time, half of the parent atoms (green) will have decayed into the daughter product (purple). The amount of time it takes for this to occur is called the *half-life*. After one half-life, there are an equal number of parent and daughter atoms.

3. After a time equal to another half-life has passed, half of the remaining parent atoms have decayed into daughter atoms. After two half-lives, 3/4 of the parent atoms have decayed and 1/4 remain.

4. This table summarizes the radioactive decay shown in the figures above. If the number of parent atoms was initially 100, half of the parent atoms (50) will have decayed to atoms of the daughter product after one half-life. After two half-lives, only 25 parent atoms remain, alongside 75 daughter atoms.

	Before Any Decay	After One Half-Life	After Two Half-Lives
Atoms of Parent	100	50	25
Atoms of Daughter	0	50	75

5. Decay rates are different for different radioactive elements, but for any given isotope, the decay rate is always the same, predictable, and measurable in the laboratory. Geologists, therefore, can calculate the age of a rock by measuring the ratio of parent atoms to daughter atoms in the rock. Dating rocks using radioactive decay is called *isotopic dating*.

Measuring and Calculating Isotopic Ages

Geologists, working alongside chemists and physicists, use an instrument called a *mass spectrometer*, shown below, to measure the ratio of parent isotopes to daughter product in the rock or the mineral to be dated.

When some minerals form, they incorporate atoms of the parent isotope, especially if this element fits in the mineral's crystalline structure. The mineral typically does not contain daughter atoms, which have a different atomic size and atomic charge than the parent atoms. Over time, radioactive decay converts parent atoms into daughter atoms, producing a specific and predictable proportion of parent and daughter atoms.

Geologists prepare a rock or mineral sample and place it in a mass spectrometer, where the sample is ionized and propelled down a tube toward a very strong electromagnet. The magnet pulls atoms with heavier atomic weights in one direction and lighter atoms in another. The strength of the magnet can be altered by adjusting the amount of electric current passing through it. With the proper settings, only atoms of the desired atomic weight reach a collector at the end of the tube, which counts the number of arriving atoms. Mass spectrometers measure ratios of isotopes more easily than absolute amounts of isotopes, so most results and calculations use ratios between isotopes. The results are calculated using equations and are commonly plotted on a graph.

Isotopes that decay quickly are used to date young rocks and archaeological artifacts. They are not appropriate for older materials because all of the parent element disappears quickly. Isotopes that decay very slowly are used to date ancient rocks. They cannot be used to date young materials because only minute amounts of daughter products will have formed. The dating process involves many potential complications and assumptions, so geologists select the correct isotope, and they consider and evaluate each assumption before applying the determined age to a rock.

09.04.mtb1

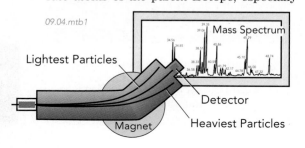

Lightest Particles

Mass Spectrum

Detector

Heaviest Particles

Magnet

B What Isotopic Systems Do We Use to Determine Ages of Rocks and Events?

A number of elements have radioactively unstable isotopes, and geologists and chemists use these different isotopic systems to measure different types of ages on rocks, minerals, sediment, and other materials. Which isotopic system is used depends on what datable materials are present, the likely age of the materials, and the geologic history of the unit.

One isotope of potassium (K) decays to the noble gas argon (Ar, which is on the far right side of the periodic table), and to an isotope of calcium. We use K-Ar dating and the related Ar-Ar (pronounced argon-argon) method to date volcanic rocks and the cooling of deep rocks brought to the surface. These methods are most useful for dating Cenozoic and Mesozoic rocks, but they are also used for older rocks.

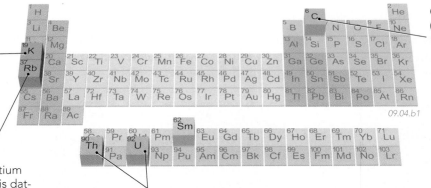

09.04.b1

One isotope of carbon (C), carbon-14, is used to date wood, charcoal, bones, shells, and carbon-rich rocks and water. Carbon-14 has a relatively short half-life that makes it useful only for dating materials that are hundreds to thousands of years old.

An isotope of rubidium (Rb) decays to strontium (Sr), providing the basis for Rb-Sr dating. This dating method provided some of the first ages for old granites and metamorphic rocks, and was key in demonstrating that some rocks on Earth and some meteorites were billions of years old.

A number of thorium (Th) and uranium (U) isotopes decay through a series of steps to different isotopes of lead (Pb), element number 82. We use these isotopes to date many kinds and ages of rocks, including grains in sediments. Samarium (Sm) decays to Neodymium (Nd), and we use Sm-Nd dating mostly to date old rocks and to investigate sources of magma.

C What Can Isotopic Ages Tell Us?

We use different isotopic systems for isotopic dating, and all isotopic ages do not provide the same kind of information. Some record when the rock incorporated the parent isotope, whereas others record later cooling.

We date volcanic units using a variety of isotopic systems, like K-Ar. Volcanic rocks form on the surface and cool rapidly, so an age of the rock is typically the *age of eruption.*

Hot plutons lose certain isotopes, so we determine the age of such bodies using only those minerals that retain isotopes and provide the *age of solidification.* Today, we mostly use U-Pb dating of the mineral zircon.

Some minerals, such as biotite mica in plutons, tell us *when a rock cooled* through a specific temperature, as when it was being uplifted to the surface.

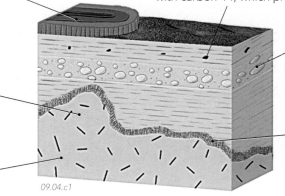

09.04.c1

Black pieces of charcoal incorporated into recent sediment can be dated with carbon-14, which provides an age for *deposition* of the sediment.

Dates from individual boulders, cobbles or even sand-sized grains in a sedimentary rock help infer the age of the *source rocks* from which the sediment was eroded. The oldest ages ever measured, more than 4 billion years old, are for individual grains in sedimentary rocks from Australia.

We investigate the *age of a metamorphic event,* like baking next to the pluton, using minerals that formed during metamorphism or minerals that record certain metamorphic temperatures.

▶ In many cases, geologists use different methods on a single rock to obtain information about different parts of the rock's geologic history. For granite, a U-Pb age on zircon, a uranium-bearing mineral, can provide the age when the magma solidified, and a K-Ar age on biotite provides the time when the rock cooled below 300°C. By using different dating methods on the same rock, geologists can show when the rock formed and how fast it cooled through time, as plotted here.

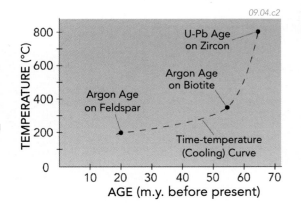

09.04.c2

Before You Leave This Page Be Able To

✓ Explain how to determine how many half-lives have passed based on the ratio of parent to daughter atoms.

✓ Describe the different ways that isotopic dating is used for dating geologic events.

✓ Describe how a mass spectrometer is used to determine isotopic ages.

9.4

9.5 What Are Fossils?

ROCKS CONTAIN FOSSILS—EVIDENCE OF ANCIENT LIFE. Rocks of appropriate age and type preserve shells, coral, bone, petrified wood, leaf impressions, dinosaur tracks, and features created by burrowing animals. What kinds of creatures left these traces of past life, and how are they preserved in the rock record?

A What Are Fossils and How Are They Preserved?

Fossils are any remains, traces, or imprints of a plant or animal that are preserved in a rock or sediment. Fossils can be of different types depending on what type of life is involved, in what environment the plant or animal lived, and how the remains were buried and preserved.

09.05.a1

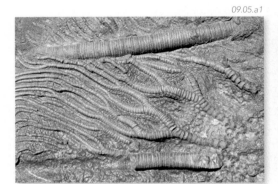

09.05.a2 Dinosaur NP, UT

09.05.a3 Petrified Forest NP, AZ

Most fossils found in the field are preserved *hard parts*, or parts that have been replaced by hard minerals of marine organisms, including shellfish and coral. The photograph above shows heads and stems of animals called *crinoids.*

Vertebrate animals have hard parts, most significantly *bones*, that can be preserved. Most bones are found as fragments instead of complete skeletons because of the destruction and dispersal caused by scavengers, weathering, erosion, and transport.

Some fossils are preserved because the original organic material is replaced by silica, pyrite, or some other material. One example is wood from trees that is replaced by fine-grained silica, forming *petrified wood.*

09.05.a4 El Paso, TX

09.05.a5

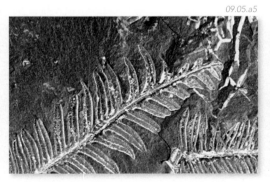

09.05.a6

Another type of fossil forms when an animal is buried and decays. This leaves a cavity in the rock that mimics the animal's shape. The cavity is a *mold* if unfilled, as shown here, and is a *cast* if it is later filled by minerals.

After burial, some carbon-rich plants and animals become *thin films* of carbon or other materials that preserve the original shape of the plant or animal. This fossil fern is almost 300 million years old.

Fish and other soft creatures can be preserved as *impressions*, especially when the remains come to rest in quiet waters of a lake or deep sea. Such fossils can preserve amazing details, including fins and scales.

09.05.a7

09.05.a8 Johannesburg, South Africa

▶ Animals can become fossils in other ways. Insects become trapped in tree sap, which through time hardens into golden-brown *amber.* Such preservation can preserve fragile features of the animal.

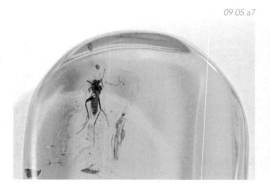

◀ Some fossils do not preserve the actual organism, but instead represent something that the organism constructed. This mound-like feature, called a *stromatolite*, was built by ancient microscopic algae.

B   What Traces Do Creatures Leave in the Rock Record?

In addition to preserved remains of organisms, rocks contain other features made by animals that moved across the surface or burrowed into soft sediment. Geologists call these features *trace fossils*.

Creatures that walk on land, such as reptiles, or on the sea bottom, like crabs, can leave *footprints* behind, such as this one from a dinosaur. Most footprints are indentations in sediment that are filled by later sediment. A trail of related footprints is called a *trackway*, from which geologists can infer how the creature moved, how much it weighed, and whether it traveled alone or in a group.

09.05.b1 Moenave, AZ

09.05.b2 Grand Canyon, AZ

Worms and other creatures wriggle, dig, or tunnel into mud, forming cavities that can be filled by a different kind of sediment, producing a trace fossil. This type of trace fossil is a *burrow*, also called a *worm burrow*. The creatures were too soft to be preserved but still left behind a record.

C   What Determines Whether a Fossil Is Preserved?

Most creatures are never preserved as fossils because fossil preservation requires certain favorable circumstances. The most important factors include the existence of hard body parts and rapid burial after death.

Hard Parts—Preservation as a fossil is much more likely if a creature, like this crinoid, has a shell, bones, teeth, or some other hard part. Only 30% of modern animals have hard parts, but such animals are overrepresented in the fossil record compared to animals like insects or jellyfish that lack hard parts. Soft parts of creatures can be eaten by scavengers, crushed during sediment compaction, dissolved by chemical reactions, or otherwise destroyed. Some ancient creatures with only soft parts may nowhere have been preserved.

Rapid Burial—A fossil cannot be preserved unless it is buried. If a creature's remains are left on the surface, whether on land or in the sea, they can decompose due to exposure to the atmosphere and water, or can be scavenged by other creatures. Rapid burial means less opportunity for destruction. Preservation is easier beneath the sea than on land because burial is generally more rapid, and because a lower content of oxygen in the deep sea slows decay.

09.05.c1

What Features Look Like Fossils but Are Not

Some natural geologic features look like fossils but are not fossils. These features form through *inorganic* processes and do not represent the remains or traces of any organism.

The most common features mistaken for fossils are the dark, branching mineral growths shown here. These growths, called *dendrites*, typically consist of manganese-oxide minerals that grow in branching patterns along joints and between sedimentary layers.

Spherical features, called *concretions*, which grow in sediment during cementation, are also commonly mistaken for fossils.

09.05.mtb1

These weather out of sediment as small spheres or oddly shaped objects that can look organic. Formation of concretions can involve some biologic processes, but concretions are not fossils.

Before You Leave This Page Be Able To

☑ Describe the different ways in which a plant or animal can be preserved as a fossil.

☑ Describe two types of commonly encountered trace fossils.

☑ Describe the two main factors that influence whether a creature is preserved as a fossil.

☑ Describe a feature that can be mistaken for a fossil.

9.5

9.6 How and Why Did Living Things Change Through Geologic Time?

DIFFERENT FOSSILS OCCUR IN DIFFERENT ROCK UNITS. Some of these differences reflect variations in sedimentary facies (depositional environment), for example between reefs and rivers, but most reflect the systematic way that living things and their fossils varied over geologic time. Why did these changes in fossils occur, and what do they tell us about how life on our planet has changed through time?

A How Do Fossils Vary with Age?

Early geologists recognized that fossils change upward from older layers of sedimentary rock to younger layers. This systematic change of fossils with age, called *faunal succession*, helped geologists identify time periods defined by major changes in life on Earth. Using the principles of relative dating and faunal succession, geologists subdivided geologic time into four major chapters, each with subdivisions. Later, results from isotopic dating provided numeric ages, in millions of years before present, for when each chapter started and ended.

1. The *Cenozoic Era,* meaning *recent life,* spans the last 65 million years. It is called the *age of mammals* because mammals, such as this (fossilized) mammoth (▼), became a dominant type of life on Earth.

09.06.a2

Cenozoic

65 Ma

Mesozoic

251 Ma

Paleozoic

542 Ma

Precambrian

(started at 4,500 Ma)

09.06.a1

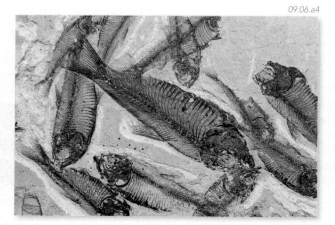

2. The *Mesozoic Era (middle life)* is known as the *age of dinosaurs* because dinosaurs (▼) rose to dominance during this era. The end of the Mesozoic Era, at 65 Ma, is marked by the extinction of dinosaurs.

09.06.a3

3. *The Paleozoic Era (ancient life)* was dominated by several major groups of marine animals, including coral, creatures like clams that had hard shells, and various types of fish (▼). Plants, insects, and amphibians also colonized the land during this era. The end of the Paleozoic Era is marked by a major time of extinction called the *Great Dying.* This extinction killed off many species of animals in the seas and on land.

09.06.a4

09.06.a5

4. The *Precambrian (before the Cambrian Period)* comprises nearly 90% of geologic time. For most of this time, only simple life forms existed, such as bacteria and algae that formed stromatolites like those shown here. ▶

B What Factors Determine Whether a Species Survives or Becomes Extinct?

Boundaries between the major chapters of the geologic timescale are defined either by the emergence of new life-forms, by massive extinctions, or both. Almost all species that ever existed are now extinct. What factors influence whether a species survives or becomes extinct?

09.06.b1 Eastern Africa

Environmental Setting—Animals, plants, and other organisms have certain ways they live, and certain survival needs. Some organisms thrive because they developed along with other plants or animals that provide them with essential food, habitat, or other needs.

09.06.b2 Amboseli NP, Kenya

Climate—Changes in climate, resulting in loss of water and food sources, along with other changes in critical habitat, can threaten a regional population of animals. Environmental stresses, including disease, can eliminate entire classes of animals.

09.06.b3 Western AZ

Reproductive Strategy—Different plants and animals reproduce in different ways. These plants flower and produce seeds, whereas the cacti grow new versions from small parts of the original plant. Some reproductive strategies will be more successful than others.

09.06.b4 Galápagos Islands

Adaptability—The more adaptable a species is, the more likely it will survive changes in the environment, such as increases of temperature or salinity of water. These marine iguanas started as land animals but developed the ability to also forage in the sea.

09.06.b5 Maasai Plains, Kenya

Competition—If two or more species are competing for the same sparse resources, there are likely to be winners and losers. Competition between members of a single species can also be a problem if it means that needed resources are in short supply for all.

09.06.b6

Predators and Prey—Being a food source for some other creature is never a good survival strategy. The opposite is also true—if an animal relies on only one kind of food, survival becomes problematic if that food source becomes scarce or even disappears.

Evolution: Observed Changes and Possible Explanations

The term *evolution* is used in two ways. First, it refers to *observed changes* in the fossil record or documented changes in more recently living animals. This is commonly called the "fact of evolution." Second, evolution refers to the *theories* that help *explain* the observed changes.

Observed changes in the fossil record over time are well documented and can be verified by anyone who studies fossil-bearing rocks from different geologic times. For more than a hundred years, geologists have used fossils to compare life-forms from rocks of different ages around the world. These comparisons are supported by many isotopic ages.

The *theory of natural selection* originated with Charles Darwin to explain the birds and other animals of the Galápagos Islands. Using this and other evolutionary theories, paleontologists try to explain how a Paleozoic fish developed front fins strong enough to support its weight on land, a mutation many paleontologists accept as having eventually led to amphibians. One evolutionary hypothesis, called *punctuated equilibrium*, explains how new organisms, or new characteristics of an existing organism, appear rather suddenly in geologic terms, instead of evolving more gradually. Proponents argue that new and favorable mutations are more likely to succeed in small, isolated populations than in large populations. After a favorable change develops fully in a small group, the group may rejoin the larger population and out-compete the other individuals, causing an observed evolutionary change.

Before You Leave This Page Be Able To

✓ Describe the four chapters of Earth history and how the boundaries are defined.

✓ Describe some factors that affect survival and extinction.

✓ Describe the difference between observed fossil changes (evolution) and evolutionary theory.

9.6

9.7 How Are Fossils Used to Infer Ages of Rocks?

THE DOCUMENTED FAUNAL SUCCESSION from older rocks to younger ones in a rock sequence provides a powerful tool for assessing the age of fossil-bearing units. Using fossils, we can judge the age of one rock relative to another, compare the ages of two sections of rocks from different places, and assign a rock to a specific time unit of the geologic timescale. Geologists use fossils to infer the ages of rocks and events, and to help interpret the environment in which a sedimentary layer accumulated.

A How Do Fossils Change Through a Sequence of Sedimentary Rocks?

In almost any fossiliferous sequence of sedimentary rocks, the types of fossils change upward through the section, from older rocks to younger ones. Some species survived unchanged for long times, and so left similar fossils in many rock layers. Other species existed for shorter times, and their fossils are restricted to a layer or two. The figure below shows how some fossils can be more useful for determining geologic ages than others. The vertical bars show the layers that contain a given fossil. Read the text from bottom to top, from the oldest layers to the youngest ones.

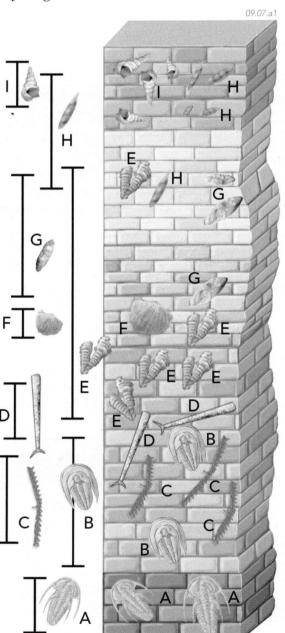

09.07.a1

9. Fossils that are large, widespread, and distinctive (fossil I) are more useful than those that are small, found only locally, and lack diagnostic characteristics.

8. The overlap between the ranges of two fossils within a sequence may tightly constrain the age of that part of the section. Microscopic fossils G and H have moderately wide ranges but occur together only within a narrow interval of rock and time.

7. Some rocks do not include large fossils, so microscopic fossils may be the only way to evaluate the age of this part of the section. Microscopic fossils G and H have been enlarged so that they are visible.

6. Some types of fossils, like the coral in fossil E, occur through a wide range of the section. These fossils represent organisms that lived essentially unchanged for a long time. Such fossils are less useful for assigning ages than fossils that are restricted to a narrower time range.

5. Some types of fossils (such as fossil F) are restricted to a small part of the sequence. These fossils are extremely useful for assigning ages to rocks and so are called *index fossils*. In addition to having a short time range, an index fossil must be abundant (commonly found) and have a widespread areal distribution.

4. The boundary between two varieties of the same fossil type, or between different types of fossils, can be abrupt, as between fossils C and D. Such boundaries may have ages that are known within narrow limits and so may be useful time markers. Moving upward, we would examine the section of rocks and look for the place where fossil C disappeared and fossil D appeared.

3. Most rocks contain more than one type of fossil; that is, they contain an *assemblage* of fossils. Fossils B and C occur in some of the same layers because both types of organisms were living at the same time. Finding both fossils may help us refine the age of the layers.

2. Different varieties of the same general type of fossil may exist and may be distinguished on the basis of shape and other characteristics. One type of trilobite (fossil B) is different from the trilobite (fossil A) found at the base of the section. The age ranges of these two varieties do not overlap, so finding one or the other is very useful for assigning ages.

1. For some time periods only a few types of fossils are useful for assigning ages. For studying the earliest part of the Paleozoic Era, a fossil called a *trilobite* (fossil A) is the most useful. It is the main fossil in the base of this section.

B How Do We Use Fossils to Correlate Two Sequences of Rocks?

The different types of fossils in rocks of different ages indicate that the types of animals changed through time. The systematic change of the types of fossils through time, faunal succession, is an important principle for comparing the ages of two different sections of rock. Such a comparison is called a *correlation* and uses rock type, interpreted environment in which the units formed, and, where possible, fossils, especially index fossils.

Comparing Two Sections That Represent the Same Age Range

This figure shows two sections of rocks that contain some of the same fossils as those shown on the previous page. These two sections, however, do not include all of the fossils because some parts of the geologic record are missing along disconformities, shown with darker, squiggly lines along contacts. We can compare fossils in the two sections to provide a basis for correlating different parts of the sections. Dashed lines connect rocks or contacts of the same age in the two sections.

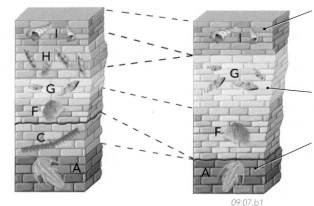

09.07.b1

The tan limestones at the top of the sections have matching rock types and fossils (I) and so are the same age.

The gray limestone with fossil H in the left section is not present in the right section but its expected position is bracketed by fossils G and I. Perhaps the environment at the time did not allow creature H to exist in the right section.

A dark gray limestone at the bases of the two sections can be correlated, but the limestone with fossil C is missing in the right section because of erosion along a disconformity.

Comparing Two Sections That Have Only Partially Overlapping Ages

We often try to compare two sequences of rocks that have only partially overlapping ages. We use faunal succession and the principle of superposition to compare the sections.

Units 4, 5, and 6 in these sections can be correlated based on similar rock types and fossils (not shown). Note that the thicknesses of these units are not exactly the same. Such thickness changes may be due to several different factors, including changes in depositional environment (facies) and differences in the rate of deposition.

Units 1, 2, and 3 are present only in the left section. These rock units are older than the rocks represented in the right section.

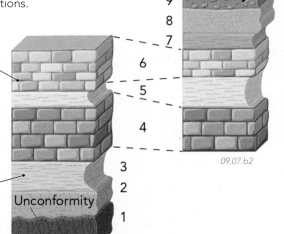

09.07.b2

Units 7, 8, and 9 are present only in the right section. They are younger than any rocks in the left section.

After correlating the two rock sections, we can mentally envision a taller section that represents the entire sequence, with units 1 through 9. We could choose average thicknesses for units 4, 5, and 6, which have thicknesses that vary between the two sections. Using this approach, we can reconstruct an entire section of rocks by observing and comparing smaller sections of rocks, correlating similar units, and by then arranging the appropriate parts of each section on top of one another. This is how geologists construct a complete stratigraphic section for a region.

How Faunal Succession Was Discovered

Most historians credit the discovery of the *principle of faunal succession* to an Englishman named William Smith. Smith also produced the first geologic map of England, Wales, and parts of Scotland, called "the map that changed the world." While surveying along canals, Smith collected fossils from the rocks, noting particularly which fossils were common in which layers. Where the canals traversed faults and folds, Smith encountered sections of the same rocks he saw elsewhere. Smith discovered that the same fossils occurred in the same layers irrespective of their location. He also noted that the fossils changed systematically up through the section of rocks. He recognized that such changes could be used to correlate rocks in different parts of England and Wales, establishing the principle of faunal succession and allowing him to map the units.

Before You Leave This Page Be Able To

✓ Describe how fossils can change through a section of rocks. Provide examples of using index fossils, abrupt boundaries between fossils, and fossil overlaps to precisely infer an age of a rock layer.

✓ Describe or sketch the ways we use fossils and rock types to correlate two rock sequences.

✓ Briefly summarize the meaning of faunal succession and how it was discovered.

9.8 How Was the Geologic Timescale Developed?

GEOLOGISTS DEVELOPED THE GEOLOGIC TIMESCALE to help them correlate rock units across countries and continents and to have a standard vocabulary for describing geologic time. The timescale was devised by using fossils or by noting the absence of fossils. Geologists commonly establish boundaries between units at places in the rock section where fossils record major changes in the types of life. Later, geologists and chemists assigned numeric ages to the timescale by using carefully calculated isotopic ages at key localities.

A What Are the Main Subdivisions of the Geologic Timescale?

After it was established that fossil assemblages change upward through sections of rock, geologists recognized that two different sites that have matching fossil assemblages were the same age. They recognized sequences of related layers across Europe and in North America and named different geologic time periods after places where rocks of that age are well exposed. The largest time intervals are *eras*, and include the Paleozoic, Mesozoic, and Cenozoic Eras, from oldest to youngest. Boundaries between eras are marked by major changes in the fossils, specifically the disappearance (extinction) of many species and families of creatures. Such major extinctions are referred to as *mass extinctions*. Geologists subdivided each era into several *periods*, shown below with the derivation of the name of each period shown to the right of the column.

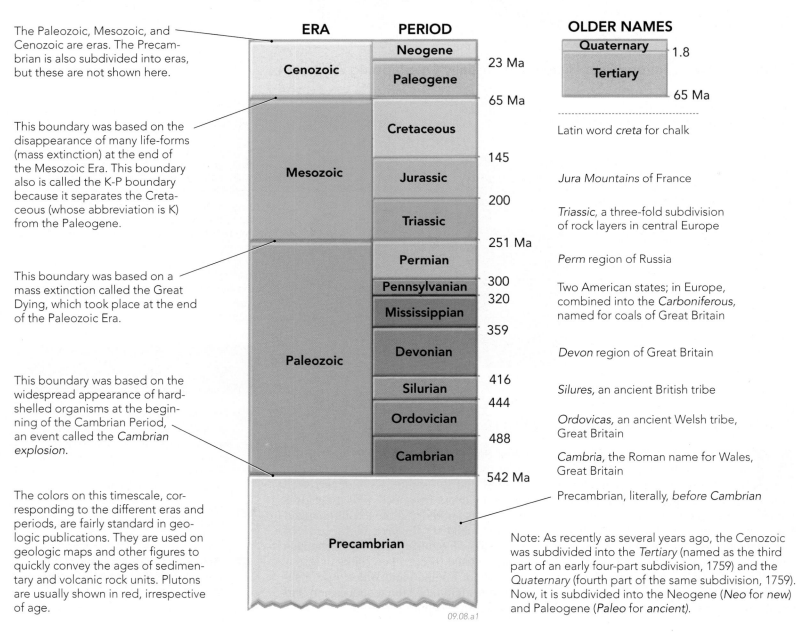

The Paleozoic, Mesozoic, and Cenozoic are eras. The Precambrian is also subdivided into eras, but these are not shown here.

This boundary was based on the disappearance of many life-forms (mass extinction) at the end of the Mesozoic Era. This boundary also is called the K-P boundary because it separates the Cretaceous (whose abbreviation is K) from the Paleogene.

This boundary was based on a mass extinction called the Great Dying, which took place at the end of the Paleozoic Era.

This boundary was based on the widespread appearance of hard-shelled organisms at the beginning of the Cambrian Period, an event called the *Cambrian explosion*.

The colors on this timescale, corresponding to the different eras and periods, are fairly standard in geologic publications. They are used on geologic maps and other figures to quickly convey the ages of sedimentary and volcanic rock units. Plutons are usually shown in red, irrespective of age.

ERA | **PERIOD** | **OLDER NAMES**

Cenozoic — Neogene — 23 Ma — Paleogene — 65 Ma

Quaternary 1.8 / Tertiary 65 Ma

Mesozoic — Cretaceous — 145 — Jurassic — 200 — Triassic — 251 Ma

Latin word *creta* for chalk

Jura Mountains of France

Triassic, a three-fold subdivision of rock layers in central Europe

Paleozoic — Permian — 300 — Pennsylvanian — 320 — Mississippian — 359 — Devonian — 416 — Silurian — 444 — Ordovician — 488 — Cambrian — 542 Ma

Perm region of Russia

Two American states; in Europe, combined into the *Carboniferous*, named for coals of Great Britain

Devon region of Great Britain

Silures, an ancient British tribe

Ordovicas, an ancient Welsh tribe, Great Britain

Cambria, the Roman name for Wales, Great Britain

Precambrian, literally, *before Cambrian*

Precambrian

Note: As recently as several years ago, the Cenozoic was subdivided into the *Tertiary* (named as the third part of an early four-part subdivision, 1759) and the *Quaternary* (fourth part of the same subdivision, 1759). Now, it is subdivided into the Neogene (*Neo* for *new*) and Paleogene (*Paleo* for *ancient*).

09.08.a1

B How Are Numeric Ages Assigned to the Geologic Timescale?

Geologists have assigned numeric ages to geologic periods and their subdivisions by studying localities where isotopically dated igneous rocks have a clear relationship to fossil-bearing layers.

◄ The isotopic age of a volcanic layer interbedded with fossil-bearing sedimentary rocks can be used to assign a numeric age to the geologic period during which these types of fossils formed. Also, the age of a fossil-bearing bed can be bracketed by dating volcanic units above and below it; the bed is younger than a volcanic unit beneath it and older than a volcanic unit above.

Volcanic Layer
09.08.b1

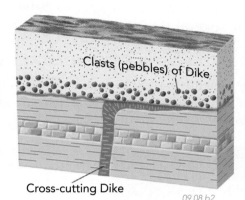

◄ Fossils must be older than the ages of dikes that cross-cut the fossil-bearing layer, like the green shale. In contrast, a tan fossil-bearing layer is younger than igneous clasts included in the layer.

Clasts (pebbles) of Dike

Cross-cutting Dike
09.08.b2

C How Is the Timescale Used to Assign Numeric Ages to Rocks and Events?

Once the ages of the periods and shorter units of geologic time were constrained, these ages could be used to estimate numeric ages of fossil-bearing units that lack datable igneous rocks.

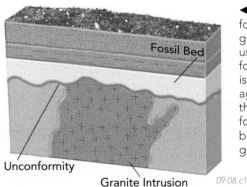

◄ The age of this unconformity, shown by the squiggly line, can be bracketed by using both isotopic ages and fossil ages. The unconformity is younger than the isotopic age of the granite. It is older than the age assigned to the fossils in the overlying bed based on their position in the geologic timescale.

Fossil Bed
Unconformity
Granite Intrusion
09.08.c1

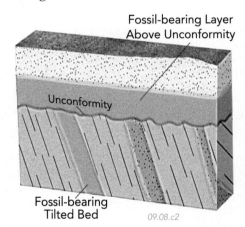

Fossil-bearing Layer Above Unconformity

◄ We can bracket the age of an unconformity by using fossils and the timescale to assign ages to rocks above and below the unconformity. The unconformity is younger than tilted sedimentary beds below, but older than rocks above.

Unconformity
Fossil-bearing Tilted Bed
09.08.c2

How to Remember the Geologic Timescale

Students have developed many techniques to help them remember the names and numbers of the geologic timescale. What do we recommend? Many students use a *mnemonic* device in which the first letters of the mnemonic words match those of the names of the subdivisions of the timescale. One mnemonic for the periods of the Paleozoic, Mesozoic, and Cenozoic Eras is the following:

Cuddly Old Sheep Dogs Make Perfect Pets; They Just Crowd People Nearby.

This mnemonic stands for Cambrian, Ordovician, Silurian, Devonian, Mississippian, Pennsylvanian, Permian, Triassic, Jurassic, Cretaceous, Paleogene, and Neogene. Envision the sheepdog and each part of the sentence. You can also use the mnemonic to help you draw a visual representation of the geologic timescale. Practice filling in the associated numeric ages until it becomes easier to draw and, thus, remember the names of the time periods and the numeric ages that mark the boundaries between the Precambrian, Paleozoic, Mesozoic, and Cenozoic.

Before You Leave This Page Be Able To

✓ Briefly summarize how the geologic timescale was developed.

✓ From oldest to youngest, list the four main geologic chapters and periods.

✓ Explain or sketch how numeric ages are assigned to the timescale and how the timescale is used to assign numeric ages to fossil-bearing rocks.

9.9 What Is the Evidence for the Age of Earth?

EARTH IS 4.5 BILLION YEARS OLD. Early geologists suspected that Earth had a long history and devised several approaches to estimate the age of Earth. The advent of isotopic dating techniques finally provided the tool needed to demonstrate that Earth is indeed very old. What evidence indicates that Earth is billions of years old and not several thousand years old?

A What Were the Early Attempts to Estimate the Age of Earth?

For centuries, scientists tried to figure out ways to date the age of Earth. Many of these ideas were reasonable for their time, but key information was missing and so early estimates of Earth's age were too young. Radioactive decay was not discovered until the late 1800s, but its use revolutionized thought about the age of Earth.

09.09.a1

09.09.a2

An early method to determine the age of Earth calculated how fast salt had accumulated in the oceans. By measuring the salinity and volume of water flowing in rivers, and estimating evaporation rates, scientists could calculate how much time it would take for ocean water to attain its present salt content. The calculation yielded an age of 90 million years, which was far too young because the estimate did not consider salt lost from the oceans to salt beds.

Lord Kelvin, a late 1800s scientist, estimated Earth's age by calculating how long a molten Earth should take to cool to its present temperature. Using thermal properties of rocks and estimates of Earth's internal temperature, he calculated an age of 100 million years. This estimate was done before the discovery that radioactive decay adds internal heat, partly explaining why Lord Kelvin's estimate is too young.

B What Is the Evidence That Earth's History Is Not Short?

Events that are seasonal and leave a physical record, such as the growth rings in trees, are easily observed. Using such features to estimate age requires only that we assume that processes that are occurring today also occurred in the past. Satellite data provide another way to check rates of geologic processes.

09.09.b1

◀ *Tree rings*, which record annual growth cycles, have textures that vary between seasons and widths that vary from dry years to wet ones. Thickness patterns in bristlecone pine rings from the American West can be correlated from living trees to dead ones to form a continuous record back to 9,000 years ago. This approach is independent of, but strongly supported by, many carbon-14 ages.

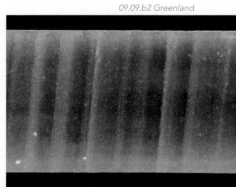
09.09.b2 Greenland

▶ *Ice cores*, cylinders of ice drilled from ices sheets in Greenland and Antarctica, contain thin layers produced by yearly cycles of the seasons. Some ice-core records are thousands of meters long, representing tens of thousands of years to over 100,000 years. Determining the age is done by counting the rings, along with determining isotopic ages for volcanic ash.

09.09.b3 Connecticut Valley, NH

◀ *Varves* are alternating light and dark sediment layers that form in lakes because of seasonal variations of sedimentation and biologic activity. Lighter layers represent increased accumulations of sand and silt during the summer, whereas darker layers record the slower deposition of mud during the winter. There are more than 4,800 varves in glacial lakes that existed in New England around 10,000 to 14,000 years ago, and the Green River Formation in Wyoming has millions of varves that represent at least several million years.

09.09.b4

▶ Current rates of plate motion can be measured precisely to be on the order of centimeters per year (using high-precision GPS data). At these rates, it would have taken the Atlantic Ocean more than 100 million years to form as Africa and the Americas moved apart. Similar rates of motion explain the ages of the Hawaiian Islands and displacement on the San Andreas fault.

C What Are the Oldest Dated Rocks on Earth?

Geologists have dated rocks from several places, including Australia, Canada, and Greenland, at more than 3.5 billion years old. The oldest date (4.4 b.y.) is on a grain in a metamorphic rock in Australia.

09.09.c1 Northwest Canada

The oldest dated rock on Earth is a metamorphic rock, called the Acasta Gneiss, in northwestern Canada. It is interpreted to have started as a granite that was later metamorphosed and deformed. Its age of nearly 4.0 billion years, when the granite solidified, was determined using uranium isotopic dating on zircon. The rock contains even older zircon crystals (4.2 b.y.) that were in the rock that was melted to form the granite magma.

09.09.c2 Grand Canyon, AZ

Many parts of North America have Precambrian metamorphic and granitic rocks beneath a younger sedimentary cover. The crystalline rocks, including those shown here, have been dated at thousands of sites and provide a systematic regional pattern of ages across North America. The oldest rocks, some more than 3 billion years old, are in crystalline rocks exposed across a broad expanse of Canada and the Great Lakes region, an area known as the *Canadian Shield*. Similarly old rocks are exposed in the uplifted mountains of Wyoming and adjacent areas. From these old centers of the continent, the ages systematically decrease to the southeast, being about 1 billion years in Texas and the southeastern United States. The pattern of ages indicates that the southern part of North America was added to an older, northern part. The consistency of the pattern is strong verification of the methodology and supporting knowledge that are the basis of isotopic dating.

Where We Get the 4.55 Billion-Year Age of the Earth

Earth and the solar system are interpreted to be the same age, as measured from the radioactive decay of isotopes in Moon rocks and in meteorites. The Moon is interpreted to have formed early in Earth's history, and meteorites are presumed to represent shattered rocky fragments formed at the same time as Earth.

Geologists and planetary scientists collect meteorites around the world, especially in Antarctica, where the dark rocks stand out on the light-colored snow and ice. They have dated various types of meteorites. The oldest ages are interpreted to represent the time just before the planets cooled. The meteorite analyses support formation of the solar system and Earth between 4.53 and 4.58 billion years ago. The meteorite shown here is the Allende meteorite, dated at 4.56 billion years old.

Nine missions to the Moon returned a limited

09.09.mtb1

number of samples of Moon rocks for isotopic dating. The oldest Moon rocks are 4.4 to 4.5 billion years old.

The age of the Milky Way Galaxy currently is estimated to be approximately 14 billion years, nearly as old as the universe. This age is based on several methods, including the stage of evolution of certain features in our galaxy compared to similar features in other galaxies.

Geologists and chemists have obtained isotopic ages for more than 70 different meteorites, several Moon rocks, and countless Earth rocks. The oldest dates converge on 4.5 billion years before present, even though the rocks are from very different places and more than one dating method was used. The slightly younger age, 4.0 billion years, for the "oldest dated rock" on Earth is expected, because erosion, deposition, and tectonic activity remove rocks, bury them, deform and metamorphose them, and

even melt them. Some geologists are surprised that such an old rock was able to survive at all. A 4.4 billion-year-old zircon grain from Australia is even closer to the age of the meteorites.

Viewed in this context, there is remarkable consistency between ages from meteorites, the Moon, and Earth. This consistency strongly supports the 4.5 billion year age for Earth, the Moon, and the solar system.

Before You Leave This Page Be Able To

☑ Describe early methods for determining the age of Earth and why they proved to be inaccurate.

☑ Describe evidence that suggests Earth has a long history, including isotopic ages on basement rocks in North America.

☑ Describe how meteorites and Moon rocks are used to interpret the age of Earth.

9.9

9.10 What Were Some Milestones in the Early History of Life on Earth?

LIFE ON EARTH BEGAN before there was much oxygen in the atmosphere. Very simple, but successful, Precambrian organisms evolved, started to photosynthesize, and eventually produced an oxygen-rich atmosphere. During the Paleozoic Era, more complex organisms appeared, eventually including fishes, plants, amphibians, reptiles, insects, and various types of marine organisms.

A What Were the Earliest Forms of Life on Earth?

Some of Earth's first inhabitants were early forms of algae, represented by fossils called *stromatolites* found in Australian rocks that are 3.5 billion years old, and rocks elsewhere that are only slightly younger. These organisms lived when Earth had little or no oxygen in its atmosphere.

09.10.a1

09.10.a2

09.10.a3

This rock contains mound-shaped *stromatolites*, the earliest non-microscopic fossils identified on Earth. The cyanobacteria that form modern stromatolites use photosynthesis to make food, and if those in ancient stromatolites did the same, they may have begun the transformation to an oxygen-rich atmosphere.

Ancient stromatolites probably had a similar appearance and structure to modern stromatolites in Australia (▲). Today's stromatolites live in an oxygen- and nitrogen-rich atmosphere, whereas the atmosphere 3.5 billion years ago had more carbon dioxide, which was produced mostly by outgassing from volcanoes.

About 2 billion years ago, cyanobacteria had produced enough oxygen through photosynthesis to increase the amount of oxygen in the atmosphere and form a protective ozone layer. Sometime after 1.5 billion years ago, organisms began to reproduce sexually, which led eventually to complex, multicellular organisms.

B What Was the Cambrian Explosion?

At the beginning of the Cambrian Period, about 542 million years ago, life became more diverse and included organisms with hard protective coverings. Creatures with shells had an obvious advantage over earlier soft-bodied organisms. Over a period of about 20 to 30 million years, many new shelled organisms appeared on Earth. This period of rapid evolutionary change is called the *Cambrian explosion*.

09.10.b1

09.10.b2

09.10.b3 Burgess Shale, BC, Canada

Trilobites were one of the dominant organisms of the Cambrian Period. They had external skeletons, diverse appearances, and lived in a wide range of environments. Many are index fossils.

The Cambrian seas produced simple forms of marine animals related to clams, starfish, sponges, and crabs, including *brachiopods* like those shown above.

Some of the best examples of Cambrian fossils come from a shale in the Canadian Rockies. The shale preserves more than 150 species, including impressions of the soft parts of some rather odd creatures, like this one.

C What Life Existed During the Paleozoic Era?

During the Paleozoic Era, an extraordinary diversity of life evolved—both in the seas and on land. Artistic reconstructions of three times of the Paleozoic were produced by Karen Carr, whose work is featured in museums.

Early Paleozoic

In the early Paleozoic, corals, crinoids (which look like platy underwater lilies), and mollusks were anchored to the seafloor, trilobites and snails moved across the seafloor, and shelled creatures with tentacles propelled themselves through the water.

Middle Paleozoic

In the middle of the Paleozoic, corals built large reefs, and pieces of crinoid stems littered the seafloor. Fish became diverse and abundant. On land, many forms of insects appeared, and plants included ferns and seedless trees.

Late Paleozoic

Amphibians and early reptiles evolved during this time, with a dramatic rise of reptile groups and a continued diversity of marine life. Land plants, insects, and marine life continued to diversify until a major extinction (the Great Dying) at the end of the Paleozoic.

Possible Causes of the Great Dying

The end of the Paleozoic Era marks Earth's greatest extinction, called the Great Dying. On land, about 70% of all species, including many invertebrates, amphibians, and reptiles, went extinct. The event took a huge toll in the oceans, extinguishing almost 90% of marine species, including trilobites. Geologists are still actively investigating a number of possible causes.

A great outpouring of lava occurred at the end of the Paleozoic. Large volumes of basalt erupted in northern Asia, in a region called the *Siberian Traps* (*trap* is an old word used to describe basalt). Such eruptions expel volcanic ash and gases, including water vapor, carbon dioxide, and sulfur dioxide. The ash and gases have the potential to warm or cool the planet and possibly cause other catastrophic effects, such as changing circulation patterns in the oceans.

There is some evidence, currently being debated, for a large meteorite impact at the end of the Paleozoic. Geologists have proposed that an impact can explain unusual carbon molecules found in rocks of this age. Geologists have found several suitably large impact craters, but none directly tied to the extinction. A huge impact could have triggered the massive eruptions of the Siberian Traps, but this connection remains conjectural.

Throughout most of the Paleozoic, continents were separated by warm, shallow seas. By Permian time, the supercontinent *Pangaea* had formed. Its formation closed seas that had once nourished Paleozoic life. The supercontinent became more arid, and vast evaporite deposits formed and could have changed the salt concentrations in seawater. These and other effects of the formation of Pangaea may have helped kill off specialized organisms and set the stage for a more dramatic event.

An alternative explanation is that conditions in the atmosphere and oceans led to a massive overturn of ocean water, causing deep, oxygen-poor water to be brought to the surface. This could have caused a dramatic change in shallow ocean temperatures and in the amount of CO_2 in the atmosphere, leading to sudden and catastrophic climate changes. Such changes could affect the entire planet, resulting in a mass extinction on the land and in the oceans. This theory, like the others, is unproven, and the Great Dying remains an unsolved mystery.

Before You Leave This Page Be Able To

✓ Describe the environments of early life and some important evolutionary events that took place during Earth's early history.

✓ Briefly describe what happened during the Cambrian explosion.

✓ Explain four possible causes for the Great Dying, the largest extinction event in Earth history.

9.11 What Were Some Milestones in the Later History of Life on Earth?

MASS EXTINCTION AT THE END OF THE PALEOZOIC ERA provided evolutionary opportunities for new life-forms. The organisms that repopulated the early Mesozoic seas and lands were very different from Paleozoic organisms. Diverse life existed during the Mesozoic Era, including dinosaurs. The end of the Mesozoic Era is defined by another major extinction event, which gave rise to yet another evolutionary chapter, the ascent of mammals during the Cenozoic Era. The artwork of Karen Carr provides us with one interpretation of the scenes represented by the bones, shells, leaves, and other fossils.

A What Life Was Abundant During the Mesozoic Era?

Diverse life existed during the Mesozoic Era, but it is known as the *age of dinosaurs*, the best known creatures of this time. The Mesozoic has three periods: Triassic, Jurassic, and Cretaceous, from oldest to youngest.

Early Mesozoic: Triassic
09.11.a1

During the Triassic Period, small and nimble dinosaur-like creatures and mammals appear beneath the seed-bearing conifer forests. In the seas, shallow-sea niches left open by the Permian extinction were occupied by coiled ammonites and other marine animals.

Middle Mesozoic: Jurassic
09.11.a2

Dinosaurs diversified and many new species appeared during the Jurassic Period, including *Stegosaurus* with plates on its back and the huge plant-eating *Apatosaurus*. Carnivorous predators, like *Allosaurus*, stalked the landscape. The Jurassic Period also featured *Archaeopteryx*, an early bird. The seas flourished with many diverse creatures, including ammonites, starfish, and large marine reptiles.

Late Mesozoic: Cretaceous
09.11.a3

09.11.a4

During the Cretaceous Period, dinosaurs remained diverse, and included various plant-eating dinosaurs that walked on four or two legs, as well as predators like the raptors lurking in the bushes. Flying reptiles and birds graced the skies. Not shown is the fearsome *Tyrannosaurus rex*. For the first time, flowering plants, called *angiosperms*, became abundant on land. Insects remained a vibrant and diverse group, and most mammals continued a rather low-key existence.

During the Cretaceous Period, animals similar to those of the Jurassic thrived in the seas, including fish of many kinds, straight and coiled nautiloids, large marine reptiles, and turtles. Not shown because of their tiny size are countless floating and free-swimming organisms called *plankton*.

B What Were Dinosaurs and What Caused Their Demise?

Dinosaurs evolved from Permian ancestors and existed on Earth for 165 million years, throughout the Mesozoic. By the middle of the Mesozoic, they dominated the land, but they and many other animals went extinct at the end of the Mesozoic Era, at what has traditionally been called the *K-T extinction*. With changes in names on the geologic timescale, it has been called the *K-P extinction*, because it separates the Cretaceous (K) from the Paleogene. Geologists and other scientists have proposed numerous hypotheses to explain the extinction.

09.11.b1

1. There were two types of dinosaurs, differing in their hip structure. One group of dinosaurs had a hip structure similar to lizards and included a diverse group of carnivores, such as *Tyrannosaurus rex*, and herbivores, such as *Apatosaurus*. Some walked slowly on four legs; others walked and ran on two legs. Another group of dinosaurs had a birdlike hip structure (but were not related to birds) and were herbivores. Some like *Stegosaurus* and *Triceratops* walked and grazed on four legs. Others, like duck-billed dinosaurs, could move on two legs.

2. A well-known hypothesis for the K-P extinction involves a huge comet or asteroid striking Earth, sending massive amounts of dust and gas into the atmosphere and blocking sunlight. Earth's surface would have been cold for decades. Many geologists conclude that the impact site, 65 million years ago, was the Chicxulub crater on the Yucatán Peninsula in Mexico (shown by the red circle).

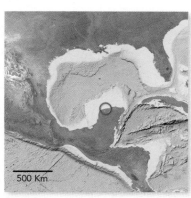
500 Km
09.11.b2

3. Another possible cause of the extinction was massive outpourings of basalt in the *Deccan Traps* in India (not shown). Huge eruptions could have put enough sulfur dioxide gas into the atmosphere to cause a winter that lasted decades.

C What Life Appeared During the Cenozoic Era?

The Cenozoic Era is also called the *age of mammals*. After dinosaurs went extinct, mammals were able to diversify rapidly and fill many niches left behind by the K-P extinction.

09.11.c2

1. By early Cenozoic time, the ancestors of modern mammals, including bats, rodents, primates, sloths, whales, hoofed animals, and carnivores were abundant and lived in a variety of habitats. Marsupial mammals, represented by modern kangaroos, thrived on the isolated southern continents of South America and Australia.

▼ 2. Although they lived 20 million years ago, many of the mammals shown here may be familiar to you because they are fairly similar to their modern descendants. Each type of mammal, however, underwent many changes between then and now. Horses, for example, changed dramatically in size. These changes are well recorded by bones and teeth of different species of horses found at thousands of sites around the world.

09.11.c1

▶ 3. Late in the Cenozoic, during the Ice Ages, a number of large mammals roamed the continents. Many of these animals, like the mammoth, saber-toothed cats, and giant beaver, went extinct as the Ice Ages ended and humans spread across the globe. The first humans (Homo sapiens) appeared before 300,000 years ago, based on fossil evidence. Human-migration data are still controversial, but by at least 50,000 years ago Homo sapiens populated several parts of the planet, having left their sites of origin in Africa. The details of human history are refined by discoveries of new archeological sites and even older ancestors.

Before You Leave This Page Be Able To

☑ Contrast the kinds of organisms that lived during the Mesozoic Era with those that lived during the Cenozoic Era.

☑ Describe some of the variety observed in dinosaurs, and summarize two theories for why dinosaurs became extinct.

9.11

9.12 How Do We Reconstruct Geologic Histories?

WE RECONSTRUCT THE SEQUENCE OF GEOLOGIC EVENTS by using the various strategies of relative dating, correlation of rock units, isotopic dating, and other geologic principles. Geologists commonly start by studying a single section of rocks and determining the sequence of events it represents. Understanding the causes and contexts of the geologic environments and events requires correlating several rock sections.

A How Do We Correlate Units and Events in Two Sections of Rocks?

There are various strategies for matching—or correlating—two sections of rocks. The general approach is to find units that match and develop a logical explanation for why other parts of the sections do not match. Read each principle and its example, and then compare that part of the two sections.

Principle

Lateral Continuity—The surest form of correlation is to be able to physically trace a unit through the landscape from one place to another.

Distinctive Rock Type—A unit may have distinctive characteristics that enable it to be matched between two sections.

Fossils—If two units contain the same assemblage of index fossils, they have the same age and are correlative.

Similar Sequence of Rocks—Two sections of rock may contain a similar sequence of layers.

Record the Same Event—Two units may record the same event, such as a change in sea level, even if they express the event in different ways.

Isotopic Age—If datable units, such as volcanic layers, yield the same numeric age, they may be time correlative.

Magnetic Signature—Earth's magnetic field has reversed direction through time. Some rocks record these changes, and we can use the resulting patterns to correlate rocks.

Position in Sequence—Two different rock types may correlate if they are in the same position in the sections.

No Correlative Unit—In some cases, a unit in one section has no correlative unit in the other section.

Relation to Unconformities—Two units may correlate if they have a similar relationship to the same unconformity.

Section 1   Section 2

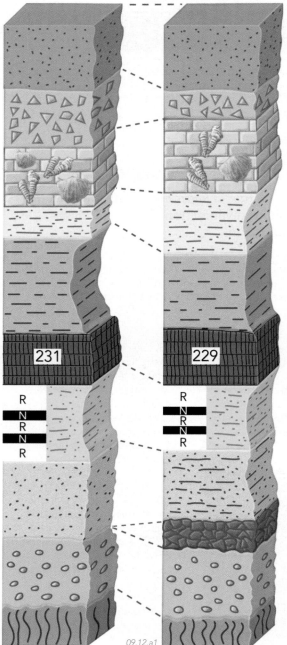

09.12.a1

Example

Both sections are capped by a reddish sandstone that can be traced through the landscape between the two sections.

Both sections have breccia that contains large blocks of gray granite.

The gray limestone in both sections contains index fossils of late Triassic age.

Both sections contain a yellowish mudstone, overlain by gray limestone and underlain by gray shale.

Gray shales in both sections record rising sea level, but different thicknesses of clay accumulated in the two areas.

Basalt flows in both sections give similar isotopic ages (~230 Ma).

Two units recorded reversals in Earth's magnetic field, from normal (N) to reversed (R), and display similar patterns of reversals. This allows but does not demonstrate a correlation.

The beach sand in section 1 is in the same position as a tidal-flat-related mudstone in section 2.

A local landslide deposit in section 2 did not extend far enough to be present in section 1.

Conglomerate at the base of both sections overlies metamorphic rocks along a regional unconformity.

B Why Do Some Rock Units Change from One Section to Another?

When investigating causes of events, geologists seek to understand how and why the rock units change laterally. Even though two units are deposited at the same time, they may be different rock types. There are several explanations why the sequence of layers changes from one place to another.

Facies Change—The type of sediment deposited at the same time can be different in two different places because the sedimentary facies (depositional environment) changed laterally, such as from a shallow marine environment to a delta.

Restricted Event—A unit may not be present in nearby sequences because it simply was not deposited there. Many units are formed by a relatively small event and have a restricted aerial distribution, like these sand dunes.

Change in Thickness—Accompanying some facies changes are variations in the thickness of sediment that is deposited. Thickness changes can also reflect variations in topography over which a unit is deposited, such as river deposits that are thickest in the center of the valley.

Eroded Away—Another explanation for why a unit is not present or is thinner is that it has been partly or completely eroded away in one place but is preserved in another. Such erosion should be marked by an unconformity of some type.

09.12.b1

C What Are Some Approaches to Investigating Geologic History?

Many geologists study geologic relationships using geologic maps, cross sections, and block diagrams. Nearly all geologic problems, including reconstructing histories, involve geometric relationships in three dimensions. The figure below shows a geologic terrain with various rock units and other geologic features, numbered in the order in which they formed, from oldest (1) to youngest (13). Examine this figure and identify the reasons why the units and features are interpreted to have formed in the relative order reflected by the numbering.

1. Cliffs expose horizontal rock layers, capped by a gray limestone (7). At the base of the cliff is a fault scarp (11) that downdrops rocks to the south, forming a valley that is covered with fairly recent sediment (9).

2. The front of the block depicts a sequence of layers (4–7), some of which are the same layers exposed in the cliff. The lowest layer in the series is the oldest (4), and the highest layer, the recent sediments (9), is the youngest. The contact between the limestone and the sediments is a disconformity (8).

09.12.c1

3. At the base of the block are metamorphic rocks (1), the oldest rocks in the area. The contact between the metamorphic rocks and layer 4 (a conglomerate) is depositional, with pieces of metamorphic rocks in the conglomerate. The contact between the steep metamorphic rocks and conglomerate is an angular unconformity (3).

5. A river valley cuts through the cliffs, forming a canyon (12). In the valley, the river contains a thin veneer of river gravels (13). The fault scarp (11) does not offset the gravels, so the scarp is older.

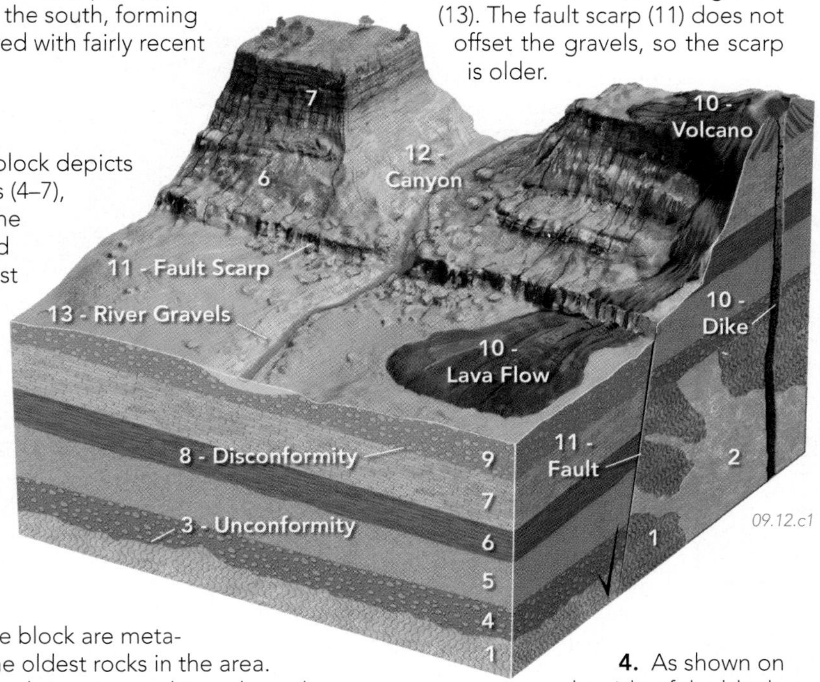

4. As shown on the side of the block, the fault cuts, and so is younger than, the metamorphic rocks, layers 4–7, and the uppermost sediments (9).

6. A recent-looking scoria cone (10) has erupted lava (also numbered 10) that flowed to the south. The lava flow poured over the cliff, but is offset by the latest movement along the fault scarp (10). Some fault movement probably predated the eruption, in order to produce the cliffs. The cone and lava flow are somewhat weathered and eroded.

7. The subsurface conduit for the scoria cone is a now-solidified dike (10). The dike cuts across all rock units it encounters on the side of the block.

8. In the subsurface, a granite (2) cuts across the metamorphic rocks (1), but is truncated by the lower angular unconformity (3). The dike (10) cuts the granite.

Before You Leave This Page Be Able To

☑ Describe or sketch the principles by which two sequences of rocks can be correlated.

☑ Describe or sketch why layers can change from one sequence to another.

☑ Reconstruct the sequence of events from a cross section or block diagram.

9.12

9.13 Why Do We Investigate Geologic History?

INVESTIGATING WHEN AND IN WHAT ORDER geologic events occurred has practical value for our modern society. Geologic history helps us evaluate the potential for geologic hazards, explore for resources, comprehend the physical world around us, and understand changes in life and the environment over time.

A How Do Geologic Ages Help Us Evaluate Geologic Hazards?

When assessing the potential for volcanic eruptions, earthquakes, and other geologic hazards, we are interested in knowing what types of processes are involved with the hazard and *when* these events last occurred. We determine when these hazardous events occurred by applying principles of relative dating, landscape development, and numeric dating, in order to estimate when such activity last occurred and when it might happen again.

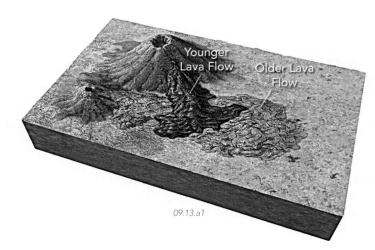

09.13.a1

Volcanic Eruptions

Earth's surface contains many volcanoes and exposures of volcanic rocks. Which of these are most dangerous, and which ones are not dangerous at all? Knowing when a volcano last erupted, and which volcanic units are the most recent, is clearly one of the most important aspects to know. In this image, a dark, recent lava flow overlies a lighter colored, older one. When did each eruption occur, and what is the likelihood of another eruption? Evaluating such volcanic hazards involves dating the relative ages of the rocks using principles of relative dating, like crosscutting relations and superposition. Next, we could estimate the ages of lavas and volcanic cones based on how much they are weathered and eroded or on the development of rock varnish. Finally, we could also use isotopic methods, like Ar-Ar dating or carbon-14 dating of charcoal from trees buried by the lava, to precisely determine the ages of each unit in thousands or millions of years before present.

Earthquakes

Earthquakes cause destruction from ground shaking and from secondary effects, like landslides and loss of soil strength. Most earthquakes, especially the large ones, result from slip along faults. Assessing the hazard for earthquakes, therefore, depends greatly on determining when a movement along a fault last occurred. Geologists use crosscutting relations to determine which units are cut by the fault and which ones are younger than any fault movement. Faults that break the surface form fault scarps, which tend to be steep and uneroded when first formed but are degraded over time from weathering, slope failure, and other erosion. In the scene shown here, the landscape is cut by two earthquake-related fault scarps. The upper scarp is recent and not eroded, and the lower one is partly eroded and covered by a lava flow. The ages of the earthquakes that formed these scarps can be investigated by examining soils and rocks that predate and postdate faults, by dating sediments associated with faulting, and by isotopic dating of the age of the lava flow.

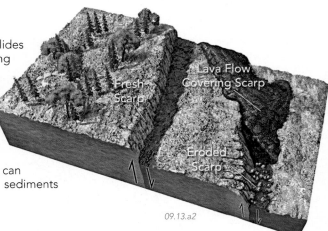

09.13.a2

Flooding

09.13.a3

Flood potential is evaluated from records of stream flow, but these records may only cover the last 100 or so years. Inferring the recurrence of larger, less frequent floods relies on geologic evidence preserved in the landscape. Rivers have an active channel, which contains moving water all or part of the year. Slightly above the active channel in elevation is the floodplain, a low, nearly flat area that gets flooded when there is too much flowing water to be contained within the active channel. The excess floodwater spills out onto the floodplain, and many floodplains are flooded nearly every year or several times every century. Above the floodplain are river terraces, which were formed by the river or stream, but are high enough to be flooded less frequently, if at all. A key strategy in assessing flooding potential, therefore, is to determine when the floodplain and terraces were last flooded. We do this using the degree of soil development, carbon-14 ages on charcoal, surface-dating methods on stones deposited by the river, and even the age of human artifacts, like bottles and cans of a certain vintage.

B How Do Geologic Ages Help Us Explore for Natural Resources?

Understanding geologic history is an essential part of evaluating an area's potential for important natural resources, especially mineral, energy, and water resources.

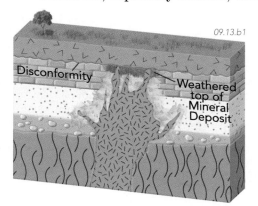

09.13.b1

Disconformity

Weathered top of Mineral Deposit

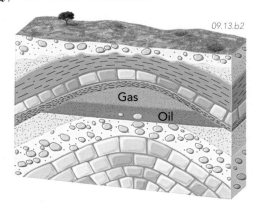

09.13.b2

Gas

Oil

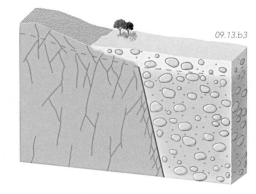

09.13.b3

Here, a granite released metal-rich fluids that formed adjacent copper and gold deposits, shaded in gray. Exploring for mineral deposits involves knowing the ages of events that contributed to mineralization, such as dating the granite or minerals deposited by the fluids. Also important are crosscutting relations between the granite, mineralized fractures, and earlier or later rocks and structures.

Oil and gas accumulate in the subsurface, commonly near the tops of anticlines. Exploration for oil and gas involves a thorough investigation of the sequence of rock units, as determined by relative dating and fossils within the sedimentary rocks. Later events, including folding and erosion, play a key role in determining whether oil is trapped at depth or could escape to the surface.

Sedimentary basins contain abundant groundwater. However, granite may only contain water in fractures. We assess the formation age of sedimentary basins and their water-rich sediments using fossils, relative dating, and isotopic ages on interbedded volcanic rocks. The age of some groundwater can also be dated using isotopes to understand when the groundwater formed.

C How Does Geologic History Help Us Investigate Our Origins?

Geologists investigate recent historical events and the origins of humans using many of the same strategies and techniques used to reconstruct the history of ancient rocks and structures. The record of historical events of the last several thousand years is commonly investigated by working with archeologists who excavate the ruins of ancient cities and other archeological sites.

09.13.c1 San Juan River, UT

Archeologists and geologists use the same relative-dating principles applied to older rocks. The oldest ruins generally are on the bottom, having been covered over by successive generations of younger habitations. Pieces of an older wall can also be contained within a younger wall if the prehistoric builders reused preexisting materials. In this cliff dwelling, built by ancient peoples of the Southwest, the walls incorporate large pieces of rock, so the pieces are older than the wall. A join in the wall reflects two different ages, of construction with slightly different craftsmanship. With time, exterior walls in such structures acquire a coating of desert varnish, with older walls having a darker varnish than younger walls. Structures that were constructed recently, within the last several hundred years, will not have visible desert varnish.

09.13.c2 Woodlands Culture, Iowa

Investigations of early human history rely heavily on the input of geologists. Bones from early human ancestors have been found in sedimentary units ranging from tens of thousands of years old to five or six million years old. These sequences are dated by interbedded volcanic rocks or by fossils of small mammals or other creatures. Events that are hundreds to tens of thousands of years old can be dated using carbon-14 techniques on bones, wood, and charcoal preserved at a site. The relative positions of dated samples become an important check for consistency.

Before You Leave This Page Be Able To

✓ Describe or sketch how geologic ages help evaluate geologic hazards.

✓ Describe or sketch how geologic ages help evaluate mineral, energy, and water resources.

✓ Discuss dating techniques used to investigate early human sites.

9.13

What Is the History of the Grand Canyon?

GEOLOGICALLY, THE GRAND CANYON HAS IT ALL. It contains some of the best exposed and studied, as well as the most beautiful, rock sequences in the world. It is discussed in almost every geology class because it so clearly expresses a history of geologic events over the last 1.7 to 1.8 billion years.

This north view of the Grand Canyon region shows the location of a geologic cross section from A to B. The Colorado River, which formed the canyon, flows from right to left, exits the canyon through high cliffs and enters Lake Mead.

The Grand Canyon is eroded into the Colorado Plateau, a region of broad plateaus, mesas, and deep canyons, which expose a mostly flat-lying sequence of Mesozoic and Paleozoic sedimentary rocks.

The river flows southwest across the area, cutting across nearly horizontal to locally tilted layers. The deepest part of the canyon is where the Colorado River erodes through the uplifted Kaibab Plateau.

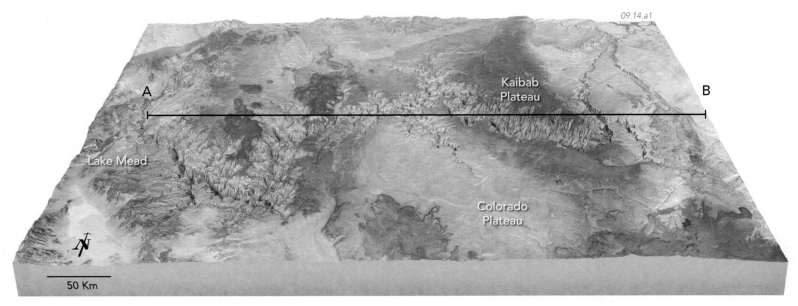

09.14.a1

Basalt flows cap some plateaus and predate formation of the main canyon. They are dated by K-Ar methods to be 8 million years old.

Large faults, like the Hurricane fault, cut across the region, downdropping rocks to the west. These faults cut basalt flows that are less than 1 to 2 million years old.

Some basalt flows flowed down into the already-carved canyon, demonstrating that much of the canyon is older than 4 to 5 million years.

Paleozoic sedimentary layers cap most plateaus and are warped over a few broad folds (monoclines).

Mesozoic sedimentary rocks are preserved on the downfolded sides of monoclines and contain famous dinosaur tracks and petrified wood in the Painted Desert.

09.14.a2

The colorful walls of the canyon expose a flat-lying sequence of, from top to bottom, late, middle, and early Paleozoic rocks. There are disconformities within the Paleozoic section, each representing tens of millions of years of missing time.

The oldest rocks are metamorphic rocks and granites that are 1.7 billion years old, with the granites being slightly younger. These rocks are exposed in the bottom of the canyon.

The near-vertical metamorphic rocks are overlain by tilted late Precambrian sedimentary and volcanic rocks, shown in purple. The contact is an angular unconformity, and is called the *lower unconformity*. Where this unconformity overlies granite plutons, it is a nonconformity.

A separate, angular unconformity marks where gently dipping Paleozoic layers, shown in blue and red, overlie the moderately tilted late Precambrian layers, shown in purple. This is called the *upper unconformity*, and to the west cuts across the lower unconformity

The Grand Canyon exposes all three types of unconformities: angular unconformity, nonconformity, and disconformity. Photographs of each type from the Grand Canyon are in Section 9.3.

Sequence of Rocks

09.14.a3 Grand Canyon, AZ

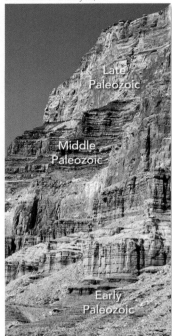

09.14.a4 Grand Canyon, AZ

Geologic History and Key Age Constraints

09.14.a5

The sequence of events in the Grand Canyon has been reconstructed using relative dating, fossils, and many different isotopic dating methods. The geologic history resulting from these studies is summarized below, which should be read from bottom to top (oldest to youngest).

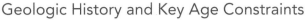

7. *Deformation, Uplift, and Erosion*—The Paleozoic strata largely have escaped deformation and remain nearly flat, except near a few faults and folds, including some monoclines visible in cross section A-B. The monoclines are bracketed, using relative-dating methods, to between 80 and 40 million years ago. The region was uplifted by some amount at this same time, but the modern canyon was not carved until much later, mostly within the last 5 million years. Some faults, like the Hurricane fault, remain active.

6. *Deposition of Late Paleozoic Layers*—Overlying sedimentary layers (shown in red, pink, tan, and blue-green) record a wide range of environments, including shallow marine, shorelines, rivers, and a dune-covered desert. These rocks are dated with marine and nonmarine fossils as late Paleozoic (Pennsylvanian and Permian). Disconformities (mostly not shown) separate some of the formations and represent time when the region was above sea level.

5. *Deposition of Early and Middle Paleozoic Units*—After erosion carved the upper unconformity, seas covered the land and deposited sandstone, shale, and limestone (shown in brown and blue). These deposits are dated by trilobites and other fossils as early and middle Paleozoic (Cambrian, Devonian, and Mississippian). Later, the seas left and in several instances formed disconformities within the limestones.

4. *Tilting and Upper Unconformity*—Layers in the Late Precambrian rocks were gently to moderately tilted and then beveled by erosion. This produced the *upper unconformity*. As this unconformity is followed west, it truncates the *lower unconformity* beneath the Kaibab Plateau (see the cross section A–B). To the west, these combined unconformities represent even more missing time (from 1.7 billion years to 540 million years, or more than 1.1 billion years); it is appropriately called the *Great Unconformity* and can be followed eastward to the Great Lakes region.

3. *Late Precambrian Rocks and Lower Unconformity*—In the Late Precambrian, sedimentary and volcanic rocks were deposited in horizontal layers across the upturned basement layers. This formed the *lower unconformity* (above the metamorphic rocks). The lower parts of these late Precambrian rocks are dated by several isotopic methods at 1.1 billion years. Since the underlying basement rocks are 1.7 billion years old, the lower unconformity represents 600 million years of time not recorded by any rocks!

2. *Uplift and Erosion of the Basement*—After the metamorphism, the basement rocks cooled as they were uplifted and eroded over a period that lasted for hundreds of millions of years. Erosion beveled across the steep metamorphic layers.

1. *Basement Rocks*—Metamorphic and plutonic rocks in the bottom of the canyon represent the oldest events. They were formed, metamorphosed, and deformed to near-vertical orientations, all between 1.76 and 1.70 billion years ago.

The Percentage of Geologic Time That the Canyon Records

Although the canyon is a classic geologic locality with a thick sequence of formations, it represents a relatively small amount of geologic time. The oldest rocks are "only" about 1.7 billion years old, so the area contains no record for 2.8 billion years of Earth history (4.5–1.7 billion years). Next, the two unconformities together cut out another 700 to 800 million years of history.

Even the Paleozoic sequence is missing more time than it records! The formations only represent five out of the seven geologic periods (rocks of the Ordovician and Silurian Periods are not present), none of the formations span an entire period, and there are major disconformities. Mesozoic and Cenozoic rocks are largely absent in the canyon, so yet more time is not represented by rocks in the canyon walls.

Before You Leave This Page Be Able To

☑ Describe examples of how different methods of dating events and rocks were used to reconstruct the geologic history of the Grand Canyon.

☑ Describe why the canyon does not represent all of geologic time.

9.14

9.15 What Is the Geologic History of This Place?

This terrain exposes various geologic relationships that have been documented in the field and recorded as descriptions. Samples collected from the area were analyzed either for their isotopes or their characteristic fossils. You will use this information to reconstruct the sequence and ages of events that produced features exposed in the landscape today.

Goals of This Exercise:

- Observe the distribution of different rock types exposed in the terrain to characterize the sequence of rocks and the geologic features that are present.
- Use descriptions of units and of key contact relationships, along with fossils, to infer the relative sequence of events.
- Calculate isotopic ages for key igneous rocks to help constrain when important events occurred.

Procedures

Use your observations to complete the following steps. Your instructor may provide you with rock or fossil specimens.

1. Observe the terrain to understand the overall pattern of rocks. Based on this pattern, use the associated descriptive text to determine in what order the units formed and where in that sequence different geologic features, such as a fault and dike, developed.

2. Examine the six fossils in the table below, and the geologic period to which each is assigned; complete the stratigraphic section on the worksheet, listing the units in the order in which the units formed, from bottom to top in the section.

3. Use the table of isotopic measurements below to calculate the age of a sample of granite and a sample of the dike.

4. Summarize the geologic history by arranging the different events in their proper order on the worksheet or online.

Field Notes

The units and features are described below. Each unit or feature has a letter assigned to it, but these do not reflect the order in which the features formed. Some letters were skipped so that some features would have letters that were easy to remember, such as V for the volcano.

Unit A—Tan sandstone with land fossils, including plants of Permian age.

Unit B—Greenish shale with marine fossils, including Ordovician trilobites. The top of the unit was weathered and eroded prior to deposition of unit A, but the layers in the two units are parallel to each other and to their mutual contact.

Unit C—Coarse sandstone and beach conglomerate that contains Cambrian trilobites. The base contains clasts derived from the underlying granite (G).

Unit D—Finely crystalline dike that has baked units A, B, C, and G.

Feature F—Fault that cuts units B, C, and G. Some units are not near the fault.

Unit G—Coarse granite that is weathered near the contact with unit C.

Unit K—Gray limestone with marine fossils of Cretaceous age.

Units L and V—Unweathered lava flow (L) associated with a volcano (V).

Feature N—Narrow canyon.

Unit R—Partly consolidated river gravels with a thick, well-developed soil. Contains land mammals of middle Cenozoic age.

Unit S—Reddish and pinkish sandstone that was deposited by rivers and in lakes. It contains Jurassic dinosaur bones.

Identification of Fossils

Rock Unit	Fossil	Period
R	Mammals	Cenozoic
K	Fish	Cretaceous
S	Dinosaurs	Jurassic
A	Plants	Permian
B	Trilobites	Ordovician
C	Trilobites	Cambrian

Table of Isotopic Measurements

Rock Unit	Half-Life of Isotope	# Parent Atoms	# Daughter Atoms
G	1 Billion Years	250	750
D	50 Million Years	500	500

This view shows a landscape with various rocks and features. There is a central plateau (high flat area) flanked by several mountains, an obvious volcano, a canyon, and a number of lines and curved features that cross the landscape. The geology in the subsurface is shown on the sides of the block. Any type of unconformity is shown with a squiggly line, reflecting some topographic relief along the erosion surface represented by the unconformity. Normal depositional contacts are shown by thin lines, and a fault is marked by a thicker line.

1. A section of layers forms a series of cliffs and slopes on three corners of the block. These were encountered first and so are lettered A, B, and C. Unit A is a brown sandstone that was deposited on land and contains Permian plant fossils. Unit B is greenish marine shale and contains Ordovician trilobites. Unit C is a coarse sandstone and beach conglomerate that contains Cambrian trilobites.

2. A dark dike (D) forms a linear wall across the landscape. It mostly is uninterrupted by other geologic features, except for one obvious gap near a belt of some tan-colored soils (associated with unit R). The dike consists of dark basalt and was dated by isotopic methods.

3. An older series of river channels (R) cross the plateau and form low troughs in the topography. One channel goes all the way to the edge of the canyon, where it stops abruptly, evidently having been cut off. Along their lengths, the channels are partially filled by river gravels and are characterized by well-developed, tan soils. They contain bones of small horses and other fossils from the middle Cenozoic.

4. The top of one mountain in the area (right corner of this figure) exposes higher layers than are preserved elsewhere. There is a red sandstone (S) that contains bones of Jurassic dinosaurs. The sandstone is overlain by a gray limestone (K) that has fish and other marine fossils from the last part of the Mesozoic (Cretaceous).

09.15.a1

5. There is a cone-shaped volcano (V) surrounded by a black lava flow (L). Neither the volcanic deposits (scoria) on the volcano nor the lava flow have developed any soil.

6. A fault (F) forms an obvious line across parts of the area, but is not continuous. It is also shown in cross section on the side of the block. It has not formed a fault scarp, but is expressed in the topography because it is the boundary between rock types that erode in slightly different ways. In a nearby area, the fault cuts the main sequence of layers, including layers C, B, A, S, and K.

7. The lowest unit in the area is a gray granite (G). Geologists determined an isotopic age on a sample of the granite, and these results are in the table on the previous page.

8. A narrow canyon (N) cuts through the area. The canyon is especially narrow in one segment where dark lava flows (L) have poured from the plateau and into the already formed canyon.

9. Reconstruct the history using superposition, crosscutting relationships, and the relationship of different features to the landscape. Be systematic, focusing your attention on any pair of objects that are in contact. For example, does the dike crosscut the fault or vice versa? Is unit A above or below unit B? Some objects may not be in direct contact with each other, but their relative age can be determined by comparing their ages relative to some other feature.

The Seafloor and Continental Margins

MUCH OF EARTH'S SURFACE IS OCEAN. Beneath the oceans is an underwater landscape that includes broad plains, submarine mountains, and deep trenches and canyons. What clues do these features provide about how our planet operates? This chapter is about the surface and subsurface of the seafloor, how we study the seafloor, and what the various features tell us about Earth processes.

Beneath Monterey Bay, off the coast of central California, the seafloor displays a puzzling feature—a great submarine canyon. In this image, satellite data are shown for land, and computer-shaded and colored data show seafloor depths.

What features are present on the seafloor, and how do we explore the depths, rock types, and structures of the seafloor?

A broad continental shelf flanks the coast, with relatively shallow water (less than about 100 m) extending out kilometers to tens of kilometers from shore. The area is a prized marine ecosystem and is the site of the Monterey Bay National Marine Sanctuary.

What is a continental shelf and how does it form?

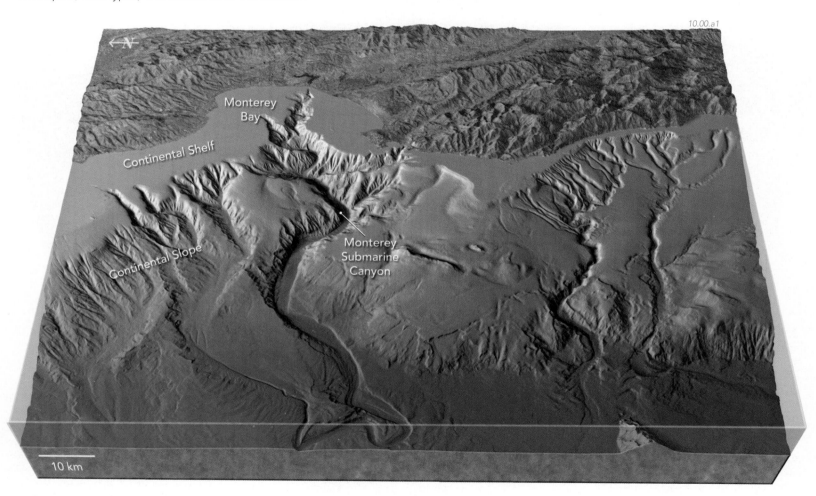

Monterey Submarine Canyon is enormous. It is similar in scale to the Grand Canyon. The canyon bottom is as much as 1,800 m (nearly 6,000 ft) below the rim and, in this deep segment, the canyon is 20 km (12 mi) wide. It resembles many valleys on land; it curves, goes from higher to lower areas, and has smaller side valleys that merge with the main channel.

What processes carve submarine canyons?

The continental margin near Monterey Canyon is heavily studied. Surveys done using ship-borne instruments provide detailed information about the canyons and other geologic features, including landslide material, bedrock ridges, and linear fault scarps.

What processes occur on the seafloor, and what types of features do they produce?

TOPICS IN THIS CHAPTER

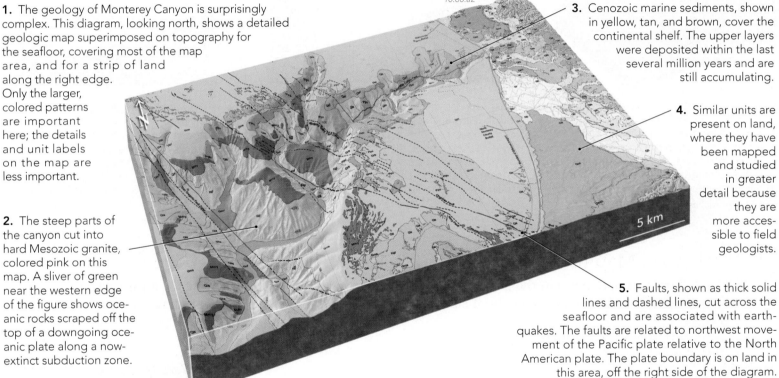

10.00.a2

1. The geology of Monterey Canyon is surprisingly complex. This diagram, looking north, shows a detailed geologic map superimposed on topography for the seafloor, covering most of the map area, and for a strip of land along the right edge. Only the larger, colored patterns are important here; the details and unit labels on the map are less important.

2. The steep parts of the canyon cut into hard Mesozoic granite, colored pink on this map. A sliver of green near the western edge of the figure shows oceanic rocks scraped off the top of a downgoing oceanic plate along a now-extinct subduction zone.

3. Cenozoic marine sediments, shown in yellow, tan, and brown, cover the continental shelf. The upper layers were deposited within the last several million years and are still accumulating.

4. Similar units are present on land, where they have been mapped and studied in greater detail because they are more accessible to field geologists.

5 km

5. Faults, shown as thick solid lines and dashed lines, cut across the seafloor and are associated with earthquakes. The faults are related to northwest movement of the Pacific plate relative to the North American plate. The plate boundary is on land in this area, off the right side of the diagram.

Origin of Monterey Canyon

We do not expect to find huge canyons beneath the sea. When and how did Monterey Canyon form, and what processes are going on today in and around the canyon? Scientists explore the submarine canyon by bouncing sound waves off the seafloor, dredging and drilling rock samples from the bottom, and diving to the bottom in small submarines.

The formation and evolution of the canyon reflect the complicated plate tectonic events that have affected California during the last 20 million years. Geologists have concluded that the upper part of the canyon was originally carved by rivers when its granitic base was above sea level, before 10 million years ago. Strike-slip motion between the North American and Pacific plates shaved off this granitic slice and transported it northward up the coast of North America. During this movement, the canyon was submerged below sea level and filled by sediments, which were later eroded by landslides and underwater currents.

For the past several million years, dense slurries of sediment-rich water, called *turbidity currents*, have flowed down the canyon, scouring the channel and undercutting the canyon walls. The canyon widens as the steep, unstable walls collapse downward in underwater landslides and debris flows. The turbidity currents carry sediment more than 200 km (120 mi) down the canyon and into deeper water, where the sediment is deposited in a broad feature called a *submarine fan*. The lower part of the canyon, like many submarine canyons, was never above sea level and has been carved entirely by turbidity currents and landslides. The position of the lower channel has shifted over time, as segments of the canyon have been offset by strike-slip faulting or buried by landslides.

10.0

10.1 How Do We Explore the Seafloor?

EXPLORING THE SEAFLOOR presents different challenges than mapping and studying geology on land. Mapping the oceans requires *remote* observation of the sea bottom by bouncing sound waves off the seafloor and by other methods. Geologists and oceanographers collect samples of rocks on and below the seafloor by going down in small submarines or by using ships to drill holes through the sediment and rock.

A How Do We Map and Investigate the Seafloor?

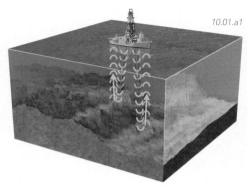

10.01.a1

10.01.a2

10.01.a3

Scientists map parts of the seafloor by transmitting sound waves from a ship and then timing how long the waves take to bounce off the seafloor and return to sensors on the ship. The longer this takes, the deeper the seafloor. Using this technique, called *sonar*, scientists can direct sound waves straight down, as shown here, or at an angle to the seafloor.

Scientists visit the seafloor in small submarines, called *submersibles*, capable of carrying two or three people. Submersibles allow direct observation of geologic features and phenomena. Scientists can take photographs for later study and can collect samples of rocks, seawater, and life-forms. Smaller robotic versions of submersibles are operated remotely from ships.

Specially equipped research vessels allow geologists to drill holes into the seafloor. We can retrieve samples of the sedimentary and volcanic rocks that make up the upper part of the oceanic crust. The layers preserved in drill cores allow geologists to reconstruct the sequence of events, the ages of the rocks, and the variations in seawater chemistry over time.

B How Do We Map the Seafloor from Space?

Most satellites can only observe the seafloor in very shallow water. They are primarily used for surface measurements, including measurements of surface-water temperature or of the height of the sea surface.

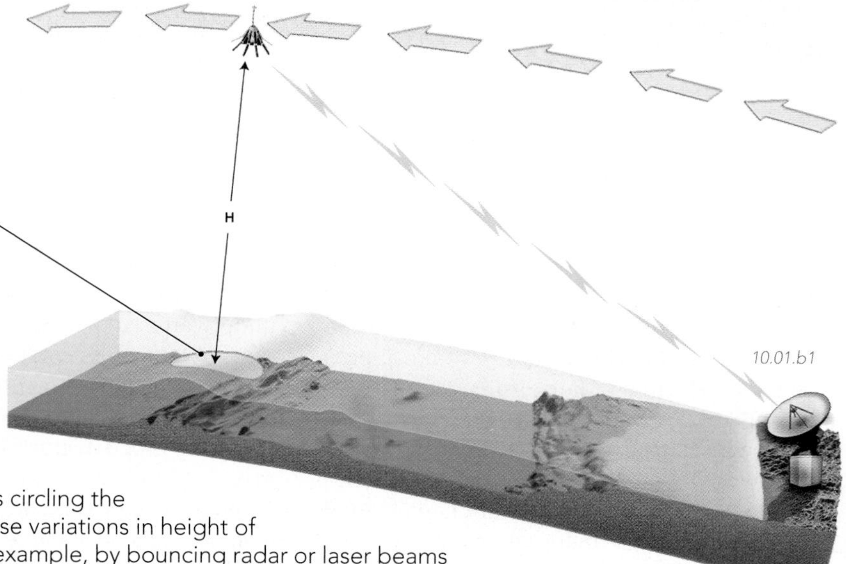

In some areas, the sea surface can be tens of meters higher than average sea level. These high areas are caused by the gravitational forces associated with submarine mountains and mid-ocean ridges. These forces attract seawater and cause it to mound up. Satellites circling the earth can detect these variations in height of the sea surface, for example, by bouncing radar or laser beams off the sea surface. In this manner, we use the satellites to estimate the topography of the underlying seafloor. In fact, we used this type of data to produce most maps of the seafloor in this book.

10.01.b1

Before You Leave These Pages Be Able To

✓ Describe the four methods we use to explore the topography and rocks of the seafloor (sonar, submersibles, drilling, and satellites).

✓ Describe the kinds of information we can obtain from cores drilled in the seafloor.

✓ Describe the seismic-reflection method for mapping the geometry of geologic units beneath the seafloor.

✓ Describe manganese nodules, and where and how they form.

C What Can We Learn from Ocean Drilling?

Geologists and oceanographers drill holes into the seafloor to retrieve samples for later study. The drilling process yields cylinder-shaped samples, called *drill core*, which provide many types of data.

10.01.c1

10.01.c2

1. *Type of Sediment or Rock*—Geologists cut open the drill core to identify the type of sediment or rock and to observe layers and other features. These observations allow interpretations of ocean-floor processes, environments, and past events.

2. *Fossils*—Microscopic and larger fossils within the drill core help geologists assign sediment and rock layers to different parts of the geologic timescale. They also provide constraints on the environments in which the sediment formed. These microscopic fossils are called *Foraminifera*.

3. *Isotopic Ages*—Small samples of the core can be crushed in order to separate minerals for isotopic dating. This is mostly done on volcanic units, especially basalt flows, when studying the seafloor. The ages of the layers, combined with measurements of layer thickness, yield the rates at which the layers accumulated:

$$rate\ of\ deposition = thickness\ /\ time\ span$$

4. *Other Measurements*—Geologists analyze core samples in other ways to answer specific questions about past climates and seawater chemistry. Analyzing for different oxygen and carbon isotopes in a series of layers yields a detailed record of how ocean chemistry and other conditions varied over the past thousands to millions of years. Many such changes were global, so these measurements can help track global climate and correlate layers between different parts of the world.

D How Do We Image What Is Below the Ocean Floor?

To investigate the geometry of rock units beneath the sea, a ship tows a device that bounces sound waves off the seafloor and off rock layers in the subsurface. We record sound waves by devices, called *geophones*, that are towed behind the ship. Sound waves reflected from shallower layers arrive back sooner than sound waves reflected by layers deeper in the subsurface.

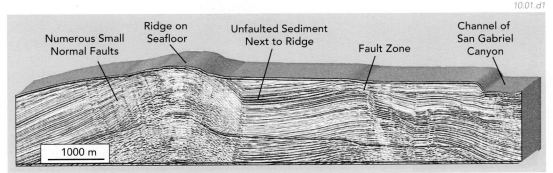

10.01.d1

Numerous Small Normal Faults — Ridge on Seafloor — Unfaulted Sediment Next to Ridge — Fault Zone — Channel of San Gabriel Canyon

1000 m

Geologists process the data using sophisticated computer programs that model the passage of the sound waves through the layers and back to the geophones. They plot the data in a type of cross-section view called a *seismic-reflection profile* (▲). This profile from offshore of Southern California shows tilted, folded, and faulted layers. The *seismic-reflection technique* is widely used in exploring for oil and natural gas beneath the seas and on land.

Diving to the Deepest Parts of the Ocean

Ocean exploration is similar in many ways to exploring space, but not quite as expensive. Getting to the deepest parts of the ocean requires specialized submarines that can only accommodate a few passengers. Such travel is quite dangerous because of the high pressure in the deep oceans. Nevertheless, humans have explored very deep regions using remotely guided probes and by diving in submersibles. One submersible can take

10.01.mtb1

humans to depths of 6,500 m (more than 21,000 ft).

Among the features observed on the seafloor are *manganese nodules*, shown here. They form when manganese precipitates out of seawater, forming baseball-sized spheres. These are an important potential source of manganese and other metals, but geologists are still investigating the logistics and environmental issues associated with remote mining on the deep seafloor.

10.1

10.2 How Is Paleomagnetism Used to Study the Ocean Floor?

PALEOMAGNETISM IS THE ROCK RECORD OF PAST CHANGES in Earth's magnetic field. The magnetic field is strong enough to orient magnetism in certain minerals, especially *magnetite*, in the direction of the prevailing magnetic field. Magnetic directions preserved in volcanic rocks, intrusive rocks, and some sedimentary rocks provide an important way to investigate the origin of the seafloor.

A What Causes Earth's Magnetic Field?

Earth has a metallic iron core, which is composed of a solid inner core surrounded by a liquid outer core. The liquid core flows and behaves like a *dynamo* (an electrical generator), creating a magnetic field around Earth.

1. The inner core transfers heat and less dense material to the liquid outer core. This transfer causes liquid in the outer core to rise, forming *convection currents*. These convection currents are limited to the outer core and are not the same as those in the upper mantle.

2. Movement of the molten iron is affected by forces associated with Earth's rotation. The resulting movement of liquid iron and electrical currents generates the magnetic field.

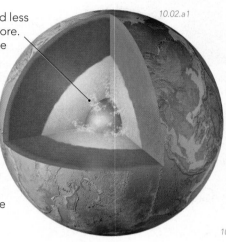

10.02.a1

10.02.a3

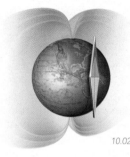

▶ **3.** Earth's magnetic field currently flows from south to north, causing the magnetic ends of a compass needle to point toward the north. This orientation is called a *normal polarity*.

10.02.a2

◀ **4.** Many times in the past, the magnetic field has had a *reversed polarity*, so that a compass needle would point south. The switch between normal polarity and reversed polarity is a *magnetic reversal*.

B How Do Magnetic Reversals Help Us Infer the Age of Rocks?

The north and south magnetic poles have switched many times, typically remaining either normal or reversed anywhere from 100,000 years to a few million years. Geologists have constructed a *magnetic timescale* by isotopically dating sequences of rocks that contain magnetic reversals. This *geomagnetic polarity timescale* then serves as a reference to compare against other sequences of rocks.

10.02.b1

1. Geologists measure the direction and strength of the magnetism preserved in rocks with an instrument called a *magnetometer*. With this device, geologists can tell whether the magnetic field had a normal polarity or a reversed polarity when an igneous rock solidified and cooled or when a sedimentary layer accumulated.

▶ **2.** This figure shows the series of magnetic reversals during the last 10 million years, the most recent part of the Cenozoic Era. This time period is within the *Neogene Period*. In older geologic literature, this time period included the *Quaternary* and part of the *Tertiary*.

3. The timescale shows periods of normal magnetization (N) in black and those of reversed magnetization (R) in white. Variability in the spacing and duration of magnetic reversals produced a unique pattern through time. Geologists can measure the pattern of reversals in a rock sequence and compare this pattern to the magnetic timescale to see where the patterns match. This allows an estimate of the age of the rock or sediment. Geologists use other age constraints, including isotopic ages or fossils, to further refine the age of the magnetized rocks. The magnetic timescale is best documented for the last 180 million years because seafloor of this age is widely preserved.

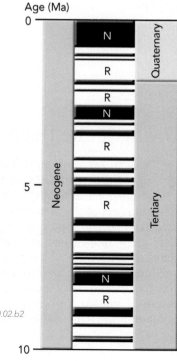

10.02.b2

C How Are Magnetic Reversals Expressed at Mid-Ocean Ridges, and How Do Magnetic Patterns on the Seafloor Help Us Study Plate Tectonics?

In the 1950s, scientists discovered that the ocean floor displayed magnetic variations in the form of matching stripes on either side of the mid-ocean ridge. Geologists Frederick Vine, Drummond Mathews, and Lawrence Morley interpreted the patterns to represent a magnetic field that had reversed its polarity. This Vine-Mathews-Morley hypothesis led to the theory of plate tectonics. Magnetic patterns allow us to estimate the ages of large areas of seafloor and to calculate the rates at which two diverging oceanic plates were formed.

As the oceanic plates spread apart at a mid-ocean ridge, basaltic lava erupts onto the surface or solidifies at depth. As the rocks cool, the orientation of Earth's magnetic field is recorded by the iron-rich mineral magnetite. In this example, the magnetite records normal polarity (shown with a reddish color) at the time the rock forms.

If the magnetic field reverses, new rocks that form will acquire a reversed polarity (shown in white). Rocks forming all along the axis of the mid-ocean ridge will have the same magnetic direction, forming a stripe of similarly magnetized rocks parallel to the ridge. Once the rocks have cooled, they retain their original magnetic direction, unless they are heated significantly or altered by certain types of fluids. In most cases, the magnetic polarity is preserved by the seafloor.

The magnetic poles switch many times, and continued seafloor spreading produces a pattern of alternating magnetic stripes on the ocean floor. This pattern is strong enough to be detected by magnetic instruments called *magnetometers* towed behind a ship or a plane.

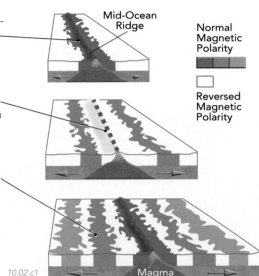

10.02.c1

As magnetic instruments are towed behind a ship, the strength of the magnetic field is measured and plotted. Stronger measurements plot high on the graph and are called *positive magnetic anomalies*. The magnetic signal is weaker over crust that was formed under a reversed magnetic field, because the magnetic direction in such rocks is opposite to and works to counteract the modern magnetic field. The reverse magnetization of the rocks slightly weakens the measured magnetic signal and will plot low on the graph, forming a *negative magnetic anomaly*.

The seafloor patterns are compared with the patterns on the geomagnetic polarity timescale to assign ages to each reversal. Geologists simplify and visualize these data as reversely and normally magnetized stripes on the seafloor, as shown in this cross section.

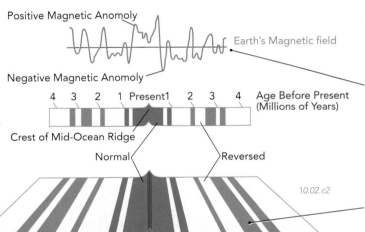

10.02.c2

We can calculate rates of seafloor spreading by measuring the width of a specific magnetic stripe in map or cross-section view and then dividing that distance by the length of time the stripe represents:

rate of spreading for stripe = width of stripe / time duration

If a magnetic stripe is 60 km wide and formed over 2 million years, then the rate at which spreading formed the stripe was 30 km/m.y. This rate is equivalent to 3 cm/year. Spreading added an equal width of oceanic crust to a plate on the other side of the mid-ocean ridge, so the total rate of spreading across the ridge was 60 km/m.y. (6 cm/year), a typical rate of seafloor spreading.

The magnetic patterns on the seafloor, in addition to magnetic measurements on sequences of rocks and sediment on the seafloor and on land, demonstrate that Earth's magnetic field has reversed many times. Scientists are currently debating the possible causes of the reversals, with most explanations attributing reversals to chaotic flow in the molten outer core, which add to or subtract from the patterns caused by the dynamo, disrupting the prevailing magnetic field and causing a reversal.

Before You Leave This Page Be Able To

✓ Describe how Earth's magnetic field is generated.

✓ Describe how magnetic reversals help with determining the age of rocks.

✓ Describe or sketch how magnetic patterns develop on the seafloor.

✓ Calculate the rate of seafloor spreading if given the width and duration of a magnetic stripe.

10.2

10.3 What Processes Occur at Mid-Ocean Ridges?

MID-OCEAN RIDGES FORM where two oceanic plates diverge. Magma ascending from the mantle erupts onto the seafloor or solidifies at depth, making new oceanic crust. Heat associated with the hot rocks and magma produces undersea vents of hot water that nourish unique life-forms on the seafloor.

A What Happens When Plates Spread Apart?

As two oceanic plates move apart, solid rock and magma rise from the mantle to occupy the space between the plates. The cooling and solidifying magma forms new oceanic crust, which then gets transported away from the mid-ocean ridge as new plate is formed along the spreading center. Slices of this rock sequence can be scraped off and preserved on land, allowing scientists to study examples of oceanic crust without diving to the bottom of the ocean.

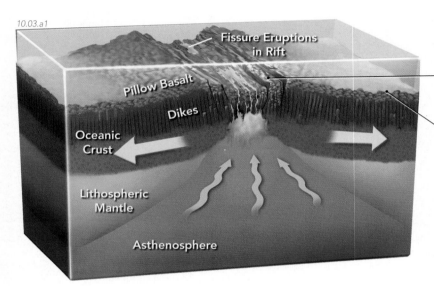

10.03.a1

As oceanic crust stretches apart, basaltic lava erupts within the rift, forming *pillow basalts* on the seafloor. Some magma solidifies within large chambers and in magma-filled fissures parallel to the mid-ocean ridge (perpendicular to plate movement). The magma-filled fissures solidify into dikes.

At many mid-ocean ridges, normal faults allow blocks of crust to be displaced down and inward toward the center, forming a fault-bounded rift.

As the cooled crust moves away from the ridge, it is progressively covered with deep-sea sediment. Over time, the sediment tends to smooth over the rough topography formed in the rift. As a result, older oceanic crust tends to have relatively smooth topography.

Below a depth of about 4,500 to 5,000 m, sediment is dominated by clay and silica-rich materials because at these depths and greater, carbonate minerals dissolve into seawater as fast as they accumulate. The depth at which this occurs is the *carbonate compensation depth* (CCD).

B What Accounts for Variations in the Shape of Mid-Ocean Ridges?

Many mid-ocean ridges possess the typical features shown above, but ridges vary in their width, ruggedness, and overall shape. These variations reflect differences in the rates of spreading, magmatism, and faulting. Compare the two topographic profiles below, each of which shows a detailed view of the center of a ridge.

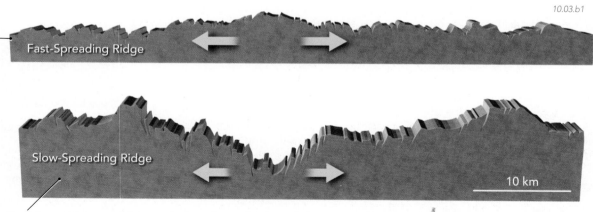

10.03.b1

1. Some ridges, including parts of the East Pacific Rise, are broad and do not have a large, well-developed rift in the center. Such ridges are spreading apart at relatively fast rates (10 cm per year). These ridges are broad because rapid spreading allows the new oceanic crust to move far from the spreading center before it cools and subsides. Furthermore, such ridges are interpreted to have more underlying magma, which rapidly pours onto the surface out of fissures, rather than forming a large, fault-bounded rift.

2. Other mid-ocean ridges, including the Mid-Atlantic Ridge, have well-defined rifts that are 1 to 3 km deep and are bounded by normal faults that dip inward toward the rift. These ridges have slower spreading rates (1 to 2 cm per year). This allows rocks near the ridge to cool and strengthen enough to form large faults. Other ridges are intermediate in character between the two end members shown. They lack a high central area or a deep rift, and are intermediate in spreading rate, breadth, roughness, and degree of faulting.

C   What Are Black Smokers and How Do They Form?

10.03.c1

◀ Mid-ocean ridges contain features called *black smokers*, shown here in a photograph taken from a submersible. Black smokers are *hydrothermal vents*, where hot water from within the rock jets out into the cold seawater. As the hot water cools, metals, sulfur, and other elements dissolved in the hot water form small crystals that make the water black and cloudy.

10.03.c2

Sulfur-bearing minerals precipitate around the vent, forming a hollow, circular column called a *chimney*. Some chimneys are more than 5 m (16 ft) high and a meter across, and can grow tens of centimeters per day. Black smokers and sulfide-rich chimneys are interpreted to have formed on mid-ocean ridges and other submarine volcanoes in Earth's geologic past, forming mineral deposits rich in copper, zinc, and other valuable elements, like these metal-rich sulfide layers. ▶

10.03.c3 Kidd Creek, Ontario, Canada

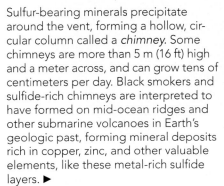

▲ Black smokers form when water in rocks above a magma chamber is heated and rises toward the surface. As the water rises, seawater from nearby areas flows in to take its place. This seawater heats up and leaches metals and other chemical elements from rocks through which it passes, becoming rich in dissolved chemicals. The heated seawater rises toward the surface along faults and other pathways, eventually venting in a black smoker. The water is very hot, commonly over 350°C, but it does not boil because of the pressure exerted by the deep water.

Life at Hydrothermal Vents

Deep-sea hydrothermal vents associated with black smokers support a unique and only recently discovered community of unusual creatures. Scientists are actively exploring the ecosystems of these vents, in part because such sites may have been where life originated on Earth. The photograph included here was taken by

scientists using a submersible to investigate these vents and their unusual inhabitants.

Sunlight is the energy source for green plants, which provide the bulk of food for animals living on Earth's surface. No sunlight reaches the deep seafloor. Instead, life around the hydrothermal vents uses a completely different energy source. Here, life is dependent on somewhat unusual bacteria that are able to break down hydrogen sulfide (H_2S), one of the chemical compounds common within black smokers. These bacteria produce sugars, which feed giant (meter-long) red tube worms. The worms can tolerate the hot water and live close to the vents, where the bacteria are abundant. In fact, many bacteria live within the worms' tissues. The worms in turn form the main food for an assembly of scavenging animals, including fish and white crabs, shown in this photograph. Large clams also live around hydrothermal vents and draw nutrients by extracting small bits of material from the water and from the bacteria. Fos-

sils of tube worms in ancient hydrothermal vent deposits show that such communities have existed for millions of years.

10.03.mtb1

Before You Leave This Page Be Able To

✓ Describe or sketch the processes that accompany the formation of new oceanic crust at mid-ocean ridges.

✓ Describe or sketch the differences between fast-spreading and slow-spreading mid-ocean ridges.

✓ Describe black smokers, how they form, and where the hot water originates and how it gets heated.

✓ Describe the type of life that exists around hydrothermal vents and where the different creatures derive their food.

10.3

10.4 What Are Major Features of the Deep Ocean?

BENEATH THE WORLD'S OCEANS lie rugged mountains, active rifts, gentle plains, broad plateaus, and deep trenches. The seafloor varies in depth and in thickness of sediment cover, largely because different parts of the seafloor have different ages. What is the topography of the deep seafloor and how do the various types of features form?

Topography of the Deep Seafloor

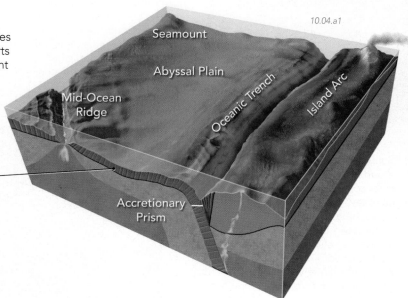

10.04.a1

Much of the ocean floor, called the *abyssal plain*, has a gentle slope and lies at depths below about 4.5 km (2.8 mi). The abyssal plain and other old parts of an oceanic plate generally have a smooth topography because sediment evens out most original irregularities. The abyssal plain contains isolated mountains, called *seamounts*, which vary from gentle submarine hills to steep mountains, many of which are volcanoes.

The elevation of the seafloor decreases from mid-ocean ridge to abyssal plain because oceanic lithosphere cools, becomes denser, and subsides as it moves away from the ridge. Subsidence slows or stops when the oceanic crust reaches an equilibrium temperature.

Trenches are the deepest parts of the ocean but comprise a very small area. They are the surface expression of a subduction zone, where an oceanic plate is flexed (bent) as it plunges beneath another plate. Within the trench, sediment and basaltic rocks are scraped off the top of the subducted plate and incorporated as slices into the *accretionary prism*.

Sediment Thicknesses

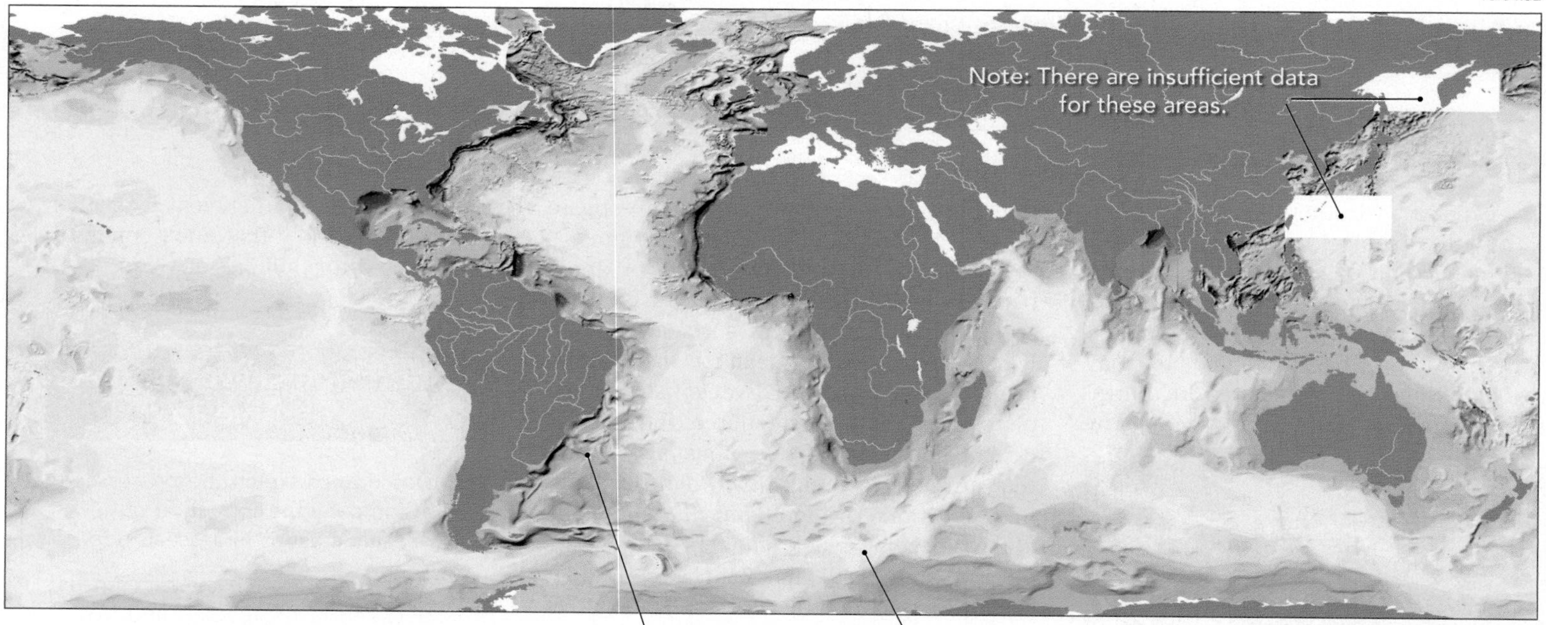

10.04.a2

Note: There are insufficient data for these areas.

This map shows sediment thickness on the seafloor. It ranges from light blue, where thickness is less than 200 m, to orange and red, where sediment is over 5 to 10 km (3–6 mi) thick. White colors indicate locations where there are insufficient data to show on this map. The white lines on the continents are rivers. What patterns do you observe on this map? Where is sediment thickest, where is it thinnest, and how does thickness relate to major rivers?

The thickest sediment is along continental margins, especially those that were formed by rifting. Seafloor sediment is also thickest near the mouths of rivers or where the oceanic crust is relatively old (see maps on the next page).

There is virtually no sediment cover over the youngest crust at the mid-ocean ridges, which here are along the belts of light blue.

Before You Leave These Pages Be Able To

☑ Sketch or describe some features of the deep seafloor.

☑ Describe how the age of the seafloor relates to mid-ocean ridges, depths of seafloor, and sediment thicknesses.

Depth and Age of the Seafloor

This map shows the depth of the seafloor. Dark gray areas are continents, and the light to dark purple and blue regions are oceans and seas. Darker purple shows the deepest areas and light purplish gray the shallowest. Letters identify some of the ridges (R), trenches (T) along convergent boundaries, and passive margins (P) that are not plate boundaries.

The deepest part of the oceans is at trenches (T) along active margins, where one plate subducts beneath another plate. The shallowest parts of the seafloor are on continental shelves, many of the widest of which are along passive margins (P). Mid-ocean ridges (R) are intermediate in depth.

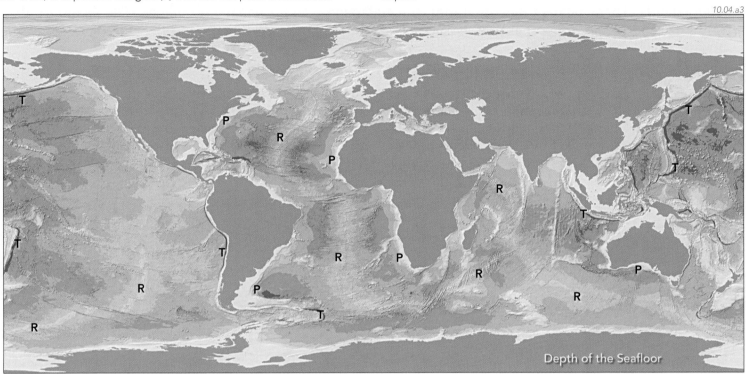

10.04.a3

Depth of the Seafloor

The map below shows the age of the seafloor. Purple represents the oldest areas (about 180 million years), and the darkest orange represents very young oceanic crust. Compare this map with the one above.

The youngest oceanic crust is near mid-ocean-ridge spreading centers (R). These areas are also higher than most of the ocean floor.

The oldest oceanic crust in any ocean is the most distant from mid-ocean ridges. None is older than about 180 million years, because all older oceanic crust has been subducted back into the mantle. The oldest seafloor is much younger than the oldest continental rocks.

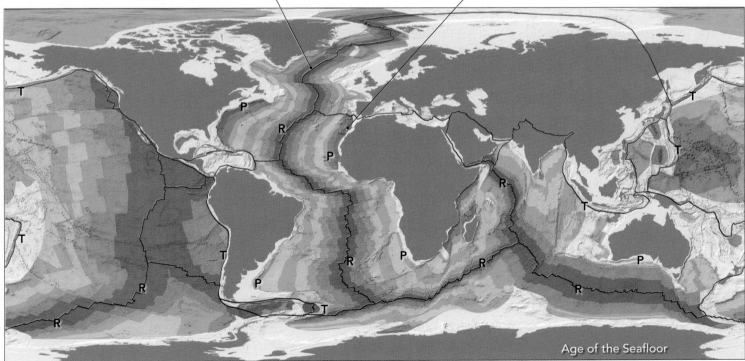

Age of the Seafloor

10.4

10.04.a4

10.5 How Do Oceanic Islands, Seamounts, and Oceanic Plateaus Form?

SUBMARINE MOUNTAINS, called *seamounts*, rise above the seafloor. In some places, they reach the surface and make islands. These islands include Tahiti, the Galápagos, and other exotic places. The seafloor also has relatively high and broad areas that are *oceanic plateaus*. How are seamounts and oceanic plateaus formed?

A How Do Some Oceanic Islands and Seamounts Form?

Most oceanic islands are made of mafic (basalt) to intermediate (andesite) volcanic rocks and are formed by a series of volcanic eruptions onto the seafloor. Many of these islands are not associated with an island arc, but form linear chains or irregular clumps of islands and seamounts, some of which are related to a *hot spot*. The direction of movement of a plate over a hot spot tends to remain constant for tens of millions of years, and this results in linear chains of islands and seamounts, which are distinct from the curved shapes of island arcs.

10.05.a1

10.05.a2

10.05.a3

Magmatism caused by an underlying hot spot begins building a submarine volcano by eruption of lava onto the seafloor. Magmatism related to hot spots is usually basaltic in composition.

Continued eruptions build up the volcano until it may eventually rise above the sea as an island. Once magmatism ceases, perhaps when an island moves off a hot spot, the oceanic plate cools and subsides.

The top of the mountain is leveled off by wave erosion and continues subsiding, becoming a submarine, flat-topped mountain. Over time, it is covered by layers of marine sediments.

B How Do Oceanic Plateaus Form?

Some large regions of the seafloor rise a kilometer or more above their surroundings, forming *oceanic plateaus*. These plateaus are largely composed of flood basalts and, like the seafloor in general, are mostly late Mesozoic and Cenozoic in age (mostly 130 million years ago to the present).

1000 km

10.05.b1

10.05.b2

10.05.b3

10.05.b4

This perspective shows the Kerguelen oceanic plateau, which rises above the surrounding seafloor in the southern Indian Ocean. The plateau is several thousand kilometers long, but it only reaches sea level in a few small islands. The small sliver of land showing in the lower right corner is part of Antarctica.

Geologists interpret oceanic plateaus as forming at hot spots, above rising mantle plumes. The plumes travel through the mantle as solid masses, not liquids.

When the top of a plume encounters the base of the lithosphere, it causes widespread melting. Submarine flood basalts pour out onto the seafloor through fissures and central vents.

Immense volumes of basalt (as much as 50 million cubic kilometers) erupt onto the seafloor over millions of years. This volcanism creates a broad, high oceanic plateau.

C What Is the Distribution of Hot Spots, Linear Island Chains, and Oceanic Plateaus?

Hot spots have created many Pacific islands that we associate with tropical paradises and exotic destinations. Hawaii is the most famous island chain formed by movement of a plate over a hot spot, but several other linear island and seamount chains, in both the Atlantic and Pacific, formed in the same manner.

On this map, red dots show the locations of likely hot spots, many of which are located at the volcanically active ends of linear island chains. There is great debate, however, about which areas really are hot spots and how hot spots form.

The dark gray areas in the oceans represent linear island chains, clumps of islands, and oceanic plateaus, similar to this high area around Iceland, which is over a hot spot.

10.05.c1

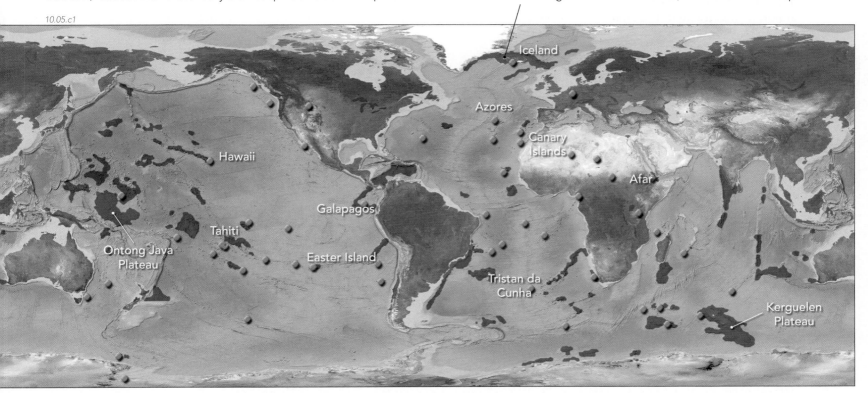

The *Ontong Java Plateau* is the largest oceanic plateau on Earth, covering millions of square kilometers, nearly 1% of Earth's surface area. It formed in the middle of the Pacific Ocean 120 million years ago and is no longer near the hot spot that produced it.

Volcanic islands near *Tahiti* define northwest-trending chains that are forming over several hot spots. In each chain, the islands to the northwest are older than those in the southeast, indicating that the Pacific plate is moving to the northwest relative to the underlying source of magma.

The *Galápagos* is a clump of volcanic islands west of South America. The western islands, shown in the satellite image to the lower left, are volcanically active and have erupted within the past several years. Eruptions build shield volcanoes and smaller scoria cones, both of which are shown in the photograph below.

Tristan da Cunha, a volcanic island in the South Atlantic Ocean, marks a hot spot just east of the Mid-Atlantic Ridge. Volcanism associated with the hot spot created a large submarine ridge (shown in gray) that tracks the motion of the African plate over the hot spot.

The *Kerguelen Plateau*, in the southern Indian Ocean, is the second largest oceanic plateau in the world. It mostly consists of basalt and was formed in several stages during the late Mesozoic (between 115 and 85 million years ago).

10.05.c2

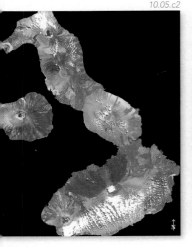

10.05.c3

Before You Leave This Page Be Able To

☑ Describe or sketch how a mantle plume is interpreted to form oceanic islands and seamounts, providing several examples.

☑ Describe how oceanic plateaus are interpreted to have formed, and provide some examples.

10.5

10.6 What Processes Form Island Arcs?

MANY ISLANDS OCCUR IN LONG ARCS that cross the seafloor. Island arcs are associated with deep ocean trenches and dangerous volcanoes. How do island arcs form, why are they curved, and what processes occur in front of, within, and behind them?

A How Do Island Arcs Form?

Island arcs form where one oceanic plate *subducts* beneath another. Subduction creates a trench and generates magma that forms an arcuate belt of volcanic islands, such as the Aleutian Islands and Java.

1. An *oceanic trench* forms where a subducted oceanic plate flexes downward beneath the overriding plate. Many island arcs are in the open ocean, away from large landmasses. In such settings, the trench receives most of its sediment from volcanic eruptions and erosion of the adjacent volcanoes.

2. As the oceanic plate subducts, it heats up, causing metamorphic reactions that release water from the minerals. This water promotes melting in the asthenosphere above the subducted plate. The asthenosphere-derived magma rises into the overriding lithosphere, erupting onto the surface or solidifying in the crust. These magmatic additions thicken the crust beneath the arc over time.

3. As a new volcano begins to grow, volcanic eruptions first occur in deep water on the seafloor. Over time, the eruptions may construct a mountain that rises above the sea. The crust becomes transitional in character and in thickness between oceanic crust and continental crust. As mantle-derived magmas interact with this thicker crust, the magmas become intermediate in composition (andesite) and form dangerous composite volcanoes. Submarine mountains and ridges form a ridge between the islands. If enough magma erupts over a long enough time, eruptions can build a longer landmass, capped by a string of volcanoes. Examples include the Indonesian island of Java and the Alaska peninsula.

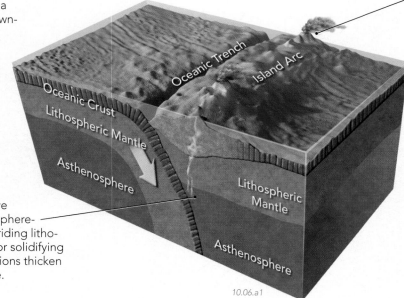

10.06.a1

B What Happens in Front of and Behind an Island Arc?

Island arcs are not fixed in position. They can migrate across the surface of Earth over millions of years, depending on what happens in front of the arc (at the trench) or behind the arc.

1. As dense oceanic plate subducts into asthenosphere, it sinks downward and tends to bend or roll back away from the island arc. The trench, the surface expression of the bend in the downgoing plate, follows the rollback of the slab, a process called *trench rollback*. Subduction continues during trench rollback, but the position of the subduction zone moves over time.

2. As the subducting slab and trench migrate, the island arc follows them because the position of the volcanoes is determined by the location of the subducting plate. As the arc and trench both migrate, stretching of the crust can cause rifting within or behind the arc. The arc can be rifted (split) into two parts, separated by a spreading center. Over time, rifting can form a new *back-arc basin*, which can be hundreds of kilometers wide. The oceanic crust in the back-arc basin typically has ages that overlap with ages of volcanic rocks within the island arc, because back-arc rifting occurred at the same time as subduction.

10.06.b1

Before You Leave These Pages Be Able To

☑ Describe the processes that occur within, in front of, and behind island arcs.

☑ Describe why island arcs and their associated trenches are curved.

☑ Describe some examples of island arcs.

C Why Are Island Arcs Curved?

1. In map view, island arcs have a distinctly curved or arcuate shape. This view of the Aleutian arc of Alaska shows the curved shape of the island arc and of the associated trench that lies in front of it.

10.06.c1

Siberia
Alaska
Aleutian Island Arc
Aleutian Trench

500 km

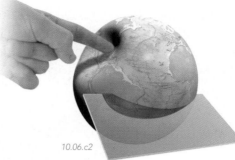

10.06.c2

10.06.c3

▲ **2.** In thinking about why island arcs are curved, we need to consider that plates are interacting on a spherical Earth, not a flat plane. Cutting into a globe (like the red plane) or depressing the surface of a globe creates a curved feature (arc).

▲ **3.** Also, on a sphere, there is more surface area on the outside than at depth. A plate that is subducted into the interior becomes buckled as it is forced to fit into a smaller width. The arc and trench have an arcuate shape because the downgoing slab does too.

D Where Are the Main Island Arcs of the World?

In *Japan* and the adjacent *Mariana arc*, the Pacific plate subducts westward beneath the islands. Bending of the subducted plate forms deep trenches, including the world's deepest trench south of the Mariana. Volcanoes in the *Philippines* are largely related to subduction of a small plate that is part of a back-arc basin west of the Mariana.

The *Aleutian arc* extends from mainland Alaska westward to the Kamchatka peninsula of Asia. It is formed where the Pacific plate is subducted northward beneath the North American plate.

The *Lesser Antilles arc* forms the eastern edge of the Caribbean Sea. It includes the islands of Montserrat, site of recent eruptions, and Martinique, site of the deadly 1902 eruption of Pelée. A small island arc in the *Aegean Sea*, east of mainland Greece, was the site of a destructive eruption on the island of Thera.

10.06.d1

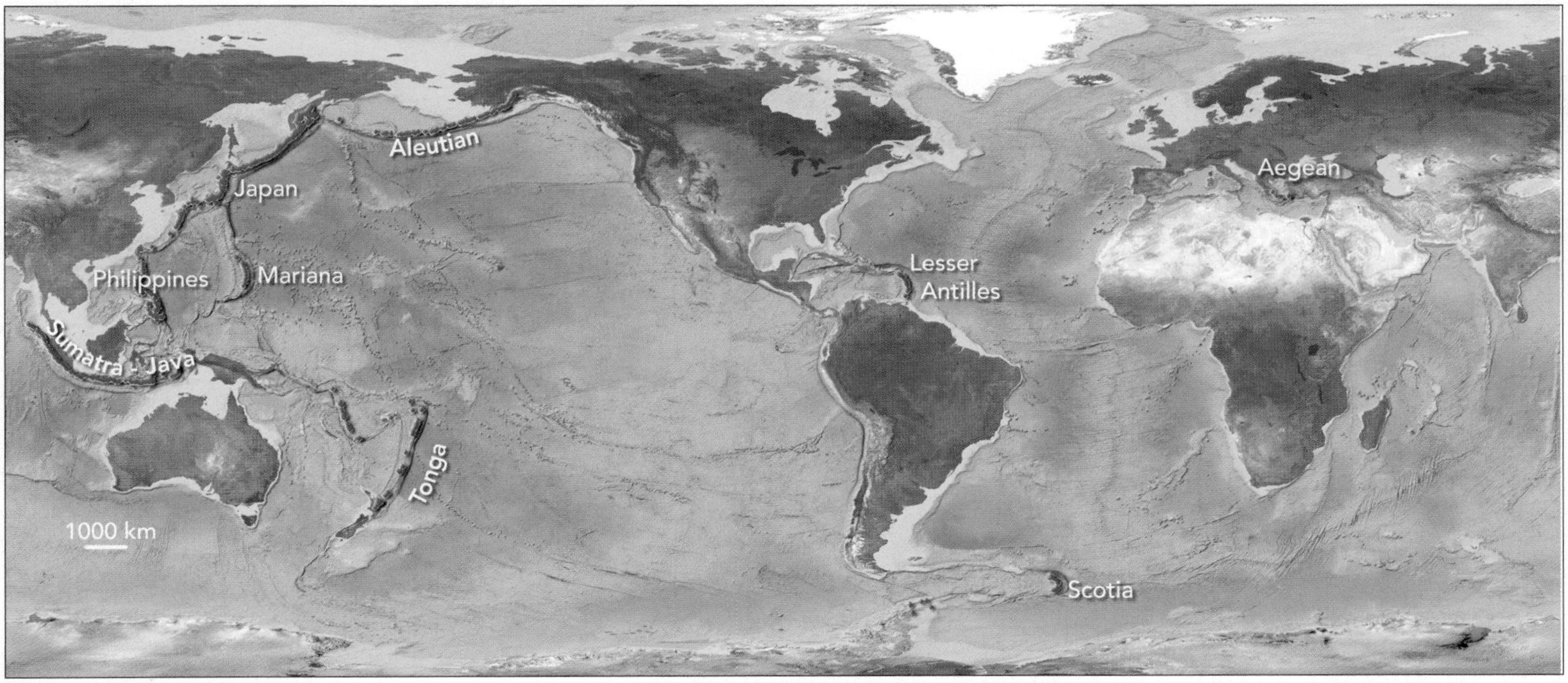

Aleutian
Japan
Aegean
Philippines   Mariana
Lesser
Antilles
Sumatra-Java
1000 km
Tonga
Scotia

The *Sumatra-Java arc* is a typical island arc in the east, but in the west it lies upon a promontory of Asian continental crust. It is located where the Indian plate is subducted northward. This subduction zone caused the deadly 2004 Indian Ocean tsunami.

Along the *Tonga trench* and island arc, the Pacific plate subducts to the west. Spreading west of the arc has created several small back-arc basins. The Tonga subduction zone is the site of many large earthquakes each year.

The *Scotia arc* is a small island arc between South America and Antarctica. Beneath the arc, an oceanic section of the South American plate subducts beneath another oceanic plate. Back-arc spreading occurs behind (west of) the arc.

10.6

10.7 How Did Smaller Seas of the Pacific Form?

A SERIES OF SMALL SEAS exist around the edges of the Pacific Ocean. They are separated from the main Pacific basin by chains of islands and slivers of continents. These include the Sea of Japan and the Gulf of California. Each sea has it own unique and interesting history, and together they illustrate the most important ways in which smaller seas in the Pacific formed.

1. The *Sea of Japan* is a moderately deep basin between Japan and mainland Asia. Before 20 million years ago, Japan was part of a volcanic arc along the coast of mainland Asia. Rifting within the arc split Japan away from Asia. This led to back-arc seafloor spreading, which formed the Sea of Japan. ▶

2. The *China Sea,* between China and the Korean Peninsula, is relatively shallow because it is mostly underlain by continental crust. Recall that continental crust floats higher on the mantle than does oceanic crust, which is thinner and more dense.

Origin of the Philippine Sea

3. The *Philippine Sea* lies between the Philippines and the Mariana island arc. It contains several distinct basins separated by long, submarine ridges, and is an example of how features on the seafloor reflect the geologic history of an area. The Mariana arc is active and is flanked by the Mariana trench, which contains the deepest seafloor in the ocean (nearly 11 km below sea level).

4. Seafloor spreading currently is forming a back-arc basin directly behind the Mariana arc. Submerged ridges farther to the west (left) represent pieces of the arc that were rifted away by different episodes of back-arc spreading.

5. The *Java Sea* of Indonesia and Malaysia is shallow and is part of a continental platform between the larger islands of the region.

6. South of Indonesia, oceanic portions of the Indian plate (bottom left on this map) are subducted northward beneath the Asian plate, forming a trench and the *Sumatra-Java island arc.* The continuation of this subduction zone to the northwest caused the huge earthquake and deadly tsunami that devastated coastlines around the Indian Ocean in 2004.

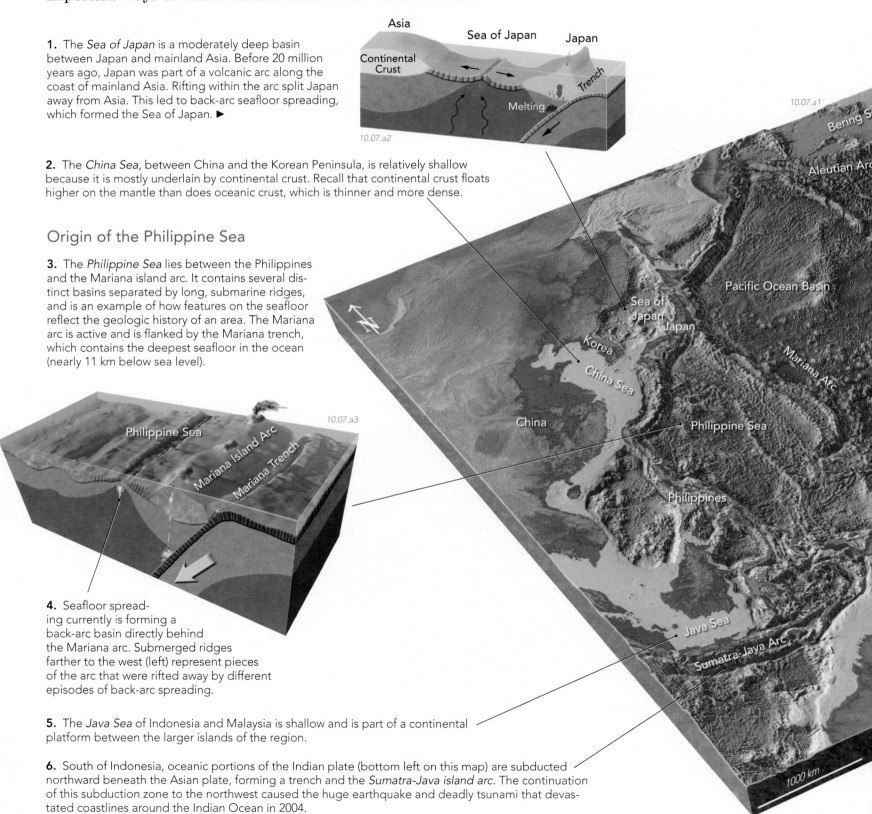

10.07.a2

10.07.a1

10.07.a3

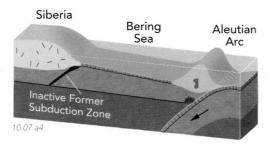

10.07.a4

▲ **10.** The *Bering Sea* lies between mainland Alaska and the Aleutian island arc. The eastern part of the sea, near Alaska, is shallow and is underlain by continental crust. The western part of the Bering Sea is deeper because it is underlain by oceanic crust. The oceanic crust is part of the North American plate (which includes Alaska), so there is no plate boundary between Alaska and the oceanic crust beneath the Bering Sea. Instead, the western edge of the Alaskan mainland is a passive margin. In the Mesozoic, oceanic plates subducted directly beneath coastal Alaska and Siberia, but the site of subduction zone migrated offshore, trapping some old oceanic crust between the new Aleutian arc and the mainland.

Origin of the Gulf of California

11. Prior to 10 million years ago, Baja, California, was part of the mainland of western Mexico, and an oceanic plate subducted eastward beneath the land.

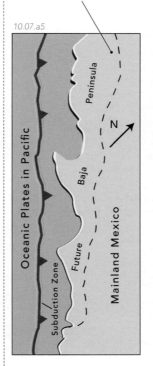

10.07.a5

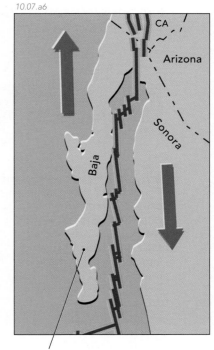

10.07.a6

12. As North America, including Mexico, moved over the East Pacific Rise spreading center, the plate boundary became a transform boundary and migrated inland, splitting Baja, California, from the mainland and shifting it northward along the coast. As Baja moved northward away from the mainland, the *Gulf of California* formed in the place Baja vacated. The gulf has long transform faults linking short spreading centers. This spreading is a continuation of the East Pacific Rise, the major spreading center that runs north-south across the eastern Pacific Ocean. Baja (and the coast of California west of the San Andreas fault) is now part of the Pacific plate and continues to move northward relative to the North American plate.

9. At the *Tonga trench*, the Pacific plate subducts westward beneath oceanic crust east of Australia. The subduction zone forms the trench and associated island arc of the Tonga Islands. It is very active, being associated with numerous large, deep earthquakes.

8. The seafloor east of Papua New Guinea and northeast of Australia is unusually complicated. It contains small basins, trenches, and island arcs, reflecting complex interactions between a number of small oceanic plates. From a plate-tectonic perspective, it is the most complex area of oceanic crust in the world.

7. The shallow seas between Australia and Papua New Guinea are underlain by a continuation of Australian continental crust.

Before You Leave This Page Be Able To

☑ Describe or sketch the different ways in which smaller seas formed in the Pacific Ocean, providing an example of each.

☑ Describe the history of the Gulf of California and how it is related to the boundary between the Pacific and North American plates.

10.7

10.8 How Did Smaller Seas Near Eurasia Form?

A NUMBER OF SEAS FLANK Europe, Asia, and Africa. These include the Black Sea, North Sea, and Mediterranean Sea. The Arabian Peninsula, between Africa and mainland Asia, has the Red Sea to the west and the Persian Gulf to the east. Several seas were formed by present or past plate-tectonic activity. Others were valleys and low areas flooded by rising sea levels after the last Ice Age.

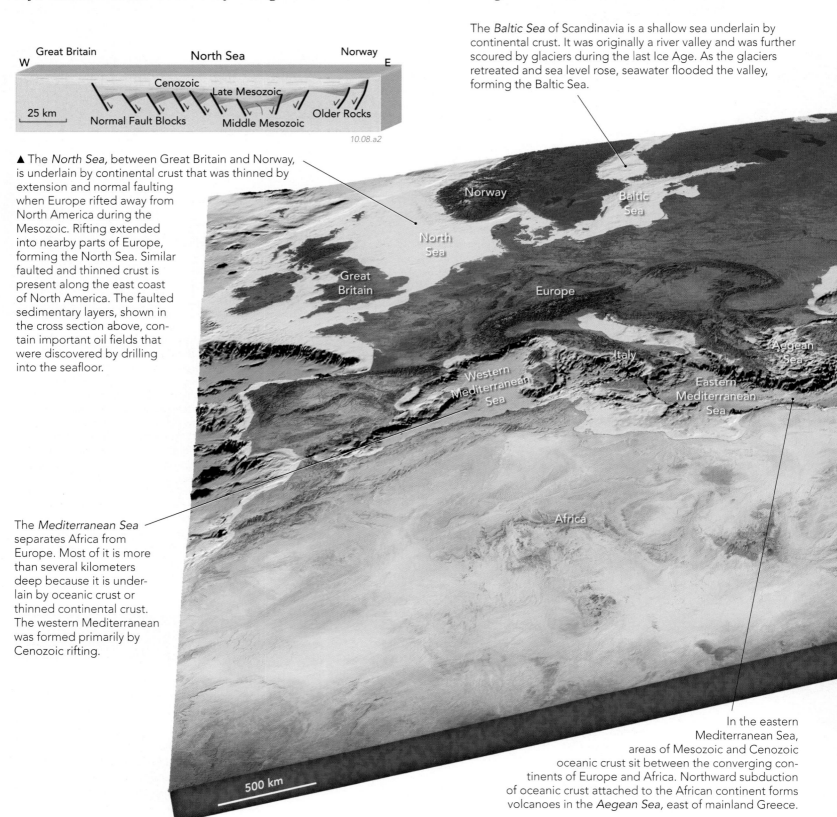

The *Baltic Sea* of Scandinavia is a shallow sea underlain by continental crust. It was originally a river valley and was further scoured by glaciers during the last Ice Age. As the glaciers retreated and sea level rose, seawater flooded the valley, forming the Baltic Sea.

10.08.a2

▲ The *North Sea*, between Great Britain and Norway, is underlain by continental crust that was thinned by extension and normal faulting when Europe rifted away from North America during the Mesozoic. Rifting extended into nearby parts of Europe, forming the North Sea. Similar faulted and thinned crust is present along the east coast of North America. The faulted sedimentary layers, shown in the cross section above, contain important oil fields that were discovered by drilling into the seafloor.

The *Mediterranean Sea* separates Africa from Europe. Most of it is more than several kilometers deep because it is underlain by oceanic crust or thinned continental crust. The western Mediterranean was formed primarily by Cenozoic rifting.

In the eastern Mediterranean Sea, areas of Mesozoic and Cenozoic oceanic crust sit between the converging continents of Europe and Africa. Northward subduction of oceanic crust attached to the African continent forms volcanoes in the *Aegean Sea*, east of mainland Greece.

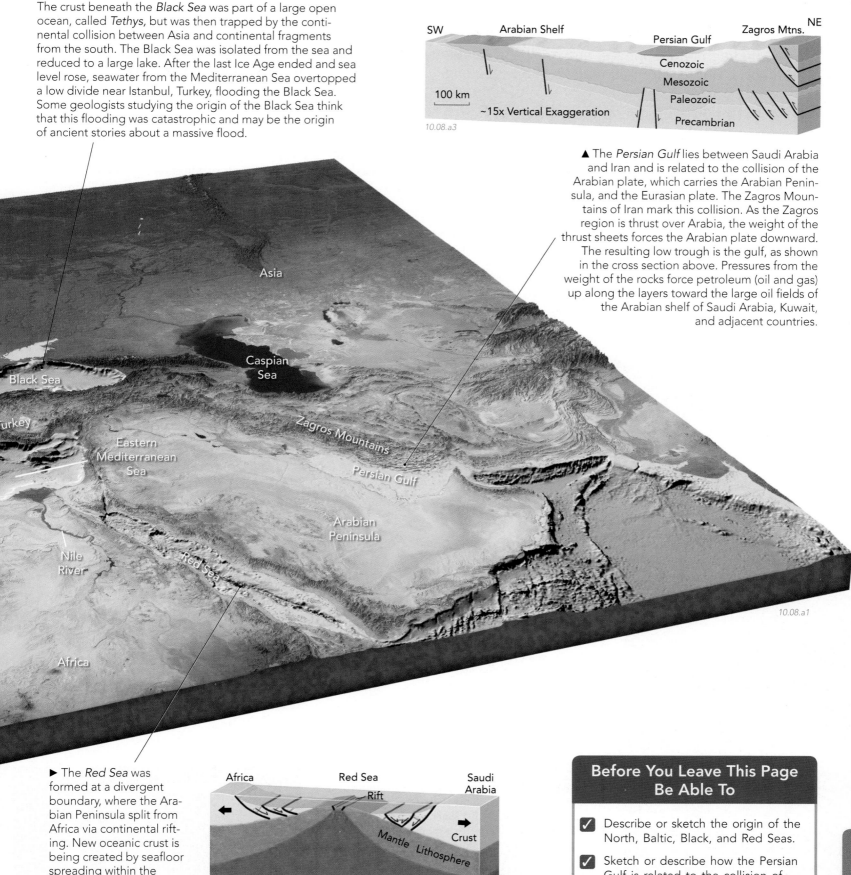

The crust beneath the *Black Sea* was part of a large open ocean, called *Tethys*, but was then trapped by the continental collision between Asia and continental fragments from the south. The Black Sea was isolated from the sea and reduced to a large lake. After the last Ice Age ended and sea level rose, seawater from the Mediterranean Sea overtopped a low divide near Istanbul, Turkey, flooding the Black Sea. Some geologists studying the origin of the Black Sea think that this flooding was catastrophic and may be the origin of ancient stories about a massive flood.

10.08.a3

▲ The *Persian Gulf* lies between Saudi Arabia and Iran and is related to the collision of the Arabian plate, which carries the Arabian Peninsula, and the Eurasian plate. The Zagros Mountains of Iran mark this collision. As the Zagros region is thrust over Arabia, the weight of the thrust sheets forces the Arabian plate downward. The resulting low trough is the gulf, as shown in the cross section above. Pressures from the weight of the rocks force petroleum (oil and gas) up along the layers toward the large oil fields of the Arabian shelf of Saudi Arabia, Kuwait, and adjacent countries.

10.08.a1

▶ The *Red Sea* was formed at a divergent boundary, where the Arabian Peninsula split from Africa via continental rifting. New oceanic crust is being created by seafloor spreading within the southern Red Sea as the two plates move apart.

10.08.a4

Before You Leave This Page Be Able To

✓ Describe or sketch the origin of the North, Baltic, Black, and Red Seas.

✓ Sketch or describe how the Persian Gulf is related to the collision of Asia and Arabia.

10.8

10.9 How Do Reefs and Coral Atolls Form?

REEFS ARE SHALLOW, MOSTLY SUBMARINE FEATURES, built primarily by colonies of living marine organisms, including coral, sponges, and shellfish. Reefs can also be constructed by accumulations of shells and other debris. Corals thrive in many settings, as long as the seawater is warm, clear, and shallow.

A In What Settings Do Coral Reefs Form?

Corals are a group of invertebrate animals that form calcium carbonate structures. To thrive, corals require nutrients, warmth, sunlight for photosynthesis, and water that is relatively free of suspended sediment. Too much sediment partially blocks the sun, can bury the coral, or can clog openings in the tiny organisms. Coral reefs form in shallow tropical seas with relatively clear water. Large waves batter many reefs, producing carbonate sediment.

1. Some reefs occur along the edges of continents, forming *barrier reefs* offshore from the main coastline. Reefs and islands protect a continent from large waves. They enclose a lagoon on the landward side but have an open ocean that is exposed to large waves and storms. Erosion of the reefs can form low, sandy islands with beaches covered by white sand produced by erosion and reworking of pieces of reef, shells, and other carbonate materials.

2. Reefs and other carbonate accumulations can form broad, shallow *platforms*, like the Bahama Islands east of Florida. In some cases, older reef deposits and dunes rise slightly above sea level. Between most islands, the water is shallow and the seabed is composed of white, carbonate-rich sand derived from wave erosion of reefs and the land.

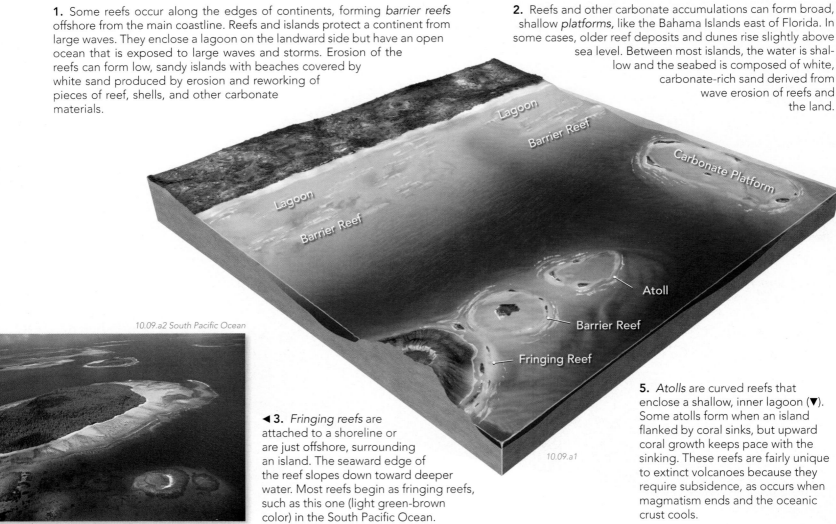

10.09.a1

10.09.a2 South Pacific Ocean

◄ 3. *Fringing reefs* are attached to a shoreline or are just offshore, surrounding an island. The seaward edge of the reef slopes down toward deeper water. Most reefs begin as fringing reefs, such as this one (light green-brown color) in the South Pacific Ocean.

5. *Atolls* are curved reefs that enclose a shallow, inner lagoon (▼). Some atolls form when an island flanked by coral sinks, but upward coral growth keeps pace with the sinking. These reefs are fairly unique to extinct volcanoes because they require subsidence, as occurs when magmatism ends and the oceanic crust cools.

10.09.a3 Great Barrier Reef, Australia

◄ 4. The *Great Barrier Reef* is along the eastern coast of Australia and has a unique history. Its base was formed along the edge of a shallow platform during the last Ice Age (17,000 years ago) when sea levels were lower. As sea levels returned to normal and began to drown the platform, the corals grew upward, keeping themselves in shallow water. Over time, the reef formed the largest organic buildup on Earth, one that is easily visible from space.

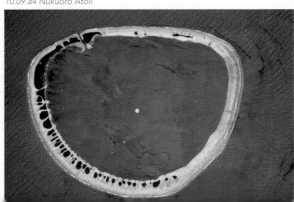

10.09.a4 Nukuoro Atoll

B How Do Atolls Form?

Charles Darwin proposed a hypothesis for the origin of atolls after observing a link between certain islands and atolls during his research aboard the ship *Beagle* from 1831 to 1836. According to his model, shown below, atolls form around a sinking landmass, such as a cooling or extinct volcano. Another model (not shown) interprets some atolls as being the result of preferential erosion of the less dense center of a carbonate platform.

10.09.b1–3

Stage 1: A volcanic island forms through a series of eruptions in a tropical ocean, establishing a shoreline along which corals can later grow and construct a fringing reef.

Stage 2: After volcanic activity ceases, the new crust begins to cool and sink. Coral reefs continue building upward as the island subsides, forming a barrier reef some distance out from the shoreline.

Stage 3: The volcano eventually sinks below the ocean surface, but upward growth of the reef continues, forming a ring of coral and other carbonate material. This forms an *atoll*, with a central, shallow lagoon.

C Where Do Reefs Occur in the World?

Most of the world's reefs are in tropical waters, located near the equator, between latitudes of 30° north and 30° south. Reef corals are more diverse in the Pacific, probably because many species went extinct in the Atlantic during the last Ice Age. The map below shows coral reefs as red dots.

10.09.c1

Reefs in the *Philippines* cover an estimated 25,000 square kilometers and consist of fringing reefs with several large atolls. Reefs also flank *Indonesia* and nearby *Malaysia*.

The *Great Barrier Reef*, along the northeastern flank of Australia, is the largest reef complex in the world. The world's second largest reef is in *New Caledonia*, a series of islands east of Australia and south of Micronesia.

The central and southwestern Pacific, including *Polynesia* and *Micronesia*, has many atolls and reefs, including a wide variety of barrier and fringing reefs. Farther north, Hawaii is also warm enough for reefs.

Well-known reefs are present throughout much of the *Caribbean* region, including *Florida*, the *Bahamas*, and the *Lesser Antilles*. The longest barrier reef in the Caribbean extends some 250 km (150 mi) along the Yucatan Peninsula, from the north of Belize, southward to Honduras.

Reefs occur along the continental shelf of *East Africa*, such as in Kenya and Tanzania. Other reefs encircle islands in the Indian Ocean and the shoreline of the Red Sea.

Before You Leave This Page Be Able To

☑ Describe the different kinds of reefs and where they form.

☑ Describe the stages of atoll formation.

☑ Name some locations with large reefs.

10.9

10.10 What Is the Geology of Continental Margins?

THE EDGES OF MOST CONTINENTS ARE HIDDEN beneath the seas, some distance from the shoreline. The edge of a continent marks the transition between continental and oceanic crust, but this transition is typically concealed by thick layers of sediment. What features are present along the edges of continents, and how do these features form?

A What Features Are Typical of Continental Margins?

Some continental margins are *active plate boundaries*, such as the western coast of South America, where oceanic crust subducts beneath the edge of the continent. Many continental margins are not plate boundaries, and instead are *passive margins*. Both active and passive margins share some features.

A *continental shelf* is a gently sloping surface that surrounds nearly all continents. On passive margins, it can extend from the shoreline as far as 1,500 km (930 mi) seaward, but it is typically narrow along active margins. The gentle slopes of most shelves, such as this one along the northeastern United States, are thought to have developed during the last Ice Age, when sea level was lower.

A *continental slope* connects the shelf with the truly deep ocean. Here the ocean floor slopes down at angles typically between 5° and 25°. The slopes are greatly exaggerated in this figure.

The *continental rise* is farther out from the continental slope. Sediment transported off the continental slope accumulates here, forming a broad, gently sloping underwater plain.

The continental shelf and slope are locally cut by submarine canyons, including Baltimore Canyon, shown here, offshore of New York and New Jersey. The Monterey Submarine Canyon offshore of California, discussed in the opening pages of this chapter, is a classic example.

100 Kilometers

10.10.a1

The transition from continent to deep ocean reflects progressively thinner continental crust and an abrupt change to oceanic crust. The thinned crust along most passive margins records rifting apart of the continent. Sediment on the continental margin varies greatly in thickness across the shelf, slope, rise, and abyssal plain.

1. Sediment is generally thinnest near the shoreline and on nearby parts of the continental shelf, which is underlain by continental crust with a close-to-normal thickness.

2. There are normal faults farther out, beneath the continental shelf and slope. These formed during the initial continental rifting that formed the margin. Normal faulting helped thin the crust, leading to deeper seafloor.

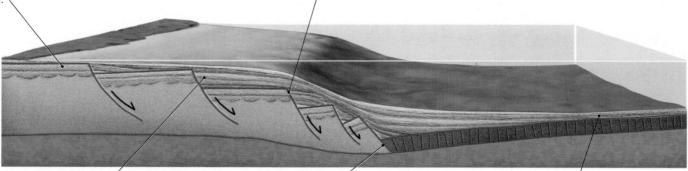

10.10.a2

3. Thick sediment accumulated over the downdropped fault blocks beneath the shelf and slope. The sedimentary layers can host important oil and gas resources.

4. The continental slope marks the abrupt change from thinned continental crust of granitic composition to even thinner oceanic crust composed of basalt and gabbro.

5. The abyssal plain is farther from land and sources of land-derived sediment. It has a thin sediment cover composed of small particles of clay and other fine-grained material.

B What Settings Lead to Underwater Slope Failure?

Continental slopes are blanketed by sediments, most of which are unconsolidated and weak. The combination of weak materials and a relatively steep angle causes some slopes to fail due to the force of gravity. Failure may be triggered by earthquakes, large storms, or overloading by newly deposited sediments.

Turbidity Currents

◄ As sediments collapse during a slope failure, they can break up and incorporate seawater between the grains. This forms a dense mixture of water and sediment (mostly clay, silt, and sand), such as this mass produced in a laboratory. These mixtures are more dense than normal seawater and flow downslope as fast-moving slurries, or *turbidity currents*. Turbidity currents have destructive potential and are capable of eroding rock, even underwater.

► The dense, cloud-like slurry of a turbidity current travels through the water until the current slows and the grains progressively settle, larger grains first, forming *graded beds*. When this happens more than once, the result is a sequence of alternating coarser and finer sediment with graded beds.

10.10.b1

10.10.b2

Submarine Canyons

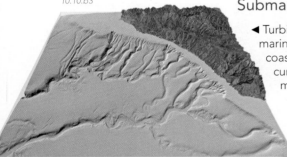

◄ Turbidity currents and other submarine movements can erode submarine canyons into the continental slope. This example is off the coast of central California, just south of Monterey Canyon. As the currents flow downhill, they erode the floor and walls of a canyon, making it larger over time.

► The upper parts of some submarine canyons, such as this one on the continental shelf near the Hudson River in New York, were carved by rivers when sea levels were more than 100 m (330 ft) lower than today. Once submerged by rising sea levels, such canyons can carry turbidity currents, which scour and enlarge the canyon.

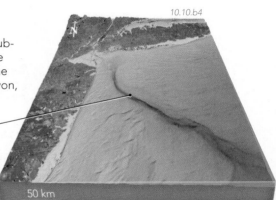

10.10.b3

10.10.b4

5 km

50 km

Submarine Fans and Submarine Landslides

This diagram illustrates a turbidity current, shown in gray, beginning on the continental slope and flowing down a submarine canyon.

As a turbidity current exits the steep canyon, it spreads out and slows down. Sand grains and any pebbles can no longer be suspended by the turbulence and settle out. As the current slows further, it deposits silt followed by clay particles. This process forms graded beds (coarser sediment at the base and finer sediment at the top).

10.10.b5

As the turbidity current slows and spreads out across the continental rise, it deposits its load of sediment in a fan-shaped deposit, or a *submarine fan*. A submarine fan can be hundreds to more than a thousand kilometers wide and typically consists of mud and other deep-marine sediment that alternate with sandy turbidite deposits with graded bedding.

Underwater slopes can also fail as *submarine landslides*. A landslide mass can contain large, fairly coherent blocks or can come apart as it detaches from the slope and moves downhill. A landslide commonly forms distinctive lumps on the seafloor and may leave behind a ragged scar on the slope above. Repeated large landslides were key to forming and widening some submarine canyons.

> ### Before You Leave This Page Be Able To
>
> ✓ Describe or sketch the features of a continental margin, such as the continental shelf, slope, and rise.
>
> ✓ Describe or sketch the rocks, sediments, and structures that occur along a typical continental margin.
>
> ✓ Explain turbidity currents, submarine canyons, submarine fans, and landslides.

10.10

10.11 How Do Marine Salt Deposits Form?

SALT DEPOSITS OF MARINE ORIGIN occur along many continental margins, forming layers, irregularly shaped masses, and structural domes. Marine salt deposits form only in specific geologic settings, especially sites where seawater evaporates. They are important sources of salt, sulfur, and petroleum. How and where do marine salt deposits form?

A How Does Salt Occur Along Continental Margins?

Natural salt is mostly composed of the sodium chloride mineral *halite* (NaCl), the mineral that makes up common table salt. Halite is associated with *gypsum* (a calcium sulfate mineral), *sylvite* (a potassium salt), and other minerals that have high solubilities in water. Most salt deposits form layers, as is typical for any sedimentary rock. Salt layers can be thinner than a centimeter or can comprise a layered sequence several kilometers thick. Salt outcrops are very soluble and so are relatively uncommon at Earth's surface. ▶

10.11.a1

10.11.a2 Canyonlands NP, UT

Salt is a very weak geologic material, flowing easily when subjected to the stresses associated with deep burial and tectonics. It is much less dense than other kinds of rocks, and so it commonly flows as solid but soft masses, like the folded layers of gypsum and salt shown here (◀). Upward-flowing masses of salt and gypsum can push up overlying layers, forming folds and domes, or in some cases even pierce through the overlying layers. Folds and other structures formed by moving salt are commonly sites where oil and gas accumulate in significant quantities, in addition to deposits of sulfur minerals.

B How Does Salt Form Near Continental Margins?

Many salt deposits form when seawater evaporates, leaving behind a residue of salt that was dissolved in the water. Such evaporation is especially efficient in warm, dry climates and in water bodies with limited connection to the oceans. Salt accumulations formed in such marine settings are *marine salt deposits*.

Marine salt deposits form along continental margins, where seawater can spill onto low areas next to the sea. They also form in narrow seas, especially those formed during continental rifting, when the two pieces of continent first separate, and during the early stages of seafloor spreading.

Smaller coastal bodies of water can receive input from rainfall, runoff from the land, and inflow of seawater from an adjacent ocean. If enough of this water evaporates, it deposits salt along the shoreline and on the floor of the water body. Inflow of water from the land and sea can effectively replace the water lost to evaporation, permitting evaporation and salt deposition to continue over a long time.

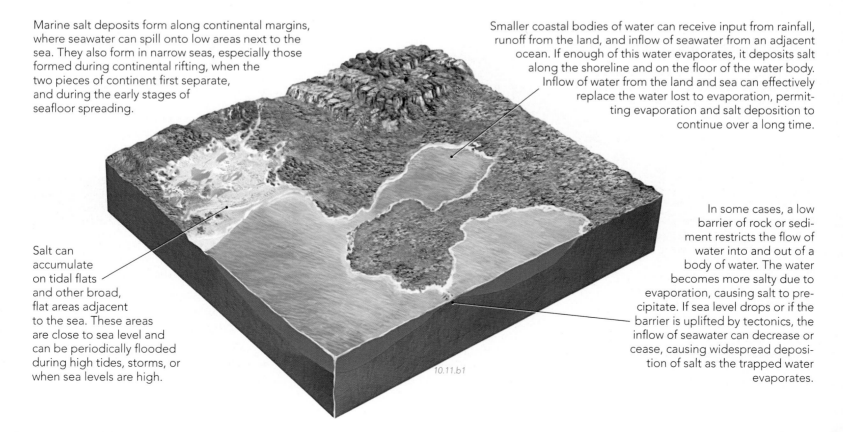
10.11.b1

Salt can accumulate on tidal flats and other broad, flat areas adjacent to the sea. These areas are close to sea level and can be periodically flooded during high tides, storms, or when sea levels are high.

In some cases, a low barrier of rock or sediment restricts the flow of water into and out of a body of water. The water becomes more salty due to evaporation, causing salt to precipitate. If sea level drops or if the barrier is uplifted by tectonics, the inflow of seawater can decrease or cease, causing widespread deposition of salt as the trapped water evaporates.

C What Structures Do Salt Deposits Form?

Because it is a weak rock, salt can form its own unique kinds of geologic structures. It also can greatly influence how faults and folds develop in overlying rocks.

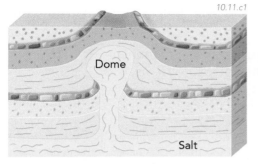

10.11.c1

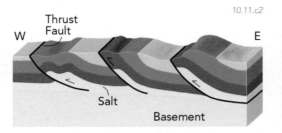

10.11.c2

10.11.c3 Iran

Salt is less dense than most rocks; when buried, it can buoyantly flow toward the surface in steep, pipe-like conduits. The resulting structure is a *salt dome.*

When a region containing a thick salt layer is deformed, the salt can slip and flow, allowing overlying rocks to fold and fault. This cross section shows part of the Jura Mountains near the French-Swiss border. The folds and faults are underlain by a weak layer of salt.

Where salt reaches the surface, such as in a salt dome or an anticline, it can flow downslope under the influence of gravity and form a *salt glacier,* such as this one.

D What Salt Structures Occur Along the Gulf Coast of the United States?

The Gulf Coast of the southern United States is world famous to geologists because it contains many salt structures, both on land and offshore. The salt structures have played a key role in the formation of the region's large oil fields and provide important sources of salt and sulfur minerals. For these reasons, they have been extensively studied by seismic surveys and by expensive drilling, sometimes in thousands of meters of water.

1. This diagram shows the land and seafloor in the Gulf of Mexico offshore of the Texas-Louisiana coast. An interpretation of the subsurface geology, drawn on the sides of the block, is based on studies by many geologists and billions of dollars of drilling.

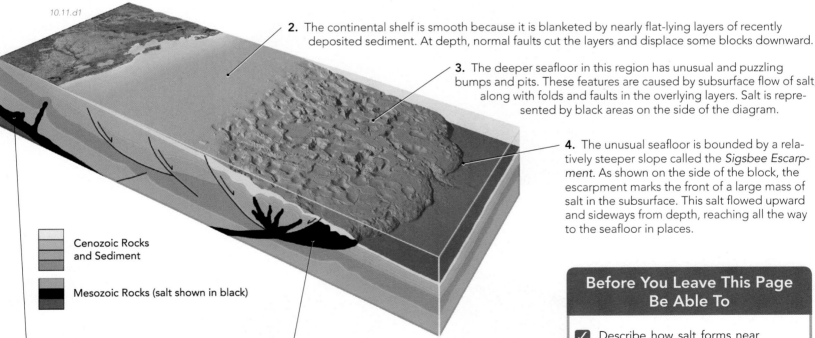

10.11.d1

2. The continental shelf is smooth because it is blanketed by nearly flat-lying layers of recently deposited sediment. At depth, normal faults cut the layers and displace some blocks downward.

3. The deeper seafloor in this region has unusual and puzzling bumps and pits. These features are caused by subsurface flow of salt, along with folds and faults in the overlying layers. Salt is represented by black areas on the side of the diagram.

4. The unusual seafloor is bounded by a relatively steeper slope called the *Sigsbee Escarpment.* As shown on the side of the block, the escarpment marks the front of a large mass of salt in the subsurface. This salt flowed upward and sideways from depth, reaching all the way to the seafloor in places.

Cenozoic Rocks and Sediment

Mesozoic Rocks (salt shown in black)

5. The salt originally was deposited in a thick layer (shown here in black) when continental rifting during the Mesozoic formed narrow basins. The basins at times had limited connection with the sea, causing evaporation of seawater and deposition of the salt layer. The salt was later buried by sediments, shown in light green, orange, and yellow.

6. As the salt was buried and subjected to increased pressure, it flowed sideways and rose up through the overlying sedimentary layers. Movement of the salt folded and domed the layers. In places, it formed steep, pillar-shaped salt domes, shown here as finger-like black masses. The salt domes and associated folded rock layer trapped oil and gas, for which the Gulf Coast is well known.

Before You Leave This Page Be Able To

☑ Describe how salt forms near continental margins.

☑ Describe how salt can occur in salt domes, some folded mountain belts, and salt glaciers.

☑ Describe how salt structures are expressed in the Gulf Coast region.

10.11

10.12 How Did Earth's Modern Oceans Evolve?

EARTH'S CONTINENTS AND OCEANS have changed over time, and their present configuration is the most recent snapshot of a longer evolution. Before 200 million years ago, the continents were joined together in a supercontinent called *Pangaea*. This huge landmass has since separated into discrete pieces, forming the modern continents and oceans.

These artistic renditions by geologist Ronald Blakey depict the breakup of a supercontinent and the movement of the continental fragments during the last 200 million years. He used an oval map that can show the entire world. It is not just one side of a globe.

200 Ma (Early Jurassic): End of Pangaea

Before 200 million years ago (200 Ma), all the continents were joined in the supercontinent of *Pangaea*, which was surrounded by an enormous ocean. Pangaea was assembled in the Late Paleozoic and, as shown here, had begun to break up via continental rifting by 180 to 200 Ma (in the Early Mesozoic).

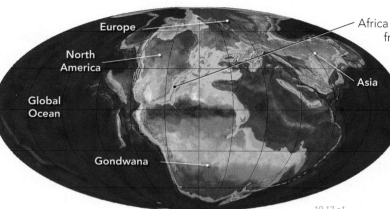

10.12.a1

Africa and South America began to rift apart from North America, forming rift valleys and a narrow sea along what is now the eastern United States. Faulting during rifting thinned crust under the East Coast and adjacent continental shelf.

As North America rifted away, the landmasses of Africa, South America, India, Australia, and Antarctica remained linked in a southern supercontinent called *Gondwana*.

150 Ma (Late Jurassic): New Oceans Open

By about 150 Ma, in the middle of the Mesozoic Era, the breakup of Pangaea was well underway. The Central Atlantic Ocean had formed as North America separated from Africa. The Gulf of Mexico formed by rifting along the southern edge of North America.

Continental rifting began along the future borders of Africa and South America, but seafloor spreading had not yet started to form the South Atlantic Ocean.

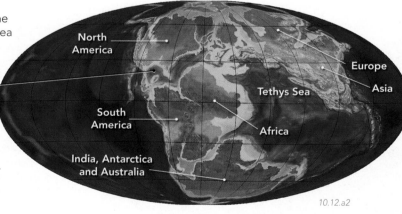

10.12.a2

The *Tethys Sea*, a tropical ocean, was a large, wedge-shaped extension of the main global ocean. To the north, a series of collisions started to consolidate Asia into a larger continent.

The landmasses south of Tethys started to rift apart and began to resemble the familiar shapes of South America and Africa. The rest of Gondwana remained mostly intact, for now.

120 Ma (Early Cretaceous): Central and South Atlantic Oceans Open

During the Cretaceous Period (120 Ma), North America, Europe, and Asia were still mostly connected. The North Atlantic Ocean, between North America and Europe, had yet to fully open.

The southern parts of the Atlantic Ocean opened as South America began to separate from Africa. The rest of Gondwana began to rift apart, as India separated from Australia and Antarctica, forming the early stages of the Indian Ocean.

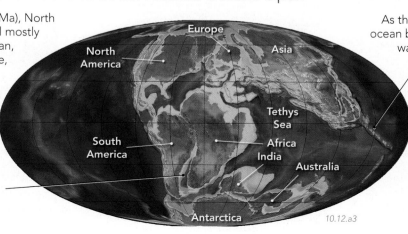

10.12.a3

As the new oceans grew, the large global ocean began to shrink because ocean crust was lost by subduction beneath North and South America, Asia, and island arcs within the ocean. As the newly formed oceans continued opening, their waters affected climate patterns, helped stabilize and moderate land temperatures, and produced many areas of shallow-marine environments. This led to an incredible diversity of life in the sea and on land.

90 Ma (Late Cretaceous): Atlantic Ocean Fully Opens

The Central Atlantic Ocean was fully open between North America and Africa, with a spreading center down the middle of the ocean. The North Atlantic, between North America and Europe, had not yet rifted open. The opening of the South Atlantic Ocean separated Africa and South America, isolating their land animals.

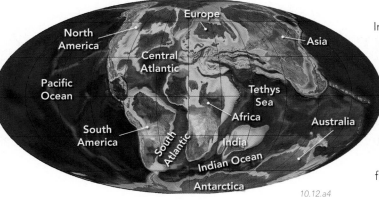

10.12.a4

India was fully separated from the rest of Gondwana and was an isolated landmass. It was headed northward across the Tethys Sea toward an eventual collision with the southern flank of Asia.

Antarctica rifted apart from Africa, which allowed the South Atlantic to connect to the southern Indian Ocean. Australia rifted apart from Antarctica, fully ending the existence of Gondwana.

30 Ma (Paleogene): Closing the Tethys Sea

Greenland and the rest of North America began rifting apart from Europe at about 80 Ma, opening the North Atlantic Ocean. The Pacific Ocean contained spreading centers (the belts of lighter blue on the seafloor), but continued to grow smaller over time as its oceanic plates subducted beneath the Americas, Asia, and many island arcs.

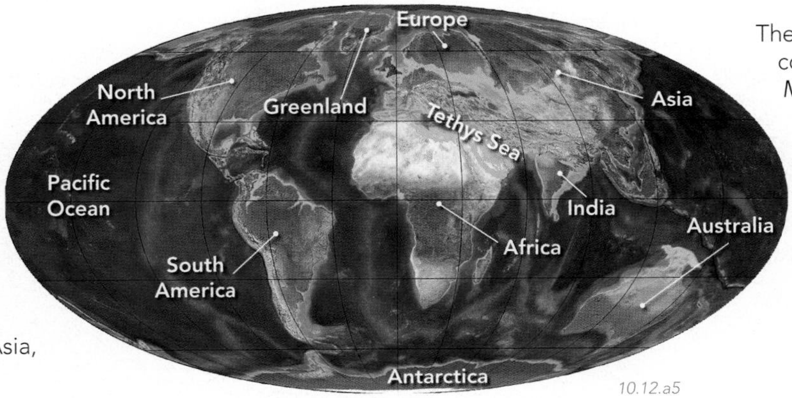

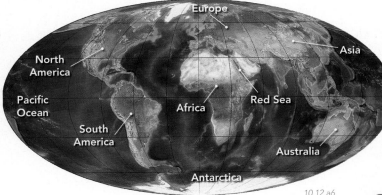

10.12.a5

The Tethys Sea was nearly closed as India collided with Asia to form the *Himalaya Mountains*, and Africa and nearby continental fragments converged with southern Europe to form the *Alps* and other ranges.

Australia was completely isolated, allowing its collection of marsupials and other unusual animals to thrive and evolve. Antarctica remained over the South Pole.

Present Day

Today, the Atlantic Ocean continues to grow because it has a spreading center that adds to the oceanic plate but does not have major subduction zones to consume any oceanic material. It has grown to its present size at the expense of the Pacific Ocean, which is the last remnant of Pangaea's global ocean.

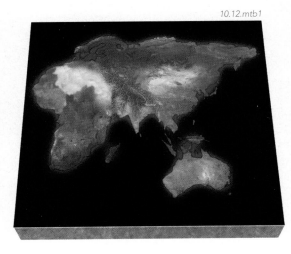

10.12.a6

Convergence between the Indian and Eurasian plates continues to form the Himalaya mountain chain north of India. It may also be starting to form a new plate boundary further south, within the Indian Ocean.

The Red Sea is a developing rift and may continue to grow at the expense of another ocean. It is uncertain whether the nearby East African Rift, south of the Red Sea, will split off yet another piece of Africa.

The Future of the Oceans

What will the oceans look like in 50 million years? Geologists calculate the likely future locations of the oceans and continents by using current plate velocities and by making assumptions about how plates act during collisions. One prediction is that Africa will collide with Europe and Asia, closing the Mediterranean Sea and forming a very large supercontinent, shown to the right. The Pacific will continue to shrink as spreading in the Atlantic Ocean pushes the Americas (not shown) farther to the west. Our present situation is not final. It is just one scene in a very long movie.

10.12.mtb1

Before You Leave This Page Be Able To

☑ Describe the major changes in Earth's oceans since 180 million years ago, including approximately when the Central Atlantic, South Atlantic, North Atlantic, and Indian Ocean formed and which continents rifted apart to form each ocean.

☑ Describe or sketch why growth of the Atlantic Ocean must have caused the Pacific Ocean to shrink over time.

10.12

10.13 How Did the Gulf of Mexico and the Caribbean Region Form?

THE GULF OF MEXICO AND CARIBBEAN SEA display an island arc, several deep troughs, and many islands and small ocean basins. The present setting and recent geologic history of the region provide an opportunity to examine various aspects of how continental margins are formed and how ocean basins evolve over time. Examine the map below, which shows seafloor depths and plate boundaries.

Present Setting

The *Gulf of Mexico* is nearly enclosed by Florida, the Gulf Coast of the United States, and Mexico. It is deepest (darker blue color) in the center and is flanked by broad continental shelves offshore of the United States and the Yucatan Peninsula.

Shallow seafloor, underlain by continental crust, flanks the Florida Peninsula and Bahama Islands (a carbonate platform). Deeper seafloor separates this region from the island of Cuba to the south and from the Yucatan Peninsula to the southwest.

A trench (sawtooth line below) curves around the outside of the Lesser Antilles island arc. The trench and island arc are the result of westward subduction of Atlantic oceanic lithosphere beneath the Caribbean plate.

10.13.a1

A deep trench marks where oceanic plates in the Pacific, including the Cocos plate, are subducted northeastward beneath Central America. Volcanoes and earthquakes are common in the overriding plate.

An east-west-trending escarpment, the *Cayman Trough*, cuts across the seafloor and the southern end of Cuba. It is a transform boundary along the northern edge of the small Caribbean plate.

The seafloor in some parts of the Caribbean plate is shallower than expected. Here, the oceanic crust is anomalously thick (up to 20 km thick) and is composed of thick sequences of basalt.

Jurassic History (~200 to 145 Ma)

By the Jurassic Period, North America had begun to rift apart from Africa and South America. Continental rifts were partially filled with sediment and salt, and the thinned continental crust became continental shelves.

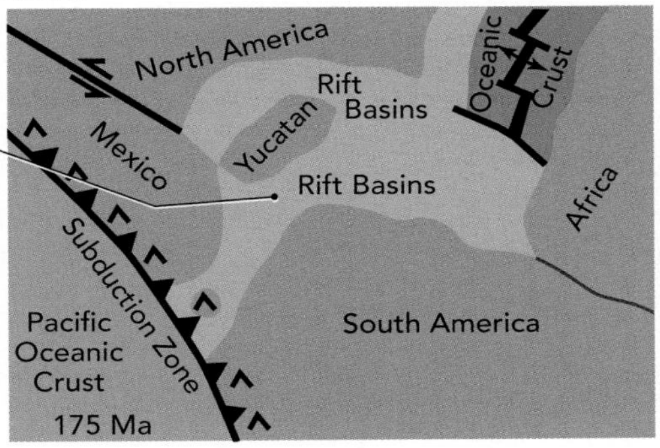

10.13.a2

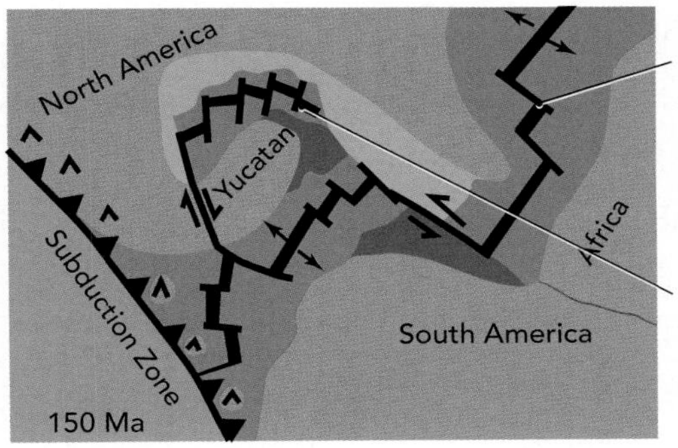

By the Late Jurassic, the continents had truly rifted apart as seafloor spreading produced new oceanic crust.

10.13.a3

Spreading formed the Gulf of Mexico when the Yucatan pulled away from the Gulf Coast.

Cretaceous History (145 to 65 Ma)

In the Cretaceous Period, spreading in the Gulf of Mexico ceased, but sediment deposition continued.

Seafloor spreading moved North America farther from Africa and South America.

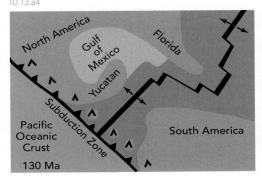

10.13.a4

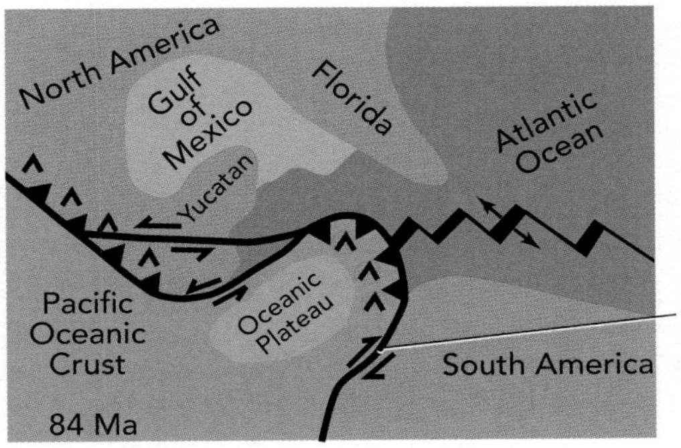

In the Late Cretaceous, an oceanic plateau from the Pacific was moving to the northeast, pushing the subduction zone and island arc in front of it.

Transform faults bounded the sides of the plate on which the oceanic plateau rode.

Tertiary History (65 to 5 Ma)

In the early Tertiary Period, the island arc and oceanic plateau collided with Florida near Cuba.

A new volcanic arc formed between South and Central America but did not connect these lands until about 5 Ma.

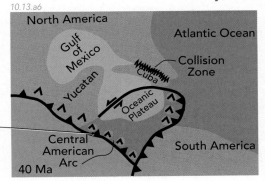

10.13.a6

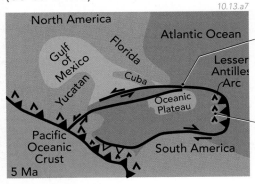

10.13.a7

By 5 Ma, long transform faults allowed the Caribbean plate to continue moving eastward.

The Lesser Antilles island arc formed above an east-dipping subduction zone.

How Geologists Reconstructed This History

The history summarized above was not easy to piece together, especially since much of the geologic record is undersea. The first studies were done on land, mapping the geology, determining the ages and sequences of rock units, and locating volcanoes, faults, and other structures.

An understanding of the undersea geology was largely obtained by seafloor mapping and geophysical surveys conducted for petroleum exploration on the continental shelves. Such surveys were followed by very expensive drilling through the sedimentary layers in order to retrieve rock samples, calibrate the surveys, and determine whether petroleum was present. The local geology was then interpreted in the context of global reconstructions of plate motions, largely derived from paleomagnetism, and calculated rates of spreading.

Geologists are actively investigating and debating many aspects of this geologic history. Perhaps the most controversial topic is whether the thick basaltic sequence in the Caribbean originated as an oceanic plateau in the Pacific, as shown above, or was formed locally, between North and South America.

Before You Leave This Page Be Able To

☑ Describe the main physical features of the Caribbean and Gulf of Mexico, describing how they relate to modern plate-tectonic boundaries.

☑ Briefly summarize the main events that shaped the Gulf and Caribbean.

☑ Describe how the geologic history of the region was studied, both on land and beneath the sea.

10.13

10.14 How Did These Ocean Features and Continental Margins Form?

The terrain below contains various features on the seafloor, as well as parts of three continents. Some general observations of each feature provide clues about what that feature is. You will use this information to interpret how each feature formed, what the area was like in the past, and how it will look in the future.

Goals of This Exercise:

- Observe the terrain and make observations about the shape, size, and character of each feature.
- Use the general descriptions to determine which features are present in different parts of the terrain.
- Interpret how each feature formed and use this information to infer the present-day plate tectonics of the area.
- Use all the information to reconstruct what the area probably looked like 20 million years ago and what it will look like 20 million years into the future.

1. This figure shows a region approximately 1,000 km (600 mi) wide. The seafloor is shaded according to depth, with lighter blue colors indicating shallower areas. Numbers indicate the isotopic ages of volcanic rocks in millions of years before present (labeled *Ma*). The lowest seafloor is part of the abyssal plain, and the only age on it is 140 Ma, in the middle of the ocean.

2. A broad oceanic plateau rises from deep water and locally forms small islands. Samples collected by drilling and dredging are mostly basalt and are dated at 40 million years. A linear chain of islands and seamounts extends from the oceanic plateau toward the southeast. The islands and seamounts are shaped like volcanoes and consist of volcanic rocks, mostly basalt. The ages of the volcanic rocks decrease to the southeast. A shield volcano at the southeast end of the chain is still active.

3. A curved belt of volcanic islands flanks a deep oceanic trench on the east. Most of the volcanoes consist of andesite and show evidence of recent explosive eruptions. Most islands have been volcanically active for more than 35 million years, as shown by the age range for volcanic rocks in several islands.

4. There is a narrow sea between the volcanic islands and a continent to the west. In the center of the sea is a low ridge, whose axis contains a rift valley and evidence of active submarine eruptions of basalt. The axis of the ridge has jogs along some type of fractures.

5. The western continent contains a narrow shelf offshore. There is no trench, evidence of recent volcanoes, earthquakes, or mountain building along this edge of the continent. The oldest oceanic crust next to the continent is 20 million years old.

10.14.a1

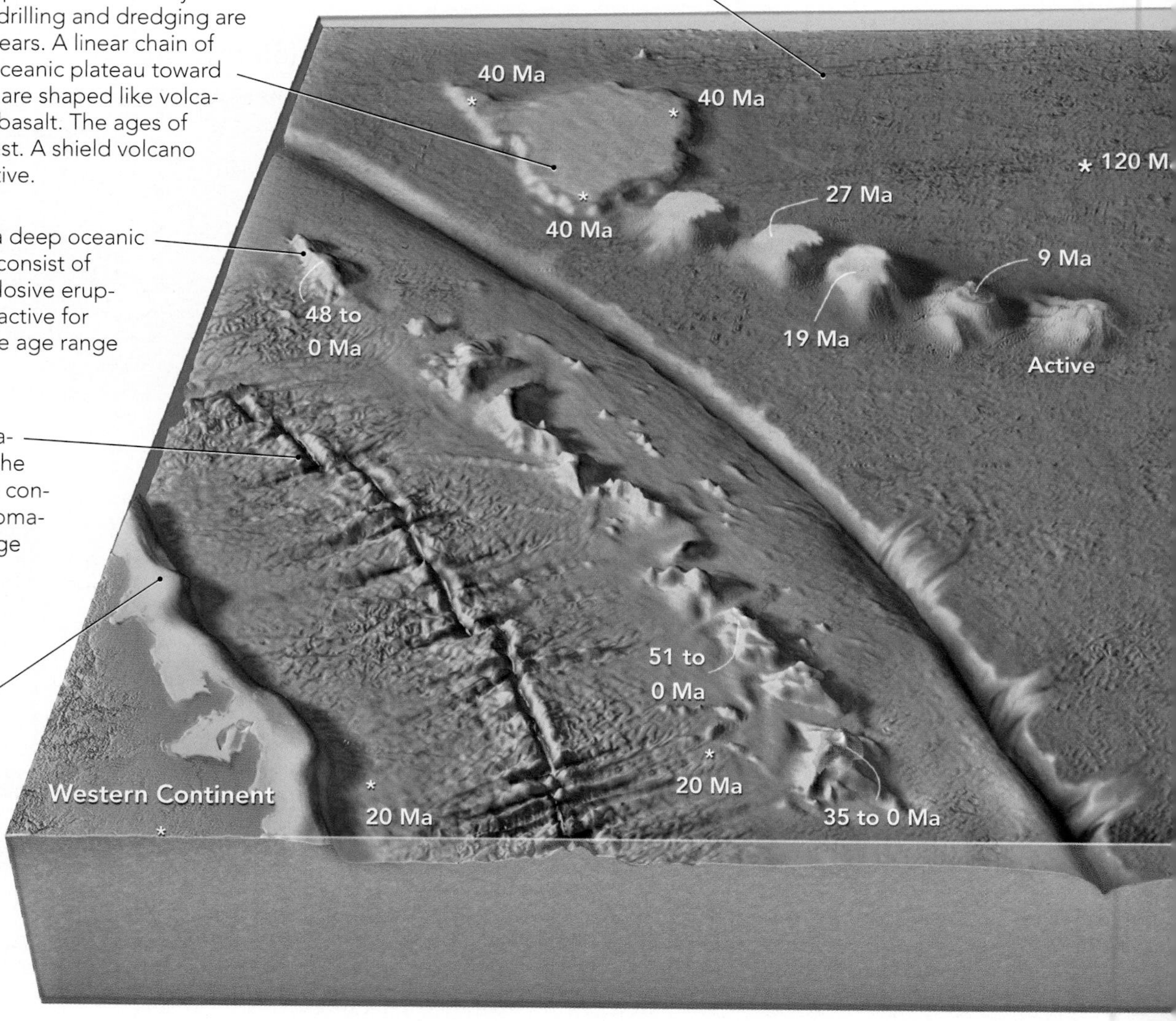

40 Ma
40 Ma
40 Ma
120 Ma
27 Ma
9 Ma
48 to 0 Ma
19 Ma
Active
51 to 0 Ma
20 Ma
Western Continent
20 Ma
20 Ma
35 to 0 Ma

Procedures

Use your observations of this region to complete the following steps, entering your answers on the worksheet or online.

1. Observe the terrain and determine which types of features are shown (e.g., mid-ocean ridge, island arc, etc.).

2. Based on the descriptions, briefly describe or identify how each feature probably formed.

3. Identify the main geologic features shown on the cross section along A–A'.

4. In the appropriate place on the worksheet, draw a cross section along the front of the terrain. Show your interpretations of the plate geometries and different types and thicknesses of crust and lithosphere.

5. Describe what the area might have looked like 20 million years ago based on the ages and relative motions of the plates. Draw a very simplified map of your interpretation on the worksheet.

Your instructor may also ask you to complete the following steps. If so, enter your answers in the appropriate tables on the worksheet.

6. Interpret whether adjacent features are related to one another using their relative positions and ages.

7. Predict what the area will look like 20 million years into the future. Draw a simplified map of your interpretation on the worksheet.

6. The shelf surrounding the central continent is broad and shallow, extending several hundred kilometers out from the shoreline. The edge of the shelf shows no evidence of earthquakes or active faulting. Several large canyons are cut into the shelf and lead down to large piles of sediment on the abyssal plain. The continent has fairly subdued topography.

7. To explore for oil, geologists used seismic surveys to investigate the shelf of the central continent. A geologic cross section summarizing these results is presented below for the line A–A' (shown on the map). All sedimentary layers are Cenozoic (younger than 65 million years).

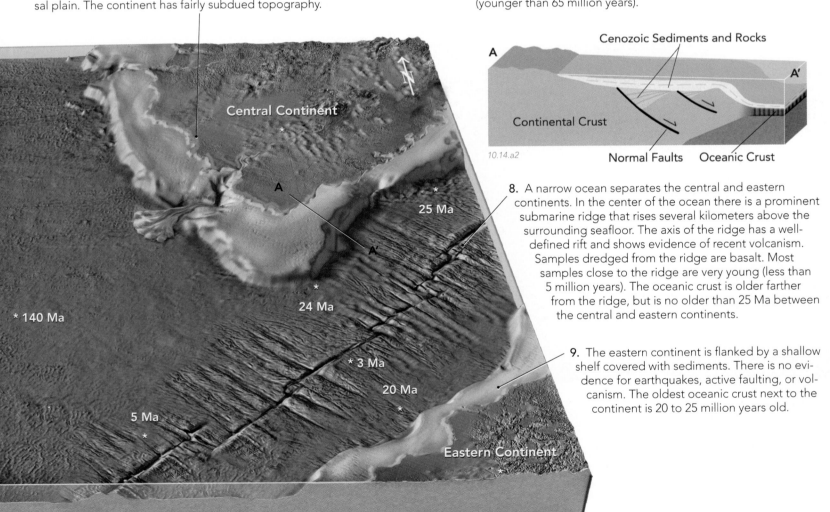

Cenozoic Sediments and Rocks

A

A'

Continental Crust

10.14.a2

Normal Faults   Oceanic Crust

8. A narrow ocean separates the central and eastern continents. In the center of the ocean there is a prominent submarine ridge that rises several kilometers above the surrounding seafloor. The axis of the ridge has a well-defined rift and shows evidence of recent volcanism. Samples dredged from the ridge are basalt. Most samples close to the ridge are very young (less than 5 million years). The oceanic crust is older farther from the ridge, but is no older than 25 Ma between the central and eastern continents.

9. The eastern continent is flanked by a shallow shelf covered with sediments. There is no evidence for earthquakes, active faulting, or volcanism. The oldest oceanic crust next to the continent is 20 to 25 million years old.

Central Continent

A

A'

25 Ma

24 Ma

* 140 Ma

* 3 Ma

20 Ma

5 Ma

Eastern Continent

10.14

Mountains, Basins, and Continents

THE SURFACE OF THE EARTH contains mountains and high plateaus, as well as basins, which are low areas where sediment accumulates. At a larger scale, Earth's surface also contains continents that have grown, rifted apart, and collided through time. How do mountains, basins, and continents form, and what factors control their elevations?

This view, looking north, shows satellite imagery superimposed on topography for the region around the Tibetan Plateau of southern Asia. A topographic profile across the region is on the next page.

What regional features can you observe in this perspective view and on the topographic profile?

The Tibetan Plateau is the largest, highest, and flattest plateau on Earth. Its average elevation is 5 km (over 15,000 ft), which is higher than any peak in the United States, except for some mountains in Alaska.

Why is the Tibetan Plateau so high, and what controls the elevation of a region?

The Tarim Basin is a large desert north of the plateau. It is 3,000 m lower than the plateau and is partially filled by sediment derived from the adjacent highlands.

How do basins form, and why are they lower than their surroundings?

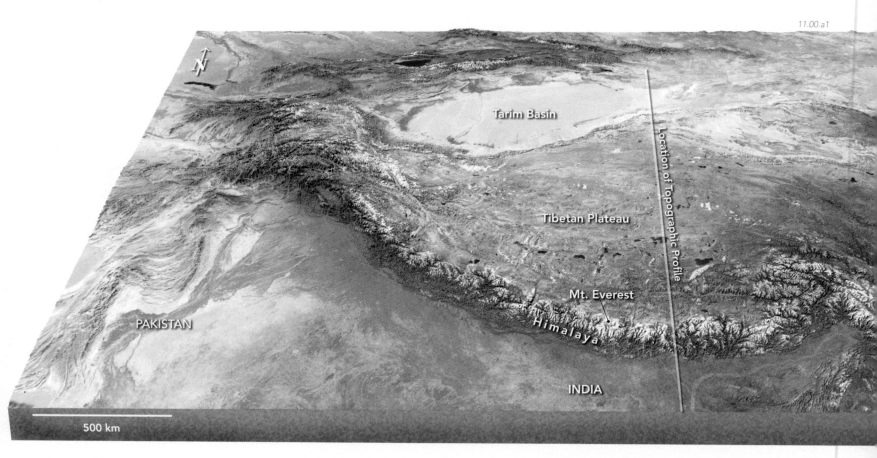

11.00.a1

The Himalaya is a spectacular mountain range that rises along the southern edge of the Tibetan Plateau. It is the world's highest mountain range, with many peaks more than 8 km (>26,000 ft) above sea level.

Why is this mountain range so high compared to all others on our planet?

Mount Everest is the world's highest mountain, rising 8,850 m (29,035 ft) above sea level. It straddles the border between Nepal (to the south) and Tibet (to the north), and climbers can approach the mountain from either side.

What geologic processes form mountains, and what controls which areas have mountains versus which ones do not?

Most of India, to the south of the Himalaya, has much lower elevation and relief, and is tectonically stable away from the mountain front. Its oldest rocks are approximately 2.7 billion years old, representing the earliest period of Earth's history.

When did the first continents form, and how do continents change over time?

TOPICS IN THIS CHAPTER

11.00.a2 Himalaya Mtns., Tibet

◀ This photograph, looking south, shows peaks of the high Himalaya. Part of the less rugged, but still high-elevation Tibetan Plateau is in the foreground.

CHINA

11.00.a3 Himalaya Mtns., Tibet

▲ From the Tibet (north) side, Mount Everest is a rugged, imposing mountain. This view is taken from one of the base camps where climbers begin their arduous and dangerous climb to the top.

▼ The topographic profile below shows the high Tibetan Plateau viewed to the west. The high mountains on the left edge of the Plateau are the Himalaya. To depict the topographic features at this regional scale, the topography is vertically exaggerated by 10 times.

Investigating the Timing of Uplift of a Region

Tibet and the Himalaya are high in elevation now, but when did they become so? Geologists use several approaches to determine when a region was uplifted. These include Global Positioning Systems (GPS), isotopically dating minerals that record the uplift history of the rocks, and examining the types, thickness, and ages of sediment in adjacent basins.

Observations using a Global Positioning System (GPS) provide a direct measurement of uplift, and indicate that parts of the Himalaya are rising a few centimeters per year. Another approach is to find rocks at high elevations that were deposited at low elevation. The top of Mount Everest contains a faulted slice of Paleozoic limestone with marine fossils; the limestone was deposited at sea level and later uplifted along with the mountain range.

Isotopic dating methods are an important way to determine the age of uplift. As deep rocks are uplifted toward the surface and uncovered by erosion, they cool, locking in daughter products from radioactive decay. Certain dating techniques tell us when deep rocks arrived to within 2 to 4 km of the surface and so indicate the age of uplift. In the Himalaya, such methods yield ages as young as several million years, indicating recent and ongoing uplift.

Uplift and erosion of mountains and other high regions contribute clasts to adjacent sedimentary basins. We can therefore infer the age of uplift by determining when clasts derived from a mountain were added to the sedimentary sequence. Sediments along the foothills of the Himalaya indicate that debris originating from the mountain range first appeared around 45 million years ago.

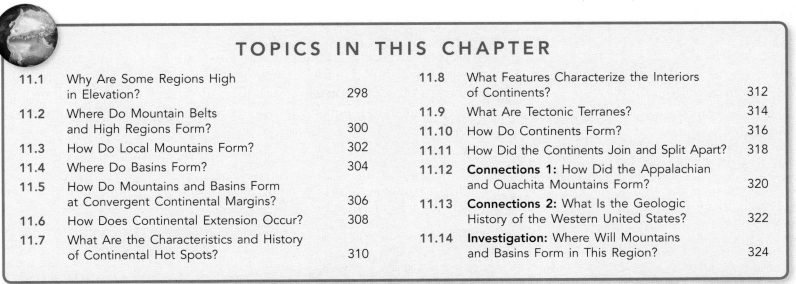

India   Himalaya   Tibetan Plateau   Tarim Basin

100 km

11.00.a4

11.1 Why Are Some Regions High in Elevation?

SOME REGIONS ARE MUCH HIGHER THAN OTHERS. Many mountains are not only steep, but are high in elevation. Elsewhere, huge regions of land are barely above sea level. What accounts for these differences? Regional variations in elevation primarily reflect the tectonic processes that occurred in the region, and the nature of the crust and mantle at depth. A change in the subsurface can cause a region to be uplifted (rise in elevation) or to subside (drop in elevation).

A What Controls Regional Elevation?

Regional elevations are controlled primarily by the thickness of the crust, but they can also be influenced by the temperature and density of materials in the crust and upper mantle.

11.01.a1

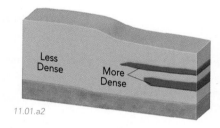

11.01.a2

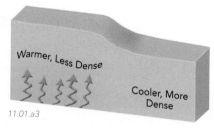

11.01.a3

Regions with thick crust are higher than those with thinner crust. In other words, mountain ranges have deep crustal roots.

Regions underlain by less dense crust will be higher in elevation than areas with a similar thickness of more dense crust.

Temperature of the crust and mantle also affects elevation. Warm rocks are less dense than cooler rocks, so areas with warm rocks are higher than areas with cool rocks.

B What Causes Variations in Crustal Thickness?

Differences in crustal thickness between regions reflect differences in their geologic histories. Such differences include whether the crust is continental or oceanic, and whether it has been deformed, eroded, or buried.

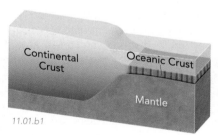

11.01.b1

Continents have relatively thick crust, generally 30 to 50 km thick, and so are higher than ocean basins, which are underlain by oceanic crust that is much thinner, typically about 7 km thick. ▲

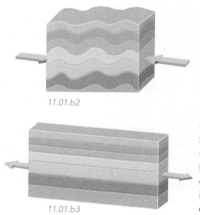

11.01.b2

◄ Crust thickens if compressed from the sides. It can respond to compression by folding or faulting.

Crust thins if it is stretched in a horizontal direction (◄), either by ductile stretching at depth or by normal faulting in the upper crust.

11.01.b3

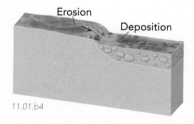

Erosion
Deposition

11.01.b4

Crust that loses material to erosion will become thinner, whereas crust that gains material by deposition of sediment or volcanic rocks will become thicker. ▲

C How Is Regional Elevation Decreased?

Normal faulting can thin the crust by displacing higher rocks off lower ones. This decreases crustal thickness and causes a region to subside.

Crustal thickness can be reduced if material is eroded from the top, as is common in many mountain belts.

Rocks contract when they cool, so cooling of large regions of the crust or mantle causes subsidence.

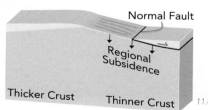

11.01.c1

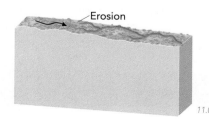

11.01.c2

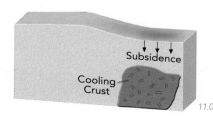

11.01.c3

D How Is Regional Elevation Increased?

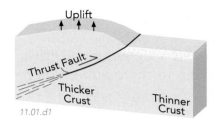

11.01.d1

Crust that is compressed and shortened by thrust faults also thickens. This thickening causes the region to be uplifted. The thrust fault can also uplift rocks, forming a mountain.

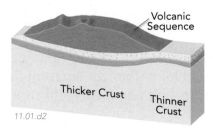

11.01.d2

Crust can thicken by material added to the surface, whether it is sediment or volcanic units, perhaps lava in huge volcanic fields. The added material builds up the surface and pushes down the crust a small amount in response to the weight.

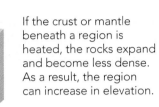

11.01.d3

If the crust or mantle beneath a region is heated, the rocks expand and become less dense. As a result, the region can increase in elevation.

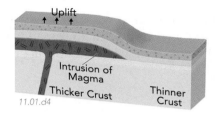

11.01.d4

Magma can add to the crust at depth, and this addition of material thickens the crust. Several processes may operate together: magma can add material and heat up the crust.

E What Is the Influence of the Thickness of the Lithosphere?

The lithosphere is, on average, about 100 km thick, but varies in thickness from nearly zero at mid-ocean ridges to more than 150 km beneath some ancient continental interiors. These variations greatly influence elevation because the mantle part of the lithosphere is more dense than the asthenosphere.

A region with thin lithosphere, such as a mid-ocean ridge, will be higher than an adjacent region with thicker lithosphere, even if they have the same type and thickness of crust.

As the new oceanic plate moves away from the ridge, the asthenosphere cools enough to become lithosphere. The plate, therefore, thickens, becomes more dense, and subsides as it cools.

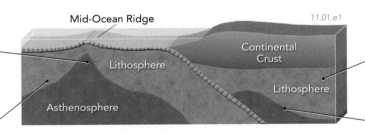

11.01.e1

Lithosphere is generally thicker in the central, ancient parts of continents, far away from modern plate boundaries, but these regions have moderate elevation because of relatively thick continental crust.

Continental lithosphere can be thinned near plate boundaries by heating and other plate activity. The affected region can rise in elevation because dense lithosphere is replaced by less dense asthenosphere.

The Discovery of Isostasy

Isostasy is the principle that regional elevations adjust to the types and thicknesses of rocks at depth. It was discovered through observations made by George Everest while surveying India around 1850. Surveyors at the time understood that a weight suspended on a line (to level the surveying equipment) was deflected from vertical a very small amount by the gravitational attraction of nearby mountains. When taking this into account, Everest noted an unexplained discrepancy in positions on his survey. He found that the deflection of the weight from vertical was less than predicted.

To explain the discrepancy, a mathematician calculated the expected gravitational attraction of the Himalaya. Astronomer George Airy then used an analogy with float-ing icebergs and other common objects to suggest that higher mountains had thicker crustal roots. By this model, lower density crustal material in the roots attracts the suspended weight less than would the denser mantle material that the crustal root has displaced.

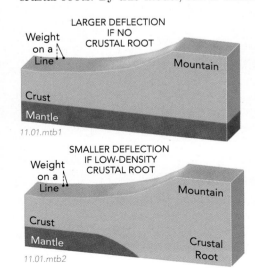

11.01.mtb1

11.01.mtb2

11.1

Before You Leave This Page Be Able To

✓ Summarize or sketch the factors that control regional elevation.

✓ Summarize or sketch what causes variations in crustal thickness.

✓ Summarize several ways to increase elevation and to decrease elevation.

✓ Explain the observation that led to the discovery of isostasy.

11.2 Where Do Mountain Belts and High Regions Form?

MOUNTAIN BELTS AND OTHER HIGH REGIONS generally owe their high elevation to thick continental crust. Less commonly, a region is higher than its surroundings due to processes originating in the mantle. Where are the world's main mountain belts and why did mountains form in these places?

A In Which Tectonic Settings Do Regional Mountain Belts Form?

Regional mountain ranges are hundreds or thousands of kilometers long. They are large enough that they can only be explained by major variations in the thickness and temperature of the crust and lithosphere. Most ranges occur near convergent plate boundaries or where there has been large-scale movement of material in the mantle.

11.02.a1

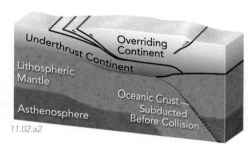

11.02.a2

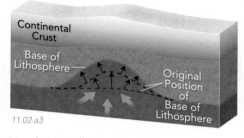

11.02.a3

Subduction Zones—Convergent margins are high in elevation largely because the crust is thickened by magmatic additions from the subduction zone and by crustal shortening. Also, in these regions, lithosphere is heated and replaced by less dense asthenosphere.

Continental Collisions—Collision zones have high elevations due to an increase in crustal thickness as one continent is shoved over another. In these settings, crustal thickening occurs by thrusting, folding, and other forms of deformation.

Mantle Upwellings—Less dense asthenosphere can move upward into the lithosphere, causing regional uplift. This occurs near hot spots, plate boundaries, and in some other settings, and is partly responsible for uplift in some parts of the western United States.

B What Causes These Regions to Have High Elevation?

Western Canada has been a convergent margin for most of the last 100 million years. Its mountain ranges overlie crust thickened from major thrust faulting, from magmatic additions, and from collisions with island arcs and pieces of continental material.

The *Alps mountain range* of southern Europe is high because it has thick crust due to collisions between Europe and smaller continental blocks that came from the south.

The *Tibetan Plateau* and the *Himalaya* are extremely high because of very thick crust that resulted when the Indian continent collided with, and was partly shoved beneath, Asia.

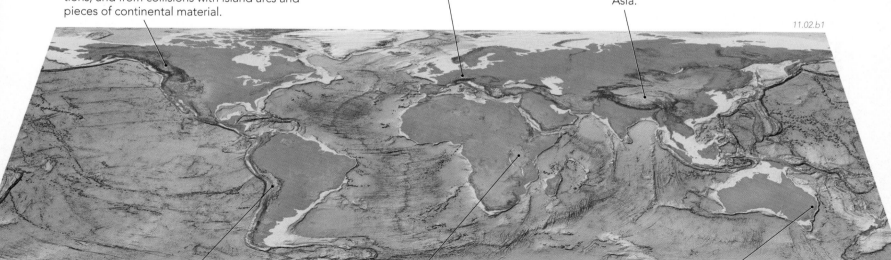

11.02.b1

The *Andes* of South America are above a subduction zone. The underlying crust is hot and thick because of magmatic additions and crustal shortening.

The *East African Rift* is higher than most of Africa because of magmatic heating of the crust, thinning of the lithosphere, and the presence of a hot spot leading to mantle upwelling.

The *Great Divide Range* forms the eastern flank of Australia. There is currently no plate boundary here, and geologists currently investigate the age and cause of uplift.

C What Happens During the Erosion of Mountain Belts?

Mountains, once formed, are subjected to weathering and erosion. These processes wear mountains down but are countered by uplift related to *isostasy*. Uplift is driven by buoyancy due to the root of underlying thick crust.

Early Mountain Building

As a mountain belt forms, uplift is commonly faster than erosion, and the mountain becomes higher and more rugged over time. A high mountain belt results from uplift that is faster than erosion.

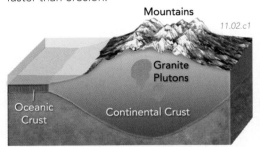

As soon as it starts forming, weathering and erosion begin to wear down a mountain, contributing sediment to streams and rivers. Sediment will be transported to adjacent low areas, perhaps in nearby oceans or other types of basins.

Erosion and Isostatic Rebound

As material erodes from a mountain belt, there is less weight holding down the thick crustal root. The buoyant crust can uplift, a process called *isostatic rebound*.

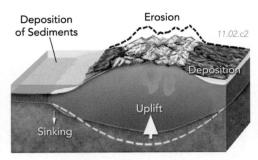

Sediment derived from the mountain is deposited in nearby basins, on both the sea and continental sides. The added weight of the sediment depresses the crust in these regional basins, making room for more sediment.

Late Stages of Evolution

Erosion and isostasy cause rocks deep in the crust to be uplifted and exposed at the surface. As a result, many mountain belts expose metamorphic and plutonic rocks.

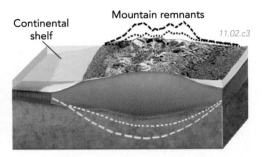

Through simultaneous erosion and isostasy, the mountain is eroded down and the thick crustal root is gradually reduced in size. Material eroded from the mountain ends up in adjacent basins, increasing the crustal thickness beneath the basins.

D What Controls Regional Elevations in North America?

The vertically exaggerated topographic profile below illustrates how elevations vary from east to west across the United States. It does not show the full thickness of crust, only the elevation of the land and depth of seafloor.

Western North America is high mostly because crust was thickened along a convergent margin. The moderately high elevation of the Basin and Range is largely due to very thin lithosphere.

Compression and shortening within the North American plate thickened crust in the Rocky Mountains. Additional uplift is due to a locally thin lithosphere and upwelling asthenosphere associated with rifting.

Elevation decreases from the Great Plains toward the Mississippi River because the lithosphere is cooler and thicker to the east.

The Appalachian Mountains were once a region of thick crust, due to the collision between North America and Africa. Much of this thickness has been lost due to erosion, and so the range has lost elevation over time.

Rule of Thumb for Elevations

Regional elevations are relatively low for regions with thinner crust, and relatively high for regions with thicker crust, but by how much? A rule of thumb is that increasing the thickness of the crust by 6 km will result in an increase in elevation of 1 km (~3,300 ft). Here is an example from Arizona.

Phoenix sits at an elevation of 300 m (1,000 ft), whereas Flagstaff is at more than 2,100 m (7,000 ft). This difference is about 2 km, so the crust beneath Flagstaff should be 12 km thicker than the crust beneath Phoenix (2 × 6 = 12). Geophysical measurements show that the crust beneath Phoenix is about 28 km thick, whereas crust beneath Flagstaff is about 40 km thick. The difference is 12 km, the value we would predict.

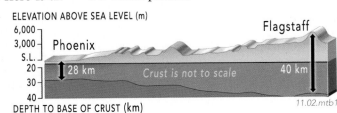

Before You Leave This Page Be Able To

- ☑ Sketch and explain the main tectonic settings of high regions, providing an example for each setting.
- ☑ Summarize the settings of the world's high mountains and plateaus.
- ☑ Explain how erosion and isostasy help expose deeply formed rocks in eroded mountain belts.
- ☑ Summarize differences in regional elevation across North America.

11.2

11.3 How Do Local Mountains Form?

THE DISTINCTION BETWEEN LOCAL MOUNTAINS and regional mountain ranges is important. Regional mountain ranges are hundreds to thousands of kilometers long, contain many peaks, and typically involve uplifted, thickened crust. Other mountains are *local* features, too small to be accompanied by regional increases in crustal thickness. Instead, such mountains simply rest upon—and are supported by—the crust.

A How Does Volcanism Form Local Mountains?

A local mountain may be formed by a volcanic eruption that piles lava, ash, and scoria onto the crust. Such mountains vary in size from small scoria (cinder) cones to large shield and composite volcanoes.

11.03.a1 Flagstaff, AZ

11.03.a2 San Pedro volcano, Guatamala

11.03.a3 Mount Kilimanjaro, Tanzania

Volcanism creates mountains by piling volcanic materials on a preexisting surface. Some of the smallest volcanic mountains and hills are scoria cones. They are clearly local features, not requiring regional changes in the thickness of the underlying crust.

Composite volcanoes consist of lava flows, variably compacted volcanic ash, and debris in mudflows and landslides. They commonly make lofty and steep mountains that have a typical volcano shape, like the one shown here.

Prolonged volcanism can build even larger mountains. *Mount Kilimanjaro*, an active volcano in Africa, is over 5,800 m (19,000 ft) high. It was built in the last 2 million years from eruptions totalling 4,200 cubic kilometers (a cube of rock 10 miles on a side).

B How Do Faults Build Mountains?

Local mountains can also arise through faulting. Thrust faults create mountains by thrusting one fault block up and over another. Normal faults also form local mountains, even though they stretch and thin the crust in a region.

Mountains Formed by Thrust Faulting

Thrust faulting will make a mountain if the overthrust block is uplifted faster than it is eroded, or if it is composed of erosion-resistant rocks like granite and other crystalline rocks.

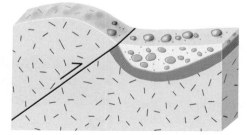
11.03.b1

Mountains Formed by Normal Faulting

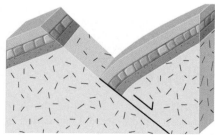
11.03.b2

During normal faulting, one block slips down, forming a basin. The other block remains high or is moved upward, and it can form a local mountain if it is not eroded away.

11.03.b4 Death Valley, CA

11.03.b3 Denali, AK

Mount McKinley (also called Denali), the tallest peak in North America, consists of granite. It was uplifted along a thrust fault that formed by localized compression along a nearby major strike-slip fault. Uplift of the range is recent and rapid, mostly occurring since 6 Ma.

Normal faulting along the eastern side of Death Valley, California, forms rugged local mountains. Faulting displaced the valley floor down relative to the mountains, forming a basin (Death Valley) that traps sediment eroded from the ranges. The floor of Death Valley is locally below sea level, but it is not connected to the sea.

C How Does Folding Build Mountains?

Another way to make local mountains is by folding. Folding can warp and uplift Earth's surface as well as the underlying rock layers. Uplift and erosion of a folded, hard layer can create a topographical high.

1. Folding can form mountains and hills by deforming the land surface and near-surface rocks, as is happening near Los Angeles, California.

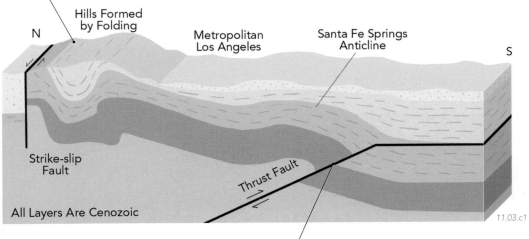

11.03.c2 Dinosaur NP, CO/UT

▲ 3. Some mountains, including this one in Dinosaur National Park in Colorado and Utah, owe their existence to folding followed by erosion. In this area, folding ended more than 45 million years ago. Erosion downcut through the rocks until it encountered these folded layers of hard, light-colored sandstone. Soft rocks underlie the valley and so were eroded away more easily, leaving the sandstone as a mountain.

2. Some oil-well pipes near Los Angeles were being crushed and bent beneath a fold, called the Santa Fe Springs anticline. No one knew why until the large 1994 Northridge earthquake revealed thrust faults in the area, including one newly discovered fault beneath L.A. This fault was breaking and bending the pipes as it folded and uplifted sedimentary rocks.

D How Can Differential Erosion Form a Local Mountain?

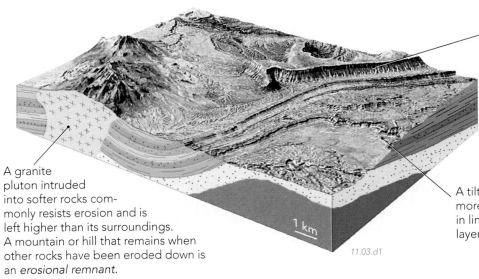

A resistant rock layer can protect softer rocks beneath from erosion, forming a local hill or mountain. Such a feature, if it has a nearly flat top, is a *mesa*. ▶

11.03.d2 Moab, UT

A granite pluton intruded into softer rocks commonly resists erosion and is left higher than its surroundings. A mountain or hill that remains when other rocks have been eroded down is an *erosional remnant*.

A tilted resistant layer, for example a sandstone, can erode more slowly than adjacent softer layers, like shale. This results in linear or curved ridges and valleys that follow the tilted rock layers across the landscape.

11.03.d1

11.03.d3 Stone Mtn., GA

Stone Mountain in Georgia consists of granite that solidified at a depth of 10 km. It was then uncovered by erosion, which removed the overlying and flanking softer rocks.

Before You Leave This Page Be Able To

✓ Describe how volcanism forms mountains.

✓ Sketch and describe how thrust faulting, normal faulting, and folding can each build mountains.

✓ Describe some ways that erosion can result in a mountain, ridge, or mesa.

11.3

11.4 Where Do Basins Form?

BASINS ARE LOW RELATIVE TO THEIR SURROUNDINGS and commonly trap sediment and water. They form in many tectonic settings, both on land and beneath the oceans, and they can accumulate different kinds of sediment, depending on their geologic environments.

A In What Tectonic Settings Do Basins Form?

Basins form on both oceanic and continental plates and along plate margins. Some basins are as large as an ocean, while others are smaller, depressed areas, commonly near a local fault or fold.

Passive Margin

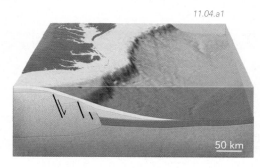

11.04.a1

50 km

The largest type of basin is a *passive margin*, a continental margin that is not a plate boundary. A passive-margin basin includes the continental shelf (lightest blue in the image above), continental rise, and continental slope, and generally is underlain by thin, previously rifted crust. It receives sediment from the continent and provides shallow-water environments for diverse life, such as offshore of North Carolina.

Continental Rift

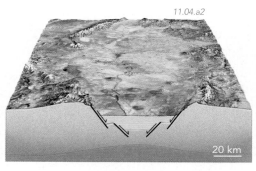

11.04.a2

20 km

Continental rifts form when forces try to pull apart a continent, forming normal faults that downdrop some fault-bounded blocks. The downdropped blocks can accumulate coarse, continental sediment, fine-grained lake beds, and evaporite deposits. If the rifting progresses to seafloor spreading, a continental rift evolves into a passive margin. The rift shown here is similar to the Rio Grande Rift that runs north from Texas and New Mexico into Colorado.

Normal Fault Blocks

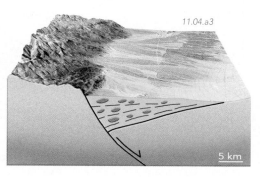

11.04.a3

5 km

Normal faulting can downdrop a block, forming a basin that fills with sediment. Steep topography along the faulted mountain front produces coarse sediment that is delivered to the basin by debris flows, landslides, and steep, rocky streams. Finer grained sediment and evaporites can accumulate in lakes. *Normal-fault basins* can occur on land, for example Death Valley, or along rifted margins, like those that flank the Atlantic Ocean.

Reverse and Thrust Faults

11.04.a4

50 km

A *foreland basin* occurs when crust (either continental or oceanic) is depressed by the weight of thrust sheets. The basin develops as a depression in front of the thrusts because the extra weight causes the crust to warp downward. The Persian Gulf is a foreland basin. Thrust faults are also common in the accretionary prism (not shown) between a magmatic arc and trench, and a basin in this setting is a *forearc basin* (in front of the arc).

Strike-Slip Faults

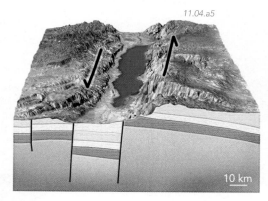

11.04.a5

10 km

Basins can develop along a strike-slip fault if motion along the fault downdrops one block relative to another. Such downdropping is most common where the fault takes a bend across the surface, as shown here for a basin similar to the Dead Sea of the Middle East. Downdropping can also occur where strike-slip motion along several nearby faults causes the crust to pull apart, dropping a block in between as a *pull-apart basin*.

Regional Subsidence

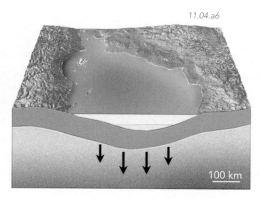

11.04.a6

100 km

Huge basins, hundreds or thousands of kilometers wide, form due to *regional subsidence*, where a broad region drops in elevation. Causes of subsidence include regional cooling of the crust and mantle, lateral or vertical movement of the underlying crust and mantle, or conversion of less dense minerals in the lower crust and upper mantle to more dense ones. A large basin in Michigan, the Michigan Basin, is an example of regional subsidence.

B What Formed These Basins?

North America contains many basins, some regional in size and others kilometers across. Some basins are currently active and accumulating sediment, whereas others are ancient and obvious only when geologists compare the thicknesses and types of sedimentary rocks from one area to another. The map below shows basins that contain more than 5 km of sedimentary and volcanic units. It color-codes basins as a function of age.

A basin sits between the Cascade volcanic arc and an offshore trench that marks subduction beneath the continent. This *Cascade forearc basin* receives abundant sediment from major rivers, like the Columbia River, that drain into the sea. The *San Joaquin Basin* of central California is an older version of a forearc basin, and formed in front of the Sierra Nevada magmatic arc.

In southern California, small, but locally deep basins formed along the San Andreas fault, a complex zone of mostly strike-slip movement. Some basins are pull-apart basins, and others are related to local thrusting or normal faulting where the fault takes a bend.

The interior of the western United States contains a passive margin formed during the Paleozoic by rifting of the western edge of North America. It locally accumulated more than 10 km (6 mi) of sediment.

The Michigan and Illinois Basins formed within the continent, probably due to Paleozoic collisional tectonics in the Appalachians and from other deep processes.

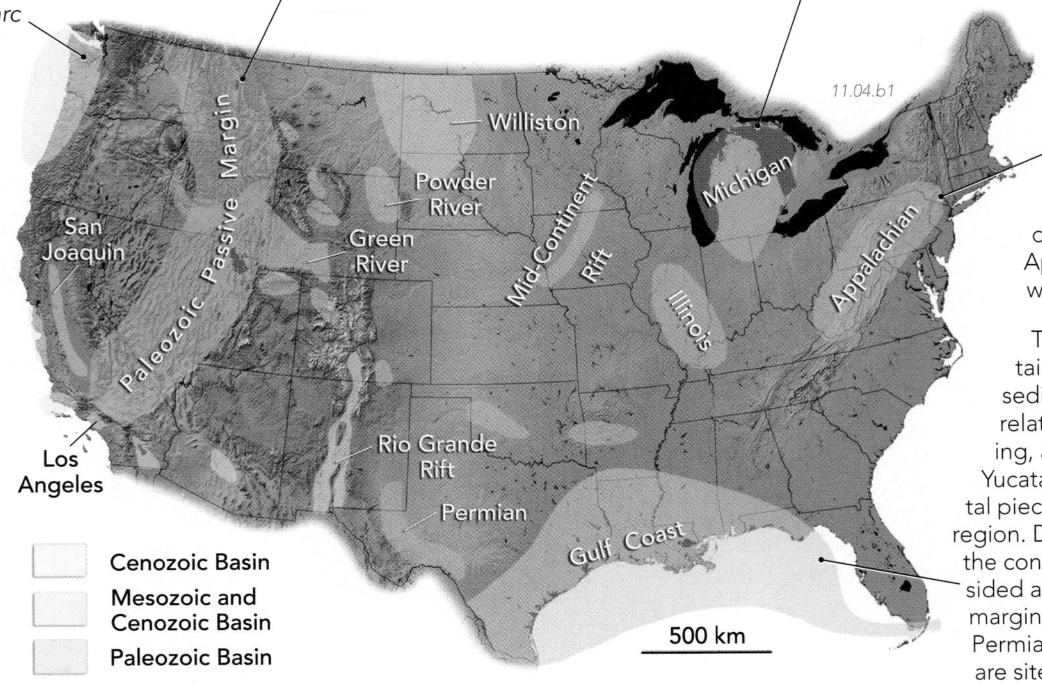

The Appalachian Mountains and nearby areas contain thick sedimentary sequences deposited along the Paleozoic continental margin and in other basins before the Appalachian Mountains were formed.

The Gulf Coast contains thick sequences of sedimentary rocks, mostly related to Mesozoic rifting, as South America, the Yucatan, and other continental pieces rifted away from this region. During and after rifting, the continental margin subsided and became a passive margin. The Gulf Coast and the Permian Basin in west Texas are sites of important oil and gas resources.

11.04.b1

☐ Cenozoic Basin
☐ Mesozoic and Cenozoic Basin
☐ Paleozoic Basin

500 km

Map labels: Williston, Powder River, Green River, Mid-Continent Rift, San Joaquin, Michigan, Illinois, Appalachian, Los Angeles, Rio Grande Rift, Permian, Gulf Coast, Paleozoic Passive Margin

The Michigan Basin

A deep basin beneath Michigan contains a fairly complete column of sedimentary rocks deposited during the early and middle parts of the Paleozoic Era. On the geologic map shown here, rock layers form a bull's-eye pattern around the roughly circular basin, with the youngest layers (yellow and green) occurring in the center of the basin. A geologic cross section across the basin (below) shows that the layers are thicker in the center of the basin. This indicates that the basin was subsiding during deposition of the sediments. The origin of the basin is somewhat enigmatic and possibly involves several causes. The basin probably formed during an episode of continental rifting, but it may also have subsided partly because of flow, thinning, and cooling of the hot lower crust.

11.04.mtb1

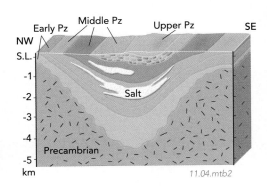

11.04.mtb2

Before You Leave This Page Be Able To

☑ Describe the different ways in which a basin can form.

☑ List some basins in the United States and describe what caused each to form.

☑ Describe the Michigan Basin and possible causes of subsidence.

11.4

11.5 How Do Mountains and Basins Form at Convergent Continental Margins?

AT SUBDUCTION ZONES BENEATH CONTINENTS, various processes create mountains and basins. Magmatic additions to the crust, along with crustal compression, cause thickening of the crust and the formation of a central mountain belt. Basins can form in front of, within, and behind the mountain belt.

A What Processes Accompany Ocean-Continent Convergence?

Along *ocean-continent* convergent boundaries, an oceanic plate subducts beneath a continental plate. Subduction causes melting in the mantle beneath the continent and also leads to compression and thickening of the continental crust. Such margins are generally dominated by a regional mountain belt.

1. As the oceanic plate approaches the convergent margin, it flexes and bends downward into the inclined subduction zone. An *oceanic trench* forms as a result and acts as a deep oceanic basin that traps sediment eroded from the adjacent mountain belt. The area between the trench and the mountain front is close to or below sea level because it is underlain by thin continental crust and oceanic material sliced off the downgoing plate.

2. An *accretionary prism* forms along the upper parts of the subduction zone as sediment is contributed by the adjacent continent and scraped off the downgoing slab. It is a structurally complex zone of faults, folds, and rocks under various metamorphic conditions. As more material is stuffed under the prism, the prism thickens and is uplifted, but generally remains below sea level.

3. Convergence of the two plates causes horizontal compression within the continent. This results in thrust faults and other structures that thicken the crust and cause further uplift of the mountain belt. A *fold and thrust belt* can form behind the main mountain belt as rocks are thrust over the interior of the continent.

4. The weight of the thrust sheets causes the continent to flex downward, forming a basin in front of the thrust belt. This basin is called a *foreland basin* because it occurs in front of the mountain belt. It receives sediment from the mountain belt and from other parts of the continent.

5. Magma generated along the subduction zone rises into the crust. It thickens the crust by erupting as volcanic rock on the surface and by solidifying at depth. The highest parts of most subduction-related mountain belts are near the areas with the greatest volcanic activity.

Oceanic Trench

Accretionary Prism

Fold and Thrust Belt

Foreland Basin

Continental Crust

Lithospheric Mantle

Subducting Plate

11.05.a1

B What Determines If the Overriding Plate Is Shortened or Extended?

Subduction is not always accompanied by compression and thrust faulting. Several factors influence whether the plate above a subduction zone experiences compression or extension, including the factors presented below.

Compression and horizontal shortening are common in subduction zones where the continental plate moves toward the subduction zone relative to the asthenosphere. This movement pushes against the subducted slab, which is difficult to move sideways through the solid mantle. As a result, the continent experiences compression, as is occurring in parts of the Andes of South America.

Extension is common when the overriding plate is not moving toward the slab relative *to the* asthenosphere, or is even moving away. The slab tends to pull back by itself, and the continent extends as its edge is pulled toward the ocean by the sinking slab. This is occurring along subduction zones in the western Pacific near Japan and the Philippines.

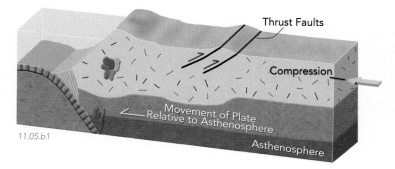

Thrust Faults

Compression

Movement of Plate Relative to Asthenosphere

Asthenosphere

11.05.b1

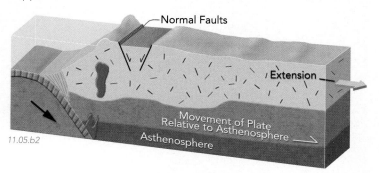

Normal Faults

Extension

Movement of Plate Relative to Asthenosphere

Asthenosphere

11.05.b2

C What Features Accompany Continental Collisions?

Continental collisions involve the convergence of two tectonic plates that each carry continental crust. A continent generally is too buoyant to be subducted deeply, so one continent is shoved beneath the edge of the other continent, and the whole region is uplifted. The collision transmits large stresses to the plates on either side, forming thrust faults and thickened crust.

During a collision, one continental plate is shoved, or *underthrust*, beneath another plate. A foreland basin forms in front of the collision zone, and the basin sediments can be overridden by or incorporated into the thrust faults.

Collisions form high mountain belts composed of faulted, folded, and cleaved rocks. Uplift and erosion bring metamorphic and intrusive igneous rocks up to the surface. In some cases, the collision forms a high continental plateau, such as in Tibet.

Behind the collision zone, rocks can be folded and thrust away from the mountain belt. The weight of the thrust sheets pushes down adjacent crust, forming sedimentary basins in front of the thrust sheets.

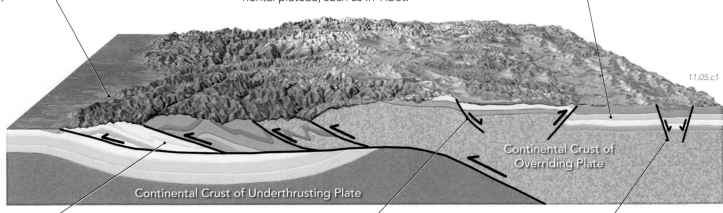

11.05.c1

Continental Crust of Underthrusting Plate

Continental Crust of Overriding Plate

A series of thrust faults forms along the collision zone and thickens the crust by shoving one slice of crust over another.

If the crust gets too thick or too hot, it may begin to spread under its own weight, flowing sideways. At the surface, such spreading can form normal faults and associated basins.

Stresses associated with the collision can cause other types of tectonic features to form hundreds to thousands of kilometers away from the actual plate boundary.

An Ancient Basin in the Eastern United States

A well-known, thrust-related basin once covered parts of what is now New York, Pennsylvania, West Virginia, Ohio, and adjacent states. The basin is of middle Paleozoic (Devonian) age and contained a feature called the *Catskill Delta*. It was related to a collision between eastern North America and a continental fragment that moved westward across an ancient ocean that lay to the east. During the collision, stresses caused thrust faulting within the Appalachian Mountains. As the thrust sheets pushed westward toward the interior of North America, their weight downwarped the crust, forming a foreland basin. The basin was originally larger than the area shown on this map, but parts of the basin have been eroded away due to uplift in the Appalachian Mountains,

northern New York, and western Ohio and Kentucky.

As the basin subsided, sediment was eroded from the mountains and was transported westward into the basin, forming a series of deltas. The sedimentary layers are coarser (mostly sandstone) to the east, closer to the source. They are thicker to the east because this part of the basin subsided

more than areas to the west. The distribution of different types of sediment indicates that thrusting occurred in the mountains during deposition in the basin. Later, coarse sediments from the mountains reached farther to the west, as faulting slowed and the basin was filled. Much later, the sedimentary basin was uplifted, tilted, and eroded to form the scenery of this region.

11.05.mtb1

[map showing NY, PA, OH, WV, VA, MD, DE, NJ, Lake Erie, with Sandstone and Shale labels]

Before You Leave This Page Be Able To

✓ Summarize how mountains and basins form in an ocean-continent convergent margin.

✓ Summarize one factor that favors shortening versus extension in a plate above a subduction zone.

✓ Summarize how mountains and basins form in a continental collision.

✓ Summarize the setting of the Catskill Delta during the Paleozoic.

11.5

11.6 How Does Continental Extension Occur?

DURING CONTINENTAL EXTENSION, continental crust is thinned and stretched horizontally, typically causing the region to subside. Continental extension also breaks the crust into faulted blocks, forming *local* mountain ranges and sedimentary basins. By studying sedimentary sequences, geologists can determine when a basin was active and how fast its sediments accumulated.

 ## How Do Continents Accommodate Crustal Extension?

When continental crust is extended, the upper part responds by breaking into discrete blocks bounded by normal faults. If the fault blocks do not rotate during extension, only a small amount of extension can occur. If the blocks and faults rotate, greater amounts of extension can take place.

Non-Rotating Fault Blocks

In some extended areas, adjacent normal faults dip in opposite directions and cut the crust into wedge-shaped fault blocks. ▶

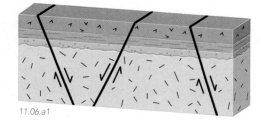

11.06.a1

Movement along the faults downdrops some blocks, forming sedimentary basins. These can be thousands of meters deep and tens of kilometers wide. ▼

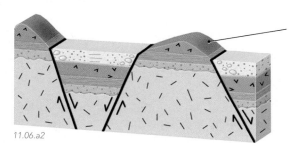

11.06.a2

The upthrown blocks become mountains bounded on both sides by faults. Erosion of the mountain contributes sediment to the basins.

Over time, the basins fill with sediment unless rivers carry most of it away.

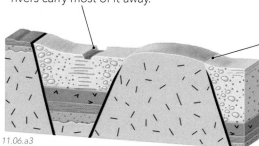

11.06.a3

The mountains are gradually eroded down, and basin sediments may overlap the edges of the range.

Rotating Fault Blocks

In other extended areas, adjacent normal faults dip in the same direction and cut the crust into book-shaped fault blocks. ▶

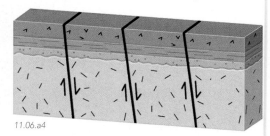

11.06.a4

During fault movement, the blocks and faults both rotate, like books sliding on a shelf. ▼

The corner of a block that is rotated down becomes a basin.

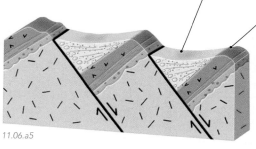

11.06.a5

The corner that is rotated up becomes a mountain or ridge. The mountains and ridges commonly are linear, following the strike of the layers.

As faulting and extension continue, units are tilted to steep dips. The oldest layers dip more steeply than more recent layers.

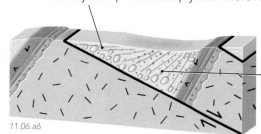

11.06.a6

Faults are rotated to gentle dips and can have kilometers of displacements, allowing large amounts of crustal extension.

11.06.a7 Death Valley, CA

◀ Death Valley in eastern California is a classic example of extension and normal faulting within a continent. This photograph, taken looking north from an overlook called Dante's View, shows the basin with mountain ranges on both sides. The white and gray units in the center of the basin are salt deposits, formed by evaporation of water within this hot, closed basin. A still-active normal fault runs along the steep mountain front in the foreground, downdropping a fault block to form the basin. The mountain in the distance is a corner of the same fault block that has been rotated upward and uplifted, as in the rotating fault blocks depicted above.

B What Happens When Extension Accompanies Subduction?

Some regions experience crustal extension and rifting in spite of being near a convergent boundary. In these cases, the region may be fairly low in elevation, except for the large volcanoes. Rifting, if it continues, can form a small ocean basin behind the arc.

1. Extension can accompany subduction of one oceanic plate beneath another oceanic plate or beneath a continental plate. In some subduction zones, extension occurs in front of the arc, causing the crust to thin by normal faulting. Thinning of the crust helps the region stay below sea level, forming a forearc basin between the arc and the trench.

2. Extension can occur behind or near the arc, where the crust is hot and weak. This causes normal faulting and thins the crust. The region subsides to lower elevations (near or below sea level) than is typical for a continental arc.

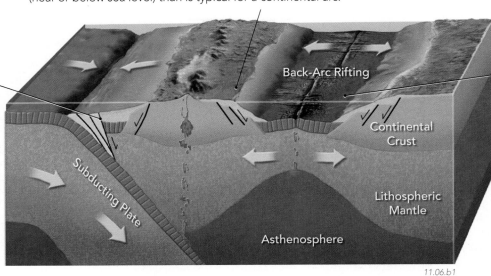

11.06.b1

3. Extension behind the arc may result in normal faults and downdropped blocks, and large amounts of extension will form a new ocean basin behind the arc. This *back-arc basin* will contain land-derived sediment along its margins and normal deep-ocean sediment in its center. Upward flow of underlying mantle continues to bring heat and material to the region, allowing the extension to continue. A well-developed back-arc basin has a generally small-scale version of a mid-ocean ridge.

C How Do We Determine the Age of a Basin?

Geologists use a variety of techniques to determine when a basin formed. They describe and measure layers in the basin, perform isotopic dating of volcanic rocks, or find key fossils. The age, thickness, and character of sediments record when and how fast a basin, like the one below, formed.

3. A unit *younger* than a basin may lie flat and may overlap the edge of the basin and its faults. It shows that the basin had stopped forming by the time the unit was deposited.

2. Units deposited *during* formation of a basin may be very thick and may contain coarse sediments that record steep slopes along the flanks of the basin.

1. Units *older* than a basin typically have the same thickness across the area because the basin did not yet exist. These older units were then tilted and faulted when the basin formed.

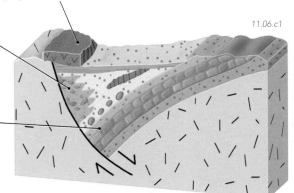

11.06.c1

4. We can calculate the rate of deposition for each unit by dividing the thickness of the unit by the time during which the unit was deposited. This plot (▼), for units in the deepest part of the basin, shows that sediment accumulated most rapidly after 15 million years ago. This indicates that the basin began forming about 15 million years ago.

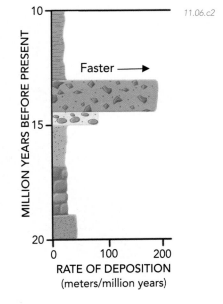

11.06.c2

MILLION YEARS BEFORE PRESENT

Faster ⟶

RATE OF DEPOSITION
(meters/million years)

Before You Leave This Page Be Able To

✓ Describe and sketch the formation of non-rotating and rotating fault blocks.

✓ Summarize where extension can occur in a plate above a subduction zone.

✓ Describe or sketch how we can determine the age of a basin.

11.6

11.7 What Are the Characteristics and History of Continental Hot Spots?

A HOT SPOT WITHIN A CONTINENTAL PLATE is marked by high elevations, abundant volcanism, and continental rifting. Hot spots can facilitate complete rifting and separation of a continent into two pieces and can help determine where the split occurs. Several continental hot spots are active today.

A What Features Are Typical of Continental Hot Spots?

Hot spots are volcanic areas interpreted to be above rising mantle plumes. Continental hot spots are associated with certain characteristics, including high elevations, volcanism, and the presence of rifts. Two examples are the Afar region of East Africa and the Yellowstone region of the western United States.

Afar Region, East Africa

Continental hot spots have high elevations largely because of heating and thinning of the lithosphere by a rising plume of hot mantle. Many geologists interpret the Afar region of eastern Africa to be located above a hot spot that is currently active.

The East African Rift is within the African plate. It may or may not evolve into a full rift that fragments the continent into two parts and that leads to seafloor spreading.

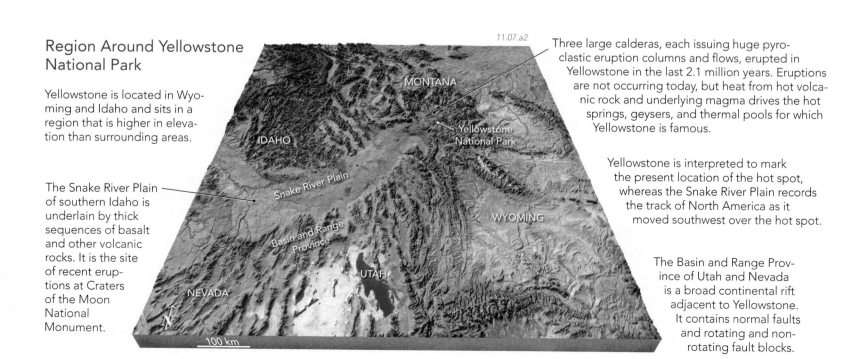

Near the hot spot, the Arabian Peninsula has pulled away from Africa along the Red Sea and the Gulf of Aden. Beneath these seas, seafloor spreading generates new oceanic crust.

The Red Sea, Gulf of Aden, and East African Rift come together in the Afar region, branching off like three spokes on a wheel. The Afar region is among the most volcanically active areas on Earth and has experienced recent volcanic eruptions. Volcanism has been so prolific here that it has created a triangular area of new land in the corner of Africa from which the Arabian Peninsula pulled away.

Region Around Yellowstone National Park

Yellowstone is located in Wyoming and Idaho and sits in a region that is higher in elevation than surrounding areas.

The Snake River Plain of southern Idaho is underlain by thick sequences of basalt and other volcanic rocks. It is the site of recent eruptions at Craters of the Moon National Monument.

Three large calderas, each issuing huge pyroclastic eruption columns and flows, erupted in Yellowstone in the last 2.1 million years. Eruptions are not occurring today, but heat from hot volcanic rock and underlying magma drives the hot springs, geysers, and thermal pools for which Yellowstone is famous.

Yellowstone is interpreted to mark the present location of the hot spot, whereas the Snake River Plain records the track of North America as it moved southwest over the hot spot.

The Basin and Range Province of Utah and Nevada is a broad continental rift adjacent to Yellowstone. It contains normal faults and rotating and non-rotating fault blocks.

B How Do Continental Hot Spots Evolve?

Many continental hot spots underwent a similar sequence of events. They started with doming and ended with the formation of a new continental margin and a new ocean formed by seafloor spreading.

1. Hot spots mark where a mostly solid, hot mass rises, probably from the lower mantle, and encounters the base of the lithosphere. The rising material melts due to decompression and also causes melting of nearby lithosphere.

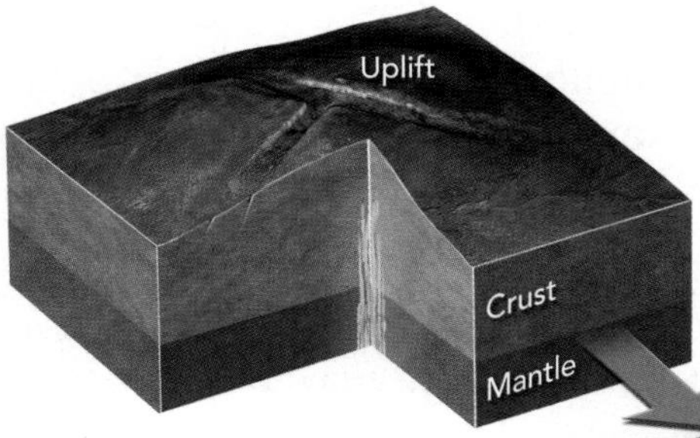

2. As the upper mantle and crust heat up, a broad, domal uplift forms on the surface. Doming is accompanied by stretching of the crust, which commonly begins to break apart along three rifts that radiate out from the hot spot.

3. Some mantle-derived magma escapes to the surface and erupts as basalts, commonly voluminous flood basalts. More felsic magma (more granitic in composition) forms where mantle-derived magma causes melting of the crust.

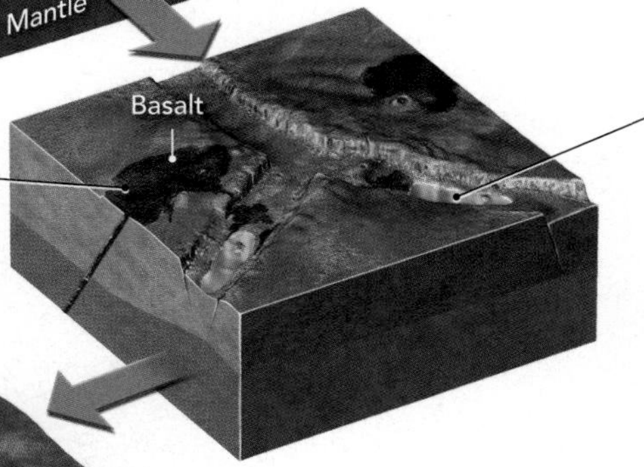

4. All three parts, or arms, of the rift are bordered by normal faults, which downdrop long fault blocks. The downdropped blocks form basins that contain lakes and are partially filled by sediment and rift-related volcanic rocks.

5. Complete rifting of the continent occurs along two arms of the rift. This results in a new continental margin and seafloor spreading in the new ocean basin. At the onset of spreading, the edge of the continent is uplifted because the lithosphere is heated and thinned due to the rifting.

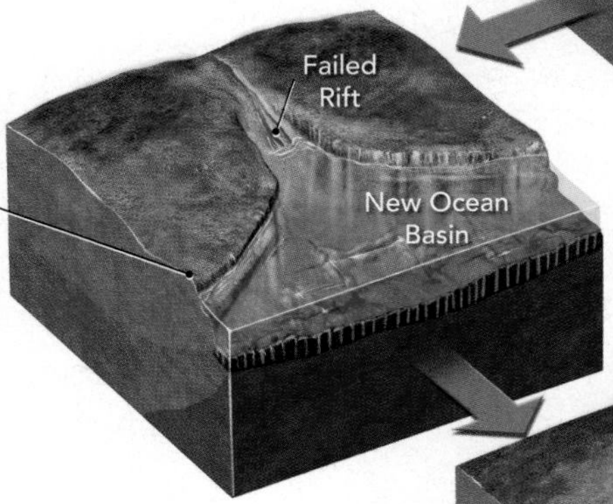

6. The third arm of the rift begins to become less active and fails to break up the continent into more pieces. This *failed rift* is lower than the surrounding continent and commonly becomes the site of major rivers.

7. As seafloor spreading continues, the generation of new oceanic lithosphere causes the mid-ocean ridge to move farther out to sea. The continental margin cools and subsides and is covered by marine sediment on the newly formed continental shelf. This continental margin is no longer a plate boundary and is now a passive margin.

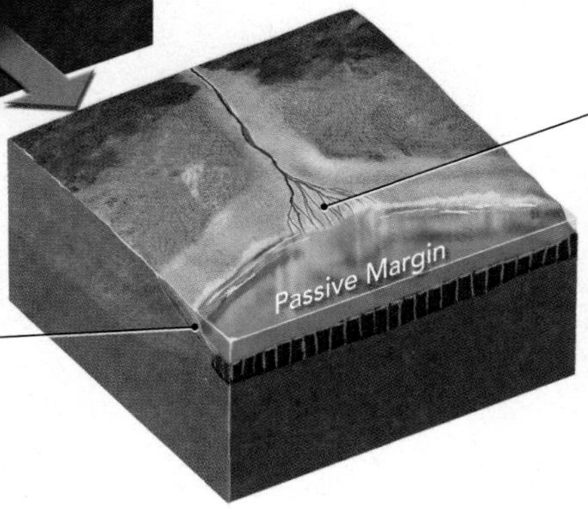

8. Sediment transported by rivers down the failed rift will form a delta at the bend in the continent. This is currently occurring along the west coast of Equatorial Africa at the large inward bend in the coast (see the figure and text below).

Hot Spots and Continental Outlines

Geologists conclude that hot spots have helped define the outlines of the continents by shaping the boundary along which continents separate from one another. The best example of this is the inward bend of the west coast of Africa. This bend occurs at the intersection of three arms of a rift, two of which led to the opening of the South Atlantic Ocean. The third *failed* *arm* cuts northeastward into Africa and is the site of several major rivers. Large eruptions of basalt (flood basalt) occurred along the rifts, and active volcanism near the failed rift may mark the location of a hot spot. This figure shows what the area may have looked like 110 million years ago, after the continents started to rift apart.

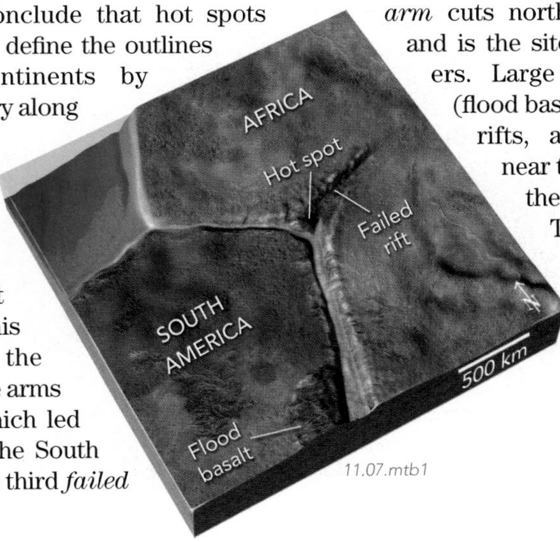

Before You Leave This Page Be Able To

✓ Summarize the features that are typical of continental hot spots, providing an example of each type of feature.

✓ Summarize or sketch how continental hot spots evolve over time.

✓ Describe or sketch how hot spots influence continental outlines, providing an example.

11.7

11.8 What Features Characterize the Interiors of Continents?

THE INTERIORS OF CONTINENTS tend to be tectonically stable, largely because they are far from plate boundaries. Sedimentary rocks formed within continental interiors contain an important record of ancient rivers, lakes, wetlands, deserts, sea level variations, and changes in global and regional climate.

A What Features Are Common in Continental Interiors?

Many continents display a similar pattern, with a central region of complexly deformed, older rocks surrounded by a relatively thin veneer of younger, nearly flat-lying sedimentary layers.

1. Many continents, including North America, have a central region called a *continental shield*. A shield consists of relatively old metamorphic and igneous rocks, commonly of Precambrian age. The crystalline (metamorphic and igneous) rocks exposed in the shield represent the kinds of rocks that underlie much of the continent, and are called the *crystalline basement*.

2. Surrounding the shield is a broad region, called the *continental platform*. It is characterized by nearly horizontal sedimentary rocks that were deposited on top of the basement. The sedimentary layers commonly contain broad basins and uplifts. Erosion across the gently dipping layers on the flanks of these structures exposes higher and lower rocks at the surface from place to place.

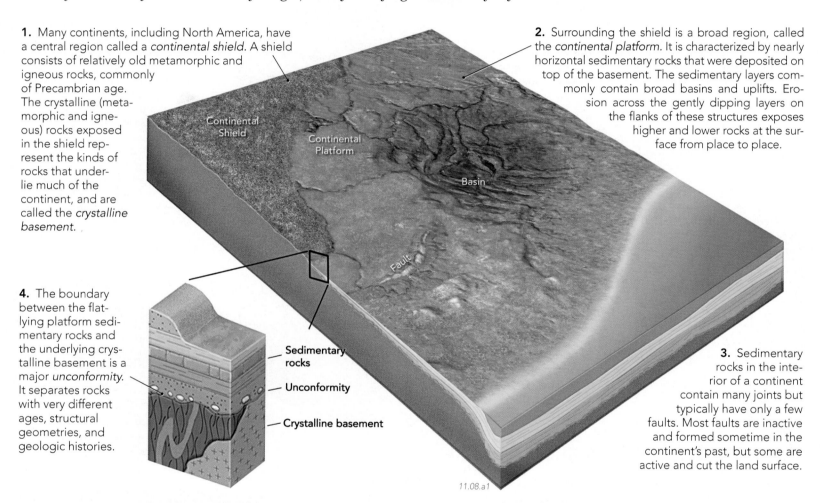

4. The boundary between the flat-lying platform sedimentary rocks and the underlying crystalline basement is a major *unconformity*. It separates rocks with very different ages, structural geometries, and geologic histories.

— Sedimentary rocks

— Unconformity

— Crystalline basement

3. Sedimentary rocks in the interior of a continent contain many joints but typically have only a few faults. Most faults are inactive and formed sometime in the continent's past, but some are active and cut the land surface.

11.08.a1

Cross Section Across Ohio

▶ 5. This geologic cross section across the state of Ohio is typical of the geology of central North America. In this area, Paleozoic sedimentary layers dip gently off the flanks of a dome, called the Findlay Arch. The section is vertically exaggerated, so true dips are less than shown here, and the thicknesses of the layers are greatly exaggerated.

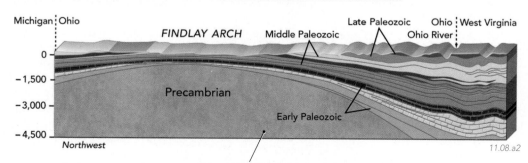

6. Igneous and metamorphic rocks of the Precambrian crystalline basement rest beneath the sedimentary layers and come within less than 800 m of the surface, but are not exposed. They have been encountered in deep drill holes.

B What Regional Effects Influence the Geologic History of Continental Interiors?

The interiors of continents are relatively stable, but they can be affected by tectonic events along distant plate boundaries. Continental interiors are strongly influenced by global environmental fluctuations, such as climate change and the rise and fall of sea level.

Tectonic activity along the edges of a continent can cause broad uplifts and basins within the continent, commonly in response to the loading (extra weight) of sediments along the margin. Additionally, thrust sheets emplaced onto the continent can create basins close to the thrusts and form uplifts farther inland as the continental plate flexes under the load.

Changes in global climate can cause sea level to rise and fall. Many continents have very low topographic relief, and their edges are barely above sea level. A sea level rise of tens of meters, therefore, can cause significant flooding, in some cases well into the interior of a continent. Such changes in sea level dominate the history of many continental interiors.

Some stresses are transmitted from plate boundaries and from distant mountain belts to the interiors of continents. If the stresses are large enough, they may cause movement on ancient faults in the crystalline basement, forming folds, faults, domes, and regional basins in the overlying sedimentary rocks.

The climate of a continent can change in response to global effects, such as global cooling, or from regional effects, including the rise of mountains along the coast. A continent also is subjected to different climates as the plate upon which it rides changes latitude (moves north or south).

11.08.b1

11.08.b2 Morgantown, WV

◄ Many continental interiors contain flat-lying layers of sedimentary rock. These rocks were originally deposited in various environments, including shallow seas, beaches, rivers, lakes, and sand dunes. The layers in the rocks are nearly flat lying today because they were deposited in nearly horizontal layers and have remained largely undeformed because they were far from plate boundaries.

11.08.b3 Pikes Peak, CO

▲ In many continental interiors, flat-lying sedimentary layers overlie Precambrian basement rocks along an unconformity. The older rocks were formed at depth, uplifted toward the surface, uncovered by erosion, and overlain by sediment.

Mountain Ranges in the Middle of Continents

Some mountain ranges occur in the middle of a continent, not along the edges. The Ural Mountains of central Russia, shown in this satellite image, are one such range, occurring far from any continental edge or plate boundary. Why is there a mountain range here?

Most mountain ranges, including the Ural Mountains, originally formed near a plate boundary along the edge of a continent. Subsequent collision between two continents can cause the continents to join, trapping the mountains within the center of a new, larger continent. In the case of the Ural Mountains, part of Europe collided with Siberia 200 to 300 million years ago.

Ural Mountains

11.08.b4   250 km

Before You Leave This Page Be Able To

☑ Summarize or sketch features that are common in continental interiors.

☑ Describe what types of regional or global effects can influence the geology of a continental interior.

☑ Describe one way that a mountain can exist in the middle of a continent.

11.8

11.9 What Are Tectonic Terranes?

EMBEDDED WITHIN CONTINENTS are pieces of crust that have a different geologic history than adjacent regions. These exotic pieces, called *tectonic terranes*, originate in a variety of tectonic settings. Many are structurally added to the edges of continents during tectonic collisions.

A How Do We Recognize a Terrane and Where Do Terranes Originate?

A tectonic terrane is defined as being bounded by faults and having rocks, structures, fossils, and other geologic aspects that are unlike those in adjacent regions.

The boundaries between a terrane and the adjacent regions are major faults or shear zones. The fault-bounded nature of a terrane, such as this volcanic terrane, means that the terrane has no continuous link with the rocks around it.

A tectonic terrane has a different sequence of rocks than adjacent regions. The terrane on the left has pillow basalt, overlain by shale, limestone, and conglomerate, but the continental rocks to the right have none of these units. Adjacent terranes usually also have different ages of rocks and different types of structures. These discrepancies imply that the two pieces of crust had different geologic histories.

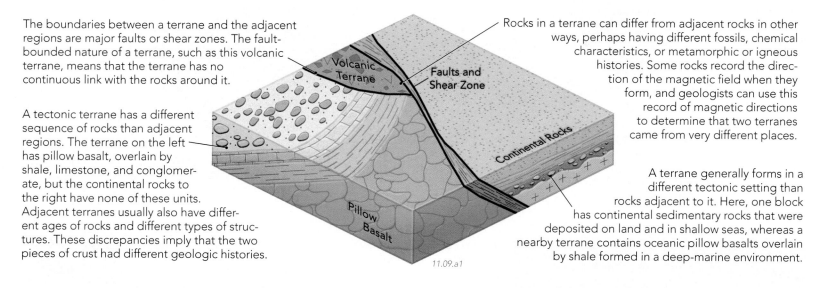

Rocks in a terrane can differ from adjacent rocks in other ways, perhaps having different fossils, chemical characteristics, or metamorphic or igneous histories. Some rocks record the direction of the magnetic field when they form, and geologists can use this record of magnetic directions to determine that two terranes came from very different places.

A terrane generally forms in a different tectonic setting than rocks adjacent to it. Here, one block has continental sedimentary rocks that were deposited on land and in shallow seas, whereas a nearby terrane contains oceanic pillow basalts overlain by shale formed in a deep-marine environment.

11.09.a1

Some Original Settings of Terranes

Some terranes contain pillow basalt, deep-sea sediment, and other attributes that indicate they originated as oceanic crust. Such terranes must be later added to the continent, or we would not see them today.

Many terranes consist of andesitic volcanic rocks and volcanic-derived sedimentary rocks that formed as island arcs. Island arcs make ideal terranes because they may move across the ocean until they collide with, and become part of, another landmass.

Some terranes have more continental characteristics, specifically thick granitic crust and continental sediment and rocks. Such terranes generally represent pieces that were sliced or rifted off another continent and then tectonically transported to their present positions.

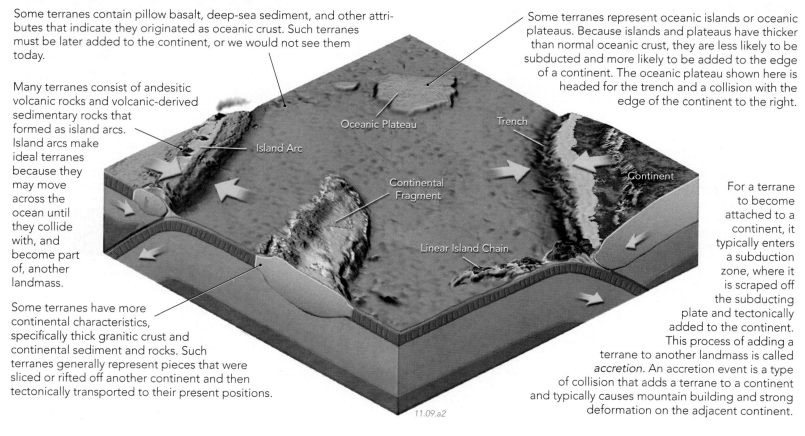

Some terranes represent oceanic islands or oceanic plateaus. Because islands and plateaus have thicker than normal oceanic crust, they are less likely to be subducted and more likely to be added to the edge of a continent. The oceanic plateau shown here is headed for the trench and a collision with the edge of the continent to the right.

For a terrane to become attached to a continent, it typically enters a subduction zone, where it is scraped off the subducting plate and tectonically added to the continent. This process of adding a terrane to another landmass is called *accretion*. An accretion event is a type of collision that adds a terrane to a continent and typically causes mountain building and strong deformation on the adjacent continent.

11.09.a2

B How Do We Infer the Origin of a Terrane and the Timing of Terrane Accretion?

Geologists investigate the origins of terranes, especially the tectonic setting in which a terrane formed and when the terrane was accreted to another landmass.

1. The origin of a terrane is revealed by the kinds of rocks it contains. For example, pillow basalts generally imply an oceanic origin. Geologists analyze the chemistry of volcanic rocks in order to compare them to rocks from modern-day mid-ocean ridges, oceanic islands, island arcs, and other volcanic environments.

2. If rocks on opposite sides of a major fault are the same age but are otherwise dissimilar, it is likely that the rocks were not close to each other when they formed. In this example, the pillow basalts on the left are the same age as the metamorphic rocks on the right, but the terranes were probably not in close proximity when the two very different types of rocks formed.

3. An intrusion that invades two terranes, or that crosscuts their boundary, indicates that the terranes were already together at the time when the intrusion invaded the crust.

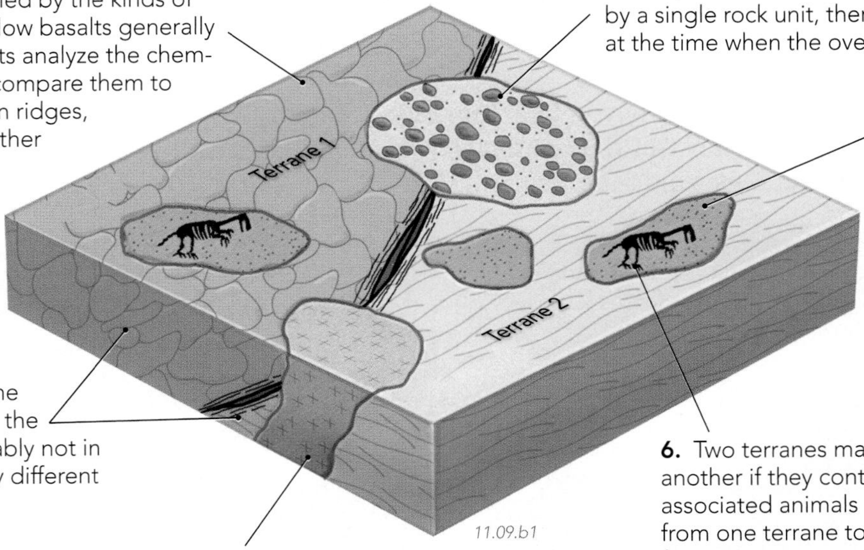

4. If two terranes and their boundary are overlain by a single rock unit, then they were already together at the time when the overlapping unit was deposited.

5. Sediment derived from one terrane can be deposited on top of adjacent rocks. The patches of brown sediment shown here contain cobbles of pillow basalt. This indicates that the two terranes were close to each other when the sediment was deposited.

6. Two terranes may have been adjacent to one another if they contain the same fossils and if the associated animals could not have swum or flown from one terrane to another. Conversely, if rocks from two terranes are the same age but have different fossils, then they probably originated in different settings and locations.

11.09.b1

Terranes of Alaska

Alaska, like most of western North America, is a mosaic of terranes. Some cover huge regions, but others are only kilometers long.

Terranes in Alaska, as simplified on the map here, are interpreted to have formed in many different tectonic settings and places. On this map, the light gray area in the northeast is part of stable North America. Blue-colored terranes represent parts of North America and its continental margin that were sliced off and transported some distance. Purple and green terranes represent slices of oceanic crust and accretionary prisms that were accreted to the continent during the Mesozoic and Cenozoic. The pink and red terranes were island arcs or continental magmatic belts. Yellowish areas depict rocks that *overlapped* the terranes after they were attached to the continent.

A famous terrane, named *Wrangellia* for the Wrangell Mountains of

southern Alaska, is red on this map. Geologists interpret it to have originated during the Late Paleozoic and Early Mesozoic (Triassic) as one or more island arcs that probably started south of the equator. It then moved northward until it collided with the West Coast, after which pieces were sliced

off by strike-slip faults and dispersed northward along the coast. Pieces of Wrangellia are scattered from western Idaho northward to Alaska, but are considered to have been part of the same terrane because they have similar ages and sequences of rocks (see below).

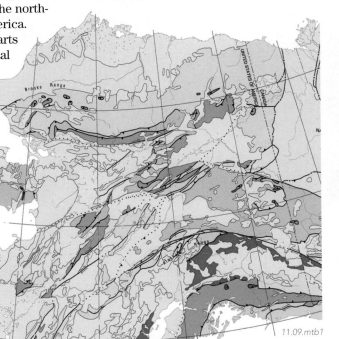

11.09.mtb1

Alaska — Western Idaho
Triassic Basalt
Permian Limestone
Late Paleozoic Arc Volcanics

11.09.mtb2

Before You Leave This Page Be Able To

✓ Summarize the characteristics used to recognize a terrane.

✓ Describe a few of the main tectonic settings in which terranes originate and how terranes are added to crust.

✓ Summarize or sketch how we determine when two terranes were apart or were brought together.

11.9

11.10 How Do Continents Form?

TECTONICS CONSTANTLY RESHAPES CONTINENTS. Pieces can be removed by rifting or added by accretion of tectonic terranes. Continents are also internally rearranged as areas are shortened during compression, stretched during extension, and shifted horizontally by strike-slip faults.

A How Old Is North America?

The crust of North America varies widely in age. The oldest parts of the Precambrian shield formed as early as 3.8 to 4.0 billion years ago, but most of the continent was added later as a series of terranes, mostly during Precambrian, Paleozoic, and Mesozoic time.

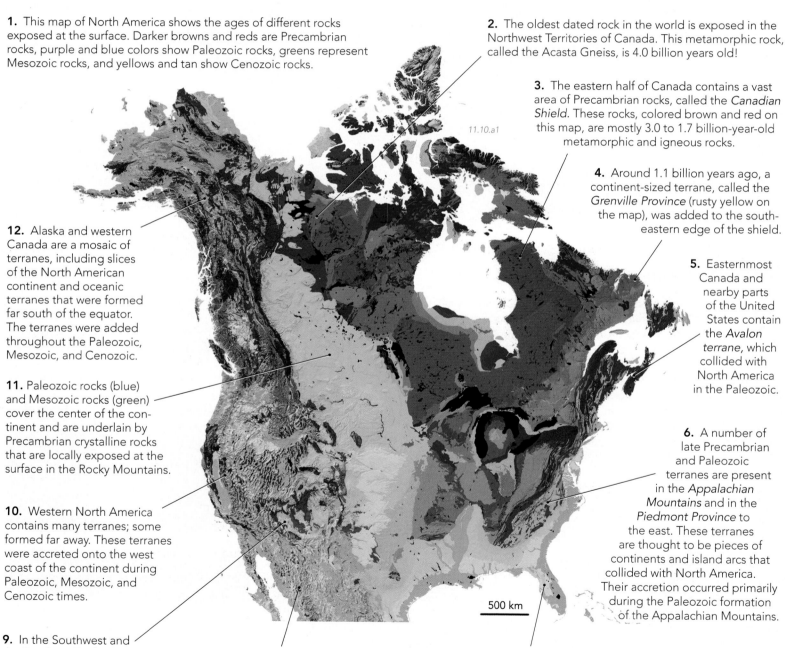

1. This map of North America shows the ages of different rocks exposed at the surface. Darker browns and reds are Precambrian rocks, purple and blue colors show Paleozoic rocks, greens represent Mesozoic rocks, and yellows and tan show Cenozoic rocks.

2. The oldest dated rock in the world is exposed in the Northwest Territories of Canada. This metamorphic rock, called the Acasta Gneiss, is 4.0 billion years old!

3. The eastern half of Canada contains a vast area of Precambrian rocks, called the *Canadian Shield*. These rocks, colored brown and red on this map, are mostly 3.0 to 1.7 billion-year-old metamorphic and igneous rocks.

4. Around 1.1 billion years ago, a continent-sized terrane, called the *Grenville Province* (rusty yellow on the map), was added to the southeastern edge of the shield.

5. Easternmost Canada and nearby parts of the United States contain the *Avalon terrane*, which collided with North America in the Paleozoic.

6. A number of late Precambrian and Paleozoic terranes are present in the *Appalachian Mountains* and in the *Piedmont Province* to the east. These terranes are thought to be pieces of continents and island arcs that collided with North America. Their accretion occurred primarily during the Paleozoic formation of the Appalachian Mountains.

7. The tan, yellow, and green areas along the southern and southeastern edge of North America represent the *Coastal Plain*. This low-lying region is covered by Late Mesozoic and Cenozoic sediments that were deposited after early Mesozoic rifting thinned the crust and blocked out this edge of North America.

8. Mexico largely consists of terranes added to North America from the Paleozoic onward. The largest terranes are Paleozoic and Mesozoic island arcs that collided with the west coast of the Mexican mainland during the Mesozoic.

9. In the Southwest and the Southern Rockies, several large Precambrian provinces were added onto the southern edge of North America between 1.9 and 1.6 billion years ago.

10. Western North America contains many terranes; some formed far away. These terranes were accreted onto the west coast of the continent during Paleozoic, Mesozoic, and Cenozoic times.

11. Paleozoic rocks (blue) and Mesozoic rocks (green) cover the center of the continent and are underlain by Precambrian crystalline rocks that are locally exposed at the surface in the Rocky Mountains.

12. Alaska and western Canada are a mosaic of terranes, including slices of the North American continent and oceanic terranes that were formed far south of the equator. The terranes were added throughout the Paleozoic, Mesozoic, and Cenozoic.

11.10.a1

500 km

B What Are the Ages of the Other Continents?

The other continents are similar to North America in that they contain Precambrian shields that are flanked by belts of successively younger rocks and deposits. The younger rocks consist of either sedimentary and volcanic material deposited over the Precambrian basement or terranes added to the edges of the shields.

On this map, Precambrian rocks are brown, with darker browns representing the oldest rocks. The other colors are blue for Paleozoic, green for Mesozoic, and yellow for Cenozoic.

Northern Europe contains a Precambrian shield (in Scandinavia), but the eastern and southern parts of the continent were added in the Paleozoic or in the Mesozoic and Cenozoic. The youngest additions are along the Mediterranean Sea and accreted during the Cenozoic collisions that formed the Alps.

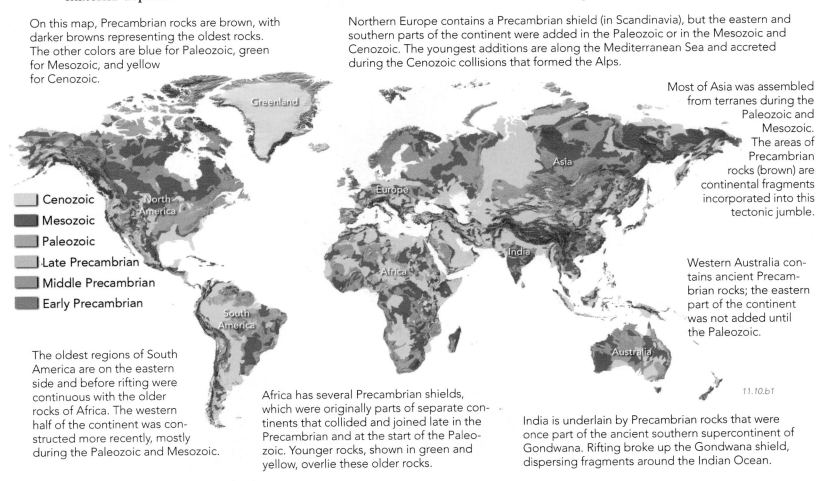

Most of Asia was assembled from terranes during the Paleozoic and Mesozoic. The areas of Precambrian rocks (brown) are continental fragments incorporated into this tectonic jumble.

Western Australia contains ancient Precambrian rocks; the eastern part of the continent was not added until the Paleozoic.

- Cenozoic
- Mesozoic
- Paleozoic
- Late Precambrian
- Middle Precambrian
- Early Precambrian

The oldest regions of South America are on the eastern side and before rifting were continuous with the older rocks of Africa. The western half of the continent was constructed more recently, mostly during the Paleozoic and Mesozoic.

Africa has several Precambrian shields, which were originally parts of separate continents that collided and joined late in the Precambrian and at the start of the Paleozoic. Younger rocks, shown in green and yellow, overlie these older rocks.

India is underlain by Precambrian rocks that were once part of the ancient southern supercontinent of Gondwana. Rifting broke up the Gondwana shield, dispersing fragments around the Indian Ocean.

11.10.b1

In Suspect Terrain

California is the area many geologists think about when they study how continents grow from the accretion of tectonic terranes. John McPhee's popular books, *In Suspect Terrain* and *Assembling California*, provide an accessible account of how terranes were recognized and how they added new real estate to North America.

This map shows the various types of terranes added to the continent. The terranes include slices of Paleozoic and Mesozoic oceanic crust and sediment, Mesozoic island arcs, and an accretionary prism. The prism consists of oceanic material scraped off oceanic plates that were subducted eastward beneath the

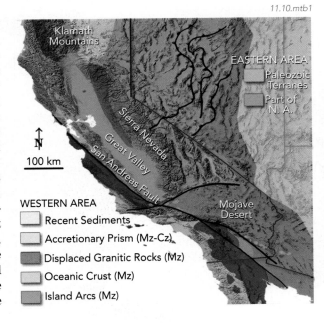

11.10.mtb1

EASTERN AREA
- Paleozoic Terranes
- Part of N. A.

WESTERN AREA
- Recent Sediments
- Accretionary Prism (Mz-Cz)
- Displaced Granitic Rocks (Mz)
- Oceanic Crust (Mz)
- Island Arcs (Mz)

continental margin. The map does not show units that formed in place, such as granites in the Sierra Nevada.

Before You Leave This Page Be Able To

☑ Identify the oldest (Precambrian) parts of North America and some areas that were added as terranes in the Paleozoic, Mesozoic, or Cenozoic.

☑ Briefly describe why different parts of a continent can be different ages.

☑ List the types of terranes added to or displaced in California.

11.10

11.11 How Did the Continents Join and Split Apart?

CONTINENTS SHIFT THEIR POSITIONS over time in response to plate tectonics. They have rifted apart and collided, only to rift apart again. Where were the continents located in the past, and which mountains resulted from their motions? The story of the movement of the continents is the same story as the origin of the modern oceans presented in chapter 10. But here we emphasize which continents were joined and how they separated. We start with 600 million years ago and work forward to the present.

600 Ma: The Supercontinent of Rodinia

The images on these pages show one interpretation of where the continents were located in the past. Geologist Ron Blakey created the artistic renderings of the continents, mountains, and oceans. For most time periods, he created two views, one focused on the western hemisphere (image on the left) and one on the eastern hemisphere (image on the right), generally with some overlap. We begin here with a single image, centered on the South Pole.

Before the Paleozoic, in the last part of the Precambrian, all of the major continents were joined. This supercontinent is called *Rodinia*. Nearly all of the other side of the globe is a huge ocean.

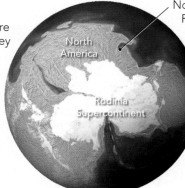

11.11.a1

North America was in the initial stages of rifting from Rodinia. This rifting outlined the western margin of North America, but geologists are not certain which continent was adjacent to North America. Options include Australia, Antarctica, and Asia.

Large parts of Rodinia were near the South Pole. There is evidence of widespread glaciation in Rodinia, but geologists are debating the extent and timing of glaciation. Geologists are also actively investigating how the continents were arranged during this time by trying to more precisely match the ages and sequences of rocks between different continents.

500 Ma: Dispersal of the Continents

At 500 Ma, in the early part of the Paleozoic, North America and Europe were separate, moderate-sized continents that had not yet joined.

Antarctica, Australia, South America, and Africa were joined in the Southern Hemisphere, together forming the southern supercontinent of *Gondwana*, which was mostly located in the Southern Hemisphere (mostly out of view on these figures). Gondwana was separated from the northern continents by some width of ocean.

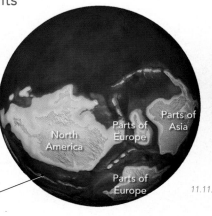

11.11.a2

With the continents still mostly clustered together, the other side of Earth was a single large ocean, much larger than the size of the present-day Pacific.

Island arcs surrounded Europe and parts of Asia. Some of the arcs would later collide with the continents, adding tectonic terranes.

370 Ma: Before Pangaea

In the middle of the Paleozoic, at 370 Ma, Europe and North America were joined but were not connected with Asia, which lay to the north. North America had collided with a microcontinent called *Avalonia*. This created mountains in what is now the northern Appalachians.

North America was approaching Africa and South America along a convergent margin. The continents were on a collision course as the intervening ocean became narrower over time.

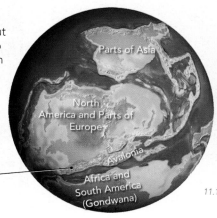

11.11.a3

Gondwana remained mostly intact, except for some slices of continental crust that probably were rifted away from the larger supercontinent. Avalonia was probably an example of one of these rifted pieces, but it had broken away sometime before 370 Ma, the time pictured here.

280 Ma: The Supercontinent of Pangaea

In the late Paleozoic, around 280 Ma, a continental collision between North America and the northern edge of Gondwana (South America and Africa) formed the Appalachian Mountains along the East Coast and the Ouachita Mountains (not labeled) in the southeastern United States.

After this collision and a series of smaller collisions, all the continents were joined in a supercontinent called *Pangaea*.

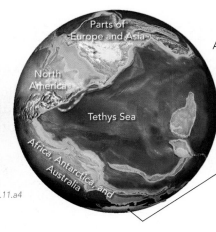

11.11.a4

A wedge-shaped ocean, the *Tethys Sea*, separated Asia from landmasses farther to the south. Southern Africa, Australia, and Antarctica were close to the South Pole and so at this time were partly covered with ice.

150 Ma: Gondwana and Laurasia

At 150 Ma, in the late Jurassic, North America had separated from Africa and South America. The Atlantic Ocean now existed, and continents on either side of the Atlantic were moving away from each other due to seafloor spreading. The left globe is rotated so that the central Atlantic Ocean is in the center of the image.

During this time, North America was still joined with Europe and Asia, forming the northern supercontinent of *Laurasia*. South America had not yet rifted away from Africa.

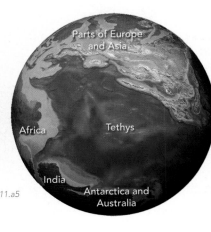

11.11.a5

Antarctica was still attached to the southern tips of Africa and South America. India was attached to the northern edge of Antarctica. These continents were soon to be rifted apart, which would mark the end of *Gondwana*.

Present

This view shows the present-day configuration of continents and oceans. Examine these globes and think about the present-day plate boundaries, envisioning which way the continents are moving relative to one another. Use the relative motions to predict where the continents were likely to have been in the past. Check your predictions by examining each set of previous globes as you step backward in time.

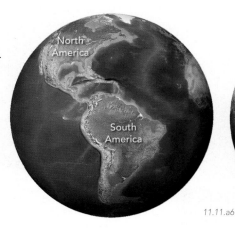

11.11.a6

When you are done working backward in time, start at 600 Ma and track the position of every continent you can and watch each ocean open. By viewing this sequence, you are observing how our present-day world came to be.

Finally, think about how the position of the continents will change if present-day plate motions continue into the future. What collisions are yet in store for North and South America?

Before You Leave This Page Be Able To

☑ Briefly summarize the general positions of the continents in the past, especially since 280 Ma.

☑ Identify times when the continents were joined in the supercontinents of Gondwana, Laurasia, Pangaea, and Rodinia.

11.11

How Did the Appalachian and Ouachita Mountains Form?

THE APPALACHIAN AND OUACHITA MOUNTAINS lie inboard of the East Coast and Gulf Coast of the United States. Unlike the settings of many mountain belts, this continental margin is not currently a plate boundary. So why are there mountains there? When and how did these ranges form? This two-page *Connections* spread and the next one present a view of how geologists interpret the geologic history of two sides of North America

North America has a central region of plains, lakes, and rivers. This central region is flanked on both sides by mountains.

The presence of mountains in western North America could be predicted because this region is near plate boundaries and associated mountain building. A mountain-building event, like that which occurred to form the Rocky Mountains, is called an *orogeny*. Geologists named each orogeny for a place where that event was first recognized or is clearly recorded.

11.12.a1

The Appalachian Mountains run parallel to the East Coast of the United States. They run from eastern Canada, through New England, and southward to Alabama.

The Ouachita Mountains are an east-west range in Arkansas and eastern Oklahoma. This mountain range and the Appalachians contain folds, thrust faults, and other geologic structures that record compression.

Paleozoic Evolution of Eastern North America

Geologist Ron Blakey has portrayed the origin of the Appalachian and Ouachita Mountains with a series of maps that illustrate mountain building (orogeny) and changes in the configuration of plate boundaries and landmasses.

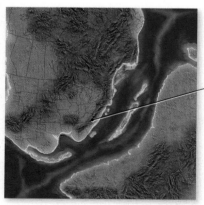

11.12.a2

◄ **1.** This map depicts the proposed tectonic setting of eastern North America at the start of the Paleozoic (around 550 Ma). By this time, the eastern edge of North America had rifted away from another landmass, perhaps the western side of South America.

11.12.a3

◄ **3.** A second mountain-building event, the *Acadian orogeny*, occurred in the middle of the Paleozoic, around 400 Ma. This event occurred when a series of landmasses, referred to as *Avalonia*, collided with the eastern coast of North America, creating the *Avalon terrane*. Further north, part of Europe collided with northern Canada.

► **2.** By 500 Ma, a subduction zone and island arc formed off the eastern coast of North America and consumed oceanic lithosphere that was attached to North America. The arc and North America moved toward each other and, eventually (at 450 Ma), the arc collided with, and was thrust over, North America, causing the *Taconic orogeny* of New England and adjacent areas.

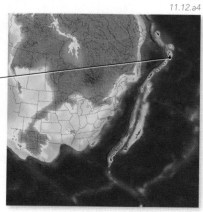

11.12.a4

► **4.** A third mountain-building event, the *Alleghenian orogeny*, occurred in the late Paleozoic (about 330 to 300 Ma). It resulted from the collision of North America with Africa and South America, which were part of the Gondwana supercontinent. This collision formed the Appalachian and Ouachita Mountains.

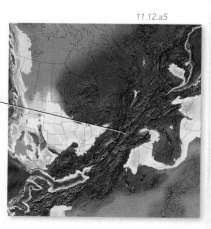

11.12.a5

Cross Sections Showing the Paleozoic Evolution of the Northern Appalachian Mountains

The evolution of the Appalachian Mountains can also be depicted with a series of cross sections, which show the inferred plate-tectonic configurations at key times. The cross sections are arranged from oldest to youngest.

1. The earliest event shown here was the initial rifting along the eastern edge of North America, probably around 600 Ma. Rifting started as a continental rift within the supercontinent of Rodinia.

2. A landmass, probably South America, was rifted away to the right. Prior to that time, North America, South America, and the other continents were joined as part of *Rodinia*.

3. By the start of the Paleozoic, around 550 Ma, the continental rift had evolved into a divergent margin. As the spreading center moved away from the land, the eastern coast of North America became a passive margin.

4. Seafloor spreading along a mid-ocean ridge had moved the two continents farther apart, creating a new ocean basin. This time corresponds to the first small map (#1) on the previous page.

5. Later, a change in plate motions formed a convergent margin where oceanic lithosphere attached to North America was subducted eastward beneath an offshore island arc, named the *Taconic Arc*.

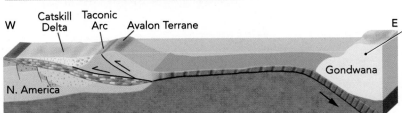

6. Over time, the arc approached the eastern coast, as shown in the second small map (#2) on the previous page. The arc eventually collided with, and was thrust over, the continental margin, causing the *Taconic orogeny* at about 450 Ma.

7. In the mid-Paleozoic (about 400 Ma), the landmass of *Avalonia* collided with and was thrust over eastern North America, during the *Acadian orogeny*. The thrust sheets warped the crust, forming a foreland basin for the *Catskill Delta*.

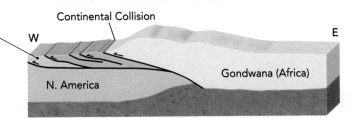

8. In the distance, the supercontinent of Gondwana approached North America by subducting oceanic lithosphere.

9. The largest mountain-building event, called the *Alleghenian orogeny*, formed the thrust faults and folds of the *Valley and Ridge Province* of the Appalachians. This event occurred when Gondwana overrode the eastern edge of North America, forming high mountains along the whole length of the Appalachian Mountains. Thrusts and folds also formed within Gondwana, but are not shown.

10. Much farther south, along the southern coast of North America, the collision between North America and the South American part of Gondwana formed the Ouachita Mountains.

Rifting Starts the Cycle Over

After the continental collision with Gondwana, the Appalachian Mountains were probably a very high mountain range, comparable to the modern-day Himalaya-Tibet region. The mountain range was progressively eroded, and rivers carried the resulting sediment away.

By 220 Ma, in the early Mesozoic, the collision zone began to be rifted apart, forming normal faults that filled with reddish sediment and basaltic lava flows, as exposed in New York, New Jersey, Pennsylvania, the Connecti-

cut River Valley, and elsewhere. Rifting led to seafloor spreading in the Jurassic (180 Ma), forming the Atlantic Ocean. With this last step, eastern and southern North America completed a cycle that started with rifting, proceeded through several collisions, and ended with another episode of rifting.

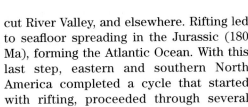

Before You Leave This Page Be Able To

☑ Summarize, using simple maps or cross sections, the main sequence of events that affected the Appalachian and Ouachita regions.

☑ Identify the causes of the Taconic, Acadian, and Alleghenian orogenies.

☑ Describe why geologists say that continents, such as North America, go through a cycle where oceans open, close, and reopen.

11.12

11.13 What Is the Geologic History of the Western United States?

THE WESTERN UNITED STATES experienced many changes in tectonic setting during its complex Precambrian to Recent history. After an early history of rifting and evolution into a passive margin, the western edge of North America was a convergent or transform margin for hundreds of millions of years.

Late Precambrian Rifting

The western edge of North America was first defined in the Late Precambrian and Early Paleozoic when the supercontinent of Rodinia began to rift apart. Geologists are as yet unsure which landmass was rifted away, but it was most likely Australia and Antarctica. This cross section shows parts of what was to become western North America (to the right) and the other unknown landmass (to the left).

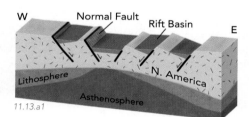
11.13.a1

Rifting thinned the crust along the western edge of North America, probably occurring mostly around 600 Ma.

Early Paleozoic Passive Margin

1. This map depicts the western United States in the early Paleozoic (about 510 Ma), some time after rifting. The white line shows the approximate location of the accompanying cross section.

2. Much of the region was a broad continental shelf, where mostly shallow-water sediment was deposited. The edge of the continent was a passive margin, not a plate boundary.

11.13.a2

3. The shoreline generally was near the boundary between thinned and rifted crust to the west and thicker, non-rifted crust to the east.

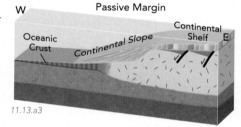
11.13.a3

4. This cross section shows the passive margin (▲), with sediment being deposited on thinned continental crust and on the adjacent oceanic crust. Deep-water sediment was deposited in more oceanic settings to the west.

Middle and Late Paleozoic Collisions

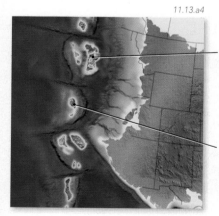
11.13.a4

1. By the middle Paleozoic (about 400 Ma), island arcs had formed within the oceanic crust to the west. Some were just offshore, probably on the same plate as the continent.

2. Other island arcs formed in open ocean and advanced toward the continent as they subducted ocean crust.

3. By the late Paleozoic (about 300 Ma), several island arcs collided with the edge of the continent.

4. In the Four Corners region, mountains called the *Ancestral Rockies* formed but were not related to west-coast events. Instead, they formed because of collisions to the east, in the Appalachian and Ouachita Mountains.

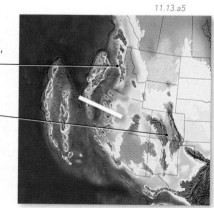
11.13.a5

The West at the End of the Paleozoic (250 Ma)

This section shows the configuration of the western United States slightly after the end of the Paleozoic. The section is located along the white line on the previous map. One or more volcanic arcs had just collided with the continent. Today, these volcanic sequences do not match rocks within the rest of North America, and they are considered tectonic terranes. ▶

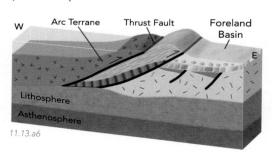
11.13.a6

Collision during the middle Paleozoic caused thrusting of deep-water sedimentary rocks (shown in light green) from the continental slope eastward over shallow-water rocks. This thrust shuffled the deeper and shallower parts of the Early Paleozoic passive margin. A foreland basin formed east of the thrust belt in eastern Nevada and adjacent areas.

Early to Late Mesozoic Convergent Margin

1. The region's setting changed in the early Mesozoic (about 210 to 150 Ma), when oceanic plates subducted beneath much of western North America.

2. A far-traveled landmass, *Wrangellia*, was offshore and approached the trench, destined for a collision with the mainland.

11.13.a7

3. Subduction caused magmatism within the overriding plate. A back-arc basin and an offshore arc ran from California northward.

4. In the Southwest, volcanoes erupted on the continent, partly within a large area of desert sand dunes (shown in orange).

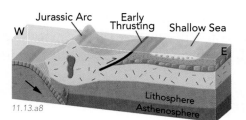

11.13.a8

5. This cross section, along the white line on the map to the left, shows back-arc thrusting and an offshore arc at 160 Ma. ◀

6. This map shows the Late Mesozoic (about 85 Ma). During this time, subduction continued beneath the western edge of North America. Inland from the coast, subduction-related magma erupted onto the surface and also solidified at depth, forming large granitic batholiths in Washington, Idaho, and the Sierra Nevada.

7. Compression associated with the convergent boundary formed thrust faults and mountains from Canada to Mexico. Inland, the continent was flexed down by the weight of the thrust sheets, allowing high sea levels at the time to flood the center of North America from Mexico to the Arctic.

11.13.a9

8. This cross section, located along the white line in the map to the left, depicts the Great Valley, Sierra Nevada batholith, and thrusting that formed a foreland basin in Utah and adjacent states. ▼

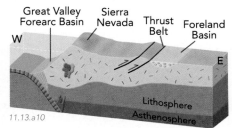

11.13.a10

Late Mesozoic and Early Cenozoic Laramide Orogeny

1. Late in the Mesozoic (about 80 Ma), the oceanic plate started to subduct at a much lower angle and to scrape along the base of the overriding lithosphere. This change in the angle of subduction moved magmatism as far east as Colorado and shut off the supply of magma to the Sierra Nevada. Large slices of crust were transported northward along the coast in the Pacific Northwest (not shown).

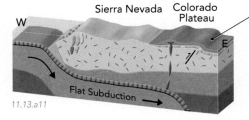

11.13.a11

2. The new subduction geometry caused compression farther into the continent, forming folds, uplifts, and basins in the Rocky Mountains and Four Corners region. This mountain-building event, called the *Laramide Orogeny*, lasted from 80 to about 40 Ma.

Middle and Late Cenozoic Extension

1. Beginning at about 40 Ma, convergence between North America and the oceanic plates slowed. This helped end the compression of the Laramide Orogeny.

2. By 15 million years ago, the Southwest had overrun a spreading center in the Pacific. This caused part of the convergent margin to be progressively converted into a transform boundary that eventually became the *San Andreas fault*.

4. Farther to the east, extension formed the *Rio Grande Rift* of New Mexico, West Texas, and south-central Colorado. Upwelling of asthenosphere beneath the rift helped cause recent uplift of the region, including parts of the Southern Rockies.

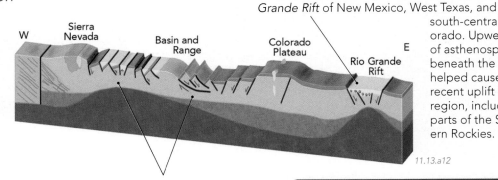

11.13.a12

11.13.a13

3. In response to changing plate settings, much of the Southwest was affected by crustal extension from about 30 Ma to the present. Extension thinned the crust and lithosphere, stretching the crust to twice its original width. It shifted the Sierra Nevada hundreds of kilometers to the west and formed the *Basin and Range Province* of the Southwest.

Before You Leave This Page Be Able To

✓ Briefly summarize or sketch the main tectonic events that affected the western United States during Paleozoic, Mesozoic, and Cenozoic times.

11.13

11.14 Where Will Mountains and Basins Form in This Region?

The figure below shows part of a continent and adjacent ocean. There are no plate boundaries now, but a subduction zone will form along the western coast of the continent, and the eastern part of the continent will be rifted away. You will use the typical patterns that form along such boundaries to predict where mountains and basins will form once the new plate boundaries are fully developed.

Goals of This Exercise:

- Observe the continent and ocean below, and read the descriptions of the types of features that will form in the future.
- Use your understanding of plate boundaries and the settings in which mountains and basins develop to predict where mountains and basins will form. Sketch your predictions on a diagram of the region.
- Predict what the regional topography will be like in different parts of the region, identifying whether an area will rise or subside, and what changes on the surface, within the crust, or in the mantle would cause this change in elevation.

1. This view shows a continent and ocean at some time, which we will call *Time 1*. The western part of the region is a typical ocean basin and has no trenches, mid-ocean ridges, or hot-spot islands.

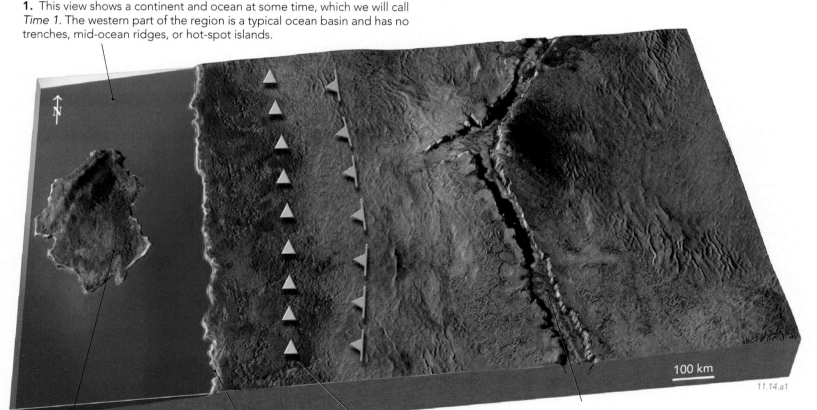

100 km

11.14.a1

2. A small piece of continent lies offshore in the middle of the ocean. When the oceanic plate begins to move, this piece of continent will be carried toward and will collide with the main continent.

3. The ocean-continent edge is currently a passive margin, not a plate boundary. It will become an ocean-continent convergent boundary, and the oceanic material will be subducted eastward below the continent.

4. Once plate convergence begins, a magmatic belt will form inland from the coast, near the position of the yellow triangles. Farther inland a thrust belt will form as shown by the blue dashed line with teeth. In the thrust belt, the western part of the continent will be thrust eastward over the central part of the continent.

5. A continental rift has formed, with three arms radiating out from a high central region, which is a hot spot marked by voluminous volcanism. This rift will split the continent into two pieces. At some later time, the piece of continent to the right will break away completely, and seafloor spreading will form a new ocean basin. Even later, at a time we will call *Time 2*, the edge of the continent will have evolved into a passive margin (not a plate boundary), and the spreading center will be out of the region.

Procedures

Use the data to complete the following steps, entering your answers in the worksheet or online.

1. Observe the regional features shown on the figure on the left page, which represents the situation at *Time 1*. Read the descriptions associated with that figure and decide what each statement implies about the *future topography* (elevations) of the area.

2. For each feature (subduction zone, thrust belt, etc.) that will form by *Time 2*, think about how that feature is typically expressed in the topography. Does it form a mountain range, a basin, or a mountain with a nearby basin?

3. On the worksheet, sketch your predictions about the area's topography for *Time 2* on the simplified figure below, which shows the same area as the figure on the previous page. The figure shows the overall shape of the continent but not the topography. Use the following letters: *O* for an oceanic trench, *A* for an accretionary prism, *M* for mountains, *V* for volcanoes in the continental magmatic belt, *B* for a basin, and *P* for a passive margin. Feel free to sketch some simple lines to portray the locations of the features. Your instructor may have you predict other features that might develop, such as a tectonic terrane **(T)** or features related to a collision **(C)**.

4. On the map below are letters A – D. A is along the coast, B is at the future position of the magmatic belt, C is within the future fold and thrust belt, and D is along the coast from which the other piece of the continent was rifted. In the worksheet, predict what will happen to the crustal thickness in each of the four locations, and identify the processes that could cause thickening or thinning of the crust or the mantle part of the lithosphere beneath each site.

Perspective of the Region in the Future (Time 2)

This boundary is now marked by a subduction zone, where the oceanic plate is being subducted beneath the continent.

By this time, subduction will result in a magmatic belt inland from the coast and a thrust belt farther into the continent.

The hot spot is no longer active, but its former position is recorded by an indentation in the continent.

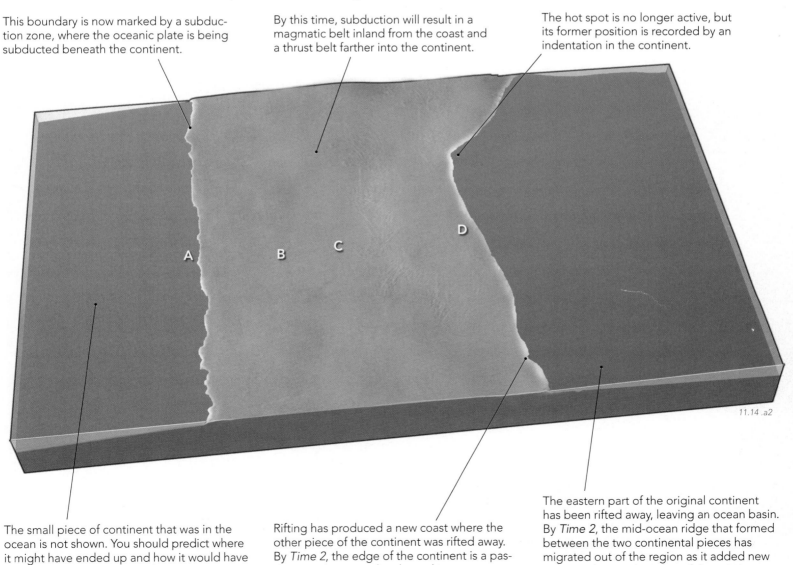

11.14 .a2

The small piece of continent that was in the ocean is not shown. You should predict where it might have ended up and how it would have changed the coastline (not shown).

Rifting has produced a new coast where the other piece of the continent was rifted away. By *Time 2*, the edge of the continent is a passive margin, not a plate boundary.

The eastern part of the original continent has been rifted away, leaving an ocean basin. By *Time 2*, the mid-ocean ridge that formed between the two continental pieces has migrated out of the region as it added new oceanic plate to the edge of the continent.

11.14

Earthquakes and Earth's Interior

EARTHQUAKES ARE AMONG EARTH'S deadliest natural phenomena. Ground shaking during an earthquake can topple buildings, liquefy normally solid ground, cause landslides and unleash massive ocean waves that wipe out coastal cities. One earthquake can kill thousands of people. What causes earthquakes, and how do we study them? In this chapter, we explore important aspects about earthquakes and Earth's interior.

The world's strongest earthquake in 40 years struck Indonesia on December 26, 2004. The magnitude 9.1 earthquake occurred west of Sumatra and was caused by movement on a fault, shown by the red line on this map. The red line shows the length of the fault that ruptured during the earthquake. The fault is part of a plate boundary where the Indian-Australian plate is subducting to the northeastward beneath the Eurasian plate. Yellow dots nearby show the locations of smaller, related earthquakes.

What causes earthquakes and where are they most likely to strike?

The earthquake occurred beneath the ocean, where the Indian-Australian plate is sliding beneath the Eurasian plate. Sudden faulting abruptly uplifted the Eurasian plate, pushing up a large region of seafloor and displacing overlying seawater. Movement on the fault started in the north and propagated south for several minutes. This caused a massive wave, called a *tsunami*, that spread across the Indian Ocean as a low wave, traveling at speeds approaching 800 km/hr (500 mi/hr)! The curved black/gray lines show a model of the wave's position by hour (numbers in circles).

What happens when an earthquake occurs under the sea, rather than on land?

The tsunami increased in height as it approached the coasts of Indonesia, Thailand, Sri Lanka, India, east Africa, and various islands. Low coastal areas were inundated by as much as 20 to 30 m of water (65 to 100 ft) in Indonesia and 12 m (40 ft) in Sri Lanka. Cities and villages were completely demolished along hundreds of kilometers of coastline, leaving more than 220,000 people dead or missing. The numbers below show casualties by location.

How does a tsunami form, how does it move through the sea, and what determines how destructive it is?

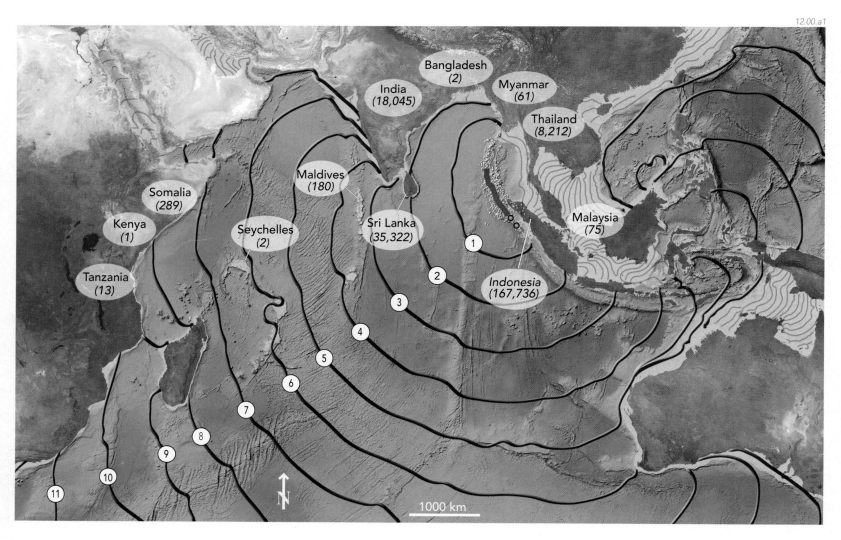

12.00.a1

TOPICS IN THIS CHAPTER

12.00.a2

The tsunami caused damage to low-lying coastlines around the Indian Ocean, reaching as far away as the eastern coast of Africa. The destructive power of the tsunami is clear from this photograph of Banda Aceh, the regional capitol of Sumatra's northernmost province. This city of 320,000 people was reduced to rubble, and nearly a third of its inhabitants were killed or are missing.

The satellite images below show Banda Aceh, before and after the tsunami. The buildings and vegetation on the "before" image (left) were stripped bare by the water's rush onto the land and the subsequent retreat back to the sea. A slightly higher area to the north (top left) was largely untouched, retaining its forest.

Which areas along a coast are most at risk for a tsunami?

12.00.a3

12.00.a4

2004 Sumatran Earthquake and Indian Ocean Tsunami

The 2004 Sumatran earthquake struck on the morning of December 26, violently shaking the region and triggering the massive Indian Ocean tsunami. It ranks as one of the two or three largest earthquakes ever recorded. The magnitude of the earthquake is variably estimated at 9.0 to 9.3, depending on how the calculations are done. Large aftershocks followed the main quake, including one with a surprisingly large magnitude of 8.7. From the seismic records of the main quake and aftershocks, it is estimated that a fault surface 1,220 km (760 mi) in length slipped by as much as 10 m during the earthquake. The earthquake lasted over 8 minutes, an unusually long duration.

The earthquake started at a depth of 30 km (19 mi) and ruptured all the way up to the seafloor. It lifted a large section of seafloor several meters in height, displacing tens of cubic kilometers of seawater, which spread out in all directions. The tsunami rose to heights of more than 30 m (100 ft) when it came ashore, and in many places it washed inland for more than a kilometer.

As a result of the earthquake, parts of the Andaman Islands northwest of Sumatra changed forever. Coral reefs, which had been undersea, were lifted above sea level. A lighthouse that was originally on land is now surrounded by seawater one meter deep. The changes to the land seem insignificant compared to the massive loss of life in this event, one of the deadliest natural disasters in world history.

12.0

12.1 What Is an Earthquake?

AN EARTHQUAKE OCCURS WHEN ENERGY stored in rocks is suddenly released. Most earthquakes are produced when stress builds up along a fault over a long time, eventually causing the fault to slip. Similar kinds of energy are released by volcanic eruptions, explosions, and even meteorite impacts.

A How Do We Describe an Earthquake?

When an earthquake occurs, it releases mechanical energy, some of which is transmitted through rocks as vibrations called *seismic waves*. These waves spread out from the site of the disturbance, travel through the interior or along the surface of Earth, and are recorded by scientific instruments at *seismic stations*.

The place where the earthquake is generated is called the *hypocenter* or *focus*. Most earthquakes occur at depths of less than 100 km (60 mi), and some occur as shallow as several kilometers. Earthquakes in subduction zones occur at shallow depths to as deep as 700 km (430 mi). The *epicenter* is the point on Earth's surface directly *above* where the earthquake occurs (directly above the hypocenter). If the seismic event happens on the surface, such as during a surface explosion, then the epicenter and hypocenter are the same.

12.01.a1

Seismic waves, once generated, spread in all directions. The curved bands show the peaks of waves radiating from the hypocenter. The intensity and duration of waves are measured by seismic stations (locations 1 and 2). Seismic stations closer to the hypocenter (station 1) will detect the waves sooner than those farther away (station 2).

B What Causes Most Earthquakes?

Most earthquakes are generated by movement along faults. When rocks on opposite sides of a fault slip past one another abruptly, the movement generates seismic waves, while materials near the fault are pushed, pulled, and sheared. Slip along any type of fault can generate an earthquake.

Normal Faults

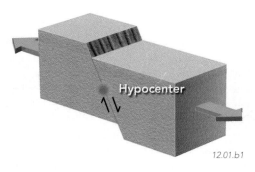

12.01.b1

In a *normal fault*, the rocks above the fault (the hanging wall) move down with respect to rocks below the fault (the footwall). The crust is stretched horizontally, so earthquakes related to normal faults are most common along divergent plate boundaries, such as oceanic spreading centers, and in continental rifts.

Reverse and Thrust Faults

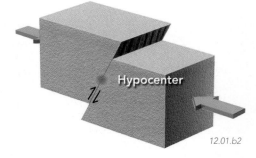

12.01.b2

Many large earthquakes are generated along *reverse faults*, especially the gently dipping variety called *thrust faults*. In thrust and reverse faults, the hanging wall moves up with respect to the footwall. Such faults are formed by compressional forces, like those associated with subduction zones and continental collisions.

Strike-Slip Faults

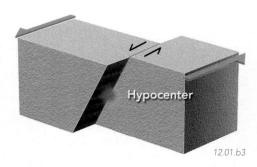

12.01.b3

In *strike-slip faults*, the two sides of the fault slip horizontally past each other. This can generate large earthquakes. Most strike-slip faults are near vertical, but some have moderate dips. The largest strike-slip faults are transform plate boundaries, like the San Andreas fault in California.

C How Do Volcanoes and Magma Cause Earthquakes?

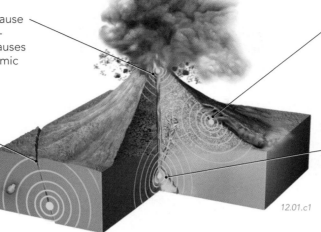

Volcanoes generate seismic waves and cause the ground to shake through several processes. An explosive volcanic eruption causes compression, transmitting energy as seismic waves (shown here with yellow lines).

Volcanoes add tremendous weight to the crust. This loading can lead to faulting and earthquakes. The fault shown here, which caused an earthquake at depth, has faulted down the volcano relative to its surroundings.

Many volcanoes have steep, unstable slopes underlain by rocks altered and weakened by hot water. The flanks of such volcanoes can fall apart catastrophically, causing landslides that shake the ground as they break away and travel down the flank of the volcano.

As magma moves beneath a volcano, it can push rocks out of the way, causing a series of small, distinctive earthquakes. In some cases, the magma causes earthquakes as it opens space by uplifting Earth's surface.

12.01.c1

D What Are Some Other Causes of Seismic Waves?

Landslides

Catastrophic landslides, whether on land or beneath water, cause ground shaking. On the Big Island of Hawaii, lava flows form new crust that can become unstable and suddenly collapse into the ocean. Seismometers at the nearby Hawaii Volcanoes National Park often record seismic waves caused by such landslides.

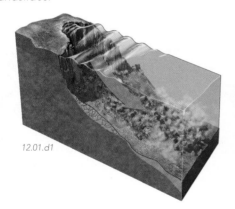

12.01.d1

Meteoroid Impacts

Ground shaking accompanies the impact of meteoroids on Earth's surface. The 100 km-wide Manicouagan ring lake in Canada is one of Earth's largest meteoroid impact sites. The impact occurred about 200 million years ago, and would have resulted in an earthquake much larger than any recorded in history.

12.01.d2

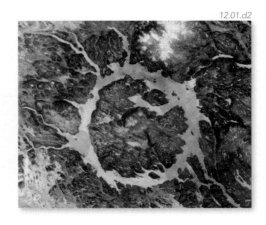

Explosions

Mine blasts and nuclear explosions compress Earth's surface, producing seismic waves measurable by distant seismic instruments. Monitoring compliance with nuclear test-ban treaties is done in part using a worldwide array of seismic instruments. These instruments recorded a nuclear bomb exploded by India in 1998. Seismic waves generated by a blast such as this are more abrupt than those caused by a natural earthquake.

12.01.d3

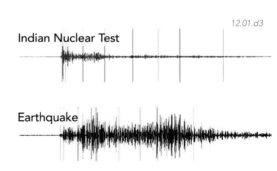

Indian Nuclear Test

Earthquake

Earthquakes Caused by Humans

Humans can cause earthquakes in several ways. Reservoirs built to store water fill rapidly and load the crust, which responds by flexing and faulting. After Lake Mead behind Hoover Dam in Nevada and Arizona was filled, hundreds of moderate earthquakes occurred under the reservoir between 1934 and 1944. Similarly, very shallow (less than 3 km deep) earthquakes occur near Monticello Reservoir in South Carolina. In China, there were fears that the filling of the Three Gorges Dam, the world's largest hydroelectric project, would trigger earthquakes in this seismically active area.

Humans have also caused earthquakes by injecting waste water underground into a deep well at the Rocky Mountain Arsenal northwest of Denver. This caused more than a thousand small earthquakes and two magnitude 5 earthquakes, which caused minor damage nearby. When the waste injection stopped and some waste was pumped back out of the ground, the number of earthquakes decreased.

Before You Leave This Page Be Able To

☑ Explain what a hypocenter and epicenter each represent.

☑ Sketch and describe the types of faults that cause earthquakes.

☑ Describe how earthquakes and seismic waves are caused by volcanoes, landslides, meteoroid impacts, and humans.

12.1

12.2 How Does Faulting Cause Earthquakes?

MOST EARTHQUAKES OCCUR because of movement along faults. Faults slip because the stress applied to them exceeds the ability of the rock to withstand the stress. Rocks respond to the stress in one of two ways—they either flex and bend, or they break and slip. Breaking and slipping causes earthquakes.

A What Processes Precede and Follow Faulting?

Before faulting, rocks change shape (i.e., they *strain*) slightly as they are squeezed, pulled, and sheared. Once stress builds up to a certain level, slippage along a fault generally happens in a sudden, discrete jump. Faulting reduces the stress on the rocks, allowing some of the strained rocks to return back to their original shapes. This type of response, where rocks return to their original shape after being strained, is called *elastic behavior*.

Pre-Slip and Elastic Strain

▶ **1.** An active strike-slip fault has modified the appearance of a streambed for hundreds of thousands of years, causing a linear trough along the fault. Some segments of the stream follow the fault. At the time shown here, the strength of the fault is greater than the tectonic forces working to slide the blocks past each other. The rocks strain and flex, but the stresses are not great enough to make the rocks break. The sizes of the yellow arrows represent the current magnitude of the stress that is building along the fault.

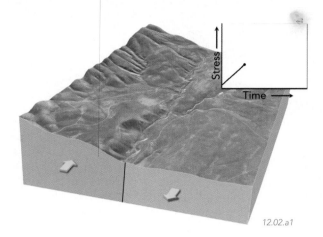

12.02.a1

◀ **2.** Stress increases with time in the rocks along the fault as depicted in this graph. In response the rocks may deform *elastically*, changing shape slightly without breaking. The fault might not be obvious at the surface because it is beneath the stream or covered with loose rocks, sand, and soil. One clue that the fault exists is its expression on the landscape, in this case a break in slope along the hillside.

Slip and Earthquake

▶ **3.** Over time, stress along the fault (represented by the yellow arrows) becomes so great that the fault slips and the rocks on opposite sides of the fault rapidly move past each other. A large earthquake occurs, generating seismic waves (not shown) that radiate outward from the fault.

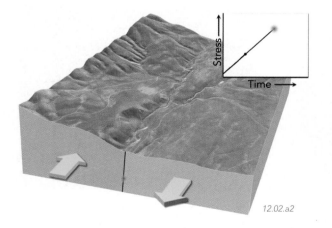

12.02.a2

◀ **4.** In the stress-versus-time graph, the point at which the earthquake occurred is shown as an orange dot. At this point the rocks were no longer strong enough and there was not sufficient friction along the fault surface to prevent movement. The built-up stress will be relieved almost instantly as the fault slips.

Post-Slip

▶ **5.** With the stress partially relieved, the rocks next to the fault relax by elastic processes and largely return to their original, unstrained shape. The movement that has occurred along the fault, however, is permanent. It is not elastic and is recorded by a new break in the topography. After the earthquake, stress again begins to slowly build up along the fault (as represented by the smaller yellow arrows). The new break along the straight part of the stream is a clue that something happened here.

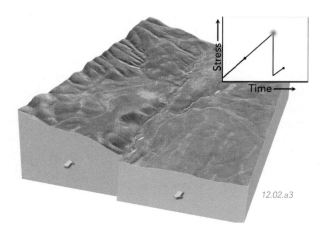

12.02.a3

◀ **6.** In the stress-versus-time graph, the release of stress after the earthquake is only temporary. The black dot at the end of the line is the current state of stress, and the cycle of stress buildup and release will continue. In this way, the rock strains elastically before the earthquake, ruptures during the earthquake, and mostly returns to its original shape afterwards. This sequence is called *stick-slip behavior* because the fault sticks (does not move) and then slips.

B How Do Earthquake Ruptures Grow?

Most earthquakes occur by slip on a preexisting fault, but the entire fault does not begin to slip at once. Instead, the earthquake rupture starts in a small area (the hypocenter) and expands over time.

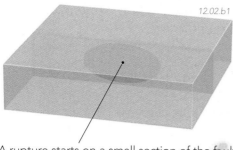

12.02.b1

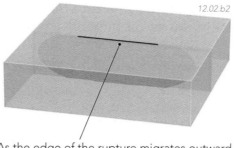

12.02.b2

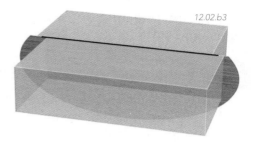

12.02.b3

A rupture starts on a small section of the fault below Earth's surface and begins to expand along the preexisting fault plane. Some rocks break adjacent to the fault, but most slip occurs on the actual fault surface, which is weaker than intact rock.

As the edge of the rupture migrates outward, it may eventually reach Earth's surface, causing a break called a *fault scarp.* Seen from above, the rupture migrates in both directions, but it may expand farther in one direction than in the other.

The rupture continues to grow along the fault plane and the fault scarp lengthens. The faulting relieves some of the stress, and rupturing will stop when the remaining stress can no longer overcome friction along the fault surface. At that point, the earthquake stops.

Earthquake Ruptures in the Field

12.02.b5 Borah Peak, ID

12.02.b6 Southern MT

12.02.b4 Mojave Desert, CA

◄ The Landers earthquake of 1992 ruptured across the Mojave Desert of California, forming a fault scarp. In this photo, the scarp is cutting through granite. The fault had mostly strike-slip movement, with some vertical movement.

► Movement along a normal fault offset the land surface during a 1983 Borah Peak earthquake, forming this fault scarp.

▲ The 1959 Hebgen Lake earthquake in southern Montana just outside Yellowstone National Park formed a several-meter-high fault scarp. The earthquake and fault scarp were generated by slip along a normal fault.

Buildup and Release of Stress

When a fault slips, it relieves some of the stress on the fault, causing the stress levels to suddenly drop. Gradually, the stress rebuilds until it exceeds the strength of the rock or the ability of friction to keep the fault from slipping. The figure below shows a conceptual model of how the amount of stress changes over time.

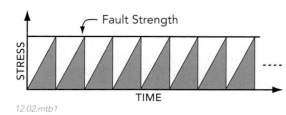

12.02.mtb1

On this plot, the magnitude of the stress imposed on the fault builds up gradually. When the *amount of stress* equals the *strength of the fault,* the fault slips, and the stress immediately decreases to the original level. In this manner, the amount of stress on a fault forms a zigzag pattern on the graph. It increases gradually (sloping line), and then decreases abruptly (vertical line) when an earthquake occurs. This process is called the *earthquake cycle,* and is one explanation for why some faults apparently produce earthquakes of a similar size. The time between repeating earthquakes is called the *recurrence interval.*

Before You Leave This Page Be Able To

☑ Describe how the buildup of stress can strain and flex rocks, leading to an earthquake.

☑ Describe or sketch how a rupture begins in a small area and grows over time and ruptures Earth's surface.

☑ Describe some characteristics of fault scarps and ruptures.

☑ Sketch and describe how stress changes through time along a fault according to the earthquake-cycle model.

12.2

12.3 Where Do Most Earthquakes Occur?

MOST EARTHQUAKES OCCUR ALONG PLATE BOUNDARIES, and maps of earthquake locations outline Earth's main tectonic plates. There are some regions, however, where *seismicity* (earthquake activity) is more widespread, reaching far away from plate boundaries and into the middle of continents. Where do earthquakes occur, and how can we explain the distribution of earthquakes across the planet?

A Where Do Earthquakes Occur in the Eastern Hemisphere?

This map and the one on the next page show the worldwide distribution of earthquake epicenters, colored according to depth. Yellow dots represent shallow earthquakes (0 to 70 km), green dots mark earthquakes with intermediate depths (70 to 300 km), and red dots indicate earthquakes deeper than 300 km. Examine these two maps and observe how earthquakes are distributed. Note how this distribution compares to other features, such as edges of continents, mid-ocean ridges, sites of subduction, and continental collisions.

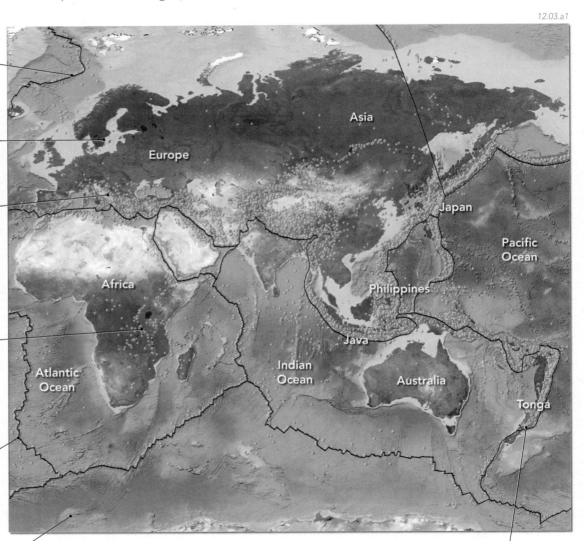

12.03.a1

1. Most earthquakes occur in narrow belts that coincide with plate boundaries. The belt of earthquakes north of Iceland marks a divergent plate boundary along a mid-ocean ridge.

2. Between the belts of earthquakes are some large regions with relatively few earthquakes, like the northern part of Europe.

3. A seismically active zone stretches along the southern part of Europe and continues eastward into Asia. This activity follows a series of mostly convergent boundaries, including continental collisions, that are occurring from the Mediterranean Sea to Tibet.

4. A diffuse zone of seismic activity cuts across eastern Africa, following the East African Rift, a region of elevated topography, active volcanism, and faulted blocks. This region is a continental rift within Africa.

5. Mid-ocean ridges, such as the one south of Africa, only have shallow earthquakes (only yellow dots on this map). In these locations, rifting and spreading of two oceanic plates produces faulting and magmatic activity, which in turn both cause earthquakes.

6. Large regions of the ocean lack significant seismicity because they are not near a plate boundary. Some seismicity in the oceans occurs away from plate boundaries and is mostly related to volcanic activity or to minor faulting that accompanies cooling and subsidence of the oceanic lithosphere.

7. Seismicity is concentrated in the western Pacific, with the main zones of seismicity being associated with oceanic trenches and volcanic islands near Tonga, Java, the Philippines, and Japan. These zones run parallel to oceanic trenches, have shallow, intermediate-depth, and deep earthquakes, and mark subduction zones. Deep and intermediate-depth earthquakes are common only along subduction zones, where there is a consistent pattern of shallow earthquakes close to the trench and progressively deeper earthquakes farther away. What do you think causes this pattern? We address this topic on the next page.

B Where Do Earthquakes Occur in the Western Hemisphere and Atlantic Ocean?

The map below shows the Western Hemisphere, including North and South America and adjacent parts of the Pacific and Atlantic Oceans. Observe the distribution of earthquakes, especially how earthquakes compare to the edges of continents, mid-ocean ridges, and sites of subduction.

1. A belt of strong seismic activity occurs along the southern part of mainland Alaska and the Aleutian Islands to the west. This belt parallels an oceanic trench and contains shallow and intermediate-depth earthquakes. It marks a subduction zone where the oceanic Pacific plate subducts beneath Alaska. This belt is a continuation of the activity in Japan and the western Pacific (i.e., the Pacific Ring of Fire).

2. Earthquakes follow the west coast of North America and extend into the mountains of the West. These earthquakes reflect diverse types of faulting (strike-slip, normal, and thrust faulting), as well as volcanism.

3. Intense seismic activity follows the western coasts of Central America, including Mexico, and South America. Included in this activity are deep and intermediate-depth earthquakes along subduction zones, especially the one beneath western South America. Shallow earthquakes are closer to the trench, and deep ones are farther away.

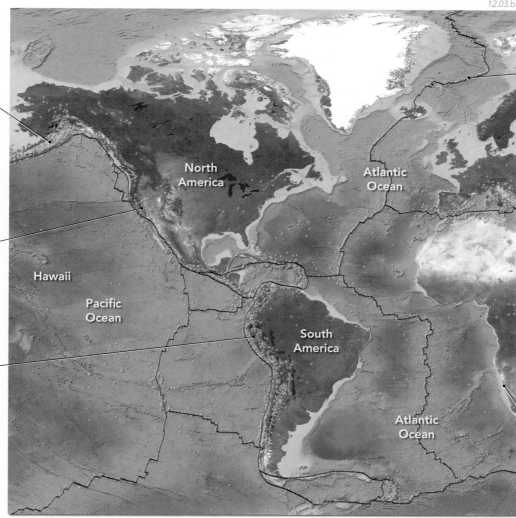

12.03.b1

6. A belt of shallow earthquakes follows the Mid-Atlantic Ridge, a mid-ocean ridge formed where the North and South American plates spread westward from the Eurasian and African plates. The pattern of earthquakes mimics the shape of the flanking continents, and the shape of the mid-ocean ridge is largely inherited from the time when these continents rifted apart in the Mesozoic.

7. Note the relative lack of seismicity along the west coast of Africa and east coasts of North and South America.

12.03.b2

4. A deep oceanic trench flanks the western coast of South America, marking a subduction zone where oceanic plates subduct beneath the western side of the continent. Observe the pattern of earthquakes for this area on the large map above before examining the figure to the right.

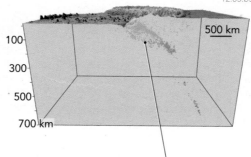

12.03.b3

5. In a side view, subduction-related earthquakes, shown as dots, are shallower to the west (near the trench) and deeper to the east, recording the descent of the oceanic plate. This pattern follows, and helps define, the position of the subducted slab, which is inclined from the shallow to the deep earthquakes.

Before You Leave This Page Be Able To

✓ Summarize some generalizations about the distribution of earthquakes, especially the relationship to plate boundaries.

✓ Sketch and explain how you could recognize a subduction zone from a map showing earthquakes colored according to depth, and how you could infer which way the subduction zone is inclined.

12.3

12.4 What Causes Earthquakes Along Plate Boundaries and Within Plates?

DIFFERENT TECTONIC SETTINGS have different types of earthquakes. Earthquakes formed along a plate boundary generally record the relative movement along this boundary (divergent, convergent, or transform) or reflect other processes, such as magmatism, associated with the boundary. Other earthquakes occur in the middle of plates, for example during continental rifting.

A How Are Earthquakes Related to Mid-Ocean Ridges?

Earthquakes are common along mid-ocean ridges, where two oceanic plates spread apart. Most of these earthquakes form at relatively shallow depths and are small or moderate in size. Some earthquakes reflect spreading of the plates, whereas others record motion as the two plates slide by one another on transform faults.

Seafloor spreading forms new oceanic lithosphere that is very hot and thin. Stress levels increase downward in Earth, but in mid-ocean ridges the rocks in the lithosphere get very hot at a shallow depth, too hot to fracture (they flow instead). As a result, earthquakes along mid-ocean ridges are relatively small and shallow, with hypocenters less than about 20 km (12 mi) deep.

Many earthquakes occur along the axis of the mid-ocean ridge, where spreading and slip along normal faults downdrop blocks along the narrow rift. Numerous small earthquakes also occur due to intrusion of magma into dikes.

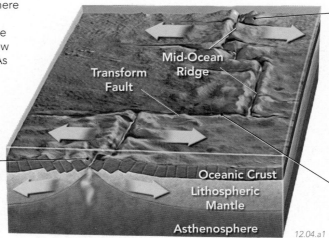

As the newly created plate moves away from the ridge, it cools, subsides, and bends. The stress caused by the bending forms steep faults, which are associated with relatively small earthquakes.

Strike-slip earthquakes occur along transform faults that link adjacent segments of the spreading center. Largely because of the typically thin lithosphere, earthquakes along these oceanic transform faults are small and shallow.

B How Are Earthquakes Related to Subduction Zones?

A subduction zone, where an oceanic plate underthrusts beneath another oceanic plate or a continental plate, undergoes compression and shearing along the plate boundary. It can produce very large earthquakes.

1. As the oceanic plate moves toward the trench, it is bent and stressed, causing earthquakes in front of the trench.

2. Larger earthquakes occur in the accretionary prism as material is scraped off the downgoing plate. Shearing within the prism causes slip and earthquakes along numerous thrust faults.

3. Large earthquakes occur along the entire contact between the subducting plate and the overriding plate. The plate boundary is a huge thrust fault called a *megathrust*. Earthquakes along megathrusts are among the most damaging and deadly of all earthquakes.

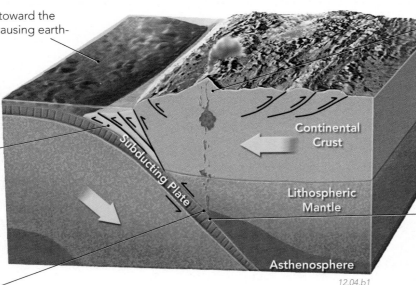

5. Earthquakes can also occur within the overriding plate due to movement of magma and from volcanic eruptions. Compressive stresses associated with plate convergence can cause thrust faulting behind the magmatic arc.

4. The downgoing oceanic plate continues to produce earthquakes from shearing along the boundary and from downward-pulling forces on the sinking slab. Subduction zones are typically the only place in the world producing deep earthquakes, as deep as 700 km (430 mi). Below 700 km, the plate is too hot to behave brittlely or to cause earthquakes.

C How Are Earthquakes Related to Continental Collisions?

During continental collisions, one continental plate underthrusts beneath another. Collisions can be extremely complex, as different parts collide at different times and rates. Collisions cause large tectonic stresses that shear and fault a broad zone within the overriding and underthrusting plates. As a result, earthquakes are widely distributed.

Large thrust faults form near the plate boundary in both the overriding and underthrusting plates, causing large but shallow earthquakes. These earthquakes can be deadly in populated areas, such as India, Nepal, Pakistan, and Iran.

Large, deadly earthquakes are produced along the plate boundary, or megathrust, between the two continental plates.

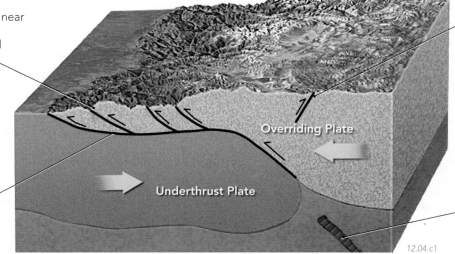

Thrust faults also form within both continental plates, causing moderately large earthquakes. The immense stresses associated with a collision can reactivate older faults within the interior of either continent, as is presently occurring in Tibet and China. Strike-slip faults and normal faults may be generated as entire regions are stressed by the collision zone or are shoved or sheared out of the way.

Any oceanic plate material that was subducted prior to the collision is detached, so subduction stops and deep earthquakes cease.

D How Are Earthquakes Generated Within Continents?

In addition to continental collisions, earthquakes occur in other tectonic settings within continents. These settings include continental rifts, continental transform faults, magmatic areas, and reactivated preexisting faults.

1. Continental rifts generally produce normal faults, whether the rift is a plate boundary or is within a continental plate. The normal faults downdrop fault blocks into the rift, causing *normal-fault earthquakes.* Such earthquakes are typically moderate in size.

2. A transform fault can cut through a continent, moving one piece of crust past another. The strike-slip motion causes earthquakes that are mostly shallower than 20 to 30 km (10 to 20 mi), but some of these strike-slip earthquakes can be quite large. The San Andreas fault of California is the best-known example of a continental transform fault, but large, destructive earthquakes also occur along continental transform faults in Turkey, Pakistan, Nicaragua (Central America), and New Zealand.

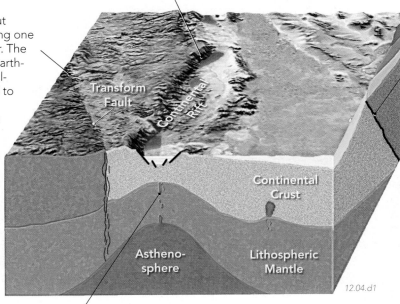

4. Preexisting faults in the crust can readjust and move as the continental plate becomes older and is subjected to new stresses, such as from distant plate boundaries. Reactivation of these structures can occur in the interior of a plate and produce large earthquakes, like those in Missouri in 1811.

3. Intrusion of magma (shown here in red) within a plate can cause small earthquakes as the magma moves and creates openings in the rock. Moving magma can produce distinctive earthquakes, which are unlike those produced by movement along faults. Heat from the magma can substantially weaken the crust, causing even more rifting and seismic activity.

Before You Leave This Page Be Able To

✓ Sketch and explain earthquakes along mid-ocean ridges, including oceanic transforms.

✓ Sketch and explain earthquakes associated with subduction zones, including earthquakes in the overriding plate.

✓ Summarize how continental collisions cause earthquakes.

✓ Describe some settings in which earthquakes can occur within a continental plate.

12.4

12.5 How Do Earthquake Waves Travel?

EARTHQUAKES GENERATE VIBRATIONS that travel through rocks as *seismic waves*. The word *seismic* comes from the Greek word for earthquake. Scientists who study earthquakes are *seismologists*. Geophysical instruments record and process information on seismic waves, and these data allow seismologists and geologists to understand where and how earthquakes occur.

A What Kinds of Seismic Waves Do Earthquakes Generate?

Earthquakes generate several different types of seismic waves. Seismologists study *body waves*, which are waves that travel inside Earth, and *surface waves*, which travel on the surface of Earth.

Shapes of Waves

Crest

Trough

12.05.a1

1. To describe seismic waves, we begin by defining waves in general. Most waves are a series of repeating crests and troughs (◄). Waves, whether moving through the ocean or through rocks, can travel, or *propagate*, for long distances. However, the material within the wave barely moves. Sound waves travel through the air and thin apartment walls, but the wall does not move much. Think of a wave as a pulse of energy moving through a nearly stationary material.

Types of Waves

2. The figure below shows an earthquake, as depicted by the red shape beneath the city shown here. Fault movement generates seismic waves. Most earthquakes occur at depth, so they first produce waves that travel through the Earth as *body waves*. The waves propagate (move outward) as shown by the circles around the earthquake.

3. When body waves reach Earth's surface, some energy is transformed into new waves that only travel on the surface (*surface waves*). It is easier to visualize processes on the surface of Earth than within it, so we begin by discussing surface waves, of which there are two kinds.

Surface Waves

4. The first type of surface wave is a *horizontal surface wave* (▼), in which material vibrates horizontally and shuffles side to side. The motion of the material is perpendicular to the direction in which the wave travels.

5. The second type of surface wave is a *vertical surface wave* (▼). It is similar to an ocean wave, in that material moves up and down in an elliptical path. These earthquake waves propagate in the direction of the large arrows, or perpendicular to the crests of the waves.

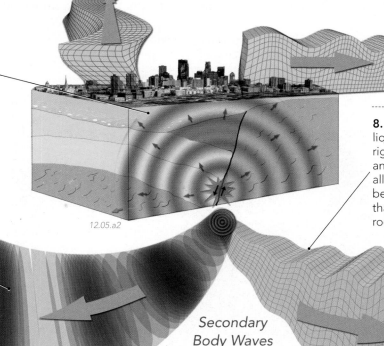

12.05.a2

8. S-waves cannot travel through liquids because liquids are not rigid (they cannot be sheared). If an area of Earth's interior does not allow S-waves to pass, then it may be molten. S-waves are slower than P-waves, travelling through rocks at about 3.6 km/s.

Primary Body Waves

6. Body waves travel through Earth and are of two main varieties. The *primary wave*, also called the *P-wave*, compresses the rock in the same direction it propagates. It is like a sound wave, which compresses the air through which it travels. P-waves can travel through solids and liquids because these materials can be compressed and then released. The P-wave is the fastest seismic wave, traveling through rocks at 6 to 14 km/s depending on the properties of the rock.

Secondary Body Waves

7. *Secondary waves*, also called *S-waves*, shear the rock side to side or up and down. This movement is perpendicular to the direction of travel. The wave to the right propagates to the right, but the material shifts up and down. It could also shift side to side, but the motion would still be perpendicular to the propagation direction of the wave.

B How Are Seismic Waves Recorded?

Sensitive digital instruments called *seismometers* are able to precisely detect a wide range of earthquakes. The recorded seismic data are uploaded to computers that process signals from hundreds of instruments registering the same earthquake. These computers calculate the location of the hypocenter and the magnitude or strength of the earthquake. From these data, we gain insight about how and where earthquakes occur.

1. A *seismometer* detects and records the ground motion during earthquakes.

2. A large mass is suspended from a wire. It resists motion during earthquakes.

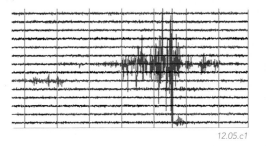
12.05.b1

3. The mass hangs from a frame that in turn is attached to the ground. When the ground shakes, the frame shakes too, but the suspended mass resists moving because of inertia. As the ground and frame move under the mass, a pen attached to the mass marks a roll of slowly rotating recording paper. As a result, the pen draws a line that records the ground movement over time.

4. This device only records ground movement parallel to the red arrows, so it only records a single direction or *single component* of motion.

5. Modern seismic detectors contain 3 seismometers oriented 90° from each other to record *three components* of motion (north-south, east-west, and up-down). From these three components, seismologists can determine the source and strength of the seismic signal.

6. Seismologists place seismometers away from human noise and vibration and bury them to reduce wind noise. Waves (in yellow) can come from any direction.

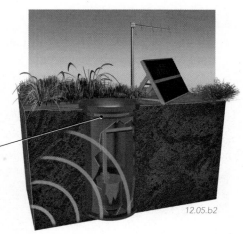
12.05.b2

C How Are Seismic Records Viewed?

1. Prior to the 1990s, seismic waveforms were mostly represented as curves on a paper *seismogram,* which is a graphic plot of the waves recorded by a seismometer. Seismologists developed this plot to better visualize the ground shaking caused by earthquakes. Today, most seismic data are recorded by digital instruments and displayed on computer screens.

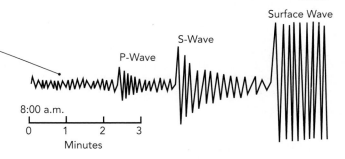
12.05.c1

2. This diagram (seismogram) shows the record of an earthquake as recorded by a seismometer. It plots vibrations versus time. On seismometers, time is marked at regular intervals so that we can determine the time of the arrival of the first P- and S-waves.

3. Background vibrations commonly look like small, somewhat random squiggles on seismograms.

4. After an earthquake, P-waves arrive first, marked by the larger squiggles. The earthquake occurred at 8:00 a.m., and the time of the P-wave's arrival was 2.5 minutes later in this example.

12.05.c2

Surface Wave

S-Wave

P-Wave

8:00 a.m.

0   1   2   3
Minutes

5. The S-wave arrives later. The delay between the P-wave and the S-wave depends primarily on how far away the earthquake occurred. The longer the distance from the earthquake, the greater the delay.

6. Surface waves arrive last and cause intense ground shaking, as recorded by the higher amplitude squiggles on the seismogram.

Amplitude and Period

Seismic waves are characterized by how much the ground moves (*wave amplitude*) and the time it takes for a complete wave to pass (*period*). Period is related to the wavelength and velocity of the wave. Both amplitude and wavelength can be measured from a seismogram. Amplitude is critical when estimating the strength and damage potential of an earthquake. The period can also be a critical component in assessing potential damage, because buildings vibrate when shaken by earthquakes. Every building has a natural period that can match, or *resonate* with, the earthquake wave. Resonance can cause intensified shaking and increased damage.

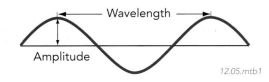
Wavelength

Amplitude
12.05.mtb1

12.5

12.6 How Do We Determine the Location and Size of an Earthquake?

EARTHQUAKES OCCUR DAILY AROUND THE WORLD, and a network of seismic instruments records these events. Using the combined seismic data from several instruments, seismologists calculate where an earthquake occurred and its magnitude (how large it was).

A How Do We Locate Earthquakes?

Seismologists maintain thousands of seismic stations that sense and record ground motions. When an earthquake occurs, parts of this network can record it. Large earthquakes generate seismic waves that can be detected around the world. Smaller earthquakes are detected only locally.

1. Seismometer Network Senses a Quake

Seismometers in the U.S. National Seismic Network (shown below) represent a fraction of all seismometers.

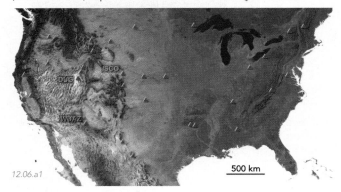

12.06.a1

On October 1, 2005, a moderate earthquake is felt in Colorado. Three seismic stations (labeled DUG, WUAZ, and ISCO) record wave arrivals and are chosen to locate the epicenter. Seismic stations are given abbreviated names that reflect their locations—for instance, ISCO for Idaho Springs, Colorado.

2. Select Earthquake Records

Records from at least three stations are compared when calculating an earthquake location. Ordinarily, records from many stations are used in an automated computer-based process.

P-waves travel faster than S-waves, and so reach a seismic station some time before the S-wave arrives. The time interval between arrival of the P-wave and S-wave is called the *P-S interval*. The farther a station is from the earthquake, the longer the P-S interval will be. Identifying the arrival of the P-wave and S-wave on these graphs is not always easy, but it can be done by seismologists or by computer.

The three seismograms show differences in the P-S interval. Based on the P-S intervals, ISCO, which has the shortest P-S interval, is the closest station to the earthquake, followed by DUG and WUAZ.

12.06.a2

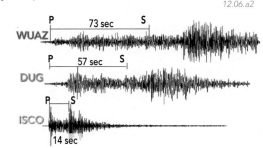

3. Estimate Station Distance from Epicenter

The P-S interval is proportional to the distance from the epicenter to the seismic station, although slightly affected by the types of materials through which the waves pass. This relationship is shown on a graph as a *time-travel curve*.

12.06.a3

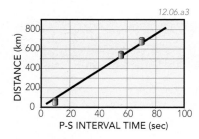

P-S intervals are measured from the seismograms shown in part 2 and then plotted on the graph. This gives the distance from each station to the earthquake's epicenter.

Station	Distance (km)
WUAZ	670
DUG	540
ISCO	65

The distance from each station to the epicenter is now known, but not the direction.

4. Triangulate the Epicenter

The distance from each station to the earthquake can be plotted as a circle on a map to find the epicenter.

12.06.a4

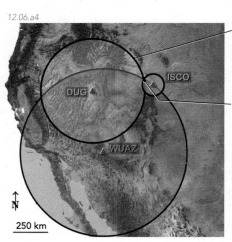

A circle is drawn around each station, with a radius equal to the distance calculated from the P-S interval.

The intersection of three (or more) circles is the epicenter of the earthquake.

We calculate the depth of the earthquake's hypocenter in a similar way, using the interval between the P-wave and another compressional wave that forms when the P-wave reflects off Earth's surface near the epicenter.

B   How Do We Measure the Size of an Earthquake?

The *magnitude* of an earthquake is a measure of the released energy and is used to compare the sizes of earthquakes. There are several ways to calculate magnitude, depending on the earthquake's depth. The most commonly used scale, called the "Richter" or "Local" magnitude (Ml) scale, is illustrated here.

Measuring Amplitude

Seismometers are calibrated so that the measurements made by two different instruments are comparable.

12.06.b1

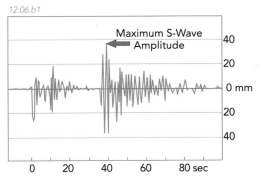

The maximum height (amplitude) of the S-wave is measured on the seismogram. It is proportional to the earthquake energy. This measure is used for shallow earthquakes.

Magnitude

The amplitude of S-waves decreases as a wave propagates. We plot the relationship between distance, earthquake magnitude, and S-wave amplitude on a graph (▶) called a *nomograph*.

For each seismic station, we draw a line connecting the distance and amplitude of the S-wave.

The earthquake's magnitude is where each line crosses the center column. These three lines for the 2005 Colorado earthquake all agree, and yield a 4.1 Ml local magnitude.

12.06.b2

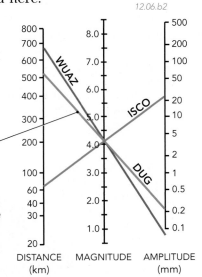

DISTANCE
(km)   MAGNITUDE   AMPLITUDE
(mm)

C   What Can the Intensity of Ground Shaking Tell Us About an Earthquake?

Some of the most damaging earthquakes occurred before seismometers were invented. For such events, we rely on reports of damage and shaking *intensity* as a way to classify the relative sizes of earthquakes.

The *Modified Mercalli Intensity Scale*, abbreviated as *MMI*, describes the effects of shaking in everyday terms. A value of "I" reflects a barely felt earthquake. A value of "XII" indicates complete destruction of buildings, with visible surface waves throwing objects into the air.

A series of very large earthquakes in 1811 and 1812 shook Missouri, Arkansas, Tennessee, and the surrounding areas. Shaking was felt over a wide region. The magnitudes on this map, numbered from III to XI, indicate what people in different areas felt and saw when the earthquake happened.

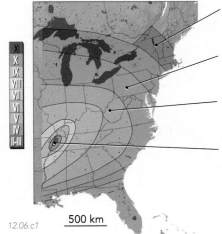

12.06.c1   500 km

III. Felt strongly by persons indoors, especially on upper floors of buildings.

V. Felt by nearly everyone; many awakened. Some dishes and windows broken. Unstable objects overturned.

VI. Felt by all, many frightened. Some heavy furniture moved. Some plaster cracks and falls. Damage slight.

X and XI. Some well-built wooden structures destroyed. Most masonry and frame structures destroyed, along with foundations. Bridges destroyed. Rails bent. Damage extensive.

Energy of Earthquakes

The Richter magnitude describes the amount of ground motion, but the scale is logarithmic. The ground motion increases by a factor of 10 from a magnitude 4 to a 5, from a 5 to a 6, and so on. The amount of energy released increases more than 30 times for each increase in magnitude, so a magnitude 8 releases more than 30 times more energy than a magnitude 7.

Another common measure of earthquake energy is *moment magnitude* or *Mw*, which is calculated from the amount of slip (displacement) on the fault and the size of fault area that slipped. Moment magnitude is useful for both large and small earthquakes. How do earthquakes compare to other energy releases with which we are familiar? An average lightning strike (*Mw* ~2) is miniscule compared to a small earthquake. However, an average hurricane is larger than the energy released by the largest historic earthquake (Chile, 1960).

Before You Leave This Page Be Able To

☑ Observe different seismic records of an earthquake and tell which one was closer to the epicenter.

☑ Describe how to use arrival times of P- and S-waves to locate an epicenter.

☑ Explain or sketch how we calculate local magnitude.

☑ Explain what a Modified Mercalli intensity rating indicates.

12.6

12.7 How Do Earthquakes Cause Damage?

MANY GEOLOGISTS SAY that "earthquakes don't kill people, buildings do." This is because most deaths from earthquakes are caused by the collapse of buildings or other structures. Destruction and collapse may result from ground shaking during an earthquake, or it can occur later due to fires, floods, or large ocean waves caused by the earthquake.

A What Destruction Can Arise from Shaking Due to Seismic Waves?

Direct damage from an earthquake results when the ground shakes because of seismic waves, especially surface waves near the epicenter of the earthquake. Damage can also be due to *secondary effects*, such as fires and flooding, that are triggered by the earthquake. In the example below most of the damage is direct damage.

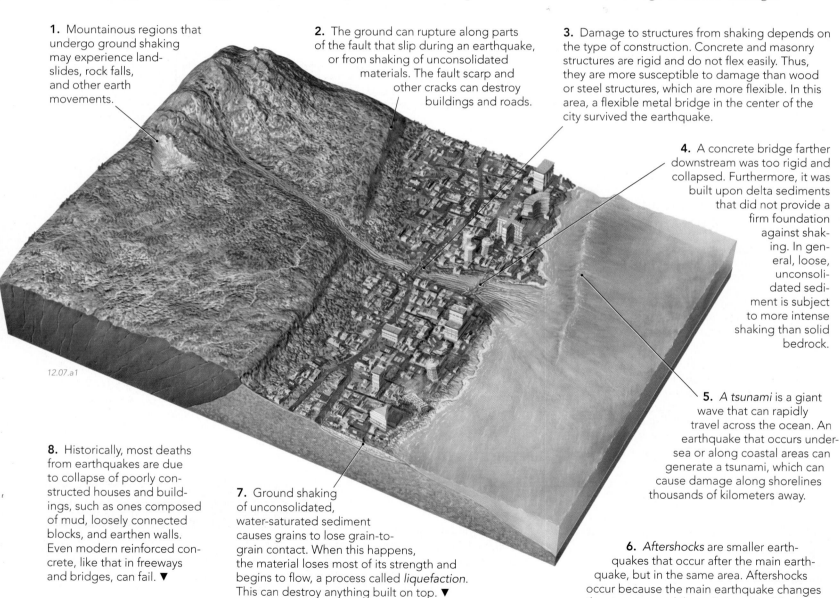

1. Mountainous regions that undergo ground shaking may experience landslides, rock falls, and other earth movements.

2. The ground can rupture along parts of the fault that slip during an earthquake, or from shaking of unconsolidated materials. The fault scarp and other cracks can destroy buildings and roads.

3. Damage to structures from shaking depends on the type of construction. Concrete and masonry structures are rigid and do not flex easily. Thus, they are more susceptible to damage than wood or steel structures, which are more flexible. In this area, a flexible metal bridge in the center of the city survived the earthquake.

4. A concrete bridge farther downstream was too rigid and collapsed. Furthermore, it was built upon delta sediments that did not provide a firm foundation against shaking. In general, loose, unconsolidated sediment is subject to more intense shaking than solid bedrock.

5. *A tsunami* is a giant wave that can rapidly travel across the ocean. An earthquake that occurs undersea or along coastal areas can generate a tsunami, which can cause damage along shorelines thousands of kilometers away.

12.07.a1

8. Historically, most deaths from earthquakes are due to collapse of poorly constructed houses and buildings, such as ones composed of mud, loosely connected blocks, and earthen walls. Even modern reinforced concrete, like that in freeways and bridges, can fail. ▼

12.07.a2 Oakland, CA

7. Ground shaking of unconsolidated, water-saturated sediment causes grains to lose grain-to-grain contact. When this happens, the material loses most of its strength and begins to flow, a process called *liquefaction*. This can destroy anything built on top. ▼

12.07.a3 San Francisco, CA

6. *Aftershocks* are smaller earthquakes that occur after the main earthquake, but in the same area. Aftershocks occur because the main earthquake changes the stress around the epicenter, and the crust adjusts to this change with more faulting. Aftershocks are very dangerous because they can collapse structures already damaged by the main shock. Aftershocks after a tsunami can cause widespread panic.

B   What Destruction Can Happen Following an Earthquake?

Some earthquake damage occurs from *secondary effects* that are triggered by the earthquake.

12.07.b1 Northridge, CA

12.07.b2 Sumatra

12.07.b3 San Fernando, CA

Fire is one of the main causes of destruction after an earthquake. Natural gas lines may rupture, causing explosions and fires. The problem is compounded if water lines also break during the earthquake, limiting the amount of water available to extinguish fires.

Earthquakes may cause both uplift and subsidence of the land surface by more than 10 m (30 ft). Subsidence, such as occurred during the 2004 Sumatra earthquake, can cause areas that had been dry land before the earthquake to become inundated by seawater, flooding buildings and trees.

Flooding may occur due to failure of dams as a result of ground rupturing, subsidence, or liquefaction. Near Los Angeles, 80,000 people were evacuated because of damage to nearby dams during the 1971 San Fernando earthquake (Mw 6.7).

C   How Can We Limit Risks from Earthquakes?

The probability that you will be affected by an earthquake depends on where you live and whether that area experiences tectonic activity. The risk of earthquake catastrophe depends on the number of people living in the region, how well the buildings are constructed, and individual and civic preparedness.

▼ **1.** Earthquake hazard maps show zones of potential earthquake damage. Near Salt Lake City, Utah, the risk is greatest (reds) near active normal faults along the Wasatch Front, the mountain front east of the city. Living away from the fault is less risky.

12.07.c2 Salt Lake City, UT

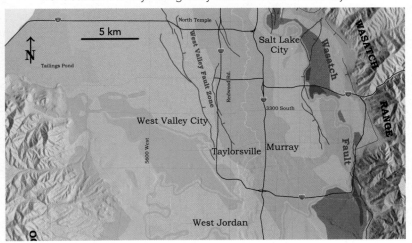
12.07.c1

2. Some utility companies and hospitals have computerized warning systems that are notified of impending earthquakes by seismic equipment. The system will automatically shut down gas systems (to avoid fire) and turn on backup generators to prevent loss of electrical power.

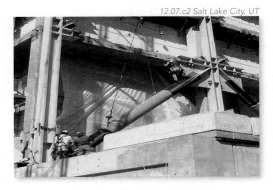

3. Earthquakes have different periods, durations, and vertical and horizontal ground motion. This makes it difficult to design earthquake-proof buildings. Some rest on sturdy wheels or have shock absorbers (▲) that allow the building to shake less than the underlying ground.

What to Do and Not Do During an Earthquake

There are actions you can take during an earthquake to reduce your chances of being hurt. If an earthquake strikes, you can seek cover under a heavy desk or table, and protect your head. You can also stand under door frames or next to inner walls, as these are the least likely to collapse. If possible, stand clear of buildings, especially those made of bricks and masonry.

During the shaking, stay away from glass bricks and heavy objects that could fall. Always keep a battery-operated flashlight handy. Avoid using candles, matches, or lighters, since there may be gas leaks. Earthquakes may interrupt electrical and water service. Keeping 72 hours' worth of food, water, and other supplies in a backpack is a prudent plan for any natural disaster.

Before You Leave This Page Be Able To

✓ Describe how earthquakes can cause destruction, both during and after the main earthquake.

✓ Describe some ways to limit our risk from earthquakes.

✓ Discuss ways to reduce personal injury during an earthquake.

12.7

12.8 What Were Some Major North American Earthquakes?

SOME VERY LARGE AND DAMAGING EARTHQUAKES have struck North America in the last several centuries. Here, we discuss seven important earthquakes chosen not because they are all the largest, but because they illustrate a range of processes, damage, and locations.

This map of the conterminous United States has yellow dots showing the locations of earthquakes with a magnitude greater than 4 that occurred during the last 15 years. The red lines on the map are faults that are interpreted to have slipped during the last 2 million years. Compare the distribution of earthquakes and these relatively young and active faults. Most active faults are in the western states, and most large earthquakes are in these same areas. Earthquakes have occurred elsewhere in the country, but most of these were too small to break the surface and form a fault scarp.

12.08.a2

Alaska, 1964

◄ A magnitude (Mw) 9.2 earthquake, one of the two or three largest earthquakes ever recorded, struck southern Alaska in 1964. It killed 128 people, triggered landslides, and collapsed parts of downtown Anchorage and nearby neighborhoods. This event was caused by thrust faults associated with the Aleutian Islands subduction zone. Most deaths and much damage were from a tsunami generated when a huge area of the seafloor was uplifted. This earthquake, like Alaska, is not shown on the map.

12.08.a1

12.08.a3

San Francisco, 1906

A huge earthquake occurred when 470 km (290 mi) of the San Andreas fault ruptured near San Francisco. The earthquake was likely a magnitude (Mw) 7.8 although not directly measured on seismometers. The earthquake ruptured the surface, leaving behind a series of cracks and open fissures. Within San Francisco, ground shaking destroyed most of the brick and mortar buildings. More than 3,000 people were killed and much of the city was devastated by fires that broke out after the earthquake.

Mexico City, 1985

A magnitude (Mw) 8.0 earthquake occurred at a subduction zone along the southwestern coast of Mexico, well west of Mexico City (not shown on this map). It damaged or destroyed many buildings in Mexico City and killed at least 9,500 people. Destruction was so extensive partly because Mexico City is built on lake sediments deposited in a bowl-shaped basin. This geologic setting amplified the seismic waves and caused intensified and highly destructive ground shaking. Surface waves, which caused the most damage, traveled 200 km (120 mi) from their source.

Northridge, Los Angeles Area, 1994

This magnitude (Mw) 6.7 earthquake was generated by a thrust fault northwest of Los Angeles. The earthquake killed 60 people and caused $20 billion in damage. A section of freeway buckled, crushing the steel-reinforced concrete slabs. The thrust is not exposed on the surface, but when it ruptured it lifted up a large section of land. Geologists are concerned about a similar fault causing a similar earthquake right below downtown Los Angeles.

12.08.a4

12.08.a5

Hebgen Lake, Yellowstone Area, 1959

◄ This magnitude (Mw) 7.3 event was generated by slip along a normal fault northwest of Yellowstone National Park. Ground shaking set loose the massive Madison Canyon slide, which buried 28 campers and formed a new lake, aptly named *Earthquake Lake*.

12.08.a6

New Madrid, 1811–1812

New Madrid, Missouri experienced a series of large (Mw 7.8-8.1) earthquakes generated over an ancient fault zone in the crust. The 1811–1812 earthquake death toll was relatively low because of the sparse population at the time. The New Madrid zone has a high earthquake risk and, as shown on the earthquake-hazard map below, is one of two areas in the eastern United States that are predicted to experience strong earthquakes in the future. Memphis lies in this zone, yet most of its buildings are not constructed to survive large earthquakes.

250 km

New Madrid

Charleston

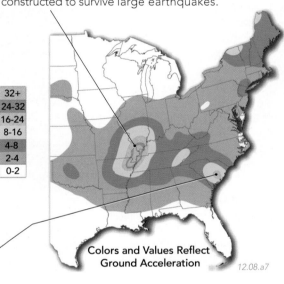

| 32+ |
| 24-32 |
| 16-24 |
| 8-16 |
| 4-8 |
| 2-4 |
| 0-2 |

Colors and Values Reflect Ground Acceleration

12.08.a7

Charleston, 1886

This earthquake occurred at the other high-risk area along the East Coast, near Charleston, South Carolina. It had an estimated magnitude of Mw 7.3, the largest ever recorded in the southeastern United States. Buildings incurred some damage (►), and 60 people died. The tectonic cause for this earthquake is still debated by geologists.

12.08.a8

Earthquakes in the Interiors of Continents

Why do large earthquakes like the ones at New Madrid, Missouri, occur in the middle of continents? Although the interior of North America is not near a plate boundary, the region is subjected to stress generated at far-off plate boundaries. In this case, the stresses are probably generated by a plate-driving force, known as *ridge push*, that originates along the Mid-Atlantic Ridge. These stresses can reactivate ancient faults that lie buried beneath the cover of sediment. In the case of New Madrid, seismic and other geophysical evidence suggest that the area is underlain by an ancient rift basin that formed about 750 million years ago during the breakup of the supercontinent of Rodinia. Modern-day stress related to the current plate configuration is interacting with the ancient faults, occasionally causing them to slip and trigger earthquakes.

Before You Leave This Page Be Able To

☑ Describe some large North American earthquakes and how they were generated.

☑ Summarize the various ways these earthquakes caused damage.

☑ Describe evidence that the eastern United States has earthquake risks.

12.8

12.9 What Were Some Major World Earthquakes?

THE WORLD HAS ENDURED a number of large and tragic earthquakes. These earthquakes have struck a collection of geographically and culturally diverse places, causing many deaths and extensive damage. Most large earthquakes occur along or near plate boundaries, especially along subduction zones.

Nicaragua, 1972

On December 23, 1972, a magnitude (Mw) 6.2 earthquake killed about 6,000 people in central America. In the capital city of Managua, wood and adobe structures were leveled and fractures opened in the street (◄). The earthquake was caused by strike-slip along a boundary of the Caribbean plate.

12.09.a2

Chile, 1960

▼ This huge, magnitude (Mw) 9.5 earthquake occurred offshore along a *megathrust* and triggered a destructive, Pacific-wide tsunami. At least 5,700 people died and $550 million of damage was done to infrastructure and buildings, such as in this city in Chile.

12.09.a3

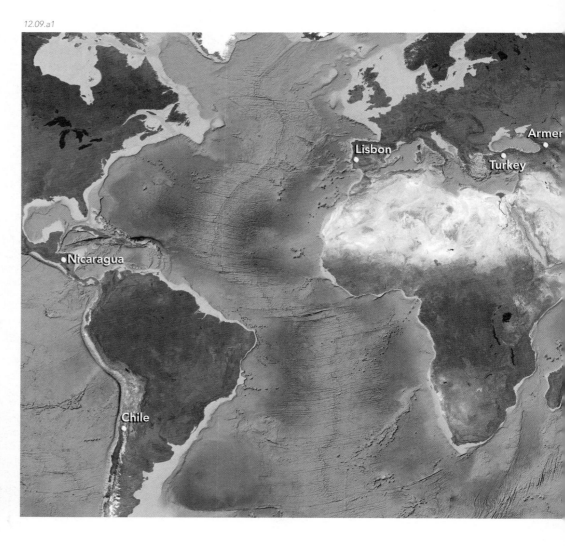

12.09.a1

Lisbon, 1755

12.09.a4

On November 1 (All Saints Day) in 1755, a large earthquake, estimated at magnitude (Mw) 8.7, shook Lisbon, Portugal. The earthquake demolished the city and triggered tsunamis that sank ships in Lisbon's famous harbor. Photography was not yet invented, but the destruction was portrayed by artists (◄). The event caused an upheaval in religious and scientific thought, as people began to think that such catastrophes must be due to natural causes, since this one struck on such a holy day. An estimated 70,000 people died.

Turkey, 1999

In 1999, a large quake (Mw 7.6) generated along a continental transform fault killed more than 17,000 people and severely impacted the economy. The earthquake destroyed many buildings, including these multi-story apartment complexes. ▼

12.09.a5

Armenia, 1988

◄ Old masonry structures were shaken apart in Leninakan, Armenia, in December 1988. This earthquake killed 25,000 people. It was caused by a continental transform fault related to lateral movement of pieces of southwestern Asia.

Kobe, Japan, 1995

► A magnitude (Mw) 6.9 thrust earthquake, also called the Great Hanshin earthquake, was the most damaging to strike Japan since the Great Kanto earthquake of 1923. The Kobe earthquake killed 5,500 and left 300,000 homeless. Damage was due largely to ground shaking and liquefaction.

12.09.a7

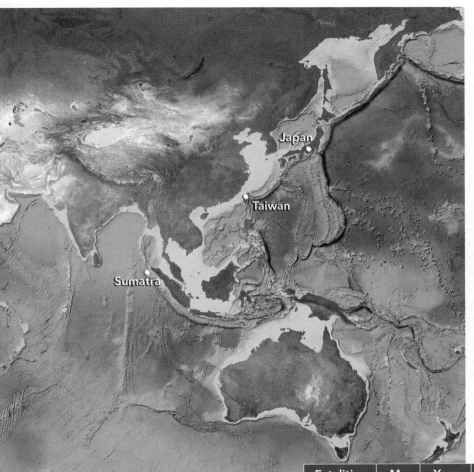

Chi Chi, Taiwan, 1999

This megathrust rupture generated a magnitude (Mw) 7.6 earthquake that was felt across Taiwan, killing 2,400 and displacing 600,000 people. The ShihKang Dam, 50 km (30 mi) from the epicenter, was breached (▼), causing flooding and shutting off the local water supply.

12.09.a8

Sumatra, 2004

This magnitude (Mw) 9.1 earthquake was along a subduction zone (megathrust) west of the island of Sumatra. It offset the seafloor and unleashed a deadly tsunami that struck coasts around much of the Indian Ocean. The earthquake left more than 220,000 people dead or missing, mostly in Indonesia. It is discussed in more detail at the start of this chapter.

Deadly Earthquakes

Earthquakes kill about 10,000 people per year on average. Most earthquake-related deaths are due to collapse of poorly built structures in cities and villages. Earthquake-generated tsunamis account for a large part of the destruction. The table to the right shows some deadly earthquake events. The highest death tolls are due to a deadly combination of high population densities, substandard construction practices, and being situated along subduction zones or other high-risk areas.

Fatalities	Mw	Year	Location
830,000	8	1556	Shaanxi, China
11,000	6.9	1857	Naples, Italy
70,000	7.2	1908	Messina, Italy
200,000	7.8	1920	Ningxia, China
143,000	7.9	1923	Kanto, Japan
200,000	7.6	1927	Tsinghai, China
32,700	7.8	1939	Erzincan, Turkey
66,000	7.9	1970	Colombia
23,000	7.5	1976	Guatemala
242,000	7.5	1976	Tangshan, China
31,000	6.6	2003	Bam, Iran
88,000	7.9	2008	Sichuan, China

Before You Leave This Page Be Able To

☑ Briefly describe some of the world's most significant earthquakes and the tectonic settings in which these deadly earthquakes occurred.

☑ Summarize some ways that these earthquakes caused deaths.

12.10 How Does a Tsunami Form and Cause Destruction?

AN EARTHQUAKE BENEATH THE OCEAN can cause a large wave called a *tsunami*, which can wreak havoc on coastal communities. Most of Earth is covered by oceans, so many earthquakes, landslides, and volcanic eruptions occur beneath the sea. Each of these events can generate a deadly tsunami.

A How Are Tsunamis Generated?

Tsunamis are waves generated by a disturbance in the sea or a lake. They are generated by abrupt changes in water level in one area relative to another, which can occur when a landslide moves into the water or when the ocean floor is unevenly uplifted or downdropped during an earthquake. Unlike typical ocean waves, which affect only the upper part of the sea, a tsunami can affect an entire body of water from top to bottom.

12.10.a1

◀ Subduction zone megathrusts can lock for long periods of time, causing the seafloor above the overriding plate to bulge, strain, and flex up or down as it accommodates the forces of convergence. This upward and downward flexing is typically most prominent near the trench.

12.10.a2

◀ When the megathrust finally ruptures in an earthquake (along the red asterisks), the bulging plate changes shape catastrophically. The water above the plate responds by lifting up from the ocean bottom toward the surface, forming a ridge of higher water. Formation of this large wave and accompanying troughs in the water can cause the ocean to retreat from the shoreline, which was observed in the 2004 Sumatra earthquake-caused tsunami and in other deadly tsunamis. A sudden retreat of the ocean along a coastline is a warning sign that a tsunami may be coming.

12.10.a3

◀ A tsunami, or a series of tsunamis, radiates away from the disturbance, traveling at speeds between 600 and 800 km/hr (370–500 mi/hr). In deep water, the wave energy is distributed over the entire water depth, forming a wave only a meter or so high but more than 700 km (435 mi) across (in wavelength). If you were in the open ocean, you probably would not notice its passing. As the wave approaches the shore, its energy concentrates in shallower and shallower water. The velocity of the front of the wave decreases to 30 to 40 km/hour (20 to 25 mi/hour), causing the following water to pile up in a higher wave. Near shore, the tsunami becomes a massive, thick wave, like the front wall of a plateau of water. It may be a series of such waves.

Tsunamis Triggered by Landslides

A large mass of rock entering the water can catastrophically displace the water, generating a tsunami that radiates outward. This has occurred repeatedly off the west side of Hawaii, where huge landslide-debris deposits (shown in green below) sit on the ocean floor.

The tsunami generated by one of these slides carried rocks and coral 6 km (3.7 mi) inland. The volume of water displaced during these events probably produced a tsunami that struck coastlines around the Pacific about 120,000 years ago.

25 km

12.10.a4

Tsunamis Caused by Eruptions

The 1883 eruption of Krakatau in Indonesia, and the collapse of its immense caldera, generated a series of huge tsunamis that killed 36,000 people. A single catastrophic volcanic explosion produced the loudest sound ever heard, and most of Krakatau Island was demolished. The tsunami was as high as 40 m (more than 130 ft), and some effects of the tsunami were recorded 7,000 km away! The painting below is a dramatic representation.

12.10.a5

B What Kind of Destruction Can a Tsunami Cause?

Tsunamis cause death and destruction along coastlines where human populations are concentrated. On May 22, 1960, the largest earthquake ever recorded on a seismometer (Mw 9.5) occurred in the subduction zone (megathrust) offshore of southern Chile. The tsunamis that followed flattened coastal settlements in Chile and traveled across the Pacific to devastate coastlines in Hawaii and Japan.

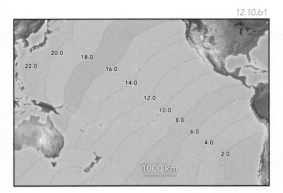
12.10.b1

Chile, May 22, 1960

◀ During this earthquake, tsunamis were generated parallel to the coast. One headed in toward the shoreline, quickly striking Chile and Peru. Another set of tsunamis swept out across the Pacific Ocean at 670 km (420 mi) per hour. Each stripe equals two hours of travel time.

▶ In Chile, the first tsunami struck 15 minutes after the earthquake. On Isla Chiloe, a 10-meter-tall wave swept over towns. The waves killed at least 2,000 people along the Peru-Chilean coast.

12.10.b2

Hawaii, May 23, 1960

About 15 hours after the earthquake in Chile, the tsunami related to the earthquake hit Hilo and other parts of Hawaii (▼). A wave 11 m (36 ft) high killed 61 people, damaged buildings, and caused $23 million in damage. Seven hours later, the tsunami killed 140 people in Japan.

Hokkaido, Japan 1993

In 1993, a magnitude 7.8 earthquake occurred off the west coast of Hokkaido. Within five minutes a tsunami struck the coastline. The tsunami killed at least 100 people and caused $600 million in property loss. It swept these boats inland across a concrete barrier built along the shoreline. ▼

Papua New Guinea, 1998

In 1998, a magnitude 7.1 earthquake and associated underwater landslides generated three tsunami waves that destroyed villages along the country's north coast, killing 2,200 people. A 10-meter-high wave destroyed a row of populated houses along the coast shown here. ▼

12.10.b3

12.10.b4

12.10.b5

Tsunami Warning System

In an international effort to save lives, the United States National Oceanic and Atmospheric Administration (NOAA) maintains two *tsunami warning centers* for the Pacific Ocean. Twenty-six nations participate in this effort. Informed by worldwide seismic networks, these centers broadcast warnings based on an earthquake's potential for generating a tsunami. After the huge loss of life from the 2004 Sumatran tsunami, the United Nations implemented a warning system in the Indian Ocean. Scien-

tists deployed warning buoys, like the one shown below, which can relay tsunami data by satellite. These buoys relay small changes in sea level detected by ocean bottom sensors as a tsunami passes overhead

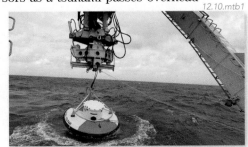
12.10.mtb1

Before You Leave This Page Be Able To

☑ Describe the different mechanisms by which tsunamis are generated.

☑ Summarize the kinds of damage tsunamis have caused.

☑ Briefly describe how tsunamis are monitored to provide an early-warning system.

12.10

12.11 How Do We Study Earthquakes in the Field?

GEOLOGISTS USE A VARIETY of tools and techniques to study evidence left behind by recent and ancient earthquakes. They examine and measure faults in natural exposures and in trenches dug across faults. Satellites and other tools allow faults to be studied in new and exciting ways.

A How Do We Study Recent Earthquakes in the Field?

Where a fault is visible at Earth's surface, it can be scrutinized in order to understand how it moves during an earthquake. Geologists investigate numerous features along a fault, some of which are shown below.

12.11.a2 Parkfield, CA

◀ **1.** When a fault ruptures the surface, geologists carefully measure its location, dimensions, and orientation. Detailed drawings and photographs are essential for documenting features along the fault. An earthquake opened up a series of ground fissures across this graduate student's thesis area on the San Andreas fault.

3. Shallow trenches dug across the fault expose what is just below the surface. Most trenches are several meters deep, allowing geologists to examine the fault zone for clues about its earthquake history. In the trench seen in this photograph (◀), a recent rupture of the San Andreas fault offsets dark layers of carbon-rich peat, which were used to date the layers and therefore date the history of earthquakes on this fault. Rocks and soils, both in natural exposures and in trenches dug to study a fault, preserve a history of motion. They give clues to the magnitudes and recurrence of past earthquakes.

12.11.a3 Pallet Creek, CA

2. When a fault moves, it can offset natural and human-made features. Streams and gullies, as well as roads, fences, and telephone lines, provide pre-earthquake reference points. Geologists can measure how much and in what direction the fault has offset these features. In this figure, a stream channel bends where it crosses the fault. The offset of the stream channels seen here and other clues tell us that along this fault (the San Andreas fault) the Pacific plate is sliding northward past the North American plate as shown by the two arrows. Can you see evidence of another location where the same stream channel was abandoned by past movements along the fault? Hint: look for a stream channel that does not seem to continue anywhere.

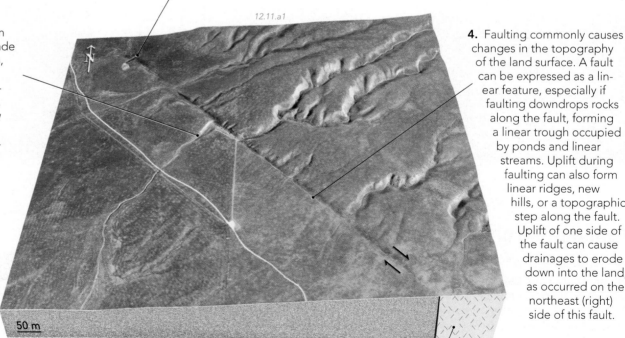

12.11.a1

50 m

4. Faulting commonly causes changes in the topography of the land surface. A fault can be expressed as a linear feature, especially if faulting downdrops rocks along the fault, forming a linear trough occupied by ponds and linear streams. Uplift during faulting can also form linear ridges, new hills, or a topographic step along the fault. Uplift of one side of the fault can cause drainages to erode down into the land, as occurred on the northeast (right) side of this fault.

5. Geologists search for distinctive rock units or other geologic features that have been cut and displaced by the fault. Such offset rocks and features provide evidence of the amount of displacement a fault has accumulated over its history, which may span many millions of years. For example, a fault may displace a granite pluton, moving part of the pluton kilometers or hundreds of kilometers along the fault. Along the San Andreas fault, geologists have matched pebbles in a conglomerate on one side of the fault with the bedrock source of the pebbles on the other side of the fault, demonstrating hundreds of kilometers of strike-slip motion on the fault over the last 10 million years.

B How Do We Study Faults with Satellites?

▶ The topography around a fault changes when the fault moves. Very small changes in elevation can be detected through laser surveying or by comparing satellite radar data sets before and after faulting. To use the satellite method, an area is mapped before and after the earthquake. The two maps are combined into an *interferogram*, which shows how Earth has deformed near the fault rupture. In this image, color bands or fringes indicate strike-slip movement associated with the 1999 Hector Mine earthquake (Mw 7.1) in southern California. The fault is cutting diagonally northwest through the view.

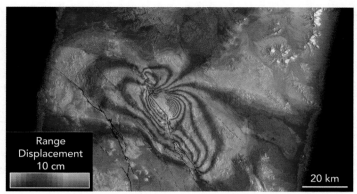

Range Displacement 10 cm

20 km

12.11.b1

C How Do We Study Faults Associated with Prehistoric Earthquakes?

Earthquake Studies Along the North Anatolian Fault, Turkey

▶ 1. In 1999, a magnitude (Mw) 7.4 earthquake ruptured over 100 km (60 mi) of the North Anatolian fault in Turkey. Soon after the earthquake, geologists conducted field studies to determine how much and how often the fault moved in the past. They used surveying equipment to precisely measure the heights of the fault scarps (▶) to determine how much the fault moved. During this earthquake, one side of the fault moved up by 1.6 m (5 ft) , but much movement was horizontal and so is not represented in this topographic profile.

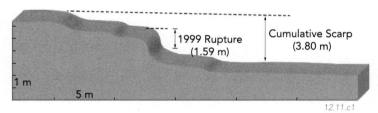

1999 Rupture (1.59 m)   Cumulative Scarp (3.80 m)

1 m   5 m

12.11.c1

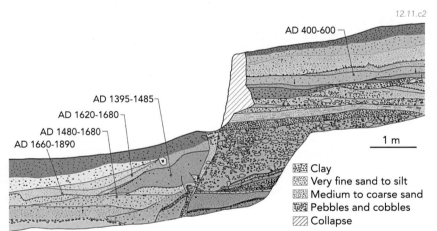

12.11.c2

AD 400-600

AD 1395-1485
AD 1620-1680
AD 1480-1680
AD 1660-1890

1 m

▦ Clay
▦ Very fine sand to silt
▦ Medium to coarse sand
▦ Pebbles and cobbles
▨ Collapse

2. Several trenches dug along the fault revealed a wealth of information about its prior history. The geologists meticulously examined the walls of the trenches and carefully mapped how the fault offset layers of sediment and soil (◀). They documented that older layers were offset by several distinct earthquake events. The colors on this figure indicate different ages of sediment. Samples of charcoal were dated by the carbon-14 method (in years AD), providing a timeline for interpreting when the fault moved.

3. From these studies, the geologists determined that a major earthquake occurs along this fault about every 200 to 300 years, and that previous events were about the same size as the 1999 event. Such earthquakes are characteristic of this fault. Determining the recurrence and likely size of earthquakes will help the people in this region plan for future earthquakes.

San Andreas Experiment

Geologists in California are engaged in a novel experiment as part of the *Earthscope* seismic project. The *San Andreas Fault Observatory at Depth* sunk a deep drill hole through part of the San Andreas fault. The drill hole is equipped with geophysical instruments to provide data on this active fault system. The scientists hope to record a large earthquake as it happens. In this figure, a drill hole crosses the fault at 3.2 km (2 mi) depth.

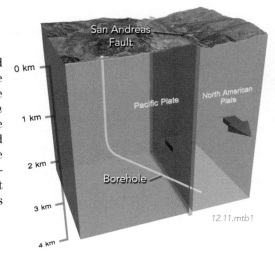

San Andreas Fault
0 km
1 km
Pacific Plate
North American Plate
2 km
Borehole
3 km
4 km
12.11.mtb1

Before You Leave This Page Be Able To

☑ Summarize the kinds of field and remote measurements geologists use to investigate recent earthquakes.

☑ Summarize the methods of investigating prehistoric earthquakes on faults, including observations within trenches dug across a fault.

12.11

12.12 Can Earthquakes Be Predicted?

EARTHQUAKES CAN BE DEVASTATING to places and people. For this reason, we have a great interest in finding ways to predict when and where earthquakes will occur. Although much is known about where earthquakes occur, there is no reliable way to predict exactly when one will strike.

A Can We Anticipate Which Areas Are Most Likely to Have Earthquakes?

We try to predict which areas will have earthquakes by understanding the (1) frequencies and sizes of historic earthquakes, (2) geologic record of prehistoric earthquakes, and (3) tectonic settings of different regions.

World Earthquake Hazard

This seismic-hazard map shows the intensity of shaking expected on land. Red areas have the highest hazard, gray areas have the lowest hazard, and yellow and green areas are considered to have a moderate to low seismic hazard.

The patterns on this map largely reflect the locations of plate boundaries. Which parts of the world have a low hazard for earthquakes, and which regions have high hazard?

Note the pattern along convergent plate margins, including the west coast of South America. The greatest hazard is from megathrust earthquakes along the coast (near the trench). Hazard decreases into the continent as the distance from the convergent boundary increases and the subduction zone becomes deep.

The Middle East is highly susceptible to earthquakes, largely because collision of the Arabian plate is causing thrust faults and strike-slip faults across the region.

Australia experiences few earthquakes, mostly because it is not along a plate boundary. Islands to the north (New Guinea) and southeast (New Zealand) straddle active plate boundaries and have higher hazards.

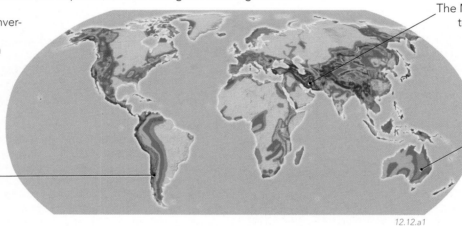

12.12.a1

United States Earthquake Hazard

These maps show the most seismically active areas of the United States, including Hawaii and Alaska. Which regions experience little damage from earthquakes, and which regions experience the most damage? Do some areas surprise you?

For the United States, the risk of earthquakes is greatest in the most tectonically active areas, especially near the plate margin in the western United States. The San Andreas fault forms the boundary between the Pacific plate and the North American plate. It is responsible for about one magnitude 8 or greater earthquake per century.

The upper Midwest and Gulf Coast areas have few active faults and very low earthquake hazards.

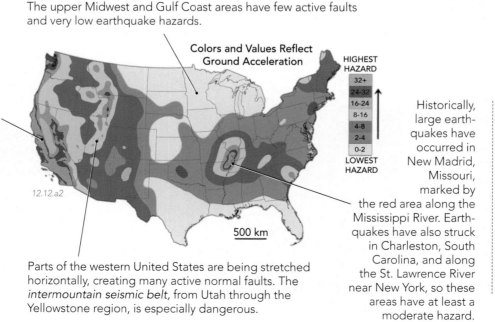

12.12.a2

Colors and Values Reflect Ground Acceleration

HIGHEST HAZARD
32+
24-32
16-24
8-16
4-8
2-4
0-2
LOWEST HAZARD

500 km

Parts of the western United States are being stretched horizontally, creating many active normal faults. The *intermountain seismic belt*, from Utah through the Yellowstone region, is especially dangerous.

Historically, large earthquakes have occurred in New Madrid, Missouri, marked by the red area along the Mississippi River. Earthquakes have also struck in Charleston, South Carolina, and along the St. Lawrence River near New York, so these areas have at least a moderate hazard.

12.12.a3

Alaska

Southern Alaska has very large subduction-related earthquakes and has a high seismic hazard.

Seismic hazard in Hawaii is higher to the southeast, closer to the most active volcanism.

Hawaii

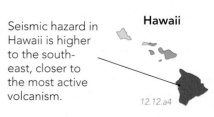

12.12.a4

B How Do We Approach Long-Range Earthquake Forecasting?

Long-term forecasting is based mainly on the knowledge of when and where earthquakes occurred in the past. Thus, geologists study present tectonic settings, geologic evidence of past events, and historical records. These studies aim to determine the locations and recurrence intervals of past earthquakes.

1. One approach to long-range forecasting is to measure patterns of seismic activity along a fault. These two cross sections show seismicity along the San Andreas fault in northern California. The top cross section shows earthquakes that occurred along the fault prior to October 17, 1989; the bottom one shows seismicity after the Loma Prieta earthquake on October 17, 1989.

2. In the top section, three segments of the fault have fewer earthquakes than other sections. These segments, called *seismic gaps*, are "locked" (not moving), and are accumulating stress. The three seismic gaps were at San Francisco, Loma Prieta, and Parkfield.

4. From various data, the USGS assigned probabilities of a magnitude 6.7 earthquake on faults of the area before 2032. The combined probability is over 60%.

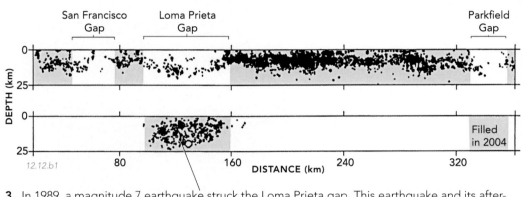

12.12.b1

3. In 1989, a magnitude 7 earthquake struck the Loma Prieta gap. This earthquake and its aftershocks, shown in the lower section, filled in the gap. The Parkfield gap was similarly filled by an earthquake and aftershocks in 2004. When will an earthquake fill the San Francisco gap?

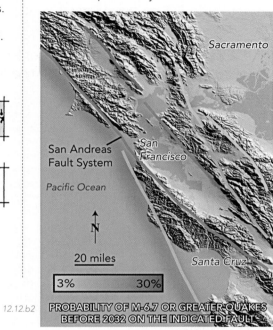

12.12.b2

PROBABILITY OF M-6.7 OR GREATER QUAKES BEFORE 2032 ON THE INDICATED FAULT

C How Successful Are Short-Term Predictions?

Short-term prediction involves monitoring the activity along an earthquake-prone fault. There are often *precursor events*, which can be gauged using sophisticated scientific equipment. The complexity inherent in fault systems means that prediction techniques are still developing, but they hold promise.

12.12.c1 Parkfield, CA

◄ **1.** Seismologists shine lasers across a fault to monitor small-scale movements that might be precursors to a larger earthquake. They can even record movement during a larger earthquake.

2. Measurements taken near active faults sometimes show that prior to an earthquake, the ground is uplifted or tilted as rocks swell under the strain building on the fault. The buildup in stress may also cause many small cracks. These can slip and produce foreshocks, small earthquakes that may advertise an upcoming main earthquake.

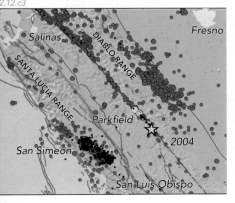

2.12.c3

◄ **3.** Prior to 2004, the Parkfield segment of the San Andreas fault, southeast of San Francisco, had six magnitude (Mw) ~6 quakes since 1857. These occurred approximately every 22 years and had similar characteristics. This situation provided an opportunity to study the short-term precursors of the next earthquake. Seismologists set up a detailed array of seismic instruments to record the region's many earthquakes, shown here as red, black, and yellow symbols.

12.12.c2

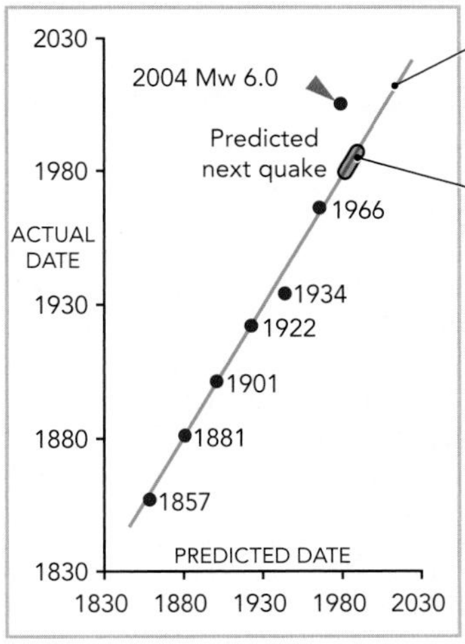

4. The blue line shows when the six large historic Parkfield earthquakes actually occurred versus when they would have occurred if they were spaced exactly 22 years apart.

5. The next big earthquake was predicted to occur between 1988 and 1993. The earthquake finally happened in 2004, 11 years later than expected.

Before You Leave This Page Be Able To

☑ Describe areas of the world that experience a high risk of earthquake activity.

☑ Summarize why certain areas of the United States experience earthquakes, while others do not.

☑ Summarize ways geologists do long-range forecasting and short-range prediction.

12.12

12.13 What Is the Potential for Earthquakes Along the San Andreas Fault?

The San Andreas Fault is the world's best-known and most extensively studied fault. It runs across California from the Mexican border to north of San Francisco, and is responsible for many destructive earthquakes. What has happened along the fault in the recent past, and what does this history say about its likelihood of causing large earthquakes? In 2008, the USGS and others forecast a 99% probability that California would have a magnitude 6.7 or larger earthquake in the next 30 years.

Recent Earthquake History of Different Segments of the San Andreas Fault and Related Faults

1. The San Andreas fault has distinct *segments* that behave differently. These segments vary in the size and frequency of earthquakes. As a result, the earthquake hazard varies along the fault. This map shows some of the major segments of the San Andreas fault that have caused earthquakes in California. Circles show epicenters of some of the more important earthquakes. The San Andreas fault accounts for the largest quakes, but there are many other recently active faults (shown in black). Some of these have caused damaging, moderate-sized earthquakes.

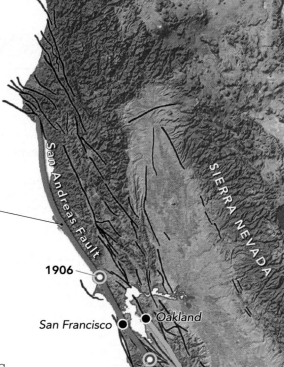

12.13.a1

12.13.a2

2. The *northern segment* of the San Andreas fault was responsible for the famous 1906 earthquake that destroyed much of San Francisco. The earthquake had a magnitude (Mw) of 7.7 and ruptured 430 km (270 mi) of the fault, from south of the city all the way to the north end of the fault (the part that ruptured is shown in red). Damage (◄) was caused by ground shaking, fires, and liquefaction of water-saturated soils in areas that had originally been part of San Francisco Bay.

12.13.a3

3. The southern part of this segment ruptured in 1989 in the magnitude (Mw) 7.1 Loma Prieta earthquake, which was centered south of San Francisco. This earthquake is famous for disrupting a World Series baseball game. Ground shaking and liquefaction collapsed buildings (◄) and parts of bridges and freeways.

4. The next segment to the south, shown in blue, is the *central creeping segment*. The two sides of the fault creep past one another somewhat continuously and slowly, rather than storing up energy for a large earthquake. Creep continues to the north along the Hayward fault, also colored blue, through Oakland. The Hayward fault was the site of a ruinous earthquake in 1868, with an estimated magnitude (Mw) of 7.

5. South of the creeping segment is the *Parkfield segment*, a short segment included here as part of a larger orange-colored segment discussed below. It produces moderate-sized earthquakes that occur, on the average, every couple of decades. The Parkfield segment receives special scrutiny from geologists and seismologists because the frequent earthquakes provide an opportunity to study the behavior of a fault before, during, and after an earthquake.

6. The San Andreas continues to the southeast through a segment (shown in orange) that last ruptured during the great Fort Tejon earthquake of 1857. This earthquake ruptured 300 km (190 mi) of the fault, from Parkfield all the way to east of Los Angeles. The earthquake was approximately magnitude (Mw) 8, but damage was limited because the area was much less populated than it is now. This part of the San Andreas commonly is called the *locked segment* because it has not ruptured since 1857. It has the potential to cause a great earthquake, commonly called "the big one."

Features Along the San Andreas Fault

The San Andreas fault generally has a clear expression in the landscape. It is marked by a number of features that are common along active faults. Some of these features can also form in ways unrelated to active faulting.

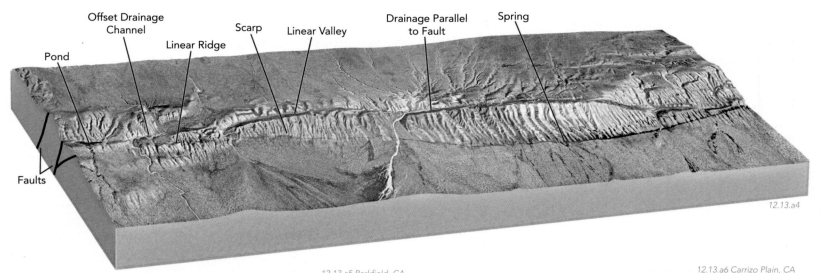

Pond
Offset Drainage Channel
Linear Ridge
Scarp
Linear Valley
Drainage Parallel to Fault
Spring
Faults

12.13.a4

12.13.a5 Parkfield, CA

▶ Geologists explore the fault to find localities that preserve a record of past faulting. Detailed studies of trenches dug across the fault help geologists unravel hundreds or thousands of years of the fault's movement history.

12.13.a6 Carrizo Plain, CA

▶ The aerial photograph to the right shows the same part of the San Andreas fault as depicted in the figure above. Can you match some of these features between the two images?

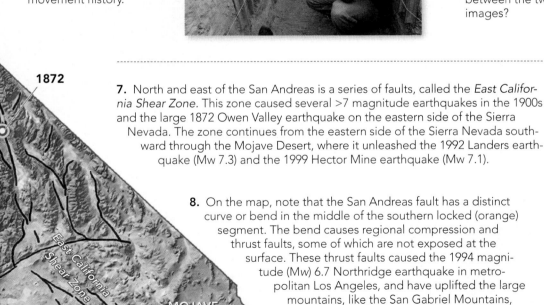

1872

East California Shear Zone

MOJAVE DESERT

Los Angeles

7. North and east of the San Andreas is a series of faults, called the *East California Shear Zone*. This zone caused several >7 magnitude earthquakes in the 1900s and the large 1872 Owen Valley earthquake on the eastern side of the Sierra Nevada. The zone continues from the eastern side of the Sierra Nevada southward through the Mojave Desert, where it unleashed the 1992 Landers earthquake (Mw 7.3) and the 1999 Hector Mine earthquake (Mw 7.1).

8. On the map, note that the San Andreas fault has a distinct curve or bend in the middle of the southern locked (orange) segment. The bend causes regional compression and thrust faults, some of which are not exposed at the surface. These thrust faults caused the 1994 magnitude (Mw) 6.7 Northridge earthquake in metropolitan Los Angeles, and have uplifted the large mountains, like the San Gabriel Mountains, north and northeast of the city.

9. East of Los Angeles, the San Andreas branches southward into several faults. Some of these experienced several moderate-sized earthquakes in the 1900s, including some close to important agricultural areas. The fault scarps for these events are colored pink and red on this map.

Before You Leave This Page Be Able To

☑ Briefly summarize the main segments of the San Andreas fault and whether they have had major earthquakes.

☑ Summarize features that might help you recognize the fault from the air.

12.13

12.14 How Do We Explore What Is Below Earth's Surface?

OUR VIEW OF GEOLOGY is typically limited to those rocks and structures that are exposed at the surface. In deep canyons we can glimpse subsurface rocks and structures, but only for hundreds to thousands of meters deep. How else do we determine what lies beneath the surface?

1. The region shown on the large figure has a few hills of granite and a dark lava flow, but is otherwise covered by soil and vegetation. There are few clues as to what types of rocks and structures lie below the surficial cover. There are two general approaches for investigating subsurface geology: obtaining samples of rocks at depth, and performing *geophysical surveys* that measure the subsurface magnetic, seismic, gravity, and electrical properties.

12.14.a2 Shiprock, NM

2. As magma rises to the surface, it can incorporate pieces of the rock through which it passes. Geologists study such pieces, called *inclusions*, to determine the types of rocks that lie beneath volcanoes. This inclusion (◄) is a piece of granite in a volcanic rock.

3. We can gain a sense of what is below the surface by examining rocks and geologic structures that have been uplifted and exposed at the surface. Geologists study rocks using microscopes and other instruments to constrain the temperature and pressure conditions under which the rocks formed and to infer the geologic processes that created the rocks at depth.

12.14.a3 Kidd Creek, Ontario

4. Mines provide a more detailed subsurface view because the tunnels provide continuous exposures of rocks and structures (◄). Some South African mines are deeper than 5 km (3 mi).

12.14.a1

5. The geometry of rock units and geologic structures can be explored by sending seismic energy (sound waves) into the ground and measuring how the waves are reflected back to the surface off boundaries between rock types. This commonly is accomplished by using large trucks that shake the ground in a controlled manner. The sound waves bounce off rock layers, faults, and other boundaries. They are then recorded using seismic receivers, called *geophones*, which are buried or stuck into the ground (such as the red-topped geophones shown on the next page). This data-collection and data-processing procedure is a *seismic-reflection survey*.

6. Seismic-reflection data are processed using sophisticated computer programs that produce the thin and commonly discontinuous lines shown in this drawing (▶). The lines represent the location of layers and surfaces that reflected seismic energy, guiding geologists who interpret the below-ground geometry of the rock units, folds, faults, and unconformities.

12.14.a4

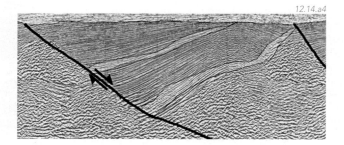

▲ 7. The geometry of the reflections, as expressed on the seismic profile, is integrated with information about the area's rock sequence and structures. We can then construct a geologic cross section (the colored zones and heavy lines) representing an interpretation of the subsurface.

8. Geologists and engineers drill holes to search for petroleum, minerals, groundwater, and scientific knowledge. Most drill holes are less than several hundred meters deep, but some reach depths of 5 km (3 mi) or more. Cylinder-shaped samples of rock, called *drill cores*, can be retrieved during the drilling process to provide samples of rocks from depth. ▼

12.14.a5 Thabazimi, South Africa

9. Instruments that measure the intensity of Earth's magnetic field can be used to determine the subsurface distributions of magnetic rocks. The equipment can be carried on foot or towed behind a plane. Earth scientists who measure and interpret magnetic, seismic, gravity, and other types of physical data are *geophysicists*. Such data are called a *geophysical survey*. Many geology graduates are involved with geophysical surveys at some point in their careers.

▶ 10. Magnetic data are generally portrayed as a map, with warmer colors (reds) representing more magnetic rocks and cooler colors (blues) representing areas with lower magnetism.

▶ 11. The red and orange areas mark the dark lava flow and hills of gray granite, which are more magnetic than the sediments that cover the rest of the area.

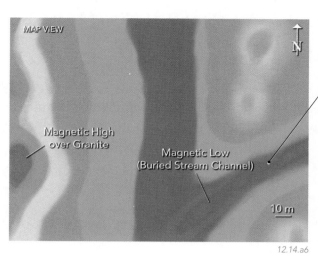

MAP VIEW

Magnetic High over Granite

Magnetic Low (Buried Stream Channel)

10 m

12.14.a6

12. A curving magnetic low, represented by the darker blue colors, coincides with a buried stream channel. In the central figure the channel forms a band of gray soil where the two teams of geophysicists are standing.

13. The strength of gravity varies slightly from one place to another on Earth's surface. This is because some rocks, such as basalt, are relatively more dense and cause a stronger pull than less dense materials, such as sediment. The variations in gravity can be measured using sensitive *gravity meters*.

14. In this area, the team of geophysicists measured gravity across the buried stream channel and plotted the data on a profile relative to the average value of gravity for the area. The plot shows a gravity minimum caused by low-density sediment within the buried channel. **▼**

12.14.a7

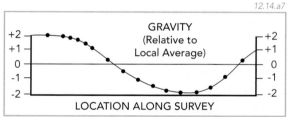

GRAVITY
(Relative to Local Average)

LOCATION ALONG SURVEY

▼ 15. From the gravity profile, computer programs can model possible density configurations that are consistent with the data.

12.14.a8

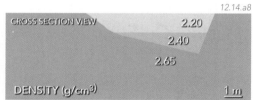

CROSS SECTION VIEW   2.20

2.40

2.65

DENSITY (g/cm³)   1 m

16. Some rocks, such as clays, conduct electrical currents better than other rocks. Rocks containing groundwater conduct an electrical current better than dry rocks. Geologists and geophysicists use these principles to explore for mineral deposits and groundwater. An electrical transmitter runs current into the ground (▼), and one or more electrical receivers some distance away measure how much current reaches the surface.

12.14.a9

17. The results of an electrical survey across the buried stream channel are plotted in cross section and contoured, with warmer colors for rocks with higher conductivity, such as those with more water. Geologists compare all the various types of data to infer the subsurface geology.

12.14.a10

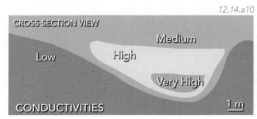

CROSS SECTION VIEW

Medium

Low   High

Very High

CONDUCTIVITIES   1 m

Before You Leave This Page Be Able To

☑ Summarize how volcanic inclusions, exposed geology, drill holes, and mines provide observations of the subsurface.

☑ Briefly summarize what is measured by the various types of geophysical surveys (seismic, magnetic, gravity, and electrical).

12.14

12.15 What Do Seismic Waves Indicate About Earth's Interior?

EARTHQUAKES, EXPLOSIONS, AND OTHER SEISMIC EVENTS generate seismic waves that can be used to interpret Earth's internal structure. The way seismic waves travel through Earth enables us to identify distinct layers and boundaries within the interior, including the crust, mantle, and core.

A How Do Seismic Waves Travel Through Materials?

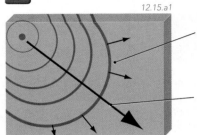

An earthquake or other source of seismic energy generates seismic waves, which radiate out from the source in all directions.

The path that any part of the wave travels is a *seismic ray*. If the physical properties of the material do not change from place to place, then a seismic ray travels in a straight line. In this case, a family of straight rays diverges outward from the source.

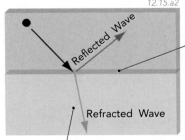

Most seismic waves encounter boundaries between materials with different physical properties, causing the waves to reflect, speed up, or slow down. Some of the energy is *reflected* off the interface as a reflected wave.

Some of the energy is bent as it crosses the boundary. This process of bending is known as *refraction*.

How Seismic Waves Refract Through Different Materials

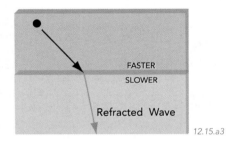

If a seismic wave passes into a material that causes it to slow down, it will be refracted away from the interface at a steeper angle.

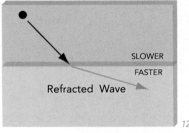

If a descending seismic ray passes from a slow material to a faster one, it will be refracted to a shallower angle.

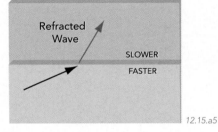

If a rising seismic ray passes from a fast material to slower one, it will be refracted upward toward the surface.

B How Do Seismic Waves Travel Through Earth's Crust and Mantle?

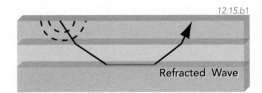

▲ 1. Refraction causes seismic waves to take curved paths through the Earth. Because rocks get denser deeper in Earth, steeply descending rays will first be refracted to shallower angles as they encounter faster and faster material at depth. Subsequently, the waves will then be bent back toward the surface as they pass back through slower, less dense material.

2. In the figures below, an earthquake sends seismic waves into the crust and mantle. Both waves are refracted back toward the surface. Waves in the mantle travel faster than those in the crust, resulting in an interesting and useful phenomenon.

3. Close to the earthquake, waves that travel through the crust arrive sooner than those from the mantle because the crustal waves travel a shorter distance.

4. Farther from the earthquake, waves that travel through the mantle arrive at the surface first because the faster velocity lets them overtake the crustal waves.

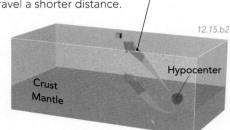

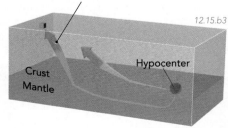

5. Seismologists observe at what distance from the hypocenter the mantle waves begin to arrive first. They then use simple computer models of velocities, crustal thicknesses, and ray paths to calculate the depth to the crust-mantle boundary.

C How Are Seismic Waves Used to Examine Earth's Deep Interior?

Seismologists recognize distinct boundaries within Earth, largely based on changes in seismic velocities. Such changes reflect the physical and chemical properties of the rock layers through which the seismic waves pass. Not all seismic waves make it through every part of Earth. Observing where particular kinds of waves are blocked helps determine which parts of Earth are molten.

1. As P-waves travel through Earth, they speed up and slow down as they pass through different kinds of material. Their velocity depends upon three factors: (1) how easily the rocks are compressed; (2) how rigid the material is; and (3) the density of the material. Based on these factors, seismologists conclude that faster velocities indicate denser rocks. The graph below (▼) plots P-wave velocity as a function of depth. Overall, P-wave velocity increases with depth in the mantle and in the core because the rocks in each part become more rigid and dense downward.

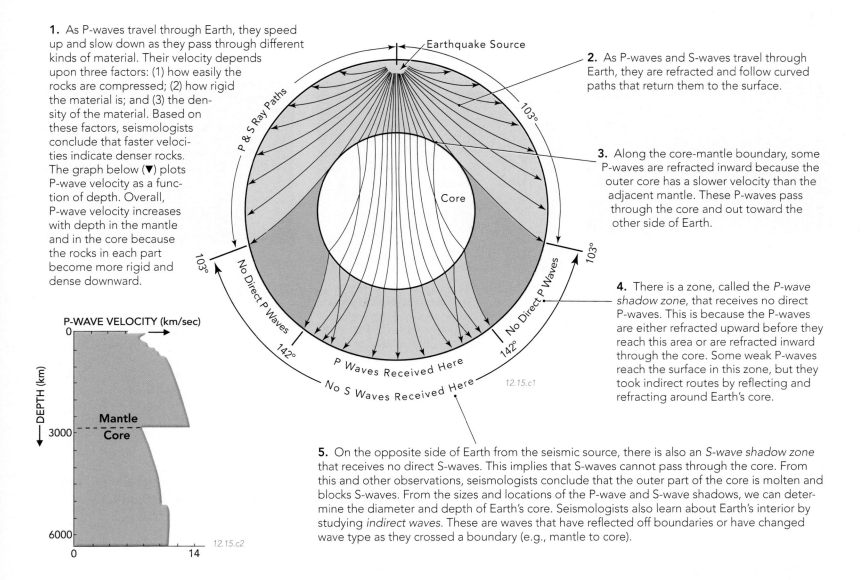

2. As P-waves and S-waves travel through Earth, they are refracted and follow curved paths that return them to the surface.

3. Along the core-mantle boundary, some P-waves are refracted inward because the outer core has a slower velocity than the adjacent mantle. These P-waves pass through the core and out toward the other side of Earth.

4. There is a zone, called the *P-wave shadow zone*, that receives no direct P-waves. This is because the P-waves are either refracted upward before they reach this area or are refracted inward through the core. Some weak P-waves reach the surface in this zone, but they took indirect routes by reflecting and refracting around Earth's core.

5. On the opposite side of Earth from the seismic source, there is also an *S-wave shadow zone* that receives no direct S-waves. This implies that S-waves cannot pass through the core. From this and other observations, seismologists conclude that the outer part of the core is molten and blocks S-waves. From the sizes and locations of the P-wave and S-wave shadows, we can determine the diameter and depth of Earth's core. Seismologists also learn about Earth's interior by studying *indirect waves*. These are waves that have reflected off boundaries or have changed wave type as they crossed a boundary (e.g., mantle to core).

The Moho

The boundary between the crust and mantle is named the *Mohorovicic Discontinuity* after the last name of the Croatian seismologist who discovered it. Most geologists simply call it the *Moho*.

Much effort is expended trying to determine the depth to the Moho because this tells us how thick the crust is. Geophysicists investigate this problem using various approaches. Some observe the arrivals of seismic waves from naturally occurring earthquakes, whereas others use mine blasts as the seismic source. We can calculate the depth to the Moho by observing whether the first waves to arrive came through the crust or the mantle, as described on the previous page. The depth to the Moho can sometimes be identified as reflections on seismic-reflection profiles. Since seismic waves travel through the crust at approximately 6 km per second, it takes 10 seconds for a wave to travel 30 km (19 mi) down to the Moho, bounce off, and travel 30 km back up. It takes less time if the crust is thin and more time if it is thick.

Before You Leave This Page Be Able To

✓ Sketch or describe reflection and refraction of seismic waves.

✓ Sketch and explain how seismic waves pass through the crust and mantle.

✓ Explain how we use seismic waves to infer the diameter of the core and to show that the outer core is molten.

12.15

12.16 How Do We Investigate Deep Processes?

ROCK PROPERTIES, SUCH AS DENSITY, temperature, pressure, and composition, change through Earth. Seismologists use seismic-wave velocities to determine how rock properties change with depth and how material moves in Earth's mantle and at the core-mantle boundary.

A How Do We Investigate Deep Conditions?

Much of what we know about Earth's interior comes from observations of rocks and our knowledge of seismic-wave velocities and how they vary within Earth's interior.

12.16.a1 Chuckwalla Mtns., CA

12.16.a2 College Station, TX

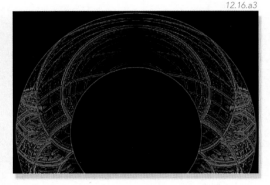
12.16.a3

One way to constrain the conditions deep within Earth is to examine rocks that have resided at great depths. Some metamorphic rocks in Norway and China contain high-pressure minerals, which indicate that they were buried at ultra-high pressures and depths of 60 to 100 km (40 to 60 mi). Documenting the minerals and structures that formed under these conditions provides insight into what processes and conditions occur at depth.

In the laboratory, we subject rocks to high temperatures and pressures in order to determine the conditions under which the rocks melt, solidify, or flow in the solid state. Many minerals change into another mineral at high temperatures, high pressures, or both. The conditions under which these changes occur are then inferred for equivalent depths and temperatures within Earth's interior.

Computers and sophisticated numerical models are used to model processes that are too deep to observe directly. Such models can illustrate how seismic waves travel through the mantle, as shown here, or how the mantle might flow upward, downward, or laterally if there are lateral variations in density. Such density variations are caused by differences in temperature and in the types of minerals that are present.

B How Does Seismic Tomography Help Us Explore the Earth?

Seismologists examine Earth using earthquakes in much the same way that medical doctors examine the internal parts of the body with CT scans and other types of imaging technologies. The technique seismologists use is called *seismic tomography*, where "tomography" means an image of what is inside.

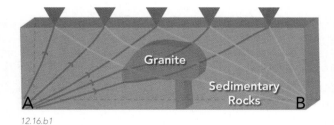

12.16.b1

12.16.b2

Seismic Observations

◄ Using seismic tomography, scientists examine earthquake waves that have passed through the same subsurface region but from different directions. In this diagram, lines called *ray paths* show the directions the seismic waves traveled. Ray paths coming from points A and B are recorded on five different seismometers, shown as triangles. If part of the crust or mantle has a higher seismic velocity than other areas, then waves passing through that area will arrive sooner than expected. Those that travel through slow regions will arrive later than expected.

Seismic Interpretation

◄ This figure models the velocities in the same region using seismic tomography. Areas that are slower than expected are shaded red and may represent areas that might be hotter than normal. Some areas, like the granite body, will be faster than expected and so are shaded blue. Fast areas might be abnormally cool or composed of stiff, dense rocks. Earthquakes do not come from every direction, so many details cannot be resolved and remain a little fuzzy.

C What Processes Are Occurring in the Mantle and the Core-Mantle Boundary?

Seismic wave velocities increase abruptly at the Moho (crust-mantle boundary), when they pass from the crust down into the mantle. The velocities vary within the mantle due to major changes in mineralogy and increasing density with depth, and because of upward and downward flow of mostly solid mantle material.

Seismic Velocities of the Lowermost Mantle

1. This globe shows computed velocities of seismic shear waves in the lowermost mantle, as modeled using seismic tomography. Red areas represent seismically slow materials, and blues represent materials that are seismically faster than average. The outlines of the continents (centered on North and South America) are shown on the surface for reference.

2. The red areas in the model are interpreted to represent rising masses of hot, mostly solid mantle material. Many, but not all, seismologists regard these rising masses as the source areas for mantle plumes and hot spots.

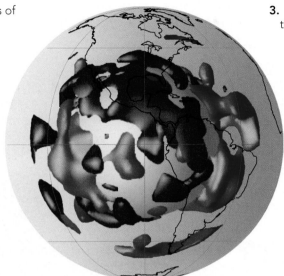

12.16.c1

3. Cooler colors (blues) are interpreted as dense plates that have been subducted into the lowermost mantle. Not all geologists agree with this interpretation.

4. Recent advances in seismic instruments, computer processing, and numerical approaches have led to the discovery of a thin layer along the boundary between the core and mantle. This boundary layer, called D" (dee-double-prime), is irregular in thickness and is interpreted to have upwellings, as shown in this model. ▼

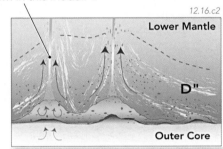

12.16.c2

Lower Mantle

D"

Outer Core

A Model of Flow Within Earth

1. Seismologists and other geologists strive to develop models for the flow of materials throughout Earth. This figure, from seismologist Ed Garnero, presents one view of the inner workings of Earth. There are many other views.

2. In this model, cold, dense material from subducted slabs sinks deep into the mantle. These slabs correspond to the blue, fast velocities in Ed Garnero's seismic tomography figure above.

3. Some cold slabs are interpreted to travel all the way down to the base of the mantle, where they pile up to form the D" layer. This figure greatly exaggerates the thickness of this layer.

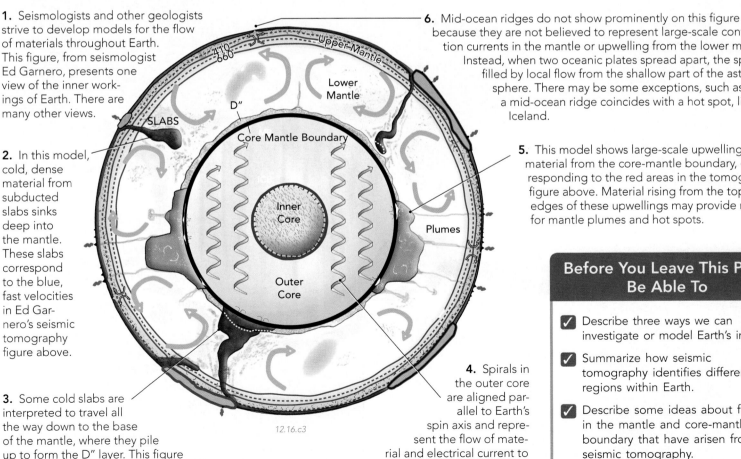

12.16.c3

6. Mid-ocean ridges do not show prominently on this figure because they are not believed to represent large-scale convection currents in the mantle or upwelling from the lower mantle. Instead, when two oceanic plates spread apart, the space is filled by local flow from the shallow part of the asthenosphere. There may be some exceptions, such as where a mid-ocean ridge coincides with a hot spot, like at Iceland.

5. This model shows large-scale upwelling of material from the core-mantle boundary, corresponding to the red areas in the tomography figure above. Material rising from the tops and edges of these upwellings may provide material for mantle plumes and hot spots.

4. Spirals in the outer core are aligned parallel to Earth's spin axis and represent the flow of material and electrical current to generate Earth's magnetic field.

Before You Leave This Page Be Able To

☑ Describe three ways we can investigate or model Earth's interior.

☑ Summarize how seismic tomography identifies different regions within Earth.

☑ Describe some ideas about flow in the mantle and core-mantle boundary that have arisen from seismic tomography.

12.16

12.17 What Happened During the Great Alaskan Earthquake of 1964?

THE SOUTHERN COAST OF ALASKA experienced one of the world's largest earthquakes in 1964. The magnitude (Mw) 9.2 earthquake, which is the strongest to have ever struck North America, destroyed buildings, triggered massive landslides, and unleashed a tsunami that caused damage and deaths from Alaska to California. This event provides an example of the causes and manifestations of an earthquake.

A What Types of Damage Did the Earthquake Cause?

The earthquake occurred along the southern coast, but was felt throughout Alaska, except for the far north coast. Ground shaking destroyed buildings and generated huge landslides of rock and soil. This dark, rocky landslide (▼) covered parts of the white Sherman Glacier.

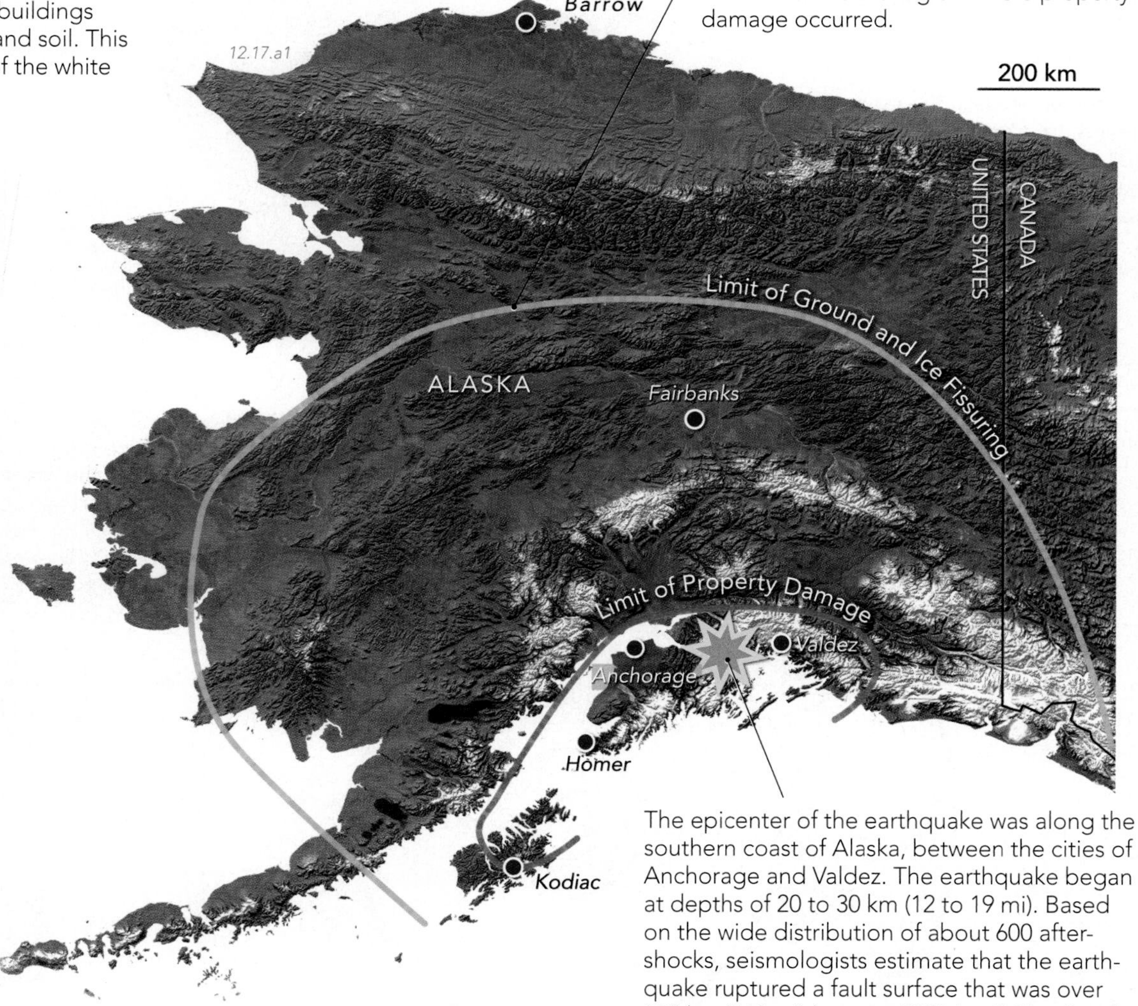

The blue line on the map marks the limit of fissuring of the ground and ice during the earthquake. The red line closer to the epicenter outlines the region where property damage occurred.

12.17.a1

200 km

UNITED STATES

CANADA

Barrow

Limit of Ground and Ice Fissuring

ALASKA

Fairbanks

Limit of Property Damage

Anchorage

Valdez

Homer

Kodiac

12.17.a2

The epicenter of the earthquake was along the southern coast of Alaska, between the cities of Anchorage and Valdez. The earthquake began at depths of 20 to 30 km (12 to 19 mi). Based on the wide distribution of about 600 aftershocks, seismologists estimate that the earthquake ruptured a fault surface that was over 900 km (560 mi) long and 250 km (160 mi) wide. The earthquake occurred on a thrust fault that dips from the Aleutian trench gently north and northwestward beneath Alaska.

12.17.a3

◄ Parts of downtown Anchorage were demolished when shaking caused the underlying land to slip and collapse. Some buildings sank so much that their second stories were level with the ground. Severe damage occurred in the Turnagain Heights area of Anchorage, where a layer of weak clay liquefied, carrying away shattered houses (►).

12.17.a4

B What Happened in the Sea During the Earthquake?

Because it occurred along the coast, the earthquake also caused (1) faulting and uplift of the seafloor, (2) huge waves from landslides, and (3) a tsunami that struck the coasts of Alaska, British Columbia, Washington, Oregon, California, Hawaii, and Japan.

▶ The main fault that caused the earthquake did not break the land surface, but two subsidiary faults did. One fault cut a notch into a mountain and uplifted the seafloor 4 to 5 m (15 ft). The white material on the uplifted (left) side of the fault consists of calcareous marine organisms that were below sea level before the earthquake. The maximum observed uplift was 11.5 m (38 ft). Other areas subsided as much as 6 m (20 ft) during the earthquake, flooding docks, oil tanks, and buildings along the coast.

12.17.b1

12.17.b2

▲ Faulting uplifted a large area of seafloor off the southern coast of Alaska, sending a large tsunami out across the sea and up the many bays and inlets along the coast. The highest tsunami recorded was 67 m (220 ft), in a bay near Valdez. The photo above shows damage done to Kodiak Island by a wave 6 m (20 ft) high. The tsunami killed 106 people in Alaska and 17 more in Oregon and California.

C How Did Geologists Study the Aftermath of the Earthquake?

Immediately after the earthquake, the U.S. Geologic Survey (USGS) dispatched a team of geologists to (1) survey the damage; (2) document the faults, landslides, and other features of the earthquake; (3) understand what happened; and (4) identify high-risk areas and devise plans to minimize loss from future earthquakes.

12.17.c1

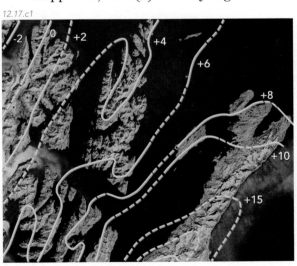

◀ 1. The USGS team investigated the coastline, measuring uplift and subsidence in hundreds of sites. They plotted and contoured the measurements (in feet) on a detailed map. Numbers are positive for uplift and negative for subsidence. Dashed lines mark extrapolated values.

12.17.c2

◀ 2. The map was used to identify broad zones of subsidence and uplift, which affected an area of over 250,000 square kilometers (100,000 square miles). The large size of the affected area reflects the huge area of the fault surface that ruptured.

12.17.c3

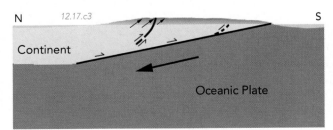

◀ 3. This huge earthquake occurred along a megathrust, where oceanic crust is subducting beneath the continent. USGS geologist George Plafker constructed a cross section showing how the uplift was explained by southward thrusting of the continent over the oceanic crust. His 1964 paper predated the idea of plate tectonics and was a key step that led to the development of the theory.

Before You Leave This Page Be Able To

☑ Summarize events associated with the Alaskan earthquake, including effects on land and sea, and how USGS studies of this area helped lead to the theory of plate tectonics.

12.18 Where Did This Earthquake Occur, and What Damage Might Be Expected?

THIS COASTAL REGION CONTAINS TWO FAULTS, an active volcano, and a steep-sided mountain prone to landslides. Any of these features could cause ground shaking. You will use seismic records from a recent earthquake to determine which feature caused the observed shaking. From this information, you will decide what hazards this earthquake poses to each of the small towns in the area.

Goals of This Exercise:

- Examine the large illustration and read the text boxes describing the types of features that are present.
- Use three seismograms to determine which feature is likely to have caused the earthquake.
- Consider potential earthquake hazards to determine what dangers each small town would face from the earthquake.
- Decide which town you think is the safest from earthquake-related hazards and justify your decision with supporting evidence.

Procedures

The area has several small towns and three seismometers, each named after the town which it is near. Seismograms recorded at each seismic station during a recent earthquake are shown at the top of the next page. Use the available information to complete the following steps and enter your answers in the appropriate places on the worksheet or online.

1. Observe the features shown on the three-dimensional perspective. Read the text associated with each location, and think about what each statement implies about earthquake hazards.

2. Inspect the seismograms for the three seismic stations to determine where the earthquake probably occurred. You can get an idea from simply comparing the time intervals between the arrivals of P-waves and S-waves for each station.

3. Your instructor may have you use the graph next to the seismograms to determine the distance from each station to the epicenter. This will allow you to more precisely locate the epicenter. Detailed instructions for this procedure are listed in topic 12.6 earlier in this chapter. A map view of the area is included with the worksheet for plotting your results.

4. From the general location of the earthquake, infer which geologic feature is likely to have caused the earthquake.

5. Use the information about the topographic and geologic features of the landscape to interpret what types of hazards the recent earthquake posed for each town. From these considerations, decide which three towns are the least safe and which two are the safest for this type of earthquake. There is not necessarily one right answer, so explain and justify your logic on the worksheet, if asked to do so by your instructor.

1. There is a deep ocean trench along the edge of the continent. Ocean drilling encountered fault-bounded slices of oceanic sediment.

2. Along one part of the coastline, there is a thin, steep beach, called *Roundstone Beach*, that rises upward to some nearby small hills. The seafloor offshore is also fairly steep as it drops off toward the trench.

3. The town of *Sandpoint* is built upon land that was reclaimed from the sea by piling up loose rocks and beach sand until the area was above sea level.

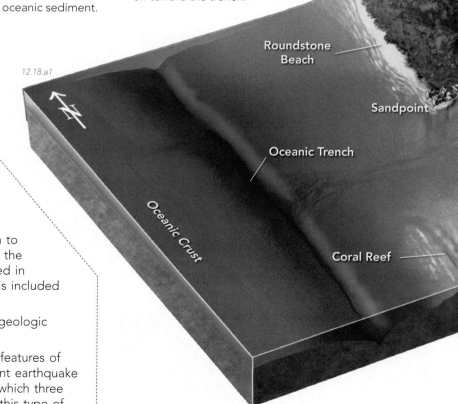

12.18.a1

12. Offshore is a coral reef that blocks larger waves, creating a quiet lagoon between the reef and the shore.

Seismograms

▶ These seismograms represent the time period from just before the earthquake to 1.5 seconds after it occurred. The first arrivals of P-waves and S-waves are labeled for each graph, along with the P-S time intervals.

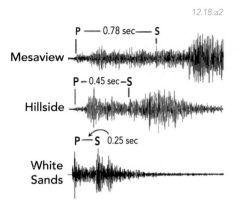

12.18.a2

Mesaview P —— 0.78 sec —— S

Hillside P — 0.45 sec — S

White Sands P—S 0.25 sec

▶ If instructed to, use this graph to determine the distance from each seismic station to the earthquake's epicenter. Find the appropriate time on the horizontal axis, follow it upward to the line, and read off the corresponding distance on the vertical axis.

12.18.a3

(Graph: DISTANCE (km) vs P-S TIME INTERVAL (sec))

---

4. A picturesque town, called *Hillside*, lies inland of some small mountains. The town is build on a flat, open area flanked by hills with fairly gentle slopes. It is a little higher in elevation than the nearby towns of Cascade Village and Riverton. The Hillside Seismic Station, shown by a triangle symbol, lies just to the east of the town.

5. In the northern part of the area, there is a flat-topped mountain, known as *Red Mesa*, surrounded by steep cliffs. A new landslide lies along the southern flank of the mountain.

6. A small town and a seismic station, both called *Mesaview*, lie between the mesa and a high volcano.

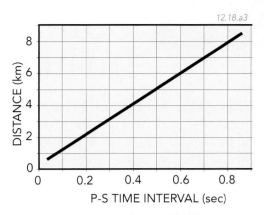

7. A volcano called *Lava Mountain* rises above the region. It has steep slopes and is surrounded by layers of volcanic ash that appear to have erupted quite recently. Every so often, the volcano releases steam and makes rumbling noises. The shaking triggers landslides down the hillsides. The small town of Ashton is on the flanks of the volcano and has a picturesque setting with huge, colorful blocks of volcanic rocks near the town.

8. The *Gray Cliffs* form a nearly vertical step in the landscape. Streams pour over the cliffs in pleasant waterfalls, each taking a jog to the left after crossing the cliffs. The small settlement of *Cascade Village* is located next to one of the waterfalls. Rocks along the cliffs are fractured and shattered.

9. The small village of *Cliffside* lies next to a gray cliff. It was built on a marshy area that was underlain by soft, unconsolidated sediments. Several streams drain into the area, but no streams are able to leave because the area is lower than the surrounding landscape. As a result, the soil is commonly very soft and people sink in as they walk.

10. *Riverton*, a picturesque town, is built near a river at the head of a sandy bay. The seafloor slopes out to the bay at a gentle angle. Muddy waters from the river prevent reefs from growing offshore in front of the bay.

11. *White Sands* is a resort town along a white, sandy beach. The sand comes from the offshore coral reef. There is a seismic station, shown by a triangle symbol, with the same name as the town.

12.18

Climate, Weather, and Their Influences on Geology

CLIMATE AND WEATHER are often mistaken for one another. Weather is a daily description of the temperature, pressure, and precipitation conditions of the atmosphere for a specific place. Climate is a longer term view of these same factors, typically taken over a period of many years. Weather and climate are responses to global circulation of water and air and help shape the landscape by eroding mountains and filling basins.

13.00.a2

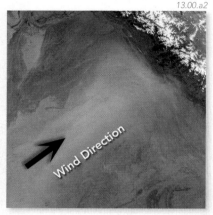

The Indian subcontinent has some of the most extreme changes in land elevation and climate in the world. India is bordered by the Indian Ocean to the east, south, and west, and so it is dominated by interactions between the land and the warm ocean water. Parts of India receive heavy rains and other parts are desert.

What causes some regions of India to be desert and others to receive seasonal downpours of rain?

13.00.a1

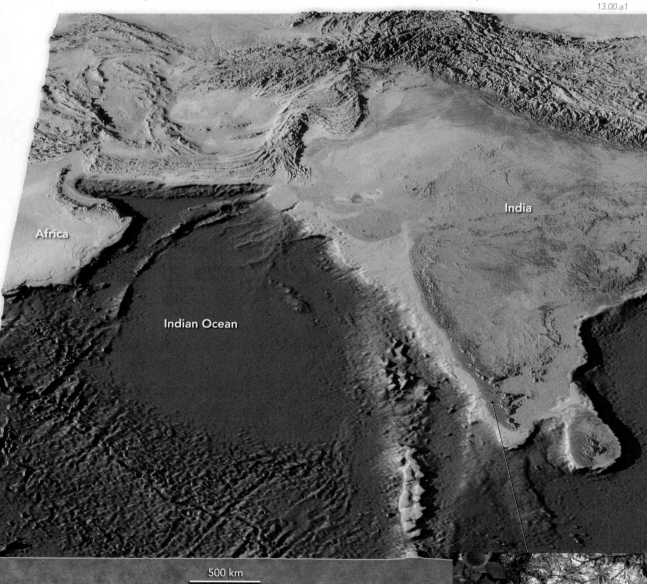

Wind Direction

▲ **This satellite image** shows a desert sandstorm and dust storm blowing across north-western India.

What types of materials does wind move and how does wind move them?

Parts of the Indian Ocean are teeming with life because of abundant nutrients. These areas are shown in green on the satellite image below.

What process brings nutrient-rich, deep ocean water to the surface?

Africa

India

Indian Ocean

Indian Ocean

Africa

13.00.a3

500 km

Rain forests and jungles with lush vegetation cover many parts of the subcontinent. ▶

What climates allow so much vegetation to be sustained?

13.00.a4

TOPICS IN THIS CHAPTER

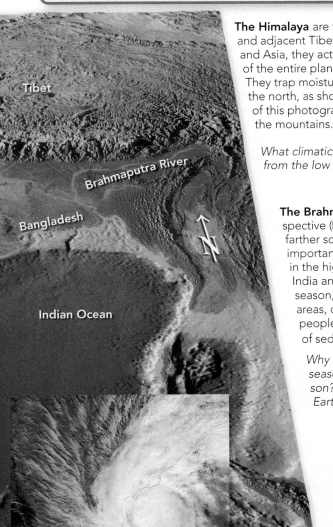

13.00.a5

The Himalaya are the highest mountains in the world. When the Himalaya and adjacent Tibetan Plateau were uplifted during the collision of India and Asia, they actually changed the atmospheric circulation patterns of the entire planet. The mountains (▶) are a major climatic boundary. They trap moisture-laden air to the south, causing a desert in Tibet to the north, as shown by the sparsely vegetated land in the foreground of this photograph. India and Nepal are on the opposite, rainy side of the mountains.

What climatic effects are caused by the extreme change in elevation from the low Indian plains to the Himalaya and Tibetan Plateau?

The Brahmaputra River, shown in this perspective (▶), along with the Ganges River farther south and west, are two of the most important rivers in the world. They begin in the high regions to the north and cross India and Bangladesh. During the rainy season, these rivers flood many low areas, displacing vast numbers of people, and carrying huge volumes of sediment.

Why do some areas have a rainy season followed by a dry season? In general, what causes Earth's seasons?

25 km

13.00.a6

Huge storms, such as the one shown here, deluge the region with torrential rainfall.

How do hurricanes and other large storms form?

13.00.a7

The Monsoons of India

The word *monsoon* is Arabic and refers to winds that reverse directions depending on the season. From June to September, India's prevailing winds blow strongly from the Indian Ocean, which lies to the south. From December to March, the winds reverse and blow southeast across the land and out to sea. The winds are caused by heating in both cases. In the summer, heating in the Himalaya and Tibet causes the air to warm and rise, generating low atmospheric pressure. Air from surrounding regions rushes in toward the low pressure. The air coming in from the Indian Ocean is rich in water vapor, which condenses when the air rises and cools over the mountains. The runoff from the torrential rain erodes the steep slopes, providing sediment to the Brahmaputra, Ganges, and other rivers. During the winter, the winds change direction, drawing dry air across the region and causing a dry season.

13.0

13.1 What Causes Winds?

THE MOVEMENT OF AIR IN THE ATMOSPHERE produces *wind*, or movement of air relative to Earth's surface. Circulation in the atmosphere is caused by pressure differences generated primarily by uneven solar heating. Air flows from areas of higher pressure, where air sinks, to areas of lower pressure, where air rises.

A What Causes Atmospheric Pressure and How Is It Measured?

Earth's atmosphere is composed of gas molecules, including nitrogen (N_2) and oxygen (O_2), that have mass and are kept from escaping to space by Earth's gravitational field. The weight of these molecules exerts a force on Earth's surface and on other objects. Pressure (force/area) is applied equally in all directions.

Pressure Decreases with Altitude

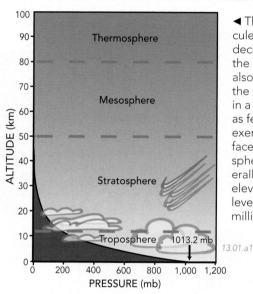

13.01.a1

◄ The density of molecules in the atmosphere decreases upward, and the force of Earth's gravity also decreases away from the surface. This results in a pressure decrease as fewer gas molecules exert force on the surfaces below them. Atmospheric pressure is generally highest at low elevations, like near sea level, and is measured in millibars (mb).

Measuring Pressure with a Barometer

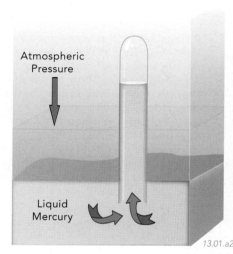

13.01.a2

◄ This barometer is a glass tube whose open end is within liquid mercury. Changes in atmospheric pressure cause the liquid level in the tube to rise or fall, allowing the measurement of relative pressure. Such barometers have units of centimeters (or inches) of mercury. Pressure is also reported in a unit called the *bar*, with one bar being approximately equal to the average atmospheric pressure at sea level. Modern digital instruments record pressure in *millibars*, which are 1/1000 of a bar.

B What Do Weather Maps Tell Us About Wind Direction and Intensity?

Weather maps contain information about high-pressure and low-pressure regions in the atmosphere as well as the location and movement of masses of warm and cold air.

A *low-pressure system* (L) has relatively low atmospheric pressure. In the northern hemisphere, winds and clouds flow counterclockwise around a low-pressure system because of the Coriolis effect, described on the next page. Areas of low pressure lift and cool moist air, causing rain and other types of unsettled weather.

A *high-pressure system* (H) has relatively high atmospheric pressure. In the northern hemisphere, air circulates clockwise around the center, and winds diverge from its center. High pressure is accompanied by sinking air and is generally associated with fair weather (i.e., not rainy or stormy). In the Southern Hemisphere, circulation directions are reversed (counterclockwise around highs and clockwise around lows).

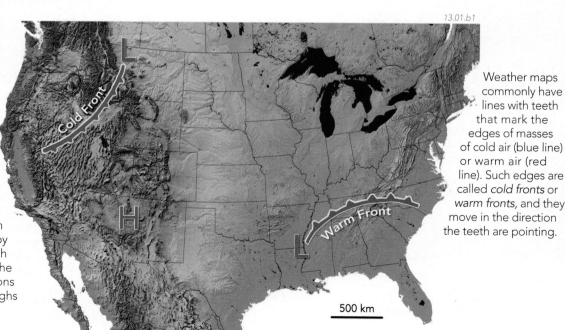

13.01.b1

Weather maps commonly have lines with teeth that mark the edges of masses of cold air (blue line) or warm air (red line). Such edges are called *cold fronts* or *warm fronts*, and they move in the direction the teeth are pointing.

C Why Do Global Wind Patterns Develop?

The Sun warms equatorial regions of Earth more than the poles, setting up a flow of warm air toward the poles, which is balanced by a flow of cold air from the poles toward the equator. Earth's rotation complicates the wind, producing curving patterns of circulating wind.

Rotation and Deflection

1. Earth is a spinning globe, with the equatorial region having a higher spin velocity than polar regions. As a result, air moving north or south is deflected sideways by the rotation, a response called the *Coriolis effect*.

2. Sunlight strikes equatorial regions more directly than it does areas closer to the poles and so preferentially heats the equatorial regions. As Earth rotates, the Sun's heat forms a band of warm air that encircles the globe and is re-energized by sunlight each day.

3. Warmed equatorial air rises and flows north and south, away from the equator. Air at the surface flows toward the equator to replace the air that rises. The Coriolis effect deflects this surface wind toward the west.

4. These flows of air combine into huge, tube-shaped cells of circulating winds, called *flow cells*. Some flow cells have surface winds flowing toward the poles. Others have winds flowing toward the equator.

Flow Cells

5. Wind direction is referenced by the direction from which it is coming. A wind coming from the west is said to be a "west wind." A wind that generally blows from the west is a *westerly*.

6. Polar regions receive the least solar heating and are very cold. Surface winds move away from the poles, carrying cold air with them. *Polar easterlies* blow away from the North Pole and are deflected toward the west by Earth's rotation.

7. *Westerlies* dominate a central belt across the United States and Europe, so weather in these areas generally moves from west to east.

8. *Northeast trade winds* blow from the northeast and were named by sailors, who took advantage of the winds to sail from the Old World to the New World.

9. *Southeast trade winds* blow from the southeast toward the equator. Near the equator, they meet the northeast trade winds in a stormy boundary called the *Intertropical Convergence Zone*.

10. *Westerlies* also occur in the Southern Hemisphere and are locally very strong because this belt has few continents, which disrupt winds.

11. *Polar easterlies* flow away from the South Pole and deflect toward the west but are mostly on the back side of the globe in this view.

Labels on globe: Polar Easterlies, Westerlies, Northeast Trade Winds, Equator, Southeast Trade Winds, Westerlies, Sunlight

13.01.c1

Why the Coriolis Effect Occurs

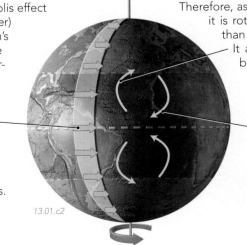

How does the Coriolis effect deflect air (and water) movement on Earth's surface? Air, like the surface, is being carried around Earth by rotation. The surface has a faster velocity near the equator than at the poles because it has to travel a greater distance in 24 hours.

Therefore, as air moves toward the poles, it is rotating faster toward the east than the land over which it moves. It appears from the surface to be deflected to the east.

The opposite occurs as air moves toward the equator and encounters areas with a faster surface velocity. The air appears to lag behind, deflecting to the west as if it were being left behind by Earth's rotation.

13.01.c2

13.1

13.2 How Does Wind Transport Material?

WIND CAN PICK UP, TRANSPORT, AND DEPOSIT MATERIAL. It moves sand and finer particles by rolling them, bouncing them, or lifting them up and carrying them. The incorporation of sediment and other material into wind results in dust storms, soil erosion, and the formation of sand dunes and other wind-related features.

A How Does Wind Transport Sediment and Other Materials?

Wind is capable of moving sand and finer sediment, as well as pieces of plants and other materials lying on the surface. It generally moves material in one of three ways.

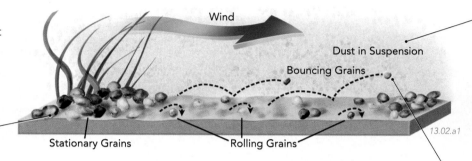

Most materials on Earth's surface are not moved by the wind because they are too firmly attached to the land (such as rock outcrops), are too large or heavy to be moved, or are both.

Wind can pick up and carry finer material, such as dust, silt, and salt, as particles drifting or sailing in air currents. This mode of transport is called *suspension*, and wind can keep some particles in the air for weeks, transporting them long distances.

If wind velocity is great enough, it can roll grains of sand and silt and other loose materials across the ground.

Very strong winds can lift sand grains, carry them short distances, and drop them. This process is akin to bouncing a grain along the surface and is called *saltation*.

B What Are Dust Storms and Whirlwinds?

Strong winds can pick up large volumes of sand and dust, forming a cloud that travels across the surface as a sandstorm, or dust storm. At a smaller scale, whirlwinds are columns of rapidly rotating air common in open fields and deserts.

In this satellite image, a massive *sandstorm* blows sand and dust west from the deserts of North Africa 1,600 km (1,000 mi) into the Atlantic Ocean. Dust storms are most common in desert areas and near cultivated fields because these areas often lack sufficient vegetation to bind the soil to the land. Low-pressure cells can lift dust high into the atmosphere, where it can drift across an entire ocean.

13.02.b2 Maricopa, AZ

The front edge of a *dust storm* can be sharp and well defined, appearing as an ominous, turbulent cloud of fast-moving dust. Such storms are common around thunderstorms in dusty regions.

13.02.b1

13.02.b3 Peru

Whirlwinds are rotating columns of wind that are typically tens to hundreds of meters across and less than 1 km high. The fast-moving winds pick up loose dust and sand, and are commonly called *dust devils*. They last only minutes, do not move much material, and do not carry material very far. Spacecraft on Mars have observed dust devils and the twisting tracks they leave on the Martian surface.

C Where Is Windblown Sediment Deposited?

Wind eventually slows down and becomes less turbulent, depositing the material it carries. Wind can form different types of sand dunes, some of which are shown below, and can also deposit windblown silt and clay as soft deposits called *loess*.

Sand Dunes

13.02.c1 Tibet

◄ Some dunes have a crescent shape, with tails pointing in the direction of the prevailing wind (to the right in this case). This type of dune is called a *crescent dune* or *barchan dune* and is very common.

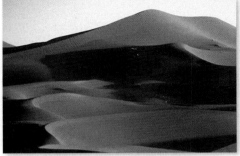
13.02.c3 Morocco

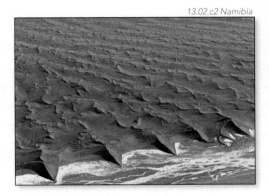
13.02.c2 Namibia

Other dunes are long and more linear or gently curved, such as the ones pictured here. Such long dunes form parallel to or perpendicular to wind direction, and can be many kilometers long. ►

Many dunes take on more irregular, complex shapes, such as the dunes shown here. Some *complex dunes*, like these in the Sahara, consist of small mountains of dunes hundreds of meters high. ▲

Loess

13.02.c4 Tibet

◄ *Loess* is wind-deposited silt and clay. Recently formed deposits can overlie topography, such as the tan material on top of this hill. They are generally loose and easily eroded, as shown here.

► Loess is very common in the Midwest of the United States, where it was the starting material for soils beneath the fertile farmlands. Most loess was blown into the area from silt and clay deposited along the front of melting ice-age glaciers.

13.02.c5 Illinois

D What Happens When the Wind Causes Erosion?

The scarcity of vegetation in desert environments and drought-stricken areas makes them susceptible to wind erosion. Strong wind can strip unprotected sand, silt, and clay-sized particles from the ground, causing locally severe soil erosion.

13.02.d1 Nebraska

Strong winds can remove so much material that the ground elevation is lowered. This wind erosion can scour away soils, causing the land to be less productive. In this photograph, wind has scoured out a depression, leaving a pillar of sediment that is capped by a resistant layer of soil.

13.02.d2 Dry Valleys, Antarctica

Wind can also bombard rocks on the surface with windblown sand, silt, or even ice, essentially sandblasting them. Many such rocks take on a smooth, polished appearance as projections and rough spots get rounded off. Others take on odd shapes, like the ones shown here.

Before You Leave This Page Be Able To

☑ Explain the ways in which the wind transports sediment and other loose material.

☑ Describe the characteristics of sandstorms, dust storms, and whirlwinds.

☑ Describe common types of windblown deposits.

☑ Describe some features formed by wind erosion.

13.2

13.3 Why Does It Rain and Snow?

PRECIPITATION TAKES THE FORM of water drops, snowflakes, and hail. It occurs when moisture-rich air is lifted or cooled in the atmosphere. Surface water evaporates, is carried in the atmosphere, and then returns as precipitation. This movement of water transfers energy from place to place and shifts energy between the ocean and atmosphere. This regulates our climate.

A What Controls the Physical State of Water in the Atmosphere?

Water occurs in three phases in the atmosphere: vapor, liquid, and solid. It converts between the different phases as the pressure and temperature change.

Evaporation and Condensation

Evaporation occurs when individual water molecules escape as gas from liquid water. These molecules mix with the nitrogen, oxygen, and other gases in the atmosphere. Evaporation at the ocean's surface is the primary contributor of water vapor to the atmosphere.

Condensation occurs when water vapor molecules in air bind together to form liquid water droplets, such as in clouds. Condensation usually occurs when air cools, as when air rises and expands.

13.03.a1

Droplet and Snowflake Formation

As water vapor condenses, it forms microdrops, which can join to form larger falling drops. As they fall, drops flatten due to wind resistance. Some drops may break apart when they become too large. Others evaporate while still in the air, providing vapor molecules that can be incorporated into other drops or into crystals of snow and ice (hail).

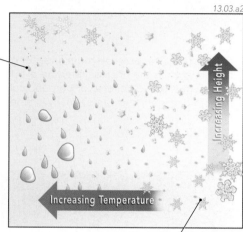

13.03.a2

Increasing Height

Increasing Temperature

The solid phase of water consists of ice crystals in snowflakes and in hail. Snow and ice form at cooler atmospheric conditions. Once formed, ice and snow can fall toward Earth's surface or remain suspended in clouds.

Water: Vapor, Liquid, and Ice

Each of the three phases of water can exist at Earth's present-day surface temperatures. This graph shows the conditions under which each phase is stable in terms of temperature and atmospheric pressure. A pressure of 1.0 atmosphere (atm) is the average atmospheric pressure at sea level.

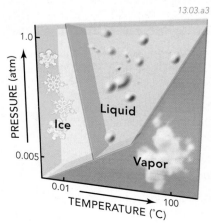

13.03.a3

PRESSURE (atm)

1.0

0.005

Ice

Liquid

Vapor

0.01

100

TEMPERATURE (°C)

Ice occurs at low temperatures, whereas liquid and water vapor are favored by higher temperatures. Higher pressure acts to hold the water molecules within the liquid rather than allowing them to escape into the air.

Cooling and Heating the Air

When water in the atmosphere changes phase, it releases or takes in thermal energy, heating or cooling the surrounding air. This diagram illustrates the change in air temperature during phase changes. Red arrows indicate that the air heats up, and blue arrows indicate that the surrounding air must provide heat to the phase change and so the air cools.

Liquid

Ice

Vapor

13.03.a4

Energy must be released into the surroundings for water to go from a less ordered to a more ordered molecular state. Heat is released (red arrows) when water vapor forms droplets or ice crystals, or when liquid water freezes.

Heat is taken in from the surroundings (blue arrows) when ice melts or water evaporates, or ice sublimates directly into vapor. This cools the air.

B What Happens When Water Evaporates or Condenses?

The evaporation and condensation of water are linked processes that occur continuously in the atmosphere. The atmosphere contains about 1% water vapor.

1. Sunlight heats an ocean or a lake, raising its surface temperature and causing liquid water to evaporate and mix with the air. The solar-heated air, with its supply of water vapor, then rises through cooler air layers. As the air rises, it creates a low-pressure zone, drawing in air from surrounding regions (not shown).

2. Prevailing wind moves the vapor-enriched air across the water or, in some cases, carries the moisture over the land.

3. As air rises over mountains, it expands and cools. This causes the water molecules to combine and form liquid droplets or ice crystals. The small water droplets and crystals form clouds. If the drops and ice crystals become large enough, they fall toward Earth, or precipitate as rain, snow, hail, or sleet.

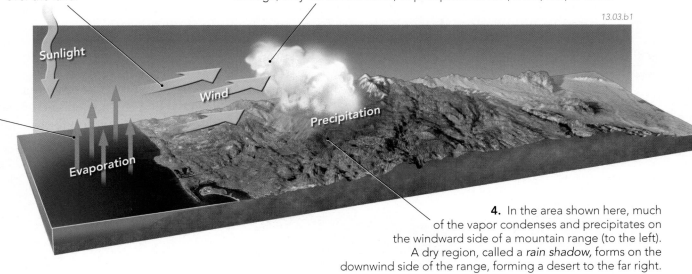

13.03.b1

4. In the area shown here, much of the vapor condenses and precipitates on the windward side of a mountain range (to the left). A dry region, called a *rain shadow,* forms on the downwind side of the range, forming a desert to the far right.

C How Does Large-Scale Atmospheric Circulation Affect Precipitation?

Differential solar radiation drives circulation both in the atmosphere and in the oceans. The atmospheric circulation patterns that develop draw cooler air toward zones of low pressure. These large-scale circulation patterns, or cells, result in different climatic zones encircling Earth.

The area north of 60° north, roughly corresponding to the polar region, is dominated by the northernmost circulation cell. In this cell, cold air descends at the pole and surface winds bring this cold air south. A similar situation occurs in the south polar region. Both polar regions have very cold temperatures, and most precipitation in these areas is as snow. This perspective shows the North Pole, but not the South Pole.

Most of Europe (and the mainland United States) is affected by a large circulation cell between 30° and 60° north. At the northern boundary of this cell, north-moving air at the surface meets the air flowing south from the pole, causing the air to rise and cool, promoting rain and snow along the boundary.

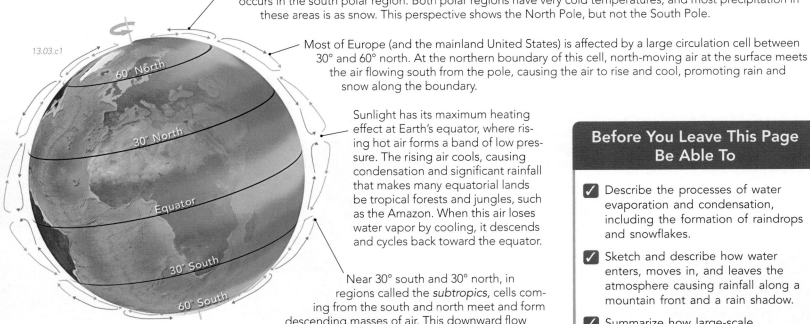

13.03.c1

Sunlight has its maximum heating effect at Earth's equator, where rising hot air forms a band of low pressure. The rising air cools, causing condensation and significant rainfall that makes many equatorial lands be tropical forests and jungles, such as the Amazon. When this air loses water vapor by cooling, it descends and cycles back toward the equator.

Near 30° south and 30° north, in regions called the *subtropics,* cells coming from the south and north meet and form descending masses of air. This downward flow dries out the air in these regions, resulting in low rainfall. Such dry regions include the Sahara and Kalahari Deserts of Africa, deserts of the Middle East, and deserts of the American Southwest.

Before You Leave This Page Be Able To

✓ Describe the processes of water evaporation and condensation, including the formation of raindrops and snowflakes.

✓ Sketch and describe how water enters, moves in, and leaves the atmosphere causing rainfall along a mountain front and a rain shadow.

✓ Summarize how large-scale atmospheric circulation affects precipitation and climate.

13.3

13.4 How Do Hurricanes, Tornadoes, and Other Storms Develop?

HURRICANES, TORNADOES, AND OTHER SEVERE STORMS are some of nature's most awe-inspiring spectacles. A *hurricane* takes days to traverse an ocean and spreads heat as it moves across the globe. Other storm events, such as *supercell thunderstorms* and *tornadoes*, are generally land-based disturbances that last from minutes to hours. How do such storms form, and how do they operate?

A What Is a Hurricane?

Tropical *hurricanes*, *typhoons*, and *cyclones* are all names for immense seasonal storms that form primarily in the warm waters of the Atlantic, Pacific, and Indian oceans, respectively. They are characterized by swirling high-velocity wind, heavy rain, and high storm surges that cause high waves and flooding ahead of the storm.

1. Hurricanes and related storms are huge, circulating masses of clouds and warm, moist air. They are zones of low atmospheric pressure that cause air to rise and condense, creating locally intense rainfall.

5. Dry air flows down the center of the storm, compresses, and evaporates clouds, forming a cylinder of relatively clear, calm air, called the *eye*.

4. Hurricanes, like other low-pressure zones, spiral counterclockwise in the Northern Hemisphere. The overall path of the hurricane is steered by air currents high in the atmosphere.

2. Warm ocean water in the hot tropics is the driving force in hurricane formation. Warm water evaporates from the sea surface, mixes with the air, and rises. As the warm, vapor-rich air rises, it cools and its vapor condenses to liquid water (rain). This heats the air, which rises higher and draws in more moisture-rich air to replace the rising air.

3. If the hurricane encounters additional warm, evaporating water as it moves, more heat is added, wind increases, and the hurricane grows in strength and size. A hurricane dissipates in strength when it passes over land or cool water.

13.04.a1

Why Do Hurricanes Rotate?

Earth's rotation deflects air and water movement on its surface in accordance with the Coriolis effect. The rotation of hurricanes is one manifestation of this effect. The Coriolis effect deflects moving air to the right of its trajectory in the Northern Hemisphere and to the left of its trajectory in the Southern Hemisphere.

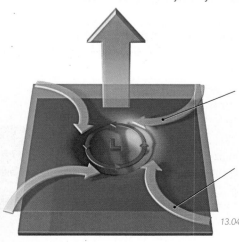

When a low-pressure zone forms by the heating of air at the ocean's surface, cool air flowing in from all directions is deflected by the Coriolis effect. This produces a vortex shape.

This deflection to the right forms the counterclockwise spin of Northern Hemisphere hurricanes.

13.04.a2

Where Do Hurricanes Occur?

Hurricanes are most common between the subtropics and the equator, and they occur seasonally as seawater warms.

13.04.a3

June-November

June-December

August-October

June-October

January-March

January-March

Hurricanes occur when sea-surface water temperatures exceed 26°C to depths of 200 m. These areas are outlined in yellow on this map. Different parts of the oceans reach these temperatures at different times.

B What Is a Supercell Thunderstorm?

Supercell thunderstorms are rare compared to ordinary thunderstorms, but they account for most of the damage caused by thunderstorm activity. These storms can generate powerful tornadoes and can last for hours.

13.04.b2

◄ **1.** *Thunderstorms* are columns of turbulent, moist air with variable amounts of lightning, thunder, rain, hail, and strong wind. Well-developed thunderstorms develop a flat-topped, anvil shape. The long point of the anvil generally points in the direction the storm is moving (left to right in the figure below).

2. Some large thunderstorms, including supercell thunderstorms, are capped by a rounded upper dome. This indicates that strong updrafts are taking moist air up higher than is typical for an average thunderstorm.

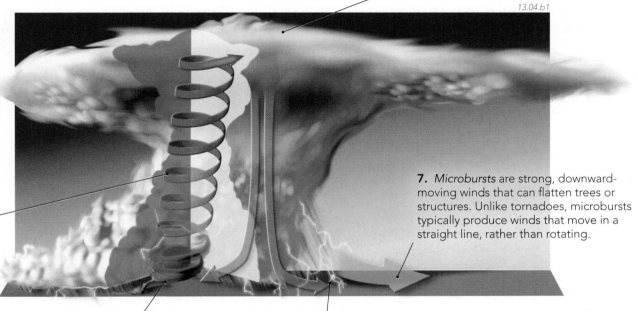

13.04.b1

3. *Supercell thunderstorms* begin as a horizontal vortex or spinning air mass that becomes vertical. All supercells have a rotating updraft.

7. *Microbursts* are strong, downward-moving winds that can flatten trees or structures. Unlike tornadoes, microbursts typically produce winds that move in a straight line, rather than rotating.

4. Intense rain and hail form within the storm and fall toward the ground, often in brief but heavy bursts (downpours).

5. Strong upward-moving winds, called *updrafts*, can rotate, spawning tornadoes as extensions of the updraft. About one-third of supercell thunderstorms form tornadoes.

6. *Lightning* results from electrical currents within the storm. It can discharge in the air or reach the ground. *Thunder* results from rapid heating and expansion of the air along the path of the lightning bolt.

C What Are Tornadoes and Where Do They Strike?

Tornadoes are violent, rotating funnel-shaped columns of air that extend down to the ground. If the funnel does not reach the ground it is simply called a *funnel cloud*. Tornadoes and funnel clouds develop in thunderstorms and along weather fronts where storm rotation is initiated. Wind speed in a tornado is commonly hundreds of kilometers per hour.

13.04.c2 Kansas

Most tornadoes have a characteristic downward-tapering funnel shape, but some are shaped like a wide cylinder. ▲

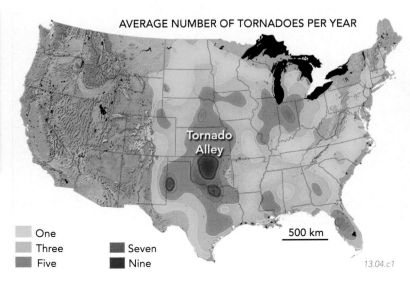

AVERAGE NUMBER OF TORNADOES PER YEAR

Tornado Alley

One	
Three	Seven
Five	Nine

500 km

13.04.c1

Tornadoes in the United States occur mostly in the central and eastern parts of the country, in what is sometimes called *tornado alley*. Tornado alley is the large, orange and red area in the center of the map.

Before You Leave This Page Be Able To

✓ Describe how hurricanes form and where they get the energy to grow.

✓ Sketch and summarize the various characteristics of a supercell thunderstorm.

✓ Describe the characteristics of tornadoes, what they represent, and where in the U.S. they are common.

13.4

13.5 How Do Ocean Currents Influence Climate?

EARTH'S OCEANS ARE UNEVENLY HEATED by the Sun. Shallow surface currents are driven mostly by the wind, whereas deep currents are driven by density differences caused by variations in temperature and salinity. A linked system, involving both deep flow and surface currents, controls many climates on Earth.

A What Are the Main Ocean Surface Currents?

Surface currents are relatively fast-moving horizontal flows of shallow ocean water. They can extend to depths of 1,000 m but are usually much shallower. They typically affect about 20% of an ocean's volume. Surface currents are driven mostly by the wind and are deflected by continents, resulting in several types of currents.

1. *Western boundary currents (in red), includ-ing* the Gulf Stream, are warm. They flow from the equator to-ward high lat-itudes and polar regions.

2. *Eastern boundary currents* are along the eastern sides of oceans. They are cold and flow from high latitudes toward the equator.

3. East- and west-flowing surface currents, shown in black, transport water between the eastern and western boundary currents, defining a circular ocean current.

4. *Northern Hemisphere currents* are combinations of two north-south and two east-west currents. These linked currents flow in a clockwise circular pattern.

5. The *North and South Equatorial Cur-rents* flow westward across all ocean basins. An eastward *counter current* between the two partially returns water to the east.

6. *Southern Hemi-sphere ocean surface currents* flow in a counterclockwise circular pattern.

7. The *Antarctic Circumpolar Current* flows from west to east around Antarctica. A west-flowing current is present closer to the continent.

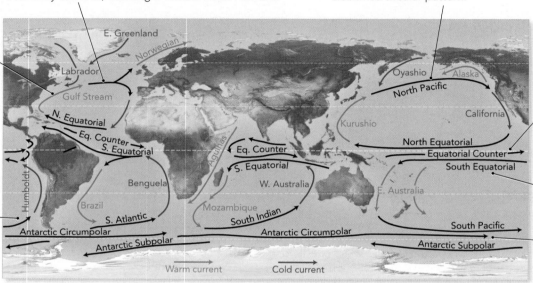

13.05.a1

B Where Do Deep Ocean Currents Flow?

Oceans have distinct currents at different depths. A deep-water current called the *thermohaline conveyor* takes deep water on a long journey. The conveyor is driven by variations in density caused by differences in tempera-ture and salinity. Water that is cold or especially saline is denser than warm or less saline water.

1. A northerly flowing current moves warm salty water up toward the cold North Atlantic. The saline water loses heat to the at-mosphere, becomes denser, and sinks.

2. The dense cold water travels south and then east along Antarctica. It mixes with very dense, salty water left over from the formation of sea ice near Antarctica.

3. The deep water flows northward, surfacing in the northern Pacific, bringing deep-sea–derived nutrients that make this area rich in marine life. After surfacing, the current loops back to the south and then west to complete the loop. The entire trip takes about 1,000 years. A second area of upwelling occurs in the Indian Ocean.

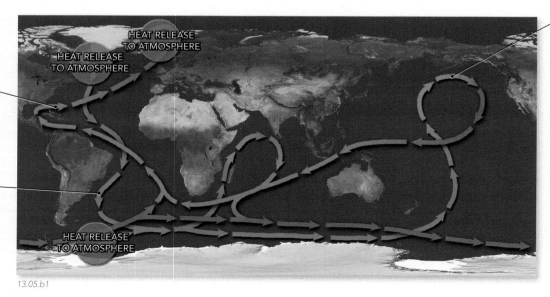

13.05.b1

How Do Ocean Currents Affect Temperatures and Precipitation on Land?

The temperatures of surface currents, combined with the prevailing wind, control many local climates. Oceans heat up and cool more slowly than land, and so they help moderate temperatures.

Surface Currents

The combination of wind patterns and surface currents has a controlling effect on climate.

13.05.c1

The warm East Australian current flows south along the continent's east side.

Prevailing easterly winds pick up moisture from the current, and then flow over Australia, causing rain along the east coast.

The driest continent on Earth would likely be even drier without the moisture associated with this warm current.

Deep Conveyor

The Gulf Stream brings waters heated from the tropics to the North Atlantic, helping warm the region.

13.05.c2

When warm waters of the Gulf Stream enter the North Atlantic, they lose some heat to wind blowing to the east.

Winds blowing from the ocean toward the land help moderate land temperatures and bring moist air over northern Europe, influencing its temperature and rainfall.

El Niño and La Niña

1. Periodically, the *Pacific Equatorial Counter Current*, which flows east across the Pacific toward South America, naturally strengthens or weakens. This significantly changes the weather conditions across the Pacific for up to a year at a time. A stronger-than-normal current causes what is called an *El Niño* condition and a weaker-than-normal current is a *La Niña* condition. These are clearly visible on maps of satellite data for sea-surface temperatures. Warm colors (red and orange) show warmer seas, whereas cool colors (blue and purple) show cooler water.

2. *El Niño* occurs when warmer-than-average ocean surface temperatures occur in the central and east-central equatorial Pacific. This image shows the pattern of increased sea-surface temperatures (red and orange) during El Niño.

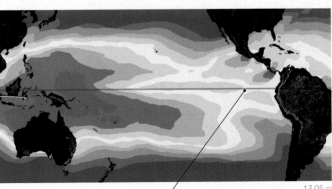
13.05.c3

3. The increased sea temperatures in the eastern Pacific Ocean during an El Niño condition lead to increased evaporation, causing more rainfall in nearby parts of South and Central America. In the United States, El Niño conditions are expressed by warmer than normal winters in the upper Midwest, cooler than normal winters in the Southeast, and increases in winter precipitation in the Southwest.

4. *La Niña* occurs when the equatorial counter current weakens, accompanied by an increase in easterly winds. High sea temperatures become restricted to the western and central Pacific and the region near Indonesia. This causes an increase in rainfall for nearby parts of Australia and Indonesia.

13.05.c4

Before You Leave This Page Be Able To

☑ Sketch and describe the main flow of surface currents in the Northern and Southern Hemispheres and how they influence sea temperatures.

☑ Describe the thermohaline conveyor.

☑ Summarize how ocean currents influence temperature and precipitation on adjacent lands, using Australia, Europe, and El Niño and La Niña effects as examples.

5. La Niña causes cooler-than-average ocean surface temperatures in the eastern equatorial Pacific. This results in relatively drier and warmer winters in the South and Southwest, and warmer and wetter winters in the Northwest.

13.5

13.6 What Causes Short-Term Climatic Variations?

THE OVERALL CLIMATE OF A REGION is controlled by its latitude (how far north or south it is), its elevation, the types of ocean currents that affect the region, the prevailing winds, and other factors. Many regions display dramatic shifts in overall climate from season to season or over years and decades. These occur in response to changes in the wind and ocean currents and to persistent, but atypical, weather patterns for that region. Such changes can bring the onset of torrential rains or can bring drought. Temperature, wind, and rainfall patterns also change during the seasons, from winter to summer or wet to dry.

A What Is a Monsoon?

The word *monsoon* signifies a change in the prevailing wind direction from one season to another. In some areas, the changing wind patterns cause torrential rainstorms.

1. The most dramatic monsoon affects India. During the summer months, the lands of Asia, such as the Gobi Desert north of Tibet, heat up from the more direct summer sunlight. The heated air rises, producing massive low-pressure zones over the land.

2. Air over the Arabian Sea and Bay of Bengal (parts of the Indian Ocean) flows inland toward the low-pressure zones, bring-ing ocean-derived moisture with it. This onshore flow of moist air causes an increase in the amount of rainfall.

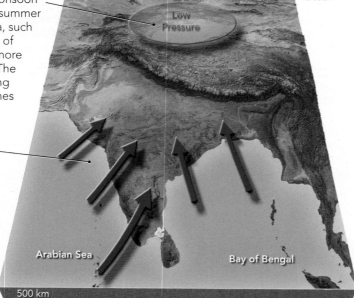

13.06.a1

3. As the moist air flows farther inland, it begins to rise over the Himalaya, where it cools, forming thick clouds (▼). These cause massive rainfall along the south flank of the range and impressive amounts of snow in the higher peaks. The precipitation is heavy because the elevation change from the low-lying plains in India to the high Himalaya is extreme.

13.06.a2 Himalaya, Nepal and Tibet

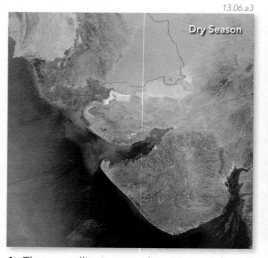

4. These satellite images show increased vegetation due to monsoon-related rains along the west coast of India. The left image is from May 2003, during the dry season, when wind patterns do not bring in moist air. The right image is from October 2003, right after the end of the mon-soon. During the monsoon, winds from the south brought abundant moisture into the area. Note the increase in plant cover (green areas) during the monsoon-caused rainy season.

Before You Leave These Pages Be Able To

✓ Explain what a monsoon is and how it affects rainfall, using India as an example.

✓ Summarize some common manifestations of drought and three causes of drought.

✓ Sketch and explain how the tilt of Earth's axis causes the seasons. Include a sketch showing the northern summer and one for the northern winter.

B What Causes a Drought?

Drought is an extended period of below-average precipitation, either a relative lack of rainfall, lack of snowfall, or both. Drought places stress on plants and animals and can cause normally green and fertile land to dry up into eroded, dusty plains. Drought can cause water shortages by drying up streams, rivers, and lakes, and by decreasing flow into groundwater. Three causes of drought are described below.

13.06.b1

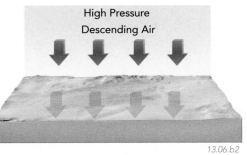
13.06.b2

13.06.b3

A persistent weather pattern or change in ocean currents can cause an unusual shift in wind direction. This can bring dry air over a region and cause drought.

Drought can be caused by weather patterns that are atypical for an area. Persistent high pressure, with its sinking air, can cause a region to be drier than normal.

A temporary change in the direction or strength of an ocean current can last multiple years, as occurs with La Niña, causing a temporary change in climate, possibly including drought.

C How Does the Tilt of the Earth Cause the Seasons?

The reason why we have summer and winter has nothing to do with being closer to the Sun during some seasons. Instead, it is because the spin axis of Earth is tilted relative to the plane in which Earth orbits around the Sun.

1. Earth's orbit around the Sun is nearly circular, when viewed from directly above the plane of the orbit (▶). The axis about which Earth spins on a daily basis, to produce day and night, is tilted relative to this plane. Over the course of a year as Earth orbits around the Sun, Earth's spin axis is tilted toward the Sun during some months and away from the Sun during other months. As a result, sunlight shines more directly on the Northern Hemisphere during some months and on the Southern Hemisphere during other months, causing the seasons.

2. At some times of the year, Earth's position in its orbit around the Sun causes the spin axis going through the North Pole to be tilted toward the Sun. This results in the situation shown below.

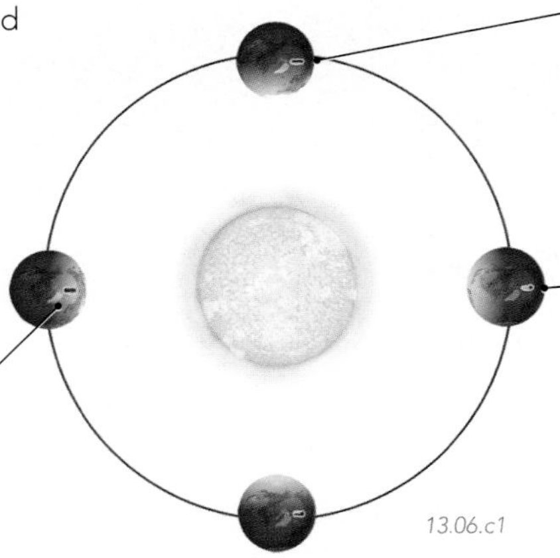

13.06.c1

4. At this position of Earth, and on the one on the opposite side of the orbit, the spin axis is tilted neither toward nor away from the Sun, and the Northern and Southern Hemispheres experience spring or fall.

5. When Earth is in this position relative to the Sun, the spin axis is tilted away from the Sun. This results in winter in the Northern Hemisphere, as shown in the figure below.

3. When the spin axis is tilted toward the Sun, sunlight strikes the Northern Hemisphere more directly, and these direct rays of sunlight pass through less atmosphere than rays striking the Southern Hemisphere. More sunlight means more infrared energy hitting the land and water, and more heat transmitted back to the atmosphere. This results in warmer temperatures, and summer, in the Northern Hemisphere. In contrast, less sunlight strikes the Southern Hemisphere and the sun rays strike the Southern Hemisphere at an angle, so the Southern Hemisphere experiences winter.

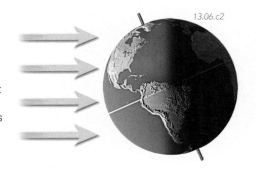
13.06.c2

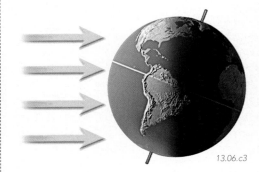
13.06.c3

6. When the spin axis is tilted away from the Sun, sunlight shining on the Northern Hemisphere is less direct and has to travel through more atmosphere than sunlight hitting the Southern Hemisphere. Less sunlight in the Northern Hemisphere means less infrared energy transmitted back to the atmosphere, resulting in colder temperatures and the northern winter. In contrast, the Southern Hemisphere now receives more direct sunlight and so experiences summer. When it is summer in the Southern Hemisphere, it is winter in the Northern Hemisphere, and vice versa. This would not be the case if the seasons were in any way related to distance from the Sun.

13.6

13.7 What Controls the Location of Rain Forests?

RAIN FORESTS CONTAIN EXTREMELY LUSH GROWTHS OF PLANTS and are one of Earth's most important ecosystems. Rain forests comprise only 2% of Earth's land surface, yet hold nearly half of terrestrial life. The majority of rain forests are tropical and lie along or close to the equator.

A What Is a Rain Forest?

A forest with high annual rainfall, no freezing temperatures, and a rich collection of plant and animal communities is a *rain forest*. Rain forests have the greatest diversity of species of any land ecosystem.

Tropical Rain Forests

Tropical rain forests are warm and humid, supporting a multitude of plant and animal species. They are the most abundant type of rain forest. The tropical rain forest below is in northwestern Australia.

Temperate Rain Forests

Temperate rain forests experience cool winters and are limited to certain coastal areas with high rainfall. Compared to tropical rain forests, they have lower plant density and less diverse vegetation and animals.

Rain Forest Structure

Treetops form a *canopy* or umbrella over the forest floor, capturing most of the sunlight. Rain forest canopies contain most of the rain forest's species.

The understory layer is shaded, so the main plants in this setting are short with broad leaves.

The forest floor is very dark, and organic debris decomposes rapidly. The soil is highly leached by high amounts of rainfall and so contains few nutrients.

13.07.a1

13.07.a2

B Where Do Rain Forests Occur?

Most of the world's rain forests lie in the tropics and are associated with the *Intertropical Convergence Zone (ITCZ)*, the sinuous line along the equator where the northeast and southeast trade winds converge. The ITCZ swings north and south seasonally, shifting low pressure and immense rainfall to different regions.

The Pacific Northwest contains almost two-thirds of the world's temperate rain forests. In this area, onshore winds bring winter rain and summer fog.

India's rain forest locations are related to the ITCZ and to inland flow of moist air during the monsoon.

The typical June position of the ITCZ is the orange line, and the December position is the blue line.

Tropical rain forests, including those in Southeast Asia, thrive along the equator, where every day has 12 hours of sunlight. Rising water-rich air within low-pressure zones brings frequent rain to promote lush plant growth

The Amazon in South America is the world's largest and most diverse rain forest. The ITCZ is located over the Amazon nearly year round.

Australia has both tropical and temperate rain forests. Rain is provided by prevailing winds that blow over warm ocean currents.

The central African rain forest is the second largest in the world. It receives heavy rainfall from the ITCZ for much of the year, but especially in December and January.

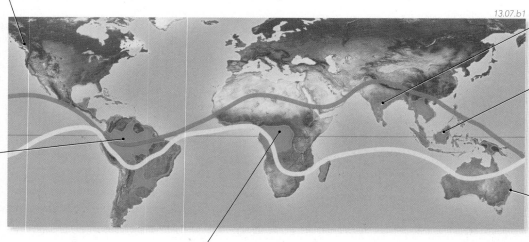

13.07.b1

C Why Are Rain Forests Disappearing?

Deforestation of rain forests is occurring at about 2.5 acres per second, or 80 million acres per year. Some ecologists estimate that most rain forests will be destroyed by 2040. The main cause is economic pressure.

13.07.c2 Brazil

13.07.c3 Central America

13.07.c1

1. Commercial logging for tropical hardwoods like mahogany consumes large areas of rain forest. Logging commonly involves *clear-cutting*, where all the trees are removed.

2. Exploitation of copper, gold, oil, and other natural resources (◄) has destroyed tracts of rain forest, especially in Africa, South America, and Indonesia.

3. Cattle ranchers clear rain forest and then plant pasture grasses. Clearing the land for whatever reason allows soil erosion. This precludes the reestablishment of indigenous forest species.

4. Major highway construction and road building in support of logging and other development destroy swaths of forest. New roads provide arteries for other types of development, leading to more loss of pristine rain forest.

5. Dam construction, such as for hydroelectric power, destroys forests in areas flooded by the reservoirs and may withhold water from rivers downstream.

6. The greatest threat to rain forests is subsistence farming, where forests are cleared by "slash-and-burn" practices (▲). This occurred in countries such as Brazil, where 3 million homeless people were encouraged by their government to farm poor soil.

D What Role Do Rain Forests Play in Ecology?

Rain forests are critical ecosystems with key ecologic niches. Extreme rainfall (150–400 cm/yr) and infiltration into the ground leaches nutrients from the soil. The rain forest ecosystem of plants, animals, insects, and microbes is nutrient recycling—organisms die, rapidly decompose, and return nutrients to the system.

Carbon Dioxide Uptake

Rain forests are responsible for about 30% of the photosynthetic activity on Earth. If CO_2 levels in the atmosphere increase because of volcanic activity (▼) or other natural or human-related causes, rain forests can increase their CO_2 uptake. Thus, they act as a buffer against climate change.

13.07.d1 Hawaii

Diversity Storehouse

Rain forests are estimated to contain at least 5 million species of plants and animals. They are the genetic storehouse for the world's ecology, including lush environments like the trees and plants below.

13.07.d2

Local Climate

Rain forests intercept and use solar energy that would otherwise strike the ground. As a result, rain forest trees and plants provide shade and help to keep the land underneath cooler and sheltered during the day.

Before You Leave This Page Be Able To

✓ Describe the characteristics and vertical structure of a rain forest.

✓ Summarize where rain forests occur and what conditions produce enough precipitation to form a rain forest.

✓ Explain threats to rain forests and why rain forests are ecologically and genetically important.

13.7

13.8 What Are Deserts and How Do They Form?

DESERTS are dry lands, often with little vegetation, that cover many parts of Earth's land. Most deserts are not barren sand dunes; they contain plants and animals adapted to life in a dry environment. What conditions create deserts, what controls their locations, and are deserts growing over time?

A What Is a Desert?

An *arid region* receives less precipitation than it loses to evaporation and other processes. Arid regions that have less than 25 cm (10 in) of rainfall per year are known as *deserts*. Vegetation is sparse in desert ecosystems, commonly covering less than 15% of the ground. Many desert areas lack permanent streams.

Sahara and Sahel

This satellite image shows the northern half of Africa. Most of the region is tan colored because it consists of sand and rock, with very sparse vegetation. This tan region is the *Saharan Desert*, stretching from Morocco to Egypt.

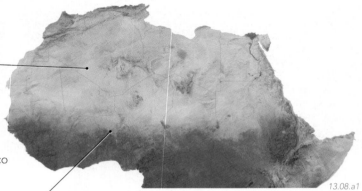

13.08.a1

South of the Sahara is the *Sahel*, a region that is relatively dry but not quite a desert. Regions intermediate between a true desert and a more humid climate are called *semiarid*. The Sahel is currently threatened by the encroachment of deserts from the north. The region has recently experienced a number of devastating droughts.

Mojave and Sonoran Deserts

▶ The Mojave Desert of Southern California has rocky mountain slopes with very few plants. The valleys can be sandy or rocky.

13.08.a2 Big Maria Mtns., CA

▶ The Sonoran Desert of Arizona receives more rain than the Mojave Desert, much of it during a summer monsoon in the Southwest. It has more cactus and other heat-adapted plants.

13.08.a3 Central AZ

B Where Do Deserts and Other Arid Lands Occur?

Examine the map below, which shows arid and semiarid regions in orange. Where are most of the world's arid lands? Do dry lands occur in certain settings? Compare the distribution of these lands with the (1) locations of atmospheric cells of rising and descending air, (2) locations and directions of ocean surface currents, (3) directions of prevailing wind, and (4) locations of mountains between the deserts and oceans.

This map of the world's deserts shows areas with that receive less than 25 cm of precipitation per year or areas that have desert biomes. A *biome* is a major community of organisms defined by the predominant types of vegetation and characterized by organisms that are adapted to that particular environment. The map does not show Antarctica, which like Greenland, is partly classified as a desert because some areas receive less than 25 cm of precipitation (mostly snow) per year.

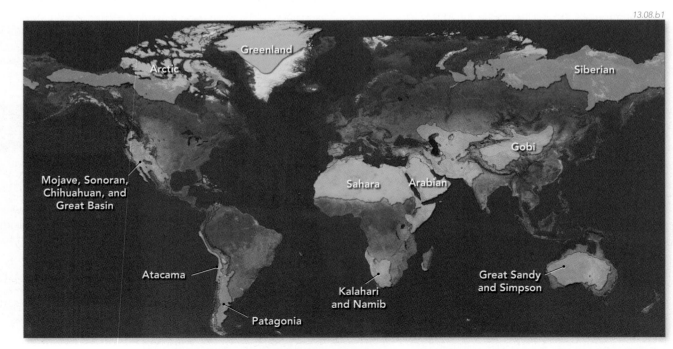

13.08.b1

C In What Settings Do Deserts and Other Arid and Semi-Arid Lands Form?

Earth's large deserts result primarily from descending air in subtropical belts and from airflow associated with cold ocean currents. Deserts also form in rain shadows associated with mountain ranges and in cold, dry polar regions.

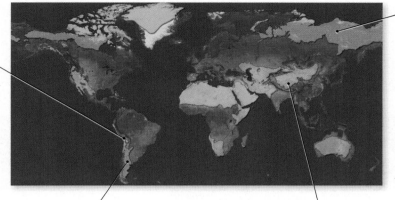

13.08.c1

Coastal deserts form where cold, upwelling ocean currents cool the air and decrease its ability to hold moisture. This applies to the Atacama Desert, one of the driest places on Earth.

The world's largest deserts, including the Sahara, Sonoran, and the Australian deserts, are *subtropical deserts*. These deserts form where the general atmospheric circulation brings dry air into the subtropics (near the Tropic of Cancer and the Tropic of Capricorn). These areas are situated beneath descending flow cells and are associated with high pressure.

Polar deserts form where cold, dry air prevails. Any available moisture is frozen for almost the entire year. Examples of a polar desert are Siberia, large areas of the Arctic, and parts of Antarctica.

When moist air rises up over mountains, it rains. As the air descends on the downwind side of the mountain, it dries, forming a *rain-shadow desert*. The Andes mountains extract moisture from westerly winds, forming the Patagonia Desert to the east. The setting of a typical rain shadow is shown below.

Continental deserts form in the interiors of some continents, far from sources of moisture, or where prevailing winds blow toward the sea. In many such settings, summers are hot and winters are very cold, like in the Gobi Desert of Mongolia. The flow of relatively dry air from a continental interior can form deserts in adjacent lands. ▼

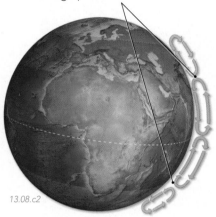

13.08.c2

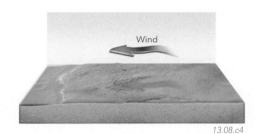

Wind   **Rainshadow**

13.08.c3

Wind

13.08.c4

D What Is Desertification?

Extended periods of drought, overgrazing by livestock, poor farming techniques, and diversion of surface water can cause soil loss and change grasslands to desert. Converting other lands to desert is called *desertification*.

This map shows the global risk of desertification. Red and orange indicate areas with high to very high risk. Areas most at risk include central Africa (the Sahel), the Middle East, and parts of the United States, including fertile farmland of the Great Plains.

13.08.d1

Plants shield the ground surface from erosion and hold the soil in place. Loss of plants and soil due to draught promotes erosion and desertification.

13.08.d2

Before You Leave This Page Be Able To

☑ Describe what deserts and other arid lands are and where they occur.

☑ Summarize how and where different kinds of deserts form.

☑ Describe desertification.

13.8

13.9 What Features Are Common in Deserts?

THE DRYNESS OF DESERTS is expressed in the sparseness of vegetation and a lack of well-developed soil on many hillsides. Deserts have a number of other characteristic landscape features, including sand dunes, dry lake beds, channels that are normally dry, and dark coatings on rock faces.

A What Landscape Features Are Characteristic of Deserts?

The low rainfall, dry atmosphere, and sparse vegetation in most deserts result in several common features. These include alluvial fans, desert washes, playa lakes, and sand dunes.

13.09.a1 Death Valley, CA

13.09.a2 Baja California, Mexico

1. *Alluvial Fans*—Loose rocks accumulate on rocky mountain slopes, are transported by streams and debris flows, and are deposited as fan-shaped aprons called *alluvial fans*. Alluvial fans form where steep, confined channels encounter more level terrain at the mountain front or where a smaller stream joins a larger one.

2. *Desert Washes*—Deserts contain sand and gravel-rich channels called *washes* or *arroyos*, which are normally dry. During intense rain, such as summer thunderstorms, a wash can rapidly fill with water draining off the land and rocky hillslopes. This causes a rapid rise in water and a *flash flood*.

13.09.a3 Badwater, Death Valley, CA

13.09.a4 Death Valley, CA

3. *Playas*—Shallow, closed basins are playas. They receive water from precipitation, runoff, or springs, but the water has no outlet. Many playas partially or totally dry up, forming *salt flats*, like those that flank Badwater, the lowest point in the United States.

4. *Dunes*—Some sand in deserts is not held down by vegetation. It will form dunes if there is a sufficient supply of sand, such as from washes, and if the wind is strong enough to pick up and concentrate material.

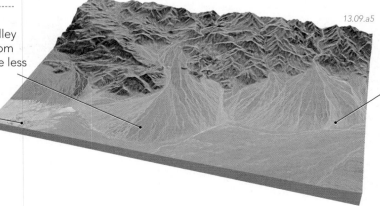

13.09.a5

5. This satellite image of Death Valley shows alluvial fans that build out from the mountain front and down to the less steep basin floor.

6. The flat parts of the basin floor contain sand dunes and lake beds, including salt, deposited when playa lakes were more extensive.

7. The lighter colored streaks running across the alluvial fans are desert washes, which branch and spread out over the fans in a network of channels. These are normally dry but contain water and muddy debris during heavy rainfall and the resulting flash floods. Older parts of the alluvial fans are darker, because of desert varnish and desert pavement (described on the next page).

B What Features Develop over Time on Desert Landscapes?

Landscapes develop a number of features, some of which are more common in deserts than in other environments. Some of these features reflect the relative lack of rainfall in deserts.

13.09.b1 Mojave Desert, CA

Pediment—Erosion and weathering of bedrock can carve a gently sloping erosion surface called a pediment. The broad dome in the distance is a gently sloping pediment. Pediments are visible in many deserts, especially where erosion has occurred over a long time, but pediments are not confined to deserts.

13.09.b2 Sonoran Desert, Western AZ

Desert Pavement—Over time, many desert surfaces become armored by rocks, forming a natural pavement. Rocks become concentrated on the surface because finer materials blow away, wash away, or move down into the soil. Desert pavement takes thousands of years to fully develop.

13.09.b3 Tempe, AZ

Caliche—Over time, soil in many environments accumulates soluble minerals, such as calcium carbonate (calcite). These minerals dissolve in rainwater, which percolates down into the soil. When the water evaporates, the dissolved components precipitate as coatings on clasts or as a distinct, hard layer, shown here, called *caliche*. In wetter environments, dissolved material is flushed completely through the soil by descending waters.

13.09.b4 Butler Wash, UT

Desert Varnish—Exposed surfaces of resistant rocks can get coated with iron and manganese oxides and other materials, forming dark, natural *rock varnish*. The material in the varnish is largely wind-derived clays, oxides, and salts. Varnish takes thousands of years to form. In the example above, Native Americans carved into the varnish to create larger-than-life petroglyphs. The larger figures carved into this varnish-coated wall are several meters high.

13.09.b5 Yampa River, UT

Natural Stains—Many desert cliffs display vertical streaks of red, brown, and black. These colors are mostly coatings of iron oxides. They were deposited by water that flowed down the rock face and evaporated. Such stains are most obvious in deserts but can develop in other environments.

Before You Leave This Page Be Able To

☑ Describe some features of deserts and how each forms. These features include alluvial fans, washes, playas, dunes, pediments, desert pavement, caliche, desert varnish, and natural stains.

13.9

13.10 What Is the Evidence for Global Warming?

OVER THE LAST 150 YEARS, people have measured atmospheric temperatures. This record, albeit short in geological terms, shows an overall increase in temperatures—*global warming*. There is currently much scientific and political discussion of this topic. What is the actual evidence that Earth is warming?

A What Is Global Warming?

Global warming means *increasing* global atmospheric and oceanic temperatures from some point in the past to the present. Scientists examine various records of Earth's temperature history to investigate whether warming has occurred and whether it is related to natural events or to human activity. In addition to direct temperature measurements, we can infer past temperatures from other types of observations, called *proxy evidence*. Most of the graphs below are from a 2006 report by the National Academy of Sciences (NAS) and indicate how temperature is interpreted to have varied, relative to an *arbitrary* mean global temperature (averaged from 1961 to 1990).

Thermometer Record

Thermometers provide a direct measurement of air temperature. This record shows an average variation in temperature for the last 140 years. According to this record, it appears that average air temperatures have increased over the last century. Before the 1940s, the data show a relatively cool period.

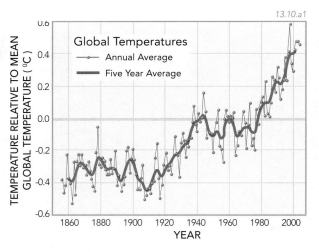

13.10.a1

Temperature data collected in the last decade show little or no warming since 1998, and possibly even a minor amount of cooling. The temperatures, however, remain above average.

Borehole Temperature

Earth's shallow subsurface is heated by two sources: heat from the atmosphere and heat from the deeper subsurface. If surface temperatures have changed over time, then the shallow subsurface temperature should reflect those changes. There are thousands of bore (drill) holes around Earth, and measurements of temperature with depth are compared to predicted subsurface temperatures to infer changes in air temperature.

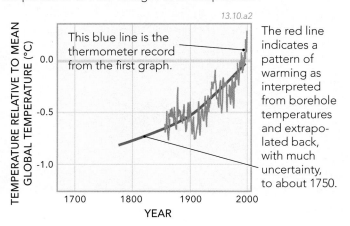

13.10.a2

This blue line is the thermometer record from the first graph.

The red line indicates a pattern of warming as interpreted from borehole temperatures and extrapolated back, with much uncertainty, to about 1750.

Glacier Length

Glaciers flow from areas of snow accumulation to lower elevations. The dynamics and energy flow of glacier movement and retreat are well understood. Most scientists interpret changes in the lengths of glaciers to be related to changes in atmospheric temperature and in the amount of precipitation. The combined data from glaciers around the world are interpreted as a warming trend beginning before the turn of the century. Note that changes in glacier length started around 1850 and mostly occurred prior to 1940.

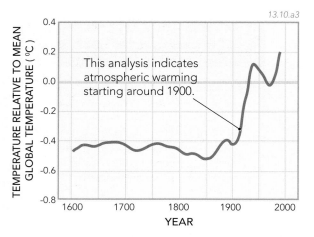

13.10.a3

This analysis indicates atmospheric warming starting around 1900.

Tree-Ring Proxy

Tree-ring growth is partly dependent on climate. Some trees can grow for 300 years or more, and thickness of tree rings of successive populations from temperature-sensitive forests can be correlated around the world. This provides a climate record going back more than 1,200 years. This record shows an increase in air temperature between 1850 and 1998.

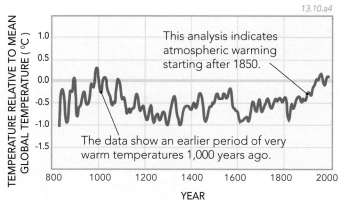

13.10.a4

This analysis indicates atmospheric warming starting after 1850.

The data show an earlier period of very warm temperatures 1,000 years ago.

Ice Core Proxy

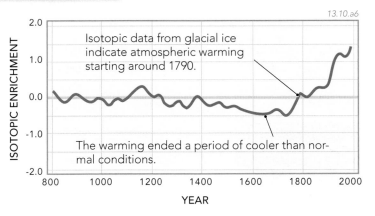

13.10.a5

Continental glaciers on Greenland and Antarctica have yearly layers that record winter precipitation and summer dust accumulations. Scientists extract ice cores by drilling, and then chemically analyze gases and ice in refrigerated laboratories (◄). Air trapped in tiny bubbles provides samples of the atmosphere back to at least 100,000 years ago. Oxygen and hydrogen isotopes in ice provide a proxy for temperature. ▼

Coral Proxy

Corals grow in yearly cycles and secrete carbonate minerals to form their skeletons. Isotopic compositions of strontium and calcium incorporated in the carbonates depend on the surface temperature of seawater. Thus, layered corals provide a temperature record for a few hundred years. Data from the Great Barrier Reef show a slight increase in sea-surface temperature from about 1920 to 1980, but they show that even warmer periods occurred earlier. These data show that the period around 1900 was relatively cool, and we are coming out of that perhaps unusual period. Otherwise, current temperatures are not unusual. ▼

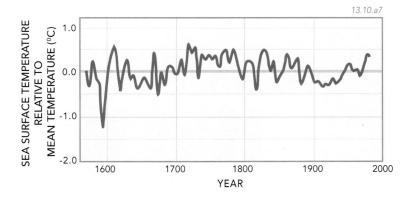

13.10.a7

Comparing Temperature Reconstructions

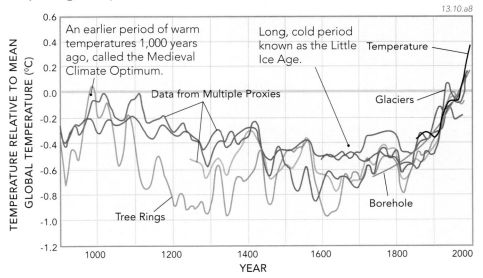

13.10.a8

The NAS published the curves in the chart to the left as part of a 2006 report to Congress. The report summarizes and compares the various temperature records shown on earlier graphs and some reconstructions produced by combining multiple types of proxies. The plot shows direct measurements, such as the temperature record, as well as the various types of proxy data, each in a different color. The curves have been smoothed to emphasize overall variations and show strong similarities in shape. This chart does not include some data, such as those from coral. Comparing the different types of data strengthens the case that some global warming has occurred. The NAS concluded that Earth's atmosphere has warmed 0.6°C in the last 100 years.

A Mystery Involving Atmospheric Satellite Data

One argument *against* warming was that it was not detected in the extensive satellite data. Microwave instruments on NOAA satellites can look into different layers of the atmosphere and detect temperatures. Climate models, as well as actual air measurements, indicated temperature increases in the troposphere, but satellite data showed no such change. The inconsistency was resolved when researchers discovered that the instruments were also measuring the colder, higher stratosphere.

Recalibration of the data now supports a temperature increase in the atmosphere, but the patterns are still being investigated.

13.10.mtb1

Before You Leave This Page Be Able To

- ✓ Describe what global warming means and how it might be measured.

- ✓ Summarize the major lines of direct measurement and proxy evidence indicating global warming in the last 100 years.

- ✓ Explain the main NAS conclusion using the combination of different temperature reconstructions.

13.10

13.11 What Factors Influence Global Warming?

MOST DATA INDICATE THAT SOME GLOBAL WARMING is occurring. Many scientists propose that human activities, including the burning of fossil fuels and the clearing of forests, contribute greenhouse gases to the atmosphere. Astronomical factors, such as Earth's orbit around the Sun and an increase in sunspot activity, can also contribute to warming. Some recent temperature changes show a clear correlation with changes in sunspot activity, and so they probably reflect changes in solar output. Other factors may lead to *global cooling*, including ash from large volcanic eruptions and an increase in certain aerosols in the atmosphere.

A What Processes Influence Atmospheric Temperature Change?

Earth's surface temperatures are dominated by energy from the Sun. Sunlight heats the oceans, land, and atmosphere, but several factors influence how much of this energy reaches the surface and how much is retained.

Interaction of Sunlight with Earth's Atmosphere, Oceans, and Land

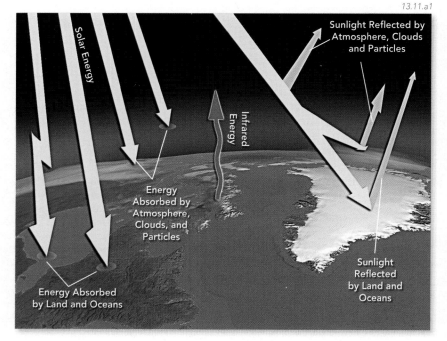
13.11.a1

1. Most of Earth's heating at the surface comes from sunlight, which heats the atmosphere, land, and oceans. Some of this energy escapes back into space. The rest is trapped by interactions with Earth, keeping the planet warm by a process called the *greenhouse effect.* The amount of solar radiation hitting Earth varies regularly due to orbital fluctuations and changes in the Sun's energy output, as expressed by changes in sunspot activity.

2. Some sunlight is *absorbed* by the atmosphere (shown as an orange disk in the figure), and some is *reflected* off the atmosphere. The reflected sunlight returns to space without heating Earth, as depicted by the blue arrows.

3. Light is absorbed by clouds, by soot from burning, and by fine particles called *aerosols*, which are produced by volcanoes, industry, and automobiles. Some of this absorbed energy radiates back into space as *infrared energy* (shown by the wavy red arrow). Clouds and particles also reflect some sunlight.

4. Some light is reflected back into space from the land surface and oceans. Ice in continental glaciers is an effective reflector. As glaciers melt, darker land or ocean is uncovered. This increases the amount of energy that is absorbed by Earth or subsequently re-radiated back to the atmosphere.

5. Some light is absorbed by the land and the oceans, both of which then radiate infrared energy back into the atmosphere. Some of this infrared energy is absorbed by atmospheric gases, such as water vapor (H_2O), carbon dioxide (CO_2), methane (CH_4), and nitrous oxides (NO_x), which are called greenhouse gases. Some portion of these gases is produced naturally and some is produced by human activities. More CO_2, for example, is produced naturally than is produced by humans.

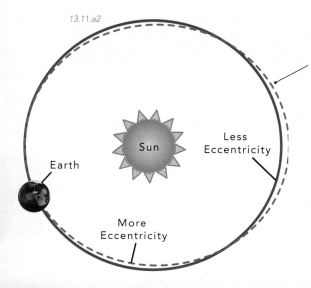
13.11.a2

Orbital Variation

Earth's orbit is affected by the Sun, Moon, and some of the other planets, causing Earth's orbit to vary from more elliptical to more circular paths over time periods of about 100,000 years. This influences Earth's climate, but the current global warming has occurred in less than 200 years. This is too short a time period for orbital changes to have caused the observed warming. The tilt of Earth's spin axis also changes, but the effects occur over thousands of years. Variations in Earth's orbit and tilt are discussed in chapter 14.

Variations in Solar Radiance

Where can we look for a record of how the Sun shined in the past? Cosmogenic isotopes in ice cores and tree rings serve as proxies. These isotopic abundances are related to the strength of the solar wind, and in a complex way, to the Sun's radiance. Using these proxies it is estimated that the Sun's energy emission has varied by about 1% over the last 2,000 years. Some data show a correlation between recent temperatures and sunspot activity, but it is currently debated whether changes in solar output are enough to cause most of the temperature changes of the late 20th century.

Greenhouse Gas Production

1. Several gases in Earth's atmosphere absorb infrared radiation emitted by Earth. This causes them to vibrate and heat up, and then to emit infrared radiation. This radiation can escape into space or be absorbed by other greenhouse gases.

13.11.a3

2. Since 1957, atmospheric scientists have collected air samples on the high peaks of Hawaii (◄). The data for Mauna Loa are plotted on this graph, which shows CO_2 content in the atmosphere as a function of time. ►

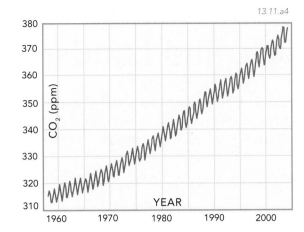

13.11.a4

3. Water vapor is the most abundant greenhouse gas, followed by CO_2. The concentration of CO_2 in air, represented by the red line, has increased by 20% in the last 40 years. Most of this increase is attributed to humans, especially the burning of fossil fuels.

Greenhouse Gases and Temperature Change Records from Ice Cores

4. Long-term records of CO_2 and other greenhouse gases are contained within ice cores and can be traced back hundreds of thousands of years. Ice cores from Antarctica have been analyzed for CO_2, isotopic temperature, and isotopic age. The data, shown on this graph (►), allow comparison between prior natural variations and changes in the last several hundred years.

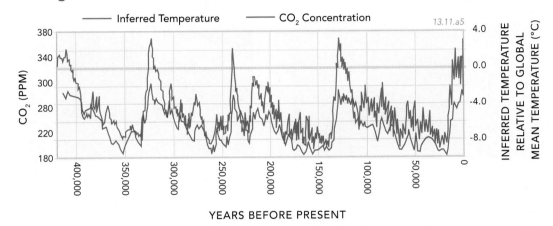

13.11.a5

5. How well do carbon dioxide concentrations in the atmosphere track temperature change? On this graph, the two trends are very similar. Graphs of other greenhouse gases, such as methane versus temperature and nitrous oxide versus temperature, show a similar correspondence. These large fluctuations in the past have natural causes, and increases in temperature generally preceded increases in carbon dioxide.

6. Scientists have used data from ice cores and other proxies to interpret how atmospheric gases changed in the past few hundred years. These data indicate that levels of carbon dioxide (CO_2), methane (CH_4), and nitrous oxide (N_2O) have increased in the last several hundred years, since the start of the Industrial Revolution of the last part of the 18th century. Many scientists infer that these increases in greenhouse gases are partly responsible for the recent increase in temperature, but there remains debate about this controversial topic.

Climate Modeling

7. The simultaneous rises of *anthropogenic* (human-caused) CO_2 and temperature may be related. Climatologists use computer models to account for the effects of the various factors that might cause warming. Some model results are consistent with observations of past climates and so may be reliable. Some models suggest that anthropogenic greenhouse gas emissions are a contributor to warming in the last century. The relative roles of different factors over the last 100 years, as predicted by one model, are shown by this bar graph (►). There is currently a healthy debate about the accuracy and validity of these models and whether the models correctly account for various factors, including coupling between the oceans and atmosphere.

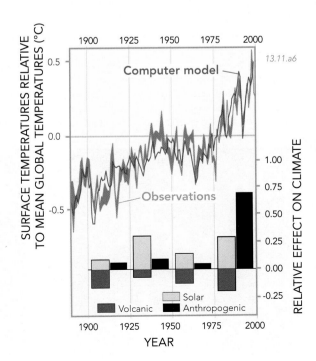

13.11.a6

Before You Leave This Page Be Able To

☑ Describe the greenhouse effect.

☑ Summarize the major factors, both natural and anthropogenic, that influence atmospheric temperature.

☑ Summarize the evidence for increased greenhouse gas concentrations in the atmosphere, and for correspondence between increases in these gases and increases in temperature.

13.11

13.12 What Is the Relationship Among Climate, Tectonics, and Landscape Evolution?

PLATE MOVEMENT AND DEFORMATION CHANGE CLIMATES by altering land elevation, pumping volcanic gases into the atmosphere, rearranging the configuration of landmasses and ocean currents, and changing sea level via changes in seafloor spreading rates.

A How Does Plate Tectonics Affect Climate?

The current arrangement and topography of the continents and ocean basins help guide wind and ocean currents that redistribute heat around Earth. Volcanic activity associated with plate boundaries releases gas and particles into the atmosphere, changing its chemistry and affecting global temperatures.

Mountains and Uplift	Seas and Subsidence	Volcanic Gas and Dust

13.12.a1 Last Hope Sound, Chile

13.12.a2 Portage, AK

13.12.a3 Augustine Volcano, AK

Mountain ranges intercept wind and water vapor, causing rain shadows, concentrated rainfall, and other climatic effects. Tectonic uplift also exposes land to chemical weathering, which removes CO_2 from the atmosphere.

Rapid seafloor spreading displaces water from the ocean basins, flooding low parts of continents. Such flooding, or tectonic subsidence (▲), changes local climates and moves CO_2 from air to rocks and soil.

Volcanic activity releases CO_2 and water vapor, which cause atmospheric warming. Volcanic ash and SO_2 gas from volcanoes reflect solar radiation, causing regional or global cooling.

B How Does Climate Change as a Plate Moves?

As a tectonic plate moves, it changes position relative to Earth's climatic zones. Plate movements open and close channels that connect oceans, thereby changing ocean currents. Tectonics can uplift mountain ranges, form new ocean basins, or bring continents together. These changes can alter a region's elevation, temperature, wind-flow patterns, and the amount and frequency of precipitation, as in these examples from North America.

In the Permian (260 million years ago), North America was part of the supercontinent of Pangaea. The newly formed Appalachian Mountains blocked easterly winds, causing a rain shadow and deserts farther west.

In the Cretaceous (75 million years ago), shallow inland seas inundated the continent during a phase of rapid seafloor spreading. The seas and ocean currents produced a warm, wet climate over most of the continent.

In the last 20 million years (Neogene), mountains along the West Coast increasingly blocked prevailing westerly winds, causing increased precipitation along the coast and a rain shadow with deserts inland.

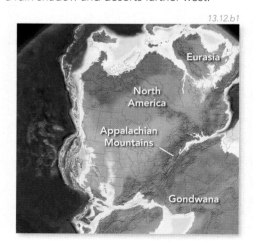
13.12.b1

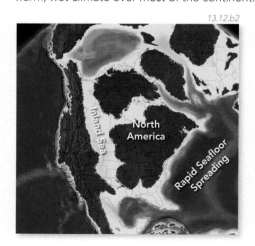
13.12.b2

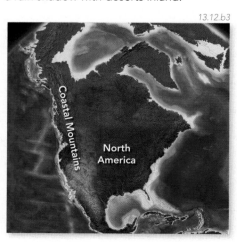
13.12.b3

How Does Climate Affect Tectonics?

Climatic conditions control the speed of erosion, which removes rocks at the surface and causes uplift of deeper parts of the mountain belt. In some mountain ranges, such as the Himalaya, erosion is thought to cause deep rocks to flow sideways and rise upward. When viewed in this way, erosion helps to maintain mountains.

Uplift and Climate

Continental collisions produce extremely high mountains, which force wind belts to flow around and over them. Even though the plate collisions continue for tens of millions of years, these mountains rise to a maximum size of about 4 to 5 km (2.5 to 3 mi) high and 120 to 240 km (75 to 150 mi) wide. This suggests that a balance is achieved between uplift and erosion when viewed at an appropriate timescale. Uplift of the mountain leads to regional climate changes, which act to lower the range through increased erosion. A balance between erosion and uplift will likely be reached, but it may take several million years.

13.12.c1

As mountains are uplifted, rocks at the surface are removed by erosion by water, ice, landslides, and the slower movement of material downhill. In some mountain ranges, erosion is almost as rapid as uplift and so limits the heights the mountains can achieve. The Olympic Mountains of coastal Washington (◄), the Southern Alps of New Zealand, and the Central Range of Taiwan are examples of mountains that are being rapidly eroded.

Episodes of Imbalance

1. Some regions of the Himalaya are estimated to be "in balance." In other areas, uplift seems to outpace erosion. Again, the scale of such observations is important. In Earth's history, countless large mountain ranges have been uplifted and then worn down.

2. Solar heating on the Tibetan Plateau causes heated air to rise, drawing in monsoon moisture toward the mountains. This climatic pattern results in intense rainfall on the south-facing mountain slopes in Nepal and Bhutan.

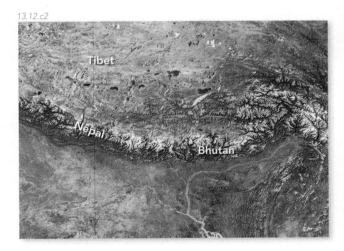

13.12.c2

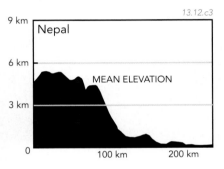

13.12.c3

3. This graph is a topographic profile, showing the slope of the land. Mountain slopes of central Nepal have a curved topographic profile that may reflect a near balance between uplift and erosion.

4. Slopes in Bhutan have straighter topographic profiles, which may indicate that the mountains are uplifting faster than erosion can wear them down.

13.12.c4

Lateral Movement of Material and Its Relationship to Climate

Geologists working in the Himalaya propose that the removal of material by erosion in Nepal and Bhutan is causing the crust to flow laterally beneath the region. According to this idea, as material is removed by erosion along the mountain front, thick, weak crust beneath Tibet flows southward (to the left), as shown by the large arrow, to replace the lost material. In other words, erosion in this area is helping the mountains to rise.

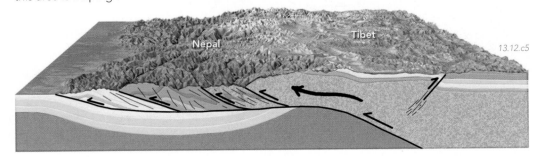

13.12.c5

Before You Leave This Page Be Able To

✔ Describe how plate tectonics affects climate.

✔ Summarize different ways that the arrangement of plates affects climate.

✔ Explain how landscapes are a product of tectonics and climate.

13.12

13.13 How Does Geology Influence Ecology?

GEOLOGY PROVIDES ENVIRONMENTS FOR LIFE by building mountains, valleys, and oceans. Rocks break down to soil and provide nutrients for organisms. Geologic factors control various aspects of climate, including the availability of water and an area's temperature and rainfall. These in turn control the *ecology*, which is the complex set of relations between living organisms and their environment.

A What Factors Affect Where Life Can Exist?

Temperature

Most life resides in environments where liquid water is stable (0 to 100°C). Life can survive in a remarkably wide range of temperatures, from freezing conditions to this 80°C hot pool in Yellowstone, where bacteria thrive. Life also exists in temperatures up to 350°C in deep-sea vents.

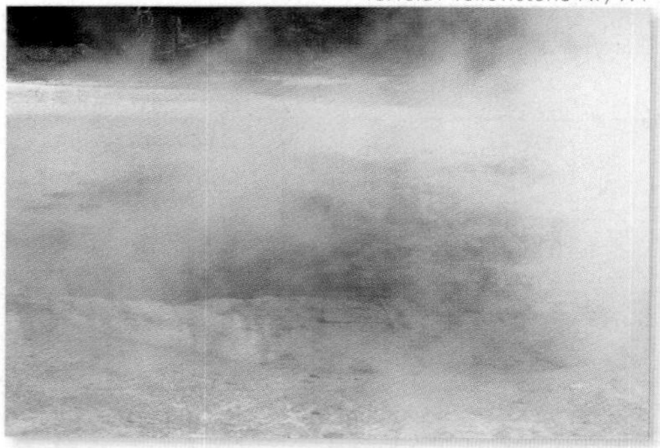

Water

Plants and animals, including humans, are composed primarily of water and cannot survive without it. Water is required to carry out metabolic processes at the cellular level. Where there is little water, there generally is little visible life, as in the dry deserts in Death Valley, California.

Nutrients and Energy

Living things require nutrients in order to produce energy. Nutrients come from several sources. Green plants process sunlight and carbon dioxide into sugars for energy, and they provide nutrients for many animals. The area shown below receives abundant sunlight and rain.

13.13.a1 Yellowstone NP, WY

13.13.a2 Death Valley, CA

13.13.a3 Kakadu World Heritage Area, Australia

B How Are Ecosystems Classified?

Climate scientists use combinations of climatic factors to classify ecological systems, or *ecosystems*. Computer techniques provide the power to analyze a variety of complex variables, which can be combined and visualized together. A technique called *ecoregion analysis* defines similar areas based on the types of ecosystems that are present and the patterns of land use, landscape type, natural vegetation, climate, and soils.

Ecosystems of the upper Midwest reflect the region's cold winters and a shorter growing season.

This map is the result of a computer analysis of nine variables important for plant growth. These fit into three categories: *elevation*, *soil*, and *climate*. The Pacific Northwest's light green color on this map reflects a combination of factors that are favorable to plant growth.

The ecosystems of the Northeast (blues) are greatly influenced by cool temperature, which controls when and where plants grow.

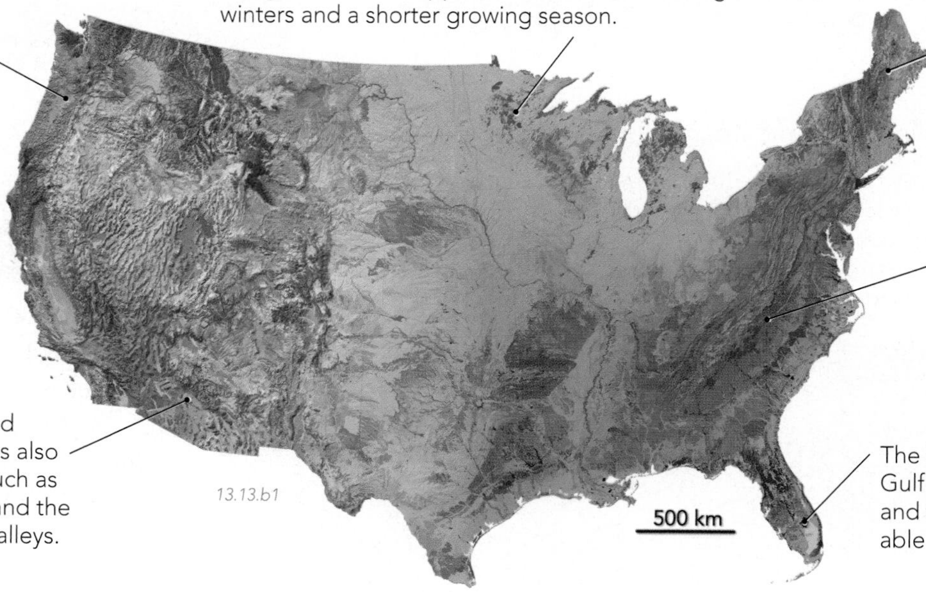

The darker green in the Southeast reflects plant growth limited by soils that are in part leached of nutrients by high rainfall.

The ecosystems of the Southwest (mostly red) are dominated by the dry climate. The patterns also reflect topographic features, such as regional changes in elevation and the alternation of mountains and valleys.

The lighter greens of Florida, the Gulf Coast, the Mississippi Valley, and the Great Plains reflect favorable plant-growing conditions.

13.13.b1

500 km

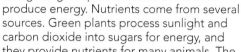

C   What Are Some Geologic Controls on Ecology?

Earth materials and geologic structures determine what kind of life can survive in a given place—*the ecology*.

13.13.c1 Banff, Alberta, Canada

Rocky mountains may catch more rainfall, but they have little soil on their steep slopes (◀). This limits the types of plants and animals that can thrive. Gentler mountains in wet climates develop thick soils that allow a greater abundance of plants and animals.

13.13.c2 Conway, NH

Topography, which is controlled by geology and climate, determines where water flows and whether it runs off or soaks into the ground. Some rocks allow more water to run off into river systems, sustaining stream-side ecosystems (◀) and lakes.

Flat areas, especially if underlain by soft rock or unconsolidated materials, allow water to soak in and develop better soils (▶). The water and soils allow more vegetation to grow. Where there are plants, there are also animals, including humans.

13.13.c3 Hudson River Valley, NY

Volcanic rocks can weather quickly, releasing nutrients such as magnesium and sulfur into soils. But some rocks, like these sulfur-mineralized volcanic rocks, release acids and other compounds that are detrimental to plants or animals (▶). Such chemicals can also contaminate streams, rivers, and lakes.

13.13.c4 Silverton, CO

D   How Does Geology Influence Agriculture?

Geologic structures and erosional processes provide environments, like broad valleys, suitable for growing crops. Decomposed rocks provide all the inorganic nutrients needed by plants except for CO_2, which plants extract from the air to make their carbon-rich leaves, stems, and wood. Examine the conceptualized terrain below.

Groundwater from these bedrock hills recharges local streams, whose water is used to irrigate crops.

Fault movement downdropped a fault block, forming a valley that filled with sediments eroded from the uplifted mountain range and adjacent areas.

This mountain range of fractured granite sheds debris rich in sodium, calcium, magnesium, and phosphorus. These are important nutrients for plants and animals.

Soils, which are composed of weathered rock and organic material, provide the foundation for crops and other plants.

This river valley contains rich soils derived in part from fine sediment deposited during floods. Sand dunes formed during the time of the Ice Ages provide well-drained soils for strawberries.

Water-vapor-rich air from warm ocean currents provides abundant rainfall and keeps plants cool during foggy, coastal summer days.

Wave action during the last glacial period eroded off this uplifted coastal bench, which is now used for cropland. The bench only exists because of uplift.

5 km

13.13.d1

Before You Leave This Page Be Able To

☑ Explain the factors that control life and distribution of ecosystems.

☑ Describe some ways that geology influences ecology, the distribution of ecosystems, and agriculture.

13.13

13.14 What Occurred During the Hurricane Seasons of 2004 and 2005?

NOAA HURRICANE FORECASTS FOR 2004 predicted that the Atlantic would see slightly above-average activity, with 6 to 8 hurricanes. Of these 2 to 4 would become major hurricanes. The conditions, however, were more favorable for hurricanes than estimated. The Atlantic had 15 hurricanes, 6 of them major. Storms and hurricanes killed more than 3,000 people and caused over $40 billion in damage. In the following year, Hurricane Katrina devastated New Orleans, Louisiana, and other communities along the Gulf Coast.

A Where Did the 2004 Hurricanes Hit and How Strong and Damaging Were They?

Hurricanes are the most energetic of common short-term Earth events. An average hurricane expends more energy than was released in the largest recorded earthquakes. Meteorologists measure hurricane intensity based on wind speed, air pressure, and surge potential. On this scale, the 2004 hurricanes were slightly above average. However, since so many made landfall in vulnerable places, they were very destructive.

During the 2004 season, four major hurricanes (Jeanne, Frances, Ivan, and Charley) made landfall along the southeastern United States. This map shows the paths they traveled.

One reason why so much damage occurred is that Florida became very popular with retirees in the period between 1966 and 2003, when only a few major hurricanes made landfall. Until 2004, the population had no experience with multiple giant storms in one season.

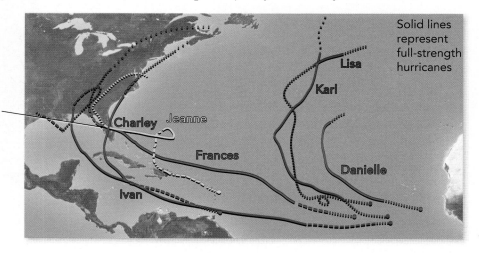

13.14.a1

Major hurricanes were born off west Africa. They traveled thousands of miles, gaining or maintaining strength before crashing onto a very small landing zone in Florida and adjacent areas. Luckily, other hurricanes turned north and missed the land.

Comparing Intensity and Damage of Hurricanes

▶ The Saffir–Simpson Hurricane Scale is a classification system based on wind velocity. It ranks hurricanes from 1 to 5, with 5 being the potentially most destructive. Ivan, shown to the left, was rated as a category 3 when it came ashore west of Florida (the land to the right of the storm).

13.14.a2

Saffir–Simpson Category	Maximum Sustained Wind Speed (miles/hour)	Minimum Central Pressure (millibars)	Storm Surge	
			(feet)	(meters)
1	74–95	> 980	3–5	1.0–1.7
2	96–110	979–965	6–8	1.8–2.6
3	111–130	964–945	9–12	2.7–3.8
4	131–155	944–920	13–18	3.9–5.6
5	155+	< 920	18+	5.6+

▶ This table compares actual damage for six recent hurricanes. The year 2004 was the worst hurricane season on record for damage until Hurricane Katrina destroyed parts of New Orleans in 2005. The damage costs have not been adjusted for inflation.

Rank	Hurricane	Year	Category	Damage (billions of dollars)
1	Katrina	2005	3	81
2	Andrew	1992	5	26.5
3	Charley	2004	4	15
4	Ivan	2004	3	14.2
5	Frances	2004	2	8.9
6	Hugo	1989	4	7

Before You Leave These Pages Be Able To

☑ Describe the Saffir-Simpson scale for hurricanes.

☑ Summarize or sketch conditions that caused 2004 to be a bad year for hurricanes.

B What Changed from Previous Seasons to Make 2004 So Damaging?

The most important factor causing a severe hurricane season in 2004 was a change in the predominant pressure system along eastern North America, which steered storms into the area around Florida.

The Atmospheric Pattern from 1995 to 2003

Previous seasons had a persistent trough of low pressure along the East Coast. It caused upper-level winds, shown with black arrows, to drop south and push hurricanes to the north and east.

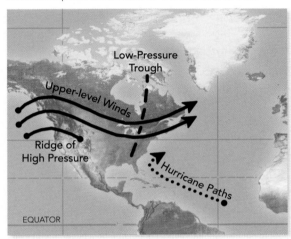

13.14.b1

◀ Hurricanes generated in the eastern Atlantic moved toward the eastern seaboard of North America, but mostly turned northeast across open ocean instead of striking land.

The Atmospheric Pattern in 2004

In 2004, a ridge of high pressure developed and persisted over northeastern North America. It caused the upper-level winds to shift farther to the north, where they could not deflect the hurricanes.

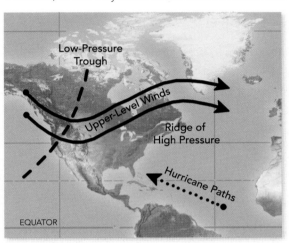

13.14.b2

◀ The high pressure ridge blocked the hurricanes from turning north, instead guiding them to the northwest toward Florida and the Gulf Region.

C What Other Factors Aligned to Produce Very Large Hurricanes?

There were exceptionally strong hurricanes in 2004, and they were more numerous than expected. There are natural cycles of increased and decreased hurricane activity, similar to the cycles for El Niño and La Niña. In 2004, perhaps as part of this cycle, there was a coincidence of sea-surface temperature and winds that helped produce a record number of storms. These storms included six major hurricanes.

1. Many Atlantic hurricanes originate as tropical disturbances near Africa and then migrate westward across the Atlantic Ocean. This map shows the atmospheric conditions that led to the hurricanes of 2004.

2. In 2004, a ridge of high pressure, shown in orange, prevented the hurricanes from turning north.

3. A persistent ridge of high pressure resides over Africa and Spain. It prevents most hurricanes and tropical storms from turning north into the eastern Atlantic and into Europe. As a result, storms travel westward across the ocean.

7. Finally, steering winds guided hurricanes across the Atlantic and toward landfall in and around Florida.

6. Pockets of warmer-than-average water off Florida helped fuel and sustain strong hurricanes.

5. A zone of rising tropical air, the Intertropical Convergence Zone (ITCZ), was positioned favorably for developing storms.

4. An upper level wind arises due to a temperature gradient between the warm Sahara and cooler coastal air in East Africa. These wind currents create disturbances, called *tropical waves,* which travel across Africa. These are the beginnings of many Atlantic storms and hurricanes.

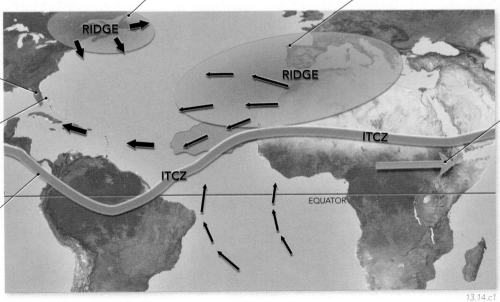

13.14.c1

13.14

13.15 What Kinds of Climate and Weather Would Occur in This Place?

PLANET M is a hypothetical replica of Earth. It has plate tectonics, oceans, and an atmosphere very similar to Earth's. You will map the climatic zones on this planet, identifying cold and warm ocean currents, prevailing winds, potential locations of rain forests and deserts, places where the climate and topographic setting would be suitable for agriculture, and sites at risk for hurricanes.

Goals of This Exercise:

- Observe the general map patterns and photographs from different parts of the planet. Note any clues about climate.
- Create a map showing probable climatic conditions, based on ocean currents and prevailing winds.
- Locate likely sites for rain forests, deserts, agricultural areas, and locations where storms will strike land.

Procedures

Follow the steps below, entering your answers for each step in the appropriate place on the worksheet or online.

1. Observe the distribution of continents and oceans on the map below and on the larger version on the worksheet. Examine the photographs of the various areas and infer the environments and climates that are represented. Each photograph has a letter that corresponds to a letter on the map. Some photographs are on the next page.

2. Carefully examine the various types of data on the next page. For each data set, examine the climatic, weather, or geologic implications for different parts of the planet.

3. Draw on the worksheet your interpretations of whether each ocean current is warm, cold, or neither, and whether prevailing winds will bring warmth or coolness, and cause dryness or precipitation. Alternatively, answer questions about these factors online.

4. Map the main climatic zones. Include the likely locations of rain forests, deserts, and areas suitable for croplands. Think about which areas are at risk for hurricanes and other major tropical storms. Be prepared to discuss your observations and interpretations.

13.15.a2

13.15.a3

13.15.a1

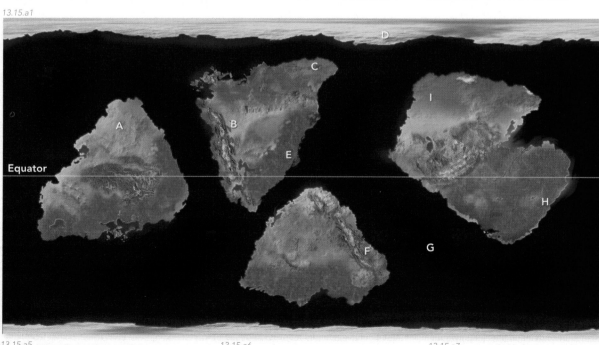

13.15.a4

13.15.a5

13.15.a6

13.15.a7

Some Important Observations About the Planet

1. The planet revolves around its axis once every 24 hours in the same direction Earth does. Therefore it has days and nights similar to Earth's. It orbits a sun similar to our own, at about the same distance. Its spin axis is slightly tilted relative to the orbital plane.

2. The process of plate tectonics is operating on this planet at rates similar to those observed on Earth. Volcanoes, mid-ocean ridges, and mountain-building processes are also similar to those on Earth. There are continents, ocean basins, and polar ice caps.

3. The oceans are likewise similar to those on Earth. They show similar variations in sea-surface temperature and have ocean currents that circulate around the edges of the main ocean basins and through gaps between the continents.

4. The planet has well-developed atmospheric currents, with wind patterns similar to those on Earth.

Preliminary Interpretation of Ocean Currents

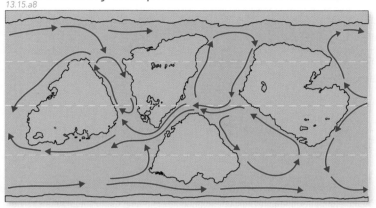

13.15.a8

This map shows the directions of ocean surface currents. From these data, identify which currents are probably cold, warm, or neither. Water temperatures will be key in inferring probable climates.

Preliminary Interpretation of Wind Data

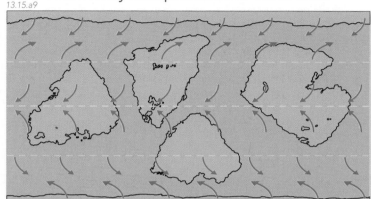

13.15.a9

These data show the directions of prevailing winds, which are similar to the patterns on Earth. Different wind directions occupy bands encircling the planet, parallel to the equator.

Ocean-Water Temperatures and Sea-Surface Pressures

This map depicts satellite data for sea-surface temperatures during late summer. Orange and red show warmer waters, whereas blue and purple show colder waters. The numbers give the temperature values in degrees centigrade (°C).

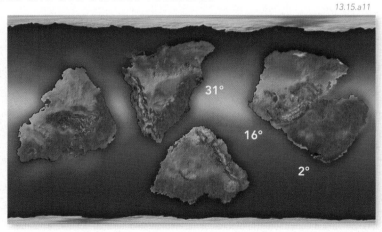

13.15.a11

31°

16°

2°

13.15.a10

I

13.15.a12

H

Climate Analysis

On the version of this map on the worksheet, show the probable locations of the following features by writing the associated letters on the map: rain forests (R), deserts (D), potential agricultural areas (A), and lands at highest risk for hurricanes (H). You can use colored pencils to shade in the extent of land associated with each letter.

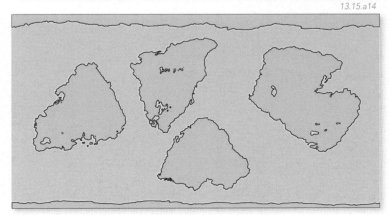

13.15.a14

13.15.a13

G

13.15

Shorelines, Glaciers, and Changing Sea Levels

SHORELINES ARE GEOLOGICALLY ACTIVE PLACES that experience dramatic changes from a single storm or from short- and long-term changes in sea level. Past sea levels have been more than 200 m (660 ft) higher and about 120 m (390 ft) lower than today, flooding the continents or exposing the continental shelves. The most rapid sea level changes are related to changes in the extent of glaciers and continental ice sheets.

The northeastern United States and adjacent Canada, shown in this shaded relief map, have a striking collection of features on land and along the shorelines. The map extends from North and South Dakota (in the northwestern corner of the map) across the Great Lakes as far east as Maine and Virginia. The northern part of the map includes southeastern Canada.

Huge, smooth troughs (each labeled with a **T** on the map) cut across the landscape. Examples are near the northwestern corner of the map, in western Minnesota and the Dakotas, and southwest of Lake Erie, in Ohio and Indiana.

What caused these smooth areas of the landscape, and is the process still occurring?

Curiously curved ridges (labeled with an **R** on the map) cross some of the smooth areas, and are especially noticeable southwest of Lake Erie and Lake Michigan.

What are these ridges, how did they form, and what do they tell us about the geologic history of this region?

14.00.a1

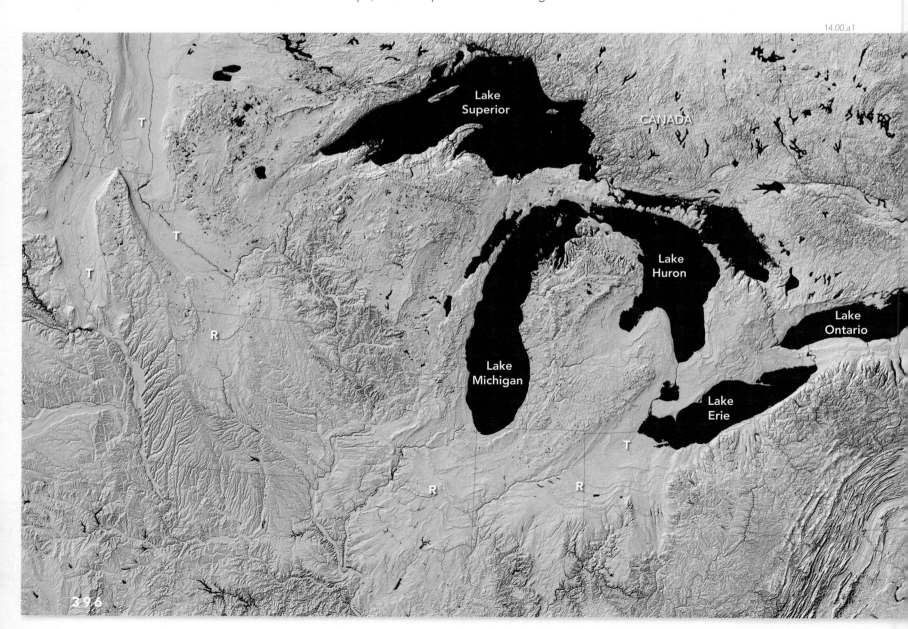

TOPICS IN THIS CHAPTER

North of the Great Lakes, the landscape of Canada is rough on a small scale, containing many lakes, more than most parts of the United States.

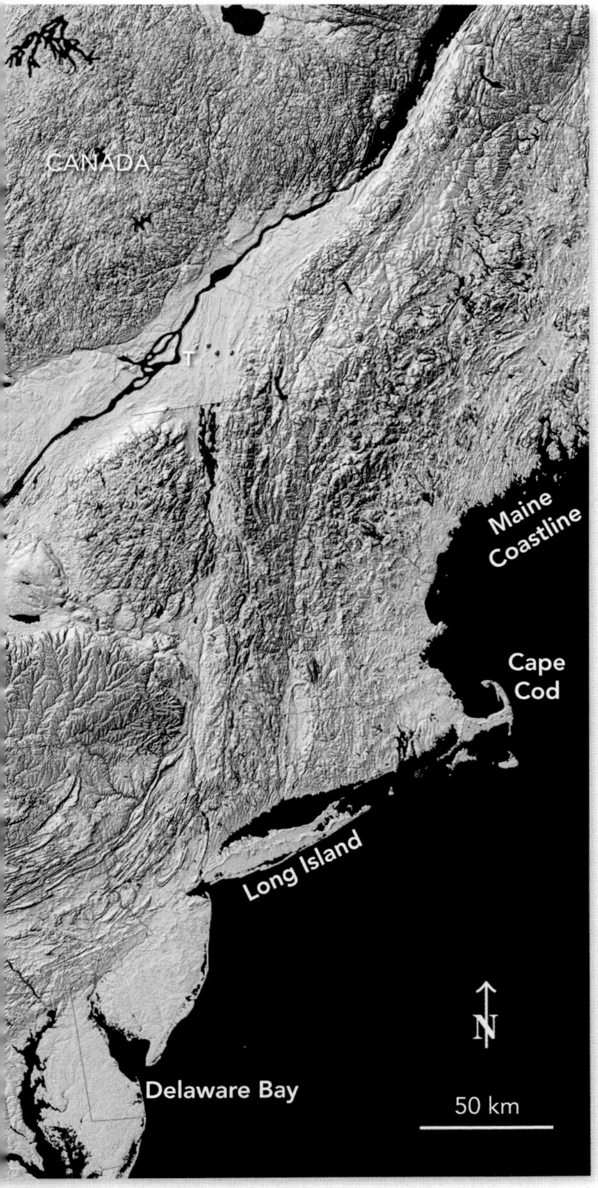

CANADA

Maine Coastline

Cape Cod

Long Island

Delaware Bay

50 km

N

Why does this region have so many lakes, and when and how did this landscape form?

The Maine coastline is very irregular, with many bays, where the ocean reaches inward like fingers into the land. Offshore are a variety of islands, for which the region is well known.

How did this coastline form, and has it always been this way?

Cape Cod, Massachusetts, protrudes into the ocean like a flexed arm or a boot with curled toes.

How did Cape Cod form, and what processes along the shoreline caused it to have this shape?

Delaware Bay and Chesapeake Bay to the south connect to the ocean, but their shapes resemble river valleys.

How did these bays form, and—if they started as river valleys—when and why did they become flooded by the ocean?

Ice Ages and Shorelines

Landscapes in the Great Lakes area contain evidence that huge ice sheets once flowed across this part of the continent—in the recent geologic past. This conclusion arises from comparing the distinctive landscape features and their associated sedimentary deposits with those observed today near currently active glaciers. With glaciers, as with most geologic features, the present is the key to interpreting the past.

For the last two million years, Earth has experienced a series of *ice ages*, during which large regions of the Northern Hemisphere, as well as Antarctica and South America, were covered year-round with ice and snow. Where the ice was thick enough, or rested on a steep enough slope, it moved downhill as a mass of flowing ice called a *glacier*. Some glaciers were small and restricted to mountain areas, whereas others covered large parts of the continents, forming *continental ice sheets*. Continental ice sheets flowed southward from Canada and smoothed off and carved grooves into the underlying landscape by grinding ice, rocks, and sand against the bedrock. The smooth troughs on this map were carved by continental glaciers that flowed southward from Canada. Some areas south of the ice had large lakes that formed in association with the glaciers.

As the climate warmed over the past 20,000 years, the ice sheets and glaciers melted and covered less area. Rocks and other sediment once carried in the ice were dropped along the front of the melting glaciers, forming a series of curved ridges south of the Great Lakes. The ice also left piles of glacially derived sediment on Long Island and Cape Cod.

Water released from melting ice carved new river valleys and flowed into the sea, causing global sea level to rise. The rising seas flooded coastlines and river valleys, forming the many inlets and bays along the Atlantic coast of the eastern United States.

14.0

14.1 What Processes Occur Along Shorelines?

SHORELINES ARE THE INTERFACE BETWEEN LAND AND WATER and so respond to processes that arise from both sides. Waves and tides affect a shoreline from the water side, while rivers, wind, and other transport agents contribute sediment from the land. Together these processes sculpt the shoreline, redistribute sediment, and present challenges for people who live along a coast.

A What Types of Processes Affect Shorelines?

14.01.a2

◀ From the water side, most shorelines are strongly affected by waves, which are near-surface features generated by wind blowing across the water. Waves typically form far from shore but can approach and break upon the shoreline, where they erode rock and loose material, deposit sand and other sediment, or simply move sediment around.

From the land side, rivers can be important contributors of sediment into the shoreline system. Silt, sand, and coarser sediment carried by rivers accumulate close to where a river meets the sea or lake, commonly forming a delta. Fine-grained sediment suspended in river water can be carried farther away from shore.

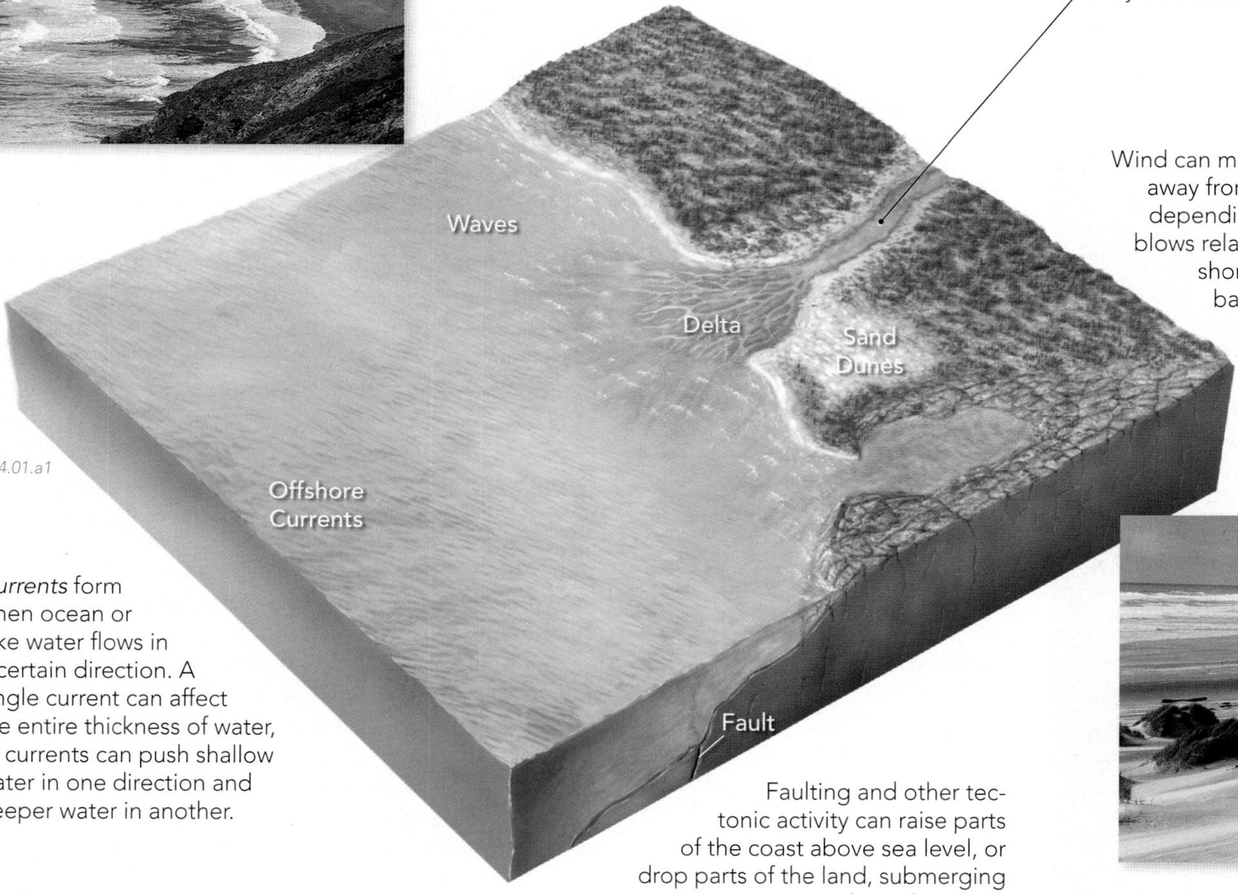

Wind can move sand and finer sediment away from, toward, or along a beach, depending on the direction the wind blows relative to the orientation of the shoreline. Many beach areas are backed by coastal sand dunes, most of which are held partially in place by some vegetation. ▼

14.01.a1

Waves

Delta

Sand Dunes

Offshore Currents

Currents form when ocean or lake water flows in a certain direction. A single current can affect the entire thickness of water, or currents can push shallow water in one direction and deeper water in another.

Fault

Faulting and other tectonic activity can raise parts of the coast above sea level, or drop parts of the land, submerging areas along the coast.

14.01.a3

14.01.a4 Mont-Saint-Michel, France

Low Tide

Changes in sea level greatly affect shorelines. In most places, tides raise and lower sea level relative to the land twice a day. Longer term changes in sea level are primarily due to changes in climate and tectonics. Tidal flats, such as the one surrounding Mont-Saint-Michel in France, shown in both of these photographs, are uncovered by low tides (◀) and flooded by high tides (▶) and storms. Such low areas could be submerged by an overall rise in sea level.

14.01.a5 Mont-Saint-Michel, France

High Tide

B   What Factors Affect the Appearance of a Shoreline?

Shorelines around the world have diverse appearances, from sandy white beaches to dark, craggy cliffs that plunge vertically into the sea, with no beach at all. A number of factors control these differences, including orientation of the coast, slope of the seafloor, hardness of the rocks, and contributions of sediment from the land.

Factors on the Water Side

The appearance of a shoreline is greatly influenced by the strength of the waves and tides that impact the shore. Stronger waves will typically cause greater erosion and move larger clasts of sediment along the shoreline.

The size and intensity of storms influence the appearance of a coast because storms bring with them large waves, strong winds, and intense rainfall. Some coasts are ravaged by hurricanes, whereas others rarely experience the erosive effects of powerful storms.

The slope of the seafloor is also a factor. Steep slopes can allow large waves to break directly against rocks along the shore, whereas more gentle slopes cause waves to break a short distance offshore.

The orientation of a coastline is also important, because waves typically approach from specific directions in response to prevailing winds. The dominant wave direction may change with the season (summer versus winter or dry versus rainy season). Also, some parts of the coast will receive less wave action because they are sheltered in a bay or are protected by an island, barrier reef, or other offshore feature. The beach below contains cobbles and boulders and is affected by strong waves, especially during storms.

14.01.b4

Factors on the Land Side

On the land side, the appearance of the shoreline reflects the hardness of the bedrock along the coast. Hard rock that resists erosion tends to form rocky cliffs (▶), whereas erosion sculpts softer sediment and rock into more gentle slopes and rounded hills.

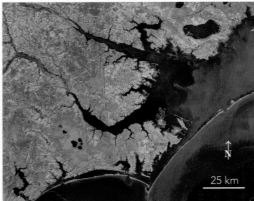

14.01.b2 Scotland

Coastal landscapes also reflect the amount and size of available sediment. A coast cannot be rocky if the only materials present are soft and fine grained. Rivers provide a fresh influx of sediment into the shoreline environment.

Coastlines undergoing uplift have a different appearance than those where the land has dropped relative to water level. A rise in sea level flooded river valleys along the North Carolina coast, producing a coastal outline marked by long, narrow inlets and bays. ▼

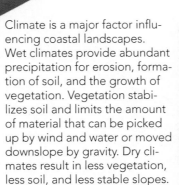

Storm
Waves
Rocky Coast
Coast Composed of Soft Sediment
River
Delta
Waves
Sheltered Bay
Flooded River Valley

14.01.b1

Climate is a major factor influencing coastal landscapes. Wet climates provide abundant precipitation for erosion, formation of soil, and the growth of vegetation. Vegetation stabilizes soil and limits the amount of material that can be picked up by wind and water or moved downslope by gravity. Dry climates result in less vegetation, less soil, and less stable slopes.

25 km

14.01.b3

Before You Leave This Page Be Able To

☑ Summarize or sketch the types of processes that affect shorelines.

☑ Summarize or sketch how different factors, from the water side and from the land side, affect the appearance of a shoreline.

14.1

14.2 What Causes High Tides and Low Tides?

THE SEA SURFACE MOVES UP AND DOWN across the shoreline, generally twice each day. These changes, called *tides*, are observed in the oceans and in bodies of water, such as bays and estuaries, that are connected to the ocean. What causes the tide to rise and fall, and why are some tides higher than others?

A What Are High and Low Tides?

Tides are cyclic changes in the height of the sea surface, generally measured at locations along the coast. The difference between high and low tide is typically 1 to 3 m, but it can be more than 12 m or almost zero.

High Tide	Average Sea Level	Low Tide

14.02.a1

14.02.a2

14.02.a3

During *high tide*, the height of water in the ocean has risen to its highest level relative to the land. At this point, the water floods into low-lying areas. In most places, high tide occurs every 12 hours and 24 minutes.

Following high tide, the water level begins to fall relative to the land—the tide is going out. At some time, water level will reach the average sea level for that location, but it keeps falling on its way to low tide.

When the water level reaches its lowest level, it is at *low tide*. Low tide in most places also occurs every 12 hours and 24 minutes. Water level begins to rise again after low tide, and the tide is coming in. Rising tide spreads water across the land.

B What Causes High and Low Tides?

Tides rise and fall largely because water in the ocean is pulled by the gravity of the Moon and to a lesser extent the Sun. As Earth rotates on its axis, most shorelines experience two high and two low tides in each 25-hour period.

1. This figure depicts Earth and Moon as if looking directly down on Earth's North Pole. It shows the Moon much closer to Earth than it would be for the size of the two bodies. Earth rotates (spins) counterclockwise in this view and, relative to the Sun, completes a full rotation once every 24 hours.

2. The Moon orbits (travels around) Earth once every 29 days, also counterclockwise in this view. Due to this motion, it takes 24 hours and 50 minutes for a point on Earth that is facing the Moon to rotate all the way around to catch up with and again face the Moon.

3. The Moon exerts a gravitational pull on Earth and its water. This pulls the water in the ocean toward the Moon, causing it to mound up on the side of Earth nearest to (i.e., facing) the Moon. Coastal areas beneath the mound of water experience *high tide*. On this figure, the thickness and mounding of the (blue) water are greatly exaggerated.

4. On parts of Earth that are facing neither toward nor away from the Moon, sea level is lower as water is pulled away from these regions toward areas of high tide. Coastal areas here experience *low tide*.

5. On the side of Earth opposite the Moon, the water is relatively far from the Moon and so feels less of the Moon's gravitational pull (recall that the force of gravity decreases with distance). The water bulges out, and the side of Earth facing directly away from the Moon, therefore, experiences *high tide*.

6. Earth rotates (spins on its axis) much faster than the Moon orbits Earth, so it is best to think of the mounds of water—but not the water itself—as remaining fixed in position relative to the Moon as Earth spins. During a complete rotation of Earth, a coastal area will pass through both mounds of water, causing most shorelines to have two high tides (and two low tides) in each 24-hour and 50-minute period.

Moon's Gravity

Rotation of Earth

14.02.b1

C Why Are Some High Tides Higher Than Others?

From week to week, not all high tides at any location reach the same level—some are higher and others are lower than average. Similarly, some low tides are very low and others are less so. Such variations are related to the added influence of the Sun's gravity and follow a predictable pattern that repeats about every 14 days.

Spring Tides

1. Like the Moon, the Sun exerts a gravitational pull on Earth and its water. The Sun is larger and more massive than the Moon, but it is farther away from Earth, so the Sun's gravitational effect on Earth is weaker than the Moon's.

2. The Sun's gravity attracts a mound of water on the side of Earth facing the Sun and causes another mound on the side facing away from the Sun. These thin mounds, shown in dark blue, are always in the same position relative to the Sun. Locations on Earth rotate through each position once every 24 hours.

Time 1: New Moon

Sun's Gravity

Moon's Gravity

Time 2: Full Moon

14.02.c1

3. The Moon orbits all the way around Earth in about 29 days. Once a month, here labeled Time 1, the Moon and the Sun are on the same side of Earth. The gravity of the Sun and Moon then pull the ocean in the same direction, causing the high and low tides to be more extreme than normal. These extreme tides are called *spring tides*, but they occur during all months of the year.

4. Two weeks later, here labeled Time 2, the Moon has moved around to the side of Earth opposite the Sun. However, the forces of the Moon and Sun are again aligned, so another *spring tide* occurs. Note that spring tides occur when the side of the Moon facing Earth is either fully illuminated (a *full moon*) or is not illuminated at all by the Sun (a *new moon*).

Neap Tides

5. Seven days after each spring tide, the Moon journeys 1/4 of the way around Earth. In this position, the Moon's and Sun's gravity are pulling at right angles to one another, and one acts to cause a high tide while the other acts to cause a low tide.

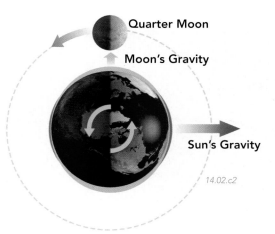

Quarter Moon

Moon's Gravity

Sun's Gravity

14.02.c2

6. At these times of the month, the effect of the Sun's gravity partially offsets the effects of the Moon's gravity, so the differences between high tide and low tide are less than average. These lower-than-average high tides and higher-than-average low tides occur every two weeks and are called *neap tides*, from an Old English word for "lacking." They happen when the Moon, as viewed from Earth, is only half illuminated by sunlight (called a *quarter moon*).

The Most Extreme Tides in the World

Some places have higher tides than others. The difference between high and low tide can be so small as to be nearly undetectable, or it can be so extreme as to be dangerous. The Mediterranean has almost no tide, and much of the Caribbean has only one tide each day. The world's highest tides are in the *Bay of Fundy* along the Atlantic coast of Canada. In this place, the geometry of the shoreline and sea bottom funnel water in and out of the bay at just the right rate to cause a tidal range of as much as 16 m (52 ft). Canadians use the large tides to generate electricity for Nova Scotia. These two photographs illustrate the extreme tidal range within the Bay of Fundy.

14.02.mtb1

High Tide

14.02.mtb2

Low Tide

Before You Leave This Page Be Able To

✓ Describe or sketch what tides are.

✓ Sketch and describe how tides relate to the position of the Moon and why.

✓ Sketch or summarize how the gravity of the Moon and Sun cause spring tides and neap tides.

14.2

14.3 How Do Waves Form and Propagate?

WAVES ARE THE MAIN cause of shoreline erosion. Most ocean waves are generated by wind and only affect the uppermost levels of the water. Waves transport and deposit eroded material along the shoreline. How do waves form, how do they propagate across the water, and what causes waves to break?

A What Is an Ocean Wave and How Is the Size of a Wave Described?

1. *Waves* are irregularities on the surface of a body of water. They vary from a series of curved ridges and troughs to more irregular bumps and depressions to breaking curls. ▼

14.03.a1

2. In the deep open water, many waves occur in sets of individual waves that are similar in size and shape to one another and follow one behind another. The lowest part of a wave is the *trough* (▶), the highest part is the *crest*, and the vertical distance between the two is the *wave height*.

3. The horizontal distance between two adjacent crests in a set of waves is the *wavelength*. The wavelength and the height of the wave can vary greatly and, in general, are greatest for high-energy waves. A typical ocean wave is several meters high, with a wavelength of several tens of meters.

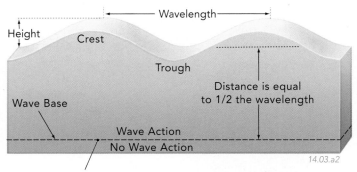
14.03.a2

4. Most waves are near-surface features. They affect only the surface of the water and depths typically down to several tens of meters. Below some level, called the *wave base*, the wave ceases to have any effect. The depth of the wave base is about half the wavelength, so if two wave crests are 20 m apart, then the wave base is about 10 m deep. The equation that expresses this relationship is:

$$\text{depth of wave base} = \text{wavelength}/2$$

B How Do Waves Propagate Across the Water?

A set of waves can travel a long distance across an ocean or other body, but the water through which the wave passes moves only a short distance.

Water waves propagate in a manner similar to seismic surface waves. Water molecules move up and down and from side to side, but they mostly stay in about the same place. Compressive forces cause water within the wave to push against the water in front of it. The three figures shown here (▶) are snapshots of a set of waves propagating to the right. Examine the motion of water at points A, B, and C within the wave. At the start, when the points are under the wave crest, they are farthest apart.

▶ Here, the waves have propagated through the water to the right, but the lettered reference points have moved only a short distance. Point A, which is close to the surface, moves more than point B, which is deeper. Point C is below the wave base and does not move at all.

▶ Later, points A and B are beneath the wave trough and have nearly returned to their positions on the reference line. As the next wave crest approaches, they will move left and upward, and then right, returning to near their starting points.

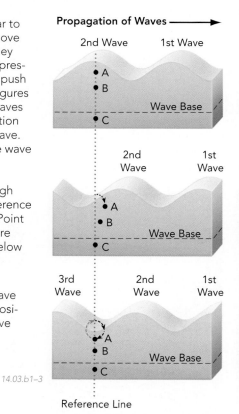
14.03.b1–3

Reference Line

▼ During the passage of an entire wave (crest to crest), points A and B will each have followed a small, circular path. The figure below shows the circular paths of 20 different points at different depths and positions along the wave.

Water particles on the surface of the water travel the most, going up and forward and then down and back, in a circular motion, as the wave passes.

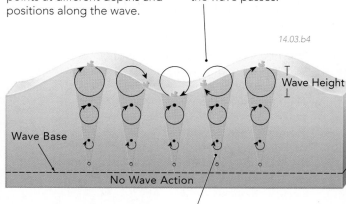
14.03.b4

Deeper within the wave, water particles travel smaller circular paths, and the paths become smaller with depth. Water that is right above the wave base barely moves, and water below the wave base does not move at all as the wave passes.

C How Do Waves Form and What Happens When They Reach Shallow Water?

As a wave moves from deep water into shallower water, it starts to interact with the bottom and changes in size and shape. A wave can also bend (refract) as one part of the wave encounters the bottom before other parts do.

How Waves Form

1. Most ocean waves are caused by wind blowing across the surface of the water.

2. When a gentle breeze is blowing, gas molecules in the air collide with the surface of the water, transferring some of their motion to the water. This forms waves that are small in height and wavelength. Once a wave forms, it catches the wind even more and so can increase in size.

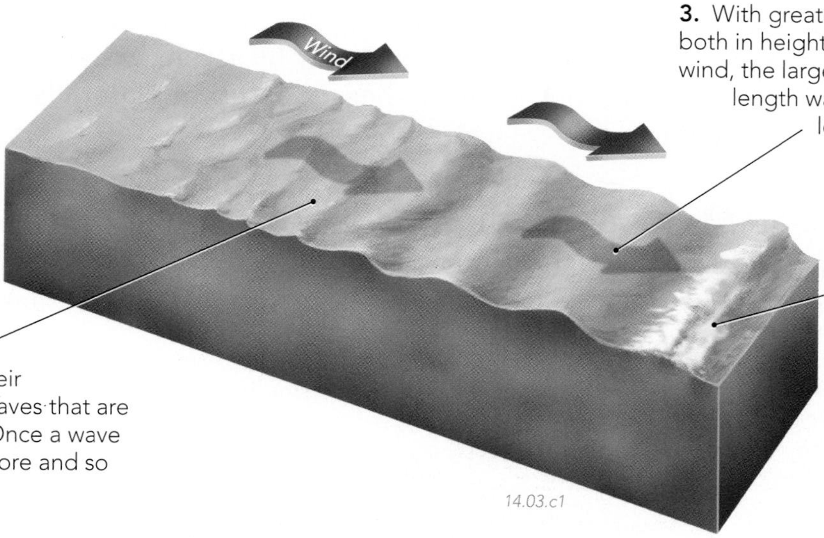

3. With greater wind speed, waves get larger, both in height and wavelength. The stronger the wind, the larger the wave generated. Long wavelength waves move faster than shorter wavelength ones and can travel farther before dying out.

4. Waves continue to move, even if the wind dies or the waves move away from the windy area. If the wind gets too strong, a wave becomes too steep to be stable and its top collapses (breaks) even if it is still out in open water. This collapse traps air and forms a white, foamy wave, a *whitecap*.

14.03.c1

How Waves Break Upon the Shore

5. As waves approach the shore, they encounter shallower water and change in a fairly systematic way. In this figure, waves are moving to the right, toward the shore.

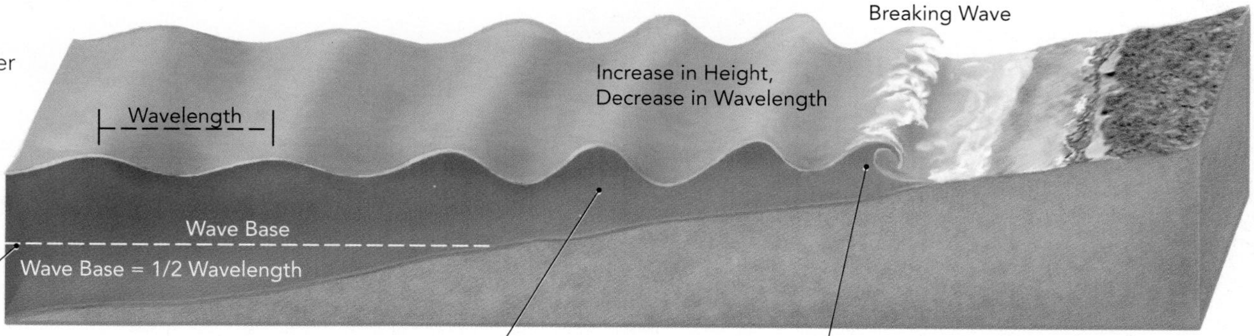

6. In deep water, waves propagate unimpeded as long as the sea bottom is deeper than the wave base. The deep-water waves shown here have a longer wavelength and lower height than waves nearer shore.

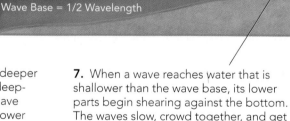

Wavelength

Wave Base
Wave Base = 1/2 Wavelength

Increase in Height, Decrease in Wavelength

Breaking Wave

14.03.c2

7. When a wave reaches water that is shallower than the wave base, its lower parts begin shearing against the bottom. The waves slow, crowd together, and get higher as their wavelengths get shorter.

8. Near shore, the wave becomes too high and steep to support itself. The top of the wave, which is moving faster than lower parts, topples over (breaks), forming a tube-shaped, plunging *breaker* or a jumble of spilling water. This area where the waves break upon the shore is the *surf zone*.

How Waves Bend If They Approach the Shore at an Angle

9. Waves almost always approach the shore at an angle, rather than straight on. In this figure, waves propagate toward the shore obliquely, from left to right. In deep water, the crests of the waves are straight or only gently curved.

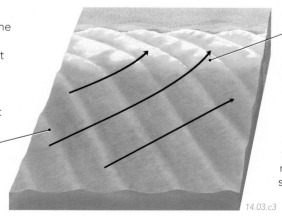

10. As the waves begin to encounter the bottom, the segments closest to the shore are slowed more than segments in deeper water. This difference in velocity causes the initially straight waves to *refract* (bend), becoming more parallel with the shore.

14.03.c3

Before You Leave This Page Be Able To

☑ Sketch and label the parts of a wave, including the height, wavelength, and wave base.

☑ Explain how the propagation of a wave differs from the motion of the water through which the wave travels.

☑ Sketch and explain why a wave rises and breaks as it reaches shallow water.

☑ Explain why waves bend if they approach the shore at an angle.

14.3

14.4 How Is Material Eroded, Transported, and Deposited Along Shorelines?

SOME SHORELINES ARE ENERGETIC ENVIRONMENTS where solid rock is eroded into loose sand, pebbles, and other kinds of sediment. Sediment can also be brought in by rivers, wind, waves, and ocean currents. Once on the beach, sediment moves in and out from the beach, and often laterally along the beach, as waves alternately transport and deposit sediment.

A How Do Waves Erode Materials Along the Shoreline?

Most erosion along shorelines is done by waves. Waves crash against rocky shores and onto beaches, breaking off pieces of bedrock that can then be reworked by more wave action and ultimately transported away.

1. When waves break directly onto a rocky coast, they cause erosion by swirling away loose pieces of bedrock and by picking up and crashing these loose pieces back against the bedrock. They also grind away the rocks by scraping sand back and forth against the bedrock.

2. Crashing and grinding water and sediment wear away at the bedrock, especially at the level where wave action is strongest. Over time, this repeated erosion in the same place may carve a *wave-cut notch* into a rocky shoreline. The notch undercuts the overlying rocks, leaving them unsupported and prone to collapsing into the sea.

3. As waves wash sand and stones back and forth across the sea bottom, they smooth off the underlying bedrock, carving a *wave-cut platform*. Knobs of resistant bedrock locally rise up above the platform, but they may eventually be worn away by the erosive action of waves.

4. Once pieces of rock are loose and within the surf zone, waves smash them together, rounding off angular corners and fracturing larger pieces into smaller ones. In this way, large angular rocks, derived from bedrock, over time become the rounded and flattened stones that dominate many beaches. With further action, stones wear down into sand.

5. Through this mechanism, waves liberate and rework pieces of bedrock and in the process create stones and sand that help the waves erode and rework even more bedrock, and break other stones into sand.

14.04.a1

B How Does the Shape of a Shoreline Influence Wave-Related Erosion?

Most coasts are not straight but have curves, bays, and other irregularities. As a result, some parts of the shoreline are somewhat protected, while other parts bear the brunt of oncoming storms and waves.

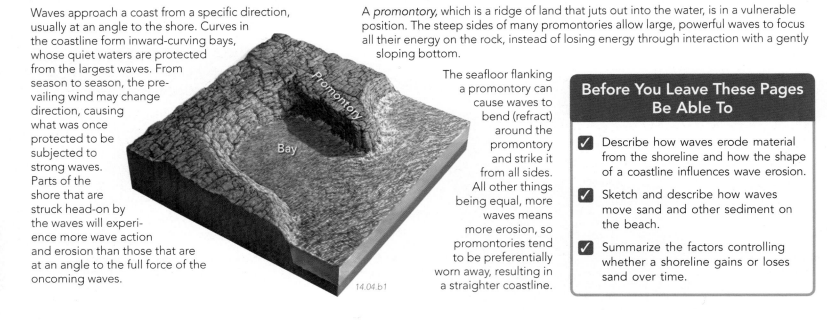

Waves approach a coast from a specific direction, usually at an angle to the shore. Curves in the coastline form inward-curving bays, whose quiet waters are protected from the largest waves. From season to season, the prevailing wind may change direction, causing what was once protected to be subjected to strong waves. Parts of the shore that are struck head-on by the waves will experience more wave action and erosion than those that are at an angle to the full force of the oncoming waves.

A *promontory*, which is a ridge of land that juts out into the water, is in a vulnerable position. The steep sides of many promontories allow large, powerful waves to focus all their energy on the rock, instead of losing energy through interaction with a gently sloping bottom.

The seafloor flanking a promontory can cause waves to bend (refract) around the promontory and strike it from all sides. All other things being equal, more waves means more erosion, so promontories tend to be preferentially worn away, resulting in a straighter coastline.

Promontory

Bay

14.04.b1

Before You Leave These Pages Be Able To

✓ Describe how waves erode material from the shoreline and how the shape of a coastline influences wave erosion.

✓ Sketch and describe how waves move sand and other sediment on the beach.

✓ Summarize the factors controlling whether a shoreline gains or loses sand over time.

C How Do Sand and Other Sediment Get Moved on a Beach?

1. During normal (non-stormy) conditions, waves wash sand and other sediment back and forth near the beach. Sediment on the sea bottom is churned up and carried toward the beach by incoming waves.

2. After a wave breaks on the beach, most water flows directly downslope off the beach, carrying sediment back toward the sea. Sediment gets reworked back and forth by the waves, but it may not be transported very far.

3. During storms, large, vigorous waves can carry sediment farther up the beach than normal, depositing it out of the reach of smaller waves that characterize more typical, less stormy conditions. Storms can also erode material from the beach, carrying it farther out to sea. Some beaches lose sand and become rocky during the winter because of the increased energy of larger winter waves, but they regain sand and become sandy and less rocky in the summer.

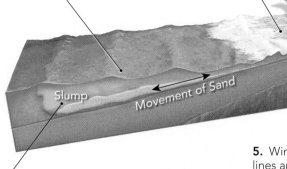

14.04.c1

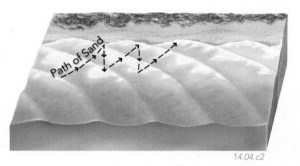

14.04.c2

4. Sediment along beaches and farther offshore can slump downhill if the sea bottom is too steep to hold the sediment or if the sediment is physically disturbed, perhaps by deep wave action during a storm, by shaking during an earthquake, or if sediment piles up too fast.

5. Wind is common along shorelines and can carry sand and finer materials long distances. Low- to moderate-strength wind cannot dislodge sand that is wet because water between sand grains tends to hold the grains together. Wind is more effective above the shoreline where the sand is dry and loose.

6. If waves approach the coastline at an angle, the sediment can be moved laterally along the coast (▲). Incoming waves move the sand at an angle relative to the coastline, and then the sand washes directly downslope when the water washes back into the sea. By this process, the sand moves laterally along the coast.

D What Determines Whether a Shoreline Gains or Loses Sand with Time?

A shoreline can gain or lose sand, depending on the rate at which sand enters the system and the rate at which it leaves. Many shorelines retain approximately the same amount of sand over time. The amount of sediment available to the system is described as the *sediment budget*, and it controls many factors of the shoreline.

1. On most coasts, sand and other kinds of sediment largely derive from erosion taking place inland. Larger volumes of sediment are produced and carried to the ocean if the land receives sufficient precipitation to generate runoff and is not overly protected by vegetation that limits the effectiveness of erosion.

2. Rivers provide an influx of sediment into the shoreline system. Deltas formed at the river–sea interface can be reworked, contributing sediment that is transported offshore or along the coast.

3. Coastal sand dunes commonly consist of sand blown onto land from the beach, representing a net loss of sand from the beach. Other dunes may derive their sand from the land and add sand to the shoreline system. Many dunes, however, simply swap sand with the beach as wind alternately blows landward and seaward, or from season to season.

4. Current flowing along the coast, a *longshore current*, transports sediment parallel to the coast. Such currents can add additional sand to the beach system, remove sand, or add as much sand as they remove (so the amount of sand is more or less in equilibrium).

6. Sediment is generated by wave erosion and associated slumping of rocks along the coast, which adds to the sand budget. Waves can bring sand in from offshore, pick up and take sand out to deeper water, or simply swash it back and forth.

14.04.d1

5. Waves erode reefs and offshore islands, especially during storms, and carry loose sediment toward the coast. Many white-sand beaches consist of calcium carbonate sand eroded from coral reefs and shells. These creatures build their shells from chemicals dissolved in the water, so they increase the amount of sand in the system.

14.4

14.5 What Landforms Occur Along Shorelines?

SHORELINES CAN DISPLAY SPECTACULAR LANDFORMS. Erosion carves some of these landforms, while sediment deposition forms others. Such landforms include wave-cut terraces, barrier islands, and various types of sandbars. Erosion is dominant in high-energy shoreline environments, whereas deposition is more common in low-energy environments. Another controlling factor is whether the coastline has been uplifted relative to ocean level or has been submerged under encroaching seas.

A What Shoreline Features Are Carved by Erosion?

In a high-energy shoreline environment, the relentless pounding of waves wears away the coastline, eroding it back toward the land. Such erosional retreat is not uniform but is often concentrated at specific locations and certain elevations. This results in some distinctive landforms along the coast.

14.05.a1 Southern Australia

◄ *Sea Cliffs*—Shorelines composed of hard bedrock can be eroded into cliffs that plunge directly into the surf or that are fronted by a narrow beach. Sea cliffs are more common in regions with active tectonism, especially where the land has been uplifted.

14.05.a2 Southern Australia

◄ *Caves and Sea Arches*—Erosion concentrates in the tidal zone where waves can undercut cliffs, forming caves. Erosion can cut through small promontories jutting out into the sea, forming arches or windows through the bedrock. Caves and sea arches are most common along uplifted coasts.

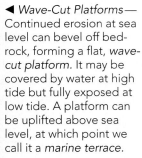

◄ *Wave-Cut Platforms*—Continued erosion at sea level can bevel off bedrock, forming a flat, *wavecut platform*. It may be covered by water at high tide but fully exposed at low tide. A platform can be uplifted above sea level, at which point we call it a *marine terrace*.

14.05.a3 Crete

◄ *Pinnacles and Sea Stacks*—Erosion along a shoreline is not uniform, and some areas of rock are left behind as erosion cuts back the shoreline. Such remnants can form pinnacles and isolated, steep-sided knobs called *sea stacks*, like these along Australia's famous Great Ocean Road.

14.05.a4 Southern Australia

Formation of a Sea Stack

A sea stack forms where a promontory extends out into the sea. As waves approach the shore, they refract, focusing erosion on the front and sides of the promontory.

If parts of the rock are weaker than others, because of differences in rock type or the relative concentration of fractures, rock behind the tip of the promontory may erode faster than the tip, forming a cave.

Continued erosion can collapse the roof of a cave and carve a passage behind the former tip of the promontory. The more resistant knob of rock becomes surrounded by the sea and is a *sea stack*.

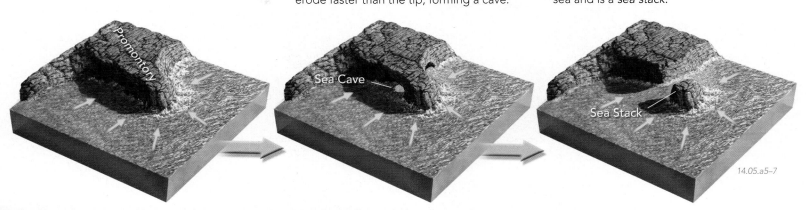
14.05.a5–7

B What Shoreline Features Result from Deposition?

As sediment moves along a coastline, it can preferentially accumulate in places where water velocity is slower, forming a variety of low, mostly sandy features.

14.05.b1 Queensland, Australia

14.05.b2 Islas Derniere, LA

14.05.b3 Baja California, Mexico

Sandbar—Offshore of many shorelines is a low, sandy area, called a *sandbar*. Bars are typically submerged much of the time and can shift position as waves and longshore currents pick up, move, and deposit the sand.

Barrier Island—Offshore of many shorelines are low islands that act as barriers, partially protecting the coast from large waves and rough seas. Many barrier islands are barely above sea level and consist of loose sand, including sand dunes, and salt-water marshes.

Spit—Along some coasts, a low ridge of sand and other sediment extends like a prong off a corner of the coast. Such a feature is a *sand spit* or a *spit* and is easily eroded, especially by storm waves. If a spit or bar blocks a bay, as shown here, it is a *baymouth bar*.

Formation of a Spit, Baymouth Bar, and Barrier Island

14.05.b4-6

A spit forms when waves and longshore currents transport sand and other beach sediment along the coast, building a long but low mound of sediment that lengthens in the direction of the prevailing longshore current.

If a spit grows long enough, it may cut off a bay, becoming a baymouth bar. This bar shelters the bay from waves and may allow it to fill in with sediment, forming a new area of low-lying land, perhaps creating a marsh.

If sea level rises, former spits and bars may become long, sandy barrier islands. Barrier islands may also form if sediment deposited by rivers when sea level is lower becomes islands when sea level rises.

Cape Cod

Cape Cod sticks out into the Atlantic Ocean from the rest of Massachusetts like a huge, flexed arm. The "curled fist" is mostly a large spit. Other features are bars and barrier islands. Much of the sediment was originally deposited here by glaciers, which retreated from the area 18,000 years ago. As the glaciers melted, global sea level rose, flooding the piles of sediment and causing them to be reworked by waves and longshore currents.

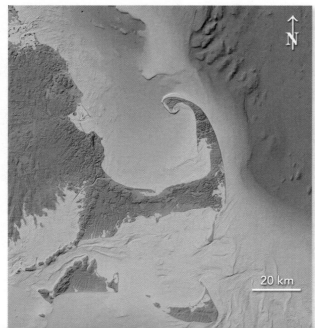
14.05.mtb1

20 km

Before You Leave This Page Be Able To

☑ Describe the different types of shoreline features.

☑ Sketch and summarize one way that a sea stack, spit, baymouth bar, and barrier island can each form.

☑ List the types of features that are present on Cape Cod, and discuss how these types of features typically form.

14.5

14.6 What Are Some Challenges of Living Along Shorelines?

SHORELINES CAN BE RISKY PLACES TO LIVE because of their dynamic nature. Destruction of property and loss of life result from waves, storm surges, and other events that are integral to the coastal environment. Beaches and other coastal lands can be totally eroded away, along with poorly situated buildings. How can homebuyers and investors identify and avoid such unsuitable and potentially risky sites?

A What Hazards Exist Along Shorelines?

Most shoreline hazards involve interactions between water and land, but strong winds also pose risks. Significant hazards accompany storms, which can produce large waves, strong winds, and surges of water onto the land.

14.06.a1

14.06.a2

14.06.a3

Waves are constantly present but are not always a threat to land, buildings, and people along shorelines. The most damage from waves occurs during extreme events, such as hurricanes and other storms. Waves can erode land and undermine hillsides, causing slopes and buildings to collapse into the water.

A *storm surge* is a local rise in the level of the sea or large lake during a hurricane or other storm. A storm surge results from strong winds that pile up water in front of an approaching storm, inundating low-lying areas along the coast. Surges are accompanied by severe erosion, transport, deposition, and destruction.

Strong winds and rain accompany storms that strike the coast. Communities right on the coast are especially susceptible to these hazards because they often lack a windbreak between them and open water. Also, many coastal lands are flat, so structures built in low-lying areas are prone to rainfall-related flooding.

Before

14.06.a4 Topsail Island, NC

After

14.06.a5 Topsail Island, NC

These images document damage caused by Hurricane Fran in 1996 along the beach on Topsail Island, North Carolina. This photograph shows the area before the hurricane. White numbers mark two houses in both photographs. Compare these two photographs to observe what happened to the two houses, and to houses nearby, during the hurricane.

This photograph, taken after the hurricane, shows the loss of beach and destruction of houses caused by waves, storm surge, and erosion. The hurricane came ashore with sustained winds measured at 185 km/hr (115 mph) and a 4-meter-high (12-foot) storm surge. It caused more than $3 billion in damage.

B What Approaches Have Been Tried to Address Shoreline Problems?

Various approaches are used to minimize the impacts of natural shoreline processes, including erecting barriers, trying to reconstitute the natural system, or simply not building in the most hazardous sites.

14.06.b1 Galveston, TX

◄ One way to limit the amount of erosion is to construct a *seawall* along the shore. Such walls consist of concrete, steel, or some other strong material. Large rocks and other debris are locally dumped along the shoreline to armor the coast, in an attempt to protect it from erosion. Material used in this way is called *rip rap*. Building a seawall commonly results in loss of beach in front of the wall.

14.06.b2 Yaquina Bay, OR

◄ Another type of wall, called a *jetty*, juts out into the water, generally to protect a bay, harbor, or nearby beach. Jetties are usually built in pairs to protect the sides of a shipping channel. In an attempt to protect one area of the coast, jetties, like many other engineering approaches to shoreline problems, can have unintended and problematic consequences. Jetties and other walls can focus waves and currents on adjacent stretches of the coast. These directed waves and currents, deprived of their normal load of sediment by the wall, erode the adjacent areas as they try to regain an equilibrium amount of sediment.

Low walls, called *groins*, are built out into the water to influence the lateral transport of sand by longshore currents and by waves that strike at an angle to the coast. A groin is intended to trap sand on its up-current side, but it has the sometimes unintended consequence of causing the beach immediately down-current of the groin to receive less sand and to become eroded.

A wall, called a *breakwater*, can be built out in the water to bear the brunt of the waves and currents. Breakwaters are built parallel to the coast to protect the beach from severe erosion and to cause sand to accumulate on the beach behind the structures. Some communities bring in sand to replenish what is lost to storms and currents. This procedure of *beach nourishment* is expensive and may last only until the next storm.

14.06.b3 Presque Isle SP, PA

Avoiding Hazards and Restoring the Shoreline System to Its Natural State

One approach preferred by some people, including many geologists, is simply not to build in those places that have the highest likelihood of erosion, coastal flooding, coastal landslides, and other shoreline hazards. Geologists can map a shoreline and conduct studies to identify the most vulnerable stretches of coastline. With such information in hand, the most inexpensive approach—in the long run—is to forbid the building of houses or other structures in those areas identified as high risk. In the wake of the destruction of New Orleans and nearby communities by Hurricane Katrina, there is a debate about whether to rebuild those neighborhoods that are at highest risk, such as those well below sea level.

In many cases, such geologic concerns are either ignored or are overruled by financial and aesthetic interests of developers, communities, and people who own the land. Beachfront property is desirable from an aesthetic standpoint and so can be expensive real estate, which some people think is too precious to leave undeveloped.

Another approach is to try to return the system to its original situation, or at least a stable and natural one, rather than trying to "engineer" the coastline. Engineering solutions can be expensive, may not last long, or may have detrimental consequences to adjacent beaches. Returning the system to a natural state may involve restoring wetlands and barrier islands that buffer areas further inland from waves and wind.

Examining the balance of sediment moving in and out of the system can help identify non-natural factors, such as dammed rivers, which if restored to original conditions would bring more sediment into the system and stabilize beaches, dunes, and marshes.

Before You Leave This Page Be Able To

✔ Summarize some of the hazards that affect beaches and other coastlines.

✔ Describe approaches to address coastal erosion and loss of sand, including not building and trying to restore systems to a natural state.

14.6

14.7 How Do Geologists Assess the Relative Risks of Different Stretches of Coastline?

UNDERSTANDING THE GEOLOGY AND DYNAMICS of a shoreline is the first step in assessing the potential risks posed by waves, currents, coastal flooding, and other shoreline processes. Geologists study coastlines using traditional field methods and new methods that involve lasers and satellites.

A What Field Studies Do Geologists Conduct Along Shorelines?

To investigate potential shoreline hazards, geologists map and characterize the topographic and geologic features of the land, shoreline, and near-shore sea bottom. They combine this information with an understanding of the important shoreline processes to identify those areas with the highest hazard.

To assess shoreline hazards, geologists document the elevation of the land surface above sea level. High areas clearly have less risk of being flooded by the sea. Coastal scientists obtain precise elevations of the land with various surveying tools, some using satellites (Global Positioning System, or GPS) or lasers that scan the ground surface from a plane. These surveys identify areas, like this high marine terrace, that are too high to be flooded, even during a hurricane.

Areas that are close to sea level may be subject to flooding by storm surges and storm-related intense rainfall that causes flooding along coastal rivers. Vulnerable low-elevation areas may extend far inland, in this case along a low river valley.

Mapping the bedrock geology, as well as the loose sediments along the beach, guides geologists when assessing how different areas will erode. Coasts backed by resistant bedrock, as along a cliff, will be less likely to be eroded by strong waves and currents. Parts of a cliff may fail over time, however, as they are undercut by constant wave action.

Geologists also map the distribution and height of coastal dunes. Dunes, especially those that are large or are stabilized by vegetation, decrease the risk inland for storm surge and associated erosion. Marshlands, like those on a delta, also help buffer areas farther inland from waves, storm surges, and strong coastal winds.

Geologists map the location and height of sandbars, islands, reefs and other offshore barriers. These barriers can protect the coast from wave action. In the photograph below, a barrier reef in the distance produces a shallow, quiet-water lagoon. The reef and shallow waters of the lagoon generally protect the shoreline from large, potentially damaging waves, except during strong storms, when a storm surge can raise water levels and deepen water close to shore.

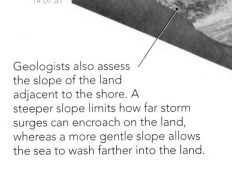

14.07.a1

Dunes

Offshore Island

Current Direction

Geologists also assess the slope of the land adjacent to the shore. A steeper slope limits how far storm surges can encroach on the land, whereas a more gentle slope allows the sea to wash farther into the land.

Geologists document the width of beaches. An area with a wide beach between the shoreline and houses is generally less risky than an area where houses sit right behind a narrow beach. Seawalls, groins, and other constructed features can greatly affect beach width and therefore potential risk. A seawall can limit the amount of landward erosion and may protect buildings from storm surges, but sand in front of the seawall may be lost. A groin affects the width of the beach differently on either side. It may decrease the risk of storm erosion to the beach on the up-current side, which gains sand and becomes wider, but increase the risk of storm erosion to the beach on the down-current side, which loses sand and becomes narrower. Beach width, wave size, and potential hazard are also affected by barrier islands and reefs, which can protect the coast.

14.07.a2 Grand Cayman

B What Can New High-Resolution Elevation Data Tell Us About Shorelines?

Satellite data (GPS) and other new methods of mapping elevation now allow scientists to more accurately characterize shoreline regions and track in detail how a shoreline changes during storms. One new method is *LIDAR*, which is an acronym for LIght Detecting And Ranging.

In the LIDAR method, a laser beam is bounced off the ground, detected back at the airplane, and timed as to how long it takes to travel to the surface and back to the instrument. The shorter the time the beam takes to reach the ground and return, the shorter the distance between the plane and the ground—and therefore the higher the elevation of the land. LIDAR elevations are accurate to within about 15 cm (6 in.) and can be quickly collected over larger areas than is practical to cover with conventional surveying.

As the plane flies forward, mirrors direct the laser toward different areas beneath the LIDAR sensor. Thousands of data measurements are recorded each second in a narrow belt, called a *swath*, across the land. The plane flies back and forth over the area, overlapping adjacent swaths to ensure that there are no gaps in the data. A GPS unit mounted on the plane allows technicians to accurately register the LIDAR data with geographic map coordinates and to match the data to features on the ground.

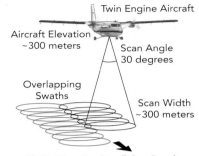

Twin Engine Aircraft
Aircraft Elevation ~300 meters
Scan Angle 30 degrees
Overlapping Swaths
Scan Width ~300 meters
Flight Direction Parallel to Beach
14.07.b1

Mapping Hurricane-Related Changes in the Shoreline of Alabama

Coastal Alabama has been hit by a series of powerful hurricanes, most recently by Hurricane Ivan in 2004 and Hurricane Katrina in 2005. The U.S. Geological Survey (USGS) and other government agencies have used LIDAR to investigate the changes that such large hurricanes inflict on the coastline. One detailed study was of Dauphine Island, an inhabited barrier island along the Gulf Coast of Alabama.

The three images to the left show perspective views of detailed LIDAR elevations taken at three different times (before Hurricane Ivan, after Ivan, and after Hurricane Katrina). Red arrows point to the same house in all five images. The first image shows a central road with houses (the colored "peaks") on both sides. In the second image, the storm surge from Hurricane Ivan has washed over the low island from left to right, eroding the left beach, covering the road with sand, and redepositing some of the sand on the right.

The aftermath of Hurricane Katrina, bottom left, is even more dramatic. All but a few houses were totally washed away. Both the width and height of the island decreased, leaving the remaining houses even more vulnerable to the next storm surge.

The two images to the right show the calculated changes caused by each storm. Each image was produced by comparing the *before* and *after* data sets and computing the difference. Features shown in reds and pinks represent losses due to erosion, green areas show where deposition occurred, and whitish features were unchanged by that storm. From all these data, how risky do you think this place is?

14.07.mtb1–3

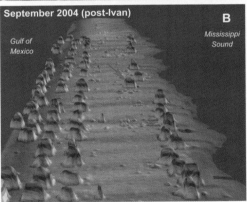

May 2004 (pre-Ivan)   A
Gulf of Mexico
Mississippi Sound

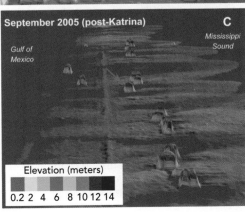

September 2004 (post-Ivan)   B
Gulf of Mexico
Mississippi Sound

September 2005 (post-Katrina)   C
Gulf of Mexico
Mississippi Sound
Elevation (meters)
0.2 2 4 6 8 10 12 14

14.07.mtb4-5

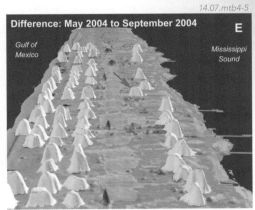

Difference: May 2004 to September 2004   E
Gulf of Mexico
Mississippi Sound

14.07.mtb5

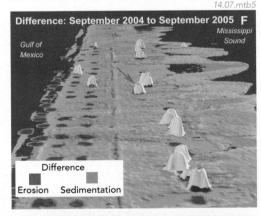

Difference: September 2004 to September 2005   F
Gulf of Mexico
Mississippi Sound
Difference
Erosion   Sedimentation

Before You Leave This Page Be Able To

✓ Describe how studying the geologic features along a coast can help identify areas of highest hazard.

✓ Summarize how LIDAR data are collected, and provide an example of how they can be used to document changes in a shoreline.

14.7

14.8 What Happens When Sea Level Changes?

SEA LEVEL HAS RISEN AND FALLEN many times in Earth's history. A rise in sea level causes low-lying parts of continents to be inundated by shallow seas, whereas a fall in sea level can expose previously submerged parts of the continental shelf. Changes in sea level form certain features along the shoreline and farther inland, and can deposit marine sediments on what is normally land. Changing sea level produces two kinds of coasts: *submergent coasts* and *emergent coasts.*

A What Features Form if Sea Level Rises Relative to the Land?

Shorelines adjust their appearance, sometimes dramatically, if sea level rises or falls relative to land. A relative change in sea level can be caused by a global change in sea level or by tectonics that causes the land to subside or be uplifted relative to the sea. Distinctive features form when sea level rises relative to the land.

Submergent coasts form where the land has been inundated by the sea because of a rise in sea level or subsidence of the land.

After the land is inundated, flooded river valleys give the coast an irregular outline, featuring branching *estuaries* and other embayments.

The shape of the land exerts a strong control on how the coastline will look after it is flooded by rising sea level. Examine this figure and imagine what will happen to different features and which areas will get flooded first.

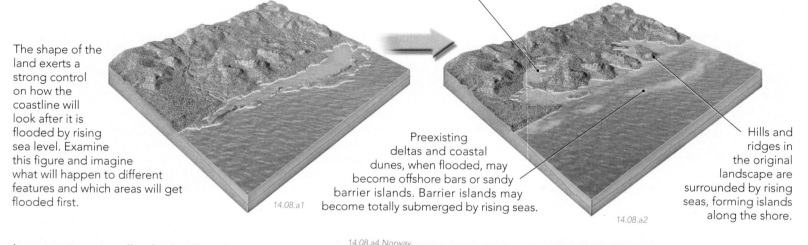

14.08.a1

Preexisting deltas and coastal dunes, when flooded, may become offshore bars or sandy barrier islands. Barrier islands may become totally submerged by rising seas.

Hills and ridges in the original landscape are surrounded by rising seas, forming islands along the shore.

14.08.a2

An *estuary* is a river valley that has been flooded by the sea, allowing freshwater from the land to interact with salt water from the sea. Water levels in the estuary and the balance between fresh and salt water are affected by tides and by changes in the amount of water coming from the land. The satellite image below shows the Chesapeake Bay estuary, with a branching form. The bay was a valley originally carved by rivers but was flooded when sea level rose at the end of the last ice age. ▼

14.08.a3

Chesapeake Bay

14.08.a4 Norway

◀ The coasts of Norway, Greenland, Alaska, and New Zealand all feature narrow, deep embayments called *fjords.* Fjords are steep-sided valleys that were carved by glaciers and later invaded by the sea as the ice melted and sea level rose.

14.08.a5 North Carolina

◀ Many barrier islands are interpreted to have been formed by rising sea level. Some barrier islands began as coastal dunes or piles of sediment deposited by rivers. As sea level rose, the rising water surrounded the piles of sediment, resulting in new islands.

B What Features Form if Sea Level Falls Relative to the Land?

Some features reflect a fall in sea level, or uplift of the land by tectonics or by isostatic processes. A fall in sea level can expose features that were submerged and can greatly affect what happens on the adjacent land.

Emergent coasts form where the sea has retreated from the land due to falling sea level or due to uplift of the land relative to the sea.

After sea level drops, erosion incises valleys into the land. Emergent wave-cut notches form topographic steps on the land, and wave-cut platforms form a series of relatively flat benches, known as *marine terraces*.

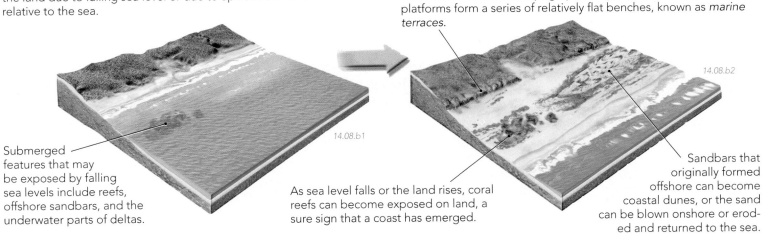

14.08.b1

14.08.b2

Submerged features that may be exposed by falling sea levels include reefs, offshore sandbars, and the underwater parts of deltas.

As sea level falls or the land rises, coral reefs can become exposed on land, a sure sign that a coast has emerged.

Sandbars that originally formed offshore can become coastal dunes, or the sand can be blown onshore or eroded and returned to the sea.

14.08.b3 California

Wave-cut platforms form within the surf zone along many rocky shorelines and, when exposed above sea level, form relatively flat terraces on the land. The surface of such marine terraces may contain marine fossils and wave-rounded stones.

14.08.b4 Galápagos

Coral reefs and other features that originally formed at or below sea level can be exposed when seas drop relative to the land. These coral reefs, now well above sea level, provide evidence of relative uplift of the land.

14.08.b5 Crete

Wave-Cut Notch

A wave-cut notch is an originally horizontal recess eroded into rock by persistent wave erosion at sea level along a shoreline. In this photograph, a horizontal wave-cut notch is now several meters above sea level because of tectonic uplift of the coast.

Before You Leave This Page Be Able To

✓ Summarize what a submergent coast is and what types of features can indicate that sea level has risen relative to the land.

✓ Summarize what an emergent coast is and what types of features indicate that sea level has fallen relative to the land.

14.8

14.9 What Causes Changes in Sea Level?

SEA LEVEL HAS VARIED GREATLY IN THE PAST. Global sea level has been more than 200 m higher and more than 120 m lower than today. What processes caused these changes in sea level? Large variations in past sea level resulted from a number of competing factors, including the extent of glaciation, rates of seafloor spreading, and global warming and cooling.

A How Does Continental Glaciation Affect Sea Level?

The height of sea level is greatly affected by the existence and extent of glaciers and regional ice sheets. At times in Earth's past, ice sheets were more extensive than today, and at other times, they were less extensive or absent.

The ice in glaciers and continental ice sheets accumulates from snowfall on land. When glaciers and ice sheets are extensive, they tie up large volumes of freshwater, causing sea level to drop.

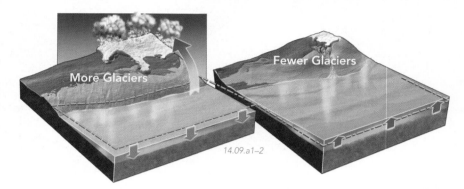

When glaciers and ice sheets melt, they release large volumes of water that flow back into the ocean, causing sea level to rise.

The growth and shrinkage of ice sheets and glaciers is the main cause of sea level change on relatively short timescales (thousands of years).

B How Do Changes in the Rate of Seafloor Spreading Affect Sea Level?

At times in Earth's history, the rate of seafloor spreading was greater than it is today, and at other times it was probably lower. Such changes cause the rise and fall of sea level.

The shape and elevation of a mid-ocean ridge and adjacent seafloor reflect the rate of spreading. As the plate moves away from the spreading center, it cools and contracts, causing the seafloor to subside, creating space for seawater.

If seafloor spreading along a ridge is slow, the ridge is narrower because the slow-moving plate has time to cool before getting very far from the ridge. Slow seafloor spreading and narrow ridges leave more room in the ocean basin for seawater. So over time, a decrease in the spreading rate causes sea level to fall.

If seafloor spreading along a ridge is relatively fast (10 cm/year or faster), the ridge is broad because still-warm parts of the plate move farther outward before cooling and subsiding. So an increase in seafloor spreading rate is accompanied by broader ridges that displace water out of the ocean basins, causing sea level to rise. In other words, faster spreading yields more young seafloor, and young seafloor is less deep than older seafloor, raising sea level.

C How Do Changes in Ocean Temperatures Cause Sea Level to Rise and Fall?

Sea level is also affected by changes in ocean temperatures, which cause water in the oceans to slightly expand or contract. Such effects result in relatively moderate changes in sea level.

Water, like most materials, contracts slightly as it cools, taking up less volume. The amount of contraction is greatly exaggerated in this small block of water.

Water expands slightly when heated, taking up more volume. Again, the amount of expansion is exaggerated in this small block of water.

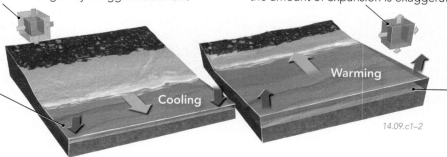

When ocean temperatures fall, water in the ocean contracts, causing sea level to fall.

When ocean temperatures increase, water in the ocean expands slightly, causing a small rise in sea level. The percentage of expansion is small, but it can cause a moderate rise in sea level.

D How Does the Position of the Continents Influence Global Sea Level?

Continents move across the face of the planet, sometimes being near the North or South Poles and at other times being closer to the equator. These positions influence sea level in several ways.

Glaciers and continental ice sheets form on land and so require a landmass to be cold enough to allow ice to persist year round. This occurs most easily if a landmass is at high latitudes (near the poles) or is high in elevation.

At most times in Earth's past, widespread glaciation occurred when one or more continents were near the poles. By allowing glaciers to exist, continents at high latitudes can cause a drop in global sea level.

14.09.d1

At times in Earth's past, the larger continents were not so close to the poles. This lower latitude position minimized or eliminated widespread glaciation, tending to keep sea levels high.

E How Do Loading and Unloading Affect Land Elevations Relative to Sea Level?

Weight can be added to a landmass, a process called *loading*. A weight can also be removed, a process called *unloading*. Loading and unloading can change the elevation of a region relative to sea level.

Weight loaded on top of a region imposes a downward force that, if large enough, can downwarp the land surface beneath the load and in adjacent areas. Loading, such as by continental ice sheets, lowers the loaded region relative to sea level. This can allow sea-water to inundate regions near the ice sheets. The ice in this figure and the amount of subsidence are very stylized and vertically exaggerated.

14.09.e1

If the weight is unloaded from the land, the region flexes back upward, a process known as *isostatic rebound*. The uplifted, rebounding region rises relative to sea level.

Unloading and isostatic rebound can occur when continental ice sheets melt. Rebound begins as soon as significant amounts of ice are removed, but it can still be occurring thousands of years after all the ice is gone.

14.09.e2

Ongoing Isostatic Rebound of Northeastern Canada

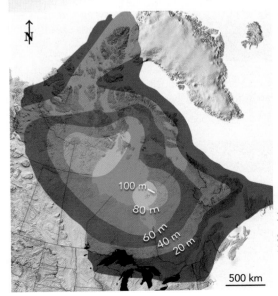

14.09.mtb1

The northern part of North America has been covered by glaciers off and on for the last two million years. The weight of these continental ice sheets loaded and depressed this part of the North American plate. When the ice sheets melted from the area, beginning about 15,000 years ago, unloading caused the land, especially in Canada, to begin to isostatically rebound upward.

The amount of rebound has been measured both directly and indirectly. We can measure uplift directly by making repeated elevation surveys across the land and then calculating the amount of uplift (rebound) between surveys. Rates of rebound are typically millimeters per year, which is enough to detect with surveying methods. Satellite measurements (GPS) are also sensitive enough to measure such changes. The amount and rate of rebound can also be inferred more indirectly by documenting how shorelines and other features have

been warped and uplifted. Contours on this map to the left indicate the amount of rebound interpreted to have occurred in northeastern North America over the last 6,000 years. Some areas have experienced more than 100 m of uplift.

Before You Leave This Page Be Able To

✔ Summarize how continental glaciation, rates of seafloor spreading, ocean temperatures, and position of the continents affect sea level.

✔ Summarize how loading and unloading affect land elevations using the example of northeastern Canada.

14.9

14.10 What Are Glaciers?

GLACIERS ARE MOVING MASSES OF ICE, ranging in size from huge ice sheets that cover large regions to smaller glaciers that are restricted to a single mountain or valley. Most glaciers are primarily ice and snow, but they typically contain significant amounts of rocks and finer sediment that were incorporated into the glacier as it flowed from higher elevations to lower ones.

A What Are the Characteristics of Glaciers?

Glaciers form where snow and ice accumulate faster than they melt, so many glaciers begin in snowfields in higher elevations or at higher latitudes. Glaciers do not form if an area is unable to sustain cold temperatures that keep ice and snow from melting faster than they accumulate.

Ice can cover broad areas, but glaciers can become confined within valleys as they flow from higher elevations to lower ones. As adjacent ice-filled valleys merge, so do the glaciers, producing a wider and commonly thicker mass of flowing ice.

As the snow gets buried, it compresses into ice, turning blue in the process. As a result, the icy parts of glaciers have a distinctly blue appearance. Most glaciers also have lines (grooves, ridges, and sediment-rich streaks) formed by flow within the glacier. These are fairly straight or gently curved if the glacier has a simple pattern of flow, but they are contorted and folded if the glacier experienced more complex patterns of flow.

Whether a glacier forms depends partly on the slope of an area. Areas with gentle slopes allow snow and ice to accumulate to sufficient thicknesses, whereas some mountain slopes are so steep that snow and ice slide downhill instead of piling up.

10 km

14.10.a1

14.10.a2 Andes, South America

◄ This glacier of blue ice flows down a steep valley and ends in a lake of meltwater from the glacier. As a glacier moves, internal stresses cause its upper surface to break, forming fractures, each of which is called a *crevasse*. Crevasses are especially abundant and well developed where a glacier flows around a curve, like around a bend in a valley, or where the land beneath the glacier changes slope, either from a steep slope to a more gentle one or from a gentle slope to a steeper drop. The glacier shown here breaks apart, opening up crevasses as it flows over a steep drop-off.

14.10.a3 Swiss Alps

◄ As a glacier moves past bedrock, it plucks away pieces of rock, and its surface may be covered by rock pieces derived from nearby steep slopes. It grinds up some rock into a fine rock powder. The glacier carries away this material, coarse and fine, depositing it where the glacier melts. This glacier has dark fringes of rocky material on both sides and at its end.

Before You Leave These Pages Be Able To

✓ Describe the characteristics of glaciers, including the three main types (ice sheets, valley glaciers, piedmont glaciers).

✓ Summarize which places have ice sheets and glaciers.

B What Are the Types of Glaciers?

The largest accumulations of ice are in *ice sheets*, regionally continuous masses of ice like those covering nearly all of Antarctica and Greenland. In this panoramic photograph, an ice sheet in the background spills over a cliff, forming a steeply flowing mass of ice called an *icefall*.

14.10.b1 Antarctica

▶ Glaciers that flow down valleys are called *valley glaciers* or *Alpine glaciers* (named after the Alps). Valley glaciers tend to be fairly narrow (several kilometers wide) but can flow down valleys for tens of kilometers.

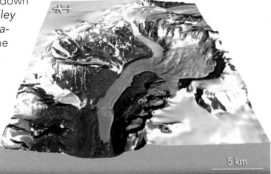
14.10.b2 Dry Valleys, Antarctica

▶ As some valley glaciers or ice sheets flow out of the mountains into broader, less confined topography, they can spread out, forming a *piedmont glacier* (piedmont means foot of the mountain).

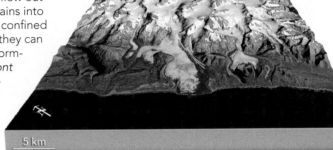
14.10.b3 Southeastern AK

C Where Are Most Ice Sheets and Glaciers?

Glaciers form where snow and ice accumulate faster than they melt, so they form in cold climates. Most glaciers are therefore in high latitudes (closer to the North or South Pole), at high elevations, or some permissible combination of latitude and elevation.

In the Southern Hemisphere, glaciers occupy high peaks of the Andes, especially in Patagonia, the most southerly part of the mountain range.

The largest ice mass on Earth is on the continent of Antarctica, which sits squarely over the South Pole. Ice and snow cover about 98% of the continent, mostly in the form of huge ice sheets. Valley glaciers form where the ice sheet flows into valleys.

Glaciers cover many of the highest parts of the Tibetan Plateau and the Himalaya, the highest mountain range on Earth, even though this region has a fairly low latitude (not close to the poles). Glaciers and ice sheets are present elsewhere in Asia, especially in islands and peninsulas along the Arctic Ocean.

Large ice sheets and smaller glaciers occupy 80% of Greenland and large areas of the neighboring islands, including Iceland.

In the main part of North America, glaciers are present in Alaska, northern Canada, the Rocky Mountains of the United States and Canada, the Coast Range of British Columbia, and on the larger volcanoes and other high peaks of the Cascade Range.

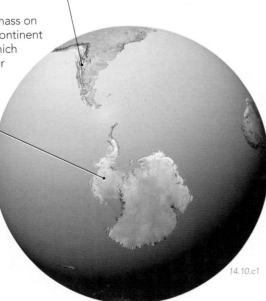
14.10.c1

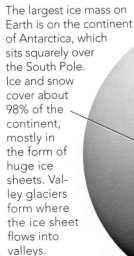
14.10.c2

14.10

How Do Glaciers Form, Move, and Vanish?

GLACIERS FORM, MOVE DOWNHILL, AND EVENTUALLY MELT AWAY. How does a glacier form? Once formed, how does a glacier move across the landscape, and what happens to it as it flows downhill, toward warmer areas with more melting and generally less snowfall?

A How Do Snow and Ice Accumulate in Glaciers?

Glaciers, including the one below, form by the accumulation of snow and ice. The snow is derived from snowfall, but can be moved around by the wind or by avalanches, which are masses of snow and ice that fall, slide, or flow downhill.

14.11.a1 Swiss Alps

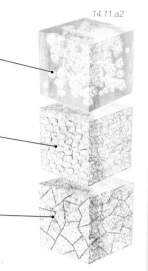

14.11.a2

Snow falls as individual flakes. Once on the ground, flakes get pressed together by the weight of other snowflakes on top. Loose snow can contain 90% air between the flakes.

As more snow accumulates on top, snowflakes farther down are compressed, forcing out more than 50% of the air. The snowflakes become compressed into small, irregular spheres of more dense snow.

With increasing depth and pressure, the snow begins to recrystallize into small interlocking crystals, forming solid ice. Ice is a crystalline material and is considered to be a type of rock. Crystalline ice contains less air and commonly has a bluish color.

B How Does a Glacier Form and Change as It Moves Downhill?

Glaciers form when the amount of snow and ice accumulating from snowfall exceeds the amount lost by melting and other processes. In this situation, the snow and ice pile up and may start to move as a glacier.

1. The upper part of the glacier or ice sheet, where snow and ice are added faster than they melt, is the *zone of accumulation*. Gravity, working on the weight of accumulating snow, causes the glacier to flow downhill.

2. As the glacier moves downhill, it loses more and more ice and snow by melting, by wind erosion, and by loss of ice molecules directly to the air, a process referred to as *sublimation*. At some point along the glacier, the losses of ice and snow exactly balance the amount of accumulation; this boundary is called the *equilibrium line*. The equilibrium line is sometimes, but not always, marked by a gradational boundary between snow-covered ice upslope on the glacier and exposed bluish ice downslope. The bluish ice formed at depth and became exposed at the surface as upper levels of ice and snow were removed. In some cases, the entire length of a glacier may be covered with snow, but blue ice can be observed at depth in fractures (crevasses) that cut the glacier's upper surface.

Zone of Accumulation

Snow

Equilibrium line

Blue Ice

Zone of Ablation

Land

1 km

Sea

14.11.b1

14.11.b2 Morteratsch, Swiss Alps

4. At lower elevations, ice melts away faster than it can be replenished by downward movement of ice within the glacier and by snowfall. This causes the glacier to end or terminate, either on land or in the sea. The end is called the *terminus*.

3. The valley glacier in the photo to the right has an upper, snow-covered part (zone of accumulation) and a lower area of blue ice below the equilibrium line.

C How Do Glaciers Move?

Glaciers move downhill because the ice is not strong enough to support its own weight against the relentless downward pull of gravity. As glaciers spread downward, they move by internal shearing and flow of the solid ice, by simply sliding across the bedrock, or by some combination of these two mechanisms.

As gravity pulls the ice downhill, friction along the base of the glacier causes the bottom of the glacier to lag behind the upper, less constrained parts. The upper part of a glacier (▼) therefore flows faster than the lower part, causing internal shearing within the glacier.

The rates at which glaciers move are extremely variable. Many glaciers move about a meter per day, but some move centimeters per day. The fastest ones move more than 30 meters per day.

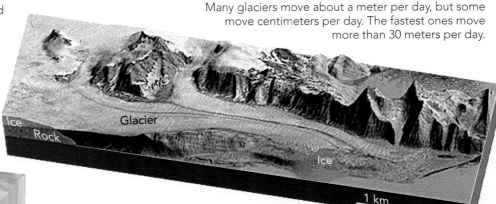

14.11.c1

14.11.c2

If the interface between the glacier and the underlying bedrock is very irregular and is relatively dry, the base may become locked to the bedrock and not move at all. Only the coldest glaciers are completely frozen at their bases.

◄ If the bedrock-glacier interface is less irregular (i.e., smoother) or contains water from melting ice, the glacier may be able to slide over the bedrock. Such glaciers can move relatively rapidly.

14.11.c3

D What Happens When a Glacier Encounters the Sea or a Lake?

When a glacier reaches the ocean or a lake, it may float on the water if the sea or lake is deep enough. Ice, even the dense blue variety within glaciers, floats because it is less dense than either freshwater or salt water.

As ice along the leading edge of a glacier floats, it tends to spread or be pulled apart, forming large crevasses within the ice. These allow large blocks of ice (►) to collapse off the front of the glacier, a process called *calving*.

14.11.d2 Glacier Bay, Alaska

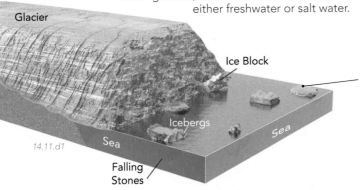

14.11.d1

As the blocks of ice fall into the water, they float, forming *icebergs*. As much as 90% of an iceberg is beneath the water. As icebergs melt, rocks and other sediment within them drop into the water. Some ice sheets and glaciers flow into the sea with such large quantities of ice that they form a large *ice shelf* that floats on seawater (►). These can be hundreds of kilometers wide.

14.11.d3 Antarctica

Glaciers, Snowfields, and Sea Ice

Not every large mass of ice on Earth's surface is a glacier. Some accumulations of snow and ice never move, and these are simply called *snowfields*. Large masses of ice also form when the upper surface of a lake or the sea freezes. In the ocean, such ice is called *sea ice*. In all but the coldest places, like parts of the Arctic Ocean, sea ice freezes in the winter and thaws in the spring or summer. Freezing excludes most salt from the crystalline structure of ice, so sea ice melts to form water that is largely fresh (not salty). In the photograph below, broken sheets of sea ice surround a rocky island in Antarctica.

14.11.mtb1 Antarctica

Before You Leave This Page Be Able To

☑ Sketch and describe how snow is transformed by pressure into ice.

☑ Summarize or sketch the differences in a glacier above and below the equilibrium line.

☑ Describe how glaciers move and what happens when they encounter a lake or the sea.

14.11

14.12 What Happens When Glaciers Erode into the Landscape?

AS GLACIERS MOVE, THEY SCOUR UNDERLYING ROCK and unconsolidated materials, picking up the pieces and carrying them toward lower elevations. In mountainous areas, glaciers pluck rocks from peaks and ridges, producing some distinctive landforms that we can use to recognize landscapes that are glacially carved. Glaciers and ice sheets grind into the underlying land surface, wearing down hills and other topographic high points, and locally polishing smooth surfaces onto bedrock.

A What Happens at the Base and Sides of a Glacier?

Ice is not a hard material, but glaciers contain rocks and other material that can scrape and grind away at the underlying land surface, smoothing off rough edges and removing rocks and other sediment, which can become incorporated into the ice and carried away.

14.12.a2 Duluth, MN

14.12.a1 Swiss Alps

◄ This view is looking down a valley to the terminus of a glacier. Piles of poorly sorted sediment liberated by the glacier flank the valley on both sides, left behind when the glacier melted. The bottom of the valley is also covered with unsorted sediment, which is being reworked by the meltwater stream emanating from the end of the glacier. Looking down past the end of the glacier shows a valley with a rounded U-shaped profile.

U-shaped Valley

► When ice sheets flow across the surface, they smooth and polish rocks over broad areas. These polished surfaces and scratch marks are evidence that a glacier once moved across the area.

Polished Surface

B How Does Glacial Erosion Modify Landscapes?

Glaciers occupy and modify landscape features that existed before glaciation, and imprint into the landscape clues that glaciers were once there. The two scenes below show one view of how glaciers shaped a modern landscape.

The San Juan Mountains of Colorado were intermittently covered by glaciers over the last two million years. The figure below depicts the possible extent of ice as glaciers began to melt away.

This is the same area as it exists today, mostly free of ice. Note the high and steep, glaciated peaks, the narrow, sharp ridges, and several prominent U-shaped valleys in the center of the image.

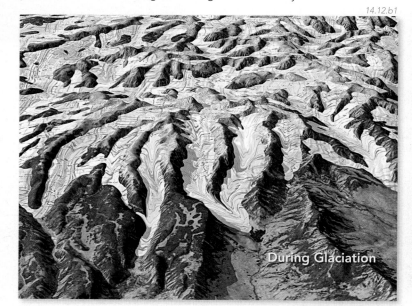

14.12.b1

During Glaciation

14.12.b2

U-shaped Valley

After Glaciation

C   What Landforms Form by Glacial Erosion?

In mountains, glaciers produce distinctive landforms, including bowl-shaped basins flanked by steep ridges and U-shaped valleys carved by the moving ice. We can use these features to recognize landscapes carved by glaciers, even after the ice melts away. The three-dimensional terrain below shows many of these features from the aptly named Glacier National Park in Montana, and photographs of examples are below the main figure.

Near the uppermost end of a mountain glacier, snow and ice accumulate in less steep areas. The ice plucks pieces from the bedrock, excavating a bowl-shaped depression, called a *cirque*. When the ice melts, it exposes the cirque.

Hard bedrock ridges that flank cirques are commonly narrow, sharp, and jagged, like the ridges shown below. Such a ridge is an *arete*, and it is jagged because it has been glacially eroded from both sides and because physical weathering is very intense in such settings.

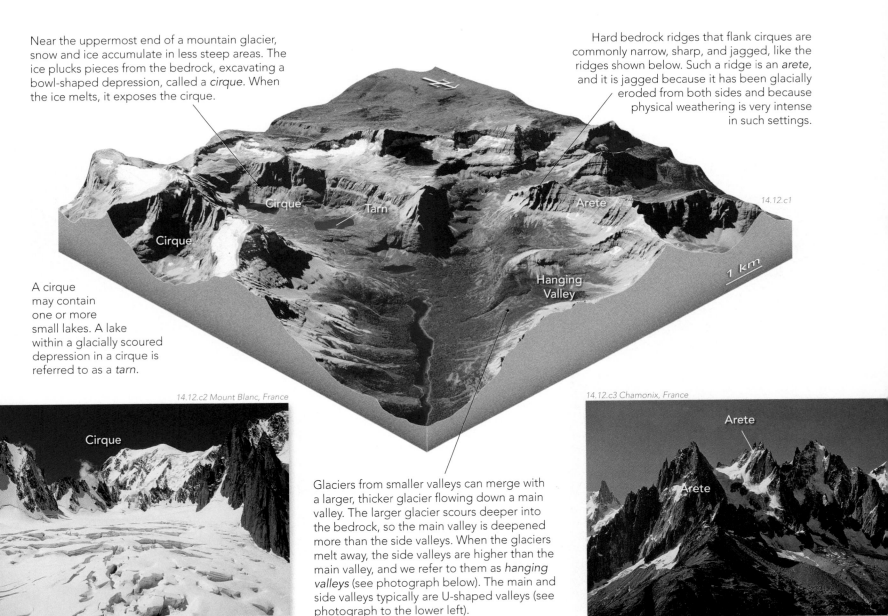

14.12.c1

A cirque may contain one or more small lakes. A lake within a glacially scoured depression in a cirque is referred to as a *tarn*.

14.12.c2 Mount Blanc, France

14.12.c3 Chamonix, France

Glaciers from smaller valleys can merge with a larger, thicker glacier flowing down a main valley. The larger glacier scours deeper into the bedrock, so the main valley is deepened more than the side valleys. When the glaciers melt away, the side valleys are higher than the main valley, and we refer to them as *hanging valleys* (see photograph below). The main and side valleys typically are U-shaped valleys (see photograph to the lower left).

14.12.c4 Norway

14.12.c5 Alaska

Before You Leave This Page Be Able To

✓ Summarize what happens at the base and sides of a glacier.

✓ Describe the origins of landforms formed by glacial erosion.

14.12

14.13 What Features Do Glacial Sediments Form?

GLACIERS AND CONTINENTAL ICE SHEETS PRODUCE LANDFORMS by depositing sediment as the ice melts. Where does this sediment come from, what are its characteristics, and what types of features result from deposition of the sediment? In other words, what sedimentary deposits and features do glaciers and larger ice sheets leave behind?

A Where Do Glaciers Carry and Deposit Sediment?

Valley glaciers pick up debris along their sides and bases, but flow within the glacier and merging of adjacent glaciers distributes the sedimentary debris throughout much of the glacier. Continental ice sheets mostly pick up sediment along their bases, by eroding into the land. When they melt, both types of glaciers deposit this sediment.

14.13.a1 Mendenhall Glacier, AK

14.13.a2 Glacier Bay, AK

A glacier carries an unsorted assemblage of large, commonly angular rocks mixed with finer sediment, as shown in the two photographs above. The sides of most glaciers contain especially abundant sediment because they receive loose materials from the slopes of hills and mountains that flank the glacier. Also, streams flowing toward the glacier bring an additional supply of sediment to the sides. The base of a glacier is also relatively rich in sediment because it plucks away pieces of bedrock and any loose materials over which the glacier moves. Sediment carried by and deposited by a glacier is *moraine*, and we classify moraines into different types according to where they form. The photograph to the left shows moraine still being carried by a glacier, whereas the three-dimensional perspective below shows deposits of moraine left behind after the glaciers melted away.

◄ A *lateral moraine* forms along the sides of the glacier and is expressed as a dark fringe of rocks and other debris. When the glacier melts, lateral moraines commonly form low ridges (►) along what were the edges of the glacier.

◄ A *medial moraine* is a sediment-rich belt in the center of the glacier. A medial moraine forms where two glaciers join, trapping their lateral moraines within the combined glacier. Medial moraines may not be well preserved, like in the area to the right.

◄ A *terminal moraine* forms at the termination of a glacier and generally marks the glacier's farthest downhill extent. Some terminal moraines are large and conspicuous (►), in many cases because the end of the glacier deposited sediment in the same place for some time.

Lateral Moraine

Medial Moraine

Terminal Moraine

14.13.a3 Mt. Spur, AK

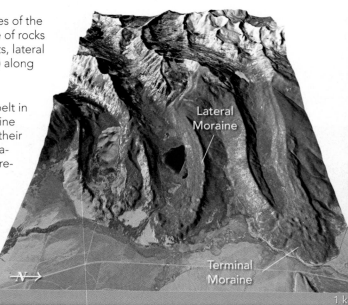

Lateral Moraine

Terminal Moraine

N →

1 km

14.13.a4 Sawtooth Mtns., ID

B What Features Do Continental Ice Sheets and Large Glaciers Leave Behind?

Continental ice sheets occur in Antarctica and Greenland, but they have also covered large parts of Europe, Asia, and North America. Continental ice sheets, like other glaciers, contain a mixture of ice, rocks, and pulverized, fine-grained sediment. Rocks and other sediment are intermixed throughout the ice sheet but are most abundant along the base and sides of the ice sheet. An ice sheet or large glacier leaves behind a suite of unique landforms.

As the ice sheet or glacier melts, it deposits its load of sediment in sheets, piles, and ridges. Such glacial sediment is *till* in all these forms, but we name the associated landform for its shape and inferred mode of origin.

▶ Some curiously shaped hills of till resemble teardrops, each with its pointed end in the direction the ice flowed. Each hill, called a *drumlin*, forms as the moving glacier sculpts soft materials into a streamlined shape.

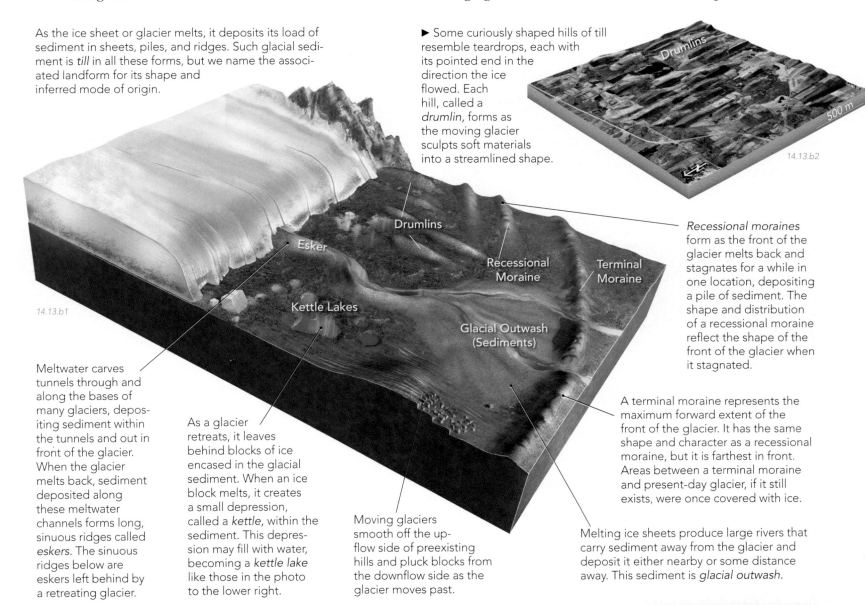

14.13.b1

14.13.b2

Recessional moraines form as the front of the glacier melts back and stagnates for a while in one location, depositing a pile of sediment. The shape and distribution of a recessional moraine reflect the shape of the front of the glacier when it stagnated.

Meltwater carves tunnels through and along the bases of many glaciers, depositing sediment within the tunnels and out in front of the glacier. When the glacier melts back, sediment deposited along these meltwater channels forms long, sinuous ridges called *eskers*. The sinuous ridges below are eskers left behind by a retreating glacier.

As a glacier retreats, it leaves behind blocks of ice encased in the glacial sediment. When an ice block melts, it creates a small depression, called a *kettle*, within the sediment. This depression may fill with water, becoming a *kettle lake* like those in the photo to the lower right.

Moving glaciers smooth off the up-flow side of preexisting hills and pluck blocks from the downflow side as the glacier moves past.

A terminal moraine represents the maximum forward extent of the front of the glacier. It has the same shape and character as a recessional moraine, but it is farthest in front. Areas between a terminal moraine and present-day glacier, if it still exists, were once covered with ice.

Melting ice sheets produce large rivers that carry sediment away from the glacier and deposit it either nearby or some distance away. This sediment is *glacial outwash*.

14.13.b3 Iceland

14.13.b4 Alaska Range, AK

Before You Leave This Page Be Able To

- ✓ Summarize where glaciers carry and deposit sediment, explaining the three main types of moraine.

- ✓ Sketch and describe the features associated with continental ice sheets, and explain how each type of feature formed.

14.13

14.14 What Features Are Peripheral to Glaciers?

REGIONS WITH COLD CLIMATES exhibit other features that are either related to glacial processes or are simply related to the freezing of the ground. Some other features reflect the cool, wet climates that accompany continental glaciation but formed in regions too warm for glaciers.

A What Types of Deposits Are Related to Glacial Episodes?

Glaciers produce an abundance of sediment and water, so glacially derived sediment can accumulate over wide regions and can be transported far from the actual glaciers by streams, rivers, wind, and waves.

14.14.a1 Tibet

14.14.a2 Tibet

Glacially produced sediment of all sizes mixes with the abundant glacial meltwater to form large rivers whose cold waters are loaded with sediment. The rivers and streams deposit sediment on broad *outwash plains* in front of the glaciers.

Glaciers pulverize the rocks they carry and the rocks they move over, producing abundant silt-sized material that can be blown away by the wind. Accumulations of windblown silt are called *loess*, and many loess deposits are glacially derived, such as these soft, tan deposits.

B What Is Permafrost and Where Does It Occur?

In cold regions, water in the uppermost part of the ground can remain permanently frozen, a condition called *permafrost*. The uppermost parts may partially thaw during some summers, but the remainder of the soil stays frozen throughout the year.

14.14.b1 Denali, AK

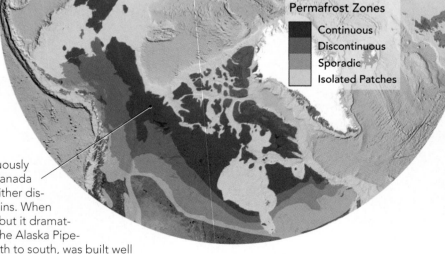
14.14.b2

◄ Some ground in this photograph is permanently frozen and so is *permafrost*. During warm months, the top few meters of the ground may thaw, allowing some vegetation to grow. Permafrost commonly does not allow trees to grow, and treeless permafrost areas are also-called *tundra*.

► In North America, large areas of continuously frozen ground are restricted to northern Canada and Alaska. In other areas, permafrost is either discontinuous or occurs in high, cold mountains. When frozen, permafrost is a very hard material, but it dramatically weakens if it thaws. For this reason, the Alaska Pipeline, which crosses central Alaska from north to south, was built well above the ground to avoid having the oil-filled pipes heat and thaw the permafrost and weaken the pipeline's supports.

Permafrost Zones
- Continuous
- Discontinuous
- Sporadic
- Isolated Patches

C What Types of Lakes Were Associated with Glacial Times?

The cool, wet climates that favor an increase in precipitation also favor the formation and maintenance of lakes. In the western United States, huge lakes existed at times during the recent ice ages, but largely dried up when the climate changed about 15,000 years ago. Evidence of the extent and height of these lakes is still visible.

An ice-age lake filled low, interconnected basins in the Rocky Mountains of western Montana. This lake, named *Lake Missoula*, caused catastrophic flooding, as described in more detail below. Shorelines from this lake were etched as horizontal lines into the hills surrounding Missoula, Montana. ▼

14.14.c2 Missoula, MT

A large ancestral lake, named *Lake Lahontan*, filled the low basins of western Nevada. The lake was up to 240 m (790 ft) deep about 13,000 years ago, and some modern lakes in the area are remnants of this larger ice-age lake.

14.14.c1

250 km

14.14.c3 Bonneville Salt Flats, UT

The Great Salt Lake of Utah is a remnant of a much larger ice-age lake named *Lake Bonneville*. As the large lake dried up, it left the Bonneville Salt Flats (▲), home to rocket testing, land-speed records, and many miles of salt.

Smaller lakes formed in closed basins across much of the American Southwest. In places, they left salt flats and fine-grained lake deposits.

Lake Missoula and the Channeled Scablands

A famous story among geologists is the history of Lake Missoula and a peculiar topographic region in eastern Washington known as the *Channeled Scablands*. The scablands are so named because the area is crossed by many gorges, which curiously do not contain streams or rivers large enough to have carved the gorges.

In the 1920s, geologist J. Harlan Bretz proposed a hypothesis to explain this mystery, but it took decades to be accepted by the larger geologic community. According to this hypothesis, on more than one occasion, glacial Lake Missoula breached the glacial dam holding back its waters, and catastrophic torrents of water raced across the landscape to the west, carving the scablands. The huge floods carried gigantic boulders, carved smooth depressions (pot

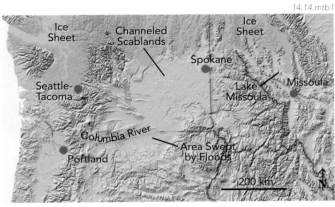

14.14.mtb1

14.14.mtb2

holes) into the bedrock (▼), and formed enormous ripples hundreds of feet high that dwarf nearby houses (above right).

14.14.mtb3 Dry Falls, WA

Before You Leave This Page Be Able To

☑ Describe the characteristics of different deposits related to glacial episodes and how each type forms.

☑ Describe permafrost and where it occurs.

☑ Describe several large ice-age lakes and some of the features they formed, either while full or while emptying.

14.14

14.15 What Is the Evidence for Past Glaciations?

SEVERAL KINDS OF EVIDENCE CONFIRM that huge ice sheets once covered the land. This evidence is directly expressed as features and deposits within the landscape, and is recorded indirectly in unusual marine deposits and in isotopic compositions of marine fossils in sediment deposited during glaciations.

A What Features Indicate the Former Presence of Glaciers?

To be able to recognize past glaciations, geologists visit active glaciers and ice sheets to observe what types of landforms and deposits are diagnostic of glaciation. Geologists observe modern-day examples to help them explain prehistoric features—another example of "the present is the key to the past."

The most obvious evidence of glaciers includes landscapes that contain features diagnostic of past glaciations, including cirques, aretes, tarns, and U-shaped valleys. Each of those features is visible in this scene. Can you find them?

14.15.a1 Absaroka Mtns., WY

14.15.a2 Plum Island, MA

Other landscape features are diagnostic of glaciers, including ridges of moraine, eskers, and, as shown above, a drumlin. The presence of these features across much of New England and the Great Lakes region demonstrates that glaciers once covered these areas, leaving behind clues to their former presence after the glaciers had totally disappeared.

14.15.a3 Kimberly, South Africa

◄ As glaciers advance and melt back across the landscape, they deposit layers or patches of poorly sorted till. Ancient examples, like as this one from Kimberly, South Africa, have lithified into rock, but they still provide evidence for ancient glaciations, in this case near the end of the Permian Period, about 250 million years ago.

► When glaciers and ice sheets flow across the land, they smooth and polish underlying rocks and deposit glacial sediments (till) on the polished surface. This site has a polished surface that displays scratch marks left behind by ancient glaciers. The glaciers deposited glacial till, now represented as sedimentary rocks in the foreground. Consolidated till is *tillite*.

14.15.a4 Kimberly, South Africa

14.15.a5 Yellowstone NP, WY

◄ Glaciers can carry huge rocks, some as big as a house, and leave them scattered about the landscape. Glaciers may transport large blocks hundreds of kilometers, taking them to places where such rock types are not present in the bedrock. Such an out-of-place block, like the one shown here, is a *glacial erratic*.

► An unusual feature of some marine and lake sediment is the presence of scattered stones in an otherwise fine-grained, clastic sediment. We call these *dropstones* because they have been carried within floating icebergs and then dropped into fine sediment on the seafloor or lake bottom.

14.15.a6

B How Do We Determine Where and When the Most Recent Ice Age Occurred?

From diverse lines of evidence, geologists determine which areas were once covered by ice and which ones were not. They then use fossils and isotopic dating methods to determine when glaciers were most widespread, a time called a *glacial period* or *glacial maximum*. A time during an ice age when glaciers are melting and retreating is an *interglacial period*. A glaciation, because it affects sea level and the influx of freshwater into the ocean, can also be investigated by examining the nature and chemistry of marine fossils and sediment.

1. During evaporation of seawater or freshwater, heavier isotopes of an element preferentially remain in the water, while lighter isotopes more easily escape into the water vapor. In the case of oxygen isotopes, evaporation causes the water to become enriched in the heavier isotope, oxygen-18 (^{18}O) while enriching the water vapor in the lighter isotope, oxygen-16 (^{16}O).

14.15.b1

2. As the water vapor condenses into clouds and precipitation (rain, snow, or hail), the water, snow, or ice contains the higher proportion of lighter isotopes that was in the water vapor. If snow and ice accumulate on land, they tend to keep the light isotopes from returning to the sea. As a result, an increase in the amount of snow and ice on land, as during a glacial event, will cause seawater to be more enriched in heavy isotopes.

3. As glaciers and ice sheets melt, they release their water, which was relatively enriched in lighter isotopes. Streams, rivers, and melting icebergs return these light-isotope-enriched waters back to the sea. A decrease in glaciation, therefore, causes seawater to shift toward lighter isotopic compositions, just the opposite of an increase in glaciation. As a result, isotopic compositions of ice on land and on water in the sea are indications of increases and decreases in the amount of snow and ice on land.

14.15.b2

5. As they grow, some marine organisms build shells of calcium carbonate by extracting the necessary chemicals from seawater. As the chemistry and temperature of the water change, so does the chemical composition of the shells formed in that water. Geologists analyze oxygen and carbon isotopes in fossils to infer changes in seawater temperature and chemistry over time. We can then use such changes to infer the times of glaciation or times when melting of glaciers released freshwater into the ocean.

14.15.b3

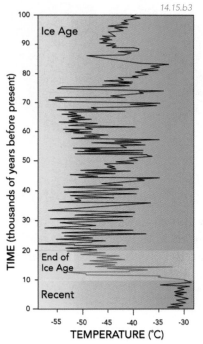

◄4. This graph shows temperatures inferred from oxygen-isotope compositions of ice from part of a 3 km-deep hole that scientists drilled into the ice sheet of central Greenland. The data show how scientists interpret temperatures in central Greenland (a very cold place) to have varied over the last 100,000 years. Points to the right indicate that temperatures were warmer, as they are today, and glaciers were less widespread. Data that plot to the left indicate that glaciers were more widespread. From these and other data, geologists infer that glaciers decreased and increased in extent many times during the last 100,000 years.

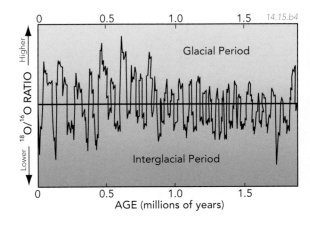

14.15.b4

◄6. This graph shows oxygen-isotope compositions of marine shells, which record the isotopic composition of seawater when the animal was alive. Higher ratios of ^{18}O/^{16}O mark times of glaciation, when snow and ice on land tied up the lighter isotope (^{16}O), increasing the relative amount of the heavier isotope (^{18}O) in seawater and the shells. These data indicate that glaciers increased and decreased in volume many times over the past two million years.

Before You Leave This Page Be Able To

✓ Describe evidence used to infer that glaciers once covered a landscape.

✓ Discuss how glaciations can be expressed in ice and the ocean, and how we can use this record to interpret when glaciation occurred.

14.15

14.16 What Happened During Past Ice Ages?

OVER THE PAST TWO MILLION YEARS, huge ice sheets and glaciers intermittently covered large areas of the Northern and Southern Hemispheres. The ice sheets and glaciers increased during some time periods we call *glacial periods*, and decreased during other times, called *interglacial periods*. Overall, the past two million years had a marked increase in glacial periods compared to most other times in Earth's geologic history. As a result, geologists call this time period the *Ice Ages*. The Ice Ages coincide with the Pleistocene Epoch.

A What Parts of the Earth Were Covered with Ice During the Ice Ages?

By examining landscapes for glacially polished surfaces, moraine, and other indications of glaciation, geologists can infer which areas were covered by ice in the past and which ones were not. We also can use sedimentary records from lakes, determining whether a lake existed at some time or was still covered by ice. By considering other factors, including elevation and latitude, we can extrapolate the actual observations to produce maps that show the interpreted extents of glaciers and ice sheets at different times during the Ice Ages.

This figure, centered on Greenland, shows the present-day distribution of ice in the Northern Hemisphere. It also shows sea ice that covers much of the Arctic Ocean between North America and Siberia (northeastern Asia). Note that the only large land areas covered by ice are Greenland, adjacent islands in northern Canada, and parts of Alaska and westernmost Canada.

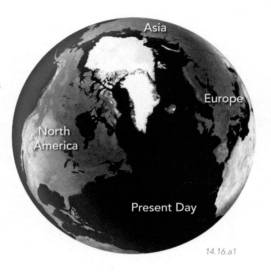

14.16.a1

14.16.a2

Approximately 28,000 years ago, ice sheets and glaciers were much more extensive on Earth, including in the Northern Hemisphere. Note that a large ice sheet covered nearly all of Canada and much of the northern parts of the United States. Ice sheets and glaciers also covered much of northern Asia, northern Europe, and the Alps of southern Europe.

B What Parts of North America Were Covered by Ice and When Did the Ice Retreat?

The maps below show interpreted ice cover for North America at two times. The left map shows a time when the ice cover was close to a maximum, approximately 20,000 years ago, whereas the map on the right shows the position of ice at about 10,000 years ago, after the ice sheets had retreated and as the Ice Ages were ending.

Most of Alaska was not covered by ice. This allowed people and animals from Asia to migrate into North America.

The center of the ice sheet was in northern Canada. Here, the ice is interpreted to have been several kilometers thick.

Ice sheets extended from Canada into the northern parts of the United States, covering New England, the Great Lakes region, and the upper Midwest. Glaciers also occupied parts of the Rocky Mountains but are not shown.

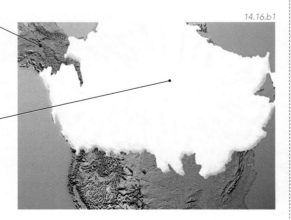

14.16.b1

When we say that the ice sheets *retreated*, we mean that ice within the sheets was still flowing forward, but the front of the ice sheet moved back (retreated) because ice melted faster than it could be replenished.

14.16.b2

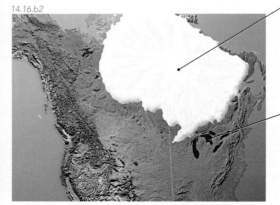

By 10,000 years ago, the ice sheets had melted back, covering only part of north-central Canada.

As the ice retreated, the northern United States emerged from beneath the ice, the Great Lakes formed, and river systems like the upper Mississippi began to develop.

C | What Record Did Past Glaciations Leave in the United States?

Glaciers once covered northern parts of the conterminous United States, leaving behind evidence of which areas they covered and which ones they did not. The map below shows the area south of the Great Lakes, highlighting the locations of ridges of moraine left behind as ice sheets retreated from the region.

A continental ice sheet once covered much of the area south of the Great Lakes, in the upper Midwest of the United States. The largest features formed by glaciation are smooth troughs that in this area trend from northeast to southwest, or locally north to south. These smooth areas were once covered by ice sheets, which smoothed off the underlying landscape as they moved southwest and south out of Canada.

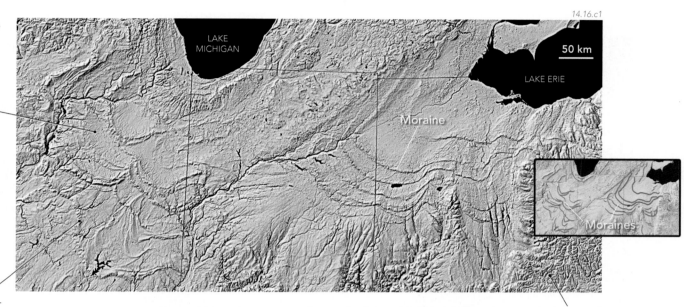

14.16.c1

As the ice sheets melted away, they left a veneer of glacial sediment on the land-scape. Ridges, representing piles of glacial sediment, mark the position of the front of the ice sheet as it melted back. Most of these ridges, highlighted in red on the inset map, are recessional moraines.

Areas with rougher topography south of the troughs, including those in the lower right corner of this map, were never glaciated. They are called *driftless areas* because they were not glaciated and so do not have a covering of glacial drift (glacial sediment).

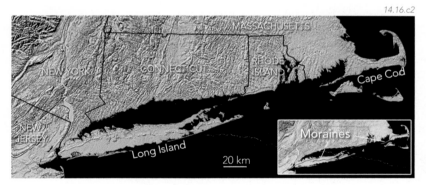

14.16.c2

▲ Glaciers also covered New England. They piled up glacial sediment, forming ridges of moraine, here shaded red, on Long Island, Rhode Island, Cape Cod and the islands of Martha's Vineyard and Nantucket.

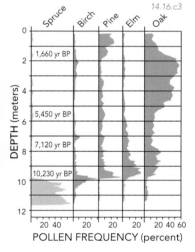

14.16.c3

Glaciers left behind other evidence in the Northeast and Great Lakes area, including outcrops that display glacial scratches and polish. As the glaciers retreated, meltwater collected in low areas, forming numerous small and large lakes. Sediment accumulating in lakes and in other settings contain a record, in the form of pollen, of the types of plants that grew at different times. These pollen records (◄) document a dramatic shift from spruce trees to leafy trees as glaciers retreated and the climate warmed.

The Ice-Age Hypothesis

The idea that huge ice sheets covered the land was not intuitive, but arose to explain an ever-growing number of hard-to-understand observations made in Europe and North America. As discussed in A. Hallam's book *Great Geological Controversies*, these observations included bones of reindeer and Arctic birds in southern France, large out-of-place boulders (erratics) scattered across much of Europe, and the presence of scratched and polished bedrock far beyond the existing glaciers in the Alps. European naturalists of the time debated how to explain these curious features. In the 1820s, they hypothesized that widespread, prehistoric glaciations had occurred in the Alps and northern Europe. Today, we call this interval of time the *Ice Ages*, and we recognize that glaciers grew and shrank many times during the last two million years.

Before You Leave This Page Be Able To

✓ Describe what parts of the Northern Hemisphere and North America were covered with ice during the Ice Ages.

✓ Discuss evidence for past glaciations in the United States.

14.16

14.17 What Starts and Stops Glacial Episodes?

TO UNDERSTAND THE REASONS FOR GLACIAL AND INTERGLACIAL TIMES, we need to further explore what causes global changes in climate. Because human history is short compared to the timescales on which global climate change occurs, we do not completely understand all the causes.

A What Variations in Earth's Rotation and Orbit Influence Global Climate?

Short-term (up to 100,000 years) variations in climate are likely controlled by the amount of solar radiation reaching Earth. Milutin Milankovitch, a Serbian astronomer and geophysicist, recognized that Earth's climate could be influenced by changes in Earth's tilt and orbit shape. We call such changes *Milankovitch cycles.*

Changes in Earth's Tilt

Over time, Earth's axis of rotation changes its tilt relative to its plane of orbit around the Sun. About every 40,000 years, the tilt changes from 22.5° to 24.5° and back again. The amount of tilt toward the Sun affects Earth's climate because tilt affects how much summer sunlight strikes higher latitudes. The diagrams below show the planet at times of the year when it is winter in the Northern Hemisphere.

The *maximum tilt angle* of Earth's rotation axis is 24.5°. This amount of tilt increases the effects of the seasons. When combined with other climatic factors, it can lead to a decrease in glacial activity because warmer summer temperatures melt more polar ice.

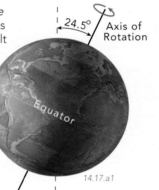

14.17.a1

The present-day position of Earth's tilt is 23.5°. Earth's tilt is currently adjusting back from the maximum 24.5° tilt, which occurred near the end of the last ice age.

14.17.a2

When Earth's tilt is at its *minimum tilt angle* (22.5°), high latitudes receive less direct sunlight during the summer, causing cooler summers and an increase in glaciers.

14.17.a3

Wobble of Rotation Axis (Precession)

Precession is similar to what happens when a spinning top slows down and wobbles. As Earth wobbles, its spin axis changes from pointing at the North Star (Polaris) to pointing at the star Vega and back again.

If the effects of precession are added to other astronomical factors, they could affect global climate. The precession cycle lasts about 23,000 years.

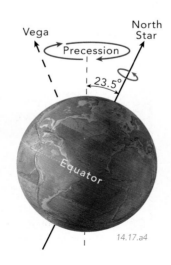

14.17.a4

Shape of Orbit (Eccentricity)

Eccentricity is the term for the noncircular shape of Earth's orbit around the Sun. Earth's orbit changes over long timescales, sometimes being more circular and sometimes slightly more elliptical. These changes cause variations in the amount of solar radiation reaching Earth.

This change from a more circular to a less circular orbit is thought to only slightly affect climate, but when added to other astronomical cycles, its effect might be significant. The eccentricity cycle lasts about 100,000 years.

Scientists have computed the effects of each of these factors (tilt, precession, and eccentricity) and then combined the effects to investigate how they interact, both in the past and in the future. When used to try to reconstruct past events, the calculated effects can be compared with records of past climate, like the isotopic composition of ice cores. Such comparisons support the hypothesis that Milankovitch cycles can explain many past climatic variations, including the waxing and waning of glaciations. Computer models can also be used to predict future Milankovitch-related climate changes, but these results are currently being debated by the scientific community.

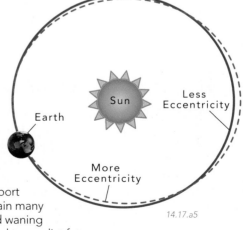

14.17.a5

B Can Variations in Solar Heating Cause Glacial Episodes?

The amount of energy given off by the Sun is not constant. As a result, different amounts of solar energy reach Earth at different times. The composition of Earth's atmosphere is not constant, so the amount of energy that can pass through and reach Earth is variable. Changes in the atmosphere's composition can lead to a warmer climate or to a cooler climate.

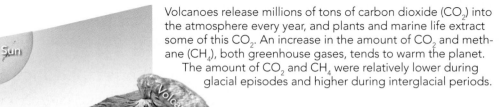

14.17.b1

Every 11 to 14 years, the level of *sunspot activity* on the Sun increases, producing small but significant changes in solar energy output. This energy fluctuation can influence Earth's climate system, affecting temperature, precipitation, and other things. When sunspot activity declines, evidence shows that Earth's climate cools slightly.

During glacial episodes, snow and ice cover more of Earth's surface, and cloud cover also increases. Both of these increase the reflectivity, also called *albedo*, of Earth so that more sunlight gets reflected off Earth's surface and is lost to space. This loss of light and heat makes the climate cooler. A cooling climate can result in more snow, ice, and clouds, leading to more cooling. In this way, the system reinforces itself.

Volcanoes release millions of tons of carbon dioxide (CO_2) into the atmosphere every year, and plants and marine life extract some of this CO_2. An increase in the amount of CO_2 and methane (CH_4), both greenhouse gases, tends to warm the planet. The amount of CO_2 and CH_4 were relatively lower during glacial episodes and higher during interglacial periods.

Large, explosive volcanic eruptions can add significant quantities of volcanic ash and dust to the atmosphere. The ash and dust reflect solar radiation back into space, allowing less sunlight to reach Earth's surface. A major volcanic eruption will increase the amount of ash and dust in the atmosphere, perhaps resulting in global cooling. Whether a huge volcanic eruption causes warming or cooling depends on the strength of the counteracting effects of gases (warming) versus ash and dust (cooling).

C What Is the Role of Ocean Currents and Continental Positions on Glaciations?

Ocean currents transport water of different temperatures from one part of the ocean to another. Such currents can warm or cool land areas, helping to increase or decrease glaciation on land.

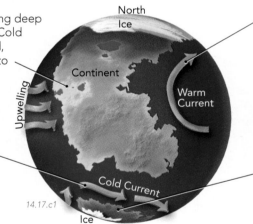

14.17.c1

Upwelling currents can bring deep cold water to the surface. Cold currents help cool the land, perhaps allowing glaciers to form, if there is sufficient precipitation.

Cold currents can also inhibit the growth of glaciers because they put less moisture into the atmosphere, leading to less snowfall.

Warm currents bring tropical waters northward that have been heated by the Sun. These currents can warm adjacent parts of a continent, inhibiting glaciations. Warm currents also bring with them warm, moist air, increasing precipitation, which can allow more ice to accumulate if the temperature is cold enough.

The position of continents affects the geometry of ocean currents, deflecting them in certain directions or blocking the connection between different oceans. In this way, continental positions influence ocean currents, which influence regional and local climates.

A continent located at or near the North or South Pole provides a large landmass on which continental ice sheets can form.

Causes of Ice Ages

The period from 2.5 or 3 million years ago to 10,000 years ago saw many *glacial* and *interglacial* periods. This time span lies mostly in what has traditionally been called the *Pleistocene* epoch, which ended with the end of the last major glaciation about 10,000 years ago. Major glacial and interglacial periods commonly lasted about 40,000 to 100,000 years. Changes between the two conditions apparently could occur rapidly in geologic terms, in some cases over less than several thousand years.

The cause of ice ages remains controversial, but four factors are considered significant: (1) changes in solar activity; (2) changes in Earth's tilt and orbit around the Sun; (3) the amount of CO_2 and methane in the atmosphere; and (4) the position of the continents, which provides a place for snow and ice to accumulate to appreciable depths.

Before You Leave This Page Be Able To

✓ Describe how variations in Earth's rotation and orbit influence global climate.

✓ Describe how global climate can be affected by atmospheric gases, volcanic ash, and the amount of snow, ice, and cloud cover.

✓ Describe the role of ocean currents and continental positions on glaciations.

14.17

14.18 What Would Happen to Sea Level if the Ice in West Antarctica Melted?

WEST ANTARCTICA HAS THE POTENTIAL TO CAUSE a dramatic rise in sea level if its glaciers melt. It contains a huge volume of ice that is especially vulnerable because it is in direct contact with the sea. If the area's ice sheets melt, rising sea level would pose a great hazard for the world's coastlines.

A What Is the Setting of Glaciers, Ice Sheets, and Ice Shelves in West Antarctica?

The continent of Antarctica, centered over the South Pole, is a frozen world mostly covered by snow and thick sheets of ice. It contains 90% of the world's ice and 75% of the world's freshwater.

1. West Antarctica, like the rest of the continent, is mostly covered by ice, with bedrock in mountains and along the coast. This large iceberg (▼) in the foreground is floating in the sea off West Antarctica.

14.18.a1 West Antarctica

▶ **2.** The Transantarctic Mountains, a major mountain range more than 4,500 m tall, divides West Antarctica from East Antarctica. West Antarctica is much smaller than East Antarctica and consists of a central landmass that leads to a peninsula extending toward South America. West Antarctica contains 11% of the ice in Antarctica.

3. The land is flanked by three large ice shelves, where glacial ice from the land has pushed out into, and is now floating on, the ocean. These three shelves are the *Ross, Ronne-Filchner,* and *Larsen* ice shelves. ▶

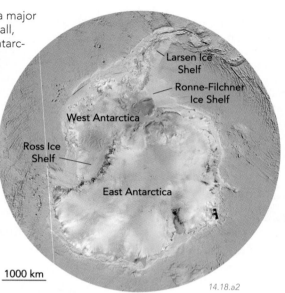

Larsen Ice Shelf
Ronne-Filchner Ice Shelf
West Antarctica
Ross Ice Shelf
East Antarctica

1000 km

14.18.a2

4. The central part of West Antarctica contains an ice sheet as thick as 3,500 m (~11,500 ft). The ice sheet accumulates from snowfall on land and feeds rapidly moving glaciers that carry massive amounts of ice toward the sea and the ice shelves. ▶

5. The base of the ice sheet is below sea level. The central part of the ice sheet is resting on solid bedrock, but outer parts are floating on the sea.

14.18.a3

Ice Sheet

6. Each ice shelf loses large volumes of ice every year by calving of icebergs and by melting of the underside, which is in contact with seawater. One part of the coast that has been studied in detail loses an average of 250 km³ of ice each year.

B What Could Happen to West Antarctica if Global Sea Levels Rise?

Dramatic events can reshape ice shelves. In early 2002, much of the Larsen Ice Shelf collapsed, breaking into millions of icebergs that floated out to sea and melted. It is possible that larger, more disastrous melting events could occur, including loss of large parts of the West Antarctic Ice Sheet.

These satellite images, taken a month apart, show the collapse of part of the Larsen Ice Shelf. The left image shows the ice shelf before the ice-loss event.

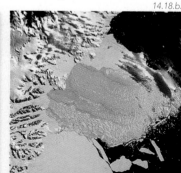

14.18.b1   14.18.b2

An area of 3,250 km² was lost, which is much larger than the entire state of Rhode Island. Other events have sprung single icebergs that were 70 km by 25 km (1,750 km²).

One possible scenario, much debated at this time, is that rising global sea level could float more of the West Antarctica ice sheet, detaching it from the underlying bedrock. If this occurred, the collapsed parts would melt, raising global sea level by some amount. But by how much? We might want to know this.

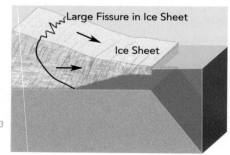

Large Fissure in Ice Sheet
Ice Sheet

14.18.b3

C   How Do We Calculate the Rise in Sea Level if West Antarctica's Ice Melts?

To evaluate how melting of ice sheets would affect our shorelines, we can make some simple calculations to determine how much sea level would rise in an unlikely scenario—melting of all the ice from West Antarctica.

1. Examine the situation below. A rectangular tub of water has one block of ice floating in it and two blocks on land that will add water to the tub if the blocks melt. The ice blocks and the grids on the side of the tub are 10 cm on a side for easy measuring.

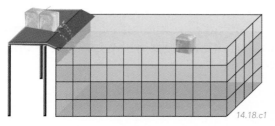

14.18.c1

2. The block floating in the water is 10 cm on all sides, or 10 cm by 10 cm by 10 cm. We simply multiply these three dimensions to get the volume of the block, which is 1,000 cm³ (10 cm × 10 cm × 10 cm = 1,000 cm³). The two blocks on the table total 20 cm (two blocks wide) by 10 cm (one block deep) by 10 cm (one block) high. If we multiply 20 cm × 10 cm × 10 cm, we get 2,000 cm³.

3. The Floating Block—Most of the floating block is below the surface. As ice melts, it yields a volume of water that is less than the volume of ice, because water is more dense than ice. As a result, melting ice that is floating in freshwater does not appreciably raise the level of the water. It does cause a slight rise in a body of saltwater because melting ice yields freshwater, which is less dense, but we'll ignore this factor to simplify things.

4. Blocks on the Table—If the blocks on the table melt, all of the water helps raise the level in the tub. To see how much, we need only to worry about the surface area of the water, not how much water is already there at depth within the tub. Also, a volume of ice produces about nine-tenths that volume of water, or a ratio of 0.9 (volume water produced/volume ice melted).

5. To get the surface area of a rectangle of water, we multiply the dimensions of its two sides. The tub is 100 cm long by 40 cm wide, yielding a surface area of 4,000 cm². To calculate how much the melting blocks will raise water level in the tub, we spread our volume of water over this surface area. The calculation is as follows:

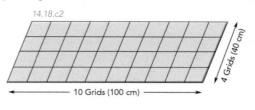

14.18.c2

4 Grids (40 cm)

10 Grids (100 cm)

| 2,000 cm³ (volume of the ice blocks on table) | × | 0.90 (to convert the ice to water) | ÷ | 4,000 cm² (surface area of the water) | = | 0.45 cm (rise in level of water) |

6. So melting an ice block floating in the water does not appreciably change sea level, but melting ice on land does. The larger the amount of ice on land that is melted, the larger the rise. But the larger the surface area of the tub, the smaller the rise. For West Antarctica and our modern seas the calculation is:

| 3,000,000 km³ (volume of all the ice) | × | 0.90 (ice to water) | ÷ | 361,000,000 km² (surface area of the world's oceans) | = | 0.0075 km (rise in sea level) |

To get meters, we multiply 0.0075 km × 1,000 m/km = 7.5 m (25 ft)

7. This calculation does not take into account that as we add water and raise sea level, the ocean spreads out over the land and so the surface area increases. The number calculated when considering this factor is more like 6 m (20 ft). Recall that this is a worst-case scenario that would occur only under a huge change in climate.

D   What Impact Would Raised Sea Levels Have on the East Coast?

Think about all the shoreline photographs in this chapter and imagine those areas if sea level were 6 m (~20 ft) higher. To plan for such contingencies, the USGS conducted a detailed assessment of the relative risk of sea-level rise for each part of the East Coast of the United States. For each segment of coast on the map shown here, geologists investigated various geologic factors, including elevation, slope of the land, hardness of the rocks, barrier islands, and various other aspects described in this chapter. From this analysis, geologists assigned each area a risk, from low to very high. The most vulnerable settings include the eastern coast of Florida, the barrier islands of Virginia and North Carolina, especially Cape Hatteras, and coastlines around Maryland, Delaware, and New Jersey. How risky is your favorite part of the East Coast?

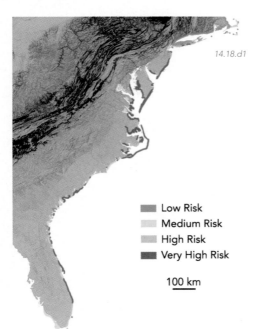

14.18.d1

■ Low Risk
■ Medium Risk
■ High Risk
■ Very High Risk

100 km

Before You Leave This Page Be Able To

☑ Briefly summarize the settings where ice occurs in West Antarctica.

☑ Calculate how much melting a block of ice will raise water levels in a tub, if you know the dimensions of the block and tub.

☑ Discuss why calculations about West Antarctica are important to people living along coastlines, including the East Coast of the United States.

14.18

14.19 How Could an Episode of Global Warming or a Glacial Period Affect North America?

This shaded-relief map of North America colors the land surface and seafloor according to elevation above and below the present sea level. These elevations represent possible levels to which the sea could rise or fall if the climate dramatically warms or cools and causes a change in the extent of glaciers. You will use estimates of the amount of ice that could be lost or gained to calculate how much sea level could rise or fall, and then evaluate the implications for the economy, transportation, and hazards for some major cities of North America. The scenarios presented in this investigation are extreme end members, not situations that will happen, except perhaps after thousands, millions, or even hundreds of millions of years.

14.19.a1

Goals of This Exercise:

- Observe the shaded-relief map of North America to identify areas that are close to sea level (above it and below it).

- Use estimates of the current amounts of ice on the planet to calculate how high sea level would rise if all the ice melted.

- Use estimates of the amount of ice that was present during the last glacial maximum (20,000 years ago) to calculate how much sea level would drop if these conditions returned.

- Use your results to identify how such rises and falls in sea level would affect some major cities of North America.

Data

Listed below are data about the present surface area of the oceans and estimates for the amount of ice that (1) is present today and (2) was present when glaciers were at a maximum 20,000 years ago. Use these data to complete the calculations on the next page.

1. The present surface area of the oceans is 361,000,000 km².

2. The total amount of ice (ice sheets, ice caps, glaciers) currently present on the planet is estimated to be 32,000,000 km³.

3. During the last glacial maximum, 20,000 years ago, the amount of ice is estimated to have been 52,000,000 km³ more than is present today. Your calculations will determine how much water is used to make this additional amount of ice, and how much this would lower sea level.

4. When ice melts, the volume of water produced is about 0.9 times the volume of the ice. That is, the volume of water produced is only 90% of the volume of ice.

Procedures

Follow the steps below, entering your answers for each step in the appropriate place on the worksheet.

1. Calculate how much water would be released if all the ice on the planet melted (an extremely unlikely scenario). For this calculation the equation is: the volume of water gained = the volume of ice × 0.9 (volume water / volume ice).

2. Calculate the volume of water that would be tied up in ice if glaciers returned to the same volume as 20,000 years ago. The equation is: the volume of water lost = the volume of *additional* ice 20,000 years ago × 0.9 (volume water / volume ice).

3. Calculate how much sea level would rise for the water volume gained in step 1 or how much sea level would fall for the water volume lost in step 2. Ignoring many important complications, the much simplified equation is: the change in sea level = change in water volume / surface area of the oceans.

4. Examine how each city shown on this map would be affected by the two extremes. Would it be flooded, not flooded but much closer to the shoreline, or much farther from the shoreline? Discuss how such changes would affect a typical city's transportation, vulnerability to coastal flooding, economic livelihood, and any other factors you can think of.

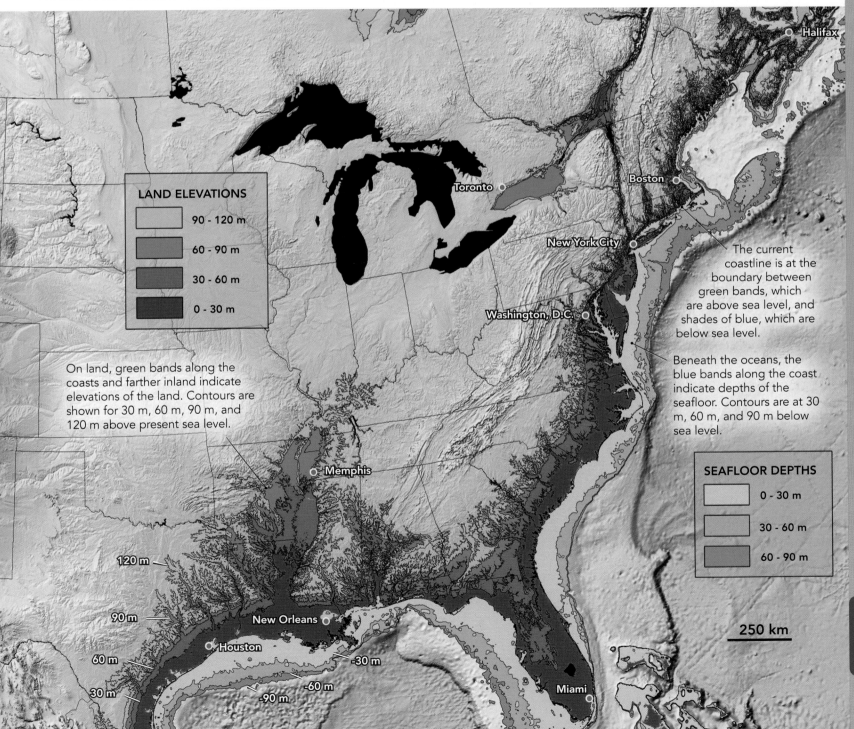

LAND ELEVATIONS

- 90 - 120 m
- 60 - 90 m
- 30 - 60 m
- 0 - 30 m

On land, green bands along the coasts and farther inland indicate elevations of the land. Contours are shown for 30 m, 60 m, 90 m, and 120 m above present sea level.

The current coastline is at the boundary between green bands, which are above sea level, and shades of blue, which are below sea level.

Beneath the oceans, the blue bands along the coast indicate depths of the seafloor. Contours are at 30 m, 60 m, and 90 m below sea level.

SEAFLOOR DEPTHS

- 0 - 30 m
- 30 - 60 m
- 60 - 90 m

Halifax

Toronto

Boston

New York City

Washington, D.C.

Memphis

120 m

90 m

New Orleans

60 m

Houston

30 m

-30 m

-60 m

-90 m

Miami

250 km

14.19

Weathering, Soil, and Unstable Slopes

SLOPES CAN BE UNSTABLE, leading to slope failures that can produce catastrophic landslides or mudslides involving thick slurries of mud and debris. Such events have killed tens of thousands of people and destroyed houses, bridges, and entire cities. Where does this loose material come from, what determines if a slope is stable, and how do slopes fail? In this chapter, we explore slope stability and the origin of soil, one of our most important resources.

The Cordillera de la Costa is a steep 2 km-high mountain range that runs along the coast of Venezuela, separating the capital city of Caracas from the sea. This image, looking south, has topography overlain with a satellite image taken in 2000. The white areas are clouds and the purple areas are cities. The Caribbean Sea is in the foreground.

In December 1999, torrential rains in the mountains caused landslides and mobilized soil and other loose material as debris flows and flash floods that buried parts of the coastal cities. Some light-colored landslide scars are visible on the hillsides in this image.

How does soil and other loose material form on hillslopes? What factors determine whether a slope is stable or is prone to landslides and other types of downhill movement?

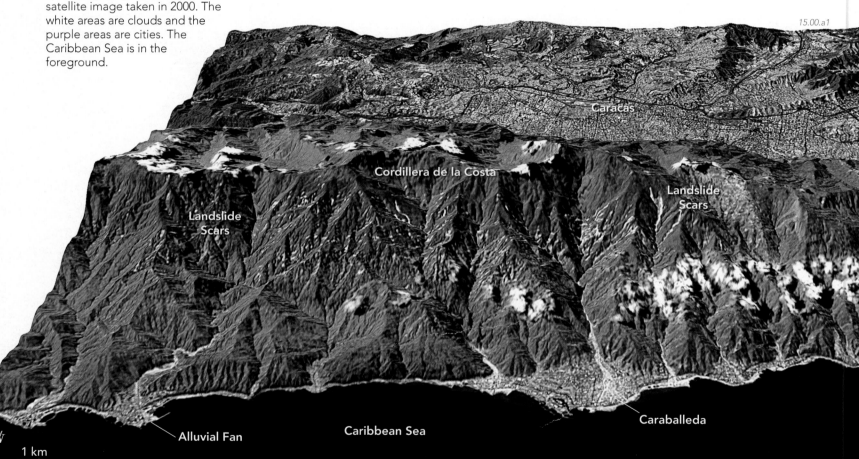

15.00.a1

The mountain slopes are too steep for buildings, so people built the coastal cities on the less steep fan-shaped areas at the foot of each valley. These flatter areas are alluvial fans composed of mountain-derived sediment that has been transported down the canyons and deposited along the mountain front.

What are some potential hazards of living next to steep mountain slopes, especially in a city built on an active alluvial fan?

The city of Caraballeda, built on one such alluvial fan, was especially hard hit in1999 by debris flows and flash floods that tore a swath of destruction through the town. Landslides, debris flows, and flooding killed more than 19,000 people and caused up to $30 billion in damage in the region. The damage is visible as the light-colored strip through the center of town.

How can loss of life and destruction of property by debris flows and landslides be avoided or at least minimized?

TOPICS IN THIS CHAPTER

Huge boulders smashed through the lower two floors of this building in Caraballeda and ripped away part of the right side (▼). The mud and water that transported these boulders is no longer present, but the boulders remain as a testament to the fury of the event.

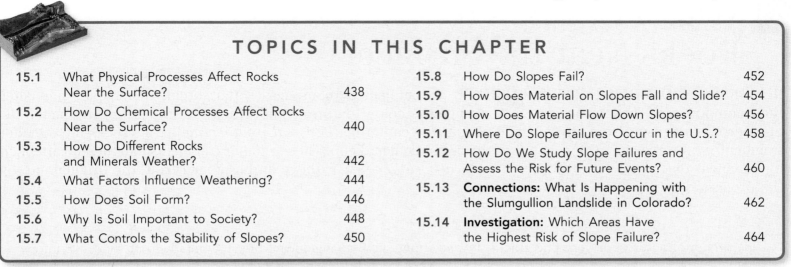

15.00.a2 Caraballeda, Venezuela

◄ This aerial photograph of Caraballeda, looking south up the canyon, shows the damage in the center of the city caused by the debris flows and flash floods.

15.00.a3

1999 Venezuelan Disaster

A *debris flow* is a slurry of water and debris, including mud, sand, gravel, pebbles, boulders, vegetation, and even cars and small buildings. Debris flows can move at speeds up to 16 m/s (36 mph). In December 1999, two storms dumped as much as 1.1 m (42 in.) of rain on the coastal mountains of Venezuela. The rain loosened soil on the steep hillsides, causing many landslides and debris flows that coalesced in the steep canyons and raced downhill toward the cities built on the alluvial fans.

In Caraballeda, the debris flows carried boulders up to 10 m (33 ft) in diameter and weighing 300 to 400 tons each. The debris flows and flash floods raced across the city, flattening cars and smashing houses, buildings, and bridges. They left behind a jumble of boulders and other debris along the path of destruction through the city.

After the event, USGS geologists went into the area to investigate what had happened and why. They documented the types of material that were carried by the debris flows, mapped the extent of the flows, and measured boulders (▼) to investigate processes that occurred during the event. When the geologists examined what lay beneath the foundations of destroyed houses, they discovered that much of the city had been built on older debris flows. These deposits should have provided a warning of what was to come.

15.00.a4 Caraballeda, Venezuela

15.0

15.1 What Physical Processes Affect Rocks Near the Surface?

ROCKS AT AND NEAR EARTH'S SURFACE are subjected to processes that break them apart and alter their components. These processes may change the color, texture, composition, or strength of the materials. Such processes result from physical and chemical weathering. *Physical weathering* breaks rocks into smaller fragments without causing any change in their chemical makeup. These smaller fragments can then be attacked by chemical reactions, the process of *chemical weathering*, or they can be moved from the original site by the process of *erosion*.

A What Is the Role of Joints in Weathering?

Joints are fractures, or very fine cracks, in rocks that show no significant offset. Joints help break rocks into smaller pieces and permit water and roots to penetrate into the rock, thereby promoting weathering.

15.01.a1

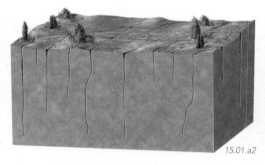
15.01.a2

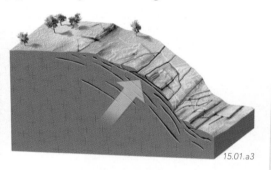
15.01.a3

Most joints form in rocks at depth and may later be uplifted to the surface. The orientation and spacing of preexisting joints and faults help determine the rates of physical and chemical weathering at the surface. More closely spaced joints promote more rapid weathering.

Some joints form as a result of expansion due to cooling or to a release of pressure as rocks are uplifted to the surface. These *expansion joints* can be difficult to distinguish from preexisting joints that formed by other processes.

As Earth is sculpted by erosion, the shape of topography influences stresses that build up when the weight of overlying rocks is *unloaded*. During unloading, expansion joints can form that mimic topography, peeling off thin sheets of rock, a process called *exfoliation*.

B How Are Joints Expressed in the Landscape?

Joints greatly influence how a landscape develops. Joints affect the strength of a rock, help control its resistance to weathering and erosion, and influence whether pieces of rock are pried loose from the landscape.

15.01.b1 Connecticut Valley, MA

15.01.b2 Arches NP, UT

15.01.b3 Yosemite NP, CA

Joints are the dominant features of this roadcut, but the amount of jointing is not uniform. The less-jointed areas are more resistant to weathering than the highly jointed ones.

The spacing and orientation of joints, along with rock type, determine how fast a rock will weather and which parts of the landscape will be most easily eroded. Joints play a prominent role in the weathering of rock layers in this photograph.

Exfoliation joints commonly mimic topography. They can shave off thin, curved slices of rock parallel to the surface, leaving curved rock faces, as shown here. They can form large, dome-shaped landforms, like famous ones in Yosemite National Park, California.

C What Other Physical Processes Loosen Rocks?

Joints, which formed by processes at depth or by expansion of rock near the surface, play a major role in weathering. Other processes may also help break rock into smaller pieces.

As rocks are heated and cooled, different minerals expand and contract by different amounts. This daily and seasonal *thermal expansion* imposes stresses on the boundaries between minerals and causes microfracturing in and along mineral grains, which physically loosens the mineral grains.

When water in a fracture freezes, it expands 8% and exerts a strong outward-directed force on the walls of the fracture. This process of *frost wedging* can widen and lengthen the fracture and pry off loose pieces of rock.

Water percolating through fractures and pore spaces may precipitate crystals of salt, calcite, and other minerals. As they grow, the crystals exert an outward force that fractures or weakens the rock. This process is called *mineral wedging*.

15.01.c1

Burrowing organisms, including rodents, earthworms, and ants, bring material to the surface where it can be further weathered and eroded. As such, these creatures are agents of physical weathering.

15.01.c2 Baja California, Mexico

Plant roots can extend into fractures and grow in length and diameter, expanding pre-existing fractures (▶). This process is *root wedging*.

D How Does Fracturing a Rock Affect Weathering?

Weathering affects rock surfaces that are exposed to air and water, so rocks weather from the outside in. Physical weathering breaks rocks into pieces, providing more surface area where chemical weathering processes can operate.

Surface Area of a Cube of Rock

If joints and other fractures in rock form a three-dimensional network, the rock may be broken into box-shaped pieces bounded by fractures, maybe like this cube. What is the total amount of exposed surface area on the sides of the cube?

2 cm

2 cm

15.01.d1

To calculate the surface area of one face (side), we multiply the height by the width of that face.

$$2 \text{ cm} \times 2 \text{ cm} = 4 \text{ cm}^2$$

There are 6 faces on a cube, so we multiply the area of one face by 6 to get the total surface area.

$$4 \text{ cm}^2 \times 6 \text{ sides} = 24 \text{ cm}^2$$

Fracturing a Cube into Pieces

What happens to the total surface area if we fracture the same cube into eight pieces? First, we calculate the surface area for each smaller cube.

1 cm

1 cm

15.01.d2

$$1 \text{ cm} \times 1 \text{ cm} = 1 \text{ cm}^2 \text{ for each side}$$

$$1 \text{ cm}^2 \times 6 \text{ sides} = 6 \text{ cm}^2 \text{ for each cube}$$

But there are eight such cubes.

$$6 \text{ cm}^2 \times 8 \text{ cubes} = 48 \text{ cm}^2$$

Therefore, this fracturing has doubled the exposed surface area, providing more surfaces where weathering can operate. The rock will therefore weather faster.

15.01.d3 Engineer Mtn., CO

Physical weathering of steep outcrops can loosen pieces that fall, tumble, or slide downhill and accumulate on the slopes below. These piles of angular blocks are *talus*, and such slopes are *talus slopes*. The largest talus blocks here are 1 m across.

Before You Leave This Page Be Able To

✓ Describe several ways that joints form.

✓ Describe how joints are expressed in the landscape.

✓ Sketch or describe physical weathering processes.

✓ Sketch or explain why fracturing aids weathering.

15.1

15.2 How Do Chemical Processes Affect Rocks Near the Surface?

CHEMICAL WEATHERING alters and decomposes rocks and minerals, principally through chemical reactions involving water. When chemical and physical weathering processes combine to break down and alter rocks, they produce minerals that are more stable in surface conditions than the original minerals. They transform rocks into clay, sand, and other materials.

A How Does Changing a Rock's Environment Promote Weathering?

Many rocks and minerals form deep within Earth. When they are brought near the surface by uplift and erosion, they encounter conditions very different from those in which they formed and so may become unstable.

Minerals that crystallize in high-temperature magmas are generally unstable when subjected to the low-temperature conditions that characterize Earth's surface. Most magma temperatures are above 700°C, whereas surface temperatures range from minus 40°C to plus 45°C (minus 40°F to plus 122°F).

During metamorphism, some minerals crystallize beneath the surface in dry, high-pressure and high-temperature environments. Once such rocks reach Earth's low-pressure and low-temperature surface, they can change to different minerals that are stable at the new, wetter, low-pressure and low-temperature conditions.

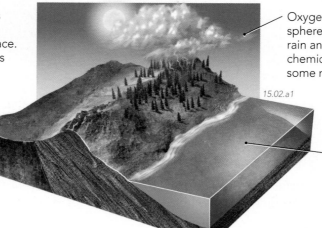

15.02.a1

Oxygen (O_2) is abundant in the atmosphere and as a dissolved component in rain and most surface water. This oxygen chemically reacts with rocks, causing some minerals to oxidize (rust).

Liquid water is more abundant on and near Earth's surface than at depth. Water, especially when it is slightly acidic, is a chemically active substance that can break the bonds in many minerals. It increases the rate of chemical weathering.

B What Happens When Rocks Dissolve?

The main agents for chemical weathering are water and weak acids formed in water, such as carbonic acid (H_2CO_3). These agents dissolve some rocks, loosen mineral grains, form clay minerals, and widen fractures.

Limestone (below) and other rocks rich in calcium carbonate or magnesium carbonate are soluble in water and in acids. They dissolve and form pits and cavities.

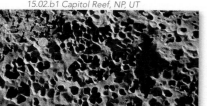

15.02.b1 Capitol Reef, NP, UT

Over time, the pits deepen, widen, and may interconnect, forming furrows (small troughs).

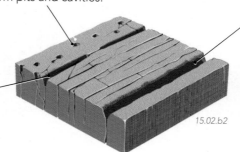

15.02.b2

Fractures can widen as water flows through them and dissolves material from the walls of the fracture, as in the limestone outcrop below. Caves can form by dissolution of limestone and other soluble rocks at depth.

15.02.b3 Austrian Alps

One Way That Calcite Chemically Dissolves in Water

Limestone is a relatively soluble rock because it is composed of calcite, which is soluble in weak acids. The most common weak acid in surface water is carbonic acid, produced when rainwater reacts with carbon dioxide (CO_2) in the atmosphere, soil, and rocks. The chemical reaction for the dissolution of calcite in carbonic acid is:

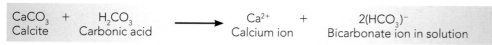

$CaCO_3$	+	H_2CO_3	$\longrightarrow$	Ca^{2+}	+	$2(HCO_3)^-$
Calcite		Carbonic acid		Calcium ion		Bicarbonate ion in solution

Acids in water produce unbonded H^+ ions, each of which is a proton without a balancing electron and is available to make other chemical bonds. H^+ ions are small and can easily enter crystal structures, releasing other ions, like calcium, into the water.

C What Happens When Rocks Oxidize Near Earth's Surface?

Oxygen (O_2) is common near Earth's surface and reacts with some minerals to change the oxidation state of an ion. This is common in iron-bearing minerals because iron (Fe) has several oxidation states.

Many mafic igneous rocks contain dark, iron-bearing minerals, such as pyroxene. Iron in pyroxene can become oxidized, producing iron oxide minerals.

Hematite consists only of iron and oxygen and is more stable than pyroxene under oxygen-rich conditions. It commonly forms during oxidation and gives oxidized rocks a reddish color.

15.02.c1 Wilson Cliffs, NV

If iron-bearing rocks become oxidized, they generally take on a red color from the iron oxide mineral hematite. Reddish rocks can lose their reddish color if they interact with fluids that have less oxygen. ▶

$$4FeSiO_3 + O_2 \longrightarrow 2Fe_2O_3 + 4SiO_2$$

| Pyroxene | Oxygen | | Hematite | Silica (in water) |

D What Happens When Minerals Chemically React with Water?

When some minerals react with water they undergo a chemical reaction where the mineral combines with water to form a new mineral. This reaction, called *hydrolysis*, converts some minerals to clay.

One kind of feldspar, containing potassium (K), is called K-feldspar. When this mineral reacts with acids (waters that have free H^+ ions), it can be converted into clay minerals by hydrolysis.

During the reaction, the H^+ ion moves into the crystalline structure, expelling the K^+ ion and silica, which both get carried away in the water.

15.02.d1 Crete

If exposed to wet conditions, many rocks convert into clay minerals. The gray limestone shown here contained impurities that weathered into clay minerals and reddish hematite that accumulated between the blocks. ▶

$$4KAlSi_3O_8 + 4H^+ + 2H_2O \longrightarrow 4K^+ + Al_4Si_4O_{10}(OH)_8 + 8SiO_2$$

| K-feldspar | Hydrogen ion | Water | | Potassium (in water) | Kaolinite (clay) | Silica (in water) |

E How Does Weathering Make the Ocean Salty?

Have you ever wondered why the ocean is salty or where the salt comes from? It turns out that most of the ocean's salts are derived from weathering and the dissolution of rocks on the land.

1. Rock, sediment, and soil on and near Earth's surface are exposed to the water and to oxygen in the atmosphere. Water that reacts with rocks and minerals can come from several sources, including from rain and other forms of precipitation.

2. Some water infiltrates into the subsurface, where it may chemically react with the materials. During weathering, hydrolysis reactions commonly produce clay minerals and also drive out positive ions (cations) from the preexisting mineral structure. The dominant cations in feldspar, a very abundant mineral, are sodium and potassium, which can form common salts.

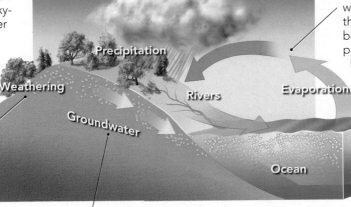

15.02.e1

Precipitation

Weathering

Groundwater

Rivers

Evaporation

Ocean

5. When seawater in the oceans evaporates, the dissolved salts remain in the seawater, increasing seawater's salt content (*salinity*). Such evaporation causes the seas to be saltier than the rivers. The seas would be even saltier if salt was not removed from some parts of the sea by deposition of salt beds. If the salt beds, once formed, are uplifted and exposed on land, they can also contribute dissolved salt to rivers and ultimately back to the sea.

3. The dissolved cations, along with negative ions like chlorine, are carried by moving water, either in rivers and streams or in the subsurface by groundwater. Much of this water eventually finds its way to the oceans, where it contributes its salts and other ions to seawater.

4. Modern oceans typically contain about 3.5% dissolved salt, and are much saltier than river water. Rivers contain some dissolved salts but only a small amount, so they are considered to be freshwater, not salt water. If the oceans got their salt from the rivers, why are the oceans saltier?

Before You Leave This Page Be Able To

✓ Describe several reasons why minerals formed at depth may not be stable at the surface.

✓ Summarize how limestone dissolves and what features are formed by dissolution.

✓ Briefly summarize the processes of oxidation and hydrolysis.

✓ Explain why oceans are salty.

15.2

15.3 How Do Different Rocks and Minerals Weather?

ALL ROCKS AND MINERALS on Earth's surface are subject to physical and chemical weathering. How a rock weathers depends on a number of factors, including the kinds of the minerals that compose the rock. These differences cause weathered rock outcrops to have a variety of distinctive appearances.

A How Do Different Rocks Respond to Weathering?

Rocks are composed of minerals. Some minerals are hard and so resist physical weathering, whereas others are weak and easily broken. Some minerals are chemically stable and resist chemical weathering, while others are chemically unstable. Some rocks are mostly a single mineral, but others contain many minerals. All these variations cause different rocks to weather in different ways.

15.03.a1 Big Maria Mtns., CA

15.03.a2 San Juan River, UT

15.03.a3 Central CO

Quartzite contains quartz grains that have grown together during metamorphism or chemical cementation. Quartz is very stable on Earth's surface, so quartzite weathers mainly by physical processes, not chemical processes. As a result, quartzite chemically weathers more slowly than most other rocks.

Limestone is composed of calcite, a relatively soft and soluble mineral. Limestone weathers rapidly in wet climates, where the calcite dissolves rapidly. In contrast, limestone can be an erosion-resistant, cliff-forming rock in dry climates. The pitted limestone shown here has been partially dissolved.

Granite consists mainly of feldspar and quartz, with lesser amounts of mica and other minerals. Each mineral weathers differently and at different rates, giving weathered granite a distinctly bumpy surface. In this rock, pink feldspar crystals weather more slowly than the mica-rich matrix.

B What Happens When Granite and Related Rocks Weather?

Granite and related igneous rocks contain feldspar, quartz, and smaller amounts of mica (biotite and muscovite), iron oxides, or amphibole. As these rocks weather, the different minerals respond in different ways.

1. Feldspar is the most abundant mineral in granites, forming the cream-colored crystals shown here. Feldspar chemically weathers by *hydrolysis* to form clay minerals. During this process sodium (Na), potassium (K), and other ions leach out of the feldspar. Some of the liberated ions are released and can be carried away by water.

15.03.b1 Baja California, Mexico

15.03.b2 Little Colorado River, AZ

15.03.b3

3. Biotite, magnetite, and other mafic minerals are present in many granites, forming the dark crystals in this rock. Some of these minerals undergo hydrolysis to form clay minerals, and others can oxidize to form hematite, an iron oxide mineral.

4. Granite is at least 25% quartz, which is the medium-gray, partially transparent mineral here. Quartz is very resistant to chemical and physical weathering. As chemically reactive minerals break down around it, quartz weathers into intact grains.

2. Clay minerals weathered from granite and similar rocks can accumulate in soil or be eroded and transported away by water and wind. Some clay particles are washed out to sea, and others are deposited in lakes, floodplains, deltas, and other muddy environments. ▶

◄ 5. Quartz grains eroded from granite typically become quartz sand, which can be transported away by water and wind. Quartz sand accumulates along rivers, in dunes, and on beaches. Feldspar can also form sand grains if the granite or other source of feldspar is not too chemically weathered, as can occur in dry climates or in areas of rapid erosion.

C How Do Mafic Rocks Weather?

Dark-colored rocks are usually composed of *mafic* (Mg- and Fe-bearing) minerals. Mafic minerals are relatively unstable at the surface and convert to more stable minerals, including clay minerals and iron oxides.

Mafic minerals generally form at high temperature and at least moderate pressures. The mafic rock shown here crystallized at depth and contains calcium-rich plagioclase feldspar and dark, mafic minerals, such as amphibole and pyroxene. Mafic minerals typically do not contain as much silica as felsic rocks, and so are not as resistant to weathering as a quartz-rich (silica-rich) rock like quartzite. Feldspar and the mafic minerals in gabbro commonly weather to clay minerals by hydrolysis.

Mafic silicate minerals, along with magnetite, an iron oxide, can also oxidize when exposed to Earth's atmosphere and water. Magnetite (Fe_3O_4), a dark gray to black magnetic mineral, can be oxidized to hematite (Fe_2O_3), which commonly is reddish in color and has a reddish streak. Oxidation has only begun to affect the rock shown here (▶).

15.03.c1 Baja California, Mexico

D What Controls How Different Minerals Weather?

Many factors determine how minerals weather, including the climate, how much time the mineral has been exposed to weathering, and the chemical composition and atomic structure of the mineral.

Chemical Bonding

15.03.d1 Cayman Islands

15.03.d2 Big Maria Mtns., CA

The bonds in some minerals allow them to be readily dissolved in water and weak natural acids, as in this dissolved limestone. Salt and gypsum also are very soluble.

Other minerals have stronger bonds that make them less soluble in water. Quartz in this quartzite has strong covalent bonds and is not very soluble in cold water.

Reactivity

15.03.d3 Durango, CO

15.03.d4 San Juan River, UT

Sandstone is composed mainly of sand-sized grains of quartz. Most quartz grains weather by physical processes, rather than chemical processes.

Limestone is very soluble and prone to chemical weathering, especially dissolution and especially in wet climates. It also weathers by physical processes.

Relative Resistance of Minerals to Weathering

1. The stability of minerals is in a very general way related to the order in which the minerals commonly crystallize from a magma. According to Bowen's reaction series, mafic minerals and Ca-rich feldspar crystallize first, followed by Na-rich and K-rich feldspar, muscovite, and quartz. In this illustration, minerals are arranged in their general crystallization order, from top to bottom.

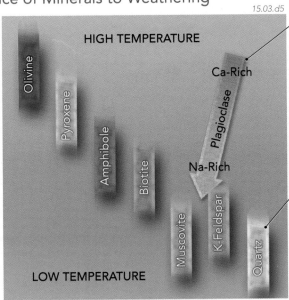

15.03.d5

HIGH TEMPERATURE

Olivine
Pyroxene
Amphibole
Biotite
Muscovite
K-Feldspar
Quartz

Plagioclase — Ca-Rich / Na-Rich

LOW TEMPERATURE

2. As a magma cools, the first minerals to crystallize do so at the highest temperatures. These minerals are typically least stable when subjected to weathering at the low temperatures of Earth's surface. Crystallization order is also accompanied by changes in the percentage of SiO_4^{-4} bonds within the mineral.

3. Quartz crystallizes late according to Bowen's reaction series and is the silicate mineral most resistant to weathering. Although Bowen's reaction series is very idealized, it is one way to think about the stabilities of minerals during weathering.

Before You Leave This Page Be Able To

- ✓ Describe differences in the weathering of quartz, feldspar, mafic minerals, and calcite.

- ✓ Explain the origin of the three main weathering products (sand, clay minerals, and dissolved ions).

- ✓ Summarize the factors that control how different minerals weather.

15.3

15.4 What Factors Influence Weathering?

DIFFERENT ROCKS WEATHER to different appearances. Some rock units remain boldly exposed or perhaps weather to large, loose blocks. Others disintegrate to smooth, soil-covered slopes or break down into clay-sized particles. What causes these variations? Weathering is largely controlled by the properties of the rock and the setting where weathering occurs.

A How Does the Character of a Rock Influence Weathering?

Differences in mineral composition, particle size, and other rock properties play an important role in how a rock responds to weathering. Equally important are fractures, bedding planes, and other discontinuities.

Composition—Weathering of a rock is influenced by the types of minerals it contains. Most sandstone, like the one in this cliff, consists largely of quartz, a mineral that is very stable on Earth's surface; it mostly weathers by physical processes. In contrast, the recesses below the cliff contain fine-grained sedimentary rocks that are more easily weathered and eroded.

Variation in Composition—Some outcrops have different parts with large contrasts in susceptibility to weathering. The more susceptible parts will weather faster than the more resistant parts. Such *differential weathering* can form alternating ledges and slopes, as shown here, or rocks with holes where less resistant material has been removed.

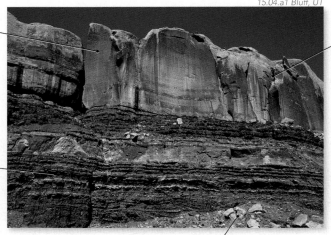

15.04.a1 Bluff, UT

Discontinuities—Fractures, bedding planes, and other discontinuities provide pathways for the entry of water into a rock body. A rock with lots of these features will weather more rapidly than a massive rock containing few such discontinuities. For example, highly fractured parts of a cliff weather faster than less fractured parts. Rocks with thin layers generally break apart and weather more readily than rocks with thick layers.

Surface Area—Rock that is already broken into small pieces, such as these loose pieces, provides more surface area on which chemical weathering can act. Solid, unfractured bedrock provides less surface area and weathers more slowly.

B What Other Factors Influence Weathering?

How rocks weather also is controlled by factors not related to the rock itself but to its outside setting. Such external factors include location, topography, temperature, and moisture.

Climate—Abundant precipitation and higher temperatures cause chemical reactions to proceed faster. Thus, warm, humid areas generally have more highly weathered rock because chemical weathering operates faster than in cold or dry climates. Elevation influences an area's climate and so is another key control.

Hillslope Orientation—The orientation of the slope on which a rock occurs is an important factor in weathering. Many mountain ranges receive more rain on one side than the other. Slopes that are more sheltered from sunlight are cooler, can better retain their moisture, and may have more plants. Moisture, soil, and plants promote chemical weathering.

Slopes facing the sun receive more light and heat than those facing away. Thus, sunny surfaces (south-facing slopes in the Northern Hemisphere) tend to be warmer and drier, to have more evaporation, and to have less chemical weathering, soil, and plants than slopes facing away from the sun.

Steepness of Slopes—On steep slopes, rainfall runs off faster, and weathering products may be quickly washed away by runoff. Soil and other loose materials can also slide down steep slopes.

On gentle slopes, weathering products can accumulate, and water may stay in contact with rock for longer periods of time, resulting in higher weathering rates. Once formed, soil and loose pieces can remain in place longer.

Time—A crucial factor in weathering is time. Depending on the other factors discussed, rates of weathering can range from rapid to extremely slow. The more time available, the more weathering will occur. The speed of weathering and the volume of material affected in a given time will depend on slope, climate, hillslope orientation, and the composition and structural condition of the rock or sediment.

15.04.b1

C Why Does Weathering Produce Rounded Features?

Weathering processes usually work inward from an exposed surface. This commonly results in rounded shapes in weathered outcrops, and weathering commonly generates loose, partially rounded blocks. The three figures below illustrate what can happen to a rock that has fractures but lacks other types of discontinuities.

Rock that is newly fractured generally has sharp, angular edges. Weathering attacks edges from two sides and corners from three sides. These edges and corners wear away faster than a single smooth surface.

15.04.c1

Over time, faster weathering of the edges and corners of the rock will begin to smooth away corners, edges, and any other parts of the rock that stick out.

15.04.c2

Weathered rocks can become moderately rounded, losing their sharp edges and angular features. Rocks dislodged from the bedrock will get smaller with time as they are weathered from all sides. Most rounding occurs while the rock is being transported.

15.04.c3

D How Is Weathering Expressed in the Landscapes?

Weathering exerts enormous control over the appearance of landscapes. Weathering helps define differences in appearance from one region to another, from one side of a hill to the other, or between different rock types.

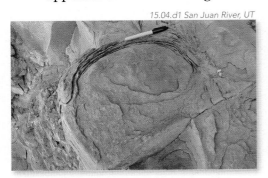
15.04.d1 San Juan River, UT

15.04.d2 Boulder MT

15.04.d3 Alice Springs, Australia

1. Weathering mostly affects rocks from the outside in, so weathered rocks have an outer weathered zone and an inner unweathered zone. The outer zone is a *weathering rind*. As weathering continues, the weathering rind thickens and can in some cases be used to infer how long the rock has been on or near the surface and exposed to weathering.

2. As weathering attacks a fractured rock, preferential weathering along the fractures can cause the intact but fracture-bounded blocks to become rounded. The outer weathered rind of the blocks splits away from the stronger, less weathered rock in the center, forming rounded shapes in a process called *spheroidal weathering.*

3. If conditions favor extensive weathering, such as exposure of a chemically reactive rock unit for a long time, thick soil can develop that obscures the underlying bedrock. In a drier climate, on a steep slope, or for a more resistant rock, weathering is slower, so soil may be less developed and hillslopes may be rockier and more barren.

4. Differential weathering and erosion of different rock units, and of different parts of a single rock unit, produce the larger landscapes we see. This image illustrates the interplay between rock type, structure, weathering, and erosion in an area loosely modeled after the area near St. George, Utah. ◄

15.04.d4

5. Less resistant rock units, or parts of units that are highly fractured, occupy the valleys, slopes, and gently rounded hills.

6. Rock layers, such as this limestone, that are more resistant to weathering may form cliffs, ledges, hills, and mountains. What can you infer about the weathering of each rock unit in this scene?

Before You Leave This Page Be Able To

☑ Summarize or sketch how weathering is affected by the properties of a rock.

☑ Summarize or sketch how weathering is affected by the setting of a rock.

☑ Describe ways that weathering is expressed in the landscape, including its role in rounding off corners.

15.4

15.5 How Does Soil Form?

SOIL BLANKETS MUCH OF THE LAND SURFACE, providing a place for plants, animals, microbes, and humans to live. Soil is affected by geologic, biologic, and hydrologic processes, and thus represents the interplay between the lithosphere, biosphere, hydrosphere, and atmosphere. The processes and factors that influence weathering also control how soil forms in different climates. What does soil contain and how does it form?

A What Is Soil?

Soil is the unconsolidated material above bedrock and contains both mineral matter and organic matter (typically decaying vegetation) along with air and water. Soil differs from sediment in that sediment is weathered rock that is transported or deposited by water, ice, or wind.

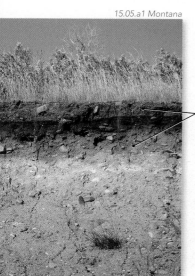

15.05.a1 Montana

◄ What do you observe in this photograph of a vertical cut through soil layers?

There are different zones or layers, called *horizons*, with rather gradational boundaries. These different layers are not the same as beds formed by sedimentation; instead each horizon forms and grows in place by weathering of rock and sediment, and by the addition of material from plants, animals, and the atmosphere.

► In this idealized soil profile, each soil horizon is assigned a letter to denote its position or its character.

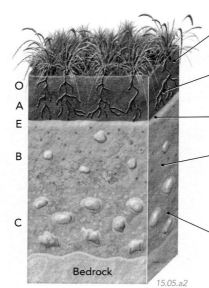

15.05.a2

O horizon is a surface accumulation of organic debris, including dead leaves, other plant material, and animal remains.

A horizon is topsoil, composed of dark gray, brown, or black organic material mixed with mineral grains.

E horizon is a light-colored, leached zone, lacking clay and organic matter.

B horizon contains little organic material, but it can have a red color due to the accumulation of iron oxide. In dry climates, the B horizon can be whitish or have whitish streaks due to calcium carbonate accumulations. It may also include gypsum and salt.

C horizon is composed of either weathered bedrock or unconsolidated sediment, and it grades downward into unweathered bedrock or sediment.

B What Processes Occur During Soil Formation?

What happens to form soil from rock and other materials? Soil forms gradually over thousands of years and involves some of the same processes as weathering, including dissolution, oxidation, hydrolysis, and root wedging. Soil formation also involves the vertical transport of dissolved material up and down through the soil profile.

Where Material Comes From

Soil material is mostly derived from weathering of underlying rock and sediment, but some material is introduced by water and wind. Sediment washes onto the surface from adjacent hillslopes or arrives as windblown dust and salts.

Soil receives several types of material from the land surface. Leaves, pine needles, twigs, and other plant parts accumulate on the surface and are worked downward into the soil. Roots emit CO_2 gas, other gases come from the atmosphere, and moisture mostly arrives as rainfall and snowmelt.

Weathering weakens and loosens underlying bedrock, providing starting material to make soil. This material can be worked up into the soil, or the soil can gradually affect deeper and deeper levels of the bedrock. Some residual material remains in place at depth.

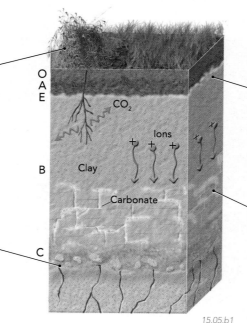

15.05.b1

How Material Moves

Soil material moves both down and up as it is carried by water, plants, animals, and gravity.

Zone of Leaching—The upper part of soil loses easily dissolved material downward. Water soaking into the soil *leaches* (dissolves and removes) soluble material liberated by chemical weathering, carrying it deeper. Clay minerals and other fine particles are carried downward by infiltrating water. Plant parts and other organic material are also worked downward into the soil.

Zone of Accumulation—Chemical ions leached from above may accumulate in an underlying zone, if the water does not carry them all the way to the water table, where they enter the groundwater system. Clay minerals, iron and aluminum oxides, salt minerals, and calcium carbonate commonly accumulate in layers, depending on how much water and oxygen pass through the soil.

C What Soil Profiles Are Typical for Different Climates?

Climate, especially temperature and moisture, strongly affects the type and rate of weathering, the abundance of plants, and the type of soil that results. Scientists and other people classify soil in many different ways, but here we limit discussion to three major soil types defined by climate. The top two types of soil (from moister climates) involve thicker sequences of soil than does the one for arid climates. The accompanying photographs only show the upper parts of the soil profiles.

Tropical Climates

1. In humid, tropical climates, there is abundant rainfall and associated plant growth. Such areas include rain forests and swamps, both of which are characterized by dense plant growth.

15.05.c1 Natal, Brazil

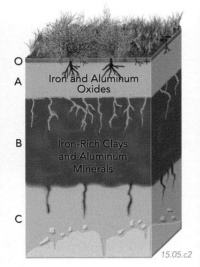
15.05.c2

2. In tropical climates, intense weathering and abundant soil moisture cause severe chemical leaching, leaving behind a soil rich in iron (Fe) and aluminum (Al) oxides, commonly giving the soil a deep red color. This extremely leached type of soil is a *laterite*.

Temperate Climates

3. Temperate climates are cooler and generally have less rainfall than tropical climates. Such areas contain savannas, grasslands, farms, or lush forests of leafy, deciduous trees or pine trees.

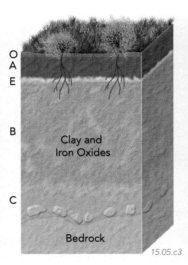
15.05.c3

15.05.c4

4. In cooler areas with moderate to high rainfall, the A and B horizons contain abundant insoluble minerals, including quartz, as well as iron oxide minerals. More soluble minerals like calcium carbonate are absent. Informal names for such soils are *grassland soil* or *forest soil*, depending on the type of vegetation.

Arid Climates

5. Arid climates are dominated by overall dryness and sparse precipitation. They can be very hot, as in subtropical deserts, very cold, as in the Dry Valleys of Antarctica, or moderate in temperature, but still dry. Plants and animals are typically sparse.

15.05.c5

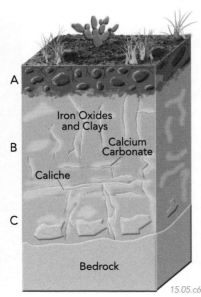
15.05.c6

6. In arid climates, there is limited vegetation so there is little or no O horizon, and usually only a thin A horizon. Clay, iron oxide, and salts, all partly derived from windblown material, accumulate at various levels in the soil. Ca^{2+} and CO_3^{2-} ions are dissolved from upper soil horizons and chemically precipitated farther down as calcium carbonate ($CaCO_3$). The amount of water passing through the soil is not enough to completely remove these ions, and so the amount of calcium carbonate increases with time, first coating clasts and eventually forming a discrete layer of caliche. Soil formed in arid climates can also be called a *desert soil*.

Before You Leave This Page Be Able To

☑ Describe what a soil is and the processes by which it forms.

☑ Sketch and describe the main soil horizons and the processes and materials that occur in each horizon.

☑ Discuss the different soils formed in different climates and the factors responsible for these differences.

15.5

15.6 Why Is Soil Important to Society?

SOIL IS ONE OF OUR MOST IMPORTANT NATURAL RESOURCES because we depend on it for food, energy, and even shelter. Soil provides a necessary foundation for grasslands, forests, and the crops that feed and clothe our growing population. The loss of soil can be catastrophic for individual communities and regions, and for the plants and animals that depend on it.

A How Do Soil and Vegetation Interact?

Soil and plants have a mutually beneficial relationship—soil permits most plants to grow, while plants contribute material to the soil and partly bind it together, helping it develop and protecting it from erosion. Bacteria, fungi, and other microbes also play an important role in the development and health of soils and plants.

15.06.a1

15.06.a2 Kansas

◄ The relationship between soil and plants is appreciated by farmers, who try to nurture and retain the soil, while harvesting crops, including grains, fruits, vegetables, and cotton.

► The grass cover of this hillslope helps the underlying soil in many ways. It protects the soil from rain and wind, captures water that helps the soil develop, and helps keep the soil from being eroded by runoff.

B What Activities Threaten Soil?

In most climates it takes 80 to 400 years to form about 1 cm of topsoil. Soil that is eroded due to poor farming practices or other detrimental activities may be lost quickly and cannot easily be replaced.

15.06.b1 Namibia

15.06.b2 Peru

◄ We use much land to raise cattle, sheep, and other livestock. Overgrazing by livestock, or by indigenous animals such as deer, removes vegetation, leaving soil more vulnerable to wind and water erosion. It also removes the nutritionally rich upper layers of soil. Overgrazing can be especially devastating in times of drought and is an extreme problem in some parts of the world.

► Soil and other weathering products can be washed away on steep slopes but can accumulate on more gentle slopes. Shaping steep terrain to provide flat areas suitable for farming, a practice called *terracing*, can better protect soil from erosion. Terracing also helps capture and retain rainfall, promotes soil formation, and provides a more level place on which to farm.

15.06.b3 Washington

15.06.b4 Colorado

◄ Cutting down forests and removing vegetation to provide lumber, grazing, or farmland can result in massive soil erosion. Severed roots rapidly decay and can no longer hold the soil in place. Eventually, the soil is unable to regenerate vegetation, which will ultimately lead to an increase in runoff, accelerated soil erosion, and possibly disasters due to floods and debris flows. The loss of soil can stop the activity (e.g., farming) for which the land was originally cleared.

► Soil can become polluted near farms that use pesticides, herbicides, fungicides, and fuel oils. Soil can also become contaminated by salt from irrigation water that has acquired a high salinity due to evaporation as it passes over croplands. Some industries use pollutants that find their way into the soil, and some mining operations contaminate soil with chemicals and with elements, such as arsenic, that occur naturally in many ores.

C What Are Some Problems Related to Soil?

In addition to being a valuable resource, soil can cause problems for people because of its low strength and how it behaves when shaken, wetted, dried, or compacted. Problematic soil can be recognized by geologists, builders, and homebuyers so that building on, or buying, such risky sites can be avoided.

15.06.c1-2

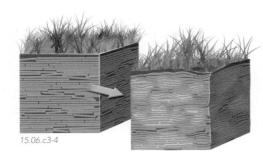

15.06.c3-4

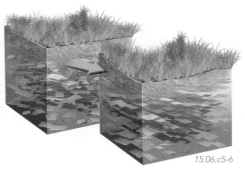

15.06.c5-6

Liquefaction occurs when loose sediment becomes oversaturated with water and individual grains lose grain-to-grain contact as water squeezes between them. Quicksand is an example of liquefaction. Liquefaction is especially common when loose, water-saturated sediment is shaken during an earthquake. The houses below, destroyed during the 1989 Loma Prieta earthquake, sank into artificial fill that liquefied during shaking.

Soil that contains a high proportion of certain clay minerals, called *swelling clays*, increases in volume when it becomes wet, expanding upward or sideways. When these clays dry out, they decrease in volume, causing the soil to shrink or compact. Repeated expansion and compaction during wet-dry cycles can crack foundations, make buildings unsafe, and ruin roads, like the one below.

In some soil, clay minerals start out arranged randomly, with much pore space between individual grains. As water infiltrates the pore spaces, the clay minerals begin to lie flat, reducing open spaces and thereby compacting the soil. Such *soil compaction* typically does not occur uniformly, because some parts of the soil have more clay than others. Differential compaction can crack walls (▼), foundations, and roads.

15.06.c7 California

15.06.c8 Boulder, CO

15.06.c9

Mineral Deposits Formed by Weathering

Weathering processes move chemical elements, leaching them from some areas and concentrating them in others. During chemical weathering, for example, a body of soluble rock can be greatly reduced in volume by dissolution and leaching. Elements that are not leached from the rock can become concentrated enough to become valuable. The most important ore of aluminum, the rock *bauxite* being mined below, forms in wet, tropical climates where

high air temperatures and abundant water produce soil rich in aluminum. The bauxite results from the breakdown by chemical weathering of clay minerals, which originally were also largely formed by weathering.

Weathering also plays a role in concentrating metals in near-surface mineral deposits. In many large copper mines, including the one shown below, weathering and down-flowing groundwater leached copper and sulfur from the top 100 m or so of the copper

deposit, reprecipitating copper- and sulfur-rich minerals farther down. In many cases, such enrichment by weathering makes the deposit rich enough in copper to mine.

15.06.mtb1

15.06.mtb2 Bisbee, AZ

Before You Leave This Page Be Able To

- ✓ Summarize activities that can threaten soil and its protective cover of vegetation.

- ✓ Describe some problems associated with certain soil types.

- ✓ Describe two ways that weathering can enrich a mineral deposit enough so the deposit can be mined.

15.6

15.7 What Controls the Stability of Slopes?

GRAVITY PULLS MATERIAL DOWNHILL, and some rocks, soil, and other loose material are not strong enough to resist this persistent force. Downward movement of material on slopes under the force of gravity is called *mass wasting*, and it occurs to some degree on all slopes. Mass wasting can proceed very slowly or very quickly, sometimes with disastrous results. Mass wasting is an important part of the erosional process, moving material downslope from higher to lower elevations, and feeding sediment from hillslopes to streams, rivers, beaches, and glaciers.

A How Does Gravity Affect Slope Stability?

The main force responsible for mass wasting is *gravity*. The force of gravity acts everywhere within and on the surface of Earth, tending to pull everything toward Earth's center.

1. On a flat surface, the force of gravity acts on a block by pushing it vertically down against the base of the block. The block will not move under this force.

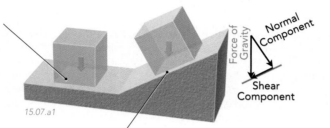

15.07.a1

2. On a slope, gravity acts at an angle to the base of the block. Part of the force pushes the block against the slope and another part pushes the block down the slope. These two parts of the force are referred to as *components*.

3. The part of the force pushing the block against the slope is the *normal component*.

4. The other component acts parallel to the slope, trying to shear the block down the slope. It is the *shear component*.

5. As the angle of slope becomes steeper, the shear component becomes larger while the normal component becomes smaller. As the slope angle steepens, the shear component becomes enough to overcome friction, and it causes the block to slide.

B How Steep Can a Slope Be and Still Remain Stable?

The steepest angle at which a pile of unconsolidated grains remains stable is called the *angle of repose*. This angle is controlled by frictional contact between grains. In general, loose, dry material has an angle of repose between 30° and 37°. This angle is somewhat higher for coarser material, for more angular grains, for material that is slightly wet, and for material that is partly consolidated. It is lower for material with flakes or rounded grains and for material that contains so much water that adjacent grains lose contact.

Dry Sand

Angle of Repose

15.07.b1

◄ **1.** Dry, unconsolidated sand grains form a pile, and the angle of the resulting slope is at the *angle of repose*. If more sand is added, the pile becomes wider and higher, but the angle of repose remains the same. If part of the pile is undercut and removed, the grains slide downhill until the pile returns to a stable slope at the angle of repose. If sand is slightly wet, surface tension between the grains and a thin coating of water enables the sand to be stable on steeper slopes.

► **2.** Loose rocks and other loose material (talus) accumulate on some slopes and at the bases of cliffs. Such talus material commonly forms slopes that are at the angle of repose for the particular sediment. The smooth *talus slope* shown here became too steep in places and so locally slid downward, forming a small pile of debris at the base of the slope.

15.07.b2 Tibet

▼ **3.** Most slopes of sand dunes reflect the angle of repose for dry sand. Slopes can be more gentle than the angle of repose, but if they begin to exceed the angle of repose then the slope fails, slumping downhill. Walking up a sand dune causes the barely stable sand to slide beneath your feet.

15.07.b3 Morocco

► **4.** The slopes of a scoria (cinder) cone reflect the angle of repose because they are typically composed of loose, volcanic scoria. The angle of repose will be steeper for coarser scoria and for material that partially fused together during the eruption.

15.07.b4 Northern AZ

C What Factors Control Slope Stability?

The main control on slope stability is the angle of repose for the material. Intact rock can form cliffs or steep slopes, but soil, sediment, and strongly fractured rock form slopes reflecting their angle of repose.

15.07.c1

The addition of minor amounts of water increases the strength of soil, but oversaturation pushes grains apart and weakens the soil. Materials with high clay-mineral content can flow downhill when they become wet.

15.07.c2

Fractures, cleavage, and bedding reduce the overall mechanical strength of a rock, and may allow rocks to slip downhill. In this illustration, rock layers oriented parallel to the slope allow material to slide.

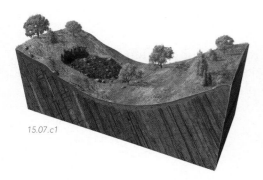

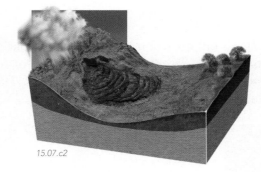

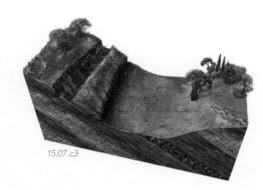

15.07.c3

D What Triggers Slope Failure?

Slope failure **occurs when a slope is too steep for its material to resist the pull of gravity. Some slopes slide or creep downhill continuously, but others fail because some event caused a previously stable slope to fail.**

Precipitation can saturate sediment, weakening an unconsolidated material by reducing grain-to-grain contact. A slope that was stable under dry conditions may fail when wet. Slopes can also fail after wildfires, which destroy plants that help bind and stabilize the soil.

Hillslopes can fail when the load on the surface exceeds a slope's ability to resist movement. Humans sometimes build heavy structures on slopes, overloading the slope and causing it to fail. Areas with gentle slopes, such as near this town, are less prone to slope failures.

Modification of a slope by humans or natural causes can increase a slope's steepness so that it becomes unstable. Erosion along river banks, like shown here, or wave action along coasts can undercut a slope, making it unsafe.

15.07.d1

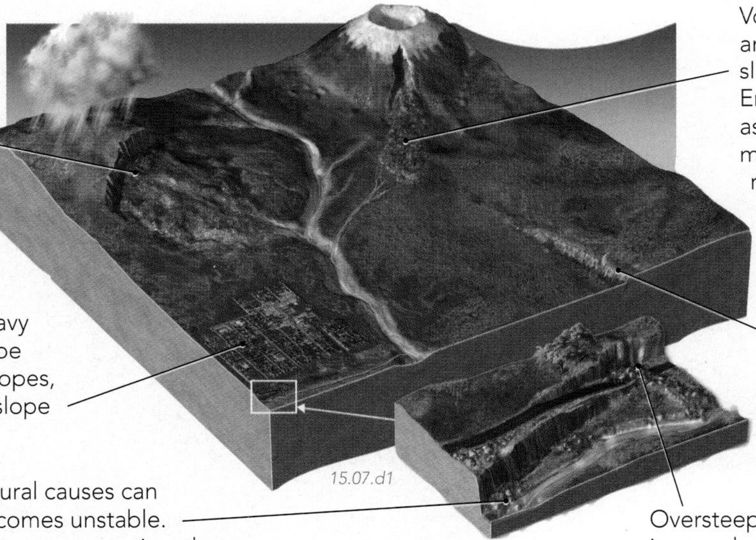

Volcanic eruptions can shake, fracture, and tilt the ground, unleashing landslides from oversteepened slopes. Eruptions can cover an area with hot ash and other loose material, causing melting of ice and snow. Melting can rapidly release large amounts of water and mobilize volcanic material in destructive debris flows.

A sudden shaking, such as tremors caused by an earthquake along this fault scarp, may trigger slope instability. Minor shocks from heavy trucks or human-caused explosions can also trigger slope failure.

Oversteepening of cliffs or hillslopes during road construction can cause them to fail, especially if fractures or layers are inclined toward the road.

Slope Stability in Cold Climates

In cold climates, water is frozen much of the year, and ice, although solid, can flow. Freeze-thaw cycles, where ice freezes and then thaws repeatedly, cause ice to flow and can contribute to slope failure.

When water-saturated soil freezes, it expands, pushing rocks and boulders on the surface upward (►). When the soil thaws, the boulders move down again. This process, called *frost heaving*, is a large contributor to the downslope movement of material in cold climates. In addition, when the upper layers of soil thaw during the warmer months, the water-saturated soil may move downslope more easily.

15.07.mtb1

Before You Leave This Page Be Able To

- ✓ Describe or sketch the role that gravity plays in slope stability.
- ✓ Describe the concept of the angle of repose and its landscape expressions.
- ✓ Describe some factors that control slope stability, and events that trigger slope failure.

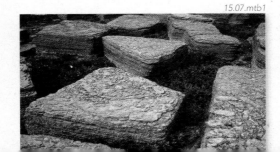

15.8 How Do Slopes Fail?

THE RAPID DOWNSLOPE MOVEMENT of material, whether bedrock, soil, or a mixture of both, is commonly referred to with the general term *landslide*. Movement during slope failure can occur by falling, sliding, rolling, slumping, or flow. We classify slope failures by how the material moves and the type of material involved.

A What Are Some Ways That Slopes Fail?

Most people have seen evidence of slope failure when hiking, driving past a road cut, or watching television news, nature shows, or movies. Slope failure can be as subtle as a small pile of rocks at the base of a hill or as dramatic as a mudflow that has destroyed a neighborhood in China or California. The photographs below show images of various types and sizes of slope failures.

15.08.a1 Yampa River, UT

15.08.a2 Denali region, AK

◄ **2.** During an earthquake, brown masses of rock and soil slid down these steep slopes in Alaska, smashed apart, and flowed as avalanches of rock, soil, and ice across a white glacier in the valley below. Parts of the avalanche flowed across the valley and partway up hillsides on the other side of the valley.

15.08.a3 Tibet

◄ **3.** On this hillside, millions of rock pieces slid off steep outcrops. Some pieces accumulated on a high talus slope, which is at the angle of repose for this material. Several scars indicate where the talus was remobilized and flowed downhill, constructing a debris fan at the base of the hill.

▲ **1.** This rocky cliff failed after being undercut by a river. Large sandstone blocks, one the size of a building, collapsed downward. The falling block detached along a prominent joint surface, which has since accumulated a brown rock varnish.

15.08.a4 El Salvador, Central America

◄ **4.** This landslide in El Salvador flowed down a steep, unstable slope and cut a swath of destruction across a neighborhood. Adjacent slopes on this hill appear to be just as steep as the part that failed, and pose a hazard to the remaining homes.

► **5.** Undercutting of hillsides by coastal erosion and highway construction has made many slopes steep and unstable. Along the California coast, rocks and soil slumped downward, covering and blocking the highway along the Pacific Palisades, California. Such slope failures commonly occur during or after intense rainstorms.

15.08.a5 Pacific Palisades, CA

B How Are Slope Failures Classified?

Classification of slope failure is imprecise because the processes commonly grade into one another, and more than one mechanism of movement can occur during a single slope-failure event. All classifications consider *how* the material moves, what *types of material* move, and the *rate of movement* of the material.

15.08.b1 Grand Junction, CO

Mechanism of Movement

◄ Geologists classify slope failures primarily by how the material moved. Rocks and other material can *fall* off cliffs, can *slip* along fractures, cleavage, or bedding planes, can *topple* over, or can do all three.

► Other slope failures involve the slow *creep* of the uppermost soil cover or the *flow* of material, as during turbulent flows of mud, rocks, and other debris. This brown mud flowed into a neighborhood in California.

15.08.b2 La Conchita, CA

15.08.b3 Monument Valley, UT

Type of Material

◄ Some slope failures involve slabs of *solid rock* or large pieces of broken rock derived from cliffs and rocky hillslopes. Such rocks can further break apart after they begin to move.

► Many slope failures mobilize *unconsolidated material* that is stripped from hillsides. Material can include soil, loose sediment, pieces of wood and other plant parts, boulders, and other types of loose debris.

15.08.b4 Grand Canyon, AZ

15.08.b5 Venezuela

Rate of Movement

◄ Fast rates—Another important factor is the *rate of movement* of the material. Some slope failures start in an instant and send material downhill at hundreds of kilometers per hour, or at least too fast for people to outrun.

► Slow rates—Other mass movements are more gradual and move downhill at rates that are imperceptible to an observer. This slow-moving mudflow carries trees, some tilted, along for the ride.

15.08.b6 Slumgullion, CO

Submarine Slope Failures

Slope failure is not restricted to the land; it can also occur on steep or even gentle slopes on the seafloor. Such *submarine slope failures* can be caused by over-loading of sediment on a slope or in a submarine canyon. They can also be triggered by shaking during an earthquake, volcanic eruption, or storm. Various types of slope failure occurred off the southwestern coast of the Big Island of Hawaii, forming the large mass of debris shown here in green.

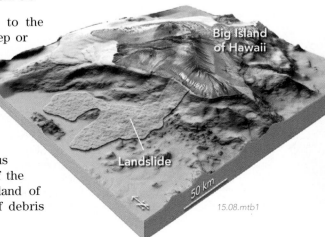

15.08.mtb1

Before You Leave This Page Be Able To

✓ Describe slope failures and some ways they are expressed in the landscape.

✓ Summarize the classification of slope failures, and describe the different types of movement, types of material, and rates of movement.

15.9 How Does Material on Slopes Fall and Slide?

SOME SLOPE FAILURES involve materials *falling* off a cliff or *sliding* down a slope. These mechanisms commonly involve rock and pieces of rock, but they can also involve materials that are less consolidated. Rocky cliffs and slopes might appear to be immune to slope failure because they consist of hard bedrock, but they can fail spectacularly and catastrophically.

A What Happens When Rocks Fall or Slide?

Rocks and other material can fall from cliffs or can slide on fractures or other weak planes. Falling and sliding rock masses may begin as relatively intact blocks, but they commonly break apart as they begin moving or when they hit the bottom of a cliff. Some slides rotate as they move. Others simply slide down the hill. One type of rock failure can lead to another because a rock fall may remove some support, causing higher parts of a cliff to slide.

15.09.a2 Eiger, Swiss Alps

Rock Falls and Debris Falls

15.09.a1

In a *rock fall*, large blocks or smaller pieces of bedrock detach from a cliff face and fall until they smash into the ground. Rock falls can be triggered by rain, frost wedging, thawing of ice that had held rocks to the cliff, an earthquake, river erosion, or human construction that undercut a cliff. Less consolidated debris, including loose sand, can also fall off a cliff. Part of the cliff in this photograph (▶) is collapsing, producing a rock fall composed of large, angular blocks. After they fell, the rocks broke apart and rolled down the hillside.

Rock Slides

15.09.a4 Naxos, Greece

In a *rock slide*, a slab of relatively intact rock detaches from bedrock along a bedding surface, preexisting fault, joint, or other discontinuity that is inclined downslope. As it slides, parts of the slab typically shatter into angular fragments of all sizes, but large blocks can remain relatively intact. In the rock slide shown to the right (▶), road construction undercut these sedimentary rocks, which had layers and fractures that were inclined downslope. At some point after construction, the rocks slid along the layers and into the road.

15.09.a3

Rotational Slides

15.09.a6 Toreva, AZ

In some slides, rock layers and other material rotate backward as they slide. Such *rotational slides* move along one or more curved slip surfaces. This type of slide, also called a *slump*, can occur in bedrock or less consolidated material. Individual slices can remain relatively intact or can break and spread apart. The rotational slide in this photograph (▶) offset and tilted the rock layers as they slumped downhill. The rock layers are nearly horizontal in the intact cliff in the background, but layers in the lower slide blocks are tilted back into the hill.

15.09.a5

B What Is the Geometry of a Rock Slide?

A combination of geological circumstances is required to detach a slab of rock and create a rock slide. It requires sufficiently steep slopes along with bedding planes, fractures, or other flaws that are inclined downslope.

1. Many rock slides occur in bedrock with discrete layers that differ in rock type and therefore in strength. Such rock layers are most common in sedimentary rocks, but they are also present in many volcanic and metamorphic sequences. In this figure, a sequence of different sedimentary layers has beds inclined downhill, toward a small stream valley.

2. At their upper end, most rock slides detach along a series of preexisting joint or fault surfaces. In other cases, as in this example, the stresses that build up in the rock before it slides are enough to form new fractures, allowing the rock slide to detach from bedrock.

3. Detachment of the base of the rock slide from underlying bedrock commonly occurs along a layer of weak rock, such as salt or shale. This weak layer may allow the overlying slab to slide fairly easily and remain partially intact as it moves downhill.

4. To be able to slide down the dip of the layers, the upper layer must have space downhill in which to slide. That is, the layers will probably not slide if they are supported by more rocks in a down-dip direction. In this example, a stream has eroded a low area, giving the sliding rock slab somewhere to go.

5. Although not shown here, rock slides can slip along joints, fault surfaces, cleavage, or some other discontinuity, rather than bedding surfaces. Preexisting faults of the proper orientation are especially susceptible to rock slides because they are planar, fairly continuous, and structurally weak.

7. Most slides leave a linear or curved scar, or *scarp*, on the hillslope, marking where the slide pulled away from the rest of the hill. This upper end is also called the *head* of the slide.

6. The leading edge of a slide, the *toe*, can overrun the land surface in front of the slide.

15.09.b1

The Vaiont Disaster, Italy

In 1960 a dam was built across the Vaiont Valley in northeastern Italy. This valley runs along the bottom of a syncline, where the rocks have been folded downward and dip into the valley from both sides. The rocks are mainly limestone but with interlayered thin beds of shale and sandstone. Some of the limestone beds contain caverns formed when groundwater dissolved the rock. Fractures in the rocks run both parallel and perpendicular to the bedding planes.

During August and September 1963, three years after the dam was completed, heavy rain fell in the area. One day in October, the south wall of the valley failed and slid into the reservoir behind the dam. The slide was 1.8 km high and 16 km wide with a volume of 240 million cubic meters. The slide moved along shale layers that parallel the bedding planes in the limestone.

As the slide moved into the reservoir it displaced an equal volume of water, forcing a surge of water 240 m above lake level onto the village of Casso on the northern side of the valley. Waves within the reservoir killed 1,000 people. Waves 100 m high swept over the dam. Although the dam did not fail, the water that overtopped the dam killed 2,000 people living in villages below the dam.

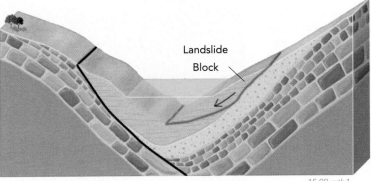

Landslide Block

15.09.mtb1

Before You Leave This Page Be Able To

☑ Sketch and describe a rock or debris fall, a slide, and a rotational slide.

☑ Sketch how the geometry of layers, faults, and other features could allow a rock slide to begin.

☑ Describe the Vaiont landslide disaster and factors that caused it to happen.

15.9

15.10 How Does Material Flow Down Slopes?

SLOPE FAILURE CAN MOBILIZE WEAK MATERIAL, including soil, sediment, broken rock, and loose debris. Material flows downhill if it is poorly attached to a hillside, is internally too weak to resist the downward pull of gravity, and the slope is steep enough to allow flow. If the material incorporates some water, it becomes weaker and better able to flow. Such material can move rapidly, as in a debris flow, or can creep down the hill at a nearly imperceptible rate. All these types of slope failure involve the *flow* of material.

15.10.a1

Creep

Creep is the very slow, continuous movement of soil, loose sediment, and weathered rock down a slope. Creep occurs on almost all slopes, but the rate varies. Evidence for creep is expressed in bent or leaning trees, warps in roads and fences, and leaning utility and fence poles (▶). Creep commonly causes bedding and other layers in the subsurface to deflect, or bend downhill. Creep can crack foundations and destroy houses constructed on creeping hillslopes.

15.10.a2 Spitsbergen Island, Norway

Debris Slides

Soil, weathered sediment, or other unconsolidated material can move downslope as a *debris slide*. Debris slides are usually less than 10 m thick and leave behind a low *scarp*. A debris slide moves downhill partly as a sliding, coherent mass and partly by internal shearing and flow. A debris slide can lose coherency as it moves, thereby evolving into an *earth flow* or a *debris flow*. The debris slide in this photograph (▶) occurred in clay-rich glacial sediment and displaced pieces of road down the mountain.

15.10.a3

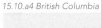

15.10.a4 British Columbia

Earth Flows

Earth flows are flowing masses of weak, mostly fine-grained material, especially mud and soil. The material moves like thick, wet concrete, generally slowly enough to outrun, but it contains enough water to be slightly fluid. Earth flows contain more mud and other fine-grained material than they do rocks, and also can be called *mudflows*. The earth flow in this photograph (▶) mobilized clay-rich altered volcanic and sedimentary rocks, as well as a surface veneer of angular rocks. The earth flow moved downhill until it reached and dammed a larger valley and completely flooded a small village.

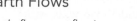

15.10.a5

15.10.a6 Thunder Mtn., ID

Debris Flows

Debris flows are wet, downhill-flowing slurries of loose mud, soil, volcanic ash, rocks, and other objects picked up along the way. Some contain only a little water, whereas others are water rich and flow like a thick soup. Debris flows, especially their mud-dominated varieties, are called *mudflows* by the media, and can move rapidly. They often result from heavy rains that saturate the soil and other loose materials. Debris flows are thick and more dense than water and so can support and carry large boulders or even houses. In this photograph (▶), debris-flow deposits from the 1999 Venezuelan disaster, discussed in the opening pages of this chapter, consist of angular blocks, some several meters across. During transport, the rocks were enclosed in a matrix of wet mud and other debris.

15.10.a8 Venezuela

15.10.a7

Rock and Debris Avalanches

Debris avalanches are high-velocity flows of soil, sediment, and rock that result from the collapse of steep mountain slopes. A debris avalanche moves down the slope, in many cases traveling considerable distances down valleys and across relatively gentle slopes. A *rock avalanche* occurs when a rock mass falls off a cliff face and shatters on contact, sending a turbulent jumble of rock fragments, some bigger than cars, flowing downhill. They can flow at over a hundred kilometers per hour, fast enough to continue flowing uphill when they encounter a topographic obstacle. Both types of avalanches are often triggered by earthquakes and volcanic eruptions, and can kill thousands of people at a time. In 1970, an earthquake shook loose a large chunk of an Andean mountain, causing a rock and debris avalanche (▲) that buried a town (not visible in this photograph) and 18,000 of its inhabitants.

15.10.a9

15.10.a10 Huascaran, Peru

Landslides and La Conchita

The coastal community of La Conchita in southern California was partially overrun by a landslide in 2005 (below right), which was a repeat of one in 1995 (far right). The landslide mobilized poorly consolidated sediment along the steep bluffs overlooking the town. Although the 1995 landslide destroyed part of the town, houses were rebuilt in the area only to be destroyed by the 2005 landslide, which remobilized parts of the 1995 landslide. Would this be a good place to rebuild again? How much should hazard insurance cost people who live here?

15.10.mtb2

15.10.mtb1

Before You Leave This Page Be Able To

✓ Sketch and describe what happens during the following: creep, debris slide, earth flow, debris flow, and debris avalanche. Compare how each of these features flows.

✓ Describe what happened at La Conchita. Was it a good idea to build, or rebuild there?

15.11 Where Do Slope Failures Occur in the U.S.?

LANDSLIDES AND OTHER SLOPE FAILURES have destroyed large parts of cities in some countries and have killed more than 30,000 people in a single event. Each year in the United States, landslides cause 25 to 50 deaths and result in more than $3 billion in damage. What slope-failure events have affected the United States? What is the likelihood they will occur again? Which areas are most at risk in the future?

A Where and How Have Large Slope Failures Affected the United States?

There have been countless landslides, debris flows, and other slope failures. The landslide that accompanied the 1980 eruption of Mount St. Helens was the largest landslide in U.S. history. It generated enough debris to fill 250 million dump trucks. Other slope failures are described below.

15.11.a1 Denali Region, AK

A large (7.9 magnitude) earthquake struck the Denali (Mount McKinley) region of south-central Alaska in 2002. Ground shaking associated with the earthquake caused huge slope failures off the region's steep mountains. Spectacular rock falls, rock slides, and debris avalanches slid down the steep slopes and flowed across and buried the Black Rapid Glacier. The region is prone to slope failures partly because tectonics has rapidly uplifted the mountains and formed an array of joints, faults, and tilted rock layers.

15.11.a2 Pacific Palisades, CA

Landslides and debris flows are common in southern California (◄), where some houses have been built on risky sites that have been undercut, causing unstable slopes to collapse downhill.

15.11.a3 Tully Valley, NY

The largest landslide to affect New York in 75 years moved through the Tully Valley, near Syracuse, in 1993. The earth flow mobilized over a million cubic meters of weak clays that had been deposited in a glacial lake.

15.11.a4 Gros Ventre, WY

The Gros Ventre slide (◄) is one of North America's largest historic landslides. During this 1925 rock slide, a slab of limestone more than 1.5 km (1 mi) long broke loose along weak bedding planes that dipped toward, and were undercut by, the Gros Ventre River of Wyoming.

15.11.a5 Thistle, UT

The most costly landslide in U.S. history devastated the small Utah town of Thistle in Spanish Fork Canyon in 1983. The landslide moved slowly downhill, reaching a maximum rate of only 1 m (3 ft) per hour. It severed railroad service between Denver and Salt Lake City, flooded two major highways, and formed a new lake where the town had been. A railroad tunnel now cuts through the light-colored landslide material.

B What Is the Potential for Landslides and Debris Flows in the United States?

All states receive some damage from landslides and debris flows, but not all areas have the same potential hazard. The potential for landslides is highest near steep mountains, such as in Colorado, the Appalachians, and other mountain areas that have weak, heavily weathered materials. It is also high in areas of recent tectonic activity.

Potential for Landslides in the United States

This USGS map portrays the landslide hazards of the lower 48 states. Red areas represent the greatest hazards, followed by yellow and then green. Other areas have less potential for landslides or are unstudied. What landslide potential exists where you live?

Many parts in the Pacific Northwest experience landslides because of the many steep mountains, heavy rainfall, and rainfall that melts snow cover.

The coastal parts of central and southern California have high landslide potential because of steep mountains, high potential for ground shaking during earthquakes, and coastal erosion that undercuts weak material along hillslopes overlooking the shoreline.

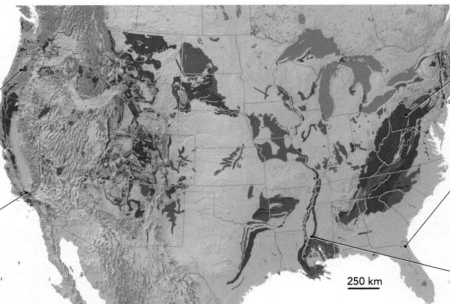

15.11.b1

In the east, landslides are common in the Appalachian Mountains, where landslides mobilize soil or occur along weaknesses in folded, faulted, and weathered rock layers.

Florida and the coastal plain of the southeast Atlantic seaboard have some of the lowest potential for landslides because the region lacks steep slopes.

Landslide hazards in the central United States are mostly along steep bluffs that flank the rivers, or in areas, such as the northern Great Plains, underlain by weak materials.

250 km

Potential for Debris Flows in the Western United States

Black areas on this USGS map show areas of the western United States that have high potential for debris flows. The USGS produced a similar map for the eastern United States.

The Pacific Northwest contains areas with high potential for debris flows, especially along slopes and valleys connected to the active Cascade volcanoes. An eruption on a snow-capped peak can unleash large debris flows.

Coastal hills and mountains of northern California produce debris flows because the rocks are weakened by deformation and intense chemical weathering in this fairly wet climate.

Debris flows are very common in the high mountains flanking Los Angeles, California. These recently uplifted mountains have very steep slopes and receive locally intense rainfall. Economic pressures on real-estate development have resulted in houses being built within and at the base of steep canyons, in areas that are repeatedly struck by debris flows. Wildfires worsen the situation.

15.11.b2

Many parts of the Rocky Mountains have a higher potential for debris flows than is shown on this map, but many areas have not been studied enough to accurately determine the risk. Mountain regions are especially susceptible to debris flows if wildfires have removed the soil-binding vegetation or if forests have been clearcut.

A belt of high debris-flow potential occurs along the Wasatch Front, the steep mountain front that flanks Salt Lake City, Provo, and most of the larger cities in Utah.

250 km

Before You Leave This Page Be Able To

☑ Briefly describe factors involved in landslides in the United States.

☑ Summarize some factors that make some areas of the United States have high risks for landslides or for debris flows.

☑ Identify whether you live in an area with a high potential for landslides.

15.11

15.12 How Do We Study Slope Failures and Assess the Risk for Future Events?

HOW CAN WE LIMIT DAMAGE CAUSED BY SLOPE FAILURES? Geologists observe and monitor active slope failures to better understand how such systems operate. They then use this information to recognize areas that either have experienced past slope failure or are susceptible to future slope failures. The most hazardous sites have dangerous combinations of geologic factors, such as steep slopes and weak materials.

A How Do We Recognize and Monitor Active or Recent Slope Failures?

To limit property damage and the potential for loss of life, geologists study and monitor active and recently active slope failures. The resulting observations and measurements provide a solid basis for investigating the processes, causes, and hazard assessment of destructive slope failures.

15.12.a1 Venezuela

After a slope-failure event, geologists conduct field studies to document the distribution, size, and other characteristics of the event, largely to understand what happened and to assess the potential for future events. This area was destroyed by debris flows.

15.12.a2 Mount Rainier, WA

Geologists use surveying equipment to document topography, especially changes in slope, of hazardous areas. From such measurements, they construct maps and cross sections showing where slope failures are most likely. Finally, geologists help devise plans to minimize losses from slope failure.

B How Do We Recognize Prehistoric Slope Failures?

Geologists identify prehistoric slope failures by observing the topography, surface and subsurface distribution of rock types, the condition of the rocks, and the geometry of layers and other geologic structures.

15.12.b1 Mount. St. Helens, WA

Landslides and other slope failures can overrun existing topography, replacing it with a random-looking assemblage of humps and pits. We call this type of landscape *hummocky topography*.

15.12.b2 Venezuela

Each slope failure leaves behind characteristic deposits. Geologists observe modern deposits and then use these characteristics to recognize deposits of past events that were similar to the modern one. These deposits were left by debris flows.

15.12.b3 Theodore Roosevelt NP, ND

Rotational rock slides can be recognized because the orientation of beds and other geologic structures is incompatible with those in adjacent areas. The evidence can be subtle, like this slightly different tilt.

15.12.b4 Mount St. Helens, WA

Debris flows, debris avalanches, and landslides are sometimes recognized because they carried distinctive rock types into areas where such rocks do not otherwise occur. Rocks in this debris pile were derived from the distant Mount St. Helens, Washington.

C How Do We Assess an Area's Potential for Slope Failure?

Geologists try to assess the likelihood that a location will suffer the disastrous consequences of slope failure by examining evidence of past slope failures in the area or in adjacent areas that have a similar setting. They also evaluate the steepness of slopes, including any recent changes, and any other factors leading to slope failure.

15.12.c1 Nepal

Evidence of Past Slope Failures

One of the best indications for potential slope failure is evidence of past failures. The more recent the failure, the more likely such an event will recur in the near future. This part of the Himalaya, with steep hillslopes scarred with slope failures, looks risky. ◄

15.12.c2 Durango, CO

Situated in Area with Known Problems

The slope angle of hillsides and mountains is clearly a key factor in the potential for slope failure but must be evaluated in the context of the types of materials and the geometry of geologic structures. ►

15.12.c3 Yellowstone NP, WY

Steepness of Slopes

A site may be at risk if its geologic setting is similar to other slopes that have failed. Part of this slope collapsed, and nearby parts have the same steep slopes, weak rocks, and position that can be undercut by the river. ◄

15.12.c4 Zermatt, Switzerland

Recent Changes in Slope

Natural processes, over time, act to adjust slopes to the appropriate, stable angle, but this equilibrium can be upset if the slope is steepened or undercut by natural or human activities. This rock slide failed in 1991, burying 31 houses. ►

15.12.c5 Capitol Reef NP, UT

Conditions of Material

Another factor is the nature of material on a slope: whether material is loose sediment or solid bedrock, and is resistant to erosion or relatively weak. The presence, spacing, and orientation of geologic structures can weaken rocks and facilitate downslope slippage of materials. ◄

15.12.c6 Augustine volcano, AK

Potential Triggers

If other factors are equal, an area is at higher risk for slope failure if there are frequent events, such as volcanic eruptions or earthquakes, that could trigger slope failure. Steep slopes with loose material on an active, shaky volcano are trouble. ►

The Blackhawk Landslide

A huge lobe of shattered rock lies in a valley north of the San Bernardino Mountains, northeast of Los Angeles, California. This mass has hummocky topography and large, shattered pieces of rock that are different than rocks in adjacent areas. This feature, known as the *Blackhawk Landslide*, formed in prehistoric times when a large segment of the mountains collapsed, shattered, and flowed as debris avalanches more than 10 km (6 mi) out into the valley.

15.12.mtb1
Landslide
1 km

Before You Leave This Page Be Able To

✓ Describe some ways that geologists investigate slope failures.

✓ Summarize characteristics used to identify prehistoric slope failures.

✓ Summarize some aspects that might indicate that an area has a high potential for slope failure.

15.12

15.13 What Is Happening with the Slumgullion Landslide in Colorado?

THE SCENIC SAN JUAN MOUNTAINS OF COLORADO contain the *Slumgullion landslide*, one of the best-studied landslides in the world. The landslide has been moving for more than 1,000 years, and part of it is still moving, allowing geologists to examine up close the processes of a relatively slow slope failure.

A What Is the Setting and Morphology of the Slumgullion Landslide?

15.13.a2 Slumgullion, CO

◄ The Slumgullion landslide is a conspicuous feature in the landscape of the San Juan Mountains of southwestern Colorado. It begins among high peaks of altered volcanic rocks, snakes downhill, and spreads out when it reaches the valley bottom. The landslide shows as a tan-fringed mass in this photograph.

The head (top) of the landslide is a steep landslide scarp within volcanic rocks. Interactions with water altered the volcanic rocks into a weak, clay-rich material. This material is too weak to support the steep slope, and so it collapsed downward, starting the landslide.

15.13.a1

The flowing material in the landslide is mostly derived from the clay-rich, altered volcanic rocks and so is weak and fine grained; it could be called an earth flow. The abundant water in the landslide reduces its strength, allowing it to flow down relatively gentle slopes. Steeper slopes at the scarp, combined with weak materials and abundant water, produced a setting favorable for starting the landslide.

When it reached the valley, the landslide blocked the Lake Fork of the Gunnison River, forming Lake San Cristobal, Colorado's second-largest natural lake. The river flows from right to left in this view.

1 km

Parts of the Landslide

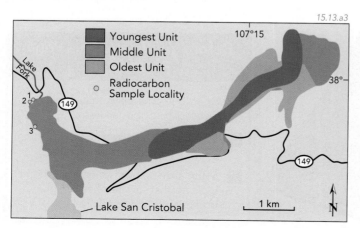

15.13.a3

Youngest Unit
Middle Unit
Oldest Unit
○ Radiocarbon Sample Locality

107°15

Lake Fork

149

38°

149

Lake San Cristobal

1 km

N

When USGS geologists investigated the landslide, they discovered that it actually consisted of three different ages of material (◄): a young, central part that was still active, and two older parts. In the active, center part, continued movement has tipped over trees and caused damage to a road that by necessity crosses the flow. ▶

15.13.a4

462

B What Structures and Other Features Are Associated with the Landslide?

To study the landslide, USGS scientists did surveying and used aerial photos to construct a very detailed topographic map. From various types of data, the geologists calculated that the weak materials in the landslide are moving down a slope of 7° to 10°, much less than a typical angle of repose for normal, dry materials. The geologists also documented that the landslide is moving quite fast. From the time the topographic map was made to when USGS geologists began to map the features within the landslide, some landmarks had already moved 18 m down the slope! The geologists used the detailed topographic map as a base to construct this map of structural features, including scarps, ridges, and other features that formed as the landslide flows downhill. ▶

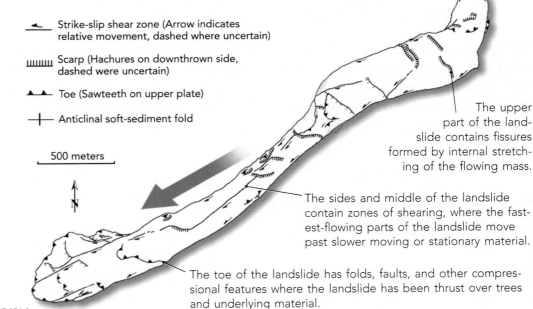

Strike-slip shear zone (Arrow indicates relative movement, dashed where uncertain)

Scarp (Hachures on downthrown side, dashed were uncertain)

Toe (Sawteeth on upper plate)

Anticlinal soft-sediment fold

500 meters

N

15.13.b1

The upper part of the landslide contains fissures formed by internal stretching of the flowing mass.

The sides and middle of the landslide contain zones of shearing, where the fastest-flowing parts of the landslide move past slower moving or stationary material.

The toe of the landslide has folds, faults, and other compressional features where the landslide has been thrust over trees and underlying material.

C What Other Studies Did Geologists Do to Understand the Landslide?

1. Geologists investigated the Slumgullion landslide not only to understand this particular landslide but to learn how landslides in general operate. They examined temperature and precipitation records to see if the landslide behaved differently during warm or wet times than during cold or dry periods. Velocity of the landslide decreased in the winter, when the formation of ice reduced the amount of water available to the landslide. The landslide also moved more slowly when conditions were drier.

2. Geologists used GPS and other surveying methods to measure how fast different parts of the landslide were moving. On this map, longer red lines indicate faster rates of downhill flow.

3. The fastest rates of movement (the longest lines on this map) are in the middle segment of the landslide, where the landslide is moving as fast as 7.8 m per year.

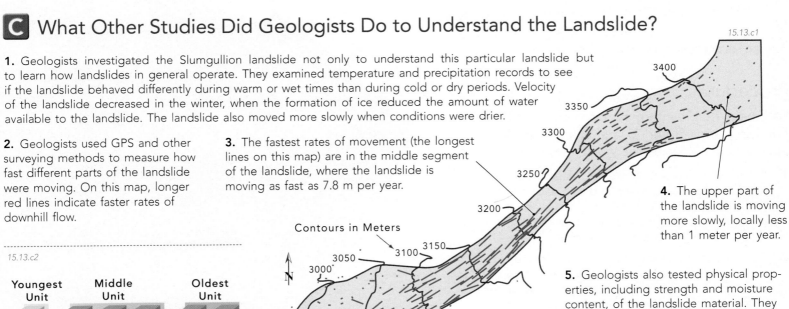

15.13.c1

3400
3350
3300
3250
3200
3150
3100
3050
3000

Contours in Meters

N

0   30 m   Direction and Amount of Displacement

500 m   Map Scale

4. The upper part of the landslide is moving more slowly, locally less than 1 meter per year.

5. Geologists also tested physical properties, including strength and moisture content, of the landslide material. They then used these data and computers to model how the landslide is moving.

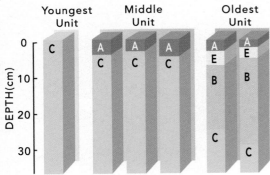

15.13.c2

Youngest Unit   Middle Unit   Oldest Unit

DEPTH(cm)

0

10

20

30

▲ **6.** To constrain the three ages of landslide material that were identified and mapped, geologists compared the soil development in the three parts. The oldest part had well-developed A and B horizons, whereas the youngest part was too young to have developed either an A or B horizon. Carbon-14 ages obtained from samples of charcoal and wood within different parts of the landslide indicate that the landslide first formed about 1,000 years ago, yet it is still very active today.

Before You Leave This Page Be Able To

✓ Describe the setting and morphology of the Slumgullion landslide.

✓ Summarize the types of studies geologists conducted on the landslide and what each type of study revealed about what was happening with the landslide.

15.13

Which Areas Have the Highest Risk of Slope Failure?

THIS GEOLOGICALLY DIVERSE PLACE has features that appear to be related to slope failure. Large, angular blocks occur in several different settings, and some of the hills may not be stable or safe. You will use descriptions and images of these features to determine what hillslope processes are occurring in different areas, and how they impact where people may live safely. The landscape is stylized and exaggerated to highlight potentially hazardous areas.

Goals of This Exercise:

- Observe the landscape to investigate the geologic setting of different areas, and interpret the geologic setting from descriptions of each location.
- Assess the hazards in different areas.
- Construct a map that shows areas that have a high risk for different types of slope failure.
- Identify locations that you think are most safe and moderately safe on which to build.

Procedures

Use the available information to complete the following steps, entering your answers in the appropriate places on the worksheet or online.

1. Observe the features shown on this landscape. Read the text boxes associated with each feature and decide what that statement implies about the geologic setting of the area and how the landscape reflects the underlying geology.

2. Think about the description of each area and consider possible types of slope failure that could occur there. Provide a reasonable interpretation of what types of slope processes are occurring and what key observations led you to that conclusion.

3. On the figure in the worksheet, draw approximate boundaries around areas that you interpret as having the highest risk for each type of slope failure. Label each area with a few words to identify the main hazard you interpret to be present.

4. Draw the letters S and M on the map for sites where you think it would be relatively safe to live. Write an (S) for one or more relatively safe places, an (M) for a moderately safe place to live. There is not a single best choice for any of these sites, so be prepared to describe your reasoning and to discuss your choice.

1. A series of small hills, referred to by local people as the *Bent Fence Hills*, contains trees that are tipped over at odd angles. Local farmers complain that they have to keep straightening their crooked fences on these hillslopes. For some reason, no one has ever built a house here.

2. A flat-topped hill, called *Flattop Hill*, is surrounded by a steep cliff formed by a resistant layer of basalt. The basalt is jointed and underlain by a weak layer of clay. Below the cliff are a series of large, angular blocks of basalt. A large, spoon-shaped scar scoops into part of the cliff.

15.14.a1

9. The *Annabelle River* cuts through the landscape, flowing from right to left. Paralleling the river on both sides are low terraces that are only a few meters higher than river level. On these low terraces are large volcanic blocks of andesite, some as big as a house. They are not present on higher areas away from the river. No one has ever seen the river with enough water to move such large blocks.

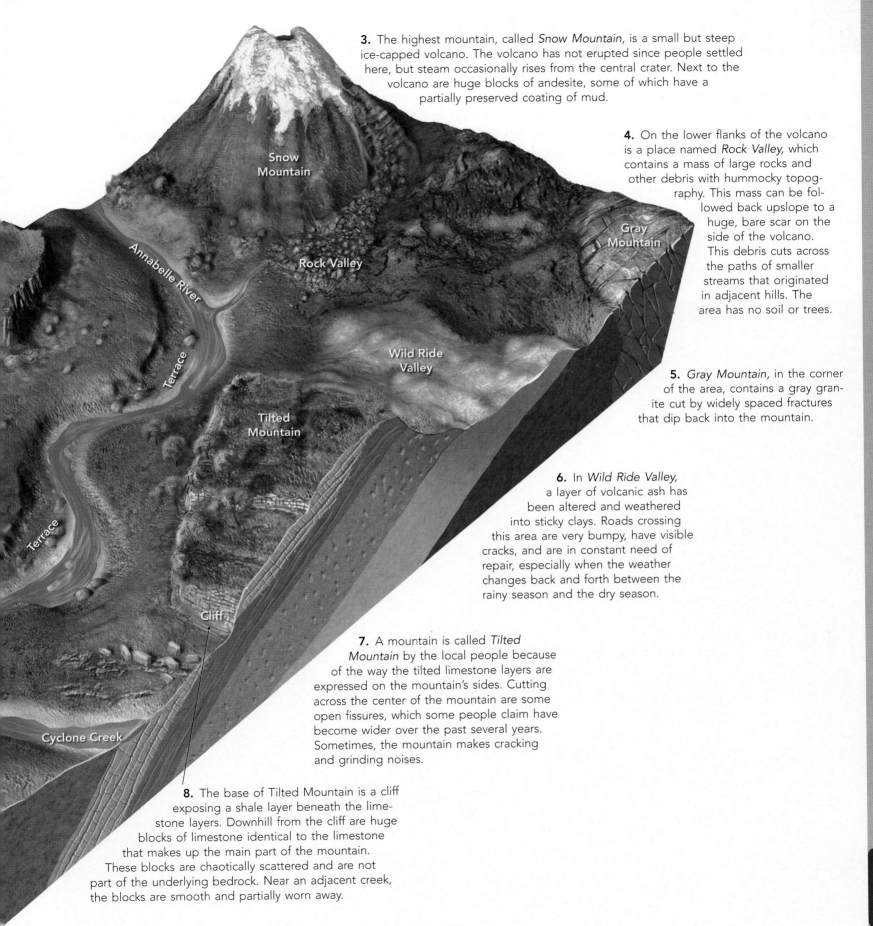

3. The highest mountain, called *Snow Mountain,* is a small but steep ice-capped volcano. The volcano has not erupted since people settled here, but steam occasionally rises from the central crater. Next to the volcano are huge blocks of andesite, some of which have a partially preserved coating of mud.

4. On the lower flanks of the volcano is a place named *Rock Valley,* which contains a mass of large rocks and other debris with hummocky topography. This mass can be followed back upslope to a huge, bare scar on the side of the volcano. This debris cuts across the paths of smaller streams that originated in adjacent hills. The area has no soil or trees.

5. *Gray Mountain,* in the corner of the area, contains a gray granite cut by widely spaced fractures that dip back into the mountain.

6. In *Wild Ride Valley,* a layer of volcanic ash has been altered and weathered into sticky clays. Roads crossing this area are very bumpy, have visible cracks, and are in constant need of repair, especially when the weather changes back and forth between the rainy season and the dry season.

7. A mountain is called *Tilted Mountain* by the local people because of the way the tilted limestone layers are expressed on the mountain's sides. Cutting across the center of the mountain are some open fissures, which some people claim have become wider over the past several years. Sometimes, the mountain makes cracking and grinding noises.

8. The base of Tilted Mountain is a cliff exposing a shale layer beneath the limestone layers. Downhill from the cliff are huge blocks of limestone identical to the limestone that makes up the main part of the mountain. These blocks are chaotically scattered and are not part of the underlying bedrock. Near an adjacent creek, the blocks are smooth and partially worn away.

15.14

Rivers and Streams

EROSION AND DEPOSITION BY RIVERS AND STREAMS are the principal sculptors of Earth's landscapes. Flowing water in rivers and streams picks up sediment, transports it to lower elevations, and deposits it in various settings. Flooding rivers deposit sediment and nutrients critical to agriculture, but they can also inundate cities and destroy structures built too close to the riverbank. How do rivers operate, and can we predict how often a flood-prone area will be flooded?

The Yukon Delta, shown in this satellite image, is a huge, fan-shaped landform formed where the Yukon River ends its 3,185 km-long journey by emptying into the Bering Sea along western Alaska. This longest of Alaskan rivers transports vast quantities of sediment eroded from the highlands of Alaska and northwestern Canada.

How do rivers and streams form, and how do they carry sediment?

Where the river meets the sea, the flowing water spreads out, slows down, and deposits its load of sediment in a *delta*. Sediment carried and deposited offshore (lighter blue) causes the delta to grow seaward with time, adding new land to the coast.

What factors determine whether and where a river deposits sediment?

16.00.a1

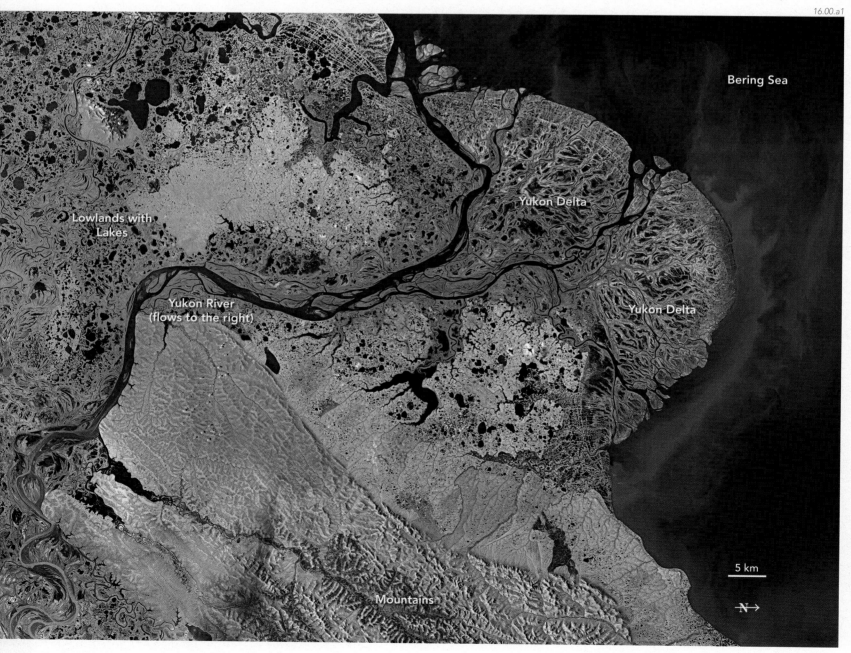

Bering Sea

Yukon Delta

Lowlands with Lakes

Yukon River (flows to the right)

Yukon Delta

5 km

Mountains

N→

TOPICS IN THIS CHAPTER

16.00.a2

◄ **The Yukon River** collects water from a large region of Alaska and Canada's Yukon Territory. It drains an area of 840,000 km² (324,000 mi²). Periodically, water volume in the river exceeds the confines of its channel, causing flooding. When winter ice on the river begins to melt and break up, it piles up in ice jams that cause additional flooding.

How is the size of a river related to the size of the area it drains, what causes a flood, and what information do we need to predict flooding events?

16.00.a3 Denali NP, AK

◄ **Many Alaskan Rivers** are full of sediment derived from weathering and erosion of the mountains and lowlands. This river in Denali National Park is choked with coarse gravel, sand, and fine sediment.

What types of sediment do different kinds of rivers carry?

16.00.a4 Yukon Delta, AK

◄ **During the summer,** lush vegetation grows on the strips of land between the delta waterways. Wetlands on the sediment-rich delta are important breeding sites for migratory birds.

What effect does vegetation have on rivers, and what effects do rivers have on vegetation?

A Variety of Rivers

Each river, like the Yukon River, has its own characteristics and history, which are specific to its geographic and geologic setting. Some rivers are steep and turbulent, moving large boulders, whereas others are slow and tranquil, transporting only silt and clay. Some rivers *meander* in huge looping turns, while others distribute their flow in a network of channels that split off and rejoin in a *braided* pattern. Certain principles govern the behavior of all rivers, such as whether a river erodes into its banks or deposits sediment. Key factors that control the behavior of a river include the steepness of the channel, the supply of sediment, the climate, and tectonic history. The processes involved with rivers cause them to change downstream and over time, producing a characteristic suite of landforms that dominate most landscapes. Rivers can flood huge tracts of land and transport enormous volumes of sediment. The Amazon River in South America (shown below) dumps millions of cubic meters of sediment-laden water into the ocean every minute.

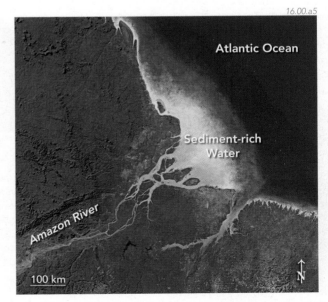

16.00.a5

Atlantic Ocean

Sediment-rich Water

Amazon River

100 km

16.0

16.1 What Are River Systems?

RIVERS ARE CONDUITS OF MOVING WATER driven by gravity, flowing from higher to lower elevations. The water in rivers comes from precipitation, snowmelt, and springs. A river drains a specific area and joins other rivers draining other areas, forming a network of rivers that drain a large region.

A What Is a River?

Rivers and streams carry flowing water through a single channel or through a number of interconnected channels. Such channels vary in size from small streams several meters wide to major rivers that are kilometers across.

1. The Brahmaputra River in India, shown in this satellite image (▼), is a main conduit for water falling on or melting off the Himalaya. The sediment load in this river is enormous, reflecting the ongoing uplift and erosion of the region.

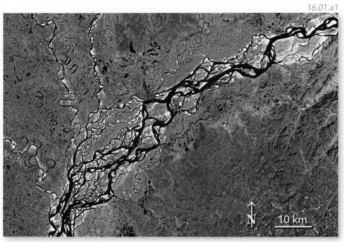

16.01.a1

2. Water flowing in rivers and streams is able to move rock fragments and dissolved minerals from high to low elevations. Note the varied sizes and styles of rivers and streams in this one image. ◄

3. The amount of water that flows through a channel over a given amount of time is the *discharge*, which has units of cubic meters per second or m³/sec. A graph showing the change in the amount of flowing water (discharge) over time (▼) is a *hydrograph*.

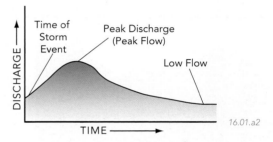

16.01.a2

4. This hydrograph (▲) shows that discharge increased and then decreased over time in response to a storm. The shape of the graph reflects how the river responds to precipitation and can tell us important information about the river and the area it drains.

B Where Does a Stream or River Get Its Water?

Each stream or river drains a naturally defined area, called a *drainage basin*. A basin slopes from higher areas, where the stream or river begins, to lower areas, toward which the stream or river flows. Runoff from rainfall, snowmelt, and springs will flow downriver and out of the drainage basin at its low point.

▼ *Drainage Basin*—In this figure, each of two adjacent streams has a drainage basin, shaded in different colors. Runoff from the red area drains into the stream on the left; runoff from the blue area drains into the stream on the right. The ridge between the two drainage basins is the boundary between water flowing into different drainage basins, and is a *drainage divide*.

16.01.b1

▼ *Basin Slope*—Overall slope of a drainage basin helps determine how fast water in the basin empties after a heavy rain or after snowmelt, as shown by the graph below.

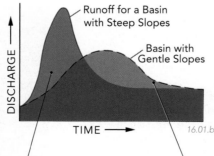

16.01.b2

Runoff from a steep drainage basin is fast, and much water arrives downstream at about the same time, yielding higher discharge values.

Runoff from a more gently sloped basin is spread out over time, leading to lower peak discharge values.

▼ *Basin Size and Shape*—A drainage basin's size and shape influence its flow response to rainfall. The plot below shows a hydrograph for a single storm event, along with a simplified map of the basin's shape.

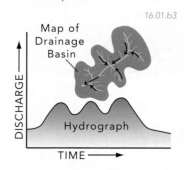

16.01.b3

Following a storm event, a hydrograph for a simple basin shows a single-peak increase in discharge with a gradual decrease, like the graph in part A. In contrast, a complex, three-part drainage basin, shown here, may show a three-peak response to a single event. Total discharge in a larger basin will be higher and more spread out because some water travels short distances and other water travels long distances.

C What Are Tributaries and Drainage Networks?

Rivers and streams have a main channel fed by smaller subsidiary channels called *tributaries*. Each tributary drains part of the larger drainage basin, but a tributary can have higher flows than the main river. The combination of tributaries and the main river forms a *drainage network*. The response of a river to precipitation is influenced by the number and size of its tributaries.

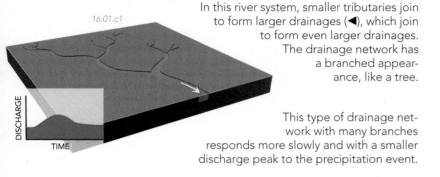

16.01.c1

In this river system, smaller tributaries join to form larger drainages (◄), which join to form even larger drainages. The drainage network has a branched appearance, like a tree.

This type of drainage network with many branches responds more slowly and with a smaller discharge peak to the precipitation event.

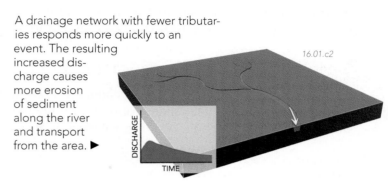

16.01.c2

A drainage network with fewer tributaries responds more quickly to an event. The resulting increased discharge causes more erosion of sediment along the river and transport from the area. ►

D How Does Geology Influence Drainage Patterns?

The patterns that river systems carve across the land surface are strongly influenced by the geology. Channels form preferentially in weaker material and so reflect differences in rock type and the geometry of folds, faults, joints, and other structural features. There are a number of drainage types, including the three shown below.

16.01.d1

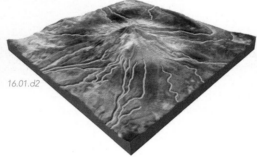

16.01.d2

16.01.d3

Dendritic Drainage Pattern—Where rocks have about the same resistance to erosion, or if a drainage network has operated for a long time, rivers can form a treelike, or dendritic, drainage pattern.

Radial Drainage Pattern—On a fairly symmetrical mountain, such as a volcano or resistant pluton, drainages flow downhill and outward in all directions (i.e., radially) away from the highest area.

Structurally Controlled Pattern—Erosion along faults, joints, or tilted and folded layers can produce a drainage that follows a layer or structure, and then cuts across a ridge to follow a different feature.

North American Drainages

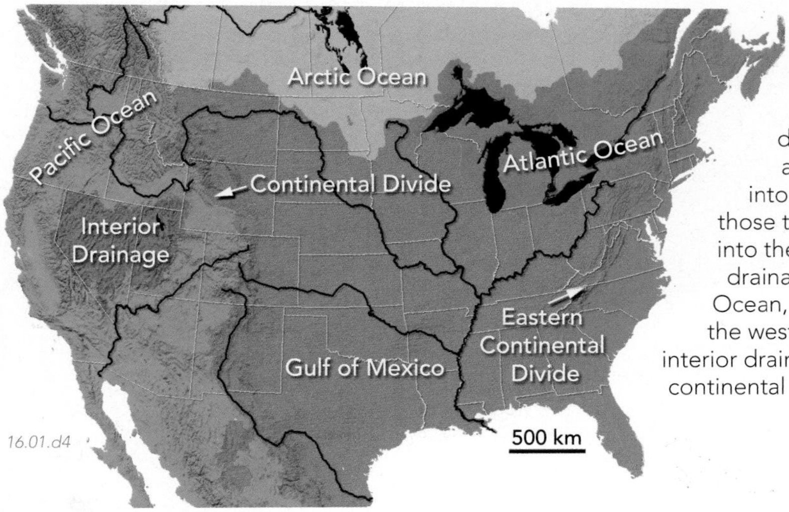

Colors on this map to the left show areas of the land that drain into different parts of the sea. Boundaries between colors are drainage divides, the best known of which is the continental divide, separating drainages that flow westward into the Pacific Ocean from those that flow east and south into the Gulf of Mexico. Other drainages flow into the Arctic Ocean, and some drainages in the western United States have interior drainage (they flow into low continental areas and do not reach the sea).

16.01.d4

500 km

Before You Leave This Page Be Able To

☑ Sketch and describe the variables plotted on a hydrograph and what this type of graph indicates.

☑ Describe how the shape and slope of a drainage basin affect discharge.

☑ Sketch or describe how the distribution of tributaries influences a river's response to precipitation.

☑ Sketch three kinds of drainage patterns, and discuss what controls each type.

16.1

16.2 How Do Rivers Transport Sediment and Erode Their Channels?

RIVERS AND STREAMS ERODE BEDROCK and loose material, transporting these as sediment and as chemical components dissolved in the water. The sediment is deposited when the river or stream can no longer carry the load, such as when the current slows or the sediment supply exceeds the river's capacity.

A How Is Material Transported and Deposited in Streambeds?

Moving water applies force to a channel's bottom and sides and is able to pick up and transport particles of various sizes: clay, silt, sand, cobbles, and boulders. The amount of sediment carried by the river, including material chemically dissolved in solution, is the *sediment load*.

1. Fine particles (silt and clay, collectively referred to as mud) can be carried *suspended* in the moving water, even in a relatively slow current. This material is the *suspended load*.

2. Sand grains can roll along the bottom or be picked up and carried down-current by bouncing along the streambed—the process of *saltation*.

3. Larger cobbles and boulders generally move by rolling and sliding, a process called *traction*, but only move during times of high flow. The largest of these clasts can be briefly picked up, but only by extremely high flows.

4. Material that is pushed, bounced, rolled, and slid along the bed of the river is the *bed load*. If the amount of sediment exceeds the river's capacity to carry it, like when velocity drops, the sediment is deposited. The balance between transport and deposition shifts as conditions change, and grains are constantly picked up and deposited again.

5. Some chemically soluble ions, such as calcium and sodium, are dissolved in and transported by the moving water. They constitute the *dissolved load*.

16.02.a1

B What Processes Erode Material in Rivers and Streams?

Moving water and the sediment it carries can erode bedrock, sediment, or other material that it flows past. Erosion occurs along the base and sides of the channel and can fragment and remove sediment within the channel. The silt, sand, and larger clasts carried by the water enhance its ability to erode.

1. Sand and larger clasts are lifted by low pressure created by water flowing over the clast tops. They can also be pushed up by turbulence. Once picked up, the grains move downstream and collide with obstacles, where they chip, scrape, and sandblast pieces off the streambed, the process of *abrasion*. Abrasion is concentrated on the upstream side of obstructions, such as larger clasts or protruding bedrock.

4. Soluble material in the streambed, such as salt, can be removed by *dissolution*. Most dissolved material in streams, however, comes from groundwater that has leaked into the stream. Other dissolved material comes from soluble rock layers along the river and from farms.

16.02.b1

16.02.b2 Adirondack Mtns., NY

◄ **2.** Concentrated erosion can also occur when water and sediment swirl in small depressions, carving bowl-shaped pits called *potholes*.

3. Turbulent flow loosens and lifts material from the streambed, especially pieces loosened by fractures, bedding planes, and other discontinuities.

C How Does Turbulence in Flowing Water Affect Erosion and Deposition?

Water, like all fluids, has *viscosity*—resistance to flow. Viscosity and surface tension are responsible for the smooth-looking surface of slow-moving streams and rivers. As the water's velocity increases, the flow becomes more chaotic or *turbulent*, and the water can pick up and move material within the channel.

1. All streams and rivers have turbulent flow to some degree. Even the smooth-looking segments of a stream have turbulence, but less than fast-moving segments. In very slow-moving streams, the viscosity limits more chaotic flow (turbulence).

2. Moving water has inertia and so tends to keep moving in the same direction unless its motion is perturbed, like where the water encounters a bend in the river. In many cases, water is able to flow smoothly over somewhat uneven surfaces.

3. As water velocity increases, viscosity is less able to dampen chaotic flow, and the water flow becomes more complex, or turbulent. As turbulence increases, swirls in the current, called *eddies*, form in both horizontal and vertical directions.

4. Fast-moving water has numerous eddies where flow is not downstream.

5. Near the bottom of the river, upward-flowing eddies can overwhelm gravitational force and lift grains from the channel. Turbulence, in general, increases the chance for grains to be picked up and carried in the flow.

16.02.c1

D How Do Erosion and Deposition Occur in Streams Confined Within Bedrock?

Many rivers and streams, especially those in mountainous areas, are carved into bedrock, and are referred to as *bedrock streams*. If the bedrock is relatively hard, the shape of the stream channel is controlled by the geology. If bedrock consists of softer material, such as easily eroded shale, then the material will have less control over the shape and character of the stream channel.

Erosion

1. The steep gradients and higher velocities typical of mountain streams (▶) erode down into the channel faster than the stream can erode down into its sides. The bed load of sand, cobbles, and boulders helps break up and erode the bedrock channel. Rapid changes in gradient, as occur along waterfalls and rapids, increase water velocity, turbulence, and the ability of the moving water to erode the channel.

16.02.d1

Deposition

16.02.d2 Grand Canyon, AZ

3. Deposition in bedrock channels occurs where the water velocity decreases, for example along the river banks during flooding or in pools behind rocks or other obstacles. Rocks and sediment constrict this river (◀), forming a pool of less turbulent water upstream. During floods, sediment is deposited in slow-moving eddies on the flanks of this pool, but such sediment is vulnerable to later erosion and is therefore very transient.

2. As a result, steep bedrock streams commonly incise deep channels (▶). They can have relatively straight sections, initially controlled by the location of softer rock types, faults, or other zones that are more easily eroded than surrounding rocks. Once formed, such hard-walled canyons make it difficult for the stream to shift its position.

16.02.d3

Before You Leave This Page Be Able To

☑ Sketch and describe how a river or stream transports solid and dissolved material.

☑ Sketch and explain the processes by which a river or stream erodes into its channel and which sites are most susceptible to erosion.

☑ Sketch and describe turbulent flow.

☑ Describe some aspects of erosion and deposition in bedrock channels.

16.2

16.3 How Do River Systems Change Downstream or Over Short Time Frames?

RIVER SYSTEMS BECOME LARGER as more tributaries join the drainage network. As a river flows downstream, it generally increases in size, discharge, and the amount of sediment it carries. A river changes over short time spans, for example after a storm, from winter to summer, and from year to year.

A How Do River Systems Change Downstream?

The character of a river changes aspects as the river flows downhill from its *headwaters*, where it starts, to its *mouth*, where it ends. The flow of a river from high elevations to lower ones is referred to as being *downstream*.

Gradient

1. The profile of most river systems is steep in the headwaters, gradually becoming less steep downstream toward the mouth. The steepness is also called the *gradient*, which is defined as the change in elevation for a given horizontal distance.

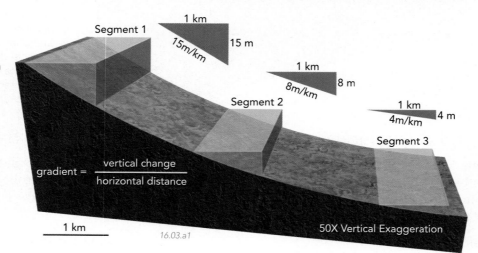

$$\text{gradient} = \frac{\text{vertical change}}{\text{horizontal distance}}$$

16.03.a1

2. This change in gradient downstream is represented by the blue triangles, which show how much the river drops for a given length of river. A steeper gradient means the river drops more over the same horizontal distance. We express gradient as meters per kilometer, feet per mile, degrees, or as a percentage (e.g., 4%). Here, gradient is calculated for three segments. It varies from 15 m/km to 4 m/km and decreases downstream. The vertical scale of the triangles is not the same as the horizontal scale.

Channel Size, Water Velocity, Discharge, and Sediment Load

3. Rivers erode bedrock and other materials and then transport the sediment down the river. Sediment can be deposited anywhere along the way or can be carried all the way to the mouth of the river. The river system shown here has a main river fed by three main tributaries, labeled T1, T2, and T3. Small graphs around the map plot how parameters change down the river, from the headwaters (H), past each tributary (T1, T2, and T3), to the start of a delta (D), and the river's mouth (M).

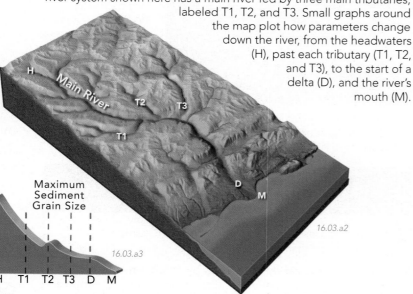

Maximum Sediment Grain Size

16.03.a2

16.03.a3

H T1 T2 T3 D M

▲ **4.** As the gradient of the river decreases from the headwaters to the mouth, the *maximum size of sediment* that the river carries decreases. Also, abrasion during transport reduces the size of clasts. For both reasons, coarse material is more common in the headwaters than it is near the mouth.

16.03.a4

Channel Size

H T1 T2 T3 D

◀ **5.** There is an increase in the overall *size of the channel*, as represented by a cross section from side to side across the channel. Specifically, size means the *cross-sectional area* of the channel, obtained by multiplying the channel's width times its average depth.

16.03.a5

Water Velocity

H T1 T2 T3 D

◀ **6.** The *velocity of water flow* increases downstream, as a higher volume of water allows the water to flow more easily and faster through the channel. Velocity looks higher in steep streams, but turbulence slows the water's overall speed.

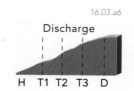

16.03.a6

Discharge

H T1 T2 T3 D

◀ **7.** Since the cross-sectional area of the channel and velocity of the water both increase, so does the total *volume of water* flowing through the river. The volume of water flowing through any part of the river per unit of time is the discharge and is calculated by multiplying the velocity times the cross-sectional area.

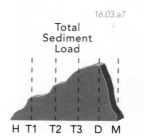

16.03.a7

Total Sediment Load

H T1 T2 T3 D M

◀ **8.** The total amount of sediment that the river is carrying, the *sediment load*, increases downstream, until large amounts of sediment begin to be deposited within the delta and near the mouth of the river.

B What Is the Relationship Between Water Flow and Transported Sediment?

A river or stream can carry sediment only up to a certain size for any particular velocity. Also, at a given flow rate, a river is capable of transporting only a certain amount of sediment, which is called its *capacity*. Normally, a river is carrying far less sediment than its capacity. As velocity decreases, so does capacity—a river is able to carry less sediment as it slows down.

▶ This graph shows the relationship between stream velocity and the size of the particles that can be carried by different modes of transport. The vertical bands of color indicate different grain sizes, and the inclined lines indicate whether sediment of that size is being carried in suspension, is being transported on the bottom of the riverbed, or is being deposited.

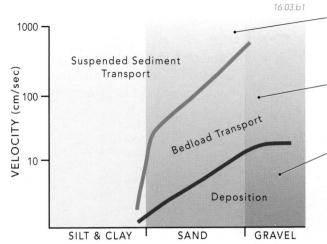

16.03.b1

At high velocities (above 100 cm/sec), clay, silt, and sand can be carried *suspended* (floating and drifting) in the water. Those grain sizes extend above the red line.

At moderate velocities (100 cm/sec), silt, clay, and fine sand remain suspended, but coarse sand and gravel slide, roll, or bounce along the riverbed, as part of the bed load.

At low velocities (below 10 cm/sec), gravel and sand remain at rest on the riverbed or are deposited if a sediment-carrying river slows down to these velocities. These grain sizes plot below the blue line. Only silt, clay, and fine sand are transported, with silt and clay in suspension and fine sand in the bed load.

C How Does River Behavior Vary Over Time?

The amount of precipitation, snowmelt, and influx from springs and groundwater varies, both during a single year and over longer timescales of decades to centuries. Here, we examine the flow of a river during a single year.

1. The amount of water flowing in a river or stream can vary throughout the year. Examine the hydrograph to the right (▶) and think about some possible explanations for this pattern. As an example, for this river and for the year shown, why would the highest discharge occur during springtime? Other rivers have other patterns than shown here, but this one is typical for rivers in some parts of the United States.

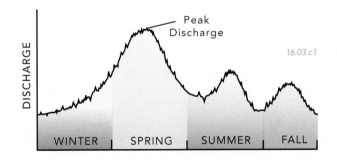

16.03.c1

2. The lowest discharge is in the winter, when some streams and lakes are frozen and precipitation occurs as snow and as gentle winter rains. Discharge is highest during spring, when the snow and ice melt. The highest value on the plot is the *peak discharge*.

3. After the most intense period of snowmelt, discharge in the river decreases throughout the rest of the spring and into the early summer. An increase in rainfall due to a yearly summer-thunderstorm season causes a short-lived increase in discharge in the middle of the summer. For this river's drainage area, the fall season typically has less precipitation than other times of the year. For the year shown, however, moisture from the tropics moved far enough north to cause a period of on-and-off rain showers.

4. This scenario is one possible explanation for a hydrograph with this shape. Try to find a hydrograph for a stream or river in your area, observe the pattern, and try to explain the increases and decreases in discharge over a year.

16.03.c2 Flagstaff, AZ

5. A stream or river that flows all year, like the one represented by the graph to the left, is a *perennial stream*. Because no place has rain all of the time to keep a stream flowing, some water in a perennial stream must be supplied by groundwater flow into springs, by a melting snowpack, by a lake, or by some combination. Some streams do not flow during the entire year, but only during rainstorms and spring snowmelt. Such a stream is an *ephemeral stream*.

Before You Leave This Page Be Able To

☑ Describe and sketch how to calculate a gradient for a river.

☑ Describe how gradient and other parameters change downstream.

☑ Describe how velocity relates to sediment size and capacity.

☑ Describe why discharge might change from season to season.

16.3

16.4 What Factors Influence Profiles of Rivers?

RIVER SYSTEMS HAVE DIFFERENT GEOMETRIES, both in map view and when viewed from the side (in profile). Rivers have diverse settings, origins, and ages, and they respond to perturbations in their environment by eroding their channels and banks, by depositing sediment, and by changing their gradient.

A What Is the Shape of a River's Profile?

Rivers are *dynamic systems* driven by precipitation and gravitational forces. They respond to many factors that influence how the river operates and how it interacts with its channel and the adjacent landscape. Over time, most rivers attain a profile that is steeper near the headwaters and is progressively less steep downstream.

1. The idealized profile of a river is represented by the side of this block. The profile is steeper (has a higher gradient) near the headwaters of the river.

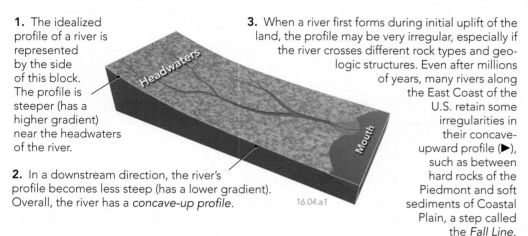

16.04.a1

2. In a downstream direction, the river's profile becomes less steep (has a lower gradient). Overall, the river has a *concave-up profile*.

3. When a river first forms during initial uplift of the land, the profile may be very irregular, especially if the river crosses different rock types and geologic structures. Even after millions of years, many rivers along the East Coast of the U.S. retain some irregularities in their concave-upward profile (▶), such as between hard rocks of the Piedmont and soft sediments of Coastal Plain, a step called the *Fall Line*.

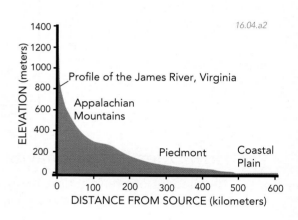

16.04.a2

B What Controls the Profiles of Streams and Rivers?

Rivers and other processes erode mountains and carry the sediment downhill, eventually depositing it in a basin or along the sea. The lowest level to which a river can erode is its *base level*. The base level controls the topography along a river, how a river develops over time, and how it responds to change.

1. This terrain shows a typical drainage system consisting of mountainous headwaters, mid-elevation foothills, and a broad, low-elevation plain, ending at a shallow inland sea.

2. High above base level, steep gradients in the mountains cause streams to erode sharply into the bedrock. The terrain appears rough and may include deep canyons cut into bedrock.

3. Foothills in front of the mountains also experience erosion but have intermediate gradients and generally appear less rough.

4. Closer to base level, rivers and streams on the broad plain have a much lower gradient, and the surrounding landscape has less relief and appears relatively smooth. This plain has low relief because either it has been eroded down or its low parts have filled with sediment. In this case, it is some of both.

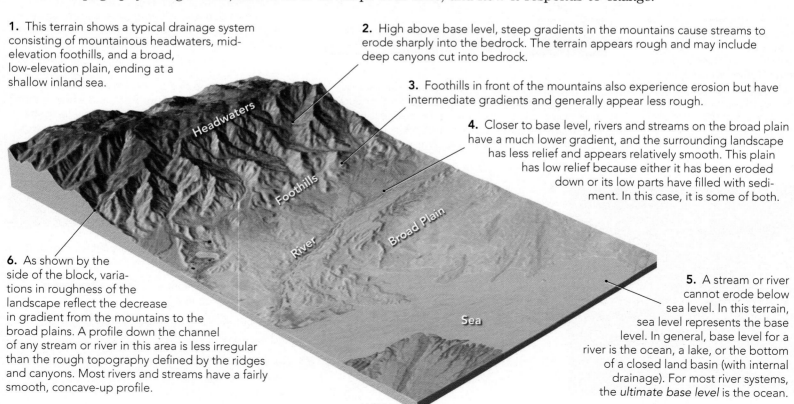

16.04.b1

6. As shown by the side of the block, variations in roughness of the landscape reflect the decrease in gradient from the mountains to the broad plains. A profile down the channel of any stream or river in this area is less irregular than the rough topography defined by the ridges and canyons. Most rivers and streams have a fairly smooth, concave-up profile.

5. A stream or river cannot erode below sea level. In this terrain, sea level represents the base level. In general, base level for a river is the ocean, a lake, or the bottom of a closed land basin (with internal drainage). For most river systems, the *ultimate base level* is the ocean.

C What Factors Influence or Change Stream Profiles?

Streams and rivers generally do not achieve equilibrium because some rocks are more difficult to erode than others and because Earth is a dynamic planet, with frequent changes in tectonics, sea level, and climate.

Rock Type

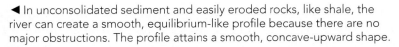

As a river flows over different kinds of rocks, its ability to erode a channel is influenced by the type of rock over which it flows. Soft rocks erode more easily than hard rocks.

◄ In unconsolidated sediment and easily eroded rocks, like shale, the river can create a smooth, equilibrium-like profile because there are no major obstructions. The profile attains a smooth, concave-upward shape.

16.04.c1

► Rocks that are more resistant to erosion will tend to form steeper slopes, with cliffs, waterfalls, steep rapids, and narrow canyons. Alternating strong and weak rocks yields a stair-stepped topography, but through time a river can smooth out its profile.

16.04.c2

Tectonics

Tectonic forces can cause *uplift* or *subsidence* of an entire region or can occur differentially, affecting one part of the region more than other parts.

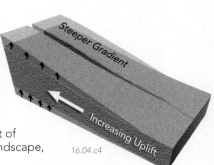

◄ Differential subsidence can flatten or steepen gradients, depending on where it occurs. In this example, subsidence occurred beneath the mountains, flattening the gradient and causing widespread deposition as stream velocity decreased and the river lost capacity.

16.04.c3

► Tectonic uplift generally causes rivers and streams to erode down into the landscape, cutting canyons and steepening the topographic relief. Here, tectonic uplift of the mountains steepened the gradient, causing erosion to cut into or *incise* the landscape, forming a narrow canyon.

16.04.c4

Sea Level

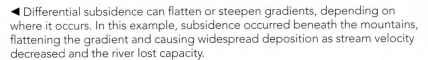

Sea level is the ultimate base level for rivers that empty into the ocean. Changes in sea level will change the location of the shoreline and the elevation of base level.

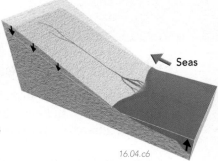

◄ If the base level is lowered, such as by a drop in sea level, the river will downcut to try to match the new base level. In this example, erosional incision begins at the coast and works its way upstream. A drop in sea level causes most rivers to downcut.

16.04.c5

► If the base level rises, such as during a rise in sea level, the river will erode inland but deposit sediment along the coastline's new position. In this manner, the river tries to achieve a new equilibrium profile.

16.04.c6

Climate

Rivers respond to changes in climate, especially an increase or decrease in rainfall. Under wet conditions, slopes will have more vegetation and so they can hold soil, but increased discharge allows streams to carry sediment away, beveling the hills more than during dry periods.

16.04.c7

Stability of Conditions

If conditions such as climate remain stable, a river may approach an *equilibrium profile*. When a river is in steady state, there is a balance between the supply of sediment and the amount the river can carry. The channel becomes stable, or nearly attains a type of *dynamic equilibrium*, neither eroding nor depositing material. We call such a stream or river a *graded stream*.

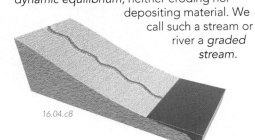

16.04.c8

Before You Leave This Page Be Able To

☑ Sketch and describe the typical profile of a river.

☑ Describe the concept of base level and how it is expressed in a typical mountain-to-sea-landscape.

☑ Summarize factors that influence a river's profile and behavior.

16.4

16.5 Why Do Rivers and Streams Have Curves?

ALL RIVERS HAVE CURVES OR BENDS, ranging from gentle deflections to tightly curved but graceful meanders. Why are rivers curved? What is inherent in the operation of a river that causes it to curve? Curves and bends are unavoidable because of processes that accompany the movement of water in a river.

A What Is the Shape of River and Stream Channels in Map View?

All rivers have curves or bends, but not all bends are the same. Some are gentle, open arcs, where the river veers slightly to one side and then the other, whereas others are tight loops. The shape of a river in map view can be thought of as having two main variables: whether there are single or multiple channels and how curved the channels are. The amount that a channel curves for a given length is its *sinuosity*.

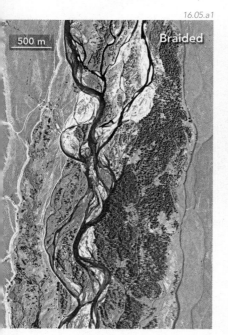

16.05.a1

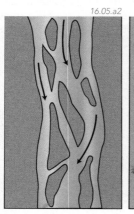

16.05.a2

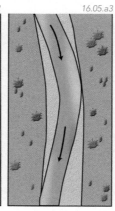

16.05.a3

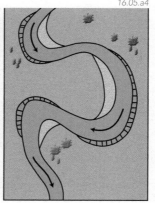

16.05.a4

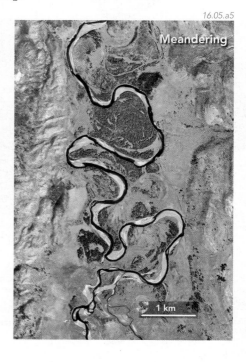

16.05.a5

▲ *Braided rivers* are characterized by a network of interweaving, sinuous channels, but the overall channel can be fairly straight.

▲ Many rivers consist of a single channel that is gently curved. This type of river is referred to as having *low sinuosity*.

▲ *Meandering rivers* have channels that are very curved, commonly forming tight loops. Such rivers have *high sinuosity*, and this type of bend is a *meander*.

B What Processes Operate When a River Meanders?

River channels in soft materials, especially rivers with low gradients, generally do not have long, straight segments. Instead they flow along sinuous paths. Curves or meanders cause differences in water velocity in the channel and reflect a balance between deposition and erosion, as illustrated below for a meandering river.

Small graphs show profiles across the channel in different locations. Arrows in the channel show relative velocity. In fairly straight segments, the channel is nearly symmetric (not deeper on one side than the other). The current is fastest in the center of the channel and slowest along the banks. In such straight segments, sediment can be deposited along the channel margins where velocity is lowest, and erosion can occur in the middle of the channel where velocity is highest.

Where the river is curved, the channel becomes asymmetric (is shallower on one side than the other). The channel is shallower and the water velocity is lower on the inside of a bend. This causes sediment to be deposited on the inside of the bend in what is called a *point bar*.

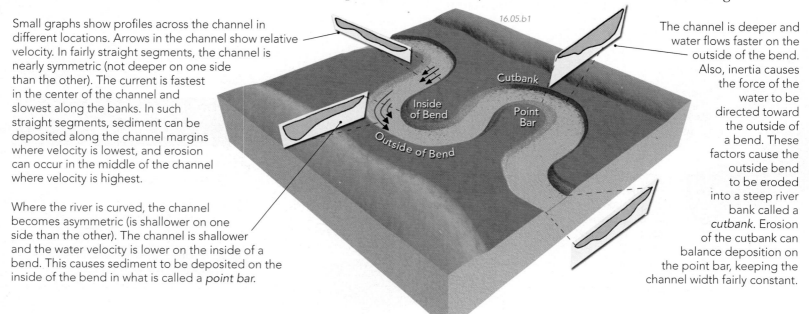

16.05.b1

The channel is deeper and water flows faster on the outside of the bend. Also, inertia causes the force of the water to be directed toward the outside of a bend. These factors cause the outside bend to be eroded into a steep river bank called a *cutbank*. Erosion of the cutbank can balance deposition on the point bar, keeping the channel width fairly constant.

C How Do Meanders Form and Migrate?

Meanders are landforms produced by migrating rivers and are extremely common in rivers that have low gradients. Meanders have been extensively studied in the field and simulated in large, sand-filled tanks. In the laboratory, water is initially directed down a straight channel in fine sand. Almost immediately, the water begins to transform the straight channel into a sinuous one, similar to the sequence shown below.

1. A curve starts to form when a slight difference in roughness on the channel bottom causes water to flow faster on one side of the channel than on the other.

2. The side of the channel that receives faster flow erodes faster, creating a slight curve. The faster moving current slightly excavates the channel bottom, deepening the outside of the bend, forming deeper areas called *pools*.

3. The overall discharge in the river is constant, so the deeper channel on the outside of a bend takes more water, leaving less water for the other side. The water on the inside of the bend becomes shallower and slower.

4. The sediment carried by the slower water on the inside of the bend is dropped and deposited on a *point bar*.

5. Erosion scours the opposite (outside) band of the channel, forming a *cutbank*.

6. Through this process, each meander begins to preferentially erode its banks toward the outside. This causes the river to migrate toward the sides and downstream, as shown by the small orange arrows.

7. Once formed, a curve continues to affect the flow by causing faster flow and increased erosion on the outside of the bend. Some secondary currents develop in the bend area and further excavate the pools, speeding flow and enhancing the cutbank.

8. As meanders migrate back and forth across the lowlands, they continuously erode and deposit the loosely bound sediment in the floodplain and channel. This is the main way in which a floodplain forms, and the old meanders remain as *scars* on the floodplain.

9. Meanders migrate until they encounter a resistant riverbank, until the volume and velocity of flow drop too low for erosion to continue, or until two parts of a meander intersect.

10. Meanders sometimes join as they migrate toward each other, in the direction of the yellow arrows. This cuts off the meander.

11. The narrow neck of a looping meander can also get cut off during a flood event, when the river rises above the channel and across the floodplain, connecting two segments of the river. In either case, the part of the meander that is abandoned is a *cutoff meander*.

12. Cutoff meanders formed in either way (10 or 11) are initially filled with water, forming isolated, curved lakes, called *oxbow lakes*.

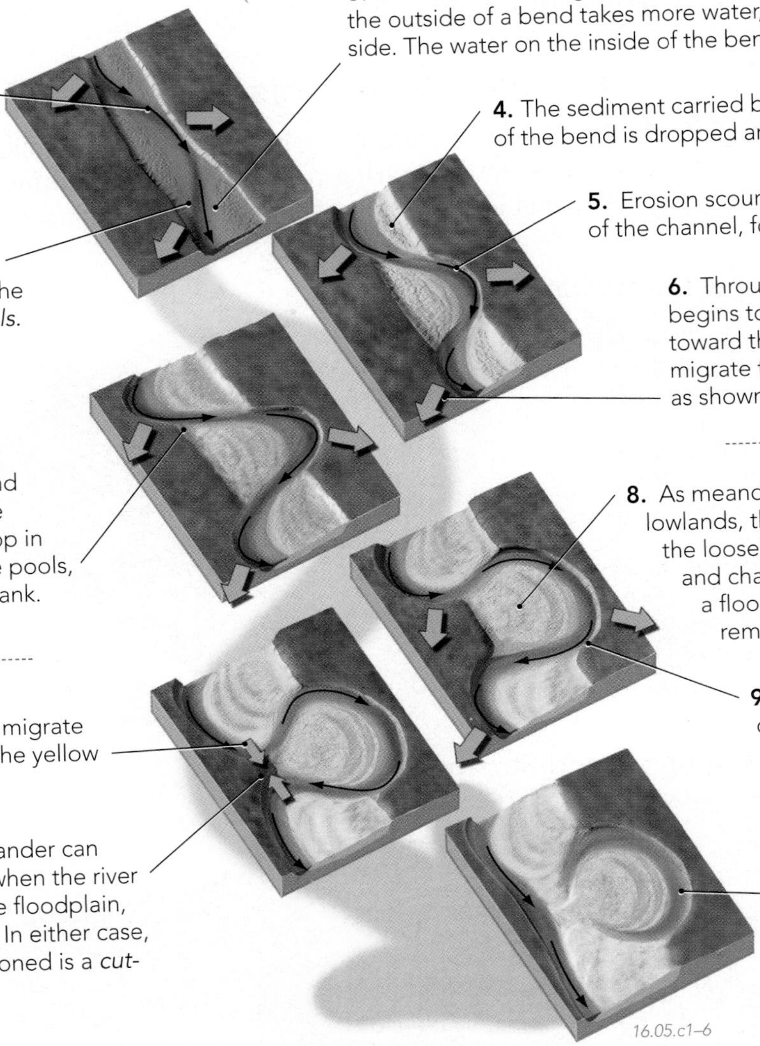

16.05.c1–6

Messing with Sinuosity

Rivers and streams have attained their characteristic sinuosity through natural processes. Their sinuosity represents the interplay between variations in channel depth, water velocity, erosion, deposition, and transport of sediment. In many cases, humans upset this balance by straightening rivers and eliminating their natural variability. These engineering solutions often cause trouble downstream because they upset the dynamics and equilibrium of the system. Rivers that have been channelized may exit the channelized segment with a higher velocity, lower sinuosity, and less sediment than is natural. Areas downstream of the channelized segment, therefore, can experience extreme erosion and destruction of riverbank property.

16.05.mtb1 Courmayeur, Italy

Before You Leave This Page Be Able To

✓ Sketch and describe the difference between braided, low-sinuosity, and high-sinuosity (meandering) rivers.

✓ Sketch or describe how flow velocity and channel profile vary in a meandering river, and what features form along different parts of bends.

✓ Sketch or describe the evolution of a meander, including how a cutoff meander forms and how it can lead to an oxbow lake.

16.5

16.6 What Features Characterize Mountain Rivers and Streams?

MOST LARGE RIVER SYSTEMS originate in mountains and are fed by rain, snowmelt, and springs. Mountain streams are steep and actively erode the land with turbulent, fast-moving water. Such erosion produces steep-sided, narrow channels and other landforms that reflect this high-energy environment.

A What Landforms Characterize the Headwaters of Rivers and Streams?

Mountain river systems begin in bedrock-dominated areas with relatively high relief and, in many cases, high elevation. In such settings, moving water is energetic, wearing rock down and sculpting the bedrock into landscapes with moderate to high relief. Steep streams and rivers are capable of carrying sediment out of the mountains.

Channel Formation

16.06.a1 Norway

◄ As water flows over the surface, it accumulates in natural cracks and low spots, such as these small channels, rather than spreading uniformly across the land.

Concentrated flow erodes or dissolves materials, especially those that are weak or loose, eventually carving a small channel or *gully*.

Once formed, a channel captures additional runoff within its small drainage basin, and the increased flow leads to further erosion and deepening of the channel.

Channels occur at all scales. Microscopic channels feed into small channels that feed into larger ones, ultimately forming a stream. The spacing and geometry of the channels is influenced by the steepness of the slope, type of material in the slope, and other factors.

16.06.a2

Landforms in the Headwaters of Rivers and Streams

The place where a river system begins is its *headwaters*. Some streams begin in high mountainous areas from rainfall, melting ice and snow, or mountain springs. Others originate in lower, flatter areas and are supplied by precipitation, lakes, springs, or the joining of small, local channels.

A *waterfall* forms when a stream's gradient is so steep that water cascades over a cliff or ledge. Cliffs and ledges typically develop where a hard, erosion-resistant rock type impedes downcutting by the stream. ▼

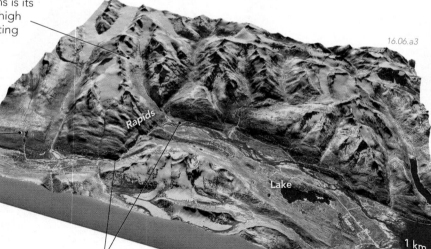

16.06.a3

Rapids

Lake

1 km

16.06.a4 Ice Lake, San Juan Mtns., CO

▲ Lakes are common in mountains where water is impounded by some obstruction, like a landslide, or where water fills a natural low spot. If a lake is created by a human-constructed dam, it is a *reservoir*.

16.06.a5 Gullfoss, Iceland

A *rapid* is a segment of rough, turbulent water along a stream or river. Most rapids develop when the gradient of a river steepens or the channel is constricted by narrow bedrock walls, large rocks, or other debris that partially blocks the channel. Many rapids form where tributaries have deposited fans of debris that crowd or clog the main channel. These obstructions cause water to flow chaotically over and around obstacles, creating extreme turbulence and big rapids. ►

16.06.a6 Grand Canyon, AZ

B What Landforms Form Along Mountain Rivers and Streams?

As mountain rivers flow toward lower elevations, they interact with tributaries and commonly decrease in gradient as they pass through foothills or mountain fronts. In response, they form other types of landforms.

16.06.b2 Buckfarm Canyon, AZ

1. Many mountain streams and rivers cut down into, or *incise*, bedrock. Early in their history, many rivers incise steep-walled notches and canyons. A canyon is narrow if downward incision is faster than widening of the canyon by erosion along tributaries and by landslides and other types of slope failure.

16.06.b1

250m

▲ **2.** This narrow canyon is cut into limestone layers that are resistant to erosion in a dry climate.

16.06.b3 Mt. Nebo, Wasatch Front, UT

▲ **5.** Where a steep, narrow drainage enters a broader, more gentle valley, coarse sediment carried by running water or by muddy debris flows piles up just below the mouth of the drainage, forming an *alluvial fan*. Deposition occurs here because of the decrease in gradient and the less confined nature of the channel, both of which decrease the velocity of moving water and mud.

16.06.b4 Grand Canyon, AZ

◄ **3.** Side tributaries play a key role in the downstream variations in the gradient, morphology, and turbulence of a mountain river. Tributaries carry sediment and deposit some of it where the tributary and main drainage meet. This sediment can constrict the channel, causing a rapid at the constriction and backing up and slowing water above the rapid, forming a pool.

4. When they reach less confined spaces, mountain streams and rivers commonly spread out in a network of sediment-filled *braided channels* (▶). These channels are not strongly incised, so the river spreads out and deposits sediment along its channel and over a broad plain.

16.06.b5 Waiapu River, New Zealand

How Do Mountain Streams Get Sediment?

Mountain rivers and streams are energetic primarily because their channels have steep gradients. Erosion dominates over deposition, forming deep *V-shaped valleys* with waterfalls and rapids. Steep valley walls promote landslides and other types of slope failure that widen the canyon and deliver material to the river for removal (▶). Soil on hillslopes can slide downhill toward the drainage. Tributaries carry debris flows and floods that scour their channels, providing more sediment. Sediment in mountain streams ranges from car-sized boulders down to silt and clay. Larger clasts start out angular but begin to round as they are transported, or as they are struck by stones and other sediment within the turbulent waters.

16.06.mtb1 Tibet-Nepal Border Region

Before You Leave This Page Be Able To

✓ Describe how channels form.

✓ Describe some of the landforms associated with the headwaters of mountain rivers and streams.

✓ Describe what conditions result in a narrow canyon.

✓ Describe why sediment is deposited along mountain fronts in alluvial fans.

✓ Describe how mountain streams get their sediment.

16.6

16.7 What Features Characterize Braided Rivers?

MANY RIVERS AND STREAMS ARE BRAIDED SYSTEMS, with a network of channels that split and rejoin, giving an intertwined appearance. Braided rivers generally have a plentiful supply of sediment and steep to moderate gradient, and typically carry and deposit coarse sediment. Braided rivers can migrate across broad plains, coating them with a veneer of sediment.

A What Conditions Lead to Braided Streams and Rivers?

Braided streams and rivers are most common in flat-bottomed valleys nestled within mountains and on broad, sloping plains that flank such mountain ranges. They can also form farther from the mountains, in areas where an abundant sediment supply nearly overwhelms the river's capacity to carry it.

Many braided rivers drain from high mountains, such as these, modeled after the South Island of New Zealand.

The Southern Alps of New Zealand are an actively uplifting and steep range. Glaciers, steep slopes, and locally heavy precipitation in the head-waters of the rivers contribute abundant sediment to the streams and rivers. ▶

16.07.a2

16.07.a1

Braided rivers deposit sediment within and beside their shallow channels and can escape their channels, especially during floods. Sediment in the riverbank is not cemented or otherwise tightly held together, so the material is easy to erode and redistribute, and the river can change position relatively easily. As the channels migrate back and forth across the broad plain, they cover the broad, low-relief area with a layer of coarse, river-deposited sediment. ▼

16.07.a3 Denali NP, AK

Braided streams form where there are steep gradients, a plentiful supply of coarse sediment, and conditions that produce variable flows. In this close-up view, individual channels are braided at various scales, but the overall path of the river channel is fairly straight. ▼

16.07.a4

B What Type of Sediment Does a Braided River Deposit?

Braided rivers are typically characterized by a wider range of sediments than is deposited in meandering rivers. Braided rivers are energetic and can carry and deposit coarse gravels and sands in addition to finer materials.

16.07.b1 New Zealand

16.07.b2 Waiapu River, New Zealand

16.07.b3 Southern Tibet

Braided rivers have numerous braided channels. They are clogged with sediment, which is constantly picked up in one place and deposited in another. As sediment is picked up, transported, and deposited, the braided channels continuously change position, width, and overall shape. They change more slowly during low flows and more rapidly during floods.

Braided rivers form when the river has a relatively high sediment load dominated by sand and larger sediment. Sand and gravels are the dominant clasts in this braided river, but some braided rivers also carry finer materials, such as mud and silt derived from glaciers and other sources. Overall, braided rivers are relatively mud-poor, especially when compared with rivers that have meanders.

This braided-river plain in Tibet contains large, partially rounded boulders, cobbles, and pebbles in a sand-dominated matrix. The river can transport these large clast sizes because it has a steep gradient and carries large amounts of turbulent water during monsoon storms and when warm temperatures cause melting of snow in the foothills of the Himalaya range (in the distance).

Making and Investigating Braided Rivers in the Laboratory

One way geologists and engineers study rivers is to make small-scale versions or models in large water tanks in a laboratory. These tanks can be several meters wide and tens of meters long, and they are sloped so that the water flows downhill. The tanks are loaded with sediment, usually sand, silt, and mud, but sometimes glass beads or other materials. Valves are opened to allow water to enter the high side of the tank and flow toward the low end. Geologists then observe the small-scale river that develops, investigating the processes that occur and the features that form. Different variables, including slope, sediment supply, and consistency of flow, can be specifically varied or controlled to isolate how each factor affects the dynamics of the river system.

The sequence of images here shows successive stages during an experiment in a 2×15 m tank at the National Center for Earth-surface Dynamics in Minneapolis. In this experiment, a braided river developed early on (far left), but became progressively less braided as alfalfa seeds embedded in the sediment sprouted and grew more dense. These experiments indicate that riverside vegetation plays a key role in stabilizing river banks, and it can actually influence whether a river remains braided. This relationship was first recognized more than 60 years ago by geologists working along rivers of the United States.

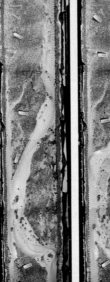

16.07.mtb1–6

Before You Leave This Page Be Able To

✔ Describe the characteristics and setting of braided rivers and streams.

✔ Describe the types of sediment that braided rivers carry and deposit.

✔ Describe how and why river processes are investigated in laboratory tanks.

16.7

16.8 What Features Characterize Low-Gradient Rivers?

IF A RIVER SYSTEM CROSSES AREAS of low relief, the gradient of its channel decreases and the river may spread out once it is no longer confined by a narrow valley. Sediments transported and deposited on low-relief plains are mostly clay to sand size but can include fine gravels. The river reworks (picks up and transports) these previously deposited sediments. The resulting landforms reflect the interaction of river velocity and sediment size with the more gentle landscape.

A What Landforms Characterize Rivers with Low Gradients?

Many rivers flow across plains that have gentle overall slopes. Such rivers reflect their environs, being dominated by the erosion, transport, and deposition of relatively fine-grained sediment. The features characteristic of these single-channel rivers occur at all scales, from those along small creeks to those along the Mississippi River. Features include meanders, floodplains, and low river terraces. On these two pages, we explore two meandering rivers: the Animas River of Colorado and the Mississippi River of the central United States.

One Main Channel

Rivers on gentle plains usually occupy a single channel rather than being braided. This single-channel characteristic is linked to the gentle downstream gradient of the river and its floodplain. Notice that the low-gradient river here occurs on a gentle plain within a mountainous region, so it is important to focus on the characteristics of the river rather than its surrounding environment. Farther upstream, this river is confined to a narrow and deep bedrock canyon.

Meanders

Rivers on gentle plains typically flow in dramatically curved paths. The degree to which the single channel is curved varies from rare, straight segments to the sinuous curves of meanders.

Meander Scars and Oxbow Lakes

Meandering rivers leave behind arcuate *meander scars*, which are exposed as low, curved ridges, lines of vegetation, or curved dry or water-filled depressions. When such depressions contain water, they are *oxbow lakes*.

Floodplain

All rivers on gentle plains have floodplains beside the channel. Floodplains represent the area covered with water when the river floods out of its channel.

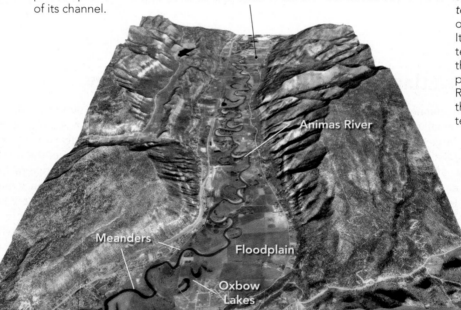

16.08.a1

River Terraces

Many rivers have older, stranded floodplains, called *terraces*, perched above and outside the current floodplain. It is common to find matching terrace levels on either side of the existing floodplain. This particular stretch of the Animas River lacks obvious terraces, but the river has well-developed terraces downstream.

Point Bars

Meandering rivers often have arcuate deposits of sand and gravel that parallel the inside bend of a meander. Point bars typically are visible as arcuate patches of bare sediment.

Scale

River channels, meanders, floodplains, and other features can occur at very different scales. Compare the two images to the right. The first is an aerial image of the same Animas River segment shown above. The second is a few meander loops on the Mississippi River. The images are at the same scale. The much smaller scale Animas River has 15 times more meanders than the Mississippi for the same downstream distance. The scale of the meanders is so different because of the huge differences in size and discharge of the rivers. The Mississippi is a huge river system with a much, much larger discharge.

16.08.a2–3

B How Do Meandering Rivers Traverse Their Floodplains?

Many major and smaller rivers meander across gentle plains, carrying large quantities of water and fine-grained sediment away from foothills or broad, low uplands. Meandering rivers, at some scale, are present in most low-relief regions.

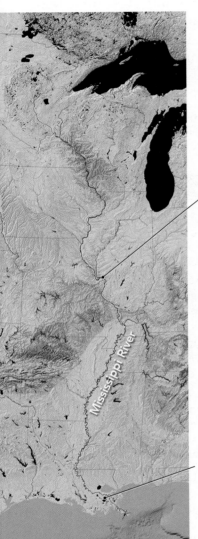

The meandering Mississippi River, shown in blue, begins in a lake in Minnesota and winds its way southward, across the center of the continent. Its length is not constant because of its shifting meanders, but it is about 3,700 km (2,300 mi) long.

From Minneapolis to the sea, a distance of about 2,900 km (1,800 mi), the river drops only 236 m (775 ft). This gives the river a very low gradient of less than 0.1 m/km.

At its mouth, the river deposits its load of sediment in a large delta southeast of New Orleans, Louisiana.

Mississippi River

100 km

16.08.b1

16.08.b2

N→
5 km

The very broad floodplain of the Mississippi River has countless crescent-shaped scars of ancient meanders, abandoned by the shifting of the river.

Many cut-off meanders are filled with water in curved oxbow lakes.

Formation of a Levee

Along the edges of many channels is a raised embankment, or *levee*. Natural levees are created by the river, and humans construct artificial levees to try to keep floodwaters from spilling onto the floodplain. It commonly works, at least for a while.

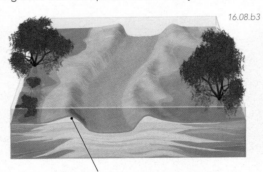

16.08.b3

16.08.b4

During flooding, sediment-carrying floodwater rises above the channel and begins to spread out. As it does, the current slows and so deposits sediment in long mounds next to and paralleling the channel.

When the flood recedes, sediment that was piled up next to the channel remains as an elevated ride or levee. Levees are barriers to water flow from the channel to the floodplain, and from the floodplain back into the channel after a flood.

Levees — Boon or Bust?

While the word *levee* likely leads to thoughts of flooding along the Mississippi, the state of California has 8,000 km (5,000 mi) of human-constructed levees that keep seasonal rainfall from inundating some of the nation's most productive farmlands. Without levees, much of this land would be permanently submerged because it has subsided and is now lower than the adjacent rivers. One problem with levee systems is that they invariably fail. It is nearly impossible to engineer an *affordable* levee system that can handle the *largest* flood events. This image shows the 1986 Linda levee failure near Marysville, California. The failure occurred nine days after the floodwaters had crested. The flood caused $400 million in damages.

16.08.mtb1

Before You Leave This Page Be Able To

☑ Sketch or describe the features that accompany low-gradient rivers.

☑ Describe the character of meander scars and oxbow lakes on the floodplains of meandering rivers.

☑ Sketch or describe how natural levees form, and describe the benefits and problems associated with levees.

16.8

16.9 What Happens When a River Reaches Its Base Level?

FOR ALMOST ALL RIVERS, base level is ultimately the ocean, where rivers slow down and drop their bed load and suspended load. Temporary base levels are established when a river is dammed by a landslide or other natural causes, or by human engineering. The new base level causes changes in the river system both above and below the obstruction. Such changes, however, are temporary—rivers win in the end.

A What Happens as a River Approaches Base Level?

Several landscape-building processes occur when a river enters the ocean, lake, or a temporary base level. Large rivers, like the Amazon and Mississippi Rivers, pump freshwater far into the ocean and carry fine sediment out to sea. They deposit coarser sediment as soon as the current slows, forming a *delta* along the shoreline.

16.09.a1

16.09.a2

1. What is a delta? This satellite view (▲) shows the green, triangular-shaped delta formed where sediment from the Nile River is deposited out into the Mediterranean.

2. A delta also forms where the Mississippi River meets the Gulf of Mexico near New Orleans. In this satellite image, the river changes from a meandering river within a broad floodplain to a series of smaller channels that branch and spread out in various directions. This branching drainage pattern is a *distributary system*.

3. Dark blue colors on this image indicate clear, deeper waters of the Gulf, whereas lighter blue areas contain suspended sediment and mostly have shallower water. Sediment from the river accumulates and builds up the delta, which is eroded by waves and by underwater slumps of the steep, unstable delta front.

4. Over the last 7,000 years, the Mississippi has created and then abandoned at least six huge mounds of sediment, each of which marks a former location of the river mouth and its associated delta; some of these abandoned deltas are labeled on the figure. A new delta, the active delta, is forming where the Mississippi River currently enters the Gulf of Mexico. Eventually, the river will shift and abandon this bird-foot-shaped delta too.

5. As a delta builds out into water, it forms new land and a characteristic sequence of sedimentary beds (▶). As the river's current slows, sand and larger particles become too heavy to be carried and are deposited in three types of beds. A set of horizontal beds forms on top of the delta.

6. A set of dipping beds forms when sediment is deposited over the front edge of the delta, moving the delta seaward.

7. Silt and clay are carried farther out into the ocean (or lake) and are deposited as nearly flat beds in front of the delta. Note the sequence of layers produced as a delta builds out into the ocean: marine clays overlain by cross-bedded sediment and then by horizontal beds deposited partly on land.

Sedimentary Beds in a Delta

16.09.a3

B What Controls the Deposition of Sediment in a Delta?

Deposition in a delta occurs where a river or stream slows, losing capacity and depositing its load of sediment. The morphology of a delta and the type of sediment deposited reflect the sediment load and discharge of the river, as well as other factors, including wave activity and the amount of vegetation or ice.

The Lena Delta of Siberia provides one of the most beautiful satellite images of Earth. This image, taken in the summer, shows a thawed East Siberian Sea and abundant vegetation on the delta. The distributary pattern of drainages is obvious. This delta nicely displays the factors that control deposition of sediment in a delta.

16.09.b1

Vegetation — The amount of vegetation and seasonal changes in vegetation affect the number and location of delta channels. Generally, deltas that have dense vegetation have fewer channels, whereas deltas with sparse vegetation have more channels. Part of the explanation is that vegetation binds the soil and stabilizes channel positions, as has been observed in sand and experimental water tanks.

Discharge — High-discharge flows tend to extend farther out into the ocean. The deposited sediment can then be affected by waves and by currents parallel to the shoreline.

Wave Erosion — Waves in the ocean or lake greatly affect the shoreline of the delta. Waves that are strong or continuous can erode and redistribute sediment along the coast or move it out to sea. Wave action is strongest and most effective during storms.

River and Ocean Ice — Seasonal changes in the amount of ice in the river and along the coast affect discharge and deposition patterns. River ice makes flow more sluggish, and sea ice tends to trap more sediment closer to shore. Since this is Siberia, known to be a very cold place, the sea around the Lena Delta freezes during the winter. The sea ice stops the action of ocean waves and causes a slowing of the river current near the north. As a result, the river, during the winter, deposits sediment along the river-ice interface.

Sediment Load — Coarser sediment, such as sand, is carried in the bed load and deposited first as the velocity drops. Finer material, carried in suspension, can be carried farther. If the river carries more sediment and is closer to its capacity, it will deposit more sediment and drop it sooner.

C What Are the Depositional and Erosional Consequences of Dams?

Human-constructed dams provide hydroelectric power generation, water storage, or flood control, but they stop a river's normal flow and transport of sediment. The reservoir behind the dam represents a *temporary base level* and so causes the river to deposit sediment behind the dam, limiting the dam's longevity.

1. When built, a dam forms a temporary base level. The river tries to achieve a new equilibrium, both upstream and downstream of the dam.

2. The change in base level causes the river to deposit sediment behind the dam in an attempt to retain its equilibrium profile. The pile of sediment builds out into the reservoir in the same way that a natural delta builds out into the sea. This sediment can eventually fill up the reservoir, shortening its lifespan.

3. Most dams release relatively clear water that is starved of sediment and that has a renewed capacity to erode. Such erosion occurs below many dams, whose clear-water releases contrast with typically muddy or sandy flows of the river before construction of the dam. For some dams, scientists and engineers are investigating ways to have the sediment in the reservoir bypass the dam, restoring the river to a more natural state.

Post-Dam Equilibrium Gradient

Pre-Dam Gradient

16.09.c1

Before You Leave This Page Be Able To

✔ Describe what happens when a river enters an ocean or lake, and what factors control deposition of sediment in a delta.

✔ Sketch and describe the stratigraphy of delta sediments and the setting in which each type of sediment formed.

✔ Describe how construction of a dam affects a river.

16.9

16.10 How Do Rivers Change Over Time?

IN GEOLOGIC TERMS, rivers come and go. Some rivers are old and others are surprisingly young. The age and history of a river are important considerations when evaluating how the river might respond to tectonic, climatic, and sea-level changes. Human activities can also evoke dramatic responses in rivers.

A How Old Are Rivers?

Rivers flow from their source to base level as long as enough water and slope are available to maintain downstream flow. A river's life can begin or end due to changes in water and sediment supply at the source, to changes in the slopes across which the river flows, or to changes in the elevation of its base level. Rivers can exist for millions of years, although their characteristics may change due to climatic, glacial, and tectonic events.

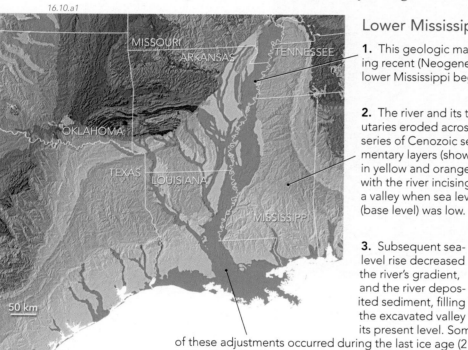

16.10.a1

Lower Mississippi River

1. This geologic map shows the Mississippi River and its tributaries in gray, representing recent (Neogene) sediments. The oldest preserved river sediments indicate that the lower Mississippi began draining the continent during Mesozoic time.

2. The river and its tributaries eroded across a series of Cenozoic sedimentary layers (shown in yellow and orange), with the river incising a valley when sea level (base level) was low.

3. Subsequent sea-level rise decreased the river's gradient, and the river deposited sediment, filling the excavated valley to its present level. Some of these adjustments occurred during the last ice age (2 million to 12,000 years ago), giving new life to an old river.

4. The river flows along a continent-scale low, the *Mississippi Embayment*, shown in this geologic cross section through Memphis, Tennessee. The embayment originated from Precambrian continental rifting, which thinned the crust and set the stage for the river's formation hundreds of millions of years later. The region has subsided well into the Neogene.

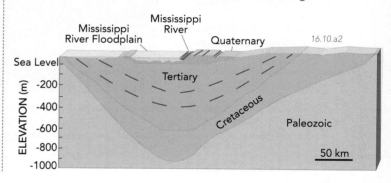

16.10.a2

Upper Mississippi River

16.10.a3

16.10.a4

The upper Mississippi River is young. It formed since the retreat of the last ice sheets, some 10,000 years ago. During the last ice age (◄), ice sheets and glaciers covered the northern half of North America, so northern rivers like the upper Mississippi did not exist. The weight of the ice sheets depressed the crust, causing large regions to slope northward (opposite of today).

◄ Melting of the ice released huge discharges of water and sediment that carved completely new river channels, including the upper part of the Mississippi.

The Fall Line

A major boundary, called the Fall Line, traces its way between the Appalachian Mountains and the east coast of the United States. The Fall Line, shown here as a red line, is marked by water falls formed along the contact between soft sediment of the coastal plain and harder bedrock in the foothills of the mountains. The Great Falls of the Potomac River, upstream from Washington, D.C., illustrate how the Fall Line developed. Before the ice age, the Potomac River occupied a broad valley. A drop in sea level during the ice age caused the river to incise deeper. Erosion proceeded upstream, stripping away the soft sediment until it encountered the harder rocks at the Great Falls. Note the concentration of cities along the Fall Line, including Washington, D.C.

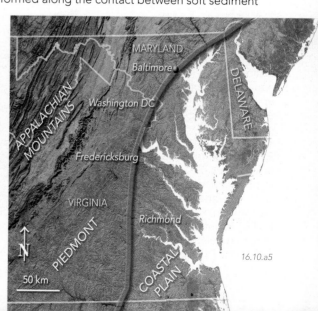

16.10.a5

B How Do River Systems Respond to Changing Conditions?

Rivers are sensitive to their environment, including local effects, like rainfall, and more distant effects, including changes in sea level. Rivers respond to changes in climate, tectonics, base level, human intervention, and the type of geology they encounter as they deposit sediment or cut deeper into the landscape.

Runoff

1. The amount of flow is the most important factor in how a river develops, and this depends mostly on the amount and timing of precipitation. Direct runoff during rainfall and delayed runoff from snowmelt supply most water to most rivers. The amount of runoff varies dramatically. The flood in the top image did this destruction (lower image) to condominiums built on an outside bend and too close to the bank.

16.10.b1 Tucson, AZ

16.10.b2 Tucson, AZ

Glacial and Sea-Level Effects

2. Global cooling and growth of ice sheets and glaciers lowers sea level. It can load and depress the crust, causing drainages to flow toward the ice sheets, where they fill large lakes.

3. Melting of the ice releases huge amounts of meltwater and sediment, creating new or larger channels. Isostatic rebound due to ice removal can tilt the land and reverse regional drainage patterns.

Loading by Ice Sheets   Melting of Ice Sheets
16.10.b3–4

Tectonism

4. Tectonism can uplift mountains, increasing slope, precipitation, and the supply of coarse sediment. The slope of a river and supply of sediment largely determine whether a river is braided or meandering.

5. Conversely, mountain uplift can create a rain shadow that decreases precipitation on the opposite side of the mountain, reducing the amount of runoff.

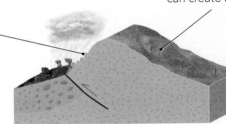
16.10.b5

Geology

6. Rivers can more easily erode unconsolidated sediment and soft rock types than hard ones. Rivers that are eroding downward may encounter rocks that have different characteristics, causing a change in the geometry of the river. The impressive Niagara Falls along the Canadian–U.S. border formed when the post-ice-age Niagara River encountered a more resistant dolostone layer underlain by less resistant shale (◄).

16.10.b6

Human Engineering

7. Dams and other flood-control structures change base level, the amount of discharge, and the supply of sediment, all of which affect the river system upstream and downstream. For example, they trap sediment in the reservoir.

16.10.b7 Glen Canyon Dam, AZ

Climate

◄ 8. When early settlers came to the American Southwest in the mid-1800s, many alluvial streams were flowing on broad valleys. Settlers built farms on the moisture-rich floodplains.

◄ 9. Climatic effects around 1880 caused streams to incise (erode down) several meters into their floodplains. This incision dried up the previous floodplain and many of the farms.

◄ 10. Around 1940, climate and other effects caused the channels to deposit sediment and begin to build up again.

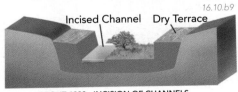
16.10.b8
Floodplain
BEFORE 1880 - STREAMS ON FLOODPLAIN

16.10.b9
Incised Channel   Dry Terrace
ABOUT 1880 - INCISION OF CHANNELS

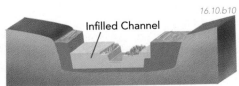
16.10.b10
Infilled Channel
ABOUT 1940 - CHANNEL FILLING BEGINS

Before You Leave This Page Be Able To

✓ Describe how rivers can be old or young, using the Mississippi River as an example.

✓ Describe how river systems respond to changes imposed by climate, tectonism, geology, and human engineering.

✓ Summarize the effect that glaciers have on river systems.

16.10

16.11 What Happens During River Incision?

RIVERS CAN INCISE INTO LANDSCAPES, as when the land is uplifted, base level drops, or the climate changes (especially an increase in precipitation). Incision by rivers forms a variety of features, including multiple levels of terraces. Rivers also carve some unusual canyons, such as those that take odd routes across the landscape, cutting right across mountains that would seem to be insurmountable obstacles. What sequence of events led to the development of these features?

A How Are River Terraces Formed?

River terraces are relatively flat benches that are perched above a river or stream and that stair-step up and out-ward from the active channel. Most terraces are composed of river-derived sediment and are essentially abandoned floodplains and alluvial plains. Other terraces are cut directly into bedrock and form by erosion. Terraces record different stages in the river's history and indicate that a river or stream has incised into the land.

16.11.a1 Jackson Hole, WY

Terraces form a series of flat to gently slop-ing benches or steps, flanked by steeper slopes. Terraces suc-cessively step up and away from the chan-nel. The highest ter-race may be tens of meters or more above the active channel. The lowest terrace commonly is only a meter or so above the channel and is often flooded, perhaps nearly every year.

This series of terraces (▼) flank the Snake River in Jackson Hole, Wyoming. The terraces are numbered from highest (1) to lowest (3). The modern floodplain also is labeled (F). Which of these ter-races formed first and which one formed last?

16.11.a2

1 km

| First Stage (oldest) | → | Second Stage | → | Last Stage (youngest) |

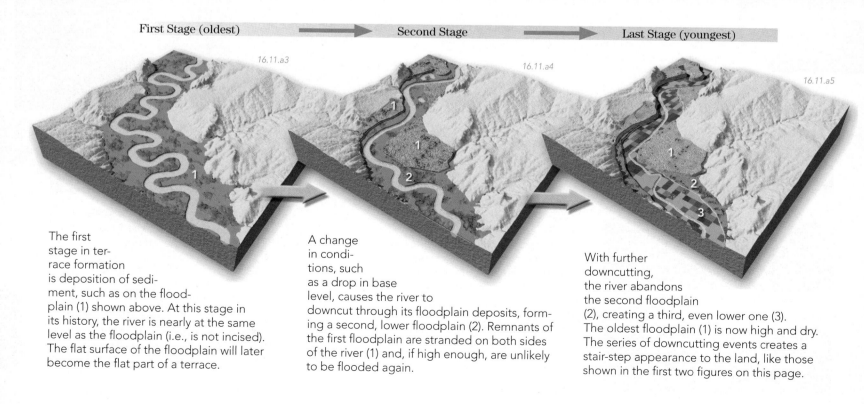

16.11.a3   16.11.a4   16.11.a5

The first stage in ter-race formation is deposition of sedi-ment, such as on the flood-plain (1) shown above. At this stage in its history, the river is nearly at the same level as the floodplain (i.e., is not incised). The flat surface of the floodplain will later become the flat part of a terrace.

A change in condi-tions, such as a drop in base level, causes the river to downcut through its floodplain deposits, form-ing a second, lower floodplain (2). Remnants of the first floodplain are stranded on both sides of the river (1) and, if high enough, are unlikely to be flooded again.

With further downcutting, the river abandons the second floodplain (2), creating a third, even lower one (3). The oldest floodplain (1) is now high and dry. The series of downcutting events creates a stair-step appearance to the land, like those shown in the first two figures on this page.

B How Are Entrenched Meanders Formed?

The landforms we know as meanders form only in loose sediments, like those on floodplains. However, in the Four Corners region of the American Southwest and areas west of the Appalachian Mountains, meanders with typical sweeping bends are deeply incised in hard bedrock, forming some puzzling canyons. What do these winding canyons, called *entrenched meanders*, tell us about the history of rivers in these areas?

First Stage (oldest)　→　Second Stage　→　Last Stage (youngest)

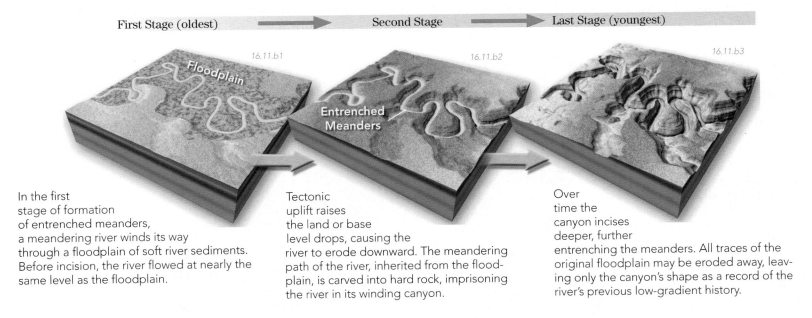

In the first stage of formation of entrenched meanders, a meandering river winds its way through a floodplain of soft river sediments. Before incision, the river flowed at nearly the same level as the floodplain.

Tectonic uplift raises the land or base level drops, causing the river to erode downward. The meandering path of the river, inherited from the floodplain, is carved into hard rock, imprisoning the river in its winding canyon.

Over time the canyon incises deeper, further entrenching the meanders. All traces of the original floodplain may be eroded away, leaving only the canyon's shape as a record of the river's previous low-gradient history.

Rivers That Cross Geologic Structures

Sometimes rivers appear to perform impossible tasks—cutting a deep canyon directly across a mountain. The Green River (below) flows across a mountain, appropriately called *Split Mountain*, as shown in the photograph to the right. This mountain ridge is an anticline of hardened sandstone in Dinosaur National Monument of northern Utah.

These odd canyons can be interpreted in at least two ways. A river may have been flowing over a region that was being actively uplifted and deformed, but the river was able to erode through the structures as fast as they were formed. Such a river is called *antecedent*, meaning it predated formation of the structure.

Alternatively, a river may establish its route when it is flowing on soft, easily eroded rocks, uninfluenced by what rocks lie at depth. As the river begins to incise, it becomes trapped in its own canyon, unable to avoid any geologic structures it encounters as it erodes down through the rocks. Such rivers are *superposed*, meaning they were superimposed on already existing features. The Green River is best interpreted as a superposed river that established a meandering course on soft rocks and then downcut into harder ones.

16.11.mtb2 Green River at Split Mountain, UT

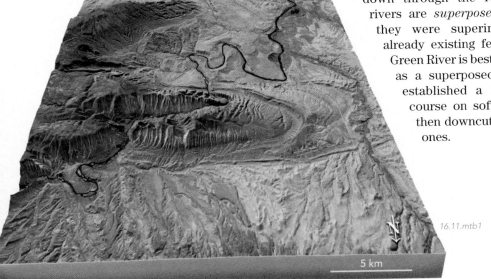
16.11.mtb1

5 km

Before You Leave This Page Be Able To

✔ Sketch and explain a series of steps showing how river terraces form.

✔ Describe one way in which entrenched meanders form.

✔ Explain how antecedent and superposed rivers are different.

16.11

16.12 What Is and What Is Not a Flood?

THROUGHOUT HISTORY, PEOPLE HAVE LIVED along rivers and streams. Rivers are sources of water for consumption, agriculture, and industry, and provide transportation routes and energy. River valleys offer a relatively flat area for construction and farming, but people who live along rivers are subject to an ever changing flow of water. High amounts of water flowing in rivers and streams often lead to flooding. In most parts of the world, flooding is a common and costly type of natural and human-caused disaster.

A What Is the Difference Between a Flood and a Normal Flow Event?

Rivers and streams are dynamic systems, and they respond to changes in the amount of water entering the system. When more water enters the system than can be held within the natural confines of the channel, the result is a *flood*.

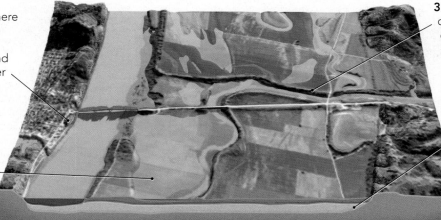

1. Flow in a channel, even when there is not a flood, may cause riverbank erosion. Such erosion can destroy structures built close to the river and make the river change position over time, turning what was floodplain into channel, and what was channel into floodplain.

2. A flood occurs when there is too much water for the channel to hold. As a result, water spills out onto the adjacent land, usually inundating parts or all of the floodplain.

3. Human-constructed levees can sometimes protect property from flooding during large flood events but trap water on the floodplain after the peak flooding ends.

4. Large floods can expand the width of the floodplain, by burying preexisting rocks and material with sediment deposited by the river. Sediment beneath the floodplain includes old channel deposits in addition to floodplain silt and clay.

16.12.a1

Normal, Bank-Full Flows

5. Normal (i.e., non-flooding) flows in rivers and streams can range from nearly dry to bank-full. Although there may be abundant water flowing down the channel, it is generally not considered a flood unless the water overflows the banks. A river's natural floodplain is an excellent place to contain excess floodwaters— as long as it remains undeveloped by society.

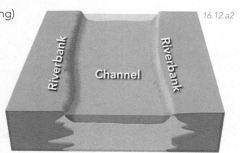

16.12.a2

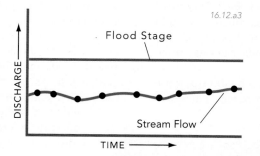

16.12.a3

6. This hydrograph shows a typical non-flood flow. The line labeled *Flood Stage* shows the amount of discharge required for the river to overtop its banks and spill out onto the floodplain (i.e., a flood). During extended times of dry conditions, or at least weather that is normal for the region, hydrographs may show little change in stream flow over time, as shown here.

Flows During a Flood

7. When the amount of water in a river exceeds the channel capacity, a flood occurs, inundating the floodplain. This hydrograph shows prolonged precipitation or snowmelt upstream that causes a flood event downstream, as represented by discharge greater than flood stage.

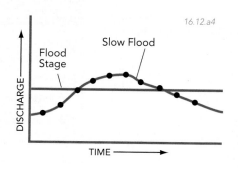

16.12.a4

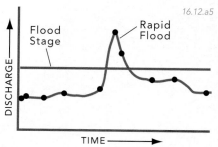

16.12.a5

8. Intense rainfall can unleash a brief *flash flood*, with a rapid rise in water levels and an increase in discharge that lasts only a short time. Similarly, rapid onsets of flooding result from failure of a natural or constructed dam, but flows last longer.

B   What Are the Causes of Flooding?

What causes discharge to exceed the channel's capacity? A simple answer is that there is more water in the channel than can be accommodated. This can be the result of natural processes or human-caused events.

Snowmelt

Flooding occurs when warming temperatures or rainfall melt snow and ice somewhere in the drainage basin.

16.12.b1 Norway

In the Northern Hemisphere, flooding from melting ice and snow occurs in the spring, from March to May. Heavy rain that coincides with melting can cause even worse flooding.

Local Heavy Precipitation

Some floods are caused by heavy rainfall over a short period of time, causing a brief, but dangerous, flash flood.

16.12.b2 Southern UT

A thunderstorm upstream of this site sent a fast-rising, muddy flash flood down this desert drainage. Vehicles attempting to cross such floods are often washed downstream.

Regional Precipitation

Regional floods occur when abnormally high precipitation falls over a large area over days, weeks, or months.

16.12.b3 Tucson, AZ

Heavy regional rains caused by moisture from a former hurricane caused this normally dry river to destroy offices built in a risky place—on loose sediment of the floodplain and on an outside bend.

Volcanic Eruption

If volcanic peaks are covered with snow when the volcano erupts, the snow will melt and cause flooding or catastrophic mudflows.

16.12.b4 Muddy River, WA

A volcanic eruption on snowy Mount St. Helens caused flooding and mudflows downstream, destroying this bridge.

Dam Failure

Dams occur as both natural and human-constructed features. Poorly engineered dams have failed, releasing floodwaters into downstream channels.

16.12.b5 Eastern ID

Catastrophic release of water during failure of the earthen Teton Dam, Idaho, in 1976 flooded towns downstream.

Urbanization

When urban growth replaces natural lands or farms, the area responds differently to precipitation and snowmelt. Foremost, urbanization increases runoff.

16.12.b6

This hydrograph shows that stream flow, for the same amount of water, became more abrupt and extreme after urbanization.

Other Manifestations of Flooding

When rivers overflow their banks, they can cause destruction to any buildings in the immediate area, but flooding can also be beneficial. Floods distribute water and sediment over large areas of land, replenishing *topsoil* and *nutrients* on agricultural land. Flooding also helps build up the elevation of the land around the river, helping keep land adjacent to the river higher than the channel and making it less prone to flooding.

In many areas, governments attempt to engineer rivers so that they will not flood. This involves building up levees, widening and deepening channels, and stabilizing the banks with concrete, soil-cement (a mixture of earth and cement), and large blocks of rock and other material. Engineering channels upstream can cause worse flooding or other serious consequences downstream.

Wetlands and other ecosystems along rivers usually develop in part from regular flooding. Allowing flooding to occur along rivers helps keep these ecosystems healthy and viable.

Before You Leave This Page Be Able To

☑ Sketch and describe a flood that overflows the channel versus a flow that stays within the channel. Include hydrographs in your sketches.

☑ Sketch the difference between a hydrograph showing a protracted flood and one of brief duration.

☑ Summarize some causes of flooding.

16.12

16.13 What Were Some Devastating Floods?

FLOODS CAN BE DISASTERS that affect millions of people and cause millions or billions of dollars in property damage. Floods occur for different reasons and over very different scales of time and area. Here we explore two kinds of floods: a regional flood on a large river and a local flash flood.

A What Happened During the 1993 Upper Mississippi River Flood?

The 1993 flood on the upper Mississippi River and other Midwestern rivers arose from heavy precipitation over several weeks. It killed 47 people and resulted in extensive property damage and economic loss. Floodwaters inundated large areas of the floodplain, including areas that were considered "safe" behind levees.

▼ 1. During June and July, the jet stream dipped south, creating a convergence zone between warm, moist air coming from the Gulf of Mexico and colder Arctic air coming from the northeast. This resulted in persistent thunderstorms in the Upper Mississippi region.

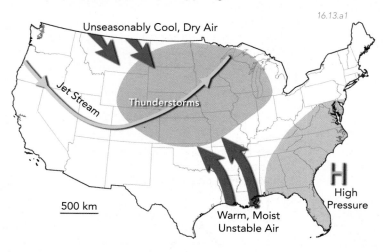

► 2. Contours on this map of the region show total rainfall (in inches) from June 1 to August 31, 1993. Some areas of Iowa received 36 inches of rain (nearly a meter).

3. High rainfall over such a large area resulted in flooding along major rivers and their tributaries. Heavy spring rains had already saturated the ground, preventing infiltration of additional rainfall during the summer storms.

Satellite Images of Three Rivers

The satellite images below both show an area at the confluence of the Mississippi, Missouri, and Illinois rivers near St. Louis, Missouri. Many homes and businesses on the modern floodplain were flooded and took months to dry out because levees, built to keep water out, trapped some water in after peak discharge.

▼ Before the flood, rivers are within their channels, and floodplains next to the channel are dry.

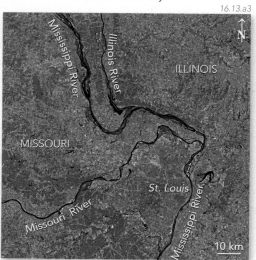

▼ During the flooding of 1993, the rivers inundated the broad floodplains, flooding places far from the river channel.

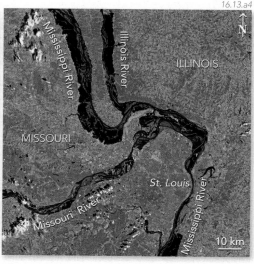

Discharge over Time

▼ As shown by this hydrograph, the Mississippi River at St. Louis, Missouri, reached 30 ft (9 m), or flood stage, on June 26 and peaked at 50 ft (15 m), 20 ft above flood stage, on August 1.

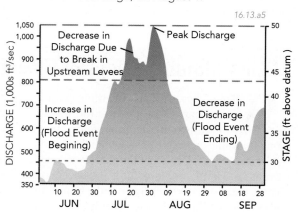

Sudden drops in discharge in mid-July correspond to breaks in the levee system upstream from St. Louis. These breaks let floodwater escape from the channel, lowering the discharge for areas downstream, like St. Louis.

B What Conditions Caused the Big Thompson River Flood in 1976?

The Big Thompson River near Estes Park, Colorado, flash flooded when as much as 12 inches of rain fell in a few hours in a small drainage basin. The flood killed 139 people and caused an estimated $16.5 million in damage.

1. This flood resulted from an unusual weather pattern. Cold polar winds converged with moist winds and pushed moist air upslope along the Colorado Front Range, forming a stationary thunderstorm. This image is an artist's depiction of the storm.

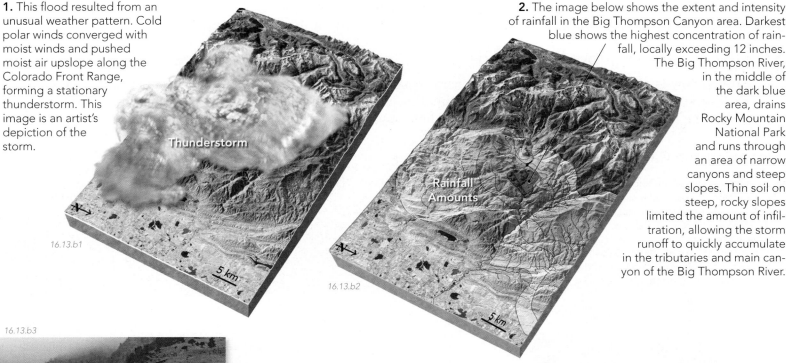

16.13.b1

16.13.b2

2. The image below shows the extent and intensity of rainfall in the Big Thompson Canyon area. Darkest blue shows the highest concentration of rainfall, locally exceeding 12 inches. The Big Thompson River, in the middle of the dark blue area, drains Rocky Mountain National Park and runs through an area of narrow canyons and steep slopes. Thin soil on steep, rocky slopes limited the amount of infiltration, allowing the storm runoff to quickly accumulate in the tributaries and main canyon of the Big Thompson River.

16.13.b3

3. The banks of the river were heavily developed with businesses, motels, campgrounds, and houses. The flood occurred in the evening, resulting in a number of deaths in a campground next to the river.

4. Houses along the Big Thompson River were totally destroyed by water and debris, which included mud, sand, and large boulders. Some houses, like this one, were ripped from their foundations and carried downstream.

16.13.b4

C What Were Some Other Notable Floods?

Flood/River	Cause of Flooding and Effect on Society
Johnstown, Pennsylvania (1889)	Failure of an earth-filled dam during heavy rains. Deadliest flood event in early-American history. Flooding destroyed the town and caused over 2,200 deaths.
Yangtze River, China (1931)	Prolonged drought followed by intense rainfall. 3.7 million people died from drowning, disease, and famine.
Fargo, North Dakota (1997)	Floods occurred along the Red River of the North in the spring of 1997. Rainfall coincided with snowmelt and a subsequent ice dam in the river. This caused the river to overflow its banks and flood a large area.
Central America (1998)	Hurricane Mitch stalled over Central America, dumping 75 inches of rain over several days. The death toll was estimated to be 11,000.
Bangladesh (2002)	Flooding occurs regularly in Bangladesh. In 2002, a combination of melting ice and snow in the Himalaya and exceptionally high precipitation filled to capacity the Ganges and Brahmaputra rivers, causing extensive flooding.

Before You Leave This Page Be Able To

☑ Describe the cause of flooding along the Mississippi River in 1993, and how this event affected the floodplains.

☑ Discuss the cause and consequences of the Big Thompson Flood of 1976.

☑ Briefly describe other circumstances that caused notable floods.

16.13

16.14 How Do We Measure Floods?

MOST FLOODS ARE A NATURAL CONSEQUENCE of fluctuating stream flow. Rivers receive most of their water from precipitation and snowmelt, and the amount of precipitation falling in any given drainage basin varies from day to day and from year to year. Stream flow, in response to rainfall and snowmelt, can vary from a trickle to a raging flood. How do we determine how big a flood was, or will be?

A How Is Stream Flow Measured?

Stream flow is measured by calculating discharge—the volume of water flowing through some stretch of a river or stream during a specified period of time. Discharge calculations help us quantify how big a flood was, determine how much water a river channel can hold, and predict the size of future floods.

Measuring and Calculating Flow

▶ The first step in calculating discharge is collecting measurements from the river at a particular site, called a *gauging station*. To collect the data, hydrologists (scientists who study water) measure cross sections of the stream at the site. The stream shown here has an overly simple stream bottom compared to natural streams. Hydrologists then measure how deep the water is and the average velocity of the water as it flows past. Many of the measurements are automated, with data being relayed by radio and computer.

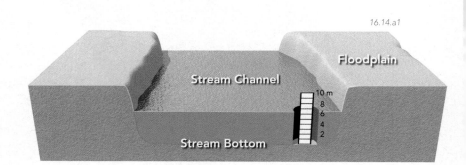

16.14.a1

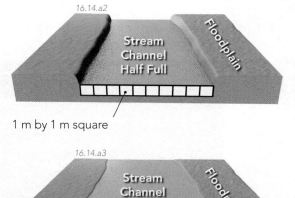

16.14.a2

1 m by 1 m square

16.14.a3

1. To calculate discharge (represented in equations by the letter Q), we need the cross-sectional area of a stream (average width x average depth) and average velocity of the current:

$$Q = stream\ depth \times stream\ width \times stream\ velocity\ (or\ Q=DWV)$$

2. If the velocity of the stream shown on the left is 1.1 meters per second, the calculation would be:

$$Q = 1\ m\ deep \times 10\ m\ wide \times 1.1\ m/sec = 11\ m^3/sec$$

3. Calculate how much discharge would be needed to fill the channel to a bank-full condition. When the river is this high, it normally flows 1.5 m/sec,

$$Q = 2\ m\ deep \times 10.5\ m\ wide \times 1.5\ m/sec = 31.5\ m^3/sec$$

or nearly three times the flow in the half-full example. The width at bank-full levels is 10.5 m wide because the bank widens a little upward.

Plotting Discharge During a Flood Using a Hydrograph

1. Before the actual flooding, this river flows at a bank-full condition. The flowing water is contained within the channel, so there is no flood. The hydrograph for a station on such a river shows a fairly constant discharge, represented by the horizontal part of the plot.

2. At some time, more water is added upstream to the river by a precipitation event (thunderstorms).

3. As the additional water reaches the station, the hydrograph shows a gradual increase. The channel can no longer contain the water, and the river floods out over its floodplain.

4. After the pulse of higher flow moves past the station, the hydrograph shows a return to the bank-full condition. The flood has passed. Most rivers will further decrease in flow, dropping well below a bank-full condition. If the river dries up completely, the discharge is zero.

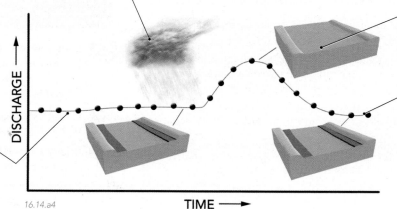

16.14.a4

B What Is the Probability That a Flood Will Occur?

Probability is the statistical description of the likelihood that an event will happen. Suppose you and a hundred other students enter a contest to win a car and each of you buys one ticket. Your probability of winning is 1 in 100 or 1%. Compare that with the probability that the Sun will rise tomorrow, which is essentially 100%. For many years, geologists and hydrologists have been collecting stream-flow data, which allows them to calculate the probability that a certain stream flow will occur in any year.

Frequency of Flows

Data used to estimate how often a river may flood comes from observations of discharge. The graph below plots peak daily average flows for the Yellowstone River from 1924 to 1998.

We plot discharge on the vertical axis and time on the horizontal axis. From this plot, it is common for this river to have peak flows around 4,000 to 7,000 m³/sec.

High flows above 7,000 to 8,000 m³/sec occur every so often.

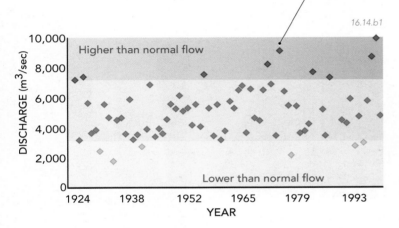

16.14.b1

Note that the highest flow event (flood) occurred only recently. To understand the river's behavior over time, and what to expect, we need data collected over a sufficiently long time period. A shorter data record means more uncertainty.

Flow Probability

Raw flow data are used to estimate probability. Hydrologists draw a rating line or curve for a river, giving the probability that a particular flow will be exceeded in any given year. This curve is for a smaller river than the Yellowstone River, with flow less than 2,000 m³/sec. To use this curve, start on either axis and follow any value to the line, and then read off the corresponding value on the other axis.

For example, the probability that a 120 m³/sec flow will occur or be exceeded in any year is about 99%.

The probability that a flow event will exceed ~2,000 m³/sec in any given year is low at about 0.5%.

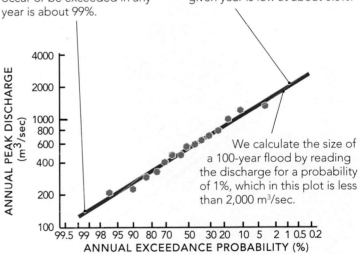

We calculate the size of a 100-year flood by reading the discharge for a probability of 1%, which in this plot is less than 2,000 m³/sec.

16.14.b2

What Probability Does and Does Not Tell Us

The probability that any particular flow from 0 m³/sec to more than 10,000 m³/sec could happen along a river in any year is estimated using graphs like the ones above. The probability for the Yellowstone River is based on a short record (since 1924) relative to the history of the river. The reliability of the mathematical estimations improves with more data. There is a very slight but real chance that floods exceeding 10,000 ft³/sec could happen three years in a row or twice in one year. The probability estimate doesn't guarantee future performance, but rather reflects what the collected data tell us about the river's past behavior. Planning for a certain size flood involves assessing how much data we have, in addition to what the existing data

predict about whether such a flood is likely to occur or not.

A term commonly used in public discussions, but less so by scientists who actually study rivers, is the concept of a *hundred-year flood*. This term signifies the size of a flood that is predicted—from the existing data—to have a 1 in 100 probability (1%) to occur in any given year. The term does not imply that such a flood will only happen every hundred years, because "100-year floods" can, and have, happened two or three years in a row along some rivers. In fact, such floods are more likely to occur in bunches, being caused by multi-year periods of abnormal amounts of precipitation and snowmelt.

Before You Leave This Page Be Able To

✓ Describe what stream discharge is and how it is measured and calculated.

✓ Sketch and describe how a precipitation event might appear on a downstream station's hydrograph.

✓ Explain how the probability of flooding is influenced by the length of time during which we have stream-gauge measurements.

16.14

How Does the Colorado River Change as It Flows Across the Landscape?

THE COLORADO RIVER SYSTEM drains a large region of the American West. The river cuts across a geologic terrain that varies from high bedrock headwaters to low, sandy valleys. It has formed a rich set of features, many of which are typical of most rivers, but some of which are unique to this river.

The large map spreading across both pages shows the drainage basin of the Colorado River. Surrounding the map are vignettes about different features, each of which is keyed to a number on the large map. The numbers begin in the headwaters. The smaller map below covers the same area as the large map and shows the Colorado River's largest tributaries.

The edge of the map is a drainage divide between the Colorado and other river systems.

16.15.a2

Reservoirs

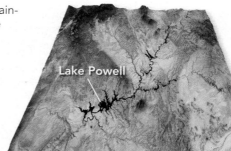

Lake Powell

◄ **6.** Dams have been constructed across the Colorado River, mostly within or bordering Arizona. The dams form large reservoirs, including Lake Powell and Lake Mead. The dams provide hydroelectric power, flood control, recreation, and water, but block sediment transport.

16.15.a3

Lakes

16.15.a4

Lava Flows

Colorado River

► **7.** Older lakes were formed by geologic events, such as lava flows that flowed into the Grand Canyon, temporarily damming the river. In geologic terms, such dams were rapidly eroded away.

Cutting Across Structures

► **8.** The Colorado River cuts across some geologic structures, such as the Kaibab Uplift in the Grand Canyon. The river may have started to cut through the uplift when a large, natural lake overtopped its rim, flooding westward across a low divide in the uplift.

16.15.a5

Kaibab Uplift

Colorado River in Grand Canyon

Nevada

California

Mexico

6

8 **7**

9

Salton Sea

▼ **9.** This large lake is located west of this area and is not shown on the large map. It filled in 1905 when a flood of the Colorado River overwhelmed canals and other structures built to divert water for irrigation in California. For two years, the river flowed into the basin, flooding 350,000 acres of land and filling a lake that had formed naturally many times in the past. Earlier lakes formed when high water volumes and high sediment load forced the river to leave its channel and flood westward into the lowlands of the Imperial Valley and ancestral Salton Sea.

Salton Sea

16.15.a6

Colorado Delta

10. As the Colorado River nears its mouth in the Gulf of California, much of its water is withdrawn for drinking and irrigation and its sediment load is blocked by dams. The delta, which has been building for hundreds of thousands of years, continues to grow but at a much slower rate because of the decrease in water volume and the sediment needed to nurture the delta's growth. The loss of water and sediment has harmed the delta's fragile ecology. ►

16.15.a7

Colorado River Delta

Gulf of California

Headwaters

16.15.a1

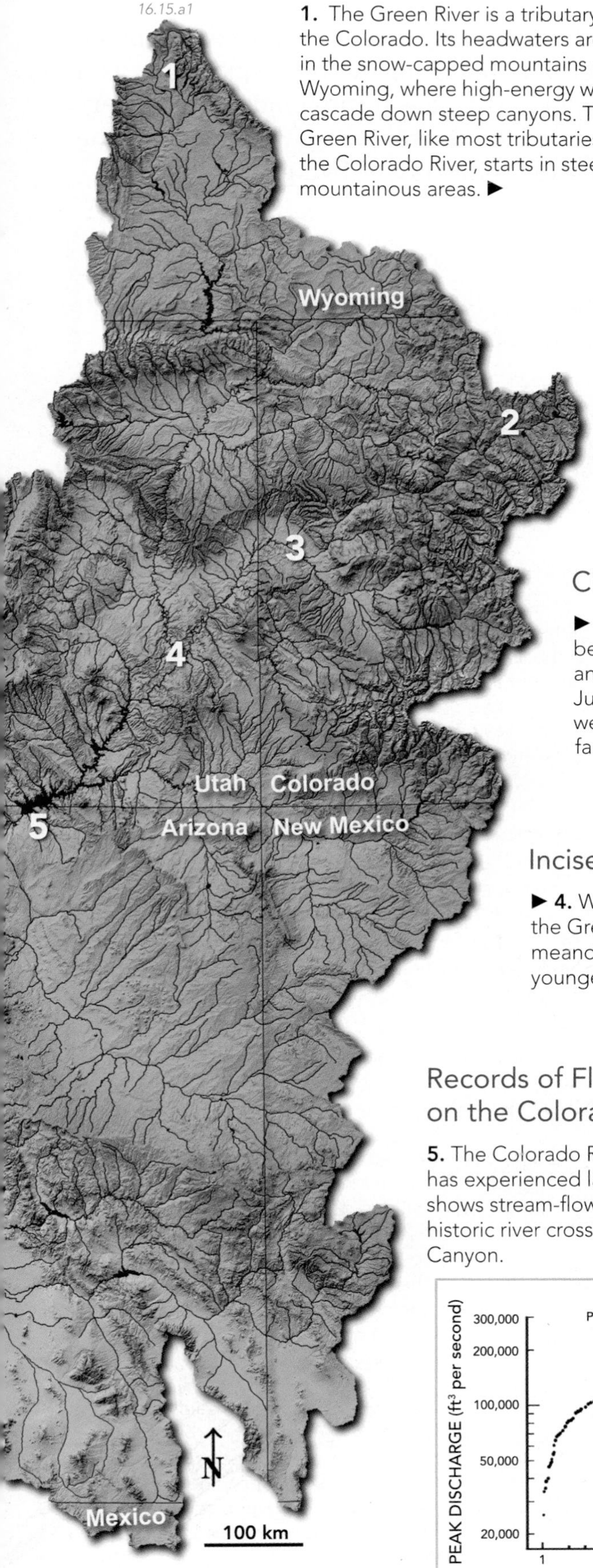

1. The Green River is a tributary of the Colorado. Its headwaters are in the snow-capped mountains of Wyoming, where high-energy waters cascade down steep canyons. The Green River, like most tributaries of the Colorado River, starts in steep mountainous areas. ▶

16.15.a8

Green River

2. The headwaters of the Colorado River (not shown by a detailed view) are in the high Rocky Mountains. Here steep mountain streams and braided rivers erode the mountains, transporting the debris to lower elevations.

Changing Conditions

▶ **3.** Where the Colorado River leaves its steep bedrock canyon, it becomes a meandering river and flows through a broad valley at Grand Junction, Colorado. Adjacent to the river is a well-developed floodplain covered with fertile farms benefiting from the Colorado's silt.

16.15.a9

Grand Junction, Colorado

Colorado River

Flood-plain

Incised Meanders

▶ **4.** Winding bedrock channels at the confluence of the Green and Colorado Rivers inherited their classic meander shapes when the river system was much younger and was flowing through softer materials.

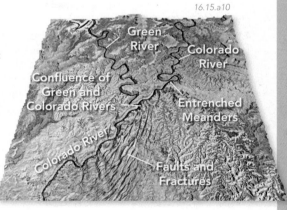

16.15.a10

Green River

Colorado River

Confluence of Green and Colorado Rivers

Entrenched Meanders

Colorado River

Faults and Fractures

Records of Flooding on the Colorado River

5. The Colorado River drains a large area and has experienced large floods. The graph below shows stream-flow data from Lee's Ferry, an historic river crossing upstream of the Grand Canyon.

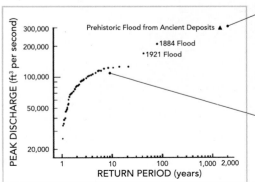

16.15.a11

Geologists investigating ancient river-flood deposits infer that a very large flood with a discharge estimated at 300,000 ft³/sec (cfs) occurred before humans were in the area. For comparison, modern, dam-controlled flows through the canyon rarely exceed 20,000 cfs.

Pre-dam flows (represented by the dots) were generally less than 130,000 cfs. During the largest flood recorded at Lee's Ferry, in 1884, the river's discharge was 220,000 cfs. Regional drought and other changes in climate greatly affect the flow of the river, and how much water is available for the rapidly growing cities of Nevada, Arizona, and California, all of which count on a steady supply of Colorado River water. What happens if flows decrease?

Wyoming

Utah | Colorado

Arizona | New Mexico

Mexico

100 km

N

16.15

16.16 How Would Flooding Affect This Place?

RIVERS PRESENT BENEFITS AND RISKS to people living along their banks. Meandering rivers provide floodplains with fertile soil and a relatively flat place to farm and perhaps build. Living on a floodplain is a hazardous proposition because it has flooded in the past, may be flooded in the near future, and owes its very existence to flooding. In this exercise, you will calculate the likelihood of flooding on two levels of the landscape and decide if potential economic and societal benefits are worth the risk of living there.

Goals of This Exercise:

- Observe and interpret features associated with a short stretch of a meandering river.
- Evaluate different locations for building a house and siting a farm, comparing and summarizing the advantages and disadvantages of each site.
- Calculate the risk of flooding for each location and discuss the risk versus the benefit.

Procedures

Use the available information to complete the following steps, entering your answers in appropriate places on the worksheet or answer questions online.

1. Observe the terrain below, in order to interpret the various parts of the landscape. Assign each landform feature or topographic level of the landscape its appropriate river term (for example, *channel*).

2. Apply your knowledge of the processes, features, and sediment associated with meandering rivers to predict what processes characterize each landform and how the landform might be affected by flow along the river.

3. Use relative elevations and other attributes to infer the order in which the features formed and the steps involved in the formation of each feature.

4. Determine which sites would be the best places to put *croplands*, considering all relevant factors, such as the flatness of the area, proximity to water, nature of the soil, what is growing there now, and possible added costs of growing crops in a specific site. You should also consider each site's vulnerability to bank erosion.

5. Evaluate the benefits of building a new house at each of the different levels of the landscape and at various locations on each level, for *both sides of the river*. Identify five homesites that are favorable, considering each site's proximity to croplands, to drinking water from the river, and any aesthetic considerations (e.g., just a nice place to live). Rank the five sites on the basis of your evaluation of their suitability.

6. Use the supplied elevation data on the profile on the next page and stream-flow data to calculate the river discharge required to flood two levels of the landscape.

7. Use the discharges you calculated and an *exceedance probability plot* for this river (provided) to estimate the probability of flooding for two levels of the landscape.

8. Evaluate the flood-risk probabilities against the other considerations (in steps 4 and 5), and describe how including the risk of flooding has changed or not changed your rankings.

Step 1: Consider the Following Observations About Different Levels Near the River

This highest flat area is a high terrace that locals call the *upper bench*. It is fairly dry and dusty, it does not contain many plants, and the soil is sandy.

The *middle bench* is a lower terrace. It has some plants and is below the dusty plain. It has a moderately good soil that could grow some crops if provided with water.

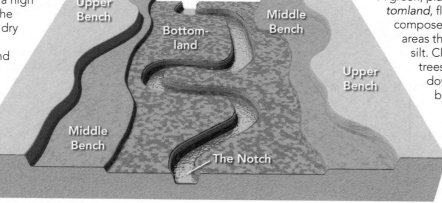

16.16.a1

A green, plant-covered, lower flat area, called the *bottomland*, flanks the river channel. It has some soil composed of silt and decayed plants, but in many areas the soil is overlain by several layers of loose silt. Close to the channel, many bushes and trees on the bottomland lean over a little in a downstream direction but were not uprooted by whatever made them lean over.

The lowest part of the valley, called *The Notch*, contains the river, whose water flows toward you in this view. When exposed during the dry season, sediment on the river bottom within the notch is loose and displays no soil development.

Step 2: Calculate Discharge for a Profile Across the River

The diagram below on the left is a profile across the river, showing the widths of The Notch and the Bottomland. You will calculate discharges along this main profile, which crosses the river near the front of the model on the right. Your instructor may provide you with a second profile (farther back in the model), because the river has different dimensions at different places. This means that the same amount of discharge may reach different heights along different segments of the river. For your profile(s), complete the following steps:

1. To calculate the discharge needed to fill the notch, first calculate the cross-sectional area of the notch in the profile: In all these calculations, we are using averages for width, depth, and velocity.

> Cross-sectional Area = Width × Depth

2. Next, calculate how much discharge is needed to fill the notch and begin to spill water out onto the bottomland. To calculate discharge, multiply the cross-sectional area of the notch by the average velocity of the river, which is 0.7 m/sec when the notch is filled:

> Discharge = Cross-sectional Area × Stream Velocity

3. Repeat the calculations, but this time determine the additional discharge needed to flood the bottomland to a height where flood-water would begin to spill onto the middle bench. The river flows faster when there is more water, so use an average water velocity of 2.0 m/sec. Enter your calculated discharges in the table on the worksheet or on a sheet of paper. You should have two discharge calculations, one to fill and overtop the notch, and another that fills up the notch and bottomland and then begins to spill out onto the middle bench.

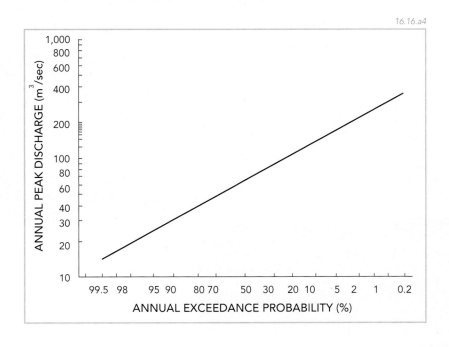

This is the location of the second (optional) profile.

This is the main profile.

16.16.a2

The steep slope between the bottomland and the middle bench is 3 m high.

The notch is 5 m high from its base.

16.16.a3

Step 3: Evaluate Flooding Risk Using Exceedance Probability

To determine the probability that each area will be flooded, compare both of your calculated discharges against the following plot, which is an *exceedance probability plot*. Follow the steps below and list in the worksheet or answer online the estimated probabilities for overfilling the notch and for overfilling the bottomland on the profile.

1. For each discharge calculation, find the position of that discharge value on the vertical axis of the plot.

2. Draw a horizontal line from that value to the right until you intersect the probability line (which slopes from lower left to upper right).

3. From the point of intersection, draw a vertical line down to the horizontal axis of the plot and read off the corresponding *chance of exceedance* (probability of flooding) on the horizontal axis. The *probability of exceedance* indicates the probability of the calculated amount of discharge being exceeded in any given year.

4. Repeat this procedure for both of your discharge calculations.

5. Consider the implications of each of these probabilities for your choice of site for cropland and a homesite. Use this information to choose final sites for cropland and a house. Explain your reasons on the worksheet or in the version online.

16.16.a4

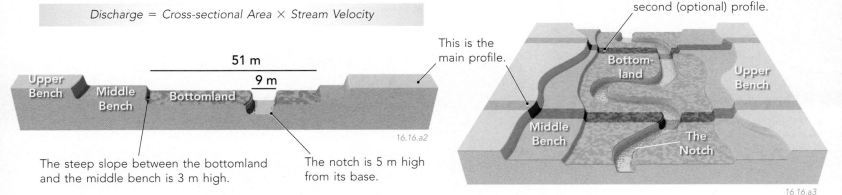

16.16

Water Resources

WATER IS THE MOST IMPORTANT RESOURCE provided by Earth—all life on Earth needs water to live and thrive. We are most familiar with *surface water*, water that occurs in streams, lakes, and oceans. Yet, the amount of freshwater in these settings is much less than the amount of freshwater that is frozen in ice and snow or that occurs in the subsurface as *groundwater*. This chapter is about surface water and groundwater and the important ways in which they interact.

17.00.a2 Shoshone, ID

The Snake River Plain, shown in this large satellite-based image, is a curved swath of low, basalt-covered land that cuts through the mountains of southern Idaho. It contains a mixture of dry, sage-covered plains, water-filled reservoirs, green agricultural fields, and recent lava flows of dark-colored basalt. Most of Idaho's population lives on the Snake River Plain near the rivers and reservoirs.

The Big Lost River, Little Lost River, and adjacent streams that enter the plain from the north never reach the Snake River. Instead, the water from the rivers and streams seeps into the ground between the grains in the sediment and through narrow fractures in the basalt. For this reason the rivers are called "lost."

Where does water that seeps into the sub-surface go?

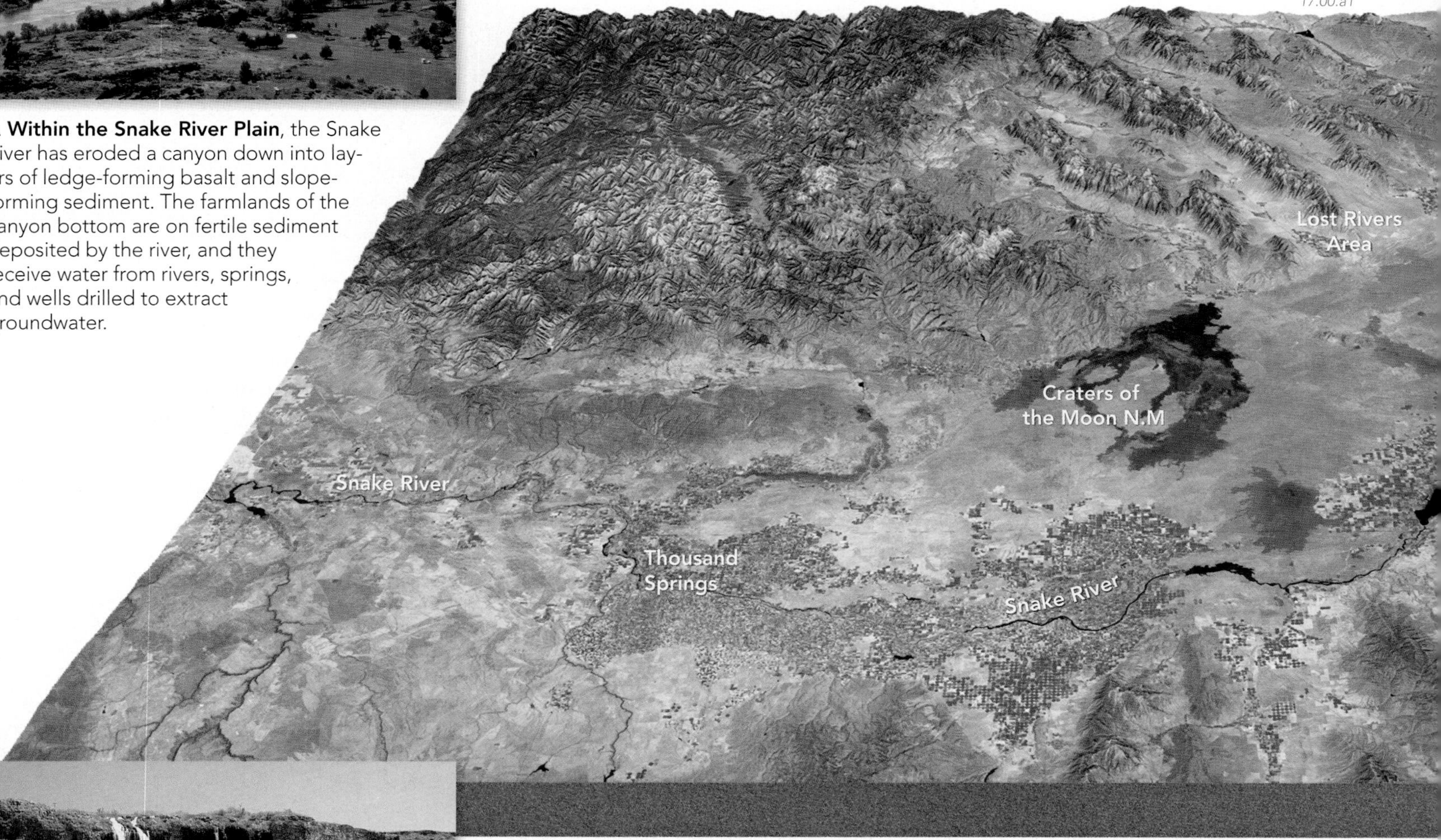

▲ **Within the Snake River Plain**, the Snake River has eroded a canyon down into layers of ledge-forming basalt and slope-forming sediment. The farmlands of the canyon bottom are on fertile sediment deposited by the river, and they receive water from rivers, springs, and wells drilled to extract groundwater.

17.00.a1

Lost Rivers Area

Craters of the Moon N.M

Snake River

Thousand Springs

Snake River

At Thousand Springs, huge springs gush from the steep volcanic walls of the Snake River Canyon (◄). The canyon includes 15 of the 65 largest springs in the United States, including those at Thousand Springs. The largest commercial trout farms in the United States use ponds fed by these springs.

What causes water from beneath the ground to flow to the surface as a spring, and where does the water in a spring come from?

17.00.a3 Thousand Springs, ID

TOPICS IN THIS CHAPTER

The Snake River winds through mountains and then flows southwest and west across the Snake River Plain.

Where does this river, flowing across such a dry plain, receive its water?

17.00.a4 Jackson Hole, WY

The river begins its journey in Jackson Hole, Wyoming (▲), from streams that drain the Tetons and nearby Gros Ventre range. The relatively higher rainfall and snowmelt in these highlands sustain the river as it flows westward across the dry plains. Farther downstream, rivers and streams entering the plain from the east and south flow directly into the Snake River, increasing its flow.

Where does the water in rivers come from, and do most rivers gain or lose water from groundwater?

Yellowstone National Park

Grand Tetons

Snake River

Snake River

Pocatello

25 km

Many lakes and farms are situated next to the Snake River. Farmers irrigate millions of acres of agriculture with surface water derived from reservoirs, lakes, and rivers, and with groundwater pumped to the surface. Chemicals used by some farms cause contamination of groundwater and surface water.

What happens if groundwater is pumped from the subsurface faster than it is replaced by precipitation and other sources? What do we do if water supplies are contaminated?

Disappearing Waters of the Northern Snake River Plain

Groundwater beneath the Snake River Plain is an essential resource for the region, providing most of the drinking water for cities and irrigation water for farms and ranches away from the actual river. Geologists and other scientists study where this water comes from, how it moves through the subsurface, and potential limits on using this resource.

Some water enters the subsurface from the Big and Little Lost Rivers, which flow into the basin from the north and then abruptly or gradually disappear as their water sinks into the porous ground. Other groundwater comes directly from the main Snake River and from tributaries that enter the basin from the south and east. Surprisingly, the largest influx of water to the subsurface is seepage from irrigated fields and associated canals.

The surface of the Snake River Plain slopes from northeast to southwest. The flow of groundwater follows this same pattern, flowing southwest and west through sediment and rocks in the subsurface. Groundwater derived from the disappearing rivers flows southwest, along the northern side and center of the basin. The groundwater does not flow like an underground river but as water between the sediment grains and within fractures in the rocks. Where the Snake River Canyon intersects the flow of groundwater, water reemerges on the surface, pouring out at Thousand Springs. This region illustrates a main theme of this chapter—surface water and groundwater are a related and interconnected resource.

17.0

17.1 Where Does Water Occur on Our Planet?

WATER IS ABUNDANT ON EARTH, occurring in many settings. Most water is in the oceans but is salty. Most freshwater is in ice and snow or in groundwater below the surface, with a smaller amount in lakes, wetlands, and rivers. Water also exists in plants, animals, and soils and as water vapor in the atmosphere.

A Where Did Earth's Water Come from and Where Does It Occur Today?

Most water on Earth probably originated during the formation of the planet or from comets and other icy celestial objects that collided with the surface. Over time, much of this water moves to the surface, for example when magma releases water vapor during eruptions.

Oceans—Of Earth's total inventory of surface and near-surface water, an estimated 96.5% occurs in the oceans and seas as *saline* (salty) *water*. The remaining 3.5% is *freshwater* held in ice sheets and glaciers, groundwater, and lakes, swamps, and other features on the surface.

Rivers—Rivers are extremely important to us and are the main source of drinking water for many areas. They contain, however, only a very small amount of Earth's freshwater.

Lakes—Water occurs on the surface in lakes of various sizes. Most are freshwater lakes, but those in dry climates are saline or *brackish* (between fresh and saline). Lakes contain a majority of the liquid freshwater at Earth's surface.

Swamps and Other Wetlands—These wet places contain water lying on the surface and water within the plants and shallow soil. They constitute about 11% of the liquid freshwater on the surface.

Atmosphere—A small, but very important, amount of Earth's water is contained in the atmosphere (0.001%). It occurs as invisible water vapor, as water droplets in clouds, and as rain, falling snow, and other types of precipitation.

Glaciers—Nearly 69% of Earth's freshwater is tied up in ice and snow in ice caps, glaciers, and permanent snow. A small amount also exists in permafrost and ground ice.

Soil Moisture—Earth's soils contain about as much water as the atmosphere (not much), but like water in the atmosphere, soil moisture is crucial to our existence.

Biological Water—Water is tied up within the cells and structures of plants and animals. It is clearly important to us but represents an exceptionally small percentage of Earth's total water (0.0001%).

Groundwater—About 30% of Earth's total freshwater occurs as groundwater. Groundwater is mostly in the open pores between sediment grains or within fractures that cut rocks. Most groundwater is fresh, but some is brackish or saline.

Deep-Interior Waters—An unknown, but perhaps very large amount of water is chemically bound in minerals of the crust and mantle. Some scientists think Earth's interior may contain more water than the oceans.

17.01.a1

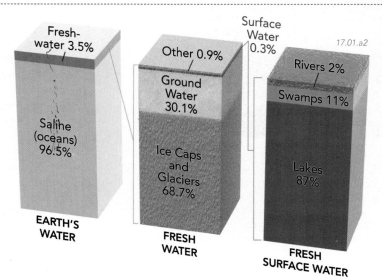

17.01.a2

► These bar graphs show USGS estimates of the distribution of water on Earth's surface and the uppermost levels of the crust. The left bar shows that the oceans contain 96.5% of Earth's total free water (water that is not bound up in minerals), but this water is saline. Only 3.5% of Earth's water is fresh. These graphs do not include water in Earth's deep interior.

EARTH'S WATER
Fresh-water 3.5%
Saline (oceans) 96.5%

FRESH WATER
Other 0.9%
Ground Water 30.1%
Ice Caps and Glaciers 68.7%

FRESH SURFACE WATER
Surface Water 0.3%
Rivers 2%
Swamps 11%
Lakes 87%

◄ The middle bar shows that most of Earth's freshwater occurs in ice caps and glaciers. Almost all the rest is groundwater. Less than 1% occurs as liquid surface water in lakes and rivers.

◄ The right bar shows where Earth's small percentage of fresh, liquid, surface water resides. Most is in lakes, followed by swamps and rivers.

B How Does Water Move from One Setting to Another?

Water is in constant circulation on Earth's surface, moving from ocean to atmosphere, from atmosphere back to the surface, and in and out of the subsurface. The circulation of water from one part of this water system to another is called the *hydrologic cycle*. From the perspective of living things, the hydrologic cycle is the critical system on Earth. It involves a number of important and mostly familiar processes. It is driven by energy from the Sun.

Evaporation—As water is heated by the Sun, some of its molecules become energized enough to break free of the attractive forces binding them together. Once free, they rise into the atmosphere as water vapor.

Condensation—As water vapor cools, like when it rises, water molecules join together. Through this process, water vapor becomes a liquid or turns directly into a solid (ice, hail, or snow). These water drops and ice crystals then collect and form clouds.

Precipitation—When clouds cool, perhaps when they rise over a mountain range, the water molecules become less energetic and bond together, commonly falling as rain, snow, or hail, depending on the temperature of the air. Precipitation may reach the ground, evaporate as it falls, or be captured by leaves and other vegetation before reaching the ground.

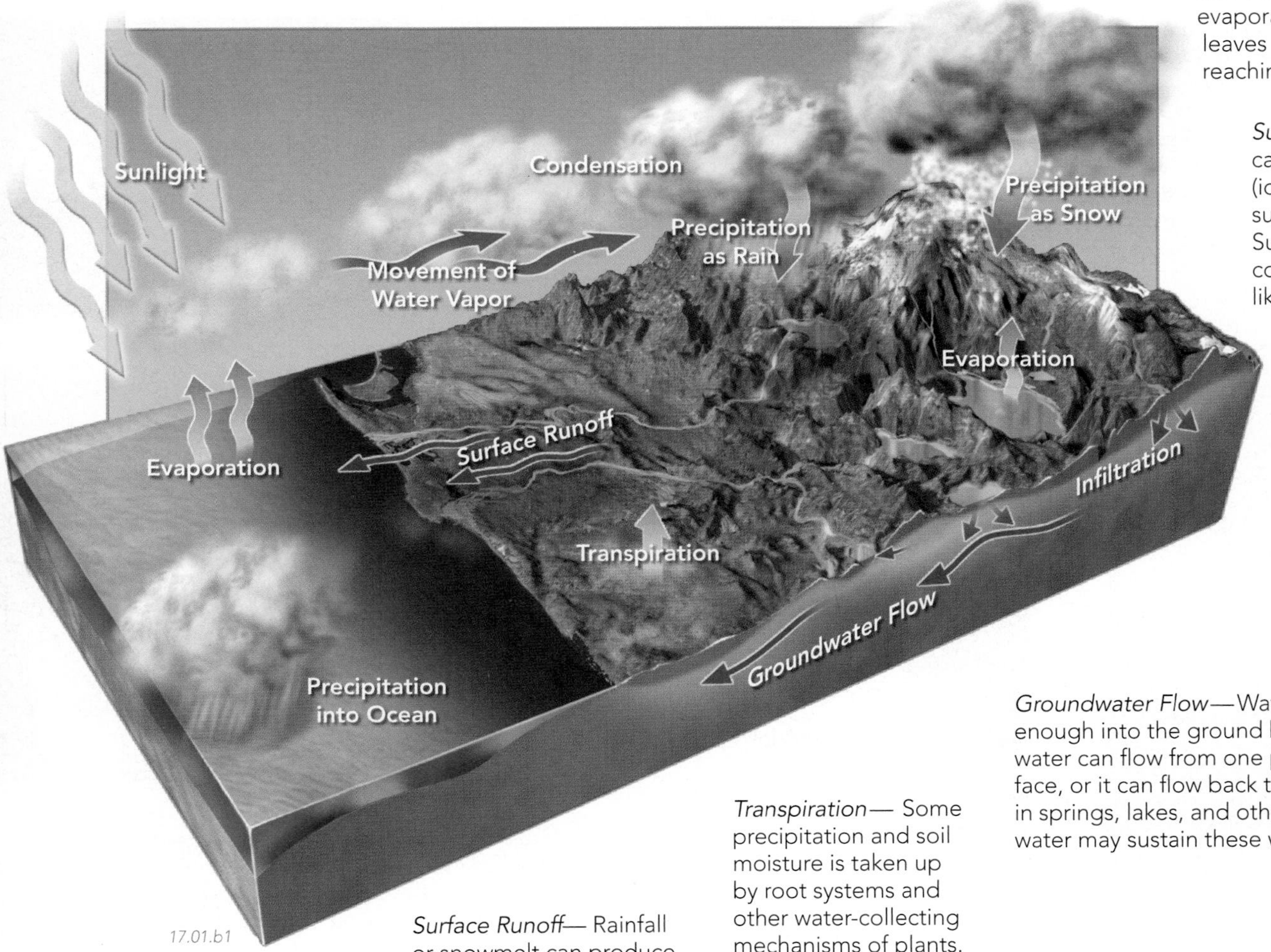

Sublimation—Water molecules can go directly from a solid (ice) to vapor, a process called sublimation (not shown here). Sublimation is most common in cold, dry, and windy climates, like some polar regions.

Infiltration—Some precipitation seeps into the ground, infiltrating through fractures and pores in soil and rocks. Some of this water becomes groundwater, some remains within the soil, and some rises back up to the surface. Water can also infiltrate into the ground from lakes, rivers, streams, canals, irrigated fields, or any body of water.

17.01.b1

Groundwater Flow—Water that percolates or infiltrates far enough into the ground becomes groundwater. Groundwater can flow from one place to another in the subsurface, or it can flow back to the surface, where it emerges in springs, lakes, and other features. Such flow of groundwater may sustain these water bodies during dry times.

Transpiration— Some precipitation and soil moisture is taken up by root systems and other water-collecting mechanisms of plants. Through their leaves, plants emit water vapor into the atmosphere by the process of transpiration.

Surface Runoff— Rainfall or snowmelt can produce water that flows across the surface as runoff. Runoff from direct precipitation can be joined by runoff from melting snow and ice and by the flow of groundwater onto the surface. The various types of runoff collect in streams, rivers, and lakes. Most is eventually carried to the ocean by rivers, where it can be evaporated, completing the hydrologic cycle.

Ocean Gains and Losses— Most precipitation falls directly into the ocean, but the ocean loses much more water to evaporation than it gains from precipitation. The difference is made up by runoff from land.

Before You Leave This Page Be Able To

☑ Summarize where most of Earth's total water resides.

☑ Describe the different settings where freshwater occurs, identifying which settings contain the most water.

☑ Describe or sketch the hydrologic cycle, summarizing the processes that shift water from one part to another.

17.1

17.2 How Do We Use Freshwater?

WE USE LARGE QUANTITIES OF WATER each day. We use water for a variety of purposes, especially power generation and irrigation of farms. How much water does each of our activities consume, and where does the water come from?

A What Are the Main Ways in Which We Use Freshwater?

The U.S. Geological Survey studied water use in the United States during 2000. The USGS compilation shows that we use freshwater in six or seven main ways, depending on how we classify the data. For the year 2000, water use in the United States was 408 billion gallons per day, most of which is from surface waters. Examine the graph below and think about the ways in which you use water.

17.02.a1 Ephrata, WA

◄ 1. *Irrigation*—Farms and ranches are one of the two largest users of freshwater, using nearly 40% of freshwater. Farms use water from groundwater, rivers, streams, lakes, and reservoirs to irrigate grain, fruit, vegetables, cotton, animal feed, and other crops. Much of this water is lost through evaporation to the atmosphere before it can be used by plants.

► 2. *Thermoelectric Power*—Electrical-generating power plants are the other large user of freshwater in the United States, also using slightly less than 40%. Such plants drive their turbines by converting water into steam and also use large amounts of water to cool hot components. Some of this water is from recycled sources. Power plants are also the largest user of saline (salty) water.

17.02.a2

▼ 3. *Public and Domestic Water Uses*—The third-largest use of freshwater is by public water suppliers and other domestic uses. We consume water by drinking, bathing, watering lawns, filling artificial lakes, and washing clothes, dishes, and cars. Much water from public water suppliers also goes to businesses. Most water for public and domestic use comes from rivers and groundwater.

17.02.a3 Phoenix, AZ

17.02.a5 West Driefontane, South Africa

FRESHWATER USAGE

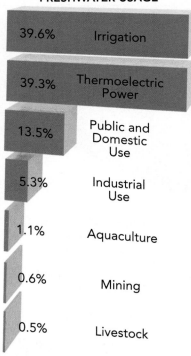

39.6%	Irrigation
39.3%	Thermoelectric Power
13.5%	Public and Domestic Use
5.3%	Industrial Use
1.1%	Aquaculture
0.6%	Mining
0.5%	Livestock

17.02.a4

▲ 4. *Industrial and Mining Uses*—Freshwater and saline water are extensively used by industries, including factories, mills, and refineries. Water is integral to many manufacturing operations, including making paper, steel, plastics, and concrete. Mining and related activities also use water in the extraction of metals and minerals from crushed rock.

17.02.a6

◄ 5. *Aquaculture*—According to the USGS study, we use approximately 1% of freshwater to raise fish and aquatic plants. Most of this use is in Idaho, near the Thousand Springs area. Such water is not totally consumed—much is released back into the Snake River.

► 6. *Livestock*—Watering of cows, sheep, horses, and other livestock accounts for only 0.5% of freshwater use, but much water is used for irrigation to raise hay, alfalfa, and other animal feed. Many ranches use small constructed reservoirs and ponds as the main water source for animals.

17.02.a7

B How Do We Use and Store Water?

17.02.b1

17.02.b2 Mississippi River

17.02.b3 Charles River, MA

Electrical Generation—The movement of surface water can generate electricity. To do this, we build dams that channel water through turbines in a hydroelectric power plant.

Transportation—We use many large waterways, such as the Mississippi River, as energy-efficient transportation systems to transport agricultural products, chemicals, and other industrial products.

Recreation—People use surface water in lakes and rivers for many types of recreation, including swimming, tubing, rafting, boating, and fishing. We also use freshwater to fill ponds, fountains, and swimming pools.

17.02.b4 Folsom Dam, CA

◄ Surface waters are commonly stored in natural lakes and in constructed reservoirs behind concrete and earthen dams. Water for drinking and other municipal uses can be stored in underground or above-ground storage tanks.

► We construct canals, raised aqueducts, and large and small pipelines to move freshwater from one place to another, especially to irrigate farms or to bring water to large cities. Pumps in water wells move groundwater to the surface.

17.02.b5 Yakima, WA

C How Do We Refer to Volumes of Water?

Water-resource studies typically report volumes of water in one of three units: *gallons*, *liters*, or *acre-feet*. Gallons and liters are familiar terms, but the concept of an *acre-foot* of water requires some explanation.

How big is an acre? An acre covers an area of 4,047 m² (43,560 ft²). If a perfect square, an acre would be 64 m (210 ft) on a side. An acre is equivalent to 91 yards of an American football field. There are 640 acres in a square mile.

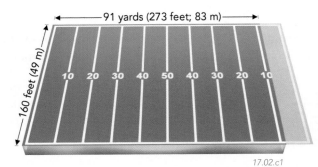

91 yards (273 feet; 83 m)

160 feet (49 m)

10 20 30 40 50 40 30 20 10

17.02.c1

An acre-foot of water is the volume of water required to cover an acre of land to a height of one foot. Imagine covering 91% of a football field (one acre) with a foot of water. An acre-foot is equivalent to about 326,000 gallons, or more than 1.2 million liters of water.

Drinking-Water Standards in the United States

The U.S. Environmental Protection Agency (EPA) sets standards for safe drinking water. Nearly all *public* water supplies in the United States meet these standards, which can be found at *www.epa.gov*. These standards set a limit on the concentrations of selected contaminants in water. Small municipalities commonly have more trouble meeting these standards than large cities because of limited budgets for water analysis, and for building and running facilities to remove contaminants. The EPA standards do not apply to private wells.

Many people prefer the taste and convenience of bottled water to public tap water, but there are generally no health reasons to buy bottled water so long as the public water provider meets all the federal, state, and local drinking water regulations. Commercially bottled water is monitored by the Food and Drug Administration (FDA) but is not as closely monitored as public water systems. The FDA requires a bottler to test its water source only once a year. Also, bottled water can cost as much as 1,000 times more than municipal drinking water.

Before You Leave This Page Be Able To

☑ Describe ways we use freshwater, and which four uses consume the most.

☑ Describe how we use and store freshwater.

☑ Describe in familiar terms how much water is in an acre-foot.

☑ Describe what a drinking water standard is, who sets the limits, and to whom they do and do not apply.

17.2

17.3 Where Is Groundwater Found?

A HUGE SUPPLY OF GROUNDWATER lies beneath Earth's surface. Groundwater occurs beneath all areas of the world but is far below the surface in some areas and very near the surface in others. Where does this water come from, and how does it find room in the solid Earth beneath us?

A What Is Groundwater?

Groundwater is free water (exists as a liquid rather than being chemically bonded in minerals) that is beneath Earth's surface. Surface and near-surface rocks and soil can be relatively dry, but deeper parts are generally saturated with water. Groundwater is present in three different settings that reflect the types of rocks, sediment, and geologic features that host the water.

In sediment or sedimentary rocks, adjacent grains do not fit together, so there is some space between the grains. These spaces, called *pore spaces*, hold groundwater. Here, the tan objects are grains, the brown represents pore spaces, and the blue indicates pore spaces that are saturated (filled) with groundwater. ▶

17.03.a1

17.03.a2

Most groundwater occurs in pore spaces and fractures, but some resides in subterranean openings and caves. Caves can be filled or partially filled with water, or they can be completely dry. The rock shown here is a soluble limestone with wide fractures, bedding planes, and small cavities. Water passing through the rock dissolved soluble materials, widening fractures and bedding planes, ultimately forming open cavities. ▶

17.03.a3

▲ All types of rocks have fractures that provide openings in which groundwater can accumulate. If fractures are interconnected, the groundwater can flow. Here, a gray rock is cut by fractures that are unsaturated (brown) or filled with water (blue).

B How Does Groundwater Accumulate?

Groundwater originates from precipitation and snowmelt that seeps from the surface down into the subsurface. The water accumulates in pores, fractures, and cavities within soil, loose sediment, and rock.

1. When rain falls on the surface or snow melts, the water can either evaporate, be absorbed by plant roots, flow downhill as runoff, or seep into the subsurface.

2. Water that soaks into the soil first passes through a part of the subsurface where most of the pore spaces are filled with air rather than water. This upper part, called the *unsaturated zone*, can be centimeters thick or can continue to depths of hundreds of meters. It can become completely dry during long periods without rain.

17.03.b1

5. The water table can be deep below the surface or can intersect the surface in lakes, streams, or swamps.

3. As water penetrates deeper into the subsurface, it eventually enters a zone where water fills nearly all the pore spaces and fractures. This zone is the *saturated zone* and is where most water occurs in the subsurface.

4. The top of the saturated zone is the *water table*, shown as a dashed red line. Below the water table, water fills and can flow through the interconnected pore spaces. Above the water table, some air remains in the pore spaces, and water within the pores can seep downward but is not connected enough to flow laterally.

C What Controls How Water Flows Through Rocks?

Water flows downhill, so the rate of groundwater flow is controlled by the steepness of the water table and two important properties of the material—*porosity* and *permeability*. These two properties are related to one another, but they are not the same thing. Porosity is a measure of how much water a rock can hold, but permeability indicates whether or how easily groundwater can flow through the rock.

Porosity

Porosity is the proportion of the volume of rock that is open space (pore space). It ranges from less than 1% to more than 50% and determines how much water a material can contain.

◄ Well-rounded and well-sorted sediment usually has higher porosity than angular or poorly sorted sediment because round grains do not fit together as tightly. This jar of marbles, analogous to well-rounded cobbles or sand grains, shows that a lot of pore space exists in such materials, provided the space is not filled with a natural cement.

17.03.c1

► Sediment that is poorly sorted tends to have less porosity because smaller grains fill the spaces between larger grains. Lower porosity also typifies sediment that has angular grains, whose corners help fill open spaces, or sedimentary grains that are held together by a natural cement, which fills in pore spaces.

17.03.c2

◄ Clay-rich sediments and sedimentary rocks, like shale, consist of small particles shaped like plates or sheets that do not fit tightly together. There is abundant open space (porosity) between them, but such pores, like the clay particles, are very small, making movement of water difficult. Clay particles can become compacted or can swell when wet, reducing porosity.

17.03.c3

► In igneous and metamorphic rocks, porosity is usually low because the minerals are tightly intergrown, leaving little free space. Some igneous rocks have less than 1% porosity. Fractures cutting any rock, however, open up narrow spaces and increase the porosity by some amount.

17.03.c4

Permeability

Permeability is a measure of the ability of a material to transmit a fluid. It is related to the size and interconnectedness of the pore spaces. Materials with low porosity usually have low permeability.

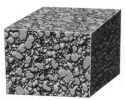

◄ Loosely cemented gravel and sand commonly contain interconnected pore spaces that allow relatively easy groundwater flow. Such materials have *high permeability* and host groundwater in many areas.

17.03.c5

► Fractures cut most rocks, opening spaces that typically represent a small volume of the rock and only slightly increase porosity. Well-connected fractures, however, allow water to flow and provide *higher permeability*. Fractures are the only significant permeability in granite and most other igneous rocks.

17.03.c6

17.03.c7

◄ When clay particles compact, they tend to become aligned parallel to one another. This decreases the porosity and causes the pore spaces to be very small. Shales and similar rocks will have very *low permeability*, or perhaps no permeability.

17.03.c8

► It is possible to have a highly porous rock with little or no permeability. A good example is a vesicular volcanic rock. The bubbles that once contained gas give the rock a *high porosity*, but most vesicles are not connected, so the rock has *low permeability*.

Below are examples of high-permeability rocks. The cobbles and sand on the left are well rounded, and the fractures on the right are interconnected. Both examples allow water to accumulate in large quantities and move easily through the material. Permeability can be measured in the laboratory or tested in drill holes, and is expressed mathematically using an equation called *Darcy's law*. Geologists and hydrologists use this equation to model the flow of groundwater.

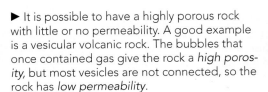

17.03.c9 Wilson Cliffs, NV

17.03.c10 Sedona, AZ

Before You Leave This Page Be Able To

☑ Sketch how groundwater accumulates and occurs in rock and sediment.

☑ Sketch and describe what the water table represents.

☑ Distinguish porosity and permeability, providing examples of materials with high and low values of each.

17.3

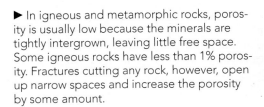

17.4 How and Where Does Groundwater Flow?

GROUNDWATER FLOWS BENEATH THE SURFACE in ways that are controlled by several key principles. The direction and rate of groundwater flow are largely controlled by the permeability of the materials, slope of the water table, and the geometry and nature of the subsurface rock. Some rock types allow easy groundwater flow, whereas others essentially preclude any significant movement.

A What Is the Geometry of the Water Table?

The water table is usually not a horizontal surface but instead has a three-dimensional shape that mimics the shape of the overlying land surface. The shape of the water table commonly has the equivalents of slopes, ridges, hills, and valleys. The shape of the water table controls which way groundwater flows.

17.04.a1

1. In most environments, the water table typically has the same shape as the overlying land surface but is more subdued. Where the land surface is high, the water table is also high. The similarity in shape between topography and the water table is less straightforward in some arid environments and in places where humans have pumped out groundwater faster than it can be replenished by precipitation.

2. The water table generally slopes from higher to lower areas. It generally is deeper below the surface under mountains than under lowlands, so its slope is less steep than that of the land surface. The shape of the water table is largely independent of the geometry of rock units through which the water table passes.

3. Groundwater just below the water table flows down the slope of the water table. In this example, it flows from left to right, from areas with a higher water table to areas with a lower water table. The red arrows show flow directions of water right below the water table.

4. Where the water table is horizontal, for example near this lake, groundwater may flow very slowly or not at all. Deeper water may flow in directions different from near-surface water.

5. The terminology used to describe features of a water table is derived from topography. A high part of the water table separating parts sloping in opposite directions is called a *groundwater divide*. Groundwater flows in opposite directions on either side of a groundwater divide.

6. Where the water table intersects the land surface, there may be lakes, wetlands, or a flowing river. The river in this figure occurs where the water table is at the surface. However, rivers do not all necessarily coincide with the water table, because some flowing rivers are underlain by unsaturated materials.

B What Controls the Rate of Groundwater Flow?

The rate of groundwater flow is typically measured in meters per day, but it can be much slower or faster. Rate is primarily controlled by permeability, which can vary by 12 orders of magnitude. The rate is controlled to a lesser extent by the steepness of the water table because flow is driven by the force of gravity. Other factors being equal, water flows faster down a steep water-table slope and slower down a more gentle one. The slope of the water table is also called the *hydraulic gradient*.

17.04.b1

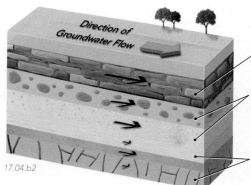

Direction of Groundwater Flow

17.04.b2

The rate of groundwater flow is strongly controlled by the *permeability* of the rock type. In this diagram, flow is fastest in highly permeable cavernous limestone.

Flow is moderately fast in a porous conglomerate or well-sorted sandstone.

Flow is slower in shale, which has small pores, and in a granite with poorly connected fractures.

C What Is an Aquifer?

An aquifer is a large body of permeable, saturated material through which groundwater can flow well enough to yield significant volumes of water to wells and springs. To be a good aquifer, a material must have high permeability, as occurs in poorly cemented sand and gravel, most sandstone, cavernous limestone, or highly fractured rocks of nearly any type.

1. The most common type of aquifer is an *unconfined aquifer* where the water-bearing unit is open (not restricted by impermeable rocks) to Earth's surface and atmosphere. Rainwater or surface water can seep unimpeded through the unsaturated zone and into an unconfined aquifer.

2. A *confined aquifer* is separated from Earth's surface by rocks with low permeability. Here, a permeable sandstone aquifer is bounded above and below by layers of low-permeability shale.

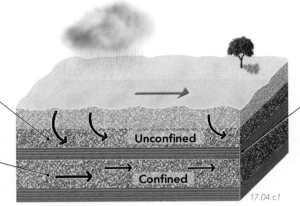

17.04.c1

3. A low-permeability unit, such as this layer of shale, can restrict flow and is referred to as an *aquitard*. An *impermeable* unit blocks flow completely. Such units are the opposite of an aquifer.

D How Are Wells Related to the Water Table?

A well is a hole dug or drilled deep enough to intersect the water table. If the well is within an aquifer, water will fill the open space to the level of the water table. This freestanding water can be drawn out by buckets or pumps.

17.04.d1

This well has been drilled from the land surface downward past the water table. The aquifer is unconfined and has filled with water to the height of the water table.

17.04.d2

In dry seasons, or during periods of high groundwater use, some wells run dry. This occurs when the water table drops and the well was not drilled deep enough into the aquifer.

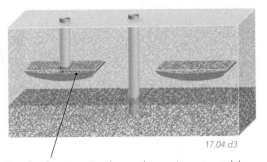

17.04.d3

Perched water sits above the main water table and generally forms where a discontinuous layer or lens of impermeable rock blocks and collects water infiltrating into the ground. Perched water bodies cam make it difficult to predict the best site to drill an adjacent well.

Artesian Wells and Water

We often hear the word *artesian*, commonly in the context of bottled water or certain beverages. What does this term imply? Does it mean that the water is better tasting, more natural, or more healthy? The short answer to these three questions is no, or at least not necessarily.

The term *artesian* means that groundwater is in a confined aquifer and is under enough water pressure that the water rises some amount within a well. The water does not have to reach the surface for the well to be called artesian, but many artesian systems have enough pressure to force the water all the way to the surface, creating a well or spring. Although it is a catchy advertising term, the term *artesian* is not indica-

tive of how the water tastes, whether it is more natural than other types of groundwater, or whether it is healthy. It only means that the groundwater is confined and under pressure, and as a result rises some amount in the well.

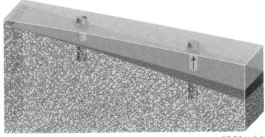

17.04.mtb1

Before You Leave This Page Be Able To

☑ Sketch and describe the typical geometry of the water table beneath a hill and a valley, showing the direction of groundwater flow.

☑ Summarize two factors that control the rate of groundwater flow.

☑ Sketch and describe the origins of perched water.

☑ Sketch and describe an unconfined, confined, and artesian aquifer.

17.4

17.5 What Is the Relationship Between Surface Water and Groundwater?

SURFACE WATER AND GROUNDWATER ARE NOT ISOLATED SYSTEMS. Rather, they are highly interconnected with water flowing from the surface to the subsurface and back again. Most groundwater forms from surface water that seeps into the ground, and some streams and lakes are fed by groundwater.

A How Does Water Move Between the Surface and Subsurface?

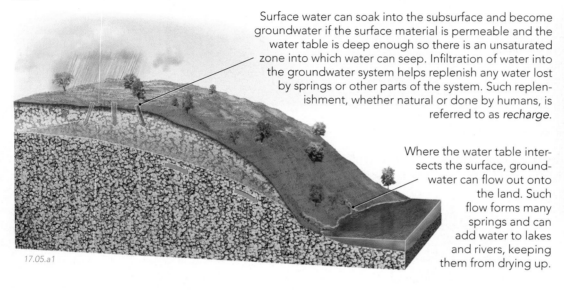

Surface water can soak into the subsurface and become groundwater if the surface material is permeable and the water table is deep enough so there is an unsaturated zone into which water can seep. Infiltration of water into the groundwater system helps replenish any water lost by springs or other parts of the system. Such replenishment, whether natural or done by humans, is referred to as *recharge*.

Where the water table intersects the surface, groundwater can flow out onto the land. Such flow forms many springs and can add water to lakes and rivers, keeping them from drying up.

17.05.a1

17.05.a2 Grand Canyon, AZ

▲ In this photograph, groundwater flows to the surface from widened fractures in limestone.

B What Causes Groundwater to Emerge as a Spring?

A spring represents a place where groundwater flows out of the ground onto the surface. At most springs, the water table intersects the surface. This can occur in a variety of geologic settings, some of which are summarized below. Some groundwater is heated by hot rocks before coming to the surface in warm springs, in hot springs, and in a *geyser*, a kind of hot spring that intermittently erupts fountains or sprays of hot water and steam.

Many springs are related to limestone aquifers. In such rocks, water can flow easily along dissolved bedding planes or through caves or fractures widened by dissolution. Where the saturated zone in the aquifer intersects the surface, water can flow out in a spring.

17.05.b1

Many springs are related to unconformities. In this example, a sedimentary unit that is a good aquifer lies above an unconformity, separating it from a less permeable crystalline rock (granite). Groundwater flows to the surface in springs along the boundary.

Unconformity

17.05.b2

Faults can serve as conduits for groundwater that feeds springs. Most faults are zones of intense fracturing and are therefore permeable. Faults can lead to springs if they juxtapose permeable against less permeable rocks, as in the example to the right.

17.05.b3

Some springs form where groundwater in permeable rock encounters a less permeable obstacle. In this example, groundwater flowing down the hydraulic gradient rises to the surface under pressure upon encountering a less permeable rock. Aquifers with such springs can be unconfined, as shown here, or confined.

17.05.b4

C How Are Lakes Related to Groundwater?

Lakes can have various relationships to groundwater. Most lakes occur where the water table intersects the ground surface, but some have a different setting. Most wetlands represent the interaction between rainfall, surface water, and groundwater and may be nourished by groundwater flow.

1. Some lakes are perched above the water table. These lakes can be transient, lasting only a short time after precipitation. A perched lake can be permanent if the inflow of water into the lake is at least equal to the amount lost by outflow to the ground, by evaporation to the air, or by other means.

2. Most lakes mark where the water table intersects and rises above the land surface. A lake can be fed entirely or partially by inflow of groundwater.

3. Many lakes are along the bottoms of valleys where groundwater is commonly close to or at the surface. Such lakes may be nearly in equilibrium with the adjacent groundwater, neither gaining nor losing water.

4. Wetlands can form peripheral to lakes, commonly at the same level as the water table. Other lakes and wetlands are perched on uplands that contain clay or other less permeable material close to the surface. The low permeability can trap precipitation and runoff, slowing the infiltration of water into the ground, forming a wetland from the ponded water.

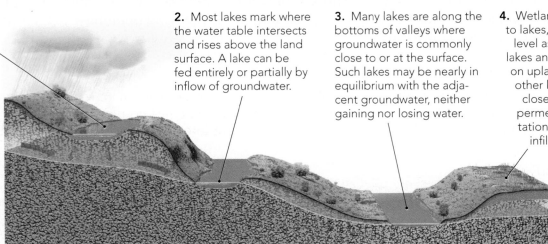

17.05.c1

D How Do Streams and Rivers Interact with the Water Table?

Water in many small rivers and streams decreases to a trickle and disappears entirely farther down the drainage. In other cases, a stream flows even when there has not been rain or snowmelt in a long time—what is the source of this water? These situations are a result of interactions with groundwater.

17.05.d1

1. Some streams and rivers are lower in elevation than the water table next to the stream, so groundwater flows into the stream or river as shown here by red arrows, which show the direction of groundwater flow below the water table. A part of a stream or river that receives water from the inflow of groundwater is said to be *gaining* or to be a *gaining stream*.

17.05.d2

2. Other channels flow across an area where the water table is at some depth below the surface. The part of the stream or river that loses water from outflow to groundwater is said to be *losing* or to be a *losing stream*. The red arrows show that groundwater below the water table flows down and away from the channel.

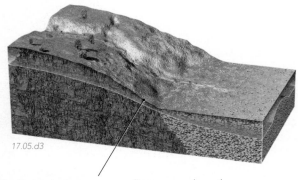

17.05.d3

3. Some losing streams disappear when they cross from hard, less permeable rocks onto softer, more permeable materials. The water seeps into the ground, where it may continue to flow at a shallow depth in the loose sand and gravel in the basin.

Before You Leave This Page Be Able To

☑ Sketch and describe how the interaction of the water table with topography causes water to flow between the surface and subsurface.

☑ Sketch or describe what is required to form a spring and possible settings where this occurs.

☑ Sketch and describe ways that lakes and wetlands relate to groundwater.

☑ Describe gaining and losing streams and how a river or stream can lose its water entirely.

17.5

17.6 What Features Form When Groundwater Interacts with Limestone?

WATER IS AN ACTIVE CHEMICAL AGENT, and groundwater can dissolve rock and other materials. Groundwater may completely dissolve limestone and other soluble rocks, leaving openings. When this happens, caves, sinkholes, and other features can form.

A How Does Groundwater Form Caves and Sinkholes?

Groundwater, especially if it is acidic, can dissolve limestone and other carbonate rocks to form caves. Cave systems generally form in carbonate rocks because most other rock types do not easily dissolve. A few other rocks, like gypsum or rock salt, dissolve too easily—they completely disappear and cannot maintain caves.

1. Limestone is primarily made of calcite (calcium carbonate), a relatively soluble mineral that dissolves in acidic water. Rainwater is typically slightly acidic due to dissolved carbon dioxide (CO_2), sulfur dioxide (SO_2), and organic material. Water reacts with calcite in limestone, dissolving it. Dissolution can be aided by acidic water coming from deeper in the earth, by microbes, and by acids that microbes produce.

3. Most caves form below the water table, but some form from downward-flowing water above the water table. Dissolution of groundwater over millions of years can form a network of interconnected caves and tunnels in the limestone. If the water table falls, groundwater drains out of the tunnels and dries out part of the cave system. If the roof of the cave collapses, the cave can be exposed to the air. Caves range in size from miniscule to huge. The Mammoth Cave System (Kentucky) is the longest in the world, over 570 km (340 mi) long.

2. Groundwater dissolves limestone and other carbonate rocks, often starting along bedding surfaces and fractures, and then progressively widening them over time. Open spaces become larger and more continuous, allowing more water to flow through and accelerating the dissolution and widening. If the openings become continuous, they may accommodate underground pools or underground streams.

17.06.a2 Winter Park, FL

4. Collapse of the roof of a cave may produce a sinkhole at the surface. The one shown here (▲) formed in 1981 and destroyed cars and buildings in Winter Park, Florida, where the underlying bedrock is limestone. The collapse resulted from lowering of the water table, which removed the water that helped support the roof of the cave in limestone beneath the area.

17.06.a1

B What Other Features Form in Dissolved Limestone Terrains?

17.06.b1 Guilin, China

17.06.b2 Havasu Creek, AZ

◀ *Karst Topography*— Many limestone terrains exhibit a distinctive topography, called karst topography. Karst is topography characterized by sinkholes, caves, limestone pillars, some poorly organized drainage patterns, and disappearing streams.

▶ *Travertine*—When limestone dissolves in groundwater, the dissolved material may later be reprecipitated when the groundwater flows onto the surface. Calcium carbonate commonly precipitates as travertine, which can build irregular terraces along a flowing stream.

C What Features Are Associated with Caves?

Caves are beautiful and interesting places to explore. Some contain twisty, narrow passages connecting open chambers. Others are immense tunnels and rooms full of cave *formations*. Caves can be decorated with intricate features formed by the dissolution and precipitation of calcite and several other minerals.

1. Most caves form by the dissolution of limestone. Certain features on the land surface can indicate that there is a cave at depth. These include the presence of limestone, sinkholes, and other features of karst topography. Collapse of part of the roof can open the cave to the surface, forming a *skylight* that lets light into the cave.

2. Caves contain many features formed by minerals precipitated from dripping or flowing water. Water flowing down the walls or along the floor can precipitate travertine (calcium carbonate) in thin layers that build up to create formations called *flowstone* or *draperies.* ▼

17.06.c2 Carlsbad Caverns, NM

17.06.c3 Kartchner Caverns, AZ

17.06.c1

6. Dissolution of limestone along fractures and bedding planes, along with formation of sinkholes and skylights, disrupts streams and other drainages. Streams may disappear into the ground, adding more water to the cave system.

5. In wet, humid environments, weathering at the surface commonly produces reddish, clay-rich soil. The soil, along with pieces of limestone, can be washed into crevices and sinkholes where it forms a reddish matrix around limestone clasts.

17.06.c4 Kartchner Caverns, AZ

◄ 3. Probably the most recognized features of caves are stalactites and stalagmites, which are formed when calcium-rich water dripping from the roof evaporates and leaves calcium carbonate behind. *Stalactites* hang tight from the roof. *Stalagmites* form when water drips to the floor, building mounds upward.

4. As mineral-rich water drips from the roof and flows from the walls, it leaves behind coatings, ribbons (▶), and straw-like tubes. The water can accumulate in underground pools on the floor of the cave, precipitating rims of cream-colored travertine along their edges.

Carlsbad Caverns

About 260 million years ago, Carlsbad, New Mexico, was an area covered by a shallow inland sea. A huge reef, lush with sea life, thrived in this warm water, tropical environment. Eventually the sea retreated, leaving the reef buried under other sedimentary layers.

While buried, the limestone was dissolved by water rich in sulfuric acid generated from hydrogen sulfide that leaked upward from deeper accumulations of petroleum. Later, erosion of overlying layers uplifted the once-buried and groundwater-filled limestone cave and eventually exposed

it at the surface. Groundwater dripped and trickled into the partially dry cave, where it deposited calcium carbonate to construct the cave's famous formations.

17.06.mtb1 Carlsbad Caverns, NM

Before You Leave This Page Be Able To

✓ Summarize the character and formation of caves, sinkholes, karst topography, and travertine along streams.

✓ Briefly summarize how stalactites, stalagmites, and flowstone form.

✓ Describe features on the surface that indicate an area may contain caves at depth.

17.6

17.7 How Do We Explore for Groundwater?

GROUNDWATER IS AN IMPORTANT RESOURCE, and much time and effort go into exploring for new sources of groundwater and gaining a better understanding of existing groundwater supplies. Geologists and hydrogeologists explore for groundwater by collecting surface and subsurface data to investigate the depth, amount, and setting of groundwater, the direction in which groundwater flows, and the quality of the water.

A What Kinds of Information Are Used to Investigate Groundwater?

Hydrogeologists are geoscientists who specialize in groundwater investigations and interactions between surface water and groundwater. They study geology on the surface and in the subsurface, and they use a variety of direct and indirect methods to constrain the subsurface geometry of rock units, sediment, and the water table.

1. Hydrogeologists usually begin a groundwater study by collecting known information, including topographic and geologic maps, reports about the geology and water resources, and information about depth to the water table, especially records of past drilling. They may also need to do new geologic field studies.

2. Because surface water and groundwater are interrelated, hydrogeologists may collect data about the flow of surface water, including the volume of water (the discharge) flowing in different stretches of streams and rivers. Such observations can indicate whether streams are gaining or losing water to the groundwater system. Additionally, sampling the chemistry of surface waters (▶) may help us understand water quality and potential threats to groundwater.

17.07.a2 Vietnam

3. Hydrogeologists and other technical staff measure the depth to the water table in existing wells by lowering an electronic device, called a water-level indicator, into the well (▼). Some water-level instruments use sound and others use electrical currents. The data are recorded for later analysis.

17.07.a1

17.07.a3

4. Geophysical surveys, involving measurements of variations in gravity, magnetism, and electrical conductivity, provide key information on the subsurface geometry of rock units and the water table. The graph below shows measurements of gravity over the edge of the basin shown above; the strength of gravity decreases slightly as the thickness of low-density sediments increases away from the mountain front and out into the basin.

5. Information about the subsurface is critical for understanding the setting and controls of groundwater flow. Hydrogeologists choose drill-hole sites that will maximize the amount of information gained. Drill holes provide direct measurements of the depth of the water table, water samples for quality analysis, samples of subsurface material, and a chance to observe the subsurface material with down-hole video cameras and geophysical instruments. The photo below shows a core of sediment retrieved by drilling.

17.07.a6

6. Geologists and hydrogeologists graphically portray the results of drilling on a *drill log* (◀), which is similar to a stratigraphic section (plotting types of rocks or sediments versus depth). A drill log commonly also includes other types of information, especially geophysical measurements that correlate with the type of material and with the presence or absence of water.

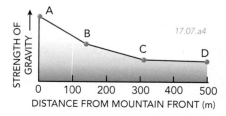

17.07.a4

STRENGTH OF GRAVITY →

A
B
C
D

0   100   200   300   400   500
DISTANCE FROM MOUNTAIN FRONT (m)

Soil
Upper gravel
Sand layer
Mud and clay
Lower gravel

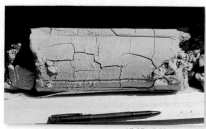
17.07.a5 Phoenix, AZ

B   How Do Hydrogeologists Depict the Water Table?

Once hydrogeologists collect the appropriate field, drilling, and geophysical data, they produce various types of maps and diagrams, especially maps showing the elevation of the water table.

The most important piece of information about ground-water is a map showing variations in the elevation of the water table. The first step in constructing such a map is to collect and plot elevations of the water table in all available wells. Each number on this map is the elevation (in meters above sea level) of the water table at a well in that location. High numbers mean the water table is higher than in sites with lower numbers.

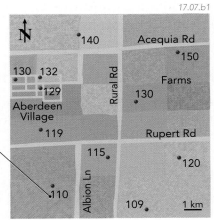

Hydrogeologists then draw contours to show the elevation of the top of the water table. The contours shown here indicate the elevation of the water table in meters. Each contour follows a specific elevation on the water table.

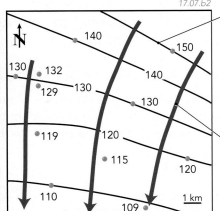

Arrows drawn perpendicular to the contours show the direction of groundwater flow, which is down the slope of the water table, from higher contours to lower ones.

Other Depictions

Hydrogeologists compare contour maps of water-table elevations to other features, including the locations of wells, rivers, farms, and other sites that may affect the ground-water, such as by taking water out of the ground. They extensively use computerized geographic information systems (GIS) to overlay and compare one data set to another and to identify patterns and relations between different types of information.

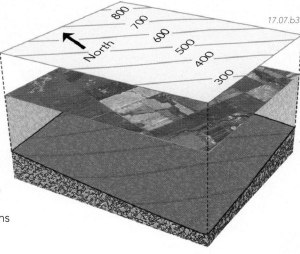

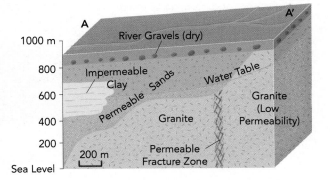

A cross section or block diagram, usually drawn with some vertical exaggeration, helps us explore how the water table relates to subsurface geology. Key considerations include the geometry and distribution of different geologic materials, especially those of different permeability, and how much of each unit is below the water table (in the saturated zone where it could yield water).

Hydrogeologists incorporate the geologic information and well data into computer programs to produce three-dimensional depictions of the water table (▶). They then model the directions and rates of groundwater flow and calculate the volumes of freshwater that will be available for drinking and other uses. The goal of the various depictions is to understand the three-dimensional geometry of the basin, rock units, water table, and topography. These factors control where and how much water accumulates, where and how it flows, and how it interacts with features we see on the surface. This computerized model shows, from top to bottom, the land surface, the base of two different sedimentary sequences in the subsurface (colored yellow and brown), and the top of hard bedrock (gray). The vertical lines are wells, color coded to show the presence of sediment saturated by groundwater (yellow). The interpreted slope of the water table is the green mesh connecting the data points (each well).

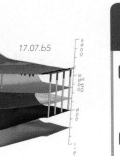

Before You Leave This Page Be Able To

✓ Summarize the types of information that hydrogeologists collect and what each indicates about the subsurface.

✓ Describe how a contour map of water-table elevations is constructed and how it would be used to predict the direction of groundwater flow.

✓ Describe factors to show in a cross section or block diagram if groundwater is the focus of the study.

17.7

17.8 What Problems Are Associated with Groundwater Pumping?

THE SUPPLY OF GROUNDWATER IS FINITE, so pumping too much groundwater, a practice called *overpumping*, can result in serious problems. Overpumping can cause neighboring wells to dry up, land to subside, and gaping fissures to open across the land surface.

A What Happens to the Water Table if Groundwater Is Overpumped?

Demands on water resources increase if an area's population grows, the amount of land being cultivated increases, or open space is replaced by industry. Groundwater is viewed as a way to acquire additional supplies of freshwater, so new wells are drilled or larger wells replace smaller ones when more water is needed.

Minor Groundwater Withdrawal

1. A simple case illustrates the problems with overpumping. The two figures below show what happens when an unconfined aquifer is pumped, first by a small-volume pump and later by a larger pump. The topography of this area is fairly flat, there are no bodies of surface water, and a single type of porous and permeable sediment composes the subsurface.

2. As people move into a nearby town, they drill a small well down to the water table to provide freshwater. The small well pulls out so little groundwater that the water table remains as it has for thousands of years, nearly flat and featureless. The well remains a dependable source of water because its bottom is below the water table.

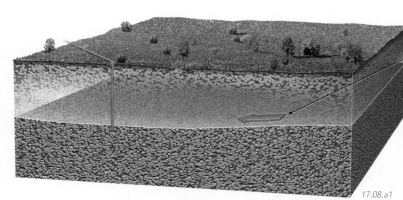

17.08.a1

3. Across the entire area, groundwater flows from right to left, down the gentle slope of the water table. The red arrow shows the direction of flow for groundwater in the saturated part of the aquifer, right below the water table. This arrow is drawn above the water table so the arrow is visible, but water above the water table is not connected enough to flow in this way.

Increased Groundwater Withdrawal

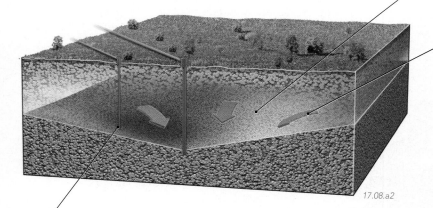

17.08.a2

4. As more people move into the surrounding area, they drill a larger well to extract larger volumes of water to satisfy the growing demand. The new, larger well pumps water so rapidly that groundwater around the well cannot flow in fast enough to replenish what is lost. This causes the local water table to drop, forming a funnel-shaped *cone of depression* around the well.

5. The direction of groundwater flow changes dramatically across the entire area. Instead of flowing in one direction, groundwater now flows toward the larger well and into the cone of depression from all directions. The change in flow direction has unintended consequences. It may cause serious safety issues, since waste-disposal sites, like landfills, are generally planned with the groundwater-flow direction in mind. The change in flow direction can bring contaminated water into previously fresh wells.

6. The original small well dries up because it no longer reaches the water table, which has been lowered by the larger well's cone of depression. A cone of depression is common around nearly all wells, but a large cone of depression, caused by overpumping of the aquifer, can have drastic consequences. It can dry up existing wells, change the direction of groundwater flow, and contaminate wells. In addition, overpumping can dry up streams and lakes, if they are fed by groundwater, or cause the roofs of caves to collapse. Effects similar to those described above can also occur in a confined aquifer.

Before You Leave These Pages Be Able To

☑ Sketch a cone of depression in cross section, describing how it forms and which way groundwater flows.

☑ Describe how a cone of depression can cause a well to become polluted.

☑ Sketch or describe some other problems associated with overpumping, including subsidence, fissures, and saltwater incursion.

B What Problems Are Caused by Excessive Groundwater Withdrawal?

Overpumping can cause the ground surface to subside if sediment within the underlying aquifer is dewatered and compacted. In certain settings, subsidence causes fissures to open on the surface.

Before Groundwater Pumping

1. Many areas have settings similar to this one: mountains composed of bedrock flank a valley or basin underlain by a thick sequence of sediment. Most water is pumped from beneath the sediment-filled basin and used by people in the valleys.

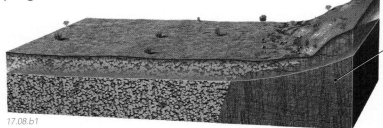

17.08.b1

2. Bedrock has interconnected fractures that give it some permeability, but it has a much lower overall porosity than sediment in the basin.

3. The water table slopes from the mountains toward the basin, across the boundary between bedrock in the mountains and sediment beneath the valley.

After Groundwater Pumping

17.08.b2 San Joaquin Valley, CA

5. As the water table drops, the upper part of the original aquifer is now above the water table and has been dewatered. Sediment within and below the dewatered zone compacts because water pressure no longer holds open the pore spaces.

4. If we overpump groundwater, the water table will drop over much of the area. In some cases it has dropped more than 100 m (~330 ft).

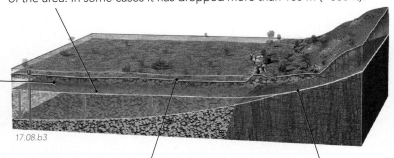

17.08.b3

17.08.b4 Chandler, AZ

6. Compaction of the sediment causes the overlying land surface to subside by several meters. Once the sediment compacts, the subsidence and loss of porosity are permanent and will not be undone if pumping stops and water levels rise again. On the left, a telephone pole in the San Joaquin Valley of California bears signs marking elevations of the land surface during subsidence—the land surface has dropped many meters! Along the coast, such subsidence could lower an area below sea level, as has occurred in New Orleans.

7. The granite cannot compact, so open fissures develop across the land surface along the boundary between land that subsided and land that did not (in the mountains). The earth fissure pictured here formed by this type of subsidence.

C How Can Groundwater Pumping Cause Saltwater Incursion into Coastal Wells?

Some wells are by necessity near the coasts of oceans and seas. These wells have a special threat—overpumping can draw salt water into the well, a process referred to as *saltwater incursion* or *saltwater intrusion*.

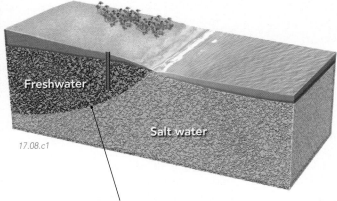

17.08.c1

Along ocean coasts, freshwater commonly underlies the land, while groundwater beneath the seafloor is salty. Freshwater is less dense than salt water and forms a lens floating on top of salt water.

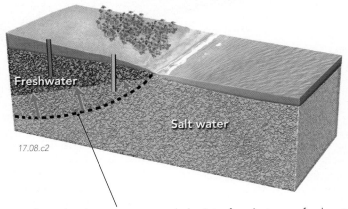

17.08.c2

When wells on land are overpumped, the interface between freshwater and salt water moves up and inland (saltwater incursion). Wells closest to the coast will begin to pump salt water and will have to be shut down.

17.8

17.9 How Can Water Become Contaminated?

CONTAMINATION OF SURFACE AND SUBSURFACE WATER SUPPLIES is a major problem facing many communities. Some contaminants are natural products of the environment, whereas others have human sources, the direct result of our modern lifestyle. What are some main sources of water contamination?

1. Examine this figure, trying to recognize every potential source that could contaminate surface water and groundwater. Then read the accompanying text blocks.

2. Water contamination can have natural causes. Weathering of rocks releases chemical elements into surface water and groundwater—some of these elements are beneficial and others are not. Rocks, especially those that have been mineralized by hot fluids, may contain lead, sulfur, arsenic, or other potentially hazardous elements. Mining activities and natural erosion move mineralized rocks away from where they formed, further spreading these contaminants.

3. We use large amounts of petroleum and coal, which have to be discovered, extracted, transported, and processed. Any of these activities potentially cause pollution. Some of the worst disasters are leaks from pipelines and supertankers, and fires at refineries and storage tanks.

4. Old landfills are the repositories for countless discarded items, many of which contain hazardous substances. Such items include diapers, lead batteries, toxic liquids from household or commercial use, compact fluorescent bulbs, old tires, and other garbage. If not properly sited and sealed from the environment, landfills can be major sources of pollution. Landfills along rivers, such as this one, can be breached by lateral erosion of channels. Supposedly impermeable linings beneath the landfill, if installed at all, can crack during settling and from daily landfill operations, allowing a toxic stew to seep into the underlying groundwater.

17.09.a1

5. One of the most basic types of contamination is human waste, which can end up in surface water and groundwater supplies if proper sanitary procedures are not followed. Contamination of this sort comes from septic tanks, accidental spills from wastewater treatment facilities, or, in less affluent parts of the world, from waste disposal in open sewers and trenches.

6. Farms, ranches, and commercial orchards are contributors of chemical and organic contamination. Chemical contaminants include fertilizers that contain nitrates, insecticides to control pests, herbicides to combat invasive weeds, and defoliants to remove leaves before harvesting crops like cotton. Irrigated fields build up salts as water evaporates, and much of this gets carried into ditches by excess irrigation water. Animal waste, which contains harmful bacteria, hormones, and feed additives, is also a potential problem.

7. Gas stations contaminate water because of leaks from underground storage tanks and spills that occur while filling vehicles. Gas stations frequently go out of business if they have to dig up leaking underground storage tanks. Spills from tanker trucks, railroad cars, and trucks delivering fuel from distribution hubs may cause water contamination if there is an accident.

8. To manufacture the items we use in our daily lives, factories use many different raw materials and chemicals. Plastic products, for example, are everywhere around us: containers for soda and bottled water, plastic bags for groceries and other purchases, and many parts of our cars. These plastics are mostly produced from petroleum, which must be refined and processed in refineries and plastic factories. Petroleum and various chemicals, along with the waste produced during the manufacturing process, can accidentally escape, as shown in this photograph, causing an industrial site to become heavily contaminated (▲). Liquid contamination may be pumped down "disposal wells," often ending up in the groundwater. Ponds intended for temporary storage can leak, contaminating surface water and groundwater. Fumes and particles emitted from smokestacks settle back to the ground or are washed down by rain and snowfall, possibly contaminating air, plants, buildings, soils, surface water, or groundwater.

17.09.a2 Russia

9. Even if a community is careful with wastes, contamination can be carried into the area by rivers that drain polluted areas upstream. Polluted surface water can seep into groundwater, and groundwater inflow can pollute streams. Soils can contaminate water, which then pollutes the next town downstream.

10. In the past, dry cleaners were sources of groundwater pollution because of the chemical solvents used to clean clothes without water. Such solvents have names from organic chemistry and commonly are referred to by their abbreviations, such as PCE for perchloroethylene ("perc" for short).

11. Houses cause water pollution during the production of the materials used to build the house, from actual construction, and from day-to-day activities that include the use of fertilizer, termite treatment, and household pesticides. Oil and gas spilled from cars and other machines, along with oil improperly disposed of during do-it-yourself oil changes, can contaminate large volumes of freshwater.

12. We may be unaware of water contamination. Subsurface rock and sediment can contain hazardous natural substances, including metallic elements and radon. We may discover the contamination only if we drill into it, often because an unusual health issue appears in a local population.

Water, Arsenic, and Bangladesh

Bangladesh, east of India, is a geologically challenged country. Much of it consists of lowlands that are flooded by storm surges in the sea and by the Ganges, one of the world's largest rivers.

One of Bangladesh's worst problems, however, is water contamination. For centuries, poor sanitation in this impoverished nation polluted the rivers and other surface-water sources with cholera, dysentery, and other diseases. To provide a new source of water, people sank more than 10 million tube wells (created by pounding tubes into the soft sediment). Unfortunately, the sediment and groundwater have a high content of naturally derived arsenic, many times the recommended limit, causing arsenic poisoning on a scale never before seen. To help solve the problem, geologists from the U.S. Geological Survey and the Geological Survey of Bangladesh, shown in these photographs, have been sampling the well waters, studying the surface and subsurface geology, drilling wells into a deeper aquifer (▼), and evaluating whether bacteria can be used to reduce the arsenic concentrations.

17.09.mtb1

17.09.mtb2

Before You Leave This Page Be Able To

✓ Describe the many ways that surface water and groundwater can become contaminated.

17.9

17.10 How Does Groundwater Contamination Move and How Do We Clean It Up?

WATER CONTAMINATION CAN BE OBVIOUS OR SUBTLE. Some rivers and lakes have oily films and give off noxious fumes, but some contamination occurs in water that looks normal and tastes normal but contains hazardous amounts of a natural or human-related chemical component. How does contamination in groundwater move, how do we investigate its causes and consequences, and what are possible remedies?

A How Does Contamination Move in Groundwater?

As contamination enters groundwater, it typically moves along with the flowing groundwater. Contamination can remain concentrated, can spread out, or can be filtered by passage through sediment and rocks.

Groundwater contamination typically moves with the groundwater down the slope of the water table.

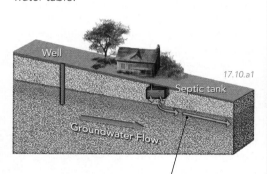

17.10.a1

Contamination from this septic tank will move to the right away from the water well. The direction of groundwater flow is clearly important in deciding where to put the septic tank relative to the well.

Contamination is drawn out parallel to the direction of groundwater flow.

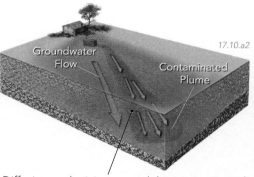

17.10.a2

Diffusion and mixing spread the contaminated zone as it migrates away from the source. Consequently, the shape of most contamination spreads out like smoke from a chimney and is called a *plume*.

Some contamination can be naturally filtered by materials through which the contaminated groundwater flows.

17.10.a3

Contamination from the septic tank on the left will be filtered by slow movement through sandstone, whereas contamination from the septic tank on the right will flow rapidly away, unfiltered, through permeable, cavernous limestone.

B How Do We Investigate Groundwater Contamination?

Hydrogeologists investigate groundwater and surface water contamination using the same approaches they use for other types of water-related geologic problems, plus a few extra strategies.

Substance	Limits	Health Issues
Arsenic	0.01 mg/L	Cancer, numbness
Cadmium	0.005 mg/L	Kidneys, liver, lungs
Chromium	0.1 mg/L	Cancer, nasal issues
Lead	0.015 mg/L	Kidneys, blood pressure
Trichloroethylene (TCE)	0.005 mg/L	Cancer, kidneys

17.10.b1 Newcastle, England

2. Water contamination is fundamentally about chemicals and hazardous microbes, so geologists collect *geochemical samples* that are analyzed either in the field or later by chemists in a laboratory. Some volatile organic compounds are detected using sensors that analyze soil gases given off by groundwater and soil.

1. Most surface-water and groundwater contamination is recognized by chemical analyses done by community water providers. In the United States, water standards are set by the Environmental Protection Agency (EPA). This table lists the EPA drinking water standards for a few of the better known or more hazardous water contaminants. Values are in milligrams per liter (mg/L), which is equivalent to *parts per million* (ppm). A standard of 0.1 mg/L for chromium means that drinking water is above the limit if it contains more than about 1 atom of chromium for every ten million molecules of water.

17.10.b2 Phoenix, AZ

3. Hydrogeologists conduct tests of an aquifer by pumping a well continuously at a specific rate and observing how that well and wells around it react during the pumping and after the pumps are turned off. This provides information about how fast groundwater and contamination might move.

C How Is Groundwater Contamination Tracked and Remediated?

Once groundwater contamination is identified, what do we do next? Hydrogeologists compile available information to compare the distribution of contamination with all relevant geologic factors. One commonly used option to clean up, or *remediate*, a site of contamination is called "pump-and-treat." Some contamination can be mostly remediated, but remediation is much more expensive than not causing the problem to begin with.

1. The first step to remediation is to properly understand the situation—what is the nature of the contamination, where is the contamination now, where did it come from, where is it going, and what are the geologic controls?

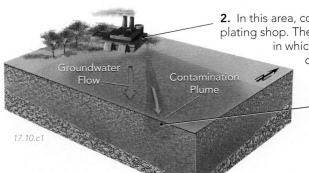

17.10.c1

2. In this area, contamination consists of chromium released by a chrome-plating shop. The water table slopes to the southeast, so this is the direction in which the upper levels of groundwater will flow. We predict that contamination will move in this same direction.

3. Chromium ions are carried away by groundwater flow and also chemically diffuse through the water, albeit at a slower rate. The combination of flow and diffusion causes the contamination to spread out like smoke from a chimney, forming a plume of contamination. There is no contamination upflow (northwest) of the shop, but the plume of contamination will spread to the southeast.

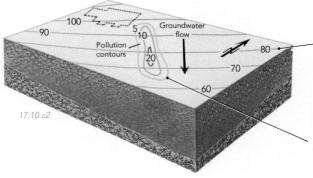

17.10.c2

4. To investigate the situation, we contour elevations of the water table to more precisely determine which way groundwater is flowing. In this case, the contours decrease in elevation to the southeast. Groundwater flows to the southeast, perpendicular to the contours (and toward lower elevation contours).

5. We draw a second set of contours based on chemical analyses of the concentration of contamination, in this case chromium. For example, areas within the 5 mg/L contour have at least 5 mg/L chromium, and those within the 10 mg/L contour have at least 10 mg/L. The EPA limit for chromium is 0.1 mg/L, so these values are well above EPA standards.

6. From these maps, we can now determine where the contamination is, which way it is moving, and where it will go in the future (down the slope of the water table). If from interviews or historical records we can determine how long ago the contamination occurred, we can use simple calculations (distance/time) to get the rate of flow. We also can use computer simulations to model past and future movement.

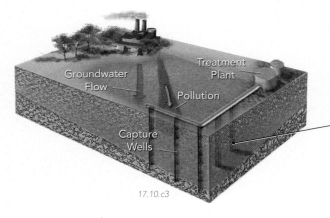

17.10.c3

7. Finally, we try to clean up the contamination. One strategy is to drill wells in front of the projected path of the contamination to contain, capture, and extract the contaminated water. Pumping brings contaminated water to the surface, where it is processed with carbon filters or other appropriate technology to separate the contaminant from the water. The cleaned water is typically reinjected into the ground, evaporates in evaporation ponds, or is channeled to flow down streams.

A Civil Action

Woburn, Massachusetts, a small town 10 miles north of Boston, was the site of a classic legal case involving groundwater contamination. The case was made famous in the book *A Civil Action* by Jonathan Harr and in a movie of the same name starring John Travolta.

The trouble began in the 1960s when the city drilled two new groundwater wells for municipal water supplies. The wells were drilled into glacial and river sediments that had filled an old valley. After the wells were installed, some residents complained that the water tasted odd and had a chemical odor. Over the next 20 years, residents began to show a high incidence of leukemia and other serious health problems. Chemical analyses showed that the groundwater was contaminated with trichloroethylene (TCE) and other volatile organic compounds. Local families filed a lawsuit against several chemical companies that were potentially responsible. The verdict remains complex, but the site is a classic example of the interaction of geology, water, health, and environmental law.

Before You Leave This Page Be Able To

☑ Sketch a plume of contamination, showing how it relates to the source of contamination and the direction of groundwater flow.

☑ Describe some ways in which hydrogeologists investigate groundwater contamination.

☑ Sketch how chemical analyses define a plume of contamination and one way a plume could be remediated.

17.10

17.11 What Is Going On with the Ogallala Aquifer?

THE MOST IMPORTANT AQUIFER IN THE UNITED STATES lies beneath the High Plains, stretching from South Dakota to Texas. It provides groundwater for about 20% of all cropland in the country, but it is severely threatened by overpumping. The setting, characteristics, groundwater flow, and water-use patterns of this aquifer connect many different aspects of water resources and illustrate their relationship to geology.

A What Is the Setting of the Ogallala Aquifer?

1. The *Ogallala aquifer*, also called the *High Plains aquifer*, covers much of the High Plains area in the center of the United States. The lightly shaded area on this map shows the outline of the main part of the aquifer. The aquifer forms an irregularly shaped north-south belt from South Dakota and Wyoming through Nebraska, Colorado, Kansas, the panhandles of Oklahoma and Texas, and eastern New Mexico.

2. The Ogallala aquifer covers about 450,000 km² (174,000 mi²) and is currently the largest source of groundwater in the country. It provides 30% of all groundwater used for irrigation in the United States. In 1980, near the height of the aquifer's use, 17.6 million acre-feet of water were withdrawn to irrigate 13 million acres of land. The water is used mostly for agriculture and rangeland. The main agricultural products include corn, wheat, soybeans, and feed for livestock.

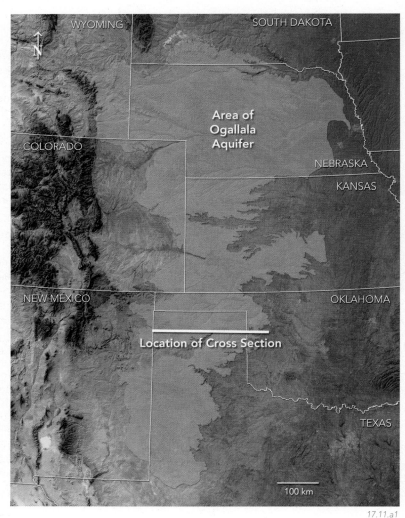

3. The aquifer is named for the Ogallala Group, the main geologic formation in the aquifer. The formation was named by a geologist in the early 1900s after the small Nebraskan town of Ogallala.

4. Much of the Ogallala Group consists of sediment deposited by rivers and wind during the last half of the Cenozoic, mostly between 19 and 5 million years ago. Braided rivers carried abundant sediment eastward from the Rocky Mountains, spreading over the landscape and depositing a relatively continuous layer of sediment. Deposition stopped when regional uplift and tilting caused the rivers to downcut and erode rather than continuing to deposit sediment. Present-day rivers continue to erode into the aquifer and drain eastward and southward, eventually flowing into the Gulf of Mexico.

17.11.a1

The Aquifer in Cross Section

5. This vertically exaggerated cross section shows the thickness of the aquifer from west to east. It shows the aquifer in various colors; rocks below the aquifer are shaded bluish gray. Note that the aquifer is at the surface and is an *unconfined aquifer*.

7. The upper part of the aquifer (shaded yellow) is above the water table and in the *unsaturated zone.*

8. Blue colors show levels of the water table for 1950 and 2000, and purple shows the predicted levels for 2050. Note that water levels in the aquifer have fallen due to overpumping. The western part is predicted to be totally depleted by 2050 (no purple).

6. The irregular base of the aquifer is an unconformity that reflects erosion of the land before deposition of the aquifer.

B Where Does Groundwater in the Aquifer Come from and How Is It Used?

Most of the water going into the aquifer is from local precipitation. This map shows the amount of precipitation received across the area, with darker shades indicating more precipitation. The western part of the aquifer receives much less precipitation (rain, snow, and hail) than the eastern part.

Areas of the aquifer that receive the least precipitation—the southwestern parts—are also those predicted to go dry by 2050.

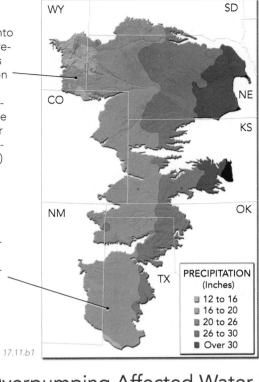

17.11.b1

PRECIPITATION
(Inches)
- 12 to 16
- 16 to 20
- 20 to 26
- 26 to 30
- Over 30

▼ This graph shows the water balance for the Ogallala aquifer. Water going into the aquifer is shown above the axis, whereas water being lost by the aquifer is below the axis. Some groundwater recharge occurs where water from precipitation seeps into the aquifer, especially in areas that receive higher amounts of precipitation, as either rain or snow.

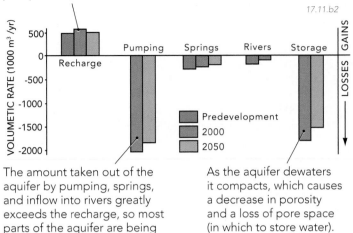

17.11.b2

The amount taken out of the aquifer by pumping, springs, and inflow into rivers greatly exceeds the recharge, so most parts of the aquifer are being dewatered.

As the aquifer dewaters it compacts, which causes a decrease in porosity and a loss of pore space (in which to store water). This cannot be undone.

C How Has Overpumping Affected Water Levels in the Ogallala Aquifer?

The USGS estimates that the aquifer contains 3.2 billion acre-feet of water. That is enough to cover the entire lower 48 states with 1.7 feet of water. How much has overpumping affected the aquifer's water levels, and what will happen to the region and to the country if large parts of the aquifer dry up?

This map shows the thickness (in feet) of the *saturated zone* within the aquifer. In some of its northern parts, more than 1,000 ft (300 m) of the aquifer is saturated with water, whereas less than 100 to 200 ft remain saturated in the southern parts.

This map shows how many feet the water table dropped in elevation between 1980 and 1995 as a consequence of overpumping. The largest drops, exceeding 40 ft, occurred in southwestern Kansas and the northern part of Texas. Compare this map to the one for precipitation.

Future Predictions—It is uncertain what will happen, but hydrogeologists are conducting detailed studies of key areas to try to predict what will happen in the next decades. Projections of current water use, combined with numerical models of the water balance, predict that some parts of the aquifer will go dry by 2050. This will have catastrophic consequences for the local farmers, ranchers, and businesses, and for people across the country who depend on the aquifer for much of their food. Subsidence related to groundwater withdrawal and compaction of the aquifer will be an increasing concern. What do you think would happen to the region if this aquifer were partly pumped dry?

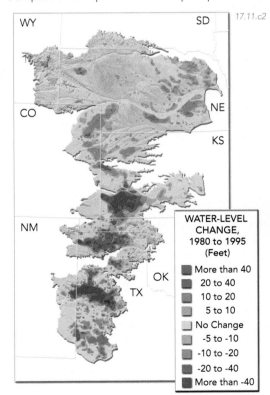

17.11.c1

SATURATED
THICKNESS
(Feet)
- 1000 to 1200
- 800 to 1000
- 600 to 800
- 400 to 600
- 200 to 400
- 100 to 200
- 0 to 100

17.11.c2

WATER-LEVEL
CHANGE,
1980 to 1995
(Feet)
- More than 40
- 20 to 40
- 10 to 20
- 5 to 10
- No Change
- -5 to -10
- -10 to -20
- -20 to -40
- More than -40

Before You Leave This Page Be Able To

✓ Summarize the location, characteristics, and importance of the Ogallala aquifer.

✓ Summarize the water balance for the aquifer and how water levels have changed in the last several decades.

17.11

17.12 Who Polluted Surface Water and Groundwater in This Place?

SURFACE WATER AND GROUNDWATER IN THIS AREA are contaminated. You will use the geology of the area, along with elevations of the water table and chemical analyses of the contaminated water, to determine where the contamination is, where it came from, and where it is going. From your conclusions, you will decide where to drill new wells for uncontaminated groundwater.

Goals of This Exercise:

- Observe the landscape to interpret the area's geologic setting.
- Read descriptions of various natural and constructed features.
- Use well data and water chemistry to draw a map showing where contamination is and which way groundwater is flowing.
- Use the map and other information to interpret where contamination originated, which facilities might be responsible, and where the contamination is headed.
- Determine a well location that is unlikely to be contaminated in the future.
- Suggest a way to remediate some of the contamination.

Procedures

Use the available information to complete the following steps, entering your answers on the worksheet or online.

1. This figure shows geologic features, rivers, springs, and human-constructed features, including a series of wells (lettered A through P). Observe the distribution of rock units, sediment, rivers, springs, and other features on the landscape. Compare these observations with the cross sections on the sides of the terrain to interpret how the geology is expressed in different areas.

2. Read the descriptions of key features and consider how this information relates to the geologic setting, to the flow of surface water and groundwater, and to the contamination.

3. The data table on the next page shows elevation of the water table in each lettered well. Use these data and the base map on the worksheet to construct a groundwater map with contours of the water table at the following elevations: 100, 110, 120, 130, 140, and 150 meters. On the contoured map, draw arrows pointing down the slope of the water table to show the direction of groundwater flow.

4. Use the data table showing concentrations of a contaminant, purposely unnamed here, in groundwater to shade in areas where there is contamination. Use darker shades for higher levels of contamination.

5. Use the groundwater map to interpret where the contamination most likely originated and which facilities were probably responsible. Mark a large X over these facilities on the map, and explain your reasons in the worksheet.

6. Determine which of the lettered well sites will most likely remain free of contamination, and draw a circle around one such well.

7. Devise a plan to remediate the groundwater contamination by drilling wells in front of the plume of contamination; mark these on the map with the letter R.

1. The region contains a series of ridges to the east and a broad, gentle valley to the west. Small towns are scattered across the ridges and valleys. There are also several farms, a dairy, and a number of industrial sites, each of which is labeled with a unique name. Geologists studied one of these towns, *Springtown*, and concluded that it is not the source of any contamination.

2. A main river, called the *Black River* for its unusual dark, cloudy color, flows westward (right to left) through the center of the valley. The river contains water all year, even when it has not rained in quite a while. Both sides of the valley slope inward, north and south, toward the river.

17.12.a1

3. Drilling and gravity surveys show that the valley is underlain by a thick sequence of relatively unconsolidated and weakly cemented sand and gravel. The deepest part of the basin has been downdropped by normal faults, one of which is buried beneath the gravel.

6. Bedrock units cross the landscape in a series of north-south stripes, parallel to the strike of the rock layers. One of the north-south valleys contains several large coal mines and a coal-burning, electrical-generating plant. An unsubstantiated rumor says that one of the mines had some sort of chemical spill that was never reported. Activity at the mines and power plant has caused fine coal dust to be blown around by the wind and washed into the smaller rivers that flow along the valley.

7. A north-south ridge is composed of sandstone, called the *lower sandstone*. *Slidetown*, a new town on this ridge, is not a possible source of the contamination because it was built too recently. A few nice-tasting, freshwater springs issue from the sandstone where it is cut by small stream valleys.

17.12.a2

Stratigraphic Section

Gravel–Unconsolidated sand and gravel in the lower parts of the valley
Upper Sandstone–Well-sorted, permeable sandstone
Upper Shale–Impermeable, with coal
Sinkerton Limestone–Porous, cavernous limestone
Middle Shale–Impermeable shale
Lower Sandstone–Permeable sandstone
Lower Shale–Impermeable shale
Basal Conglomerate–Poorly sorted with salty water
Granite–Sparsely fractured; oldest rock in area

8. The highest part of the region is a ridge of granite and sedimentary rocks along the east edge of the area. This ridge receives quite a bit of rain during the summer and snow in the winter. Several clear streams begin in the ridge and flow westward toward the lowlands.

9. A company built a coal-burning power plant over tilted beds of a unit named the *Sinkerton Limestone*, so called because it is associated with many sinkholes, caves, and karst topography. The limestone is so permeable that the power plant has had difficulty keeping water in ponds built to dispose of waste waters, which are rich in the chemical substances (including the contaminant) that are naturally present in coal.

10. The tables below list water-table elevations in meters and concentrations of contamination in milligrams per liter (mg/L) for each of the lettered wells (A–P). This table also lists the concentration of contamination in samples from four springs (S1–S4) and eight river segments (R1–R8). The location of each sample site is marked on the figure. Wells M, N, and P are deep wells, drilled into the Sinkerton Limestone aquifer at depth, although they first encountered water at a shallow depth. The chemical samples from these wells were collected from deep waters.

5. From mapping and other studies on the surface, geologists have determined the sequence of rock units, as summarized in the stratigraphic section in the upper right corner of this page. These studies also document a broad anticline and syncline beneath the eastern part of the region. Note that contamination can flow through the subsurface, following limestone and other permeable units, instead of passing horizontally through impermeable ones, like shale.

4. Based on shallow drilling, the water table (the top of the blue shading) mimics the topography, being higher beneath the ridges than beneath the valleys. Overall, the water table slopes from east to west (right to left), parallel to the regional slope of the land. All rocks below the water table are saturated with groundwater.

Well	Elev. WT	mg/L
A	110	0
B	100	0
C	105	0
D	110	20
E	120	10
F	115	0
G	120	0
H	120	50

Well	Elev. WT	mg/L
I	130	30
J	130	0
K	120	0
L	130	0
M	150	50
N	150	10
O	140	0
P	150	0

Spring	mg/L
S1	50
S2	0
S3	0
S4	0

River	mg/L
R1	0
R2	20
R3	0
R4	0

River	mg/L
R5	0
R6	0
R7	5
R8	5

17.12

CHAPTER
18 Energy and Mineral Resources

NATURAL RESOURCES ARE THE FOUNDATIONS OF MODERN SOCIETY. They provide us with the materials that sustain our way of life, including *energy resources* for electricity, transportation, and industry, and *mineral resources*, the starting materials for metals, concrete, bricks, and many other things. This chapter is about energy and mineral resources, whose study involves nearly every other aspect of geology, including plate tectonics, mineralogy, sedimentary environments, deformation, and movement of groundwater.

The Arabian Peninsula, most of which is part of Saudi Arabia, is a dry, desert land bounded on the west by the *Red Sea* and on the east by the *Persian Gulf*. The peninsula is asymmetrical: Its western edge along the Red Sea is a series of steep escarpments. On the east, it gradually decreases in elevation until it slips beneath the shallow waters of the Persian Gulf.

How did the peninsula form, why is it asymmetric, and why is the Red Sea much deeper than the Persian Gulf?

The Persian Gulf region produces a quarter of the world's oil—about 24 million barrels a day. The established oil reserves (the amount documented to be present) are more than 700 billion barrels of oil, equivalent to about 57% of the world's reserves of crude oil. The region also has 45% of the world's known gas reserves, and the United States Geological Survey estimates that the region has the greatest potential for undiscovered oil of any part of the world.

Where do oil and gas come from, and why does this region have such a large share of these critical resources?

The Zagros Mountains, which run parallel to the Persian Gulf, are also rich in oil. These mountains contain large folds that formed as the Arabian plate pushed beneath the southwestern edge of the Eurasian plate.

How are oil and gas related to folded mountain belts?

18.00.a1

▶ **This tectonic map** shows the main tectonic features of the Arabian Peninsula and adjacent regions. It shows tectonic boundaries, oil and gas fields, exposures of Precambrian rocks, and large geologic structures.

18.00.a2

The Arabian Peninsula is on the Arabian plate, which is bounded by all three types of plate boundaries (divergent, convergent, and transform). On its southwestern side, Arabia is pulling away from Africa along a divergent boundary in the Red Sea. The northwestern and southeastern boundaries of the Arabian plate are transform faults, where the Arabian plate slips past plates in the Indian Ocean and Mediterranean Sea. The northeastern boundary of the plate is a convergent boundary, where the Arabian plate pushes beneath Asia. Compression along this boundary produces folds and faults, forming structures that trapped the world's largest oil and gas resources in the Zagros Mountains and Arabian Peninsula.

TOPICS IN THIS CHAPTER

In cross section, the Arabian Peninsula is tilted—the western part rifted and uplifted, while the eastern part subsided beneath the weight of thrust sheets in the Zagros Mountains. Upper sedimentary layers thicken toward the Persian Gulf and contain folds, salt domes, and oil fields, including the largest oil field in the world in eastern Saudi Arabia.

Rifting uplifted land that flanks the Red Sea, exposing deeper Precambrian basement rocks in the western Arabian Peninsula.

Paleozoic and Mesozoic sedimentary layers near the Persian Gulf were buried and slightly heated, which converted organic material in the layers into oil and gas. Oil and gas migrated through the sedimentary layers, becoming trapped in anticlines, near salt domes, and in other structures.

Natural Resources and Our Modern Society

We use many different natural resources. Some resources, like oil and natural gas, are obvious, but others may not be noticed. It is often said—and it is true—that if something we use is not grown, it probably has a geologic origin and was found by a geologist. This is especially true of *energy and mineral resources*.

Energy resources include oil and gas, as well as coal, nuclear fuels, and energy derived from dams, wind, and the Sun. These resources are not equally distributed in every part of the world. Some areas, like Saudi Arabia and Wyoming, are rich in energy resources. Others, like South Africa and Nevada, are rich in mineral resources. Some areas have neither. Why is this so? What factors cause some areas to be rich in resources and others to have so few?

The answer, of course, is geology. Each region has its own unique geologic history, which means that some areas have thick sequences of sedimentary rocks, and others have granite and metamorphic rocks. Some areas have folds and faults; others have horizontal layers. These variations in geology lead to differences in the abundance and kinds of energy and mineral resources found in different places.

Our society, and we as individuals, use large quantities of energy, mineral, and water resources. Most people are unaware of the amounts of energy and mineral resources consumed in the United States per person. These amounts include materials used to construct roads, gypsum in wallboard, and copper in wiring. As summarized in the table included here, the National Mining Association estimates that average consumption in the United States in 2005 was 47,000 pounds of minerals per person! Nearly half of this amount was sand, gravel, and stone. We also consume large amounts of coal, natural gas, and petroleum used for fuels and to make plastics and many other items.

Finding mineral resources is essential to support our modern society, so geologists study many aspects of resources, from their general characteristics to the processes by which they form. Certain geologic processes form oil, and others form copper deposits. For many important resources, we are at the mercy of geologic events, most of which happened millions of years ago. The political and economic systems of the world must function around the geologic reality.

Material	Per Capita Consumption (pounds)
Sand, Gravel, Stone	22,060
Petroleum Products	7,667
Coal	7,589
Natural Gas	6,866
Cement	940
Iron Ore	425
Salt	400
Phosphate Rock	302
Clays	276
Aluminum (Bauxite)	77

18.0

18.1 How Do Oil and Natural Gas Form?

OIL AND NATURAL GAS SEEM LIKE ODD SUBSTANCES to find in solid rock. Where do they come from? Oil and natural gas, together called *petroleum*, form naturally when sediment rich in organic material is deposited, buried, and heated to slightly elevated temperatures. Once formed, petroleum can escape to the surface or be trapped at depth, where it can be discovered by geologists and extracted through drilling.

A What Is Petroleum and Where Does It Come From?

Naturally occurring petroleum is an organic substance, largely composed of carbon chemically bonded with hydrogen and smaller amounts of other elements. The dominance of hydrogen and carbon atoms is the reason we use the term *hydrocarbons* to refer to oil and natural gas, as well as to their refinery-produced derivative products, including gasoline and diesel fuel. The organic material that turns into hydrocarbons comes from several sources.

18.01.a1

18.01.a2

18.01.a3 Wisconsin

Reefs teem with life, including fish and other marine organisms, some of which are microscopic but build coral and other structures. Other creatures live in deeper and colder water and contribute organic material to deep-ocean sediment.

Plants, whether they are *terrestrial* (grow on land) or *aquatic* (grow in water), contribute organic material to sediment. When buried, land plants can change to make coal and methane gas, but generally they do not decompose to oil.

Most petroleum comes from *microorganisms* that occur in great variety and abundance in seas and lakes. These organisms include algae, bacteria, and other tiny organisms that live in shallow or deep water and settle to the bottom when they die.

B What Processes Turn Organic Material into Oil and Gas?

Natural organic matter comes in many forms, ranging from highly complex organisms to simple waxes or fats. When heated, these organic materials convert to a succession of other hydrocarbons, including oil. The conversion of organic material to oil, like most geologic processes, generally takes millions of years.

18.01.b1

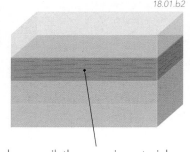
18.01.b2

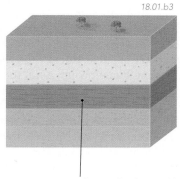
18.01.b3

The first stage in the formation of oil and gas is accumulation of organic material, perhaps in a layer of dark, organic-rich mud. A rock that contains enough organic material to produce petroleum is referred to as a *source rock*. At low temperatures on or near Earth's surface, the organic material is relatively unordered or still retains some of the structure of the animals or plants from which the material was derived.

To end up as oil, the organic material must be preserved before it can decompose. This usually involves being deposited in oxygen-poor conditions and buried under other layers of sediment. When buried to shallow depths and heated to less than about 60°C, the organic starting material is converted into *kerogen*, a thick substance composed of long chains of hydrocarbons.

Over time, source rocks can be buried by more sedimentary layers, becoming heated by the temperature increase with depth. When heated to 60°C to 120°C, long hydrocarbon chains in kerogen break down into heavy and light *oils*. At these and greater temperatures (up to about 200°C), the oily hydrocarbons convert into *natural gas*.

C Where Do Oil and Gas Go?

Once oil and gas form, what happens? Both are mobile, fluid materials and can travel along fractures and through pore spaces between grains. In many situations, they remain within, or fairly close to, the source rock where they originated. In other cases, they migrate far from where they formed.

1. As you can observe for yourself by placing several drops of any oil in a bowl of water, oil is lighter (less dense) than water. It floats on the surface of water and so will buoyantly rise through groundwater toward the surface. Water under pressure can force oil and gas upward or laterally (sideways) through the rock. Gas is even lighter than oil.

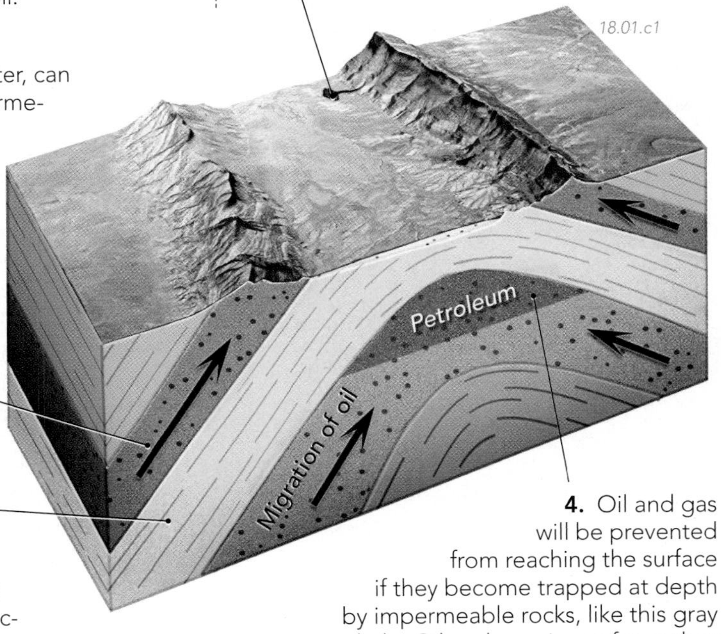

2. Oil and gas, like groundwater, can flow through rocks that are permeable. Some rocks, like many sandstones, have open spaces between the grains and along fractures, and so are relatively permeable. Oil and gas may move up through a permeable layer, such as the inclined sandstone layer shown here.

3. Other rocks are less permeable and block the flow of oil, gas, and groundwater. A rock unit can be relatively impermeable if it lacks interconnected pore spaces and through-going fractures. Rocks that are typically impermeable include (1) shale, which has very small pore spaces, (2) unfractured granite, which has crystals that generally fit tightly together, and (3) salt, which flows easily to close up any open spaces.

6. If oil flows into sandy sediments, it can form *oil sands* or *tar sands*. Large deposits in Alberta, Canada, are mined in large pits to extract the hydrocarbons from the sandy host. Oil can migrate upward until it reaches the surface, where it flows out onto the surface as an *oil seep* (▶).

18.01.c2 Southwestern UT

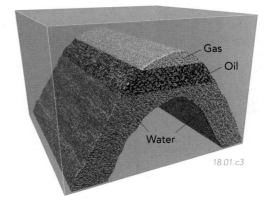

18.01.c1

Petroleum

Migration of oil

4. Oil and gas will be prevented from reaching the surface if they become trapped at depth by impermeable rocks, like this gray shale. Oil and gas rise as far as they can, floating on top of water within the rock (▶). Gas is lighter than oil, so it floats on top of the oil, which floats on top of the water.

5. To trap oil and gas at depth, a rock unit must have no through-going pores or fractures to provide an easy pathway to the surface. Severely deformed and fractured rocks, therefore, generally are less able to trap hydrocarbons than undeformed rocks. Some faults, however, effectively block the flow of fluids because the faulting has produced finely crushed rock fragments that filled open pore spaces.

Gas
Oil
Water

18.01.c3

The La Brea Tar Pits

Los Angeles, California, contains one of the world's best known fossil sites, at the *Rancho La Brea Tar Pits*. The tar formed—and is still forming—as oil seeps onto the surface, where it loses its lighter, more easily evaporated components and leaves behind a sticky, dense material called *tar*. The oil was formed at depth by the same processes that formed the many oil fields near Los Angeles, but in the case of the tar pits, the subsurface geology did not trap the oil at depth.

We have recovered from the tar pits more than one million bones of modern and ice-age animals unlucky enough to have been stuck in the tar. Among the ani-

mals are now-extinct wolves, saber-tooth cats, ground sloths, and many smaller mammals, all of which roamed the area in the last 30,000 years.

18.01.mtb1

Before You Leave This Page Be Able To

✓ Summarize where the organic material in petroleum comes from.

✓ Summarize how oil and gas naturally form by burial and heating.

✓ Sketch or describe how oil and gas move through rocks and how they can be trapped at depth or end up on the surface.

✓ Briefly describe the La Brea Tar Pits.

18.1

18.2 In What Settings Are Oil and Gas Trapped?

OIL AND NATURAL GAS CAN BE TRAPPED in the subsurface by various combinations of rock types and geologic structures. To trap oil or gas, there must be a rock in which the hydrocarbons can accumulate. Such a rock unit is known as a *reservoir*, and the area of the reservoir rock at depth that actually contains hydrocarbons is called an *oil field*, *gas field*, or *oil and gas field*. In addition to the reservoir, the oil or gas must be overlain by one or more impermeable rock units to prevent them from rising all the way to the surface.

A How Do Folded Layers Trap Oil and Gas?

The classic view of an oil and gas field is of an arch-shaped fold (anticline) that traps petroleum near its crest. Many of the world's oil and gas fields, including some of the largest, are in anticlines.

1. When rock layers are folded, they may form anticlines that arc upward (shown here) or synclines that are U-shaped (not shown). In an anticline, the layers on the flanks of the fold dip down and away from the *crest* (central high point) of the fold.

2. Oil and gas migrate (as shown by the arrows) up the flanks of the folds until they reach the crest. If there is no impermeable seal on top, the petroleum can escape to the surface, as in this brown, unconfined sandstone layer.

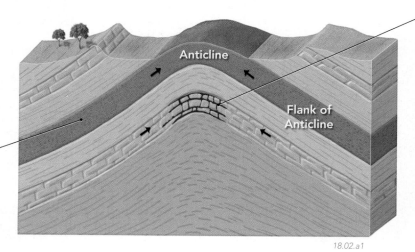

18.02.a1

3. An impermeable layer that traps oil is called a *seal*. If a reservoir rock is capped by a seal, then oil and gas (shown in black) can accumulate in the crest of the fold. In this case, a permeable gray limestone is the reservoir and an overlying greenish shale is the impermeable seal.

4. Petroleum does not form an "open pool" of hydrocarbons; instead it fills the pore spaces between grains and the narrow open spaces along fractures and beds. So, the amount of porosity (open spaces) in a reservoir rock is of great importance.

B How Do Salt Domes Help Trap Oil and Gas?

Salt is a low-density, geologically weak material that flows relatively easily when subjected to forces. Salt masses buried at depth are squeezed by the weight of overlying rocks. The salt flows to try to escape the pressure and, because salt is less dense than surrounding rocks, can rise toward the surface and create a domelike structure called a *salt dome*. Arching of rocks over and adjacent to a salt dome can trap petroleum, as along the Gulf Coast and Persian Gulf regions.

1. The weight of a sequence of sedimentary rocks presses downward on the layer of salt at the base of the diagram. The salt responds by flowing as a weak, but solid, mass. It rises upward, piercing through the overlying rocks and creating a salt dome.

2. At depth, salt flows laterally along the layer to replenish and perpetuate the rising salt mass.

18.02.b1

3. Rocks over the salt dome bend upward and can be eroded into circular or oval features on the surface. In a region where oil is known to be present, such features are targets for oil and gas exploration.

4. Petroleum can accumulate in the crest of folded layers directly above the salt dome, such as in this dome-shaped fold of a limestone reservoir rock.

5. Petroleum is also commonly trapped on the flanks of a salt dome, where the petroleum migrated upward along an uptilted layer until it encountered the central mass of impermeable salt. Many oil wells are drilled to explore the uptilted rocks that encircle a salt dome.

C What Are Other Ways That Petroleum Can Be Trapped?

Faults can trap oil and gas by juxtaposing permeable rock against impermeable rock, or by causing folding as rock layers move over bends in the fault. Other common traps are an unconformity and a trap formed by a sedimentary layer thinning or changing in character, including from one rock type to another.

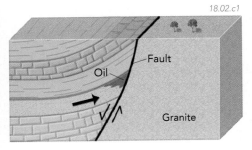

In this diagram, a normal fault displaced sedimentary layers downward against a granite. Petroleum (in this case oil) migrated up the tilted layers until it encountered the fault and granite, which stop further upward flow of the petroleum.

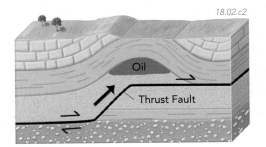

These rock layers moved up and over a bend or step in a *thrust fault*. As the layers above the fault move over the bend, they fold upward into an anticline. Petroleum migrates up the layers until it is trapped in the anticline.

An unconformity is an old erosion surface separating two sequences of rocks. Rocks below an unconformity were tilted and eroded before the layers above the unconformity were deposited. Petroleum can migrate up the tilted layers below the unconformity and be trapped by impermeable sedimentary layers along or above the unconformity. Petroleum accumulates along the unconformity and in underlying rocks.

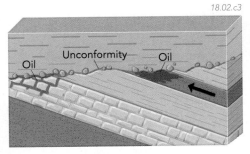

All sedimentary rock units eventually end when traced laterally, either because they decrease in thickness or because they change character into another type of rock in a facies change. The permeable sandstone bed (shown here) is encased within thick, impermeable shale. Petroleum migrating up the sandstone layer was trapped where the sandstone thinned and ended. Ancient reefs can form lenses and trap oil in similar ways.

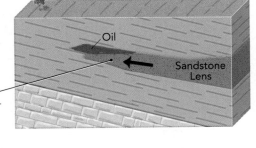

Petroleum Basins of the United States

This map shows the distribution of the main sedimentary basins that contain petroleum in the lower 48 states. Note which parts of the country have petroleum, and which do not. The top producing oil field in the United States is Prudhoe Bay in Alaska (not shown). The amount of production of oil and gas is not proportional to the size of the basin. Some small basins produce more oil and gas than larger basins.

Are there any petroleum basins where you live? If not, why do you think there are none?

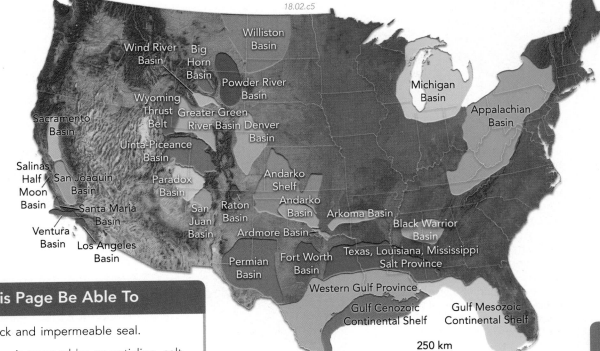

Before You Leave This Page Be Able To

✓ Describe the role of a reservoir rock and impermeable seal.

✓ Sketch and describe how petroleum is trapped by an anticline, salt dome, fault, thrust-related fold, unconformity, and facies change.

✓ Briefly summarize where petroleum basins are located in the U.S.

18.2

18.3 How Do Coal and Coal-Bed Methane Form?

COAL IS ANOTHER CARBON-BASED RESOURCE that provides energy and the raw materials to make other products. Coal forms from buried and compacted plants. There are different types of coal and different ways in which we mine, transport, and process coal, generally with the aim of producing electrical energy. Coal beds also release a type of natural gas called *coal-bed methane*.

A How Does Coal Form?

The development of coal begins with accumulation of plant matter and other organic materials on the surface. Progressive burial, compaction, and heating change the coal from one type to another, improving its quality.

Processes of Coal Formation

1. Formation of coal requires that plants accumulate on the surface in sufficient amounts so that the plant matter is much greater than the input of sand, clay, or other sediment. The most common setting for this is swamps and other wetlands.

18.03.a1

3. The plant matter must then be rapidly buried so that it is not oxidized or otherwise totally destroyed. Burial can occur because of rising sea level that covers land with water and sediment, or in other ways. The pressure that accompanies burial squeezes water and other impurities out, converting the decomposing plant material to a low-quality variety of coal called *lignite*. Lignite has less carbon than other coals, and it may be more than 50% water by weight.

5. With further burial, the increasing weight of the overlying rock layers compresses the lignite, making it more dense and compact. The thickness of the coal layer will decrease as the material is compressed into less and less space. As the coal is buried, it is also heated. The higher temperature begins to cook the coal, driving off sulfur and other chemical components that are relatively *volatile* (that readily evaporate or turn to vapor). Compaction and increased temperature convert lignite into *subbituminous coal* and then *bituminous coal*, both of which contain relatively more carbon and less water than lignite. The processes by which coal changes as it is buried and heated is *maturation*.

18.03.a3

Types of Coal

18.03.a2

2. The organic material in coal starts as compressed and partially decomposed plant matter, including *peat*, a water-soaked mass of relatively unconsolidated plant remains found in bogs. Once dried, peat can be burned. It was an important fuel resource in the past and is still used in many regions.

18.03.a4

4. Lignite is a brown, not-very-dense coal.

18.03.a6

6. Bituminous coal, a black, fairly dense coal.

7. As burial and maturation continue, the coal becomes even more compacted and, therefore, thinner and more dense. The further increase in temperature drives off even more impurities, resulting in coal with a high concentration of carbon (92% to 98%). Such coal, called *anthracite*, is the most highly prized variety because it burns cleaner and has a higher energy content for a given volume of coal. We describe the energy content of coal in terms of *calories:* the amount of heat produced by combustion. Anthracite can have more than twice the calorie content of lignite.

18.03.a5

8. Anthracite is black, dense, and shiny, and is the highest quality coal.

18.03.a7

9. During the progression from plant matter to high-quality coal, hydrocarbons released from the coal can include methane (CH_4), a colorless and odorless gas. Methane generated during the maturation of coal is *coal-bed methane*. Coal that is subbituminous and bituminous is most likely to contain coal-bed methane. Lignite generally has not matured enough, and anthracite is too mature (has been heated too much).

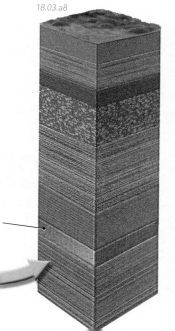

18.03.a8

B How Is Coal Mined, Transported, and Used?

To provide us with energy, coal is usually dug from the ground, transported to a power plant or other location, and then burned to convert water into steam to drive electrical generators.

▶ **1.** Most coal is mined in large open pits, where coal layers lie close to the surface. Miners remove one long strip of coal at a time, producing a *strip mine*. The entire operation moves across the coal-bearing area (▼), and it may affect very large areas.

18.03.b1 Kirtland, NM

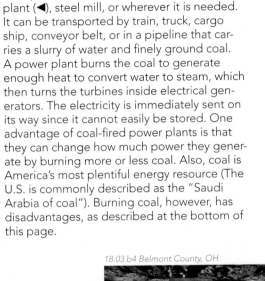

Power Plant
Coal Stockpiles
Strip Mine
Reclaimed Land

4. Whether it is mined from a strip mine or an underground mine, the coal is crushed, stockpiled, and then transported to a power plant (◀), steel mill, or wherever it is needed. It can be transported by train, truck, cargo ship, conveyor belt, or in a pipeline that carries a slurry of water and finely ground coal. A power plant burns the coal to generate enough heat to convert water to steam, which then turns the turbines inside electrical generators. The electricity is immediately sent on its way since it cannot easily be stored. One advantage of coal-fired power plants is that they can change how much power they generate by burning more or less coal. Also, coal is America's most plentiful energy resource (The U.S. is commonly described as the "Saudi Arabia of coal"). Burning coal, however, has disadvantages, as described at the bottom of this page.

18.03.b2 Witbank, South Africa

2. Huge shovels strip overlying material, called *overburden* or *waste*, off the top to expose the coal (▶). These waste materials are stored or immediately used to fill parts of the pit where the coal has already been removed. The refilled areas are smoothed and replanted with grass, trees, and other local vegetation. The aim is to make the *reclaimed land* look as undisturbed and natural as possible.

18.03.b3 Witbank, South Africa

3. We mine much coal, especially the better varieties (bituminous and anthracite), from *underground mines*. These are the typical mines we think of—small, dark, wet passageways (▶) blasted into the side of a mountain or sunk thousands of feet into Earth. Underground mines are common in the Appalachian Mountains and other places. In underground mines, miners remove relatively thin layers called *coal seams*, and then haul it to the surface, either through vertical mine shafts or up gently sloping roadways. Underground mining is more labor intensive, more dangerous, and therefore more expensive than strip mining.

18.03.b4 Belmont County, OH

The Environmental Consequences of Using Coal

Coal has some advantages over other forms of energy used for generating electricity, but the mining and burning of coal comes at a cost to people and the environment.

Coal is plentiful, widely distributed (in both the eastern and western parts of the United States), and can be fairly easily moved from mine to generator by nonpolluting conveyor belts or electric trains.

Coal also has its downside. Mining coal in underground mines is difficult and dangerous, with the potential for methane-caused explosions, cave-ins, and asphyxiation from noxious gases. Coal mining, especially as done in the past, can devastate

landscapes, leaving piles of mining waste that can contaminate water supplies. When coal is burned in power plants, in industrial furnaces, and for home heating and cooking, the incombustible material in the coal, called *ash*, rises into the air, polluting the air and settling onto the surrounding landscape. Most coal also contains sulfur, which burns to form *sulfur dioxide* (SO_2), a leading cause of *acid rain*. Some coal, called *high-sulfur coal*, contains a relatively high content of sulfur and is worse for the environment than cleaner-burning, *low-sulfur coal*. Burning coal also releases carbon dioxide and other *greenhouse gases*, which can affect our global climate.

Before You Leave This Page Be Able To

☑ Summarize how coal forms.

☑ Describe the different types of coal, ranking them from lowest quality to highest quality.

☑ Summarize or sketch how coal is mined from strip mines and underground mines.

☑ Summarize how coal is used and some of the environmental downsides.

18.3

18.4 What Are Other Types of Hydrocarbons?

NATURE STORES LARGE AMOUNTS OF HYDROCARBONS in shale and sand, in an ice-like solid beneath the ocean floor, and in other unconventional places. These resources are enormous, in several cases larger than all conventional oil and gas reserves in the world. These newly developed or potentially important sources of hydrocarbons include *coal-bed methane*, *gas hydrate*, *oil shale*, and *tar sands*.

A Where Do Coal and Coal-Bed Methane Occur, and How Is Methane Extracted?

Coal beds release *coal-bed methane*. Methane is the simplest form of natural gas (CH_4) and is a relatively clean-burning fuel that provides about 7% to 8% of natural gas used in the United States.

This map shows the general distribution of coal in the lower 48 states. Where there is coal, there is also some coal-bed methane, even if it is not extracted.

The plant matter that turned into coal accumulated at different times in different places. Some has been buried to significant depths, and some has not, because of regional differences in geologic history. Paleozoic coal of the Appalachians was buried enough so it is mostly bituminous coal and anthracite. Younger (Mesozoic and Cenozoic) coal of the western states and Gulf Coast was buried less and is lignite and bituminous coal. The largest production of coal-bed methane comes from the Colorado Plateau and southern Appalachians.

Methane sticks to surfaces within the coal or dissolves in groundwater flowing in pores and fractures in coal. Holes are drilled into the coal seam to release the pressure, allowing the gas to rise toward the surface. Collection tanks (▼) feed into local and then regional pipelines.

18.04.a1

Lignite
Bituminous and Subbituminous
Anthracite

250 km

Pacific Northwest · Northern Great Plains · Michigan Basin · Rocky Mountains · Colorado Plateau · Western Interior Province · Illinois Basin · Appalachian Basin · Anthracite Belt · Gulf Coast

18.04.a2 Farmington, NM

B What Is Gas Hydrate and Where Does It Occur?

Gas hydrate is an ice-like solid mixture of water and a natural gas, usually methane. The methane is derived from decaying organic material that is buried. It is estimated that the world's gas hydrate reserves contain twice the amount of carbon held in all other fossil fuels combined. Due to technical challenges, gas hydrate is not currently used for energy, but it could represent an important future resource as oil reserves become depleted. However, methane is a greenhouse gas that is worse for the environment than carbon dioxide.

18.04.b1

In gas hydrate, the cage-like structure of frozen water traps molecules of methane gas. Gas hydrate is called the "ice that burns." Here (▼), the methane being released from the ice is burning, not the water ice, which is incombustible. As this map shows (▶), gas hydrate occurs within ocean-floor sediments, especially in parts of the ocean that are cold and deeper than 500 m (1,600 ft). Common settings are along passive margins and trenches. Gas hydrates also occur on land, beneath the frozen arctic tundra.

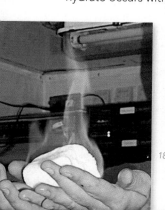

18.04.b2

1000 km

C   What Is Oil Shale and Where Is It Found?

Oil shale is another potentially important resource. Large reserves occur in the western United States and in Australia and Canada. Oil shale accumulated as clay-rich sediment that trapped abundant organic material within small pore spaces. Most formed from algal matter in lakes, lagoons, and estuaries.

18.04.c1 Green River, UT

Northeastern Utah and adjacent states have large deposits of *oil shale* in Cenozoic lake beds called the *Green River Formation* (►). The most oil-rich shale is dark brown to nearly black because of organic matter trapped between the compacted clay particles. Oil shale can be mined like coal, by open-pit or underground methods. It is then heated to 450–500°C in the presence of steam to liberate the oil. Companies have done feasibility studies of extracting the oil in place (*in situ*) by heating the rocks while still underground. Today, however, processing oil shale requires too much energy to make it an economically viable energy source.

► The United States has 60% to 70% of the world's oil shale reserves. The largest deposits are in large sedimentary basins in Colorado, Utah, and Wyoming, where several large lakes formed in early Cenozoic time. Paleozoic shale with less organic material is present in the central and eastern United States but is less likely to be developed.

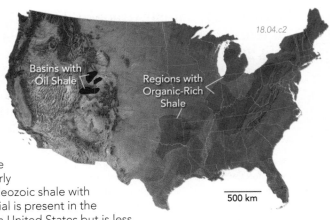
18.04.c2

Basins with Oil Shale

Regions with Organic-Rich Shale

500 km

D   What Is Tar Sand and Where Are the Largest Deposits?

Tar sand, another source of hydrocarbons, consists of thick, heavy oil intermixed with sand, clay, and other sediment. It forms when oil generated through normal processes is degraded by bacteria, which preferentially consume the lighter components of the oil, leaving a thick sludge in the sedimentary layers.

18.04.d1 Utah

◄ This tar sand has thick oil in the pore spaces. Tar sand is generally mined from open pits, but energy companies are investigating new technologies that would allow them to extract the oil by injecting steam into the deposits.

► The world's largest reserves of tar sand are in the *Orinoco Basin* of Venezuela and the *Athabasca Basin* in northeastern Alberta, Canada. Each place has one-third of the world's tar sands. Tar sand may represent two-thirds of the world's oil. Most of the petroleum that the United States imports from Canada comes from tar sands in Alberta.

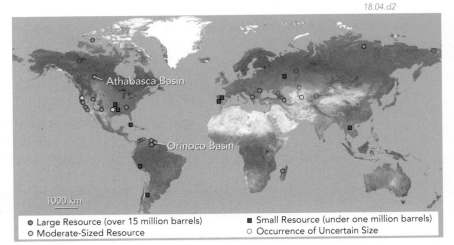
18.04.d2

Athabasca Basin

Orinoco Basin

1000 km

● Large Resource (over 15 million barrels)   ■ Small Resource (under one million barrels)
○ Moderate-Sized Resource   ○ Occurrence of Uncertain Size

Real and Potential Hazards of Gas Hydrates

Gas hydrates are potentially very hazardous, for both technical and environmental reasons. They are hazards to oil drillers because drilling into unexpected hydrate can cause a rapid buildup of gas and even an explosion that could blow steel drill pipe out of the drill hole.

Gas hydrates also appear to be responsible for large underwater landslides. Geologists think that some unusual pits and depressions on the seafloor, caused by gas escaping from the hydrates, are associated with submarine landslides. The released gas hydrate percolates through the sediment, causing it to lose strength and slide downslope.

Gas hydrates, if melted *en masse* due to warming of the oceans, could release large amounts of methane, which is a very potent greenhouse gas, perhaps causing even more warming. Analyses of methane in Greenland ice cores may have detected such events in the past.

Before You Leave This Page Be Able To

✓ For coal-bed methane, oil shale, and tar sand, summarize what the substance is, how it forms, where it is most abundant, and how it is, or could be, extracted.

✓ Describe gas hydrate, where it occurs, and its known or potential hazards.

18.4

18.5 How Do We Explore for Fossil Fuels?

OIL, NATURAL GAS, AND COAL ARE KEY RESOURCES used to power our lights, computers, industries, and automobiles. We call these three resources *fossil fuels*, because they formed in the past, there is a finite supply, and they can be depleted—they are *nonrenewable*. In essence, the energy contained in fossil fuels is solar energy initially trapped during photosynthesis, millions of years ago. Geologists use various strategies to discover new deposits of fossil fuels.

A How Do We Identify Regions That Are Favorable for Fossil Fuels?

The search for fossil fuels (oil, gas, coal, and coal-bed methane) begins by considering how each type of resource forms and then investigating whether the geologic history of a region could have permitted this resource to form and be preserved. Many regions have low potential because they do not contain the right types of rock.

Nearly all fossil fuels form in sedimentary environments, so a fundamental requirement is the presence of sedimentary rocks. Next, geologists examine the sequence of sedimentary rocks to see if they represent the right environments for the resource that is being evaluated—oil mostly forms in marine conditions, whereas coal forms from land plants and is preserved in rocks deposited on land.

18.05.a1

An area more likely contains oil and gas if it has folds, faults, salt domes, or facies changes that could trap pockets of oil and gas. If it does not have preserved sedimentary layers, there will be no coal to mine. If an area contains only granite or other igneous rock, it does not contain fossil fuels. Metamorphic rocks, too, are unlikely to contain hydrocarbons because they generally reached temperatures that drove off or destroyed any hydrocarbon-producing organic materials.

B What Studies Do Geologists Conduct When Exploring for Fossil Fuels?

Once a region is deemed favorable, or at least not prohibitive, for fossil fuels, geologists conduct detailed field investigations of key areas and outcrops. Depending on which resource is sought, geologists will tailor their investigation to observe and map the appropriate information.

18.05.b1 Capitol Reef, UT

18.05.b2 Hunters Point, AZ

18.05.b3 National Pike, PA

One of the most important aspects to investigate is the *sequence* and thickness of rock layers in the area. Geologists describe the rock types and use these characteristics to interpret the environment in which the rocks most likely formed. We call such geologists *soft-rock geologists*, because their focus is on sedimentary rocks and deposits, rather than harder igneous and metamorphic rocks (the focus of *hard-rock geologists*).

Another critical piece of the puzzle is identifying any geologic structures. Geologists document the distribution of rock units by constructing a geologic map. They measure the strike and dip of sedimentary beds within the rock layers, as well as the orientations of faults, folds, and other geologic features. From these data, geologists commonly construct a geologic cross section to determine how deep the resource-bearing beds will be.

A widely used approach for investigating the geometry of buried rock layers is the *seismic-reflection method*, where large trucks shake the ground, generating seismic vibrations that travel into the subsurface. The seismic waves reflect off boundaries between layers and return to the surface, where they can be recorded. Geophysicists computer process the data to model the subsurface structures, using the field observations as constraints.

C How Do We Determine What Is in the Subsurface?

It is one thing to document the characteristics, distribution, and attitudes of units on the surface, but exploring for geologic resources also generally requires understanding what is going on in the subsurface.

An excellent way to start understanding what rock units lie at depth is to find places where nature has exposed the sequence, such as in deep canyons or on the sides of mountains.

Observations made at the surface are then extrapolated to the subsurface. Layers can be projected downward using the thicknesses and structural attitudes measured on the surface. Confidence in such extrapolations, however, decreases with increasing depth, as the interpretation gets farther from the actual data. In this example, layers are horizontal at the surface and should continue that way for some distance downward.

Exploration for fossil fuels is challenging because of uncertainties in how the geology may change with depth and because there may be geologic structures that are not expressed on the surface. To limit the number of possibilities, exploration companies invest large sums in geophysical investigations, including gravity and seismic-reflection surveys. The ultimate test of a geologist's interpretation of the subsurface is to drill an exploration hole. The drill core, cuttings, and various physical measurements obtained from a drill hole provide a clear test of the subsurface interpretation. Such data may support the interpretation or require going back, literally, to the drawing board.

18.05.c1

D What Tools Do Geologists Use to Visualize Subsurface Geology?

18.05.d2

The main tool used to interpret and visualize the subsurface geometry of rock units and geologic structures is a *geologic cross section*. Seismic-reflection data and various techniques help geologists infer the geometry

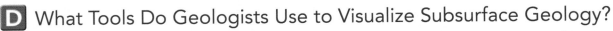

Projection of Faults into Air

18.05.d1

- 0 —
  Syncline
  Late Mesozoic Rocks
  Anticline
  Thrust Faults
- -5 —
  Late Paleozoic Rocks
- -10 —
  km
  Precambrian and Early Paleozoic Rocks   5 km

of folds and faults at depth, based on the faults and orientation of layers exposed at the surface. This cross section shows an interpretation of part of the Arctic National Wildlife Refuge (ANWR), an oil-bearing area in Alaska that has not been developed due to ecological and political considerations. The area has rocks and structures that are favorable for oil.

▲ Computer-based visualization programs and expensive visualization rooms are key components of modern exploration efforts. They help geologists integrate information from surface studies, drill holes, and geophysical surveys, especially seismic-reflection profiles.

The Costs of Exploration

The stakes in exploring for fossil fuels are high, both because of the nearly prohibitive cost and the possibility, however remote, of lucrative return on investment. Some companies employ hundreds of geologists, along with all the necessary support staff, including business people hired to obtain exploration leases from federal, state, tribal, and private landowners. Doing field work is commonly one of the least expensive parts of the operation, mostly requiring money for four-wheel-drive vehicles, fuel, accommodations, and a team of well-paid geologists.

One of the most expensive aspects involves conducting *geophysical surveys*, which may use dozens of large trucks and expensive sensors and computer gear. Drilling exploration holes is astoundingly expensive: an individual drill hole can cost tens of millions of dollars. Drilling costs, quoted in dollars per foot of depth, range from less than $100/ft for shallow wells to more than $500/ft for very deep ones—and some wells reach depths of 20,000 feet. An offshore drilling platform can cost a company more than $100,000 a day!

Before You Leave This Page Be Able To

✓ Describe aspects to consider regarding an area's potential for fossil fuels.

✓ Summarize the types of field studies geologists conduct in exploring for fossil fuels.

✓ Summarize or sketch how geologists infer what is in the subsurface, and describe the tools they use to visualize these data.

✓ Describe why exploration is so costly.

18.6 How Is Nuclear Energy Produced?

NUCLEAR REACTIONS PRODUCE ENORMOUS ENERGY that can be harnessed to power electrical generators. Currently, nuclear power is the second largest source of electricity in the United States (after coal-fired plants), supplying about 16% of the nation's electricity. Nuclear power provides nearly 80% of France's electricity and 50% of Switzerland's. How do nuclear reactions supply so much electricity, and in what geologic settings do we find deposits of uranium, the key component in the process?

A How Does Nuclear Fission Produce Energy?

Present-day nuclear power plants are based on the process of *fission*, during which an unstable isotope of a radioactive element splits into two parts. There is a significant amount of research being done to develop a reactor based on a sustained *fusion* reaction, involving atoms that collide to produce a larger and heavier element. Earth's radioactive heat arises from *fission*, whereas the Sun's energy comes from *fusion*. Here, we discuss *fission*, which is used in power plants today.

1. Uranium is a large and heavy element, with an average atomic mass of 238. The main isotopes of uranium, ^{238}U and ^{235}U, both decay by fission. ^{235}U decays more rapidly than ^{238}U, and so produces more energy, but it is much less abundant (more than 99% of uranium is ^{238}U).

2. When a uranium atom splits apart by fission, it releases relatively large amounts of energy, partly in the form of heat. In a reactor, ^{235}U atoms are bombarded by neutrons, and this induces fission. The heat produced by fission converts water to steam, which is then used to turn the turbines in electrical generators.

3. Most reactors are designed to keep the uranium and its decay products isolated from the water that is converted to steam. Only heat is exchanged between the two parts of the system, and so the steam being released does not contain radioactive materials.

18.06.a1

B In What Settings Do Uranium Deposits Form?

Uranium atoms are large compared to other elements and so have difficulty fitting into the structure of common minerals. Partly for this reason, uranium is mobile and is easily transported by groundwater and other fluids.

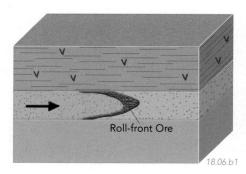

Some deposits form where uranium, carried by groundwater, encounters water with a different chemistry. Uranium accumulates along the boundary between the two waters, forming an arcuate deposit called a *roll-front*.

18.06.b1

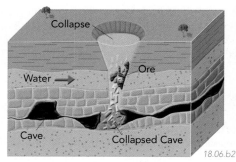

Some uranium occurs in pipe- or cone-shaped structures filled with angular fragments (breccia). Such *breccia pipes* form when limestone caves collapse and uranium is introduced later by groundwater.

18.06.b2

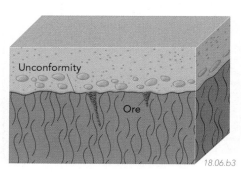

Some large deposits of uranium ore were deposited by groundwater along *unconformities* during Precambrian time. Large deposits of this type are present in Canada and Australia.

18.06.b3

Uranium is mined using underground or open-pit mining methods and then is processed in a mill that physically or chemically concentrates the uranium minerals. The relatively high price for uranium ore, typically tens of dollars per pound, makes it possible to use underground mines, which are expensive to operate. If uranium ore is close to the surface, it generally is more economical to mine the deposit via an open pit.

18.06.b4 Olympic Dam, Australia

C How Is Uranium Used to Generate Electricity?

A number of steps, some of them technically difficult, are required to obtain material that is rich enough in uranium, specifically in ^{235}U, to sustain a controlled fission reaction.

The most technically challenging part of the operation is using devices called *gas centrifuges* (◄) to preferentially *enrich* the uranium in ^{235}U, the more energetic isotope. Since ^{235}U and ^{238}U have nearly identical properties, uranium enrichment is the major obstacle for countries trying to produce nuclear materials for power generation—or for nuclear weapons. The *enriched uranium,* now with enough ^{235}U to sustain a fission reaction, is processed into pellets

18.06.c1 Piketon, OH

or rods for the reactor. Nuclear power generation requires much lower concentrations of ^{235}U than are needed for nuclear weapons, and producing weapons-grade material requires much more sophisticated technology.

18.06.c2

The rods of enriched uranium are used in one of several different types of nuclear reactors. In each type, some type of material, called the *moderator,* is placed between the rods to slow the neutrons so that they cause fission upon colliding with atoms of ^{235}U. Most commercial reactors in the United States use water as a moderator and are called *light-water reactors.* The Chernobyl disaster in the Soviet Union involved a totally different design that used graphite as the moderator. In most reactors, boron-containing fluid is mixed into the water around the rods to control the rate of the reaction. Heat from nuclear fission converts water into steam to drive electrical generators. The only emissions from nuclear power plants are heat and water vapor.

D What Are Some Environmental Issues Associated with Nuclear Energy?

Nuclear power holds the promise of generating large amounts of electrical energy without adding more carbon dioxide or methane, both greenhouse gases, but costs, environmental concerns, and political considerations have limited its wider use in the United States. No new reactors have been built since the 1970s.

18.06.d1 Moab, UT

Uranium mining and milling involve materials that are relatively rich in uranium, which can contaminate the water, land, and air around facilities if materials are handled improperly. Also, some miners and mill operators have become gravely ill from overexposure to radioactive materials. These uranium-contaminated materials in Moab, Utah, lie right next to the Colorado River, a major source of water for the West.

18.06.d2

After fission has depleted some percentage of their uranium, the fuel rods are removed from the reactor and mostly stored on-site. Such *spent fuel rods* contain high concentrations of uranium, as well as other radioactive elements, so they pose a large risk of radioactive contamination.

18.06.d3 Chernobyl, Ukraine

A steam explosion, followed by a number of tragic mistakes, led to other explosions and a reactor meltdown at the Chernobyl nuclear reactor in the Ukraine in 1986. The disaster killed at least 56 people and contaminated a very large area, requiring resettlement of more than 200,000 people.

Yucca Mountain

Yucca Mountain is a mesa capped by volcanic rocks in an isolated part of the Mojave Desert west of Las Vegas, Nevada. It has been designated as the future repository for spent fuel rods and other high-level (uranium-rich) radioactive waste from commercial reactors. We have spent more than $6 billion at the site on scientific studies, initial construction, and related activities. The scientific studies have focused on the surface and subsurface geology of the site, especially whether the region is stable and whether the deep water table will keep radioactive materials out of the environment for thou-

sands to hundreds of thousands of years. After decades of geologic study and intense scrutiny by various interest groups and government agencies, politics will decide the site's fate. Yucca Mountain, if used, cannot solve the nation's nuclear-waste problem, in part because of large amounts of nuclear waste currently stored in sites across the country.

18.06.mtb1 Yucca Mtn., NV

Before You Leave This Page Be Able To

✓ Summarize how nuclear fission releases energy and is used to generate electricity.

✓ Summarize or sketch some settings in which uranium deposits form.

✓ Describe positive and negative aspects of the use of nuclear energy.

✓ Briefly describe what material may be stored at Yucca Mountain.

18.6

18.7 How Is Water Used to Generate Electricity?

WATER IS PLENTIFUL ON OUR PLANET, and so we have found ways to use this abundant resource to generate energy. Using the movement of water to generate electrical energy produces *hydroelectric power*, which commonly employs some type of dam across a river valley but can also involve tapping the energy of moving water in ocean tides and currents.

A How Is Electricity Generated by Hydroelectric Dams?

Hydroelectric power, almost all from dams, provides about 10% of electrical energy for the United States and about 20% of the world's electricity. It is the main source of electricity for some topographically rugged western states, like Idaho, Washington, and Oregon, as well as nations including Norway and Iceland.

How Hydroelectric Dams Generate Electricity

To generate hydroelectric power, dams capture water from rivers or streams, storing it in a *reservoir* behind the dam. The dam is constructed out of concrete or compacted clay, rock, and other material. Large steel pipes or concrete tubes within the dam guide water to flow from higher elevations to lower ones, under the constant—and free—force of gravity.

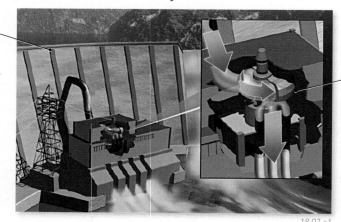

18.07.a1

When we need power, operators of the dam allow water to flow through the pipes and tubes within the dam and then down into a powerhouse, where the moving water turns turbines or blades in electrical generators. The amount of energy produced is related to the velocity and volume of the water passing through the turbines. An ideal dam, from an energy perspective, is high and has a high-volume reservoir. An advantage of hydroelectric energy is that the *potential energy* of the water can be stored until we need the electricity (which cannot be stored). If the demand for electricity goes up, like during the day, operators of the dam let out more water and generate more electricity. At night, when there is less demand, operators let out less water.

Geologic Factors Important to Hydroelectric Dams

1. Geologic factors control the suitability of a site for a dam, in many cases making a site unsuitable for a dam of any type. This figure summarizes these factors.

2. A main geologic factor is whether there is a deep-enough canyon or valley in which to build a dam. Depth of a canyon or valley is in turn controlled by the types and sequence of rocks and by the area's history of uplift, erosion, and deposition.

3. Engineers build most dams where the two walls of the canyon or valley are relatively close together. This keeps the width of the dam to a minimum, improving structural stability and keeping construction costs down.

4. Rocks beside and beneath the dam should be relatively impermeable to limit seepage of water around and under the dam. In addition to the loss of valuable water, such seepage can dissolve or loosen material in the rocks and weaken the foundation of the dam over time.

18.07.a2

9. The canyon shape determines the width and depth of the reservoir. Deep, narrow lakes lose less water to evaporation than shallow, wide ones and have greater energy potential.

8. Engineers anchor a concrete dam to the rocks on either side, the *abutments* of the dam. Rocks in the abutments need to be relatively strong and unfractured, and be composed of insoluble rock types.

7. Dams cannot safely be built in sites where the adjacent and underlying rock types, like this layer of shale, are too weak to anchor or support the dam.

6. Faults, fractures, and other zones of weakness and high permeability can be a fatal flaw, both in terms of site suitability and for public safety downstream of a dam.

5. Rocks below and adjacent to the dam should also be *chemically nonreactive*. Most limestone is unsuitable because it is soluble. Volcanic tuff can convert to clays and other weak materials if exposed to water over sufficient time. Some dams built on such altered rocks have been abandoned or else required a total redesign of the dam structure because the foundation was simply too weak to support a dam.

B How Is Electricity Generated from Ocean Tides and Currents?

Hydroelectric power can be generated using the changes in local sea level that accompany rising and falling tides. Power can also be generated by submerging propellers or turbines in the shallow ocean to be spun by the moving water in ocean currents.

1. One way to generate electricity from tidal changes is to construct a dam-like structure across a narrow, shallow inlet, like an estuary. The barrier impedes the flow of water, causing water to pile up on one side of the barrier during high tide and on the other side during low tide. An ideal location is one with a large difference between high and low tide levels. One favorable site for tidal power is the Bay of Fundy in Nova Scotia, which can have a tidal range of 16 m (52 ft).

18.07.b3

2. When the tide rises, tidal forces pull ocean water toward the land and pile it up on the seaward side of the barrier. When the water level on the seaward side is sufficiently higher than the water level on the landward side, the gate is opened and the inward-rushing water turns the turbines.

Rising Tide   18.07.b1

3. During a falling tide, water is trapped on the landward side of the barrier, while the water level on the seaward side drops. The trapped water flows through the gates toward the sea, turning the turbines. Electricity cannot be generated at all times, only when the water levels on the two sides of the barrier are sufficiently different. Tidal power generation occurs only during tidal changes, and only along coastlines with significant tides.

Falling Tide   18.07.b2

4. Alternatively, we can anchor large turbines or propellers to the shallow seafloor (▲) in an area where water flows past because of tides or prevailing ocean currents. The moving water spins a turbine or propeller blade, which turns the shaft of an electrical generator. With this technique, ocean currents can generate electricity most of the time.

Environmental Issues Associated with Dams

Dams hold the promise of a nearly constant supply of electrical power with no associated emissions of greenhouse gases or toxic contaminants. They have some advantages compared to other ways of generating electricity—the amount of electricity being generated can be changed rapidly just by increasing or decreasing the amount of water released through the turbines. This is important because the demand for electricity varies greatly between daytime and nighttime, and from hot to cold seasons, and large quantities of electricity are not easily stored. With other systems, if the amount of electricity being generated is greater than the amount being used, the excess is simply lost.

Unfortunately, dams, like *Glen Canyon Dam* in northern Arizona (shown here), and those along rivers of the Tennessee Valley, have some important negative environmental aspects. During filling of the reservoir, rising water inun-

18.07.mtb1 Glen Canyon Dam, AZ

dates the canyon or valley behind the dam. This will destroy any farmlands, houses, or even cities that are located where the reservoir will be. Flooding will destroy all the existing vegetation and animal habitat, as well as any special natural places and archeological sites.

Dams also interfere with the natural river dynamics, changing the natural flow patterns, such as spreading out large spring floods into a more consistent and managed flow. A dam traps sediment carried by the rivers and streams flowing

into the reservoir, and the reservoir will eventually fill up with sediment (called *silting up*). This blockage deprives the downstream river of sediment and associated nutrients. This can cause erosion and drastically change the downstream habitat, as clear, cold water from the depths of the reservoir replaces the warmer, muddy water that flowed down the river before construction of the dam.

Before You Leave This Page Be Able To

✓ Sketch or describe how electricity is generated by hydroelectric dams and from tides and ocean currents.

✓ Summarize how geology affects the location of a dam.

✓ Summarize some advantages and disadvantages of hydroelectric dams.

18.7

18.8 What Are Alternative Energy Sources?

FOSSIL FUELS, NUCLEAR ENERGY, AND DAMS have drawbacks, so considerable research and development have gone into exploring other ways to produce energy. The goal is to find energy sources that are friendly to the environment in terms of not producing greenhouse gases and that are *renewable resources*, meaning that their supply is essentially limitless and using them doesn't remove something irreplaceable from Earth. Such approaches include using heat within Earth, wind, and solar energy.

A What Is Geothermal Energy and What Sites Are Most Favorable for Its Use?

Geothermal energy uses Earth's natural heat as an energy source. Geothermal power plants convert natural *hot* water to steam to power electrical generators. *Naturally warm* water can be piped from the ground to places where it can be used to heat buildings and greenhouses, or to keep streets and sidewalks free of ice and snow.

18.08.a2

18.08.a1

18.08.a3 Nesjavellir, Iceland

1. In some regions, hot waters form hot springs, streaming or smoldering pools of water, or an intermittent rising fountain of hot water and steam called a *geyser*. This geyser is at Geyser, Iceland.

2. Temperature increases with depth, so water circulating through the crust can become heated at depth and then rise to the surface. This is how *hot springs* form.

3. The ideal combination of high temperatures and relatively shallow depths is most common in areas of recent volcanic activity, commonly within a collapsed caldera formed by eruption of volcanic ash.

4. Shallow magma or solidified, but still hot, magma chambers that are still hot can heat water to high temperatures, exceeding 200 to 300°C (~500°F). Although water of these temperatures would boil on the surface, it generally does not boil at depth because water pressures work against the great expansion in volume required to convert liquid water into gaseous steam.

5. To generate electricity, hot water is piped to the surface and into power plants. In the plant, the confining pressure on the overheated water is released, and the hot water flashes into steam, driving the turbines in the electrical generators, like these in Iceland. ▼

6. Regions with recent faulting can also be promising sites for geothermal energy. Faults disrupt the continuity of aquifers and provide a conduit for heated water to rise to the surface and issue from hot springs. The Iceland site shown above is within an active rift zone.

7. If rocks are hot at a shallow depth but dry, water derived from some other source can be pumped down drill holes to be heated by the rocks and then pumped back to the surface and used for heating or power generation.

B How Is Electricity Produced from the Wind?

Wind is another clean and renewable energy source. It currently provides very little of the world's and North America's power requirements, but it is one of the fastest growing sources of energy, both here and abroad.

18.08.b1

Large-scale generation of electricity from wind requires a site that has strong winds much of the time. Important geologic factors to consider are how the surface topography interacts with or controls wind, and whether the materials beneath a site are suitable for building the necessary facilities. Each wind turbine has its own small electrical generator.

An advantage of wind power is that it is renewable, is nonpolluting, and can be used in remote locations and in areas that have little other infrastructure.

One disadvantage of wind power is that winds, and the resulting power, are variable. If there is no wind, there is no electricity generation. Wind turbines affect the *aesthetics* of the site, being large and conspicuous, even if they are painted to help blend in with the environment. They can be noisy, are relatively expensive to maintain, and kill birds. The downsides are weighed against the obvious benefits of clean, renewable power.

C How Is Solar Energy Used?

Solar energy involves using the Sun's free electromagnetic energy to heat buildings and generate electricity. There are many strategies for using solar energy, including *passive solar*, *active solar*, and *photovoltaic panels*. All solar-energy approaches work best in sunny climates and in sites with unrestricted views of the Sun.

▶ In *passive solar*, light and infrared energy from the Sun enter a space through glass windows, naturally heating the inside air. Passive solar does not use any moving parts (hence the name *passive*) and is as easy as designing a house with large windows facing south (in the Northern Hemisphere) to collect the winter sun. Overhangs shield the windows in the summer.

Active solar implies that there are moving parts and some use of electrical energy, such as a fan for moving heated air or an electric pump for circulating heated fluids from the solar panel to the interior of the building.

18.08.c1 Tucson, AZ

18.08.c2

Photovoltaic panels convert sunlight directly into electricity. Such panels, although expensive to produce and install, provide nonpolluting renewable energy, even to remote locations and small sites.

D What Are Some Other Alternative Sources of Energy?

18.08.d1 California

Biomass—Energy can be produced by burning scrap wood, by burning methane released from decaying organic material in landfills, and in other ways. One downside of producing biomass and other biofuels is an increase in water usage and potential contamination by fertilizers.

18.08.d2 Indiana

Ethanol—Ethanol is a type of alcohol that can be used to fuel cars. To produce ethanol, corn, sugar cane, and other plant material is soaked in ammonia, fermented, and distilled. Recent increases in ethanol use caused steep rises in the price of corn and other foods, locally causing food riots.

18.08.d3

Fuel Cells—This potentially important technology uses electricity to break water molecules into hydrogen and oxygen, which then can be used as fuel. Some other source of energy, such as coal, must be used to generate the electricity, and using today's fuel cells does not lead to any net energy gain. ▶

Trade-offs Between Different Ways of Producing Energy

In an ideal world, our energy sources would be renewable, nonpolluting, ubiquitous, portable, cheap, and easily extracted without affecting the ecology or aesthetics of the site. All current energy sources have *trade-offs* of one sort or another—cheap and portable but polluting; renewable and nonpolluting, but expensive and requiring huge facilities. No single energy source does it all.

The decision of which type of energy to use involves identifying the *advantages* and *disadvantages* of each type, and then carefully comparing them against each other to see if the good outweighs the bad. This decision process is similar to the scientific method—we pose a question, make

observations, and collect data to better understand the variables. We then make predictions or models for several scenarios, and we test the predictions by observing or numerically modeling the system.

The final decision, however, also involves *values* of individuals, companies, and governments. Such values can involve such nonscientific aspects as aesthetics, ethics, emotions, and politics. An excellent example is the *Cape Wind Project*, where people living around Cape Cod, Massachusetts, known for their strong environmental sentiments, are resisting, for aesthetic reasons, the installation of wind turbines to generate electricity using environment-friendly wind energy.

Before You Leave This Page Be Able To

☑ Describe or sketch the surface and subsurface geologic factors favorable for geothermal energy.

☑ Summarize electricity production using the wind.

☑ Describe how we produce energy from solar power, biomass, ethanol, and fuel cells.

☑ Discuss some of the trade-offs involved in each of the various energy sources.

18.8

18.9 What Are Mineral Deposits and How Do They Form?

PHYSICAL AND CHEMICAL PROCESSES concentrate and disseminate elements and minerals, causing rocks to have a higher content of some elements and minerals, and a lower content of others. If a volume of rock is enriched enough in an element or mineral to be potentially valuable, we call it a *mineral deposit*. Materials extracted from mineral deposits provide the very foundation for our modern world.

A What Is a Mineral Deposit and What Is an Ore?

Most rocks are not considered to be mineral deposits, even though they are indeed composed of minerals. Instead, the term *mineral deposit* means the rock is especially rich in some commodity that might be valuable, and such rocks are said to be *mineralized*. If a mineral deposit contains enough of a commodity to be mined at a profit, it is an *ore deposit*, and the valuable rocks or other materials in that deposit are *ore*.

An outcrop of plain white quartz is not a mineral deposit, unless it contains flecks of gold. If rich enough in gold, like the fist-sized, gold-rich sample in the inset photograph, the piece of quartz may also be ore.

18.09.a1 Murchison, South Africa

18.09.a2

18.09.a3 Hemlo Mine, Ontario, Canada

Ore can be conspicuous, like this rock that contains shiny, brass-colored sulfide minerals. Some ore is much more subtle, being enriched in some element but otherwise looking like a typical igneous, metamorphic, or sedimentary rock.

B What Determines Whether a Mineral or Rock Is an Ore?

Many factors, some of them nongeologic, determine whether a rock or other material is considered ore. These include concentration of the commodity (valuable material) in the rock, how easily the commodity is extracted from the rock, the proximity to markets, and the economics that controls prices, especially supply and demand.

18.09.b1 Kidd Creek, Ontario, Canada

◄ *Grade of Ore*—The percentage or concentration of the valuable commodity in a rock is called the *grade*. A rock that is very rich in the commodity, like this sulfide–rich copper ore, is *high grade;* the opposite is *low grade.*

► *Type of Ore*—A commodity can occur in different types of minerals, such as copper in these blue-green copper-oxide minerals. In the case of copper, it is cheaper to extract copper from oxide minerals than from the copper sulfide minerals shown in the left photograph.

18.09.b2 Morenci, AZ

18.09.b3 Morenci Mine, AZ

Location of Deposit—A deposit that is close to markets and to infrastructure, such as railroads and electrical lines, will be more economical than one that is far from civilization or in an environmentally sensitive place. Economic factors, such as the price for which the commodity can be sold, and political factors, especially whether the area has a stable government, can determine whether a deposit can be mined. ►

◄ *Size and Depth of Mineral Deposit*—The size of a deposit determines if it is worth mining, because of the large cost of setting up a mining operation. If a deposit is small, the investment in equipment may not be worthwhile unless the ore is very rich. A shallow mineral deposit is cheaper to mine than a deeper one. A large, open-pit mine is more economical to operate than a small mine or a deep, underground one.

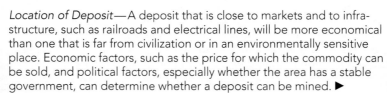 

18.09.b4 Bingham Canyon Mine, UT

C   What Processes Can Form a Mineral Deposit?

Many geologic processes can concentrate minerals or chemical elements, but it takes special circumstances to form a mineral deposit, especially one rich enough to become ore. Some ore-forming environments involve hot or deep processes and others involve low-temperature processes typical of near-surface environments.

Hot or Deep Processes

18.09.c1 Bushveld, South Africa

Igneous Crystallization—A crystallizing magma can form a mineral deposit by having one or two minerals crystallize at one time. Heavy crystals can sink to the bottom of a magma chamber, forming enriched parts, as occurred to crystallize this dark layer of iron oxide minerals.

18.09.c3 Oatman, AZ

Hydrothermal Deposition—Hot water, called a *hydrothermal fluid*, can precipitate minerals in fractures, on the surface in hot springs, or from hydrothermal vents on the seafloor. Mineralization within or along a fracture generally involves a hydrothermal fluid and is called a *vein*.

18.09.c5 Rooiberg, South Africa

Hydrothermal Replacement—A hydrothermal fluid can permeate through a rock, replacing the materials with new minerals containing chemical components that were carried by the fluid. This dark ore formed when tin-rich waters replaced a reactive limestone.

18.09.c7 Salome, AZ

Metamorphism—Metamorphism, due to increased temperatures and pressures, can convert existing minerals in a rock into new minerals that may be valuable. This blue kyanite grew during metamorphism of an initially clay-rich rock.

Surficial Processes

18.09.c2 North Silverbell, AZ

Weathering Enrichment—As mineralized rocks are exposed at the surface, valuable elements can be *leached* from the rocks by weathering, carried by groundwater, and deposited elsewhere. Leaching removed copper from this rock and enriched rocks tens of meters below.

18.09.c4 Brazil

Formation by Weathering—Some ore deposits represent *residual* materials left behind as other chemical components are leached away. Deposits enriched in aluminum (bauxite) and in nickel can form as residual soils from weathering.

18.09.c6 Kimberly, South Africa

Mechanical Concentration—As materials are transported by rivers and washed by waves, minerals can be sorted and concentrated, usually on the basis of size and density. These river gravels in South Africa were mined for diamonds, which are dense and collect in river bottoms.

18.09.c8 Gypsum, CO

Low-Temperature Precipitation—Valuable minerals are deposited by evaporating sea or lake water, or are deposited by groundwater flowing through permeable rocks. These gypsum beds were deposited by evaporation of seawater.

Before You Leave This Page Be Able To

✓ Explain the meaning of mineral deposit, mineralization, and ore.

✓ Summarize geologic and non-geologic factors that determine whether a mineralized body can be mined.

✓ Summarize the processes that can form a mineral deposit.

18.9

18.10 How Do Precious Metal Deposits Form?

GOLD, SILVER, AND PLATINUM are three members of a family of valuable metallic elements called *precious metals*. These metals are widely used for industrial and monetary purposes in addition to jewelry, but they occur only in relatively minor concentrations in Earth's crust and so are high-cost materials. Where do precious metals come from, and how do we find new deposits?

A In What Settings Do Gold- and Silver-Rich Mineral Deposits Form?

Gold and silver occur together in many geologic environments because these two elements behave similarly under many geologic conditions. Mines that produce gold usually get some silver from the ore as well. Since gold is much more valuable than silver, the discussion below focuses on gold.

Gold and Silver Veins

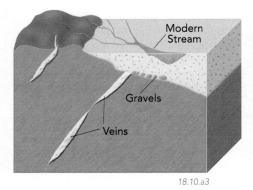

◄ Gold and silver commonly occur in narrow, steep, tabular bodies, called *veins*. Veins represent fractures through which hydrothermal fluids passed and deposited minerals. They are typically not extensive, but ore in veins can be very high grade. Many veins are associated with volcanic areas or igneous intrusions, which heat up large volumes of water.

▶ Veins are common in metamorphic rocks, and they may contain enough gold to be mined. In some cases, gold was already present in the rocks before metamorphism but was redistributed and concentrated in veins by metamorphic processes. In other cases, hydrothermal fluids introduce gold and silver.

18.10.a1

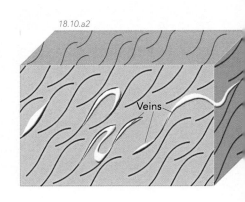

18.10.a2

Modern and Ancient Placer Deposits

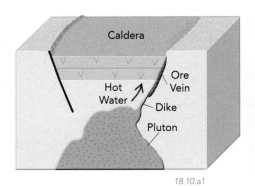

◄ Pieces of gold, liberated from bedrock by weathering and erosion, can be carried away by mountain streams and rivers. Gold is more dense than any other grains being carried by the running water, so it will be deposited at the base of gravels. Such deposits are *placer gold deposits*, or simply *placers*.

▶ If gold-bearing gravels are buried by later rocks, the unconsolidated gravels will be compacted and lithified into *conglomerate*. The famous gold mines of South Africa are in ancient conglomerate beds. Miners have followed such gold-bearing conglomerate to depths of nearly 4,000 m (13,000 ft).

18.10.a3

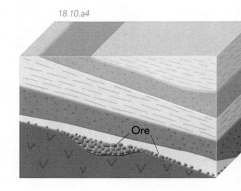

18.10.a4

Large, Low-Grade Gold Deposits

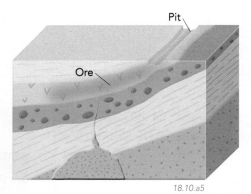

◄ If gold prices are high, it becomes profitable to mine rocks that contain low concentrations of gold. These deposits, typically hosted by volcanic or sedimentary rocks, are low grade but fairly large, and they are mined by open-pit methods. The gold typically is not visible, but it can be extracted by chemical processing.

By-Product Gold

▶ A substantial amount of gold is recovered from other types of mineral deposits, especially from large, open-pit copper mines. The ore is mined for the copper content, but the net worth of the gold recovered as a *by-product* of the copper mining can make the mine profitable during times of low copper prices.

18.10.a5

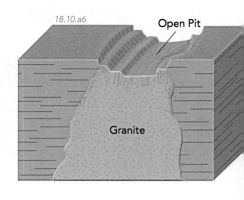

18.10.a6

B What Parts of the United States Have Large Gold Deposits?

In the United States, most gold has been mined from bedrock and rivers of the mountainous west and Alaska. In Canada, gold deposits are common in the western mountains and in Precambrian rocks of the Canadian Shield.

This map shows larger gold deposits of the lower 48 states. Gold rushes in Alaska and along the Yukon River of northwestern Canada (not shown on this map) were largely in river gravels and so are modern placer deposits. The famous California Gold Rush was touched off by flakes of gold found in river gravels at Sutter's Mill, but the gold originated within veins in metamorphic rocks of the Sierra Nevada foothills. This area produced some large gold specimens.

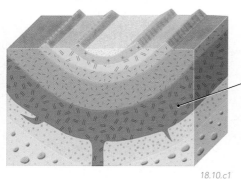
18.10.b1 Murphys, CA

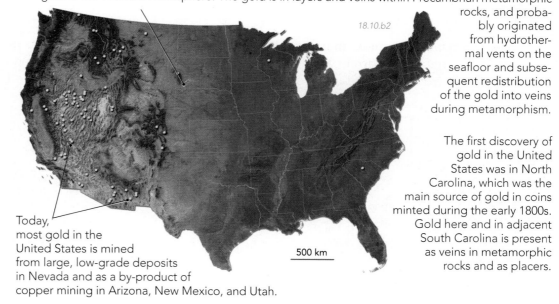

The deep Homestake Mine of the Black Hills, now inactive, is the largest historic producer of gold in the Western Hemisphere. The gold is in layers and veins within Precambrian metamorphic rocks, and probably originated from hydrothermal vents on the seafloor and subsequent redistribution of the gold into veins during metamorphism.

18.10.b2

The first discovery of gold in the United States was in North Carolina, which was the main source of gold in coins minted during the early 1800s. Gold here and in adjacent South Carolina is present as veins in metamorphic rocks and as placers.

Today, most gold in the United States is mined from large, low-grade deposits in Nevada and as a by-product of copper mining in Arizona, New Mexico, and Utah.

500 km

C Where Does Platinum Occur?

Platinum is chemically very different from gold and silver and occurs in very different settings. Most platinum deposits are associated with mafic or ultramafic igneous rocks, which rose into the crust from the mantle.

18.10.c1

South Africa is the largest supplier of platinum, typically producing 70% of the world's platinum. Most mines are within a large mafic to ultramafic intrusion that developed layers as different minerals crystallized. This intrusion, the Bushveld complex, is the largest layered intrusion on Earth.

18.10.c2 Merensky Reef, South Africa

Much of the platinum occurs in a specific zone that is less than one meter thick. The layer has centimeter-sized igneous crystals and a high chromium and platinum content. Note the black pen cap for scale.

Precious Metals and Deep Mines

Precious metals are so precious that they are priced by the ounce, specifically by the *troy ounce*, which is about 1.1 normal ounces. What do you think is the going rate for an ounce of gold, silver, or platinum?

Gold is more precious than silver, typically costing 50 times as much per ounce. The price of gold fluctuates tremendously, but when this book was written, gold was more than $900 per troy ounce. Silver was nearly $13 per troy ounce, and platinum was more than $1,000 per troy ounce.

Such high prices make it worthwhile to dig deep mines. The Homestake Gold Mine is nearly 2.4 km (8,000 ft) deep. Some gold mines in South Africa are nearly 4 km (>13,000 ft) deep—to mine a gold-rich pebbly layer (▼) less than 30 cm thick.

18.10.mtb1 Ventersdorp, South Africa

Before You Leave This Page Be Able To

☑ Describe or sketch the main geologic settings for gold and silver deposits.

☑ Identify where gold is mined in the United States, and describe the type of gold deposits in each region.

☑ Describe the geologic setting of the world's largest platinum deposits.

☑ Explain why precious metal mines can afford to be so deep.

18.10

18.11 How Do Base Metal Deposits Form?

SOME METALS ARE MUCH MORE COMMON than precious metals and, unlike gold, tarnish fairly easily in air. Such metals are called *base metals*, and they include iron, nickel, copper, lead, and zinc, which are fundamental to our daily lives. Where and how do deposits of these metallic elements form?

A How Do Iron Deposits Form?

Iron is arguably our most important metal because it is the main ingredient in steel. Iron deposits that are mined generally contain two iron oxide minerals: *magnetite*, which is magnetic, and *hematite*, which has a characteristic red streak. Iron carbonate and iron sulfide minerals are also present in some iron deposits.

We mine most iron from sedimentary sequences called *banded iron formations*, which contain many thin layers of iron-rich and quartz-rich rocks. Nearly all large banded iron formations are Precambrian (about 2 billion years old) and are interpreted to mark a time when increasing oxygen in Earth's atmosphere caused dissolved iron in the ocean to precipitate as iron minerals on the seafloor.

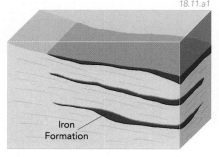
18.11.a1

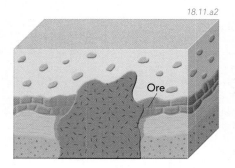
18.11.a2

Other iron mines occur along the flanks of igneous intrusions. When these intrusions invade the crust, they release metal-rich fluids that permeate adjacent rocks. Limestone and other rocks that are chemically reactive were partly or completely replaced by iron-rich minerals, producing a dark, heavy, magnetic rock. Pieces of ore are often mistaken for meteorites.

B How Do Most Copper Deposits Form?

Copper is a relatively mobile element in water and so occurs in a variety of different types of mineral deposits. The three types discussed here illustrate the wide spectrum of copper deposits that exist around the world.

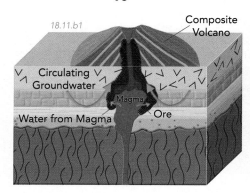
18.11.b1

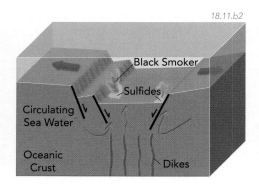
18.11.b2

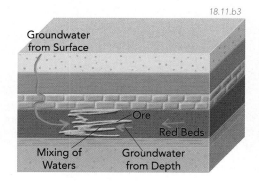
18.11.b3

Most of the world's copper ore comes from large open pits within or adjacent to intermediate to felsic intrusions. Many ore-related intrusions have a *porphyritic* texture (larger crystals in a finer matrix) and the deposits are called *porphyry copper deposits*. They are fairly low grade (<1% copper sulfides and oxides). The photograph below shows samples of weathered ore and porphyritic rocks.

Other copper deposits are much higher grade, but smaller. They contain lenses and pods mainly of sulfide minerals, and are referred to as *massive sulfide deposits*. The bronze-colored rocks below consist almost entirely of sulfide minerals. Many massive sulfides formed in association with volcanic rocks that were erupted in seawater, and represent submarine hydrothermal vents called *black smokers*.

A different type of copper deposit forms in sedimentary rocks. These *sedimentary copper deposits* are common in central Europe and in the copper belt of west-central Africa. They probably formed when copper-rich groundwater mixed with chemically different groundwater, depositing copper. The photograph below shows blue-green copper minerals replacing plant fossils in sandstone.

18.11.b4 Boulder Mtns., MT

18.11.b5 Ansil Mine, Quebec, Canada

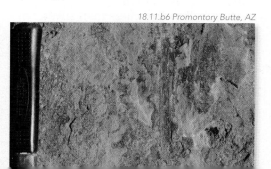
18.11.b6 Promontory Butte, AZ

C What Happens When a Copper Deposit Is Weathered?

Porphyry copper deposits are relatively low grade, and some are barely economical, especially when world markets drive down the price of copper. Weathering can increase the ore grade, making a deposit profitable.

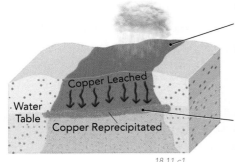

When a copper deposit is at the surface, the upper part becomes weathered and oxidized. Weathering leaches copper from the top of the deposit and leaves these rocks a reddish color.

Percolating groundwater carries the leached copper down until it reaches the water table, where it redeposits the copper, making this part of the deposit higher grade (richer in copper). The photograph to the right shows the reddish weathered top of a copper deposit and the underlying gray, enriched part.

18.11.c1

18.11.c2 Silverbell, AZ

D How Do Some Other Base Metal Deposits Form?

There are many types of base-metal deposits, with diverse modes of formation, as illustrated by two very different examples: one formed in sedimentary layers and another formed at the bottom of a large magma chamber.

Famous lead-zinc deposits of the mid-continental United States are hosted by limestone and other carbonate rocks that were deposited over hills of an older, eroded granite. Metal-rich ground-

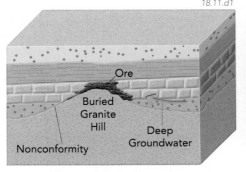

18.11.d1

water flowing through the sedimentary layers encountered the buried hills and deposited the lead and zinc minerals as replacements. This type of deposit, called a *Mississippi-Valley deposit*, contains crystals of galena (lead sulfide), zinc sulfides, calcite, and green and purple fluorite (a calcium fluoride mineral).

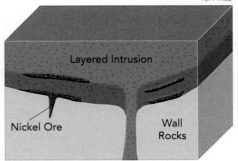

18.11.d2

Near Sudbury, Ontario, Canada, large nickel-sulfide deposits formed at the base of a mafic to ultramafic magma chamber. Most of the sulfide ore minerals crystallized from the large magma chamber. Rocks around Sudbury contain unusual features, such as these unusual cone-shaped joints (▶). This type of joint, called a *shatter cone*, is common around craters interpreted to have formed from a meteoroid impact. From this and other evidence, many geologists conclude that an ancient meteoroid-impact event fractured the crust and triggered the mantle-derived magma that crystallized to form the Sudbury mineral deposits.

18.11.d3 Sudbury, Ontario, Canada

Some Environmental Issues Associated with Mining

Extracting and processing minerals has consequences to the environment. Mines disrupt the land by leaving pits in the ground, along with large piles or *dumps* of rock and other materials that were moved to get to the ore. In some cases, ore is crushed and pulverized to extract the precious commodity, and this produces heaps of light-colored powder called *mill tailings*. Dumps and tailings are unsightly, cover whatever used to be there, and can collapse downhill, sending these loose materials into streams and even houses.

18.11.mtb1 La Plata, CO

Water that interacts with pyrite and other sulfide minerals in dumps, tailings, and exposures of mineralized rock can become a form of sulfuric acid and be contaminated with metals and other toxic chemicals. The acidic and contaminated water can flow into streams or into groundwater. Such waters kill many types of life and make the water unusable to humans. Finally, some sulfide ores must be roasted in mineral-extraction facilities called *smelters*, which release sulfur dioxide gas into the atmosphere, a major contributor to acid rain.

Before You Leave This Page Be Able To

☑ Describe or sketch two ways that iron deposits form.

☑ Describe or sketch ways that copper deposits form, and what happens when a deposit is weathered.

☑ Summarize how the Mississippi Valley lead-zinc and Sudbury nickel deposits formed.

☑ Discuss a few of the environmental issues involved with mining and processing ore.

18.11

18.12 How Do We Explore for Mineral Deposits?

MINERAL DEPOSITS ARE ESSENTIAL TO SOCIETY, and much time, effort, and money are spent exploring for them. If you wanted to find a new mineral deposit, where would you start looking? Geologists explore for new mineral deposits by first studying ones they know about—to become familiar with any diagnostic attributes that would help the geologist recognize a new deposit. Geologists then use various strategies and tools to find deposits that are partially exposed on the surface and those that are completely buried.

A How Do Field Geologists Explore for Mineral Deposits?

Much of the search for new mineral deposits occurs in the field. Mineral-exploration geologists conduct various investigations. They hike across the countryside doing geologic mapping, conducting structural studies, and collecting samples for chemical analyses. Mineral exploration takes geologists to many far-off places, including Peru, Indonesia, and Mongolia.

1. Exploration geologists use existing geologic maps or construct new geologic maps to document the geometry and distribution of rock units. The geologists pay special attention to geologic structures that provide conduits for hydrothermal fluids, because many mineral deposits are in mineral-filled veins.

2. In addition to mapping the rock units, exploration geologists also map mineral changes caused by the passage of hydrothermal fluids through fractures and other pathways. The fluids change the adjacent rocks by altering existing minerals and adding new minerals, especially quartz and pyrite. These changes, called *alteration*, can be recognized in the field and on satellite images processed to emphasize alteration.

4. Mineralizing fluids leave chemical traces in rocks through which they pass. Geologists collect samples of rocks and soils to later analyze the *geochemistry* of the samples for the commodity of interest and for other elements (such as arsenic) that are associated with this type of mineralization.

18.12.a1

5. From the various types of geologic information, exploration geologists reconstruct the geologic history of the region. They determine the sequence of rocks and structures and also when the mineralization and alteration occurred. Events that could have destroyed, preserved, enriched, or moved a mineral deposit are carefully considered. Such events include weathering and erosion that could remove or enrich a deposit, faulting that moves all or parts of a deposit, and sediment deposition, which can bury and hide a deposit. Exploring for mineral deposits involves many different aspects of geology.

18.12.a2 North Silverbell, AZ

◄ 3. These rocks show reddish alteration with iron oxide minerals and a later set of crosscutting fractures that have gray alteration. Such alteration effects can extend far beyond the limits of the actual ore body and thus can help us find a new ore body that is nearby, but perhaps partially hidden.

B How Do Geologists Use Plate Tectonics to Explore for Ore Deposits?

Plate tectonics is a critical consideration in mineral exploration because many mineral deposits are in rocks formed in specific plate-tectonic settings.

Convergent

In convergent boundaries, slices of oceanic crust are scraped off the oceanic plate, metal-rich magma invades the overlying plate, and metal-rich fluids escape from thrust zones. ▶

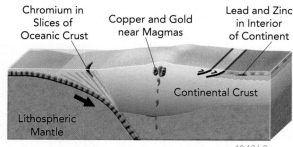

18.12.b2

Divergent

◄ Along a divergent oceanic boundary, plates spread apart, decompression causes magmatism, and heated seawater forms submarine hot springs that deposit sulfide minerals in black smokers.

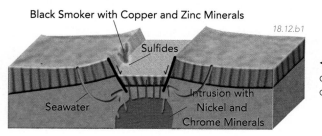

18.12.b1

C How Do We Find Buried Mineral Deposits?

Many mineral deposits are exposed on the surface, and most of these have already been found. Undiscovered mineral deposits are most likely partially or completely buried, requiring different exploration strategies.

1. To locate mineral deposits that are buried beneath younger deposits or obscured by thick soils and vegetation, geologists conduct surveys across potential exploration targets using several *geophysical techniques*.

2. This porphyry copper deposit is buried beneath younger sediment (shown in yellow), but is flanked by a zone of high-grade replacement ore that contains abundant magnetic iron minerals (magnetite) in addition to copper minerals.

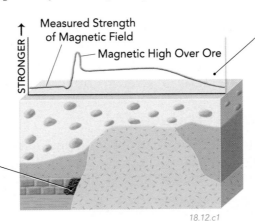

18.12.c1

3. This graph shows the strength of Earth's magnetic field across the area, as determined by a magnetic survey. The survey recorded a strong magnetic signal, called a *magnetic high*, over the ore. Exploration geologists, applying the knowledge that deposits of replacement ore commonly contain magnetic minerals, might hypothesize that such ore lies at depth. This hypothesis can be tested by drilling an exploration drill hole into the location of the magnetic high. Other types of information could be obtained by measuring variations in the strength of gravity or by running electrical current through the ground to find conductive sulfide minerals.

D How Are Mineral Deposits Extracted and Processed?

Once a new mineral deposit has been discovered, it is extensively drilled so samples can be chemically analyzed to determine the size, shape, depth, and grade of the deposit. If all these factors are favorable for development, mining geologists work with engineers to design plans to mine and process the ore.

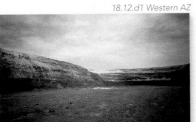

18.12.d1 Western AZ

1. *Open-Pit Mine*—If an ore deposit is shallow enough, it is cost effective to mine the ore in an open pit (◀). Ore can be blasted loose with explosives and loaded into huge ore trucks.

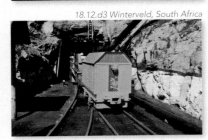

18.12.d3 Winterveld, South Africa

2. *Underground Mine*—Deeper ore deposits must be mined by more expensive underground methods. Ore is blasted loose at depth and hauled to the surface by train (◀), ore carts, elevators, in vertical shafts, or large trucks that drive through even larger tunnels.

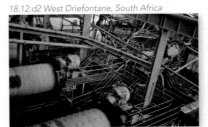

18.12.d2 West Driefontane, South Africa

3. *Mill*—From the mine, the ore goes to the mill, where it is crushed and run through various processes to separate the ore minerals from the rest of the rock. These large, rotating cylinders (◀) contain hard metal spheres that crush the pieces of ore.

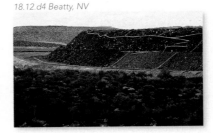

18.12.d4 Beatty, NV

4. *Smelter or Leach Pads*—Some ore is roasted in furnaces in a *smelter*. Other ore is crushed and placed on pads (◀) that are sprinkled with chemical solutions that dissolve (leach) soluble minerals so they can be recovered.

The Geologic Setting of Diamonds

Diamonds are a classic ore mineral; a little bit of diamond can be worth a lot, and it is worthwhile to mine a lot of rock just to get a little bit of diamond. Natural diamonds have an unusual origin. They form at depths of more than 100 km (60 mi) within the mantle and then are violently carried toward the surface in pipe-like volcanic conduits.

South Africa, the largest producer of diamonds, has two main types of diamond deposits. Most diamonds are mined from vertical volcanic conduits called *diamond pipes*, such as the mine shown here. Some diamonds are mined from gravels deposited by streams that carried diamonds eroded from diamond pipes.

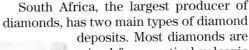

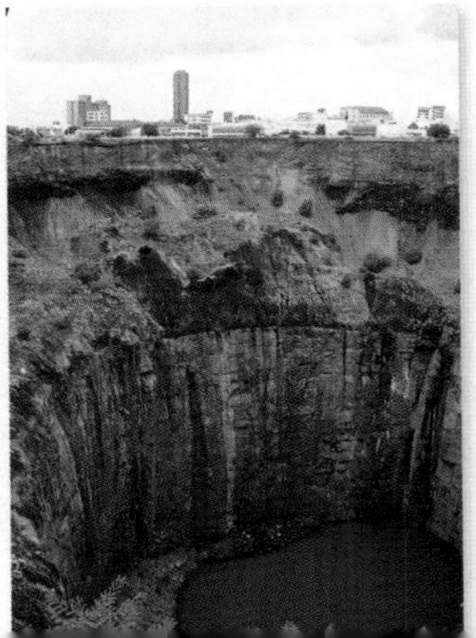

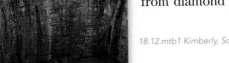

18.12.mtb1 Kimberly, South Africa

Before You Leave This Page Be Able To

- ✓ Describe how field geologists explore for mineral deposits.
- ✓ Describe why the plate-tectonic setting of a region is an important consideration for mineral exploration.
- ✓ Describe one way we find buried mineral deposits.
- ✓ Briefly summarize how minerals are extracted and processed.
- ✓ Describe the setting of diamonds.

18.12

18.13 Why Are Industrial Rocks and Minerals So Important to Society?

GOLD, PLATINUM, AND DIAMONDS get all the attention, but modern society especially relies on everyday rocks and minerals, like limestone and gravel. We use these common materials, called *industrial rocks and minerals*, to build much of the infrastructure of our civilization, such as highways, bridges, water pipes, and sewer lines. We also use them to obtain the material used to build houses and other buildings.

A How Are Cement and Concrete Produced?

Cement and concrete are everywhere in our cities. They form the foundations of our buildings, overpasses across our highways, sidewalks, curbs, and block walls. Where does this material come from?

1. The starting material for cement and concrete is *limestone*, especially one that is free of impurities, like certain detrimental clay minerals. Limestone is extracted (mined) mostly in open pits called *quarries* (▼). The price of limestone rarely justifies expensive underground mining.

18.13.a2

18.13.a1 Marana, AZ

▲ 2. Limestone is crushed and processed in a cement plant, where the crushed limestone is mixed with clays and other materials and then roasted in large kilns. The material is cooled and mixed with gypsum to make *cement*. We can also produce a different product, *lime*, by roasting limestone. We use lime to produce paper, help make steel, and treat soils.

Cement Versus Concrete

3. *Cement* is a whitish powder produced at the plant. When the powder is mixed with water (and usually other materials), complex reactions take place and the mixture sets to a solid mass.

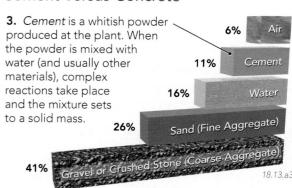

6% Air
11% Cement
16% Water
26% Sand (Fine Aggregate)
41% Gravel or Crushed Stone (Coarse Aggregate)
18.13.a3

4. *Concrete* is a mixture (▲) of cement with sand, gravel, or crushed stone, which together are called *aggregate*. Concrete is a thick, wet slurry when it is being transported and laid down. It can be mixed at the plant or on site in small batches. Cement makes up only 10% to 15% of concrete, but it is what holds concrete together.

B Where Do Sheetrock and Plaster Come From?

Sheetrock, also called wallboard, and plaster are the main materials covering the inside walls of houses and buildings. These are both produced using *gypsum*, a calcium sulfate mineral.

Most interior walls in buildings constructed in the United States in the last 40 years are made from wallboard containing gypsum. Since gypsum is not a very expensive material, the distance from mine to market is critical in determining if the mining operation is profitable. Most gypsum mines are shallow open pits, which are cheaper to operate than an underground mine. Gypsum is common enough that many communities have fairly local sources. In 2006, the leading producers of gypsum in the United States were Oklahoma, Iowa, Nevada, New York, California, and Arkansas. The total amount of gypsum mined in the United States in 2007 was 22 million tons, worth $165 million—a lot of gypsum and a lot of wallboard.

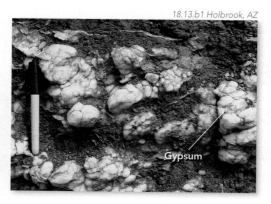
18.13.b1 Holbrook, AZ

Gypsum

◀ Gypsum is an evaporite mineral, meaning that it can be precipitated by evaporation of water. Most gypsum forms when water in a sea or lake evaporates, leaving behind gypsum, salt, and other substances dissolved in the water.

▶ Most gypsum forms in specific sedimentary layers, like these shown here. Mining operations strip off any overlying rocks, mine away the gypsum, and stop at the base of the gypsum-rich layer.

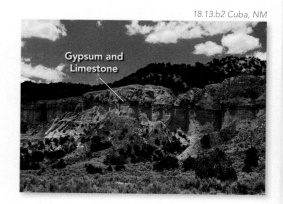
18.13.b2 Cuba, NM

Gypsum and Limestone

C How Do We Use Sand, Gravel, and Crushed Rock?

The Earth materials we use the most are *sand and gravel*. Our society uses enormous quantities of sand and gravel, along with crushed rock, to mix into concrete. We use sand to produce brick, roof tile, and glass, as well as to build up or fill in spaces beneath highways and the foundations of buildings. An average of 10 metric tons (22,000 pounds) of sand, gravel, and crushed stone is mined each year for each person in the United States.

18.13.c1

◄ The largest source of sand and gravel is rivers and streams. Some is quarried from within the active river channel or adjacent low terraces, because these sediments are unconsolidated and individual clasts are generally not highly weathered. If a modern-day channel is not available, sand and gravel can be quarried from older stream and river deposits.

► Various types of hard rocks, such as quartzite, are crushed for use as aggregate, decorative rock, road and railroad beds, and various other uses.

18.13.c2 Flagstaff, AZ

D What Are Some Other Important Industrial Mineral Deposits?

Many other industrial rocks and minerals are mined and used by society. You will recognize some of these, like salt and clay, but others will be unfamiliar. We use certain minerals as chemical filters and others for fertilizer. Here, we present a small selection of these mineral commodities.

► *Silica sand* is like a typical sand, but consists of nearly all quartz grains with few other impurities. Quartz sand this pure comes from windblown deposits and some beach deposits. We use it to make glass, ceramics, and microchips.

18.13.d1 Berkeley Springs, WV

► *Clay* refers to a size of particle and to a family of platy minerals. Clay minerals are a main ingredient in tiles and other ceramics, bricks, and expensive paper. There are several varieties of clay, each with a different use, value, and geologic origin.

18.13.d2 St. Austell, Cornwall, England

18.13.d3 Mojave Desert, CA

◄ *Salt* is a necessary commodity for nutrition and for many industrial and chemical uses. Salt is an *evaporite* and mostly comes from salt deposits that formed when a sea or lake evaporates enough to precipitate salt. Common salt is the mineral *halite* (NaCl).

18.13.d4

◄ *Phosphate rock* is a general term for rocks that are rich in the element phosphorus, which is essential for plant growth. It is mined from recent marine deposits in Florida and elsewhere and from older (Paleozoic) marine rocks in Idaho. We primarily use it for fertilizer.

Ways to Obtain Salt

Rock salt can be obtained in several ways, depending on the geologic setting where the salt occurs. Salt can be harvested by trapping salty water and allowing it to evaporate in the sunlight. Salt is harvested this way from Great Salt Lake, Utah, shown in the photograph here. We can also mine salt from underground salt layers or from salt domes, such as along the Gulf Coast of the southern United States. Another way to extract salt is to pour freshwater into underground salt bodies, pump the resulting salty water back to the surface, and then use front-end loaders to harvest the salt after the salty water evaporates and the salt dries.

18.13.mtb1

Before You Leave This Page Be Able To

☑ Summarize how limestone is used to make cement, concrete, and lime.

☑ Summarize where the gypsum in wallboard and plaster comes from.

☑ Discuss why sand, gravel, and other aggregate are our most used mineral resource.

☑ Briefly describe the origins and uses of silica sand, salt, clay, and phosphate rock.

18.13

Why Is Wyoming So Rich in Energy Resources?

WYOMING HAS A WEALTH OF ENERGY RESOURCES. It contains as many large and diverse energy resources as any region in the world. It has the most prolific coal region in the world, and the three largest coal mines in the United States. It ranks first in the country in uranium production and second in natural gas, including coal-bed methane. It shares with adjacent states the largest resource of oil shale in the world. What happened during the geologic history of Wyoming to make the state so energy rich?

A What Is the Geology of Wyoming?

Wyoming has a long and interesting geologic history, spanning more than 2.5 billion years. Nearly all of the state's energy resources, however, resulted from geologic events during the last 80 million years.

The topography of Wyoming reflects its geology, as expressed in this geologic map on a shaded-relief base. Colors reflect types and ages of the exposed rock units. On this map, large uplifted mountain ranges are surrounded by broad valleys (basins).

Heavy lines on the map represent faults, many of which are along the edges of mountains, including the Grand Tetons.

The details of the geologic units on this map are not critical here, but the colors represent the following ages of rock, from youngest to oldest:

 Cenozoic sedimentary (yellow) and volcanic (reddish orange) rocks

 Mesozoic sedimentary rocks (various shades of green)

 Paleozoic sedimentary rocks (blue)

 Precambrian metamorphic and granitic rocks (dark brown and purple)

The mountain ranges contain old (Precambrian and Paleozoic) rocks, whereas the basins are covered by younger Mesozoic and Cenozoic sedimentary rocks. The line A–A' on the map shows the location of the geologic cross section below.

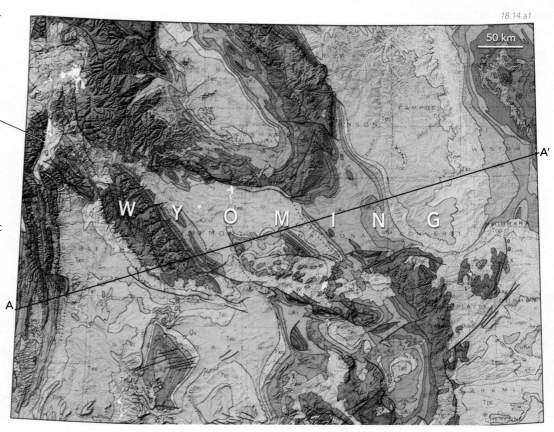

18.14.a1

50 km

▼ On a cross section (A–A') across the state, the western edge of the state is a fold and thrust belt, the Overthrust Belt, which has shuffled the Paleozoic and Mesozoic sedimentary rocks.

The cores of the mountain ranges expose ancient Precambrian rocks (brown), which include the 2.5 billion-year-old metamorphic and granitic rocks that crop out in the Wind River Range.

A thin veneer of Paleozoic and Mesozoic sedimentary rocks, such as those in the Casper Arch, sit on the Precambrian rocks or are beneath reverse faults.

The largest basin is the Powder River Basin in the eastern part of the state (the yellow area beneath the "G" in Wyoming on the map). It is a broad downward-bending fold (syncline) that preserves the youngest Cenozoic rocks in the center of the basin. The rock layers rise up toward both flanks of the basin, climbing westward toward the Casper Arch and eastward toward the Black Hills uplift of Wyoming and South Dakota.

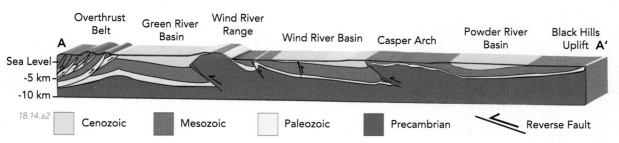

18.14.a2

| Cenozoic | Mesozoic | Paleozoic | Precambrian | Reverse Fault |

B What Is the Geologic Setting of Wyoming's Natural Resources?

The uplifts, basins, and other large-scale geologic features of Wyoming control the distribution of oil and gas fields, coal, oil shale, uranium deposits, and hot springs (potential for geothermal energy).

Oil and Gas, and Oil Shale

1. The oil fields (blue-green) and the gas fields (orange) are in the basins, where sedimentary rocks are preserved and relatively thick. There are almost no oil and gas fields in the uplifted mountain ranges, where the sedimentary layers are eroded away, exposing the underlying Precambrian crystalline rocks.

2. The southwestern part of the state, including the Overthrust Belt, has mostly gas fields instead of oil because rocks in this area were heated more than rocks farther east.

3. The world's largest oil shale deposits (light purple) are in southwestern Wyoming and nearby states, where large lakes formed in early Cenozoic time within the basins.

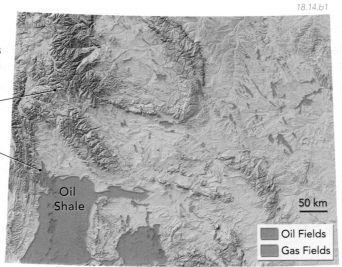

18.14.b1

4. The oil and gas are mostly derived from Paleozoic and Mesozoic sedimentary layers, because these rocks are young enough to have been organic-rich and were buried deep enough to turn the organic material into oil and gas. The reservoir rocks for the oil and gas include a Paleozoic limestone and a Mesozoic sandstone, both of which have high permeability. The main oil and gas traps are anticlines formed from 80 to 40 Ma.

Coal and Coal-Bed Methane

5. Coal fields (dark gray) are also in the basins, especially in those parts of the basins that have the youngest Mesozoic and Cenozoic sedimentary rocks. Many of the coal-bearing rocks formed in swampy, plant-rich deltas.

6. There are no coal fields in the large mountain ranges because some coal-bearing layers were eroded off the uplifts and others were only deposited in the basins.

7. Areas of coal-bed methane (light gray) are more widely distributed than the coal mines or the coal areas shown here, because coal can be mined only where the coal-bearing units are near the surface, whereas the gas wells can tap deeper layers of coal.

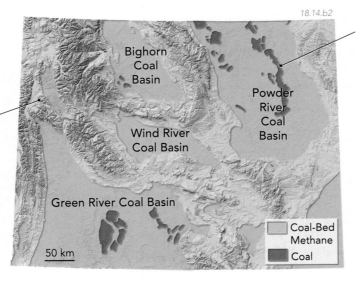

18.14.b2

8. The Powder River Basin produces more coal than any place in the world and also has the largest remaining coal reserves. In 2008, it produced 446 million tons of coal. Most of the mined coal goes to coal-fired electrical-generating plants, some of which are located close to the coal fields to avoid transporting the coal great distances. Large quantities of coal are shipped to power plants in other parts of the country.

Uranium and Geothermal Energy

9. Areas with potential for geothermal energy commonly have hot springs (shown in red) and geysers. In Wyoming, most hot springs (>100°F) are near areas of recent volcanic activity and faulting near Yellowstone National Park.

10. The largest uranium deposits (yellow) are in Cenozoic sandstone and were deposited by groundwater. Miners have extracted the uranium by traditional methods and also by pumping water through subsurface ores to dissolve the uranium.

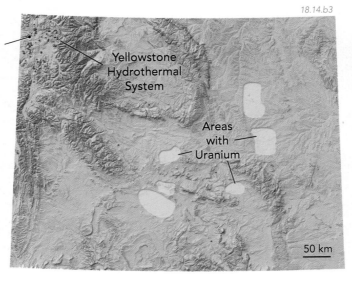

18.14.b3

Before You Leave This Page Be Able To

☑ Summarize or sketch in cross section the main types of geologic features present in Wyoming.

☑ Summarize how large-scale geologic features control the distribution of oil, gas, oil shale, coal, coal-bed methane, uranium, and geothermal energy.

18.14

Where Would You Explore for Fossil Fuels in This Place?

THIS REGION EXHIBITS CLUES that it contains several types of fossil fuels. There are oil seeps, exposed coal seams, and other features that may be related to hydrocarbons. You will use the character and distribution of different rock types on the surface, along with some subsurface information, to identify places to explore for hydrocarbon-based energy sources. This investigation involves many aspects of geology, including sedimentary environments, structural geology, groundwater, and energy resources.

Goals of This Exercise:

- Observe the landscape to understand the geologic setting of different areas, and read the descriptions of each location in order to interpret the significance for exploring for fossil fuels.
- Use a stratigraphic section and descriptions of the rocks to interpret the environment in which each rock layer formed.
- Use surface observations of geologic structures, along with the geologic section shown on the side of the diagram, to interpret the subsurface geology.
- Determine the best locations to explore for hydrocarbon-based fossil fuels, either on the surface or in the subsurface.

Procedures

Use the information to complete the following steps, entering your answers in the worksheet or online.

1. Observe the features shown on the landscape to the right. Read the text box associated with each area and consider what that statement implies about the rock types and geologic structure in that area.

2. On the stratigraphic section, read the description of each rock unit and interpret the environment in which each formed. Next, consider what implications each rock's character has for that rock's potential role in the generation, preservation, or trapping of fossil fuels.

3. Use the various types of structural information to characterize the main geologic structures that cross the area.

4. Integrate your understanding of the rock sequence and the structural geometries to predict what rocks would lie at depth beneath any area, and whether any particular rock layer will be at a shallow, medium, or great depth below the surface.

5. Draw the letters OG, C, and S any place on the map that you think has potential for oil and natural gas (OG) including coal-bed methane, coal (C), and oil shale (S). Note that not all of these types of hydrocarbons may be present. You may decide to write a letter (such as C for coal) in more than one location. If you do, label them C1, C2, C3, etc., in order of highest to lowest potential.

6. Be prepared to write or discuss a justification for each of your proposed sites, including what you think would be present in the subsurface and how you intend to extract the resource.

1. The highest feature in the area is a large ridge, *called Tan Mountain*, which is largely composed of tan-colored sandstone. The sandstone is well sorted and is reportedly a good source of groundwater (i.e., is an aquifer). The sandstone unit is named after the mountain and is called the *Tan Mountain Sandstone*. On both flanks of the mountain, the sandstone dips away from the ridge crest, defining an anticline that has nearly the same shape as the mountain.

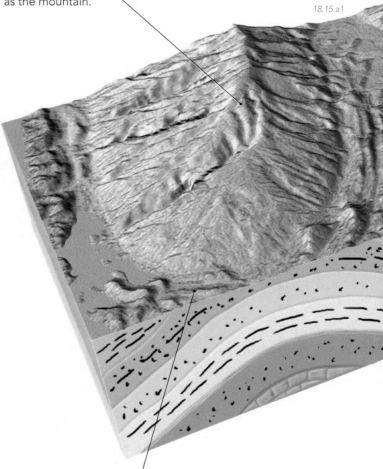

18.15.a1

2. Wrapping around the flanks of the mountain is a sequence of reddish and gray sedimentary rocks, which are shown on the cross section with a pale reddish color. Local people informally call this unit the *carbon beds* because the bottom part of the unit contains a layer of coal up to 5 m (16 ft) thick (too thin to show on the cross section). The coal is exposed only here and there, but digging beneath the surface has shown that the coal layer is fairly continuous. Miners drilling into the unit while exploring for coal experienced some minor explosions due to some type of flammable gas.

3. The two valleys in the center of the area are underlain by tan, dark gray, and brown shale, which geologists call the *upper shale*. This unit, colored brownish tan on the cross section, is the highest unit exposed in the area, except for some thin gravels along the rivers and mountain fronts. The dark-gray and brown parts of the shale emit an oily smell. Some long-time residents claim that the shale actually burns, but this has not been verified.

4. In one valley, the rock layers do not quite match up when geologists compare the rocks exposed on either side. A zone of crushed rocks marks a fault zone along one of the mountain fronts. There are some springs and small oil seeps along the fault, but overall the fault does not seem to be very permeable because some type of natural cement has filled in the pore spaces between the broken pieces of rock.

5. The eastern ridge, like the other two, is an anticline. It is more eroded than the other two anticlines, and so exposes deeper rock layers. The Tan Mountain Sandstone caps the highest peaks and overlies a thick sequence of yellowish-tan shale. This shale is called the *muddy shale*, because water will not sink into the unit when it rains, remaining on the surface and making a muddy mess.

6. Some recently deposited gravels cover the older rocks in a few places. These gravels are loose and unconsolidated.

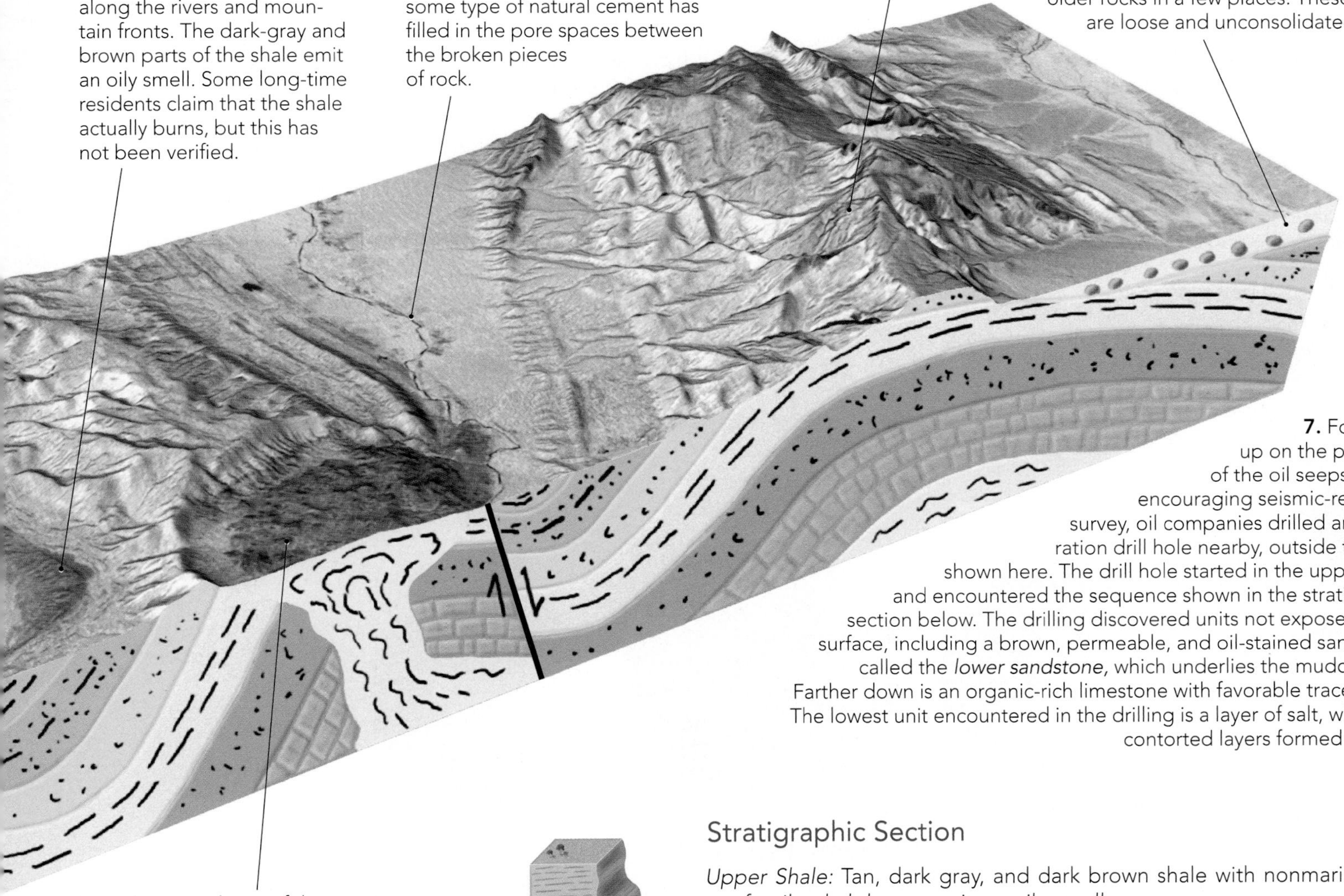

7. Following up on the presence of the oil seeps and an encouraging seismic-reflection survey, oil companies drilled an exploration drill hole nearby, outside the area shown here. The drill hole started in the upper shale and encountered the sequence shown in the stratigraphic section below. The drilling discovered units not exposed at the surface, including a brown, permeable, and oil-stained sandstone, called the *lower sandstone*, which underlies the muddy shale. Farther down is an organic-rich limestone with favorable traces of oil. The lowest unit encountered in the drilling is a layer of salt, which has contorted layers formed by flow.

8. In the central part of the area is an odd circular feature that seems out of place in the rock sequence. The feature is composed of salt and has a rough, irregular surface because the salt is flowing and spreading out in all directions. Areas of pure salt are light colored, but the entire feature is dark because the salt contains minor amounts of shale and oil intermixed with the salt. No one has been able to extract the oil from the salt, because the salt is too impermeable.

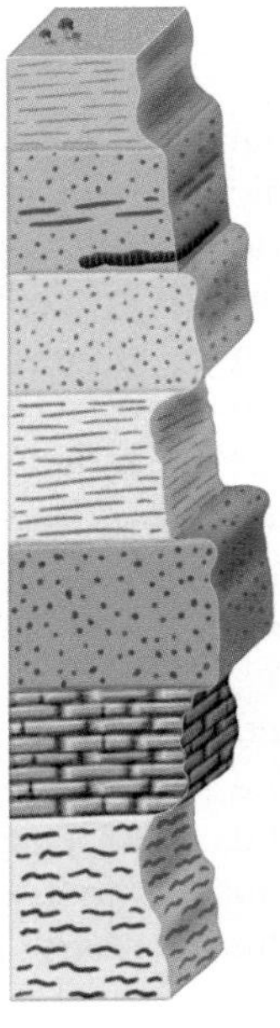

Stratigraphic Section

Upper Shale: Tan, dark gray, and dark brown shale with nonmarine fish fossils; dark layers emit an oily smell.

Carbon Beds: Reddish and gray sedimentary beds with coal near the base.

Tan Mountain Sandstone: Tan, locally cross-bedded sandstone with both marine fossils and land plants.

Muddy Shale: Light-gray marine shale that is impermeable.

Lower Sandstone: Porous and permeable, well-sorted sandstone with very little natural cement between the grains; locally oil stained.

Lower Limestone: Dark-gray limestone with abundant marine fossils; contains many open fractures and bedding planes, some of which are locally oil stained.

Salt: A mostly cream-colored to light gray salt with some thin organic-rich shales. Tests show that the salt, shale, and overlying limestone have all been buried and regionally heated to 80°C to 100°C.

18.15.a2

18.15

Geology of the Solar System

GEOLOGIC PROCESSES AND FEATURES are not restricted to Earth. Geologic features are exposed on the surfaces of our solar system's four innermost planets, on our own Moon, and on moons of Saturn and Jupiter. The planets and moons highlighted in this chapter provide a brief portrait of the most important or interesting bodies in our solar system. Compared to illustrations in this chapter, the planets and moons are vastly farther apart and are more different in size than can be shown.

The four inner planets are called *terrestrial planets* because they have solid rocky surfaces (*terra* means earth). These planets also have similar overall compositions but not similar geologic histories.

Mercury, closest to the Sun, is a small, heavily cratered planet with almost no atmosphere. It has a 650°C difference between night and day temperatures.

Why is the planet so heavily cratered?

Venus has a thick atmosphere of carbon dioxide that captures much of the solar radiation that reaches the planet. This extreme greenhouse effect causes surface temperatures to reach 450°C. The planet is shown here as if it did not have its thick atmospheric shroud.

What is the land like beneath the clouds?

Earth has plate tectonics and also a strong magnetic field caused by its rotating, molten iron outer core. Abundant surface water sustains a diversity of life.

Why is Earth so different from the other inner planets?

Mars has been explored recently, and new data show that water once flowed on the Martian surface. But Mars lost most of its atmosphere sometime in the past and now is so cold that liquid water cannot exist in large quantities on its surface.

What is the evidence for past movement of water on Mars?

Asteroids are rocky fragments concentrated in an orbit between Mars and Jupiter. They are similar in composition to certain meteorites and are interpreted to be fragments left over from the formation of the solar system, probably with some pieces of small planetary objects that broke apart.

Mercury

19.00.a1

Venus

Earth

Mars

How the Solar System Formed

The Sun formed about 5 billion years ago from the remnants of previous stars and cosmic dust, all of which had a beginning in what is called the "big bang." According to modern theories the entire universe arose from the *big bang* 10 to 15 billion years ago, so the universe was 5 to 10 billion years old before our solar system began to form. Current theories for the formation of our solar system suggest that the Sun and planets condensed from a *nebula*, a shapeless cloud of gas and dust. Particles of dust clung together to form small chunks and then larger and larger pieces, eventually ending up as planets. The Sun, meantime, continued to attract more material and became massive enough to begin atomic fusion and to emit protons and electrons in a *solar wind*. The

solar wind reached the inner planets, blowing away hydrogen, helium, and other light elements near the surface, leaving only heavy materials. Later, Earth gradually re-acquired its supply of hydrogen and other light elements. The outer planets were less affected by solar wind and had enough gravity to retain hydrogen and helium, and so remain gaseous.

During the early stages of its formation, Earth is thought to have collided with another large object that was not quite yet a planet. This catastrophic collision ripped away part of Earth, forming our Moon. It also likely knocked Earth off its original axis of spin, giving the planet the present 23.5° tilt of its spin axis. In other words, we probably have a moon and seasons because of this immense collision nearly 4.5 billion years ago.

Jupiter, the largest planet in the solar system, consists of hydrogen and helium with a small rocky core, making it compositionally more similar to the Sun than to Earth. It has a banded, swirling atmosphere and many interesting moons.

Do Jupiter's moons look like ours and are they all the same?

Jupiter

Saturn is similar to Jupiter, but it has a beautiful system of delicate, icy rings around the planet. Our spacecraft are actively exploring Saturn and its moons.

What have our spacecraft found so far?

Saturn

Outside Our Solar System

Our galaxy, the *Milky Way Galaxy*, is only one of countless immense galaxies in the universe. Each galaxy is composed of millions of stars, like the Sun, and many of these stars are orbited by planets. Astronomers have captured some amazing images of other galaxies and of nebulae, which are large accumulations of space dust and stars. Below are the Whirlpool Galaxy (top) and Eagle Nebula (bottom).

19.00.a2

Uranus

Uranus is large, but smaller than Jupiter and Saturn. It is much farther out in the solar system, being as far from Saturn as Saturn is from the Sun. Uranus has arguably the oddest moon in the solar system.

Neptune

Neptune is very similar in size and composition to Uranus. Both planets are gaseous and have a blue color.

Why are Neptune and Uranus blue?

Pluto is a tiny body with an icy surface and an unusual orbit. Once considered to be the ninth planet, Pluto is no longer classified as a planet, leaving our solar system with only eight true planets.

Pluto

19.00.a3

The four large outer planets—Jupiter, Saturn, Uranus, and Neptune—are known as the *gas giants* because of their large size and gas-rich character, which is quite different from the terrestrial planets. All four planets have their own moons and some type of rings. One way to think of the solar system is as an inner zone of rocky planets, an outer zone of giant, gaseous ones, and finally a zone of small, distinct objects, like Pluto, dominated by ice. This outward progression is related to how the solar system formed and evolved.

19.0

19.1 How Do We Explore Other Planets and Moons?

EXPLORING THE GEOLOGY OF OTHER PLANETS and moons is not as easy as studying geology on Earth, even compared to Antarctica or other remote parts of our planet. Nearly all exploration of other planets and their moons has to be done remotely (from a distance) by examining them with telescopes and other instruments, either on Earth or in orbit around Earth. We gain additional information by sending spacecraft to visit these distant objects. Our Moon is the only place other than Earth where geologists and other humans have walked on the surface, making observations and collecting rock samples.

A What Can We Observe with Telescopes on Earth and in Earth Orbit?

Historically, most investigations of other planets and moons in the solar system have used Earth-based telescopes. Astronomers still rely heavily on Earth-based telescopes but also use telescopes in orbit around Earth, or launched into space to avoid the distorting effect of Earth's unpredictable atmosphere.

▼ These images of Mars were taken several months apart from the *Hubble Space Telescope*, which orbits Earth.

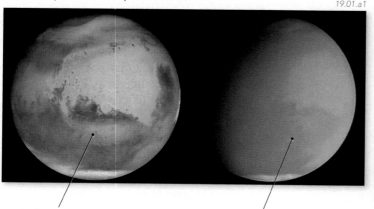
19.01.a1

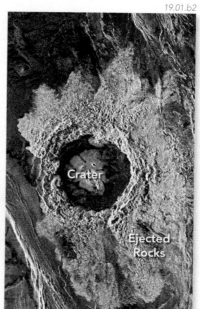
19.01.a2

In this view, Mars has an overall orange-red color, but contains dark rocky areas, as well as ice caps at the north and south poles.

Several months later, a planet-wide dust storm had covered most of Mars, obscuring its surface.

Common telescopes view the night sky using visible light, but astronomers use telescopes that observe other parts of the *electromagnetic spectrum*. These three views of Saturn from the Hubble Space Telescope show how the planet appears when imaged in visible light (center), short-wavelength ultraviolet light (top), and long-wavelength infrared light (bottom). Each type of image provides data about different aspects of an object.

B What Do Radar Observations Tell Us About a Planetary Surface?

We can observe the surfaces of other planets and moons remotely by using radar or other techniques that can penetrate the atmosphere of a planet and reveal topographic features and roughness of the surface.

1. In this technique, a satellite transmits radar waves down and sideways toward a planetary surface. A sensor on the satellite then measures the amount of the radar signal reflected back from the surface.

2. If the planetary surface is rough or slopes toward the satellite, the surface will reflect more radar waves back toward the satellite. The area will appear bright on the radar image (lots of returned energy).

3. If the planetary surface is smooth or slopes away from the satellite, most of the radar energy will bounce away rather than return to the instrument. As a result, this area will be relatively dark on the radar image.

Radar Waves

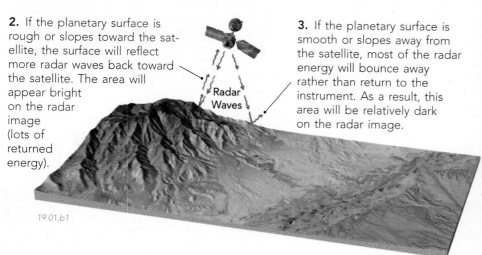
19.01.b1

19.01.b2

This radar image from Venus shows a crater produced by a meteoroid impact. Much of the interior of the crater is dark because the surface there is relatively smooth. Broken rocks ejected from the crater form an apron of ejected rock around the crater and appear bright because they form rough topography.

Crater

Ejected Rocks

C How Can We Remotely Observe Temperature and Composition of Surfaces?

Geologists, working with engineers, develop sophisticated instruments that allow us to measure the temperature of, and to infer the composition of, a planetary surface from afar. One technique measures the energy given off by rocks in the *infrared* part of the electromagnetic spectrum.

When rocks and other materials are heated by the Sun, they re-emit some of the energy as thermal energy with infrared wavelengths. By measuring how much infrared energy is given off, geologists can calculate the temperature of the surface, even at night. They infer what types of material are present by how well the material holds heat. Unconsolidated sediment and other low-density materials lose their heat faster than solid rock and materials that are relatively dense.

19.01.c1

This image of an impact crater on Mars combines visible and infrared measurements. The colors depict night-time temperatures: reds show warmer areas and blues show colder ones. The bright colors on this and similar images in this chapter are not the actual colors of the Martian surface.

The floor of the crater is bluish in this image, showing that the materials cooled relatively quickly after sunset. Geologists therefore infer that the crater contains some loose, unconsolidated material, probably sediment.

D How Have We Explored the Surfaces of the Moon and Some Planets?

In addition to observations from spacecraft, NASA and other space agencies have landed or intentionally crashed probes on several planets, moons, and asteroids. Astronauts have walked on the Moon with the expressed intent of collecting rock samples and observing other aspects of its geology.

19.01.d1

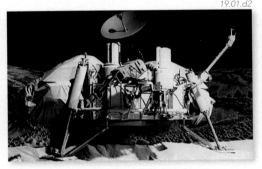
19.01.d2

19.01.d3

Astronaut and geologist Harrison (Jack) Schmitt collects a sample near the Apollo 17 landing site on the Moon. What an amazing field site in which to do geologic field work!

Humans have landed spacecraft on the surfaces of Venus, Mars, Earth's moon, a moon of Saturn, and even an asteroid. Such spacecraft, called *landers*, collect images and various types of data during their descent and after they have landed.

In recent years, NASA has explored the geology of the Martian surface using *rovers*, which are small-wheeled vehicles that drive around on the surface, following commands issued from Earth. The rovers take images, collect infrared and other data, and even drill through the surface coating of rocks.

Encounters with Asteroids

Farther from the Sun than Mars, but this side of Jupiter, is a belt of more than 90,000 large and small rocky chunks drifting in space. Some are more than 500 km (300 mi) across, others are less than 1 km. Most asteroids are thought to be debris left over from the formation of the solar system 4.5 billion years ago, but some are probably parts of objects that broke up. Several space missions have passed close enough to asteroids to take detailed photographs or even to land on the surface. NASA's *Galileo* spacecraft, on its way to Jupiter, passed asteroids within the main asteroid belt between Mars and Jupiter. This colorized photograph shows an asteroid named *Gaspra*. ▶

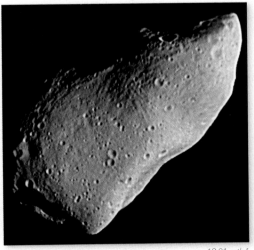
19.01.mtb1

Before You Leave This Page Be Able To

✓ Briefly explain why we put telescopes in orbit to better observe space.

✓ Sketch or describe what radar and infrared observations indicate about a planetary surface.

✓ Describe ways we collect information by landing spacecraft on an object, including those with geologists or rovers.

✓ Describe what asteroids are and where most are located.

19.1

19.2 Why Is Each Planet and Moon Different?

THE PLANETS AND MOONS OF OUR SOLAR SYSTEM exhibit a remarkable diversity of atmospheric and geologic characteristics. Some are gaseous giants with thick atmospheres, whereas others are like cratered snowballs. What processes affect the surfaces of planets and moons and what causes these differences?

A What Determines the Kinds of Materials and Features on a Planet or Moon?

The surface environments of planets and moons are governed by their position in the solar system. The distance from the Sun to a planet or moon influences what material the planet or moon contains and how much solar heating they experience. The planetary objects are equally influenced by their size, whether they have an atmosphere, and whether liquid water currently exists, or ever existed, on the planet or moon.

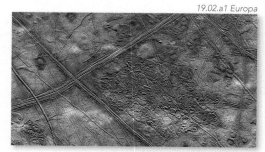
19.02.a1 Europa

19.02.a2 Jupiter

19.02.a3 Mars (Thermal Infrared Image)

Composition—The appearance and dominant processes of a planetary object are influenced by its composition, including its chemical composition, and the proportions of ice (▲), rock, liquid, and gas.

Atmosphere—The presence of a thick atmosphere can obscure the planet's surface. It can block incoming solar radiation, trap in heat, and lead to other phenomena, such as rain and erosion.

Impacts—All planetary surfaces have at least some craters formed by the impact of meteoroids and other objects. Some planets and moons contain many craters, whereas others have few preserved craters.

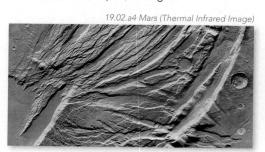
19.02.a4 Mars (Thermal Infrared Image)

19.02.a5 Olympus Mons, Mars

19.02.a6 Mars

Tectonics—Faulting and other tectonics modify planetary surfaces, causing variations in topographic relief. Only Earth exhibits our style of plate tectonics, but some other objects have tectonics.

Volcanism—Magma erupts onto the surface and may form a silicate rock, like on this basaltic shield volcano on Mars, may be sulfur rich, as on Jupiter's moon Io, or may be liquid water on an icy moon.

Erosion and Deposition—Wind, water, and gravity can erode some areas and redeposit eroded material elsewhere. In this image, sediment eroded from gullies accumulates in fan-shaped bodies (alluvial fans) at the base of steep slopes. Erosion, deposition, volcanism, and tectonics can remake a surface, a process called *resurfacing*.

B How Are Impact Craters Formed?

The dominant geologic process in our solar system is the formation of impact craters. Impact craters of all sizes are abundant on the surfaces of most planets and moons, except in areas that have been resurfaced. Impacts were more frequent early in the solar system's history because there was more debris in space.

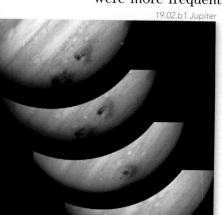
19.02.b1 Jupiter

◄ An impact crater forms when a meteoroid, comet, or other object from space strikes a planetary surface at an extremely high velocity, blasting open a crater and fracturing adjacent rocks. Pieces of the Shoemaker-Levy 9 comet collided with Jupiter in 1994, as recorded by this series of images. These collisions caused a disruption of Jupiter's gaseous surface, but did not form visible craters.

► Most impact craters are circular and surrounded by an apron of material ejected from the crater during the impact. Larger impacts can cause melting of the rocks by the intense shock waves that pass through the rocks. The impacting comet or meteoroid may be totally vaporized, leaving only a crater as a record.

19.02.b2 Ma

C | What Determines Whether an Object Has Active Tectonics and Volcanism?

Why do some planets and moons have active tectonics and volcanism, whereas others are inactive and heavily cratered? The main cause of this difference is the size of the object, which controls how fast it loses heat.

1. The size of an object affects heat loss. As the radius (r) of a sphere increases, surface area increases by r^2 (area of sphere = $4\pi r^2$), whereas volume increases by r^3 (volume of sphere = $4/3\pi r^3$). Therefore, larger objects have less surface area relative to their volume than do smaller objects. We examine three objects, which are shown at their correct relative sizes.

2. Smaller objects, like the planet Mercury, have a large surface area relative to their volume. This causes them to lose heat relatively rapidly, which in turn causes them to solidify, shutting off volcanism and tectonism.

19.02.c1   19.02.c2

3. The solar system has a few smaller objects that remain volcanically and tectonically active. Io, an inner moon of Jupiter, is active because gravitational effects of Jupiter and its other moons provide additional heat energy to Io.

4. Larger rocky planets, like Earth, have a relatively small surface area compared to their inner volume. This allows them to retain heat generated from their initial formation and from post-formation radioactive decay. As a result, larger planets remain volcanically and tectonically active longer than smaller planets.

19.02.c3

D | How Can Water and Wind Modify a Planet's Surface?

Water is currently abundant on Earth, and scientists think that water was present in the past on Mars and perhaps on other planets. Water promotes chemical and physical weathering, erosion, transportation, and deposition of sediment. Flowing water leaves telltale landforms, including drainage networks, that point to its earlier existence. Wind forms dunes and other distinctive landforms and can bury or modify parts of a planet's surface.

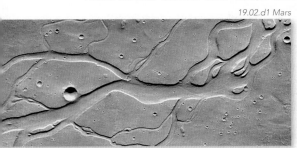
19.02.d1 Mars

19.02.d2 Mars

19.02.d3 Mars

Weathering and Erosion—These channels on Mars are interpreted to have been eroded by running water, because they are similar to stream channels on Earth. If there is no water or wind, there will be little or no erosion of an object's surface.

Deposition—Material can be moved and deposited by various processes, including running water, wind, flowing ice, and slope failure. This channel on Mars has a curved meander that is floored with sedimentary deposits that look like a typical point bar.

Wind—If a planet has an atmosphere and strong winds, sand and dust can be blown across the surface, forming distinctive dunes and covering up what lies beneath. Much of the surface of Mars is obscured by windblown dust and sand dunes, like those shown above.

Using Crater Density to Estimate the Age of a Surface

19.02.mtb1 Moon

Across the solar system, the degree to which the surface of a planet or moon is cratered varies widely. Planetary geologists use the density of craters to estimate the *age* of a planetary surface. The underlying principle is that the longer a surface is exposed, the more impact scars it receives. Surfaces that have remained undisturbed by tectonics, volcanism, and deposition for a long time will be more heavily cratered than those that have been more recently resurfaced by these processes. So a region that has a high density of craters is interpreted as being older than a region with fewer craters. Which part of this image of the Copernicus area of Earth's moon (◄) is oldest? The foreground is less cratered and is younger than the older, more heavily cratered part in the background.

Before You Leave This Page Be Able To

☑ Summarize the factors and processes that affect the appearance of a moon or planet.

☑ Summarize the characteristics of an impact crater and how one forms.

☑ Explain why smaller objects are more likely to be tectonically or volcanically inactive compared to larger objects.

☑ Explain how crater density can be used to estimate the age of a planetary surface.

19.2

19.3 What Is the Geology of the Inner Planets?

THE FOUR PLANETS CLOSEST TO THE SUN are the *inner planets*: Mercury, Venus, Earth, and Mars. These four *terrestrial planets* all have a rocky surface. The geology of these four planets differs primarily because of their different sizes and different atmospheres. For each planet, observe the images first, noting what features are most obvious, before you read the accompanying text.

A What Is on the Surface of Mercury, the Closest Planet to the Sun?

Mercury is a small and rocky planet that is relentlessly baked by its proximity to the Sun. It speeds through space at about 50 km/sec and orbits the Sun in 88 days (that is, a Mercury-year is only 88 days long).

◀ Mercury has almost no atmosphere, and the temperature can reach 460°C (860°F) during the day and an extremely cold minus 80°C (−290°F) at night.

With so little atmosphere, there is no erosion by water or wind, so the surface is not modified by these processes. As a result, Mercury's surface is covered with numerous craters produced when meteoroids and other planetary debris collided with the planet's surface. Because of its extremely thin atmosphere, impacts are more common on Mercury than on the other terrestrial planets. If plants have a thick atmosphere, meteoroids and other objects often burn up before they reach the planet's surface.

19.03.a1

19.03.a2

◀ In this close-up view of part of Mercury, the surface is covered by impact craters of various sizes. The surface appears so heavily cratered because tectonic and volcanic activity on the planet ceased early in the solar system's history, when there still was an abundance of debris to collide with the planet. The subsequent lack of volcanoes, wind, or water means that the craters have not been eroded or covered by lava or sediment.

B What Is Beneath the Atmospheric Shroud of Venus?

Venus, the second planet out from the Sun, is similar to Earth in size, mass, and composition, but has a thick atmosphere that hides the planet's surface. It has experienced volcanism and tectonism in the last billion years, but there is no evidence of active plate tectonics.

19.03.b1

◀ In this telescopic view of Venus, the surface of the planet is obscured by a thick atmosphere of clouds and toxic gases that swirl around the planet. The atmosphere consists mostly of carbon dioxide and droplets of sulfuric acid, with almost no water vapor. The atmosphere exerts a stifling amount of air pressure. It allows the Sun's energy in, but keeps the heat from escaping back to space, causing the atmosphere to heat up, like a closed greenhouse. As a result of these extreme greenhouse effects, surface temperatures are more than 450°C (840°F).

19.03.b2

19.03.b3 Venus

◀ Planetary scientists used radar to image the surface through the thick atmosphere. The radar image shows that the planet has significant topographic relief, including large, continent-sized high areas and vast, low plains. This map is shaded by elevation: blues are low regions and reds are high regions.

19.03.b4

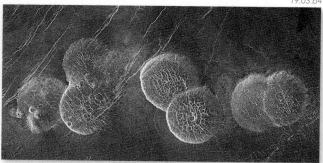

◀ This radar image shows bright areas that have a rough surface. The rough areas are interpreted as lava flows that flowed across a linear ridge formed by faulting or some other type of deformation.

▲ These 25-km-wide, pancake-shaped mountains are interpreted to be thick lava domes, demonstrating that volcanism has played an important role in the history of Venus.

C What Makes Earth Unique?

Earth is very different from its neighbors because it is just the right distance from the Sun to contain abundant water and to allow a thick, but not too thick, atmosphere to develop, along with oceans, rivers, and life.

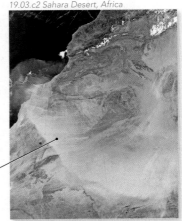

19.03.c2 Sahara Desert, Africa

1. In this computer-rendered view of Earth, the planet is dominated by its blue oceans but also shows green and brown continents and white clouds and ice. Even when viewed over the short time frame of hours or days, Earth is a dynamic place, with clouds, oceans, wind, rivers, and other features being in constant motion. Over the longer perspective of geologic time, mountains rise and are eroded down, continents break apart and travel great distances, and oceans widen or shrink with time due to seafloor spreading and subduction.

2. Earth's surface conditions allow water to exist as water vapor, liquid, or ice (in the atmosphere, oceans, and glaciers, respectively). Moving water and ice erode the land and deposit sediment, modifying or completely resurfacing the land surface.

3. On Earth, weathering produces abundant sediment, which can be transported by water and atmospheric winds. A thin veneer of sediments and sedimentary rocks covers most of the land and seafloor. This image shows a large dust storm moving many tons of sediment across northern Africa.

19.03.c1

4. Earth is big enough and generates enough heat to allow plate tectonics to operate and form the large-scale features observed today. The existence of tectonic activity on Earth, especially plate tectonics, is largely responsible for Earth's uniqueness. Tectonic-related volcanoes, like the steam-emitting, divergent rift shown here (▶), release water vapor and other gases that are an essential part of our atmosphere. Subduction removes carbon-rich crustal materials, like limestone and organic shale, from Earth's surface, allowing more CO_2 from the atmosphere to take its place. Without plate tectonics, Earth could have a thick, stifling, CO_2-rich atmosphere like that of Venus, and might be lifeless. It might not have a temperature or atmospheric chemistry that would support life as we know it.

19.03.c3 Krafla, Iceland

5. Without tectonics to dewater and degas Earth's interior, Earth might not have oceans, lakes, rivers, or other parts of the hydrosphere. Without oceans and rain, it is likely that wind and dust storms would be the dominant agents of erosion, transport, and deposition of sediment, as is the current situation on Mars.

D What Is on the Surface of Mars, the Red Planet?

Compared to Earth, Mars is smaller and has a thinner atmosphere. The color of the planet is due to an abundance of reddish, windblown dust. Mars has been orbited by spacecraft with sophisticated cameras and instruments, and has been visited by landers and rovers. It is the focus of the connection pages near the end of this chapter.

19.03.d1

Ice Cap

Dust

Bedrock

Mars has enough atmosphere to maintain ice caps on the south and north poles, as well as patches of ice in some shaded areas. The ice caps shrink and grow with the Mars seasons.

Much of the surface consists of reddish dust that has been weathered from rocks and blown around the surface by strong Martian winds. The reddish color is from iron oxide minerals, like hematite.

Geologists interpret Martian bedrock to consist of basaltic lavas, including some with recently photographed columnar joints. Other areas have layered rocks that many geologists think are sedimentary in origin.

Before You Leave This Page Be Able To

☑ Explain why the surface of Mercury is so heavily cratered.

☑ Describe why radar was required to investigate the surface of Venus, and what types of features we found.

☑ Discuss factors that make the surface of Earth so different from its neighbors.

☑ Summarize the materials and features present on the surface of Mars.

19.3

19.4 What Is the Geology of Our Moon?

OUR NEAREST NEIGHBOR IN SPACE is the Moon. It is much closer than any other object in the solar system and can be observed in detail with the simplest of telescopes or binoculars. It was the first object in the solar system to be systematically studied by geologists, who observed the Moon with telescopes. Subsequently, geologists mapped the topography, geology, and composition by sending spacecraft to orbit the Moon, and finally walked and drove on the Moon's surface, observing the features and collecting rock samples to bring back to Earth for detailed study.

A What Are the Main Geologic Features of the Moon?

Observe this large image of the Moon. The surface of the Moon is not all the same. It has lighter colored areas, with some dark, somewhat-circular patches. With binoculars, we can observe individual craters.

1. This view of the Moon shows the features on the side of the Moon that always faces Earth, called the *near side*. The other, or *far side*, cannot be seen from Earth but has been photographed by spacecraft.

2. Much of the near side and nearly all of the far side consists of light-colored material that is heavily cratered (▶). This material mostly occurs at higher elevations and is called the *lunar highlands* or the *cratered highlands*.

19.04.a2

19.04.a1

3. From samples collected by astronauts and other information, we know that the cratered highlands contain igneous rocks that are light colored because they consist almost entirely of feldspar. Rocks from the highlands (▶) have been dated to more than 4 billion years.

19.04.a3

4. The dark patches on the Moon (▼) are lower, flatter, and less cratered; they are *maria* (plural of *mare*, meaning sea). The maria have far fewer craters than the highland and so are much younger. The maria consist of dark basalt erupted as lava flows that buried and filled craters that existed in the lunar highland material.

6. The other obvious features on the Moon are *impact craters*, some of which have bright rays of material radiating outward. The rays overlie and cut across the top of the maria. Such *rayed craters* are some of the youngest features on the Moon, in some places probably being less than 100 million years old. Samples collected from lunar craters are mostly breccia containing angular rock fragments (▶) generated during impacts.

5. Samples from maria consist of basalt lava (▼), mostly dated at 3.8 to 2.5 billion years old. At these past times, the Moon retained enough heat to allow volcanism. Vesicles in the basalt record gas in the magma.

19.04.a6

19.04.a4

19.04.a5

B What Other Features Are Observed on the Moon?

The Moon is tectonically and volcanically inactive, and has been so for more than two billion years. The vast majority of craters on the Moon are due to impacts of meteoroids and other objects onto the Moon's surface, but some are volcanic in origin. These impacts and volcanoes led to several types of features similar to those on Earth.

19.04.b1

19.04.b2

After a crater is excavated by an impact, the shattered walls of the crater are commonly too steep to withstand the Moon's gravity, even though it is only 1/6 that of Earth. Slope failures are common along the walls, like the rotated slump blocks, debris avalanches, and other loose debris shown here.

Several types of troughs cut across the Moon's surface. One type, shown here, is fairly straight and is thought to represent a down-faulted block (graben), similar to fault blocks observed on Earth. The steep edges of the trough are fault scarps. The tensional forces that pulled the surface apart were probably related to adjustments after a nearby impact.

C What Causes the Phases of the Moon?

Every month the Moon appears to change its illumination on a regular schedule, going from completely dark to fully lit and back to dark again. Why is this cycle, called *phases of the Moon*, happening?

19.04.c1

19.04.c2

19.04.c3

1. At times, the side of the Moon facing Earth (near side) appears fully illuminated by sunlight. This is a *full moon*.

2. Seven days later, only half of the near side can be seen from Earth. This is a *quarter moon*.

3. Six days later, a thin sliver of the near side is illuminated. The next day, none is illuminated, which is a *new moon*.

4. Half of the Moon is always illuminated by the Sun, but from Earth we may not be able to see the entire sunlit half because of the Moon's position. The Moon orbits Earth in 28 days, and we see different amounts of the Moon's sunlit side on different nights.

5. During a new moon, the side being illuminated by sunlight is away from our view. The Moon appears dark to us, but the other side of the Moon is still completely sunlit. The yellow arrows depict the rays of sunlight, but the Sun is not shown.

6. At other times, we see half of the sunlit half of the Moon, so it is a quarter moon. When the moon is in this position, we can tell that the phases of the Moon are not in any way related to Earth's shadow.

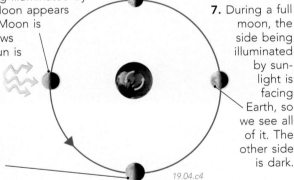
19.04.c4

7. During a full moon, the side being illuminated by sunlight is facing Earth, so we see all of it. The other side is dark.

A Model for the Formation of the Moon

Where did the Moon come from, and how did it form? Geologists and other scientists have investigated this question by examining several types of data. They calculate the age of the Moon by dating actual lunar samples. The chemical composition of these samples, including isotopic analyses, showed some unexpected similarities to rocks on Earth. This led to a hypothesis, currently favored by many scientists, that the Moon formed when a Mars-sized object collided with Earth early in the history of the solar system. The col-

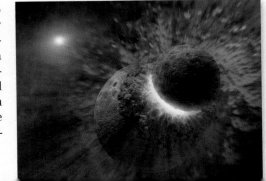
19.04.mtb1

lision ejected a huge part of Earth's mantle into space, where it later aggregated under the force of gravity and formed the Moon.

As the Moon formed and soon thereafter, it became hot enough for large parts to melt. As the magma began to solidify, heavier crystals sank downward (crystal settling), while less dense crystals, especially feldspar, floated upward. The floating crystals accumulated near the surface, forming the light-colored igneous rocks of the highlands. Early, intense impacts cratered the highlands. Later, basaltic

magmas from depth erupted, forming the dark-colored maria. Rayed craters formed even later, from more recent impacts.

Before You Leave This Page Be Able To

☑ Summarize the physical characteristics and rock compositions of the lunar highlands, maria, and craters, and explain how each feature formed.

☑ Sketch and describe what causes the phases of the Moon.

☑ Summarize one model for how the Moon and its different parts formed.

19.4

19.5 What Is Observed on Jupiter and Its Moons?

JUPITER, THE LARGEST PLANET IN THE SOLAR SYSTEM, is a gas giant more than three times farther from the Sun than Mars. Jupiter is orbited by, at present count, 49 officially named moons, including the largest moon in the solar system. To geologists, the icy and rocky moons are of greater interest than the gas-dominated planet itself because of their solid surfaces, spectacular geologic features, and wide diversity.

Jupiter is nearly 780,000,000 km from the Sun, but it is so large that we can see it on most clear nights. It is about 2.5 times more massive than all the other planets combined, and contains more than 300 times the mass of Earth. Examine the large, page-spanning image, which was computer generated by wrapping actual images of Jupiter around a sphere. What do you observe on the surface of the planet?

19.05.a1

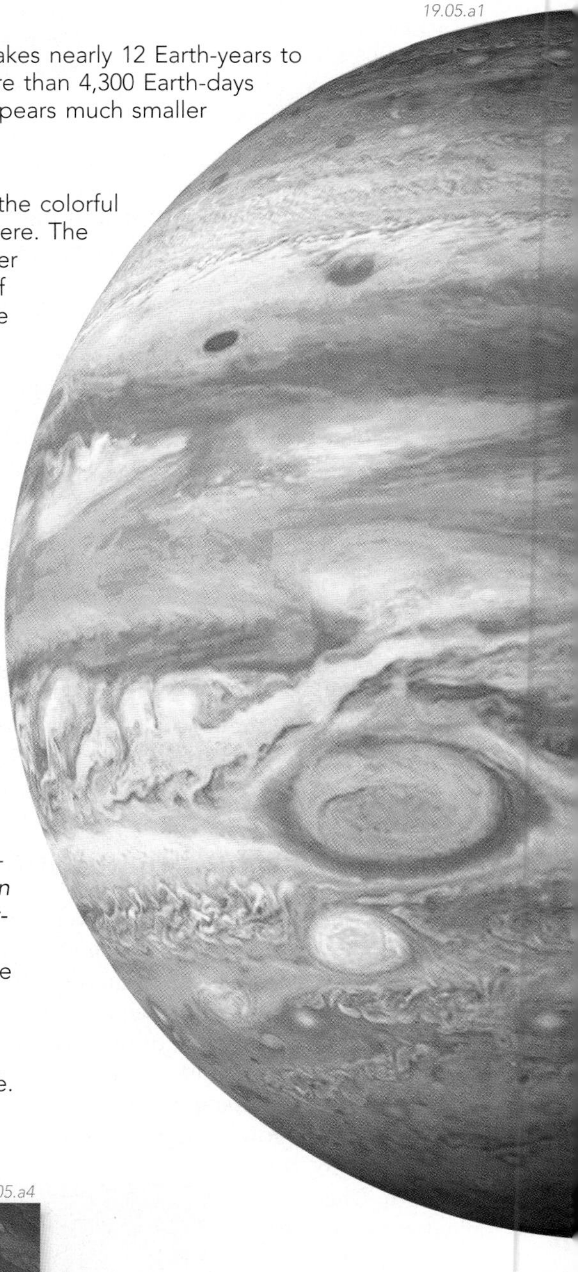

1. Jupiter is so far from the Sun that it takes nearly 12 Earth-years to complete an orbit—a Jupiter year is more than 4,300 Earth-days long. As viewed from Jupiter, the Sun appears much smaller and dimmer than it does from Earth.

2. The dominant features of Jupiter are the colorful bands and swirls of the planet's atmosphere. The atmosphere is mostly hydrogen with lesser amounts of helium, and trace amounts of methane, ammonia, and other gases. The interior of the planet is interpreted to consist of hydrogen in liquid and liquid-metallic forms, surrounding a solid core of mostly iron silicate minerals. Most of the planet is gas, so its overall density is less than that of Earth.

19.05.a2

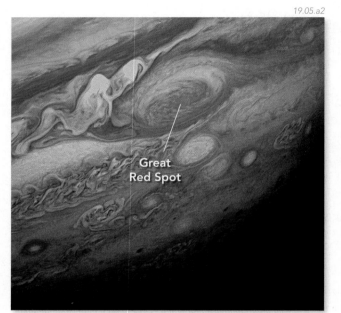

Great Red Spot

◄ 3. One of the most distinctive features in Jupiter's atmosphere is the Great Red Spot, which is a storm that has existed for at least a hundred years. This spot is three times wider than the diameter of Earth.

19.05.a3

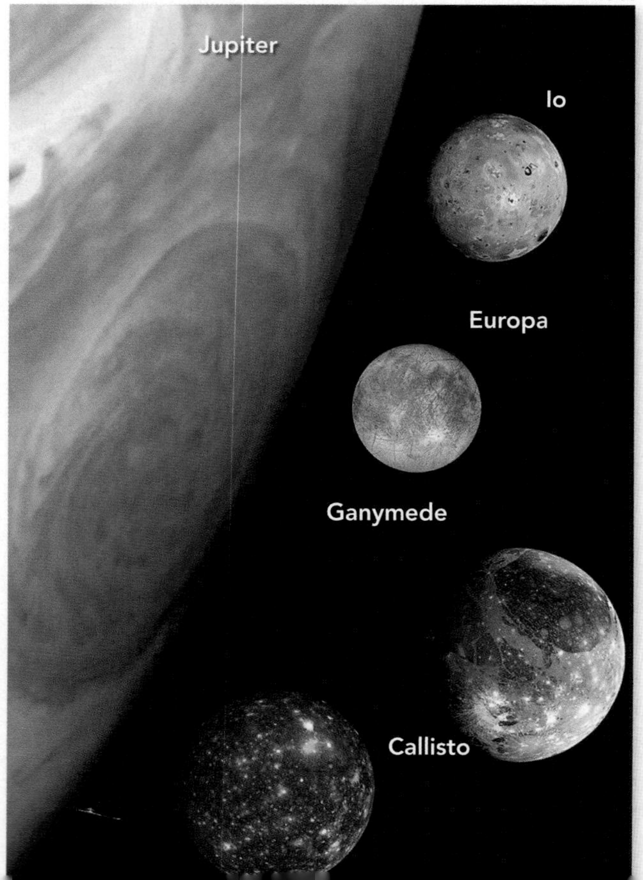

Jupiter

Io

Europa

Ganymede

Callisto

◄ 4. Jupiter's four largest moons were discovered by Galileo Galilei in 1610 when he observed the planet with a telescope. These four moons are the *Galilean moons*, and are named *Io, Europa, Ganymede*, and *Callisto*. The dramatic differences between the moons are largely due to differences in their distance from the massive gas giant around which they revolve. The moons are not as close to one another nor to Jupiter as shown here.

19.05.a4

Io

Io's Shadow

◄ 5. This image shows Jupiter's moon Io and the shadow of Io on Jupiter's surface. The image was taken by the Hubble Space Telescope, which orbits Earth.

19.05.a5

6. Of the four Galilean moons, *Io* is closest to Jupiter. It is slightly larger than Earth's moon. Because it is so close to massive Jupiter, it is subjected to extreme tidal forces that deform its land surface up and down by as much as 100 m, in the same way that our Moon moves Earth's oceans up or down a few meters, causing tides in our oceans.

7. Pulling and squeezing of rocks by tidal forces generates heat, making Io the most volcanically active object in the solar system. Sulfur-rich lava flows cover its surface. NASA's *Galileo* spacecraft photographed one such eruption of lava. ▶

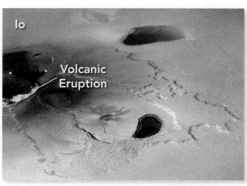
19.05.a6

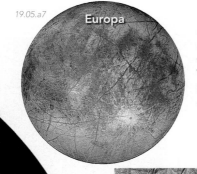
19.05.a7

◀ **8.** *Europa* is farther away from Jupiter but is still heated by the tidal forces of Jupiter and the other Galilean moons. These forces allow Europa to remain volcanically and tectonically active longer than would be merited by its size. Volcanic and tectonic processes have extensively reworked the surface, accounting for the nearly complete lack of craters. Beneath the icy crust is probably an ocean of liquid water.

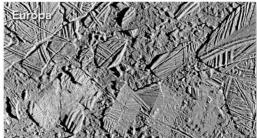
19.05.a9

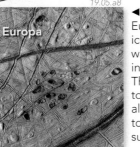
19.05.a8

◀ **9.** The surface of Europa is a crust of ice (mostly frozen water) marked by intersecting lines. These lines appear to be fissures that allowed liquid water to erupt onto the surface.

▲ **10.** Parts of Europa's surface are covered by huge blocks of ice that broke apart and then froze in place, like icebergs in a frozen sea.

19.05.a10

▲ **11.** *Ganymede*, the largest moon in the solar system, is thought to consist of a rocky core with a water-ice mantle and a crust of water-ice and rocks.

19.05.a11

▲ **12.** Ganymede's surface contains dark, cratered patches that are relatively old. Younger patches and belts crosscut the older surfaces and contain tectonic features similar to those seen on Europa, including some interpreted to be water-erupting fissures.

19.05.a12

◀ **13.** *Callisto*, the third-largest moon in the solar system, is the most heavily cratered object in the solar system. It is far enough from Jupiter's tidal forces that its surface has remained largely intact for the last 4 billion years.

Before You Leave This Page Be Able To

☑ Summarize the key characteristics of Jupiter, such as its size, internal composition, and atmospheric composition.

☑ Briefly summarize the main characteristics of each Galilean moon.

19.5

19.6 What Is Observed on Saturn and Its Moons?

SATURN IS A BEAUTIFUL PLANET, a gas giant encircled by a set of spectacular rings. Saturn is the second-largest planet in the solar system and is orbited by more than 50 named moons. Saturn's moons are quite diverse, reflecting differences in the materials and in the role of different geologic processes.

Saturn is farther from the Sun than Jupiter, in fact nearly twice as far. The two planets are very far apart, with the distance between the two planets being greater than the distance between the Sun and Mars. The large photograph was taken by the *Voyager* spacecraft, and colors have been enhanced in the small photograph to accentuate the bands.

19.06.a1

1. Saturn, like Jupiter, consists mostly of hydrogen and helium, which make up the gaseous atmosphere. The gases become liquid as they are compressed closer to the center of the planet. The center of the planet is interpreted to have a solid core of rock and metal. Saturn is more than 1.4 billion km from the Sun, and takes 29.5 Earth-years to orbit the Sun.

2. Like Jupiter, Saturn is a mini solar system, orbited by a collection of large and small moons. Our knowledge of the geology of these moons has increased dramatically due to the arrival of the *Cassini-Huygens* spacecraft in 2004. The image to the right shows Saturn, its rings, and some of its moons (small light-brown spots), four of which are discussed here: Titan, Iapetus, Enceladus, and Mimas. Titan is the largest of Saturn's moons, and the other three are included, not because they are the three next largest, but because they are geologically interesting and nicely illustrate the geologic diversity of Saturn's moons.

▼ **3.** Saturn is best known for its rings, which extend outward from the planet a distance nearly equal to the distance between the Earth and Sun. The rings consist of widely separated icy chunks floating in space. Most of the icy chunks are the size of sand, pebbles, and boulders, with some larger pieces. Close-ups of the rings display intricate details of concentric thick and thin rings separated by dark-colored, more-empty space, as viewed in the image below, which is colored by particle size. Purple indicates regions where particles are larger than 5 cm (pebble size and larger), green and blue indicate particles smaller than 5 cm, and white bands mark rings where particles blocked the radio signals used to determine size.

19.06.a2

Titan

19.06.a3

◄ 4. *Titan* is the largest of Saturn's moons and the second-largest moon in the solar system, even larger than the planet Mercury. Its surface is obscured by a thick, cloudy atmosphere of mostly nitrogen and methane, but this image generated from various types of data shows some of Titan's surface features. The surface is inferred and observed to contain solid materials (ices) and liquids, including liquid methane and other hydrocarbons.

5. The *Cassini* spacecraft released the *Huygens* probe, which parachuted through Titan's atmosphere and softly landed on the surface. On the way down, it captured images (►) of drainage networks and a lake or ocean, confirming that liquids are widespread on Titan's surface. Once on the surface, the Huygens probe sent back an image of well-rounded icy boulders, presumably rounded by transport in flowing liquid.

Titan

19.06.a4

Iapetus

19.06.a5

◄ 6. *Iapetus*, another moon, is distinctive in that most of its icy, cratered surface is light colored, but part is quite dark. The light-colored side is water ice; the darker side is interpreted to be a coating of dust that probably escaped from an adjacent moon and has been plastered on the leading edge of Iapetus as it orbits Saturn.

▼ 7. *Enceladus* is one of the lightest-colored objects in the solar system, possibly because an icy frost continuously forms on the surface.

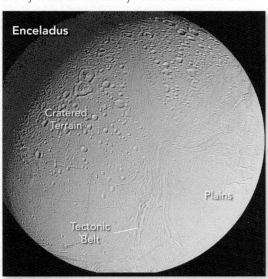

Enceladus

Cratered Terrain

Plains

Tectonic Belt

19.06.a6

8. The surface of Enceladus consists of at least three distinct types of terrain, the oldest of which is heavily cratered. Broad plains lie adjacent to the cratered terrain, and are much less cratered and are therefore younger. They are interpreted to have been resurfaced by the eruption of water onto the surface.

9. The third type of terrain consists of tectonic belts that slice through the heavily cratered material and through the plains. These belts have linear ridges and troughs probably formed as fissures through which water erupted to the surface. The *Cassini* team discovered active ice geysers fountaining from the surface. ►

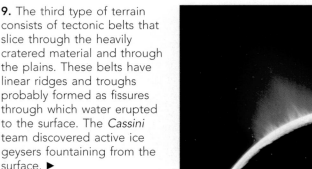

19.06.a7

▼ 10. *Mimas* is a relatively small moon whose pockmarked surface contains a large crater 130 km in diameter, with walls nearly 10 km (~33,000 ft) high. This crater formed from an impact that scientists have calculated nearly blasted the moon apart.

Mimas

19.06.a8

Before You Leave This Page Be Able To

☑ Summarize the key characteristics of Saturn, such as its size and composition.

☑ Describe what materials compose the rings of Saturn.

☑ Summarize the main characteristics of the four moons of Saturn described here and the main geologic processes expressed on the surface of each.

19.6

19.7 What Is the Geology of the Outer Planets and Their Moons?

THE OUTER PLANETS OF THE SOLAR SYSTEM and their moons are less well known than those from Saturn inward to the Sun. Many of the observations are based on images taken by the *Voyager 1* and *2* spacecraft, which flew through the outer reaches of the solar system in the late 1980s. These images provided a wealth of new information about the planets Uranus and Neptune, and some of their moons.

A What Features Characterize Uranus and Its Unusual Moons?

The next large planet out from Saturn is another large, gaseous world called *Uranus*. This planet and some of its moons have some unusual characteristics and features.

1. The planet Uranus is nearly 2.9 billion km from the Sun. The distance from Uranus to Saturn, the next planet in, is comparable to the distance from Saturn to the Sun. It takes Uranus 84 Earth-years to orbit the Sun.

2. Uranus consists largely of liquid and icy materials, including water, methane, and ammonia. The atmosphere is a mixture of hydrogen, helium, and methane. The blue-green color is caused by methane, which absorbs red light and reflects blue light. Uranus does not appear to have a solid surface.

3. Uranus has rings and at least 27 moons, named after characters from the works of William Shakespeare and Alexander Pope, including Oberon, Titania, Juliette, Puck, Ariel, and Miranda.

4. Uranus is unusual in that its axis of rotation, orientation of rings, and orbits of its moons are roughly perpendicular to those of every other planet in the solar system. That is, if Earth's Moon orbits "horizontally" around Earth, Uranus' rings and moons go around it vertically. ▼

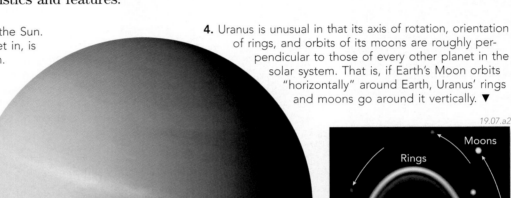
19.07.a1

19.07.a2

Ariel

▼ **5.** Ariel is a moderate-sized moon, approximately 580 km (360 mi) in diameter, and is covered by ice. The surface is cut by long fractures, some of which have been filled by upwelling liquid water. This moon is thought to be mostly inactive.

19.07.a3

Ariel

Miranda

▶ **6.** To some geologists, Miranda, a small (236 km-diameter) moon of Uranus, is the most bizarre world in our solar system. The surface of Miranda is covered with ice and displays several distinct types of terrain.

7. There is highly cratered terrain that is lighter colored and relatively old.

8. Disrupting the heavily cratered terrain are huge oval to angular features, each of which is a *corona*. The origin of these features is unresolved among planetary geologists, but they involve some normal faulting and probably upwelling of deeper materials.

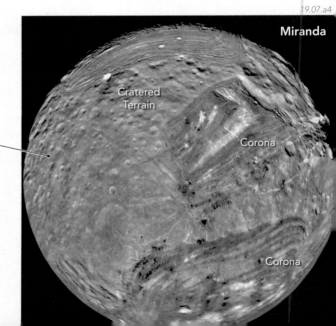
19.07.a4

Miranda

Cratered Terrain

Corona

Corona

B What Is the Geology of Neptune and Its Moon, Triton?

Neptune is the eighth and last planet from the Sun and is another gas giant similar in many ways to Uranus. The existence of Neptune was predicted mathematically before the planet was discovered by telescope.

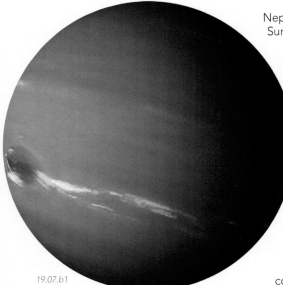

19.07.b1

Neptune is nearly 4.5 billion km from the Sun, and the distance from Uranus to Neptune is more than that between Uranus and Saturn. In other words, the distance between adjacent planets increases as one moves outward through the solar system. Neptune is so far from the Sun that it has not completed even one orbit since it was discovered. It takes 165 Earth-years to go around the Sun.

Neptune is about the same size as Uranus (~50,000 km in diameter) and has a similar composition, with ices and liquids inside and an outer atmosphere of hydrogen, helium, and methane. Its blue color is due to methane.

19.07.b2

Triton

▲ The surface of Triton, Neptune's only large moon, consists of ices of nitrogen and carbon substances. It has two distinct halves; one part appears like the surface of a cantaloupe and is interpreted to represent activity from volcanic eruptions and active geysers.

C What Do We Know About Pluto and Its Companions?

Pluto was once considered to be the ninth and outermost planet, but astronomers recently reassigned Pluto to a type of solar system object called a *plutoid*. So our solar system has eight planets, not nine.

Pluto was always an oddity compared to the eight planets. It is a relatively small, icy object, even smaller than Earth's moon. Pluto orbits the Sun in a very elliptical orbit that sometimes brings it closer to the Sun than Neptune. A circuit around the Sun takes 248 Earth-years. Pluto has a large companion named *Charon*, which is half Pluto's size, plus several very small moons.

19.07.c1

No spacecraft has yet visited Pluto, but NASA's *New Horizons* spacecraft is on the way. This spacecraft will also visit the *Kuiper Belt,* a disk-like zone of objects that lies beyond the orbit of Neptune. This belt has a number of objects that are similar to and far beyond Pluto. Some of the named Kuiper Belt objects are shown in this artist's conception. ▶

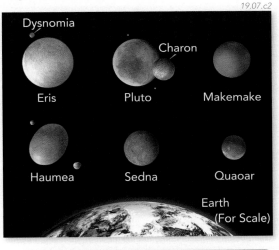

19.07.c2

Dysnomia

Charon

Eris

Pluto

Makemake

Haumea

Sedna

Quaoar

Earth
(For Scale)

Comets

19.07.mtb1

Comets are one of the more interesting spectacles that sometimes grace the night sky. Comets are small, icy and rocky objects with very elliptical orbits around the Sun. Some comets, such as Halley's Comet (▶), visit the inner solar system regularly, whereas others visit at very long intervals. Comets are thought to come from a very outer part of the solar system, well beyond the orbit of Neptune. As a comet nears the Sun, gas and dust are stripped off by the solar wind and carried outward, forming a tail that always points away from the Sun.

Before You Leave This Page Be Able To

☑ Describe some key features of Uranus and Neptune, and explain how they are similar.

☑ Describe unusual features on Ariel, Miranda, and Triton, and identify the materials that comprise the surfaces of these moons.

☑ Describe what is known about Pluto and its companions.

☑ Describe what comets are and why they have a tail.

19.7

19.8 What Have We Learned About Mars?

EXCITING DEVELOPMENTS IN PLANETARY GEOLOGY involve Mars, the *Red Planet*. Recently, Mars has been explored by orbiting spacecraft that carried sophisticated cameras and other instruments, many designed and controlled by planetary geologists. Spacecraft have landed on the planet and unleashed small, robotic rovers that travel across the surface, exploring and collecting data.

A What Have We Learned from Instruments Orbiting Mars?

Several spacecraft, including *Mars Odyssey*, *Mars Express*, and *Mars Reconnaissance Orbiter*, orbit the planet. As they pass over the planet's surface, they record images and take measurements designed to detect water and determine the composition of rocks, sediment, and ice. Using these spacecraft, we have made major discoveries.

19.08.a2 Valles Marineris, Mars

◄ **1.** Mars contains a huge canyon system, 4,000 km long—*Valles Marineris*. The canyon's length is equivalent to the width of the United States. The canyon, which began as a large rift, has been widened by inward collapse of the steep canyon walls and by other types of erosion.

► **4.** Many parts of Mars have layered rocks, which variably have a sedimentary or volcanic origin. Layers in this image are colorized for emphasis. The surface of Mars is not this color.

19.08.a7

19.08.a3 Candor Chasm, Mars

▲ **2.** This image shows *Candor Chasm*, a part of the Valles Marineris system. The steep walls of the chasm have collapsed downslope, providing some of the most spectacular examples of slope failure in the solar system. Gullies carve into the cliffs and steep slopes.

19.08.a1

► **5.** Some parts of Mars have spectacular channels, interpreted to have been formed by torrents of running water flowing on the surface some time in the past. Where the channels reached the gentle plains, they deposited piles of sediment, equivalent to deltas and alluvial fans on Earth.

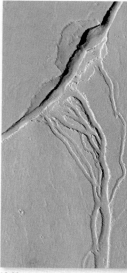

19.08.a6

19.08.a4 Victoria Crater, Mars

19.08.a5 Olympus Mons, Mars

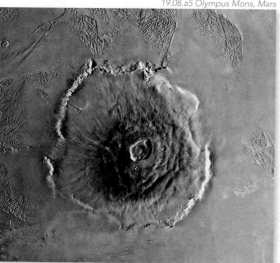

◄ **3.** Mars' atmosphere is less dense than Earth's, but the winds are strong. Lace-like sand dunes occupy the center of beautiful Victoria Crater. The crater was visited by the Mars Exploration Rover named *Opportunity*.

◄ **6.** Mars has the solar system's largest volcano, *Olympus Mons*. The volcano is 600 km across and 27 km high, nearly three times the height of Mount Everest. It is a large shield volcano like those on Hawaii, but is inactive. Large segments of the mountain collapsed in landslides and debris avalanches that moved downhill, spreading out to form areas of hummocky topography.

B   What Have We Learned from Landers and Rovers on Mars?

NASA successfully landed three spacecraft that carried small rovers to navigate and photograph the Martian landscape. The two famous recent rovers, *Spirit* and *Opportunity*, provided a wealth of new geologic information, including a few surprises.

► **1.** Each rover-bearing spacecraft landed on the surface by cushioning itself with large air bags that inflated just before the spacecraft bounced onto the surface. Then the rover, shown here in an artist's conception, rolled off to explore nearby parts of the planet.

19.08.b1

19.08.b2 Mars

◄ **2.** The rovers roll across the surface on wheels, stopping to inspect any outcrops or interesting rocks. They can spin their wheels to dig up sediment on the surface or use a tool to scratch at the rocks. They carry cameras and scientific instruments to measure composition, temperature, and other aspects.

19.08.b3 Payson, Mars

◄ **3.** This image of layered rocks was taken with *Opportunity's* camera at a Mars site named *Payson*.

19.08.b6 Mars

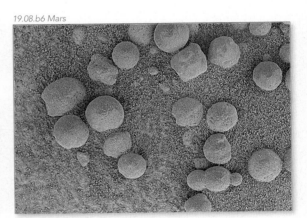

19.08.b4 Mars

19.08.b5 Mars

4. *Layers*—One of the first features that caught geologists' attention was the layers exposed in the walls of several craters. The layers are interpreted to have been formed by running water or by fast-moving debris ejected from nearby impacts.

5. *Meteorites*—On its travels across the plains, *Spirit* encountered a lone rock that seemed out of place to a geologist monitoring its travels. When the rover was redirected to investigate the rock, it discovered the first meteorite found on Mars.

6. *Blueberries*—Within the layered rocks and weathering out onto the surface are millimeter-size, spherical objects, nicknamed *blueberries*. Measurements document that these contain the mineral *hematite*, which some planetary scientists interpret formed in the presence of water.

Choosing a Landing Site

How do researchers choose where to land? For the *Opportunity* rover, they chose the site on the basis of *infrared measurements* that, to geologists, indicated the presence of abundant hematite in the area. On Earth, hematite most commonly forms under wet, oxidizing conditions. The geologists therefore concluded that if you were looking for water on Mars, this would be a good place to start. When *Opportunity* rolled off its platform to explore, it confirmed the presence of hematite, in blueberries lying on the ground and weathering out of rocks. This image shows areas with several percent hematite in blue to 20% hematite in red.

19.08.mtb1

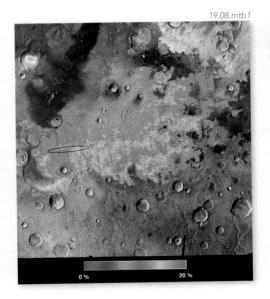

0 %   20 %

Before You Leave This Page Be Able To

- ☑ Summarize two of the ways that geologists have explored Mars.

- ☑ Describe some features found by orbiting spacecraft and what they imply about processes that have occurred on the planet's surface.

- ☑ Describe some features discovered by the rovers *Spirit* and *Opportunity*.

- ☑ Explain how *Opportunity's* discoveries were made possible by prior spacecraft measurements.

19.9 How and When Did Geologic Features on This Alien World Form?

TRAVELING THROUGH SPACE YOU ENCOUNTER AN UNKNOWN WORLD. Your spacecraft orbits the planet and takes images and measurements of the different geologic regions and of the most interesting geologic features. You will use these images and some initial observations about the features to interpret how each region or feature formed and in what chronological order they formed.

Goals of This Exercise:

- Observe the planet to identify large regions that have a similar geologic appearance.
- Examine close-up images of features and read descriptions for each, to interpret how each feature formed.
- Use several strategies to reconstruct the sequence in which the different features formed.
- Summarize geologic features and history of this planet.

Procedures

Use the available information to complete the following steps, entering your answers in the appropriate places on the worksheet or online.

1. Observe the image of the entire planet on the next page. Identify regions that have different geologic characteristics, and locate their approximate boundaries.

2. Observe each of the close-up images and read the description that accompanies each, looking for further clues about what types of geologic features are present and how each feature might have formed.

3. Determine the relative ages of the different geologic regions and features using crosscutting relationships and density of impact craters.

4. Your instructor may have you draw a simple geologic map of the planet, on which each map unit is a different type of geologic region or geologic feature. Draw a legend to accompany your map that has (1) a small box with the color or pattern you chose to depict that geologic terrain, (2) the name of the geologic terrain or feature, and (3) a brief description of less than 30 words that conveys the key characteristics of this terrain and your interpretation of the terrain's origin.

5. Write a short report or list summarizing the geology of the planet and its geologic history. Your instructor will guide you about the length and detail expected. This report should demonstrate the breadth of knowledge you have gained in this course, not just the concepts from this last chapter. In other words, use this final investigation to bring together concepts you have learned throughout the course.

The large image shows one side of the planet as illuminated by the local sun. The surface contains different types of geologic terrain as well as several obvious large geologic features. North is up in this view, south is down, west is to the left, and east is to the right. Some observations about this place are listed below, labeled with letters corresponding to the name or character of the place. Corresponding letters mark the place on the large view of the planet.

Western Terrain (W)—The western side of the planet consists of a heavily cratered terrain with many large and small craters. Samples of the rocks are very shattered and contain many angular fragments of highly weathered basalt.

Dark Terrain (D)—A dark, wide strip curves across the planet from south to northwest. Radar measurements indicate that it has a rough upper surface. A few normal faults cut across the dark material. As shown in the image below (▼), the dark material locally protrudes into the adjacent, heavily cratered terrain, covering it and filling some craters. The dark material is partly weathered basalt. The dark terrain has some small impact craters, but fewer than the western terrain.

19.09.a2

Chasm (C)—Cutting across the highly cratered terrain is a deep chasm that narrows progressively toward the south. On the image of the entire globe, the chasm has some important relationships with the dark terrain and to a reddish-brown sedimentary area (S) to the north. The close-up below (▼) shows one wall of the chasm. What features do you observe?

19.09.a3

Polar Ice (P)—The north and south poles are covered with water ice year round. The close-up image to the left (◄) shows the edge of the layered ice overlapping a crater. The ice has almost no craters.

19.09.a4

Sedimentary Terrain (S)—Adjacent to the north pole and the northern ice cap is a distinctive reddish-brown region. The unit has layers and appears to be sedimentary in origin. Along the southern edge of the terrain, the soft-looking, loose sediment is in contact with terrain that is more heavily cratered, as shown in the detailed image to the right (►). The sedimentary region has very few craters. Similar material may be present near the south pole but is not visible in this view of the planet.

19.09.a5

Valleys (V)—A few valleys or channels extend south from the sedimentary region. They appear to be filled with sediment, and there is a feature that looks like a delta or fan where one channel empties into a crater. The large crater is part of moderately cratered terrain that makes up much of the eastern part of the globe. A close-up view of one channel is shown below (▼).

19.09.a1

19.09.a6

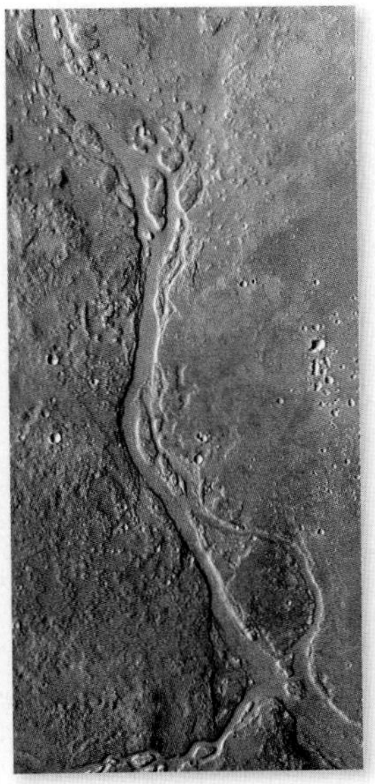

Mountains (M)—In the southeastern part of this view, three large mountains rise above the plain. The close-up below shows one of the mountains (▼). The mountain is cone shaped with a central crater. The flank is indented by small craters, and the lower part of the mountain appears to be missing on one side (upper left side in this view).

19.09.a7

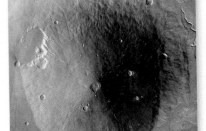

Eastern Terrain (E)—Much of the eastern hemisphere of the planet consists of a moderately dark, moderately cratered terrain. This terrain has fewer craters than the heavily cratered, western terrain, but has more craters than are present in the valleys, sedimentary terrain, chasm, or dark terrain.

19.9

GLOSSARY

Definitions in this glossary are derived largely from this textbook, with additions from the American Geological Institute *Glossary of Geology* (K.K.E. Neuendorf and others, 5th edition, 2005, Alexandria, VA) and the *McGraw-Hill Dictionary of Geology and Mineralogy* (2nd edition, 2003, New York). Numbers in parentheses after each entry refer to the chapter and section number in this textbook where that term is defined or described.

A

AA lava A blocky type of lava with a rough, jagged surface. (6.3)

abrasion The mechanical wearing away of rock surfaces by contact with sand and other solid rock particles transported by running water, waves, wind, ice, or gravity. (16.2)

absolute age *See* numeric age. (9.4)

abyssal plain A relatively flat, smooth region of the deep ocean floor. (10.4)

Acadian orogeny A middle Paleozoic mountain-building event in eastern North America; occurred when a series of landmasses, referred to as Avalonia, collided with the eastern coast of North America. (11.12)

accretion The addition of tectonic terranes and other material to an existing landmass, usually along a convergent or transform plate boundary. (11.9)

accretionary prism A prism- or wedge-shaped, structurally complex zone of faults, folds, and mostly metamorphosed rocks that form along the upper parts of a subduction zone; material derived from sediment contributed by adjacent volcanoes or a continent, along with oceanic crust scraped off the downgoing slab. (10.4)

acre-foot The volume of water required to cover an acre of land to a height of one foot. (17.2)

aerial photograph A photograph taken from the air, such as a photograph taken by a camera mounted in an aircraft. (2.3)

aerosols Very small solid particles or drops of liquid dispersed in the atmosphere. (6.6)

aftershocks Smaller earthquakes that occur after a main earthquake and in the same area. (12.7)

aggregate Construction materials, such as sand and gravel or crushed stone, used alone as in construction fill or mixed into concrete, asphalt, and other materials. (18.13)

A horizon Topsoil composed of dark gray, brown, or black organic material mixed with mineral grains. (15.5)

albedo The ratio of the amount of electromagnetic energy reflected by a surface to the amount of energy incident upon it. (14.17)

Allegheny orogeny Late Paleozoic mountain-building event that occurred during the collision of Africa and eastern North America. (11.12)

alluvial fan A low, gently sloping mass of sediment, shaped like an open fan or a segment of a cone; typically deposited by a stream where it exits a mountain range or joins a main stream. (7.8)

alluvial plain A nearly level or gently sloping plain formed during widespread deposition of sediment by a stream or river. (7.8)

alpha decay The radioactive decay process in which an unstable atom fissions, releasing a positively charged particle with two protons and two neutrons. (5.4)

alpine glacier A glacier that begins in mountainous terrain and flows down a valley as a narrow tongue of ice and rocks; *syn.* valley glacier. (14.10)

amphibole A group of silicate minerals, including hornblende, characterized by double chains of silicate tetrahedra; most common in igneous and metamorphic rocks. (4.6)

analytical dating methods Analytical methods, commonly using isotopes, to determine the ages and histories of rocks and minerals in hundreds, thousands, millions, and billions of years. (9.2)

andesite An intermediate-composition igneous rock that is the fine-grained equivalent of diorite; typically gray or greenish-gray, commonly with phenocrysts of cream-colored feldspar or dark amphibole. (5.2)

angle of repose The steepest angle at which a pile of unconsolidated grains remains stable. (15.7)

angular unconformity An unconformity (ancient erosion surface) in which the older, underlying strata dip more steeply or at a different angle than the younger, overlying strata. (9.3)

anthracite A metamorphic coal with a high carbon content, high density, and black, shiny appearance. (18.3)

anticline A fold, generally concave downward, with the oldest rocks in the center. (8.5)

aphanitic rock An igneous rock that does not contain crystals visible to the unaided eye; can consist of microscopic crystals, fine-grained volcanic ash, volcanic glass, or a combination of these. (5.1)

aquatic plant A plant that grows in water. (18.1)

aquifer A permeable body of saturated material through which groundwater can flow and which yields significant volumes of water to wells and springs. (17.4)

aquitard A low-permeability bed or other unit that retards but does not prevent the flow of water to or from an adjacent aquifer. (17.4)

arid region A relatively dry region of land that receives less precipitation than it loses to evaporation and other processes. (13.8)

arkose A sedimentary rock composed of sand-sized clasts that contain a high proportion of feldspar grains in addition to quartz and other minerals. (7.5)

arroyo Small, deep gully containing a sandy or gravel-rich channel in an arid area. (13.9)

artesian Groundwater that is in a confined aquifer and is under enough water pressure to rise above the level of the aquifer. (17.4)

asbestos A general name for the fibrous varieties of a number of silicate minerals whose common characteristic is that they tend to form fibers. (4.6)

assimilation The process in which melted wall rocks become incorporated into magma. (5.6)

asteroid One of countless rocky fragments in orbit around the Sun, mostly concentrated between the orbits of Mars and Jupiter. (19.0)

asthenosphere The area of mantle, beneath the lithosphere, that is solid, but hotter than the rock above it and can flow under pressure; functions as a soft, weak zone over which the lithosphere may move. (1.3)

atoll A curved to roughly circular reef that encloses a shallow, inner lagoon and low islands. (10.9)

atom Smallest possible particle of an element that retains the properties of that element. (4.11)

atomic number The number of protons in an atom. (4.6)

atomic mass The number of neutrons and protons in an atom. (4.11)

atomic symbol One or two letters representing the name of an element, as in the Periodic Table. (4.6)

atomic weight The sum of the weight of the subatomic particles in an average atom of an element, given in atomic mass units. (9.4)

atmosphere The gaseous envelope that surrounds the solid Earth, and includes air, clouds, and precipitation. (1.7)

atmospheric pressure The pressure, or force per unit area, exerted by the atmosphere; generally greater at sea level than in high elevations. (1.4)

axial surface A surface that connects the hinge lines in a fold. (8.5)

B

back-arc The region adjacent to a subduction-related volcanic arc, on the side opposite the trench and subducting plate. (10.6)

banded iron formation A sedimentary mineral deposit displaying layers or bands of iron-rich minerals and fine-grained, silica-rich material. (18.11)

barchan dune A crescent-shaped dune of wind-blown sand with its tails pointing in the direction of the prevailing wind. (13.2)

barometer An instrument used to measure atmospheric pressure. (13.1)

barrier island A long, narrow island adjacent to a coastline and commonly containing sandy areas, like dunes, marshes, and sandy beaches. (7.2)

barrier reef A long, commonly curved, coral reef that parallels the coast of an island or continent from which it is separated by a lagoon. (10.9)

basalt A fine-grained, dark-colored mafic igneous rock, with or without vesicles and phenocrysts of pyroxene, olivine, or feldspar. (5.2)

base level The lowest level to which a stream can erode, commonly represented by the level of the sea, lake, or main river valley. (16.4)

basement rock The oldest crystalline rocks (igneous or metamorphic) in an area, commonly underlying all other rock units. (11.8)

base metal A relatively common and chemically active metal, including copper, lead, and zinc. (18.11)

basin 1. A low-lying area or an area in which sediments accumulate with no surface outlet. 2. A structural feature toward which rock strata are inclined in all directions. (11.4)

batholith One or more contiguous plutons that cover more than 100 km². (5.12)

baymouth bar A bar of sand or gravel deposited entirely or partly across the mouth of a bay. (14.5)

beach A stretch of coastline along which sediment, especially sand and stones, has accumulated. (7.2)

beach nourishment The procedure of bringing sand to a beach to replenish any sand lost to storms and currents. (14.6)

bed A distinct layer formed during deposition, generally in sediment or sedimentary rock; can be present in volcanic tuff and some metamorphic rocks. (2.1)

bedding Layers or beds of varying thickness and character, generally in a sediment or sedimentary rock. (2.1)

bed load Material, commonly sand and larger, that is transported along the bed of a river. (16.2)

bedrock stream A stream that is carved into bedrock, commonly in mountainous areas. (16.2)

beta decay Radioactive decay that involves an atom losing a beta particle, which is an electron or a positron. (5.4)

B horizon A zone in the soil characterized by the accumulation of material, including iron oxide, clay, and calcium carbonate, depending on the climate and starting materials. (15.5)

biochemical rock A sedimentary rock resulting from chemical processes and the activities of organisms. (7.6)

biome A major community of organisms defined by the predominant types of vegetation and characterized by organisms that are adapted to that environment. (13.8)

biosphere The spherical zone that includes life and all of the places life can exist on, below, and above Earth's surface. (1.7)

biotite A typically black or brown mica (sheet silicate mineral). (4.8)

bituminous coal A dark brown to black coal that is high in carbonaceous matter and has 15% to 50% volatile matter. (18.3)

black smokers Hydrothermal vents on the seafloor in which hot water from within the rock jets out into cold seawater and forms a cloud of minerals, especially those rich in sulfur. (10.3)

block diagram A diagram that portrays in three dimensions the shape of the land surface and the subsurface distributions of rock units, folds, faults, and other geologic features. (2.4)

blueschist A metamorphic rock, commonly with a bluish color, formed by high pressure/low temperature metamorphism and associated with subduction zones. (8.9)

body waves Seismic waves that travel within Earth. (12.5)

bonding The process in which two atoms bond together by sharing, donating, or borrowing electrons from their outermost orbital shells. (4.12)

boulder A clast or rock fragment with a diameter exceeding 0.25m. (7.4)

Bowen's Reaction Series An idealized sequence in which minerals could crystallize from a magma as it cools. (5.8)

brachiopods A phylum of marine-invertebrate animals, commonly containing two valves and living on or close to the bottom of the seafloor. (9.10)

brackish Water that is intermediate in salinity between normal freshwater and seawater. (17.1)

braided stream A stream or river with an interconnecting network of branching and reuniting shallow channels. (16.5)

breakwater An offshore structure (such as a wall), typically parallel to the shore, that breaks the force of the waves to protect a shoreline or harbor. (14.6)

breccia A rock composed of large, angular fragments; typically formed in sedimentary environments, but can be formed by volcanic, hydrothermal, or tectonic processes. (7.5)

breccia pipe A pipe- or cone-shaped structure filled with breccia; mostly formed by cave collapse (18.6), but can also form by igneous and hydrothermal processes.

brittle A rock or material that deforms by fracturing and frictional processes. (8.1)

burrow A commonly tubular opening formed when creatures wiggle or tunnel into mud; can be filled with a different type of sediment to form a trace fossil. (9.5)

butte An isolated, steep-sided feature that rises above the surrounding landscape. (2.2)

C

calcite A common rock-forming calcium carbonate mineral occurring in limestone and a variety of water-related deposits. (4.9)

caldera A large volcanic depression that is typically circular to elongate in shape and formed by collapse of a magma chamber. (6.10)

caliche A soil-related accumulation of calcareous material that cements sand, gravel, and other materials, commonly forming a hard layer. (15.5)

calving The breaking off of a mass of ice from a glacier, iceberg, or ice shelf. (14.11)

Cambrian The earliest geologic period of the Paleozoic Era.

Cambrian Explosion The widespread, relatively rapid appearance of diverse types of hard-shelled organisms near the beginning of the Cambrian Period. (9.10)

Canadian Shield A vast area of mostly crystalline rocks exposed in the eastern half of Canada and the Great Lakes region of the United States. (1.1)

capacity The amount of sediment, at a given flow rate, that a river is capable of transporting. (16.3)

carbonate compensation depth (CCD) The depth in the ocean at which carbonate minerals dissolve into seawater as fast as they accumulate so that no calcium carbonate is preserved; typically 4 to 5 km deep. (10.3)

carbonate minerals A mineral containing a significant amount of the carbon-oxygen combination called carbonate. (4.9)

carbonate rock A rock composed mostly of carbonate minerals, especially calcite and dolomite. (7.11)

carbonates Rocks or minerals that contain abundant carbonate. (4.6)

Carboniferous A Paleozoic geologic period, used outside of North America, for the combined duration of the Pennsylvanian and Mississippian Periods. (9.8)

cations Positively charged atoms. (4.7)

cave A natural underground chamber, generally large enough for a person to enter. (17.6)

cement 1. A natural material precipitated in the pore spaces between grains, helping to hold the grains together. 2. A processed, whitish powder, typically derived from calcium carbonate, that sets to a solid mass when mixed with water and usually other materials. (7.5)

cementation The precipitation of a binding material (the cement) around grains in sedimentary rocks. (7.5)

Cenozoic Era A major subdivision of geologic time from 65 Ma to the present; characterized by an abundance of mammals. (9.6)

chalk A soft, very fine-grained limestone that forms from the accumulation of calcium carbonate from microscopic organisms that float in the sea. (7.6)

chemical weathering Chemical reactions that affect a rock or other material by breaking down minerals and removing soluble material from the rock. (7.3)

chert A sedimentary rock composed of fine-grained silica. (7.6)

chimney A hollow, circular column formed by sulfur-rich minerals precipitating around a submarine, hydrothermal vent. (10.3)

C horizon A zone in the soil composed of weathered bedrock or sediment, grading downward into unweathered bedrock or sediment. (15.5)

chrysotile A fibrous form of the silicate mineral serpentine; the most commonly used asbestos in the United States. (4.6)

cinder cone *See* scoria cone. (6.1)

cirque An open-sided, bowl-shaped depression formed in a mountain, commonly at the head of a glacial valley; produced by erosion of a glacier. (14.12)

clast An individual grain or fragment of rock, produced by the physical breakdown of a larger rock mass. (4.2)

clastic A material consisting of pieces (clasts) derived from preexisting rocks; usually formed on Earth's surface in low-temperature environments; *syn.* detrital. (4.2)

clay 1. Any fine-grained sedimentary particle that is finer than 1/256 mm (7.4). 2. A family of finely crystalline, hydrous-silicate minerals with a two- or three-layered crystal structure. (4.8)

cleavage 1. The tendency of minerals to break along specific orientations of closely spaced planes (4.3); 2. The tendency of a rock, especially a metamorphic rock, to split along mostly parallel planes. (8.6)

coal The natural, brown to black rock derived from peat and other plant materials that have been buried, compacted, and heated (7.6); used for heating and to generate electricity. (18.3)

coal-bed methane A type of natural gas released by coal beds. (18.3)

coal seam A layer or bed of coal, typically much thinner than its lateral extent. (18.3)

coastal desert A desert that forms adjacent to the ocean, typically where cold, upwelling ocean currents cool the air and reduce its ability to hold moisture. (13.8)

coastal dune A windblown sand dune that is inland from a beach along a sea or lake. (7.2)

cobble A rock fragment larger than a pebble and smaller than a boulder, having a diameter in the range of 64 to 256 mm. (7.4)

cold front The sloping boundary surface between an advancing mass of cold air and a warmer air mass. (13.1)

columnar joints Distinctive fracture-bounded columns of rock formed when hot but solid igneous rock contracts as it cools. (5.13)

compaction Process by which soil, sediment, and volcanic materials lose pore space in response to the weight of overlying material. (7.5)

composite volcano A common type of volcano constructed of alternating layers of lava, pyroclastic deposits, and mass-wasting deposits, including mudflows; *syn.* stratovolcano. (6.7)

compression The type of differential stress that occurs when forces push in on a rock. (8.2)

concave-up profile A river profile that becomes less steep in a downstream direction (has a lower gradient in a downstream direction). (16.4)

concrete A mixture of cement, aggregate, and water that hardens to a rocklike consistency. (18.13)

concretions A hard, compact accumulation of mineral matter in the pores of sedimentary or volcanic rocks; representing a concentration of constituents of the rock or cementing material. (9.5)

condensation The process by which a vapor becomes a liquid; the opposite of evaporation. (13.3)

conduction Transfer of thermal energy by direct contact. (5.4)

cone of depression A depression in the water table that has the shape of a downward-narrowing cone; develops around a well that is pumped, especially one that is overpumped. (17.8)

confined aquifer An aquifer that is separated from Earth's surface by materials with low permeability. (17.4)

confining pressure The type of pressure that results when the force imposed on a rock is the same amount from all directions. (8.2)

conglomerate A coarse-grained clastic sedimentary rock composed of rounded to subrounded clasts (pebbles, cobbles, and boulders) in a fine-grained matrix of sand and mud. (7.5)

contact effects Evidence of baking, passage of hot fluids, or some other manifestation of the thermal and chemical effects of a magma chamber or a hot volcanic unit, as expressed in changes to adjacent wall rocks. (9.1)

contact metamorphism Metamorphism that principally involves heating of the rocks next to a magma or hot igneous material. (8.8)

continent-continent convergent boundary A plate-tectonic boundary where two continental masses collide. *syn.* Continental collision (3.5)

continental collision A convergent plate boundary that involves the collision of two masses of continental crust. (3.5)

continental crust The type of Earth's crust that underlies the continents and the continental shelves; average granitic composition, but includes diverse types of material. (1.3)

continental desert A desert that forms in the interior of a continent, far from sources of moisture or where prevailing winds blow toward the sea. (13.8)

continental drift The concept of movement of continents and other landmasses across the surface of the Earth. (2.9)

continental ice sheet A large mass of ice, including glaciers, that covers a large part of a continent. (14.0)

continental platform A broad region that surrounds the continental shield and typically exposes horizontal to gently dipping sedimentary rocks. (11.8)

continental rift A low trough or series of troughs bounded by normal faults, especially where two parts of a continent begin to rift apart. (3.4)

continental rifting The pulling apart of a continent, forming a low, fault-bounded trough (continental rift); may lead to a divergent plate boundary that leads to seafloor spreading and splitting apart a continent. (3.4)

continental rise A gently sloping edge of a continental plate, connecting the continental slope and the abyssal plain; built up by shedding of sediments from the continental block. (10.10)

continental shelf A gently sloping, relatively shallow area of seafloor that flanks a continent and is underlain by thinned continental crust. (10.10)

continental shield A central region of many continents, consisting of relatively old metamorphic and igneous rocks, commonly of Precambrian age. (11.8)

continental slope A submarine slope that connects the continental shelf with deeper seafloor. (10.10)

contour line A line on a map or chart connecting points of equal value, generally elevation. (2.3)

convection Transfer of thermal energy by flow of a liquid or a solid, but weak material. (5.4)

convection cell The movement of material in an elliptical to roughly circular loop by the process of convection; movement is driven by differences in density, especially those caused by temperature variations. (5.4)

convection current A flowing liquid or solid material that transfers heat from hotter regions to cooler ones, commonly involving movement of material in a loop. (10.2)

convergent boundary A plate-tectonic boundary in which two plates move toward (converge) one another. (3.3)

Coriolis effect The tendency of particles in motion on Earth's surface to have an apparent deflection related to the rotation of Earth about its axis. (13.1)

covalent bond A chemical bond created when two atoms share an electron. (4.12)

crater A typically bowl-shaped, steep-sided pit or depression, generally formed by a volcanic eruption or meteorite impact. (2.11, 6.1)

cratered highlands See lunar highlands. (19.4)

creep The slow, continuous movement of material, such as soil and other weak materials, down a slope. (15.10)

crescent dune *See* barchan dune. (13.2)

crest The highest part of a wave, fold, hill, or other feature. (14.3)

Cretaceous The youngest geologic period of the Mesozoic Era. (9.8)

crinoid A marine creature anchored to the seafloor with a stem or column and capped by a starlike head. (9.5)

cross bed A series of beds inclined at an angle to the main layers or beds. (7.7)

crosscutting relations The principle that a geologic unit or feature is older than a rock or feature that crosscuts it. (9.1)

cross section A diagram representing the geology as a two-dimensional slice through the land. *See* geologic cross section. (2.4)

crust The outermost solid layer of the Earth, consisting of continental and oceanic crust. (1.3)

crystalline A mineral that has an ordered internal structure due to its atoms being arranged in a regular, repeating way. (4.1)

crystalline basement The crystalline (metamorphic and igneous) rocks that underlie sedimentary and volcanic rocks in an area; widely exposed in the continental shield. (11.8)

crystalline rock A rock composed of interlocking minerals that grew together; usually formed in high-temperature environments by crystallization of magma, by metamorphism, or by precipitation from hot water. (4.2)

crystal setting The process in which more dense minerals settle and less dense minerals rise through magma. (5.6)

cubic A common arrangement of atoms in a mineral or the tendency for a mineral to break along three perpendicular planes. (4.4)

cutbank A steep cut or slope formed by lateral erosion of a stream, especially on the outside bend of a channel. (16.5)

cutoff meander A new channel formed when a stream cuts through the neck of a meander. (16.5)

cyclone An atmospheric low-pressure system, with a closed, roughly circular wind motion. (13.4)

D

Darcy's law An equation used to describe the flow of a fluid through a porous material. (17.1)

daughter product The element produced by radioactive decay of a parent atom; *syn.* Daughter atom. (9.4)

debris avalanche A high-velocity flow of soil, sediment, and rock, commonly from the collapse of a steep mountainside. (15.10)

debris flow Downhill-flowing slurries of loose rock, mud, and other materials, and the resulting landform and sedimentary deposit; sometimes called a mudflow. (6.7)

debris slide The downslope movement of soil, weathered sediment, or other unconsolidated material, partly as a sliding, coherent mass and partly by internal shearing and flow. (15.10)

decompression melting Melting of a rock or other material due to a decrease in pressure. (5.5)

deformation Processes that cause a rock body to change position, orientation, size, or shape, such as by folding, faulting, and shearing. (8.0)

delta A nearly flat tract of land formed by deposition of sediment at the mouth of a river or stream. (7.1)

dendrite One or more minerals that has crystallized in a branching pattern. (9.5)

density A measure of how much mass is present per given volume of a substance. (2.5)

desert A dry region with a mean annual precipitation of 25 cm or less; commonly applied to areas with sparse vegetation or the presence of desert-type plants and landscapes. (13.8)

desertification The converting of a land into a desert by natural or human causes. (13.8)

desert pavement A natural concentration of pebbles and other rock fragments, mantling a desert surface of low relief. (9.2)

desert soil A soil developed in a desert or semiarid region, generally characterized by an accumulation of abundant calcium carbonate and by a relative lack of organic material. (15.5)

desert varnish A thin, dark film or coating of iron and manganese oxides, silica, and other materials; formed by prolonged exposure at the surface; *syn.* rock varnish. (9.2)

desert wash A sandy or gravel-rich channel in a desert or other arid region. (13.9)

detrital *See* clastic.

Devonian A geologic period near the middle of the Paleozoic Era. (9.8)

diamond The hardest naturally occurring mineral, commonly used as a gemstone and industrial abrasive. (4.12)

diamond pipe A steep, cylindrical to funnel-shaped volcanic conduit that contains diamonds dispersed in an igneous rock; *syn.* kimberlite pipe. (18.12)

differential stress A condition with unequal stresses from different directions. (8.2)

differential weathering Weathering of different rock units, or different parts of a rock, at different rates. (15.4)

dike A sheetlike intrusion that cuts across any layers in a host rock, commonly formed with a steep orientation. (5.13)

diorite A medium- to coarse-grained, intermediate-composition igneous rock; the phaneritic equivalent of andesite. (5.2)

dip The angle that a layer or structural surface makes with the horizontal, measured perpendicular to the strike. (8.4)

dip-slip fault A fault on which the relative movement is parallel to the dip of the fault. (8.4)

⁓charge The volume of water flowing through some stretch of a river or stream ⁓ unit of time. (16.14)

disconformity An unconformity in which the bedding planes above and below the break are essentially parallel, but recording erosion or some other interruption in the deposition of layers. (9.3)

dissolution The process by which a material is dissolved. (7.3)

dissolved load Chemically soluble ions, such as calcium and sodium, that are dissolved in and transported by moving water, as in a stream. (16.2)

distributary system The branching drainage pattern formed when a stream branches and spreads out into a series of smaller channels. (16.9)

divergent boundary A boundary in which two plates move apart (diverge) relative to one another. (3.3)

dolomite A carbonate mineral containing calcium and magnesium. (4.9)

dolostone A rock composed mostly of the mineral dolomite. (7.6)

dome 1. A circular or elliptical anticlinal structure in which the rocks dip gently away in all directions (8.5). 2. A dome-shaped accumulation of lava and other volcanic materials. *See also* volcanic dome. (6.9)

drainage basin An area in which all drainages merge into a single stream or other body of water. (16.1)

drainage divide The boundary between adjacent drainage basins. (16.1)

drainage network The configuration or arrangement of streams within a drainage basin. (16.1)

drapery A thin sheet of travertine formed by the evaporation of water dripping from the ceiling of a cave. (17.6)

driftless area A region that was never glaciated. (14.16)

drill core A cylinder-shaped sample produced by drilling into rock and sediment in the subsurface. (12.14)

drill log A log that geologists and others use to graphically portray the results of drilling. (17.7)

driving force A force that drives the movement of tectonic plates. (3.7)

dropstone A stone that was carried within a floating iceberg and then dropped into fine sediment on the seafloor or lake bottom. (14.15)

drought An extended period of below-average precipitation. (13.6)

drumlin A commonly teardrop-shaped hill formed when a glacier reshapes glacial deposits. (14.13)

ductile A rock or material that is able to flow as a solid or otherwise deform without fracturing and faulting. (8.1)

dust devil A fast-moving whirlwind that picks up loose dust and sand. (13.2)

dust storm A moving mass of dust and sand propelled by strong winds. (13.2)

dynamic equilibrium A condition of a system in approximate steady state where there is a balanced inflow and outflow of materials. (16.4)

dynamo An electrical generator. (10.2)

E

earthflow A flowing mass of weak, mostly fine-grained material, especially mud and soil. (15.10)

earthquake Sudden movement of the earth caused by the abrupt release of energy. (12.1)

earthquake cycle The gradual accumulation of stress on a fault followed by an abrupt decrease in stress during an earthquake. 12.2)

eccentricity The barely noncircular shape of Earth's orbit around the Sun. (14.17)

eclogite A metamorphic rock, commonly with certain minerals, formed at very high pressure, like in subducting slabs. (8.9)

ecology The complex set of relations between living organisms and their environment. (13.13)

ecoregion analysis A computer technique climate scientists use to define similar areas based on the types of ecosystems that are present and the patterns of land use, landscape type, natural vegetation, climate, and soils. (13.13)

ecosystem An ecologic system, consisting of organisms and their environment. (13.13)

effervescence The potential of a mineral to have a vigorous bubbling reaction when a drop of dilute hydrochloric acid (HCl) is placed on it. (4.3)

effervescing A vigorous bubbling reaction that results when a drop of dilute hydrochloric acid (HCl) is placed on a mineral like calcite. (4.3)

E horizon A light-colored, leached zone of soil, lacking clay and organic matter. (15.5)

elastic behavior The ability of a material to strain a small amount and then return to its original shape when the stress is decreased. (12.2)

electromagnetic energy Various forms of energy, including light, infrared, and ultraviolet radiation. (1.4)

electromagnetic spectrum A range of electromagnetic radiation that includes visible light, infrared, ultraviolet, X-rays, and other wavelengths. (19.1)

electron cloud The area most likely to contain the electrons within an atom. (4.11)

electronegativity The measure of an element's ability to attract electrons. (4.12)

electron A stable, subatomic particle with a negative charge. (4.11)

electron shells The different energy states of electrons arranged around the nucleus of an atom. (4.11)

element A type of atom that has a specific number of protons and chemical characteristics. (4.6)

elevation The vertical distance of an object above or below a reference datum (usually mean sea level); generally the height of a ground point above sea level. (2.4)

El Niño A condition that occurs when warmer-than-average ocean surface temperatures occur in the central and east-central Equatorial Pacific. (13.5)

emergent coast A coast that forms where the sea has retreated from the land due to falling sea level or due to uplift of the land relative to the sea. (14.8)

entrenched meander A curved canyon that represents a meander carved into the land surface. (16.11)

ephemeral stream A stream that has periods during the year when it does not flow. (16.3)

epicenter The point on the Earth's surface directly above where an earthquake occurs (directly above the focus or hypocenter). (12.1)

equilibrium line The zone in a glacier where the losses of ice and snow balance the accumulation of ice and snow. (14.11)

equilibrium profile A profile reflecting a river in an approximate steady state where deposition of sediment is balanced by erosion. (16.4)

era A main subdivision in the geologic timescale. (9.8)

erosion The wearing away of soil, sediment, and rock through the removal of material by running water, waves, currents, ice, wind, and gravity. (1.7)

erosional remnant A mountain or hill that remains when adjacent areas have eroded to lower levels. (11.3)

eruption column A rising column of hot gases, tephra, and rock fragments that erupts high into the atmosphere. (6.2)

esker A long, narrow, sinuous ridge composed of sediment deposited by a stream flowing within or beneath a glacier. (14.13)

estuary A channel where freshwater from the land interacts with salt water from the sea and commonly is affected by ocean tides. (14.8)

ethanol A type of alcohol that is produced when corn, sugarcane, and other plant material is soaked in ammonia, fermented, and distilled. (18.8)

evaporate mineral A mineral precipitated as a result of the evaporation of water. (7.2)

evaporation The process by which a substance passes from a liquid to a vapor. (13.3)

evapotranspiration The process where plants take moisture from the soil, surface water, or air, and release water vapor into the atmosphere. (1.4)

evolution The observed changes in the fossil record or in living organisms; also used to refer to theories that help explain the observed changes. (9.6)

evolutionary diagram A block diagram, cross section, or map that shows the history of an area as a series of steps, proceeding from the earliest stages to the most recent one. (2.4)

exfoliation The processes by which a rock sheds concentric plates, such as that which occurs due to the release of pressure during exposure. (15.1)

exfoliation joint A joint that forms during exfoliation and mimics topography. (15.1)

expansion joint A joint that forms as a result of expansion due to cooling or to a release of pressure as rocks are uplifted to the surface. (15.1)

external energy Energy that comes from outside Earth, especially from the Sun. (1.4)

extrusive rock An igneous rock that forms when magma is erupted onto Earth's surface; *syn.* volcanic rock. (5.6)

F

fall line An imaginary line connecting waterfalls on several adjacent rivers, especially along the boundary between the Piedmont and Coastal Plain Eastern North America. (16.4)

far side The side of the Moon that cannot be seen from Earth. (19.4)

fault A fracture along which the adjacent rock surfaces are displaced parallel to the fracture. (8.3)

fault block A block of rock bounded on at least two sides by faults. (8.4)

fault breccia A rock composed of angular fragments formed by fracturing and crushing within a fault zone. (8.12)

fault scarp A step in the landscape caused when fault movement offsets Earth's surface. (8.12)

fault zone A zone of faults and associated fracturing. (8.12)

faunal succession The systematic change of fossils with age. (9.6)

feldspar A very common rock-forming silicate mineral that is abundant in most igneous and metamorphic rocks and some sedimentary rocks. (5.2)

felsic rock An igneous rock with a felsic composition, including granite, a light-colored igneous rock that contains abundant feldspar and quartz. (5.2)

fission The process by which an unstable isotope of a radioactive element splits into two parts. (18.6)

fissure 1. A fracture or crack on the land surface, such as that which forms by differential subsidence (17.8). 2. A magma-filled fracture in the subsurface, which typically solidifies into a dike, or a linear volcanic vent erupting onto the land surface. (6.1)

fissure eruption A volcanic eruption that occurs when magma rises through and erupts onto the surface in a long fissure. (6.5)

fjord A long, narrow arm of the sea contained within a steep-sided valley, interpreted to be carved by a glacier and later invaded by the sea as the ice melted and sea level rose. (14.8)

flash flood A local and sudden flood of short duration, such as that which may follow a brief but heavy rainfall. (16.12)

flood The result of water overfilling a channel and spilling out onto the floodplain or other adjacent land. (16.12)

flood basalts Large-volume basaltic lava flows that cover vast areas. (6.5)

floodplain An area of relatively smooth land adjacent to a stream channel that is intermittently flooded when the stream overflows its banks. (16.8)

flood stage The level at which the amount of discharge causes a river to overtop its banks and spill out onto the floodplain. (16.12)

flow band Layering in an igneous rock, especially a lava flow, formed by shearing and other processes within the magma. (6.9)

flow cell A huge, tube-shaped cell of atmospheric circulation. (13.1)

flowstone Any deposit of calcium carbonate or other mineral formed by flowing water on the walls or floor of a cave. (17.6)

fluvial Pertaining to rivers or streams, including the processes, sediment, resulting rock, and landforms. (7.1)

focus The place where an earthquake is generated; *syn.* hypocenter. (12.1)

foliation The planar arrangement of textural or structural features in metamorphic rocks and certain igneous rocks. (8.6)

footwall The wall rock beneath an inclined fault. (8.4)

foraminifera A group of small to microscopic, mostly marine animals that produce shells. (10.1)

force A push or pull that causes, or tends to cause, change in the motion of a body. (8.1)

fore-arc basin A sedimentary basin that lies between the volcanic arc and the trench in a convergent plate boundary. (11.4)

foreland basin A basin that forms when crust (either continental or oceanic) is warped by the weight of thrust sheets, especially when formed between a mountain belt and continental interior. (11.4)

foreshocks Small earthquakes that occur before a main earthquake. (12.12)

forest soil A soil formed in temperate climates and in forests of deciduous trees or pine trees. (15.5)

formation A rock unit that is distinct, laterally traceable, and mappable (7.7); also used as an informal term for an eroded, perhaps unusually shaped, mass of rock.

fossil Any remains, trace, or imprint of a plant or animal that has been preserved from some past geologic or prehistoric time. (9.5)

fossil fuel A nonrenewable resource formed in the past, especially petroleum, natural gas, and coal. (18.5)

fracture A break or crack in a rock; subdivided into joints and faults. (4.3)

framework silicates A group of silicate minerals in which tetrahedra share all four oxygen atoms, forming a structure bonded well in three dimensions. (4.7)

fringing reef A reef that fringes the shoreline of an island or continent. (10.9)

frost heaving The uneven upward movement and distortion of soils and other materials due to subsurface freezing of water into ice. (15.7)

frost wedging Process by which jointed rock is pried and dislodged by the expansion of ice during freezing. (15.1)

fuel cell An electrochemical device that produces electrical energy by reacting a fuel with an oxidant, such as hydrogen with oxygen. (18.8)

funnel cloud A rapidly rotating, funnel-shaped column of air that does not reach the ground. (13.4)

fusion The combination, or fusion, of two nuclei to form a heavier nucleus, in the process releasing a large amount of energy. (18.6)

G

Ga Billions of years before present (Giga-annum). (2.6)

gabbro A medium- to coarse-grained mafic igneous rock, the phaneritic equivalent of basalt. (5.2)

gaining stream The part of a stream or river that receives water from the inflow of groundwater. (17.5)

galena A lead sulfide mineral with a high specific gravity and distinctive metallic-gray cubes. (4.9)

gamma decay Radioactive decay in which an isotope emits an energetic particle called a photon. (5.4)

garnet A fairly common silicate mineral with a distinctive shape but nearly any color (red is most common). (4.8)

gas centrifuge A mechanical device used to separate elements or isotopes by their weight, such as in the enrichment of uranium isotopes. (18.6)

gas field A volume of rock that contains natural gas, or the projection of that area to the surface. (18.2)

gas giant A large, gas-dominated planet, including the four outer planets of our solar system (Jupiter, Saturn, Uranus, and Neptune). (1.8)

gas hydrate An icelike solid mixture of water and a natural gas, usually methane. (18.4)

gauging stage A site where measurements from a stream are collected in order to calculate discharge of that segment of the stream. (16.14)

geochemical samples Samples that are collected and analyzed, either in the field or later by chemists in a laboratory, for their content of specific chemicals. (17.10)

geologic cross section The two-dimensional diagram, representing a slice through Earth, that depicts the subsurface geometry of rock units and geologic structures. (18.5)

geologic map A map that shows the distribution, nature, and age relationships of rock units, sediments, structures, and other geologic features. (2.3)

geologic time The exceptionally long period of time dealt with by geology; the time extending from the formation of Earth to the beginning of written history. (2.6)

geologic timescale A chronologic subdivision of geologic time depicting the sequence of geologic events, including those represented by fossils; ages of boundaries are assigned through numeric dating of key rock units. (2.6)

geologist A person trained in any of the geological sciences. (2.10)

geology The study of planet Earth and other solid planetary objects, including their materials, processes, products, and history. (1.0)

geomagnetic polarity timescale A chronology based on the pattern and numeric ages of reversals of Earth's magnetic field. (10.2)

geophone A portable electronic device used to record seismic waves. (10.10)

geophysical survey The use of one or more geophysical techniques to explore Earth's subsurface, including seismic, gravity, electrical, magnetic, radioactivity, and heat-flow measurements. (12.14)

geophysicist A geoscientist who measures and interprets seismic, gravity, magnetic, and other geophysical data. (12.14)

geoscientist A person trained in the geological sciences or closely related fields. (1.3)

geothermal energy Energy that can be extracted from Earth's internal heat. (18.8)

geothermal gradient The rate at which temperature increases with depth into the subsurface. (5.4)

geyser A type of hot spring that intermittently erupts fountains of hot water and steam. (12.5)

glacial erratic A rock fragment carried by moving ice and deposited some distance from where it was derived. (14.15)

glacial maximum A time when glaciers were most widespread. (14.15)

glacial outwash Glacially derived sediment that is carried and deposited by a stream. (14.13)

glacial period A time interval when glaciers were abundant. (14.15)

glacier A moving mass of ice, snow, rock, and other sediment. (14.10)

Global Positioning System (GPS) An accurate location technique that uses small radio receivers to record signals from several dozen Earth-orbiting satellites. (3.7)

global warming Increasing global atmospheric and oceanic temperatures as measured or inferred from some point in the past to the present. (13.10)

gneiss A metamorphic rock that contains a gneissic foliation defined by a preferred orientation of crystals and generally by alternating lighter and darker colored bands. (8.6)

gneissic structure A metamorphic foliation defined by a preferred orientation of crystals and generally by alternating lighter and darker colored bands representing varying percentages of different minerals. (8.6)

Gondwana Name given by geologists to the hypothetical combination of the southern continents into a single large supercontinent. (11.11)

graben An elongate, downdropped crustal block that is bounded by faults on one or both sides. (8.4)

graded bed A sedimentary or volcanic layer that displays a gradational change in grain size from bottom to top. (7.7)

graded stream A stream in equilibrium, showing a balance between its capacity to transport sediment and the amount of sediment supplied to it, thus with no overall erosion or deposition of sediment. (16.3)

gradient The change in elevation for a given horizontal distance. (16.3)

granite A coarse-grained, felsic igneous rock containing mostly feldspar and quartz. (4.1)

graphite A soft, black, greasy-feeling carbon mineral. (4.12)

grassland soil A soil formed in a temperate climate beneath a surface of grass and other small plants. (15.5)

gravity The force exerted between any two objects, such as that between the Sun, Earth, and Moon; *syn.* gravitational pull (1.4)

gravity meter An instrument that measures variation in the gravitational field from place to place. (12.14)

graywacke Sandstone containing grains of different materials, including quartz, feldspar, mica minerals, and small fragments of other rocks. (7.5)

greenhouse effect The process that occurs when infrared energy radiating upward from a planetary surface is trapped by the atmosphere, warming the planetary body; greenhouse gases include water vapor, methane, and carbon dioxide. (1.4)

greenschist A greenish metamorphic rock with a schistosity defined by crystals of green or black mica and amphibole. (8.7)

groin A low wall built out into a body of water to affect the lateral transport of sand by waves and longshore currents. (14.6)

ground deformation Changes in the height or shape of Earth's surface, such as those that take place before a volcanic eruption. (6.13)

groundwater Water that occurs in the pores, fractures, and cavities in the subsurface. (17.0)

groundwater divide A relatively high area of the water table, separating groundwater that flows in opposite directions. (17.4)

gully A small channel eroded into the land surface. (16.6)

gypsum A common calcium sulfate mineral, generally formed by the evaporation of water. (4.9)

H

half-life In radioactive decay refers to the time it takes for half of the parent atoms to decay into a daughter product. (9.4)

halide mineral One of a family of minerals that consist of a metallic element, such as sodium or potassium, and a halide element, usually chlorine or fluorine. (4.6)

halite A salt mineral (NaCl) generally formed by the evaporation of water; cleaves into cubes and has a distinctive salty taste. (4.4)

hand lens A small magnifying glass, commonly used in the field to examine a rock, mineral, or fossil. (4.1)

hand specimen A hand-sized piece of rock for study, sampling, or for inclusion in a collection. (4.1)

hanging valley A glacial valley whose mouth is higher than the bottom of a larger glacially carved valley it joins. (14.12)

hanging wall The wall rock above an inclined fault. (8.4)

hard-rock geologist A geologist who focuses on the geology of igneous and metamorphic rocks as opposed to softer sedimentary rocks. (18.5)

hazard The existence of a potentially dangerous situation or event. (6.6)

headwaters The location or general area where a stream or river begins. (16.3)

heat The transfer of thermal energy from high-temperature to low-temperature objects. (5.4)

heat flow Transfer of thermal energy that results when two adjacent masses have different temperatures, especially from depth to Earth's surface. (5.4)

heft test An approach to determining the approximate density of a mineral by simply holding a mineral and noting how heavy it feels. (4.3)

hematite An iron oxide metal that has a reddish streak and commonly forms under oxidizing conditions. (4.9)

high-pressure system An area in the atmosphere characterized by relatively high atmospheric pressure, sinking air, and generally fair weather. (13.1)

high tide The maximum height to which water in the ocean rises relative to the land in response to the gravitational pull of the Moon; also refers to the time when such high levels occur. (14.1)

hinge The part of a fold that is most sharply curved. (8.5)

hogback Any ridge with a sharp summit and one slope inclined approximately parallel to the dip of layers, resembling in outline the back of a hog. (8.12)

horizon A zone in soil that is distinct from adjacent zones, including differences in color, texture, content of minerals and organic matter, or other attributes. (15.5)

horizontal surface wave A type of surface wave in which material vibrates horizontally, from side to side, perpendicular to the direction of wave propagation. (12.5)

hornfels A fine-grained, nonfoliated metamorphic rock, typically formed by contact metamorphism of nearly any kind of starting rock. (8.7)

horst An elongate, relatively uplifted crustal block that is bounded by faults on two sides. (8.4)

hot spot A volcanically active site interpreted to be above an unusually high-temperature region in the deep crust and upper mantle. (10.5)

hummocky topography A type of chaotic landscape characterized by randomly distributed humps and pits, commonly created by a landslide or less commonly by a pyroclastic eruption. (15.12)

hundred year flood The size of a flood that is predicted, from the existing data, to have a 1 in 100 probability (1%) of occurring in any given year. (16.14)

hurricane A tropical cyclone in which the wind velocity equals or exceeds 64 knots (73 mph); *syn.* typhoon. (13.4)

hydraulic gradient The slope or gradient of the water table. (17.4)

hydrocarbon A gaseous, liquid, or solid organic compound composed of carbon and hydrogen. (18.1)

hydroelectric power Power produced by the movement of water, specifically the generation of electrical energy. (18.7)

hydrogen bond A weak bond in water that forms between one molecule's hydrogen atom and another molecule's oxygen atom. (4.13)

hydrogeologist A geoscientist who specializes in groundwater and surface-water investigations. (17.7)

hydrograph A graph showing the change in the amount of flowing water (discharge) over time. (16.1)

hydrologic cycle The cycle representing the movement of water between the oceans, atmosphere, land, rivers and other surface water, groundwater, and organisms. (1.7)

hydrolysis A decomposition reaction involving water and commonly producing clays, as in soil. (7.3)

hydrosphere The part of Earth characterized by the presence of water in all its expressions, including oceans, lakes, streams, wetlands, glaciers, groundwater, moisture in soil, water vapor, and drops and ice crystals in clouds and precipitation. (1.7)

hydrothermal rock A rock that precipitated directly from hot water, either at depth or on the surface. (1.5)

hypocenter The place where an earthquake is generated; *syn.* the focus. (12.)

hypothesis A conception or proposition that is tentatively assumed, and then tested for validity by comparison with observed facts and by experimentation. (2.7)

I

ice age A period of time in which large regions of land are covered year-round with ice and snow, especially in the last 2 m.y. (14.16)

iceberg A massive piece of ice floating or grounded in the sea or other body of water. (14.11)

icefall A steeply flowing mass of ice. (14.10)

ice sheet A mass of ice of considerable thickness and more than 50,000 km² in area, forming a nearly continuous cover of ice and snow over a land surface. (14.10)

igneous rock A rock that formed by solidification of molten material (magma). (1.5)

imbrication A sedimentary fabric characterized by disk-shaped or elongate pebbles and larger clasts inclined in a preferred direction in response to the direction of current. (7.13)

inclusion A fragment of older rock or material that is contained within another rock or material, as in a fragment of preexisting rock in a magma. (12.14)

index contour A dark line on a topographic map which helps emphasize the broader elevation patterns of an area and allows easier following of lines across the map; on most topographic maps, every fifth line is an index contour. (2.3)

infiltration Water and other fluids that seep into the ground through open pores, fractures, and cavities in soil and rocks. (17.1)

infrared energy A form of electromagnetic energy with longer wavelengths than visible light; much of the Sun's light that reaches Earth converts into this type of energy. (1.4)

inner core The solid central part of Earth's core, extending from a depth of about 5,100 km to the center of Earth (6,371 km); its radius is about one-third of the whole core. (1.3)

inner planets The four planets closest to the Sun (Mercury, Venus, Earth, and Mars); *syn.* terrestrial planets. (19.3)

inorganic Pertaining to a compound that is not produced by living organisms or that chemically contains no carbon. (4.1)

interferogram The combining of satellite mapping images taken of a place at two different times to compare changes in the land surface, for example, during an earthquake. (12.11)

interglacial period A time during an ice age when glaciers are melting, retreating, or diminished in extent. (14.15)

intermolecular force A bond that occurs when several types of weak bonds attract a molecule (a combination of atoms) to another molecule. (4.12)

internal energy Energy that comes from within Earth and includes both the heat energy trapped from when the planet formed and the heat produced by radioactive decay. (1.4)

Intertropical Convergence Zone (ITCZ) A sinuous line more or less along the equator where the northeast and southeast trade winds converge, promoting rain and the development of rain forests. (13.7)

intrusive rock An igneous rock that solidified from magma below Earth's surface; *syn.* plutonic rock. (5.6)

ion A charged atom. (4.11)

ionic bond Chemical bond formed because of the attraction of two oppositely charged ions, such as by the loaning of one or more electrons from one ion to another. (4.12)

iron formation A rock composed of millimeter- to centimeter-thick layers of iron-bearing minerals, especially iron oxide, commonly with quartz. (7.6)

island A tract of land smaller than a continent, surrounded by water of an ocean, sea, lake, or stream. (3.1)

island arc A generally curved belt of volcanic islands above a subduction zone; also used as an adjective to refer to this setting. (3.1)

isostasy The condition of equilibrium, comparable to floating, of the crust resting on the solid mantle. (1.3)

isostatic rebound Uplifting caused by the removal of weight on top of the crust, as when an ice sheet melts away or when erosion strips material off the top of a thick crustal root of a mountain. (11.2)

isotope One of two or more species of the same chemical element but differing from one another by having a different number of neutrons. (9.4)

isotopic dating The process of dating rocks using radioactive decay. (9.4)

J

jetty An engineering structure built from the shore into a body of water to redirect current or tide, for example, to protect a harbor. (14.6)

joint A fracture in a rock where the rock has been pulled apart slightly, without significant displacement parallel to the fracture. (8.3)

joint sets A set of parallel or nearly parallel joints. (8.12)

Jupiter The largest planet in the solar system. (19.5)

Jurassic A geologic period in the middle of the Mesozoic Era. (9.8)

K

karst topography Topography characterized by sinkholes, caves, limestone pillars, poorly organized drainage patterns, and disappearing streams; generally formed from the dissolution of limestone. (17.6)

kerogen A thick substance composed of long chains of hydrocarbons. (18.1)

kettle A pitlike depression in glacial deposits, commonly a lake or swamp; formed by the melting of a large block of ice that had been at least partly buried in the glacial deposits. (14.13)

kettle lake A body of water occupying a kettle. (14.13)

K-feldspar A very common feldspar mineral containing potassium (K); *syn.* potassium feldspar. (15.2)

kinetic energy Energy due to movement of an object. (1.4)

K-P extinction Refers to the extinction of the dinosaurs and many other animals at the end of the Mesozoic Era, between the Cretaceous Period (K) and the Paleogene Period (P); traditionally referred to as the K-T extinction, separating the Cretaceous Period (K) and the Tertiary (T). (9.11)

K-T extinction *See* K-P extinction. (9.11)

Kuiper Belt A zone of planetary objects beyond the orbit of Neptune. (19.7)

L

ꞁlith A bulge-shaped igneous body that has domed and tilted overlying layers ꞏhat is observed or interpreted to have a relatively flat floor. (5.13)

lagoon A shallow part of the sea between the shoreline and a protecting feature, such as a reef or barrier island, farther out to sea. (7.2)

lahar A mudflow mostly composed of volcanic-derived materials and generally formed on the flank of a volcano. (6.7)

lander A spacecraft that lands on a planetary object and collects images and other data. (19.1)

landslide A general term for the rapid downslope movement of soil, sediment, bedrock, or a mixture of these; also the material or landform formed by this process. (15.8)

La Niña A condition that results in the concentration of warm water in the western equatorial Pacific Ocean, accompanied by cooling in the eastern equatorial Pacific Ocean. (13.5)

lateral moraine Sediment carried in and deposited along the sides of a glacier. (14.13)

laterite A type of tropical soil rich in iron (Fe) and aluminum (Al) oxides, commonly giving the soil a deep red color. (15.5)

Laurasia The northern supercontinent that existed in the Mesozoic and included North America, Europe, and Asia. (11.11)

lava Magma that is erupted onto the surface, or the rock mass into which it solidifies. (6.2)

lava dome A dome-shaped mountain or hill of at least partly solidified lava generally of felsic to intermediate composition. (6.2)

lava flow Magma that erupts onto the surface and flows downhill from the vent; also the solidified body of rock formed by this magma. (6.2)

lava fountain A fountain of molten lava propelled into the air by pressure and escaping gases. (6.2)

lava tube A long, tubular opening under the crust of solidified lava and representing an active or partially emptied subsurface channel of lava. (6.3)

leaching The separation or dissolution of soluble constituents from a rock, sediment, or soil by percolation of water. (15.5)

levee A long, low ridge of sediment deposited by a stream next to the channel; some levees are built by humans to keep floodwaters from spilling onto a floodplain. (16.8)

LIDAR (LIght Detecting and Ranging) A mapping method that uses reflection and scattering of light to determine distance and other characteristics of the land. (14.7)

lightning An abrupt atmospheric discharge of electricity. (13.4)

lignite A brownish-black coal that is intermediate in quality between peat and bituminous coal. (18.3)

limbs The planar or less curved parts of a fold on either side of the hinge. (8.5)

limestone A sedimentary rock composed predominantly of calcium carbonate, principally in the form of calcite, and which may include chert, dolomite, and fine-grained clastic sediment. (7.11)

lineation A linear structure in a metamorphic rock. (8.6)

liquefaction Loss of cohesion when grains in water-saturated soil or sediment lose grain-to-grain contact, as when shaken during an earthquake. (12.7)

lithification The conversion of unconsolidated sediment or volcanic ash into a coherent, solid rock, involving processes such as compaction and cementation. (1.6)

lithosphere Earth's upper, rigid layer composed of the crust and uppermost mantle. (1.7)

lithospheric mantle The part of the uppermost mantle that is in the lithosphere. (1.3)

loading The process by which weight is added to the lithosphere. (14.9)

local mountain A mountain that is supported by the strength of the crust and is too small to be accompanied by a regional increase in crustal thickness. (11.3)

loess An essentially unconsolidated sediment consisting predominantly of silt, interpreted to be windblown dust, commonly of glacial origin. (13.2)

longshore current A current, generally in an ocean or large lake, flowing more or less parallel to a coastline. (14.4)

losing stream Part of a stream or river that loses water from outflow to groundwater. (17.5)

low-pressure system An area in the atmosphere characterized by relatively low atmospheric pressure, rising air, and commonly stormy weather. (13.1)

low tide The lowest height to which water in the ocean drops relative to the land in response to the gravitational pull of the Moon; also refers to the time when such low levels occur. (14.1)

lunar highlands High elevations on the Moon that contain a light-colored, heavily cratered material; *syn.* cratered highlands. (19.4)

luster The reflection of light from the surface of a mineral, especially its quality and intensity; the appearance of a mineral in reflected light. (4.3)

M

Ma Millions of years before present (Mega-annum). (2.6)

mafic A material having high contents of magnesium (Mg) and iron (Fe), generally accompanied by a decreased amount of silica. (5.3)

mafic mineral A generally dark-colored, silicate mineral with a high magnesium (Mg) and iron (Fe) content. (4.8)

mafic rock A generally dark-colored igneous rock with a mafic composition. (5.2)

magma Molten rock, which may or may not contain some crystals, solidified rock, and gas. (1.5)

magma chamber A large reservoir in the crust or mantle that is occupied by a body of magma. (5.12)

magma mixing Process whereby two different magmas come into contact and partially mix, forming a magma that has a composition intermediate between the two. (5.6)

magnetic reversal A reversal of the polarity of Earth's magnetic field, from normal polarity to reversed polarity. (10.2)

magnetite An iron oxide mineral that is typically black and is strongly magnetic. (4.9)

magnetometer An instrument used to measure the direction and strength of magnetism in rocks and other materials. (10.2)

magnitude A measure of the amount of energy released by an earthquake; used to compare sizes of earthquakes. (12.6)

manganese nodule Small, irregular, black to brown concretions formed on the seafloor and consisting primarily of manganese and iron minerals. (10.1)

mantle The most voluminous layer of Earth; located below the crust and above the core. (1.3)

mantle convection Movement of mantle material in response to variations in density, especially those caused by differences in temperature. (3.7)

marble A metamorphic rock composed of recrystallized calcite or dolomite. (8.7)

mare A dark, low-lying, relatively smooth area on the Moon consisting of basalt. (19.4)

maria Plural of mare. (19.4)

marine salt deposit A salt accumulation formed when seawater evaporates, leaving behind a residue of material that was dissolved in the water. (10.11)

marine terrace A platform that was cut or constructed by waves but is now elevated above sea level; commonly covered by a thin veneer of marine sediment. (14.5)

Mars Fourth planet out from the Sun. (19.3)

mass extinction The disappearance of many species and families of creatures in a geologically short period of time. (9.8)

mass spectrometer An instrument used to measure the abundance of different atoms and isotopes in a material, such as a rock or mineral to be numerically dated. (9.4)

mass wasting Downward movement of material on slopes under the force of gravity. (15.7)

matrix The finer grained material enclosing or filling the areas between larger grains, crystals, or fragments of a rock. (7.5)

maturation The process by which coal increases its carbon content and loses volatiles and other impurities as it is buried and heated. (18.3); also the process by which organic material becomes petroleum upon burial and heating. (18.1).

meander A sinuous curve or bend in the course of a stream or river. (16.5)

meandering river A river that has a strongly curved channel (with meanders). (16.5)

meander scar A crescent-shaped feature in the landscape that indicates the former position of a river meander. (16.8)

medial moraine Sediment carried in the center of a glacier, representing where two glaciers joined; also refers to the deposited sediment and resulting landform. (14.13)

megathrust A huge thrust fault, representing the boundary between the subducted slab and overriding plate. (8.9)

Mercury The closest planet to the Sun. (19.3)

mesa A broad, flat-topped and steep-sided, isolated hill or mountain. (11.3)

Mesozoic Era A major subdivision of geologic time from 251 Ma to 65 Ma; characterized by dinosaurs. (9.6)

metallic bond A chemical bond formed when electrons are shared widely by many atoms. (4.12)

metamorphic rock A rock changed in the solid state by temperature, pressure, deformation, or chemical reactions that modified a preexisting rock. (1.5)

metamorphism The mineralogical and structural changes of solid rock in response to changes in environmental conditions, especially at depth. (8.0)

metarhyolite A metamorphic rock formed through the metamorphism of rhyolite, tuff, and other felsic volcanic rocks. (8.7)

Meteor Crater A meteoroid impact crater located in northern Arizona. (2.11)

meteorite A fragment of a meteoroid that has fallen to a planetary surface. (2.8)

meteorite crater A crater formed by impact of a meteoroid onto a surface. (2.11)

meteoroid Solid object in interplanetary space; distinguished from asteroids by a smaller size. (2.11)

microburst A strong, downward-moving wind, generally associated with thunderstorms. (13.4)

microorganism An organism of microscopic size, such as bacteria. (18.1)

mid-ocean ridge A long mountain range on the floor of the ocean, associated with seafloor spreading. (3.1)

migmatite A rock composed of both metamorphic and igneous or igneous-appearing material. (8.9)

Milankovitch Cycle Periodic variations in Earth's orbit and tilt, interpreted to influence Earth's climate. (14.17)

Milky Way Galaxy The spiral galaxy in which Earth is located; only one of countless galaxies in the universe. (19.0)

mineral A naturally occurring, inorganic, crystalline solid with a relatively consistent composition. (4.1)

mineral deposit A mass of naturally occurring rocks and other materials that are especially rich in some commodity that might be valuable. (18.9)

mineralize To convert to or impregnate with mineral material, as in the processes of ore deposition and of fossilization. (18.9)

mineralogists Geologists and other scientists who study minerals. (4.4)

mineral wedging The growth of minerals that exert an outward force that can fracture rock or loosen grains. (7.3)

Mississippian A geologic period near the middle of the Paleozoic Era. (9.8)

modified Mercalli scale An earthquake intensity scale, recording the relative amount of damage and how the earthquake was perceived by people. (12.6)

Mohorovičić Discontinuity The boundary between the crust and mantle, commonly referred to as the moho. (12.15)

Mohs Hardness Scale Consists of ten common minerals ranked in order of increasing relative hardness, from 1 to 10. (4.3)

moment magnitude (Mw) A measure of the amount of energy released by an earthquake. (12.6)

monocline A fold defined by local steepening in gently dipping layers. (8.5)

monsoon Refers to winds that reverse directions depending on the season. (13.5)

moraine Sediment carried by and deposited by a glacier; also refers to the resulting landform. (14.13)

mouth The location where a stream, river, or canyon ends, such as where a river enters the sea. (16.3)

mud A mixture of silt and clay, or a general term used to refer to either silt or clay. (7.10)

mudcrack A somewhat polygonal pattern of fractures, formed by the shrinkage of mud as it dries. (7.13)

mudflow A general term for mass movement involving a fluidly flowing mass of mud and other material and the resulting landform; applied to some earthflows and debris flows. (6.7)

mudrock A general term for a rock composed of consolidated clay, silt, or a combination of these. (7.10)

mudstone A rock composed of consolidate mud (silt, clay, or especially a combination of the two). (7.10)

muscovite A light-colored, sheet silicate mineral that is part of the mica family. (4.8)

N

native mineral Rock-forming mineral that contains only a single element. (4.6)

natural gas Hydrocarbons that exist as gas or vapor at ordinary temperatures and pressures. (18.1)

natural resource A naturally occurring resource, including energy, mineral, and water resources, as well as soil and timber. (18.0)

natural selection The process by which the organism best adapted to its environment tends to survive and transmit its genetic characteristics to the population; one theory for natural evolution. (9.6)

neap tide Lower-than-average high tides and higher-than-average low tides caused when the Sun's gravity partially offsets the effects of the Moon's gravity. (14.2)

near side The side of the Moon that always faces Earth. (19.4)

nebula A shapeless cloud of gas and dust in space. (19.0)

negative magnetic anomaly A measurement of Earth's magnetic field that is lower than average. (10.2)

Neogene The youngest geologic period of the Cenozoic Era. (9.8)

Neptune The eighth and last planet from the Sun. (19.7)

neutron A subatomic particle that contributes mass to a nucleus and is electrically neutral. (4.11)

nomograph A type of graph, used in seismology, to determine the local magnitude of an earthquake. (12.6)

nonclastic rock A sedimentary rock not composed of clasts. (7.6)

nonconformity An unconformity in which the older rocks below the unconformity are not layered. (9.3)

nonsilicate A mineral or other material that does not include silicon. (4.9)

nonvesicular An igneous rock that does not contain obvious gas pockets or vesicles. (6.3)

normal fault A fault in which the hanging wall moves down relative to the footwall. (8.4)

normal-fault basin A low area that has been downdropped by one or more normal faults. (11.6)

normal polarity Refers to the current polarity of Earth's magnetic field. (10.2)

nuclear fission The breaking apart of atoms during radioactive decay, in the process releasing a large amount of energy. (1.8)

nuclear fusion The combination, or fusion, of two nuclei to form a heavier nucleus, in the process releasing a large amount of energy. (1.8)

nucleus A particle composed of protons and generally neutrons in the core of an atom. (4.11)

numeric age Geologic age of a rock, sediment, fossil, or event calculated in thousands, millions, or billions of years before present; *syn.* absolute age. (2.6)

numeric dating The process of determining ages of rocks by using analytical measurements; *syn.* absolute dating. (9.1)

O

oblique-slip fault A fault on which slip has both dip-slip and strike-slip components. (8.4)

obsidian A generally gray to black, shiny volcanic glass, usually of felsic composition. (5.3)

ocean basin The deeper parts of the ocean basins, especially those that are not underlain by continental crust. (11.4)

ocean-continent convergent boundary A plate-tectonic boundary where an oceanic plate converges with a continental plate, generally expressed by subduction of the oceanic plate beneath the continent. (3.5)

oceanic crust The type of thin, mafic crust that underlies the ocean basins. (1.3)

oceanic fracture zone Crack or step in elevation of the seafloor that formed as a transform fault along a mid-ocean ridge but is no longer a plate boundary. (3.1)

octahedron A polyhedron with eight faces. (4.4)

O horizon An upper, organic-rich soil horizon composed of dead leaves and other plant and animal remains. (15.5)

oil field The region in the subsurface that contains petroleum, especially oil; also the projection of that area to the surface. (18.2)

oil sand A sand or other porous sediment impregnated by petroleum. (18.1)

oil seep The seepage of liquid petroleum at the surface. (18.1)

oil shale A kerogen-bearing, thinly layered, fine-grained sedimentary rock that yields liquid or gaseous hydrocarbons upon heating and distillation. (18.4)

olivine A green iron-magnesium silicate mineral that composes much of the upper mantle and also occurs in mafic and ultramafic igneous rocks. (4.8)

ophiolite complex A consistent sequence of, from top to bottom, oceanic sediment, pillow basalt, sheeted dikes, and gabbro, interpreted to represent a slice of oceanic crust or an oceanic volcano. (5.9)

Ordovician A geologic period in the early part of the Paleozoic Era. (9.8)

ore A rock, sediment, or other material that can be mined for a profit. (18.9)

ore deposit A mineral deposit that contains enough of a commodity to be mined at a profit. (18.9)

original horizontality The principle that most sediments and many volcanic units are deposited in layers that originally are more or less horizontal. (9.1)

orogeny The formation of mountains or the time period during which tectonic activity causes deformation and forms mountains. (11.12)

outer core The molten outer part of Earth's core, extending from a depth of 2,900 km to 5,100 km. (1.3)

outwash plain An area in front of a glacier where streams deposit glacially produced sediment. (14.14)

overpumping Pumping of groundwater at a rate that causes severe lowering of the water table in the aquifer or other deterimental effects. (17.8)

oxbow lake An isolated, curved lake formed when a cutoff meander is filled with water. (16.5)

oxidation The chemical process during which a material combines with oxygen. (7.3)

oxide 1. Any member of a group of minerals that consist of oxygen bonded with a metallic element, like iron (4.6). 2. Mineralized rock that contains minerals, especially those of iron and copper, that formed near the surface or in other oxidizing conditions. (18.9)

P

pahoehoe A type of lava or lava flow that has a smooth upper surface or folds that form a "ropy" texture. (6.3)

paleomagnetism The rock record of past changes in Earth's magnetic field. (10.2)

Paleogene The oldest geologic period of the Cenozoic Era. (9.8)

Paleozoic Era A major subdivision of geologic time, beginning at the end of the Precambrian; from 542 to 251 million years ago. (9.6)

Pangaea A supercontinent that existed from about 300 to about 200 million years ago and included most of Earth's continental crust. (11.11)

parallel bedding A sequence of beds that are approximately parallel. (7.7)

parent atom An atom before it undergoes radioactive decay; *syn.* parent isotope. (9.4)

passive margin A continental margin that is not a plate boundary. (3.4)

passive solar The use of solar energy, involving light and infrared energy from the Sun entering a space through windows and naturally heating the inside air and mass. (18.8)

peat An unconsolidated deposit of partially decayed plant matter. (7.6)

pebble A small stone between 6 and 64 mm in diameter. (7.4)

pediment A gently sloping, low-relief plain or erosion surface carved onto bedrock, commonly with a thin, discontinuous veneer of sediment. (13.9)

pegmatite An igneous rock containing very large crystals, which may be centimeters to meters long. (5.1)

Pennsylvanian A geologic period in the latter part the Paleozoic Era. (9.8)

perched water Groundwater that sits above the main water table and generally is underlain by a layer or lens of impermeable rock that blocks the downward flow of groundwater. (17.4)

perennial stream A stream or river that flows all year. (16.3)

peridotite An ultramafic igneous rock generally containing abundant olivine, commonly with smaller amounts of pyroxene. (5.2)

period A time interval in the geologic timescale; a subdivision of an era. (2.6)

Periodic Table Table that organizes all the chemical elements according to the element's atomic number and electron orbitals. (4.6)

permafrost A condition in which water in the uppermost part of the ground remains frozen all or most of the time. (14.14)

permeability A measure of the ability of a material to transmit a fluid. (17.3)

Permian The last geologic period of the Paleozoic Era. (9.8)

petrified wood A piece of fossilized wood that has been replaced by silica and other material, preserving some of the original structure of the wood. (9.5)

petroleum A general term for naturally occurring hydrocarbons, whether liquid, gaseous, or solid. (18.1)

petroleum geologist A geologist engaged in exploration for, or production of, oil or natural gas. (2.10)

phaneritic rock An igneous rock containing crystals that are visible to the unaided eye. (5.1)

phases of the moon The monthly cycle of the Moon in which its illumination changes on a regular schedule, going from completely dark to fully lit and back again. (19.4)

phenocrysts Crystals in an igneous rock that are larger than those around them, as in a porphyritic rock. (5.1)

phosphate rock A rock that is rich in the element phosphorus. (18.13)

photon An energetic particle released from an atom, such as during gamma decay. (5.4)

photosynthesis The process by which plants produce carbohydrates, using water, light, and atmospheric carbon dioxide. (1.4)

photovoltaic panel A solar-energy device that converts sunlight directly into electricity. (18.8)

phyllite A shiny, foliated, fine-grained metamorphic rock, intermediate in grade between slate and schist. (8.7)

physical weathering The physical breaking or disintegration of rocks when exposed to the environment. (7.3)

piedmont glacier A broad glacier that forms when an ice sheet or valley glacier spreads out as it moves into less confined topography. (14.10)

pillow A rounded, pillow-shaped structure that forms when lava erupts into water. (6.4)

pillow basalt A basaltic lava flow that includes pillow structures. (6.4)

pinnacle A tall, slender tower or pillar of rock. (14.5)

placer A surficial mineral deposit in which a valuable substance, like gold or diamonds, has been concentrated, such as on rivers and beaches. (18.10)

plagioclase A very common rock-forming feldspar mineral that contains sodium, calcium, or both elements. (4.8)

planet A large celestial body that revolves around a sun in a solar system. (19.0)

planetary geologist A geoscientist who applies geologic principles and techniques to the study of planets, moons, and other planetary objects. (2.10)

plate boundary Zone of tectonic activity, including earthquakes, along the boundary between two lithospheric plates that are moving relative to one another. (3.3)

plateau A broad, relatively flat region of land that has a high elevation. (3.1)

plate tectonics A theory in which the lithosphere is divided into a number of mostly rigid plates that move relative to one another, causing tectonic activity along these boundaries. (3.3)

platinum The native metallic element (Pt) and metal alloys that contain it. (18.10)

playa A shallow, closed basin in a generally dry environment. (13.9)

plume 1. A rising mass of mostly solid mantle material, thought to be the causal mechanism of a hot spot (5.11). 2. A mass of groundwater contamination that spreads out away from the source. (17.10)

plunge The inclination of a geologic structure, commonly measured in degrees, from the horizontal. (8.5)

plunging An adjective that describes an inclined, linear geologic structure, for example, a fold whose hinge is not horizontal. (8.5)

plutoid A class of dwarf planet farther from the Sun than the orbit of Neptune. (19.7)

pluton A subsurface magma body or the mass of rock in which it solidifies; *syn.* intrusion. (5.12)

plutonic rock An igneous rock that solidified at depth rather than on the surface; *syn.* intrusive rock. (5.6)

point bar A series of low, arcuate ridges of sand and gravel deposited on the inside of a stream bend or meander. (16.5)

polar desert A desert that forms where cold, dry air prevails and available moisture is frozen most of the year. (13.8)

pore space Any open space within rocks, sediment, or soil, including open space between grains in a sedimentary rock, within fractures, and in other cavities. (7.5)

porosity The percentage of the volume of a rock, sediment, or soil that is open space (pore space). (17.3)

porphyritic An igneous texture in which larger crystals are set in a finer grained matrix. (18.11)

porphyry A porphyritic igneous rock that contains conspicuous phenocrysts in a fine-grained matrix. (18.11)

positive magnetic anomalies A measurement of Earth's magnetic field that is higher than average. (10.2)

potassium feldspar A very common silicate mineral that contains potassium; *syn.* K-feldspar. (4.8)

pothole A bowl-shaped pit eroded into rock by swirling water and sediment. (16.2)

Precambrian A very long interval of geologic time, from the formation of the solid earth to the beginning of the Paleozoic; it is equivalent to 90% of geologic time. (9.6)

precipitation Atmospheric water that falls to the surface as rain, snow, hail, or sleet. (1.4)

precious metal Gold, silver, or any minerals of the platinum group. (18.10)

primary wave (P wave) A seismic body wave that involves particle motion, consisting of alternating compression and expansion, in the direction of propagation. (12.5)

principle of superposition The concept that a sedimentary or volcanic layer is younger than any rock unit on which it is deposited. (9.1)

probability The statistical description of the likelihood that an event will occur. (16.14)

promontory A ridge of land that juts out into a body of water. (14.4)

proton Principal particle of an atomic nucleus with a positive charge. (4.11)

proxy evidence Types of observations, other than direct measurements, used to infer past conditions, including temperatures. (13.10)

P-S interval The time interval between the arrivals of the P wave and the S wave. (12.6)

pull-apart basin A basin that forms as the result of movement within a zone of strike-slip faulting. (11.4)

pumice Volcanic rock, especially of felsic or intermediate composition, containing many vesicles (holes) formed by expanding gases in magma. (5.3)

pump-and-treat A commonly used option to clean up, or remediate, a site of groundwater contamination. (17.10)

punctuated equilibrium A hypothesis that new organisms or new characteristics of an existing organism appear rather suddenly in geologic terms instead of evolving gradually. (9.6)

P wave *See* primary wave. (12.5)

P-wave shadow zone The region on Earth's surface, 103° to 142° away from an earthquake epicenter, in which direct P waves from the earthquake are not recorded. (12.15)

pyrite A common, pale bronze to brass yellow, iron sulfide mineral, commonly called "fool's gold." (4.9)

pyroclastic A volcanic eruption where hot fragments and magma are thrown into the air; also refers to a deposit or rock produced by such an event. (6.2)

pyroclastic flow A fast-moving cloud of hot volcanic gases, ash, pumice, and rock fragments that generally travel down the flanks of a volcano; *syn.* ash flow. (5.1)

pyroxene One of a group of mostly dark, single-chain silicate minerals. (4.8)

Q

quantitative data Data that are numeric and typically visualized and analyzed using data tables, calculations, equations, and graphs. (2.5)

qualitative data Data that include descriptive words, labels, sketches, or other images. (2.5)

quartz A very common rock-forming silicate mineral, consisting of crystalline silica. (4.8)

quartzite A very hard rock consisting chiefly of quartz grains joined by secondary silica that causes the rock to break across rather than around the grains; formed by metamorphism or by silica cementation of a quartz sandstone. (8.7)

quartz sandstone Sandstone composed mostly of quartz grains. (8.7)

R

radiant heat transfer Transfer of thermal energy as electromagnetic waves. (5.4)

radioactive decay The spontaneous disintegration and emission of particles from an unstable atom. (9.4)

rain forest A forest with high annual rainfall, no freezing temperatures, and a rich collection of plant and animal communities. (13.7)

rain shadow A relatively dry region on the downwind side of a topographic obstacle, usually a mountain range; rainfall is noticeably less than on the windward side. (13.3)

rain-shadow desert A desert formed in a rain shadow. (13.8)

rapid A segment of rough, turbulent water along a stream. (16.6)

recessional moraine A moraine that forms as the front of a glacier melts back and stagnates for some time in one location, depositing a pile of sediment. (14.13)

recharge The replenishment of water into a groundwater system, whether natural or done by humans. (17.5)

recurrence interval The time between repeating earthquakes. (12.2)

recrystallization The formation, essentially in the solid state, of new crystalline grains in a rock. (8.8)

reefs Shallow, mostly submarine features, primarily built by colonies of living marine organisms, including coral, sponges, and shellfish, or by the accumulation of shells and other debris. (10.9)

refraction The deflection of a ray, as in light or a seismic wave, due to its passage from one material to another of different density. (12.15)

regional metamorphism Metamorphism affecting an extensive region and related mostly to regional burial, heating, and deformation of rocks. (8.8)

regional mountain range A mountain range that is hundreds to thousands of kilometers long, contains many peaks, and typically involves uplifted, thickened crust. (11.1)

regional subsidence The process by which a region decreases in elevation, for example, subsidence due to crustal thinning. (11.4)

regression The retreat of the sea from land areas and evidence of such withdrawal. (7.12)

relative age The age of a fossil, organism, rock, geologic feature, or event as defined relative to other geologic features or events. (2.2)

relative motion The motion of tectonic plates relative to one another across a plate boundary. (3.3)

ridge push A plate-driving force that results from the tendency of an oceanic plate to slide down the sloping lithosphere-asthenosphere boundary near a mid-ocean ridge. (3.7)

relief The difference in elevation of one feature relative to another; *syn.* topographic relief. (2.4)

remediate To remedy a fault or deficiency; for example, to clean up a site of soil or water contamination. (17.10)

renewable resource A resource that has a virtually unlimited supply and does not remove something irreplaceable when it is used. (18.8)

reservoir 1. A lake that is created by a human-constructed dam (16.6). 2. A subsurface volume of rock that has sufficient porosity and permeability to permit the accumulation of oil or natural gas under adequate trap conditions. (18.1)

resisting force A force that resists the motion of an object, such as resisting the movement of tectonic plates. (3.7)

resurfacing Remaking a surface through erosion, deposition, volcanism, or tectonics. (19.2)

reversed polarity Refers to times in the past when the polarity of Earth's magnetic field was the opposite of what it is today. (10.2)

reverse fault A fault in which the hanging wall moves up relative to the footwall. (8.4)

rhomb An oblique, equilateral parallelogram, with a shape like a sheared box. (4.5)

rhyolite A mostly fine-grained, felsic igneous rock, generally of volcanic origin; can contain glass, volcanic ash, pieces of pumice, and variable amounts of visible crystals (phenocrysts). (5.2)

Richter scale A numeric scale of earthquake magnitude, devised in 1935 by the seismologist C. F. Richter. (12.6)

ripple marks Small ridges and troughs formed by moving currents. (7.13)

riprap A layer of large, durable fragments of broken rock, concrete, or other material, placed to prevent erosion by waves or currents. (14.6)

risk An assessment of whether a hazard might have some societal impact. (6.6)

river A large moving stream of water driven by gravity and flowing from higher to lower elevations. (16.1)

river terrace A relatively flat bench that is perched above a river or stream and that was formed by past deposition or erosion of the river or stream. (16.11)

rock avalanche High-velocity, turbulent flow of rock, sediment, and soil that results from the collapse of a steep mountain front. (15.10)

rock cycle A conceptual framework presenting possible paths and processes to which a rock can be subjected as it moves from one place to another and between different depths within the earth. (1.6)

rock fall A mass-wasting process whereby large rocks and smaller pieces of bedrock detach and fall onto the ground. (15.09)

rockslide A slab of relatively intact rock that detaches from bedrock and slides downhill, shattering as it moves. (7.8)

rock varnish A thin, dark film or coating of iron and manganese oxides, silica, and other materials; formed by prolonged exposure at the surface; *syn.* desert varnish. (9.2)

Rodinia A supercontinent, consisting of all the continents joined, that existed near the Precambrian-Paleozoic boundary. (11.11)

roll-front An arcuate deposit of uranium that forms where uranium carried by groundwater encounters water with a different chemistry. (18.6)

root wedging The process of plant roots extending into fractures and growing in length and diameter, expanding preexisting fractures. (15.1)

rotational slide A slide in which shearing takes place on a well-defined, curved shear surface, concave upward, producing a backward rotation in the displaced mass; *syn.* slump. (15.9)

runoff Precipitation that collects and flows on the surface, such as in streams. (1.7)

S

salinity The concentration of salt in water. (15.2)

saltation Transport of sediment in which particles are moved in a series of short, intermittent bounces on a bottom surface. (16.2)

salt dome A structure formed when buried salt buoyantly flows to the surface in steep, pipelike conduits. (10.11)

salt flat The nearly level, salt-encrusted bottom of a lake that is temporarily or permanently dried up. (13.9)

salt glacier A gravitational flow of rock salt downhill on the surface. (10.11)

saltwater incursion Displacement of fresh groundwater by the advance of salt water, usually in coastal areas; *syn.* saltwater intrusion. (17.8)

sand A grain or rock fragment smaller than 2 mm and larger than 1/16 mm. (7.4)

sandbar A low, sandy feature, possibly submerged, offshore of a shoreline or within a sandy river. (14.5)

sand dune An accumulation of loose sand piled up by the wind. (7.1)

sand spit A low ridge of sand and other sediment that extends like a prong off a coast. (14.5)

sand storm *See* dust storm. (13.2)

sandstone A medium-grained, clastic sedimentary rock composed mostly of grains of sand, along with other material. (7.9)

satellite A celestial body, natural or manmade, that revolves around a planet or other large planetary object. (19.1)

satellite image Image taken by an artificial satellite and generally depicting the types of materials on the surface of Earth or another planetary object. (2.3)

saturated zone The area in the subsurface where water fills nearly all the pore spaces. (17.3)

Saturn The second-largest planet in the solar system; the sixth planet outward from the Sun. (19.6)

scarp 1. A linear or curved scar left behind by a landslide on a hillslope, marking where the landslide pulled away from the rest of the hill (15.9). 2. A break or step in the land surface formed by movement along a fault (fault scarp). (12.2)

schist A shiny, foliated, metamorphic rock generally containing abundant visible crystals of mica. (8.6)

schistosity A metamorphic foliation representing the parallel arrangement of mineral grains, especially mica in schist or other coarse-grained metamorphic rocks. (8.6)

scoria A dark gray, black, or reddish volcanic rock that contains abundant vesicles, usually having the composition of basalt or andesite; *syn.* volcanic cinders. (5.3)

scoria cone A relatively small type of volcano that is cone shaped and mostly composed of scoria; *syn* cinder cone. (6.3)

sea arch An opening through a thin promontory of land that extends out into the ocean. (14.5)

sea cave A cave at the base of a sea cliff, usually flooded by seawater. (14.5)

sea cliff A cliff or steep slope situated along the coast. (14.5)

seafloor spreading The process by which two oceanic plates move apart and new magmatic material is added between the plates. (2.9)

sea ice Ice that forms from the freezing of seawater. (14.11)

seal An impermeable layer that traps petroleum at depth. (18.2)

seamount A submarine mountain, in some cases flat-topped, that rises above the seafloor. (10.5)

sea stack An isolated, pillar-like, rocky island or pinnacle near a rocky coastline. (14.5)

sea wall A human-constructed wall or embankment of concrete, stone, or other materials along a shoreline, intended to prevent erosion by waves. (14.6)

secondary wave (S wave) A seismic body wave propagated by a shearing motion that involves movement of material perpendicular to the direction of propagation; an S wave cannot travel through magma and other liquids. (12.5)

sediment Grains and other fragments that originate from the weathering and transport of rocks, and the unconsolidated deposits that result from the deposition of this material. (1.5)

sediment load The amount of sediment, including material chemically dissolved in a solution, carried by a stream. (16.2)

sedimentary rock Rock resulting from the consolidation of sediment. (1.5)

sediment budget The amount of sediment available to a system, such as along a shoreline. (14.4)

seismic activity *See* seismicity. (12.3)

seismicity Earth movements, either on the surface or at depth, caused by earthquakes. (12.3)

seismic energy Energy, in the form of vibrations, that is released during an earthquake, explosion, or some other ground-shaking event. (12.6)

seismic gap A segment of an active fault zone that has not experienced a major earthquake during a time interval when most other segments of the zone have experienced earthquakes. (12.12)

seismic ray The path that any part of a seismic wave travels. (12.15)

seismic-reflection profile A cross section plotting data gathered from a seismic-reflection survey. (10.1)

seismic-reflection survey A geophysical survey that uses the seismic-reflection technique. (10.1)

seismic-reflection technique A technique, widely used in exploring for petroleum, in which seismic energy (sound waves) is generated near the surface, bounces off subsurface layers, is recorded on the surface using geophones, and processed with sophisticated computers and numerical methods. (10.1)

seismic station The location of a scientific instrument (seismograph) that measures seismic vibrations. (12.1)

seismic tomography A technique using the arrival times of seismic waves to identify materials with different physical properties within the Earth. (12.16)

seismic wave Elastic waves produced by earthquakes or generated artificially. (12.1)

seismogram The record made by a seismograph, an instrument that records seismic waves. (12.5)

seismologist A scientist who studies seismic waves by analyzing when and how these waves arrive and by using powerful computers and sophisticated programs to model subsurface parameters, such as density. (12.5)

seismometer An instrument that measures ground shaking or seismic activity. (12.5)

shaded-relief map A map of an area whose relief is made to appear three-dimensional by simulating the shading on mountains, valleys, and other features. (2.3)

shale A fine-grained clastic sedimentary rock, formed by the consolidation of clay and other fine-grained material. (7.5)

shatter cone A distinctively conical fracture, usually interpreted to form during meteoroid impacts. (18.11)

shear The type of differential stress that occurs when stresses on the edge of a mass are applied in opposite directions. (8.2)

shear zone A generally tabular zone of rock that is more highly sheared and deformed than rocks outside the zone. (8.2)

sheet silicates A group of silicate minerals, including micas, that have a distinctly sheetlike crystalline structure. (4.7)

shield volcano A type of volcano that has broad, gently curved slopes constructed mostly of relatively fluid basaltic lava flows. (6.4)

silica Silicon dioxide (SiO_2), appearing either as a relatively pure form in a mineral (e.g., quartz) or as a component in more chemically complex minerals and rocks. (4.7)

silicates Minerals that contain silicon-oxygen tetrahedra; the most common mineral group on Earth. (4.6)

silicon The fourteenth element in the Periodic Table, having the atomic symbol Si. (4.7)

silicone A synthetic material in which carbon is bonded to silicon atoms to keep the material in long chains. (4.7)

sill A tabular igneous intrusion that parallels layers or other planar structures of the surrounding rock and which usually has a subhorizontal orientation. (5.13)

silt A fine-grained rock fragment or clast, 1/256 to 1/16 mm in diameter. (7.4)

silting up A filling, or partial filling, with silt of a reservoir as it receives sediment brought in by streams and surface runoff. (18.7)

siltstone A sedimentary rock composed of consolidated silt-sized particles, generally mostly quartz. (7.5)

Silurian A geologic period in the early part the Paleozoic Era. (9.8)

sinkhole A closed, circular depression, usually in a karst area, resulting from the collapse of an underlying cave. (7.11)

sinuosity The amount a river or stream channel curves for a given length. (16.5)

slab pull A plate-driving force generated by the sinking action of a relatively dense, subducted slab. (3.7)

slate A compact, fine-grained, low-grade metamorphic rock that possesses slaty cleavage. (8.7)

slope failure The sudden or gradual collapse of a slope that is too steep for its material to resist the pull of gravity. (15.7)

slump A slide in which shearing takes place on a well-defined, curved shear surface, concave upward, producing a backward rotation in the displaced mass; *syn.* rotational slide. (15.9)

smelter A mineral-extraction facility in which sulfide ores and other material are roasted in a furnace. (18.11)

snowfield A large area covered with snow and ice that, unlike a glacier, does not move. (14.11)

soft-rock geologist A geologist who focuses on the geology of sediments and sedimentary rocks as opposed to igneous and metamorphic rocks. (18.5)

soil Unconsolidated material at and near the surface, produced by weathering; includes mineral matter, organic matter, air, and water, and is generally capable of supporting plant growth. (15.5)

solar energy 1. Electromagnetic energy from the Sun (1.4). 2 Energy that uses the Sun's electromagnetic energy to heat buildings, heat water, or generate electricity. (18.8)

solar system The Sun, its eight planets, and other celestial bodies that orbit the Sun. (1.8)

solidification The process in which magma cools and hardens into solid rock, with or without the formation of crystals. (1.6)

sonar Using sound waves to determine the distance to reflecting objects, especially depth of the seafloor or a lake bottom. (10.1)

source rock A rock or sediment that contains enough organic material to produce petroleum. (18.1)

specific gravity The ratio of the density of a substance to the density of freshwater. (4.3)

spheroidal weathering A form of mostly chemical weathering in which concentric or spherical shells of decayed rock are successively separated from a block of rock. (15.4)

spit A small point or low ridge of sand or gravel projecting from the shore into a body of water. (14.5)

spreading center Divergent boundary where two oceanic plates move apart (diverge). (3.4)

spring A place where groundwater flows out of the ground onto the surface. (17.5)

spring tides Higher-than-average high tides and lower-than-average low tides caused when the Sun's gravity adds to the effects of the Moon's gravity. (14.2)

stalactite A conical or cylindrical cave formation that hangs from the ceiling of a cave and is composed mostly of calcium carbonate. (17.6)

stalagmite A conical, cylindrical, or moundlike cave formation that is developed upward from the floor of a cave and is composed mostly of calcium carbonate. (17.6)

stick-slip behavior The sequence of a rock straining before an earthquake, rupturing during an earthquake, and then mostly returning to its original shape after the earthquake. (12.2)

storm surge A local rise in the level of a sea or a lake during a hurricane or other storm. (14.6)

strain Change in shape or volume of a body as a result of stress. (8.1)

stratigraphic section A columnar diagram that shows the sequence of rock units, generally in their approximate relative thicknesses. (2.4)

stratovolcano *See* composite volcano. (6.1)

streak The color of powder a mineral leaves when rubbed against a porcelain plate. (4.3)

streak plate A piece of unglazed porcelain used to obtain a streak during mineral identification. (4.3)

stress The amount of force divided by the area on which the force is applied. (8.1)

stress field The entire array of stresses applied on a point or volume of rock. (8.2)

striation 1. A series of straight, subparallel lines on the surface of a crystal (4.8). 2. Linear features, resembling scratch marks, on a fault surface. (see 8.12.b4)

strike The direction of a horizontal line on an inclined surface. (8.4)

strike-slip fault A fault in which the relative movement is essentially horizontal, parallel to the strike of the fault surface. (8.4)

strip mine A mine, usually a coal mine, in which a long strip of material is mined at any one time, as the mining operation moves across the landscape. (18.3)

stromatolite A mound- or column-shaped feature of concentrically laminated carbonate materials, generally in ancient sedimentary rocks, interpreted to have been constructed by microscopic algae; also modern, live examples. (9.10)

subbituminous coal A black coal, intermediate in maturity, between lignite and bituminous coals. (18.3)

subduction The process along a convergent plate-tectonic boundary in which an oceanic lithospheric plate descends beneath the overriding plate. (3.5)

subduction zone A zone in which subduction takes place, either referring to the actual downgoing slab and its surroundings, or to the region, including Earth's surface, above the subducting slab. (3.5)

sublimation The process by which material moves from a solid phase directly into a vapor, as occurs when a glacier loses ice molecules directly to the air. (14.11, 17.1)

submarine canyon A submarine valley incised into the continental shelf or slope. (7.2)

submarine delta The part of a delta that is below sea level. (7.2)

submarine fan A broad, fan-shaped accumulation of sediment on the seafloor, especially below the mouth of a large river or submarine canyon. (10.0)

submarine slope failure Slope failure that occurs on the seafloor. (15.8)

submergent coast A coast that forms where land has been inundated by the sea because of a rise in sea level or subsidence of the land. (14.8)

submersible A small submarine, typically capable of carrying two to three people, that scientists use to study the ocean floor. (10.1)

subtropical desert A desert located in the subtropics, generally due to atmospheric circulation of dry, descending air. (13.8)

subtropics Geographical zones located directly north and south of the tropics, approximately centered on 30°S and 30°N. (13.3)

sulfates A group of minerals that contain sulfur (S) bonded to oxygen. (4.6)

sulfides A group of minerals containing sulfur (S) bonded with a metal. (4.6)

supercell thunderstorm An especially large and violent thunderstorm, commonly associated with damaging hail, wind, and tornadoes. (13.4)

surface currents Relatively fast-moving flows of shallow ocean water. (13.4)

surface water Water that occurs in streams, rivers, lakes, oceans, and other settings on Earth's surface. (17.0)

surface waves Seismic waves that travel on Earth's surface. (12.5)

surf zone The area where waves break and spread water upon the shore. (13.3)

suspended load Fine particles, generally clay and silt, that are carried suspended in moving water. (16.2)

suspension A mode of sediment transport in which water or wind picks up and carries the sediment as floating particles. (13.2)

S wave *See* secondary wave. (12.5)

S-wave shadow zone The region on Earth's surface (at any distance more than 103° from an earthquake epicenter) in which direct S waves from the earthquake are absent because they cannot pass through Earth's molten outer core. (12.15)

swelling clays Soil that contains a high portion of certain clay minerals and increases in volume (expands) when it becomes wet. (15.6)

syncline A fold, generally concave upward, with the youngest rocks in the center. (8.5)

synthetic Refers to material produced by humans. (4.1)

T

Taconic orgeny An early Paleozoic orogeny that occurred in the northern Appalachians and is interpreted to represent a collision between an island arc and North America. (11.12)

talus Loose rock fragments upon a steep slope, or an accumulation of such fragments. (15.1)

talus slope A steep slope composed of loose rock fragments, that is, talus. (15.1)

tar A thick, brown to black organic liquid, formed when petroleum loses its volatile components. (18.1)

tarn A small lake, especially one within a cirque, a glacially scoured depression. (14.12)

tar sand A sand or other sediment containing tar. (18.4)

tectonic activity *See* tectonics. (3.3)

tectonic plates The dozen or so fairly rigid blocks into which Earth's lithosphere is broken. (3.3)

tectonics Earthquakes, volcanoes, and other processes that deform Earth's crust and mantle; *syn.* tectonic activity. (3.3)

tectonic terrane A fault-bounded body of rock that has a different geologic history than adjacent regions. (11.9)

temporary base level Any base level, other than sea level, that limits the downward extent of erosion. (16.9)

tension The type of differential stress where stress is directed outward, pulling the material. (8.2)

tephra A pyroclastic material, regardless of size or origin, ejected during an explosive volcanic eruption; includes ash, pumice, and rock fragments. (6.2)

terminal moraine Glacially carried sediment that accumulates at the terminus (end) of a glacier and a landform composed of such material; generally marks the glacier's farthest downhill extent. (14.13)

termination The well-defined, commonly sharp end of a crystal. (4.4)

terminus The lower end of a glacier. (14.11)

terrace A relatively level or gently inclined surface, or bench, bounded on one edge by a steeper descending slope. (9.2)

terracing The practice of shaping steep terrain to provide flat areas suitable for agriculture, ranching, or to protect soil from erosion. (15.6)

terrestrial planets Our solar system's four inner planets, which have solid rocky surfaces and include Mercury, Venus, Earth, and Mars; *syn.* inner planets. (19.0)

terrestrial plant A plant that grows on land. (18.1)

tetrahedron A four-sided pyramid, as in a silica tetrahedron. (4.4)

texture The general physical appearance or character of a rock, especially the size, shape, and arrangement of minerals and other materials. (4.2)

theory An explanatory system of propositions and principles, supported to some extent by experimental or factual evidence and held to be true until contradicted or amended by new facts. (2.8)

theory of plate tectonics The theory currently accepted by nearly all geologists that Earth's lithosphere is broken into a number of fairly rigid plates that move relative to one another. (3.3)

thermal expansion The expansion of material as it is heated, such as heating and cooling that cause different minerals, or different parts of a rock, to expand and contract by different amounts. (7.3)

thermohaline conveyor A mostly deep-water ocean current driven by density differences that are caused by variations in temperature and salinity. (13.4)

thrust fault A reverse fault that has a gentle dip. (8.4)

thrust sheet The sheet of rock that has been displaced above a thrust fault. (8.4)

thunder A loud sound, resulting from the rapid heating and expansion of air along the path of a lightning bolt. (13.4).

thunderstorm A cloudy column of turbulent, moist air with variable amounts of lightning, thunder, rain, hail, and strong wind. (13.4)

tidal flat A low, gently sloping to horizontal area, commonly covered by sediment, that is flooded by ocean water during high tide. (7.2)

tide A cyclic change in the height of the sea surface, generally measured at locations along the coast; caused by the pull of the Moon's gravity and to a lesser extent the Sun's gravity. (14.1)

till Unsorted, generally unlayered sediment, deposited directly by or underneath a glacier. (14.13)

tillite A consolidated sedimentary rock formed by lithification of glacial till. (14.15)

tiltmeter A scientific instrument used to determine whether a site is being tilted by earth movements. (6.13)

time-travel curve A graph that plots the time difference between the arrivals of P waves and S waves as a function of the distance from the epicenter or other seismic disturbance to a seismic station.

topographic map A map showing the topographic features of a land surface, commonly by means of contour lines. (2.3)

topographic profile A cross-sectional view across a part of Earth's surface, showing variations in elevation or depth. (2.4)

topographic relief The difference in elevation of one feature relative to another; *syn.* relief. (2.4)

topography The general configuration of a surface, especially the land surface or seafloor, including its elevation, relief, and features; shape of the land. (2.4)

tornado A violent, rapidly rotating, funnel-shaped column of air that extends to the ground. (13.4)

tornado alley The area in the central part of the United States that experiences frequent tornadoes. (13.4)

tourmaline A group of generally dark to multicolored, prismatic silicate minerals, commonly used as semiprecious gemstones. (4.0)

trace fossils Features in rocks made by animals that moved across the surface or burrowed into soft sediment. (9.5)

traction The process by which particles roll, slide, or otherwise move on the surface, by such transport agents as streams, wind, or waves. (16.2)

trade winds Winds that blow in a generally consistent direction that can be used by wind-powered ships to cross the ocean. (13.1)

trading location for time A strategy that uses different parts of a landscape to represent different stages in the evolution of the landscape. (2.2)

transform boundary A plate boundary in which two tectonic plates move horizontally past one another. (3.3)

transform fault A strike-slip fault that accommodates the horizontal movement of one tectonic plate past another. (3.6)

transgression The advance of the sea across the land and the evidence of such an advance. (7.12)

transpiration The process by which plants, through their leaves, emit water vapor into the atmosphere. (17.1)

travertine A variety of limestone that is commonly concentrically banded and porous. (7.11)

trench A narrow, steep-sided, elongate depression of the deep seafloor, formed by bending down of a subducting oceanic plate at a convergent plate boundary; includes the deepest parts of the ocean. (3.1)

trench rollback Process by which a dense oceanic plate subducts into the asthenosphere, sinks, and tends to bend or roll away from the island arc. (10.6)

Triassic The earliest geologic period in the Mesozoic Era. (9.8)

trilobite A marine creature of the Paleozoic Era, characterized by a three-lobed external skeleton. (9.10)

tributary A secondary stream that joins or flows into a larger stream, river, or lake. (16.1)

triple junction The place where three tectonic plates, and three plate-tectonic boundaries, meet. (3.6)

trough 1. An elongate depression of the land or seafloor (3.4). 2. The lowest part of a wave. (14.3)

troy ounce A unit of weight (about 1.1 normal ounces) by which a precious metal is priced. (18.10)

tsunami A large sea wave produced by uplift, subsidence, or some other disturbance of the seafloor, especially by a shallow submarine earthquake. (12.7)

tsunami warning system An array of sensors in the ocean and computerized infrastructure to provide a warning upon formation of a tsunami. (12.7)

tuff Volcanic rock composed of consolidated volcanic ash and other tephra, commonly including pumice, crystals, and rock fragments. (5.3)

tundra A treeless, gentle plain in arctic and subarctic regions; usually has a marshy surface and is underlain by permafrost. (14.14)

turbidite A sediment or rock deposited from, or interpreted to be deposited from, a turbidity current; typically characterized by graded bedding. (7.2)

turbidity current A current in water or air that moves downward because it is more dense than the adjacent water or air, especially applied to a swift, bottom-flowing current of water and suspended sediment on the seafloor or the bottom of a lake; produces a turbidite. (7.2)

turbulent Chaotic flow of water, air, or some other fluid. (16.2)

typhoon A tropical cyclone that occurs in the western Pacific Ocean; *syn.* hurricane. (13.4)

U

ultramafic A generally dark or greenish igneous rock composed chiefly of mafic minerals rich in magnesium and iron. (5.2)

ultraviolet radiation The ultraviolet part of the electromagnetic spectrum. (1.4)

unconfined aquifer An aquifer where the water-bearing unit is open (not restricted by impermeable rocks) to Earth's surface and atmosphere. (17.4)

unconformity A boundary between underlying and overlying rock strata, representing a significant break or gap in the geologic record; an unconformity represents an interval of nondeposition or erosion, commonly accompanied by uplift. (9.3)

uniformitarianism The concept that the present is the key to the past; that is, geologic processes occurring today also occurred in the geologic past and can be used to explain ancient events and the geologic features they produced. (9.0)

unloading The process during which weight is removed from a landmass, such as by the melting of a glacier. (14.9)

unloading joints Joints formed from stresses that arise during uplift of buried rocks and that cause rocks to fracture due to reduced pressure. (8.3)

unsaturated zone A part of the subsurface where most of the pore spaces are filled with air rather than water. (17.3)

updraft An upward-moving current of air, as in a thunderstorm. (13.4)

Uranus The seventh planet out from the Sun. (19.7)

V

valley glacier A glacier that flows down a valley and tends to be narrow; *syn.* Alpine glacier. (14.10)

varves Thin, alternating light-colored and dark layers of sediment that form in lakes because of seasonal variations of deposition and biologic activity. (9.9)

vein A generally tabular accumulation of minerals that filled a fracture or other discontinuity in a rock; formed by precipitation of material from fluids, especially hydrothermal fluids. (8.1)

Venus The second planet from the Sun. (19.3)

vertical surface wave A type of surface wave in which material moves up and down, perpendicular to the propagation direction of the wave. (12.5)

vesicles Small holes found in a volcanic rock, representing gas bubbles in a magma that were trapped when the lava solidified. (5.1)

vesicular Adjective used to describe a rock containing vesicles. (5.1)

vesicular basalt A basalt that contains vesicles. (6.3)

viscous magma Magma that does not flow easily. (5.7)

viscosity A measure of a material's resistance to flow. (5.7)

volatile A chemical component that readily converts to a vapor phase under the proper conditions. (18.3)

volcanic ash Particles of volcanic tephra that are sand-sized or smaller, and accumulations of such material. (5.1)

volcanic bomb A large rock fragment representing either a large blob of magma or a solid angular block ejected during an explosive volcanic eruption. (6.1)

volcanic breccia A volcanic rock containing angular fragments in a matrix of finer material. (5.1)

volcanic dome A dome-shaped volcanic feature, largely composed of solidified lava of felsic to intermediate composition. (6.9)

volcanic field A region that contains an abundance of volcanic rocks and perhaps preserved volcanoes. (6.3)

volcanic glass A natural glass produced by the cooling and solidification of molten lava at a rate too rapid to permit crystallization. (5.1)

volcanic neck A steep, typically buttelike topographic feature composed of volcanic materials that formed in the conduit within or beneath a volcanic vent and that were more resistant to erosion than surrounding materials. (5.13)

volcano A vent in the surface of Earth through which magma and associated gases and ash erupt; also the form or structure, usually conical, that is produced by material erupted from the vent. (6.1)

W

warm front The sloping boundary between an advancing warm air mass and a cooler air mass. (13.1)

waterfall A steep descent of water within a stream, such as the place where it crosses a cliff or steep ledge. (16.6)

water table The surface between the unsaturated zone and the saturated zone, as in the top of groundwater in an unconfined aquifer. (17.3)

wave 1. An irregularity on the surface of a body of water (14.3). 2. *See* seismic wave.

wave amplitude The height between the trough and the crest of a wave, including ocean waves and seismic waves. (12.5)

wave base The depth at which the action of a wave in an ocean or lake no longer has an effect. (14.3)

wave-cut notch A notch or indentation produced in rocks or sediment by continued wave action at a specific level along a coast. (14.4)

wave-cut platform A gently sloping surface or bench produced by wave erosion. (14.5)

wave height The vertical distance between the trough and the crest of a wave. (14.3)

wavelength The horizontal distance between two adjacent crests in a set of waves. (14.3)

weathering Physical disintegration and chemical decomposition of rocks, sediment, and soil due to exposure to water and other chemical, atmospheric, and biological agents. (1.6)

weathering rind A weathered, outer crust on a rock fragment or bedrock mass exposed to weathering. (15.4)

weight A measure of how much downward force the mass of an object exerts under the pull of gravity. (2.5)

westerly A wind that generally blows from the west. (13.1)

whirlwind A rotating column of wind and usually dust. (13.2)

whitecap A white, foamy wave that forms when wind causes a wave to become too steep and to collapse. (14.3)

wind Movement of air relative to Earth's surface. (13.1)

Z

zone of accumulation The upper part of a glacier or ice sheet, where snow and ice are added faster than they are removed by melting and other processes. (14.11)

CREDITS

PHOTOGRAPHS

Unless otherwise credited: © Stephen J. Reynolds.

CHAPTER 1

Image Number: 01.00.a2: © Image Ideas/PictureQuest RF; 01.00.a3: Mike Doukas/U.S. Geological Survey; 01.00.a5: © Getty RF; 01.02.d1: "Jurassic Landscape," © Karen Carr; 01.03.a5: © Dr. Parvinder Sethi; 01.05.a4: © Getty RF; 01.05.b2: C.G. Newhall/U.S. Geological Survey; 01.05.b3: J.D. Griggs/U.S. Geological Survey; 01.05.b4, 0107.a4: Photos by Michael M. Kelly; 01.09.a2: U.S. Geological Survey; 01.09.a3: Woodbridge Williams/National Park Service; 01.09.b2: Photo by Perry H. Rahn.

CHAPTER 2

Image Number: 02.03.a2: Wendell Duffield/U.S. Geological Survey; 02.05.a1: Cyrus Read/U.S. Geological Survey; 02.05.a2: Kate Bull/U.S. Geological Survey; 02.05.a3: T.A. Plucinski/U.S. Geological Survey; 0205.b1: Karl String/U.S. Geological Survey; 02.05.b2: Photo by Ramon Arrowsmith; 02.05.b3: G. McGimsey/U.S. Geological Survey; 02.05.b4: Photo by Hillairy Hartnett; 02.05.c1: Photo by Chris Marone; 02.05.c3: Photo by Ariel Anbar; 02.07.d1: David E. Wieprecht/U.S. Geological Survey; 02.08.mtb1: Daniel Ball/Arizona State University; 02.10.a1: Photo by Paul Fitzgerald; 02.10.a2: Hawaiian Volcano Observatory/U.S. Geological Survey; 02.10.b1: P.R. Christensen/NASA/JPL/Arizona State University; 02.10.b2: Jason Leigh/Electronic Visualization Lab, University of Illinois, Chicago; 02.10.b3: Edward Garnero/Arizona State University; 02.10.mtb1, 02.11.a1: U.S. Geological Survey.

CHAPTER 3

Image Number: 03.06.b2: Robert E. Wallace/U.S. Geological Survey; 03.07.d3: Peter J. Haeussler/U.S. Geological Survey.

CHAPTER 4

Image Number: 04.00.a4: © Phototake/Alamy; 04.01.a1: © Dr. Parvinder/Sethi; 04.01.a2: © Paul Collis/ALAMY; 04.01.a4: Photo by Susanne P. Gillatt; 04.01.a6: Photo by Michael M. Kelly; 04.01.b1: © Doug Sherman/Geofile; 04.01.b2: © Corbis Royalty Free; 04.01.b3: © Doug Sherman/Geofile; 04.01.b4: The McGraw-Hill Companies, Inc./John A. Karachewski, photographer; 04.01.b5: © Vol. 30/PhotoDisc/Getty Images; 04.01.b6: © Corbis RF; 04.01.b8: © The McGraw-Hill Companies, Inc./Bob Coyle, photographer; 04.01.b9: © The McGraw-Hill Companies, Inc./Stephen Frisch, photographer; 04.01.b10: © The McGraw-Hill Companies, Inc./John Karachewski, photographer; 04.02.b6: Photo by Susanne P. Gillatt; 04.03.a4: © The McGraw-Hill Companies, Inc./Jacques Cornell, photographer; 04.03.a5–04.03.a6: © The McGraw-Hill Companies Inc./Ken Cavanagh, photographer; 04.03.a7: Photo by Dexter Perkins; 04.03.a8: © The McGraw-Hill Companies, Inc./Jacques Cornell, photographer; 04.03.a9: © Parvinder Sethi; 04.03.a10: © The McGraw-Hill Companies Inc./Ken Cavanagh, photographer; 04.03.a11: © Robert Rutford/James Carter, photographer; 04.03.a12, 04.03.b1, 04.03.b2, 04.03.b3: © Doug Sherman/Geofile; 04.03.b4: © Dr. Parvinder/Sethi; 04.04.a1: © The McGraw-Hill Companies, Inc./Bob Coyle, photographer; 04.04.a5 © The McGraw-Hill Companies, Inc./Bob Coyle, photographer; 04.05.a2: Photo by Dexter Perkins; 04.05.b2: © Robert Rutford/James Carter, photographer; 04.06.b1: © Gary Dyson/ALAMY; 04.06.b3: © The McGraw-Hill Companies, Inc./Jacques Cornell, photographer; 04.06.b5: Thomas Sharp/Arizona State University; 04.06.b6: © The McGraw-Hill Companies, Inc./Doug Sherman, photographer; 04.06.b7: © Doug Sherman/Geofile; 04.08.a1, 04.08.a2: © Robert Rutford/James Carter, photographer; 04.08.a3: © Doug Sherman/Geofile; 04.08.a4: © Charles E. Jones; 04.08.b1–b3: Thomas Sharp/Arizona State University; 04.08.b4: © Doug Sherman/Geofile; 04.08.b5, 04.09.a1, 04.09.a2: Thomas Sharp/Arizona State University; 04.08.mtb1: © Wally Eberhart/Visuals Unlimited; 04.09.a4, 04.09.a5: © Doug Sherman/Geofile; 04.09.a7, 04.09.a9: © Doug Sherman/Geofile; 04.09.a10, 04.09.a12, 04.09.a13: Thomas Sharp/Arizona State University; 04.10.a2, 04.10.a3: © Dr. Parvinder Sethi; 04.13.mtb1: Photo by Susanne P. Gillatt; 04.14.a4: © Vol. 31/PhotoDisc/Getty Images; 04.14.a5: © Brand X/Images RF; 04.14.a6: © EP100/PhotoDisc/Getty Images; 04.14.b4: © Brand X/FotoSearch RF; 04.15.a1, 04.15.a2: Thomas Sharp/Arizona State University; 04.15.a3: © Doug Sherman/Geofile; 04.15.a5, 04.15.a6, 04.15.a7: Thomas Sharp/Arizona State University.

CHAPTER 5

Image Number: 05.03.a3: © Brand X/Punchstock RF; 05.03.a7: © B.Runk/S. Schoenberger/Grant Heilman Photography; 05.07.a2: © Arctic Images/ALAMY; 05.08.b2: Thomas Sharp/Arizona State University; 05.08.b3, 05.08.b4: © Robert Rutford/James Carter, photographer; 05.08.b5: Thomas Sharp/Arizona State University; 05.08.b6: © Charles E. Jones; 05.13.a6: Photo by George H. Davis; 05.13.b2: Photo by Steven Semken; 05.13.mtb1: Anonymous/National Park Service; 05.14.a2: Photo by Allen Glazner; 05.14.a3, 05.14.a4: Photo by Michael Ort; 05.14.a5: Photo by Scott Chandler; 05.14.a6: Photo by Michael Ort; 05.14.mtb1: Photo by Allen Glazner.

CHAPTER 6

Image Number: 06.00.a1: Donald A. Swanson/U.S. Geological Survey; 06.00.a3: John Pallister/U.S. Geological Survey; 06.00.a4: Austin Post/U.S. Geological Survey; 06.00.a5: Lyn Topinka/U.S. Geological Survey; 06.01.a1: J.D. Griggs/U.S. Geological Survey; 06.01.a2: E. Klett/US Fish and Wildlife Service; 06.02.a1: J. Judd/U.S. Geological Survey; 06.02.a2: J.D. Griggs/U.S. Geological Survey; 06.02.a3: Don Swanson/U.S. Geological Survey; 06.02.a4: R. Clucas/U.S. Geological Survey; 06.02.a5: U.S. Geological Survey; 06.02.a6: M.E. Yount/Alaska Volcano Observatory, U.S. Geological Survey; 06.02.c1: John Pallister/U.S. Geological Survey; 06.02.c2: Hawaii Volcano Observatory/U.S. Geological Survey; 06.03.a1: J.D. Griggs/U.S. Geological Survey; 06.03.a3: U.S. Geological Survey; 06.03.a4: Photo by Michael M. Kelly; 06.03.a5, J.D. Griggs/U.S. Geological Surrvey; 06.03.a6: Photo by Michael M. Kelly; 06.03.a7: J.D. Griggs/U.S. Geological Survey; 06.03.c2: Photo by Michael M. Kelly; 06.03.c4: Photo by Jessica Barone; 06.03.mtb1: Photo by Michael M. Kelly; 06.04.a2: Hawaii Volcano Observatory/U.S. Geological Survey; 06.04.a3: C.C. Heliker/U.S. Geological Survey; 06.04.a4: Hawaii Volcano Observatory/U.S. Geological Survey; 06.04.b1, 06.04.b2, 06.04.b3: J.D. Griggs/U.S. Geological Survey; 06.05.a2: Photo by D.P. Schwert; 06.05.a3: Hawaii Volcano Observatory/U.S. Geological Survey; 06.06.a2: © Jacques Collet/EPA/Corbis; 06.06.b1, 06.06.b3: Photo by Michael M. Kelly; 06.06.c1, 06.06.c2: J.D. Griggs/U.S. Geological Survey; 06.06.d2: © AP Images; 06.07.a1: U.S. Geological Survey; 06.07.a4: Game McGimsey/Alaska Volcano Observatory, U.S. Geological Survey; 06.07.a5: TA Plucinski/Alaska Volcano Observatory, U.S. Geological Survey; 06.07.b1: R. Christensen/U.S. Geological Survey; 06.07.b4: C.A. Neal/Alaska Volcano Observatory, U.S. Geological Survey; 06.07.b5: R.G. McGimsey/Alaska Volcano Observatory, U.S. Geological Survey; 06.07.mtb1: © Vol. 23/PhotoDisc/Getty Images; 06.07.mtb2: International Space Station/NASA; 06.07.mtb3: Image by Jim Williams, NASA, GSFC Scientific Visualization Studio, and the Landsat 7 Science Team; 06.08.a2: © Bettmann/CORBIS; 06.08.b2: © Popperfoto/ALAMY; 06.08.c3: Harry Glicken/U.S.Geological Survey; 06.08.c4–c6: © Copyright Gary Rosenquist, 1980; 06.08.c7: Peter W. Lipman/U.S. Geological Survey; 06.09.a1: T. Miller/U.S. Geological Survey; 06.09.a2: Steve Schilling/U.S. Geological Survey; 06.09c5: Willie Scott/U.S. Geological Survey; 06.09.mtb1, 06.09.mtb2: U.S. Geological Survey; 06.10.b5: R.A. Bailey/U.S. Geological Survey; 06.10.b8: S.R. Brantley/U.S. Geological Survey; 06.10.mtb1: Mike Dukas/U.S. Geological Survey; 06.11.a1: NASA; 06.12.a1: G. McGimsey/Alaska Volcano Observatory, U.S. Geological Survey; 06.13.b1: Photo by Michael M. Kelly; 06.13.b2: U.S. Geological Survey; 06.13.c1: Maxim Sorokin/U.S. Geological Survey; 06.13.c2: Steve J. Smith, Alaska Volcano Observatory/U.S. Geological Survey/Geophysical Institute, University of Alaska, Fairbanks; 06.13.d1: M. Sako/U.S. Geological Survey; 06.13.d2: J.D. Griggs/U.S. Geological Survey; 06.13.d3: S.R. Brantley/U.S. Geological Survey; 06.13.e1: U.S. Geological Survey; 06.13.mtb1: Dan Dzurisin/U.S.Geological Survey; 06.14.a2: Lyn Topinka/U.S. Geological Survey.

CHAPTER 7

Image Number: 07.00.a2: © Digital Vision/PunchStock; 07.01.a3: Photo by Jessica Barone; 07.01.a4: © Digital Vision RF; 07.01.a6: © Peter Bowater/ALAMY; 07.01.a7: Photo by Michael M. Kelly; 07.01.a8: © Getty Royalty Free; 07.02.a2: © PhotoDisc/Getty Images; 07.02.a3: © Getty Royalty Free; 07.02.a4: © Digital Vision/Getty Images; 07.02.a5: Tim McCabe/NRCS; 07.02.a7: NASA; 07.02.a8: U.S. Geological Survey; 07.03.c2: Photo by Charles M. Carter; 07.04.a4–6: © The McGraw-Hill Companies, Inc./Bob Coyle; photographer; 07.04.a12: U.S. Geological Survey; 07.04.b4: Photo by Michael M. Kelly; 07.04.b3, 07.04.b5: Photo by Susanne Gillatt; 07.06.a1, 07.06.a5: © Digital Vision/Getty Images; 07.06.a7: © Creatas/PunchStock; 07.06.b2: © Bruce Molina, Terra Photographics/Earth Science World Image Bank, http://www.earthscienceworld .org/imagebank; 07.06.b5: © The McGraw-Hill Companies, Inc./Jacques Cornell, photographer; 07.06.b10: © Mark Schneider/Visuals Unlimited; 07.07.b1: Photo by William R. Dickinson; 07.07.b6: © Sheila Terry/Photo Researchers, Inc.; 07.07.b9: Photo by William R. Dickinson; 07.07.b10:, 07.08.b7: Photo by Michael M. Kelly; 07.08.b8: © Gerald and Buff Corsi/Visuals Unlimited; 07.09.b1: © Digital Vision RF; 07.09.d1: © The McGraw-Hill Companies, Inc./John Karachewski, photographer; 07.09.d3: © Jerome Nuefeld/The Experimental Nonlinear Physics Group/The University of Toronto; 07.09.d4: Photo by Brian Romans; 07.10.b2: © Julia Waterlow/Eye Ubiquitous/Corbis; 07.10.c1: © The McGraw-Hill Companies, Inc./John A. Karachewski, photographer; 07.11.d2: Courtesy of Terre Sans Frontieres Avions Sans Frontieres; 07.11.d4: U.S. Geological Survey; 07.13.a12: © Dr. Parvinder Sethi; 07.14.b1: Guy Gelfenbaum/U.S.Geological Survey; 07.15.b1–07.15.b3, 07.15.c2–07.15.c4: Vincent Matthews/Colorado Geological Survey; 07.15.c5–07.15.c7: Ron Blakey, Colorado Plateau Geosystems

CHAPTER 8

Image Number: 08.00.a3: © Duncan Heron; 08.01.mtb1: Spokane Research Lab/NIOSH/CDC; 08.02.mtb1: Courtesy of J.M. Logan and F.M. Chester, Center for Tectonophysics, Texas A & M University; 08.03.c6: © Dean Conger/Corbis; 08.10.c2: © Phil Degginger/Animals Animals; 08.11.a9: © Dr. Marli Miller/Visuals Unlimited; 08.14.a2: © Charles Ver Straeten/New York State Museum.

CHAPTER 9

Image Number: 09.03.a5: Photo by Michael M. Kelly; 09.05.a6: © Phil Degginger/Carnegie Museum/ALAMY Royalty Free; 09.05.a7, 09.05.b2: Photo by Michael M. Kelly; 09.06.a1: © Getty Royalty Free; 09.06.a4: © Michael Freeman/Corbis; 09.06.b1: © Digital Vision/PunchStock

17.02.b1: © Vol. 39/PhotoDisc/Getty Images; 17.02.b2: © Vol. 16/PhotoDisc/Getty Images; 17.02.b3: © PhotoDisc/ Getty Images; 17.02.b4: Bureau of Reclamation; 17.02.b5: David Walsh/Bureau of Reclamation; 17.05.a2: © The McGraw-Hill Companies, Inc./John A. Karachewski, photographer; 17.06.a2: U.S. Geological Survey; 17.06.b1, 17.06.b3: © Corbis Royalty Free; 17.06.c2: © Dr. Parvinder Sethi; 17.06.mtb1: D Luchsinger/National Park Service; 17.07.a2: © Terry Whittaker/Photo Researchers, Inc.; 17.07.a3: Dustin Reed/USGS Ohio Water Science Center/ U.S. Geological Survey; 17.08.b2: Photo by Richard O. Ireland, USGS; 17.09.a2: © Gyori Antoine/Corbis/Sygma; 7.09.mtb1, mtb2: U.S. Geological Survey International; 17.10.b1: © Colin Cuthbert/Newcastle University/Photo Researchers, Inc.; 17.10.b2: Doug Bartlett/Clear Creek Associates.

CHAPTER 18

Image Number: 18.00.a2: NASA; 18.01.a1: Photo by Cynthia Shaw; 18.01.a2: © Steven P. Lynch; 18.01.a3: Photo by Belinda Rain, Courtesy of EPA/National Archives; 18.01.mtb1: © Photo by William R. Dupre; 18.03.a2: © Dorling Kindersley/Courtesy of the Natural History Museum, London; 18.03.a4: © Scientifica/Visuals Unlimited; 18.03.a7: © Breck P. Kent/Animals Animals; 18.03.b4: © AP Image; 18.04.a2: Photo by Julia K. Johnson; 18.04.b2: Gary P. Klinkhammer/Oregon State University; 18.04.d1: Dept. of Energy; 18.05.d2: © Greg Smith/Corbis; 18.06.c2: © Tim Wright/Corbis; 18.06.d1: © Dewitt Jones/Corbis; 18.06.d2: Courtesy of Department of Energy; 18.06.d3: © Igor Kostin/Sygma/Corbis; 18.06.mtb1: Department of Energy; 18.07.b3: © Marine Current Turbines TM Ltd.; 18.08.b1: © The McGraw-Hill Companies, Inc./John Flournoy, photographer; 18.08.c1: Photo by Susanne Gillatt; 18.08.c2: © Corbis Royalty Free; 18.08.d1: Photo by K.J. Kolb; 18.08.d2: Photo by Georgi Banchev; 18.08.d3: Jet Propulsion Laboratory/NASA; 18.09.c4: © Indiapicture/ALAMY; 18.13.a2: © Vol. 39/PhotoDisc/Getty Images; 18.13.c1: © The McGraw-Hill Companies, Inc./John A. Karachewski, photographer; 18.13.d1: © Robert Garvey/Corbis; 18.13.d2: © Andrew Brown/Ecoscen/Corbis; 18.13.d3: © The McGraw-Hill Companies, Inc./John A. Karachewski, photographer; 18.13.d4: © Vol. 31/PhotoDisc/Getty Images; 18.13.mtb1: © Vol. 39/PhotoDisc/Getty Images.

CHAPTER 19

Image Number: 19.00.a2: NASA and The Hubble Heritage Team (STScI/AURA) Acknowledgement: N. Scoville (Caltech) and T. Rector (NOAO); 19.00.a3: J. Hester, P. Scowen (ASU), HST, NASA; 19.01.a1: NASA, James Bell (Cornell Univ.), Michael Wolff (Space Science Inst.), and the Hubble Heritage Team (STcI/AURA); 19.01.b2: NASA/ JPL; 19.01.c1: P.R. Christensen/NASA/JPL/Arizona State University; 19.01.d1: NASA; 19.01.d2, 19.01.d3: NASA/JPL; 19.01.mtb1: NASA/Galileo Project; 19.02.a1: JPL/NASA/ University of Arizona; 19.02.a2: NASA; 19.02.a3, 19.02.a4: P.R. Christensen/NASA/JPL/Arizona State University; 19.02.a5: NASA/JPL; 19.02.a6: NASA/JPL/University of Arizona; 19.02.b2: NASA/JPL/Malin Space Science Systems; 19.02.d1: NASA/JPL; 19.02.d2: NASA/JPL/Malin Space Science Systems; 19.02.d3: NASA/JPL/University of Arizona; 19.02.mtb1: NASA; 19.03.a1: NASA/U.S. Geological Survey; 19.03.a2: NASA/JPL/Northwestern University; 19.03.b1: NASA/JPL; 19.03.b2: NASA/U.S. Geological Survey; 19.03.b3, 19.03.b4: NASA/JPL; 19.03.c2: NASA; 19.03.d1: NASA, James Ben (Cornell Univ.), Michael Wolff (Space Science Inst.), and the Hubble Heritage Team (STScI/AURA); 19.04.a1: U.S. Geological Survey; 19.04.a2: NASA/U.S. Geological Survey; 19.04.a3, 19.04.a4, 19.04.a5, 19.04.a6, 19.04.b1, 19.04.b2: NASA; 19.04.c1, 19.04.c2, 19.04.c3: Photo by Donald Burt; 19.04.mtb1: NASA/JPS-Caltech; 19.05.a2: NASA; 19.05.a4: John Spencer (Lowell

Observatory) and NASA; 19.05.a5: PIRL/University of Arizona; 19.05.a6: NASA/University of Arizona; 19.05.a7: NASA; 19.05.a8: NASA/JPL/R. Pappalardo (University of Colorado); 19.05.a9, 19.05.a10: NASA/JPL; 19.05.a11: NASA/Brown University; 19.05.a12: PIRL/University of Arizona; 19.06.a1: JPL; 19.06.a2: NASA/JPL; 19.06.a3: NASA/University of Arizona/LPL; 19.06.a4: NASA/JPL/ ESA/University of Arizona; 19.06.a5: CICLOPS/Space Science Institute; 19.06.a6: NASA; 19.06.a7, 19.06.a8: NASA/ Cassini Imaging Team; 19.07.a3: Calvin J. Hamilton/NASA; 19.07.a4, 19.07.b2, 19.07.mtb1: NASA; 19.08.a2: NASA/JPL/ Arizona State University; 19.08.a3: NASA/STScI; 19.08.a4: NASA/JPL-Caltech/Univ. of Arizona; 19.08.a5: NASA/U.S. Geological Survey; 19.08.a6, 19.08.a7: NASA/JPL/University of Arizona; 19.08.b2: NASA/JPL/Cornell; 19.08.b3: NASA/JPL-Caltech/USGS/Cornell; 19.08b4, 19.08.b5, 19.08.b6: NASA/JPL/Cornell; 19.09.a2: P.R. Christensen/ NASA/JPL/Arizona State University; 19.09.a3: NASA/JPL; 19.09.a4, 19.09.a5, 19.09.a6, 19.09.a7: P.R. Christensen/ NASA/JPL/Arizona State University.

TEXT AND LINE ART

CHAPTER 1

01.02.c1, 01.02.c2: Martin Jakobsson/Stockholm Geo Visualisation Laboratory; 01.09.a4: p. 20 After U.S. Geological Survey HA-743.

CHAPTER 2

02.06.c1: Kentucky Geological Survey; 02.07.d2: After U.S. Geological Survey Fact Sheet 100-03; 02.12.b2: After T. Kenkmann and D. Scherler, *Lunar Planetary Science*, 2002.

CHAPTER 3

03.02.a1, 03.02.b1, 03.03.a1, 03.03.c1, 03.05.c1, 03.07.b1: Data from Michelle K. Hall-Wallace; 03.07.c1: Kreemer, C. and others, 2002.

CHAPTER 5

05.08.mtb1: Data from GEOROC/Max-Planck Society; 05.11.mtb1: Ocean Drilling Project, Site 1201, Texas A&M University; 05.12.mtb1: Modified from King and H. Beikman/U.S. Geological Survey; 05.14.a1: Modified from U.S. Geological Survey Digital Data Series 11.

CHAPTER 6

06.04.a1: Inset from Schmincke, 2004, Springer-Verlag, used with permission; 06.05.a1: After D. Swanson, *American Journal of Science*, 1975; 06.05.c4: After Chris Jenkins, Institute of Arctic & Alpine Research, University of Colorado at Boulder; 06.08.a1: After L. Gurioli, *Geology*, 2005; 06.08.c1: After J. Verhoogen/California University, *Department of Geological Sciences Bulletin*; 06.08.c2: U.S. Geological Survey; 06.10.a2: After F. Goff, *Geo-Heat Center Bulletin*, 2002; 06.11.c1: U.S. Geological Survey Fact Sheet 100-03; 06.11.c2: U.S. Geological Survey Fact Sheet 2005-3024; 06.11c3: U.S. Geological Survey Fact Sheet 100-03; 06.11.c4: U.S. Geological Survey Fact Sheet 100-03; 06.11.mtb1: U.S. Geological Survey Open-File Report 95-59; 06.12.c1: Siebert L. and Simkin T. (http://www.volcano.si.edu/world/), 2002; 06.13.a1: After E. Wolfe/ U.S. Geological Survey, 1996; 06.13.a2: U.S. Geological Survey 1996; 06.13.b3: After A. Daag/ U.S. Geological Survey 1996; 06.13.d4: After C. Wicks/ U.S. Geological Survey; 06.14.a3: U.S. *Geological Survey Bulletin 1292*; 06.14.c1, 06.14.c2: U.S. Geological Survey Open-File Report 98-428; 06.14.mtb1: U.S. Geological Survey Open-File Report 94-585.

CHAPTER 7

07.15.a1: Colorado Geological Survey; 07.15.c1: Colorado Geological Survey.

CHAPTER 8

08.00.a2: After P. King, Princeton University Press, 1977; 08.14.a1: P. King and H. Beikman/U.S. Geological Survey, 1974; 08.14.b1: R. Stanley and N. Ratcliffe, *Geological Society of America Bulletin*, 1985; 08.14.c1: Geology Department, Union College; 08.14.c2: After F. Spear, *Journal of Petrology*, 2002.

CHAPTER 9

09.04.mtb1: After McGraw-Hill.

CHAPTER 10

10.00.a1: After G. Hatcher, Monterey Bay Aquarium Research Institute; 10.00.a2: After D. Wagner and others/ California Geological Survey CD 2002-04; 10.01.d1: After M. Fisher and others, U.S. Geological Survey Professional Paper 1687; 10.02.b2: After J. Kious/U.S. Geological Survey, 1996; 10.02.c1: U.S. Geological Survey; 10.02.c2: U.S. Geological Survey; 10.03.b1: After K. MacDonald, Academic Press, 2001; 10.04.a2: After Gabi Laske, University of California, San Diego; 10.04.a3: After David Sandwell/ Scripps Institution of Oceanography; 10.04.a4: After R. Dietmar Muller/School of Geosciences University of Sidney; 10.05.c1: Don Anderson/Seismological Laboratory, California Institute of Technology; 10.06.d1: Smithsonian Global Volcanism Program; 10.07.a5, 10.07.a6: After P. Gans, *Tectonics*, 1997; 10.08.a2: After N. White, Oxford University Press, 1990; 10.08.a3: After Z. Beydoun, *Episodes*, 1998; 10.09.c1: Chris Jenkins/Institute of Arctic & Alpine Research, University of Colorado at Boulder; 10.10.a1: D.L. Divins/National Geophysical Data Center; 10.10.b3: Gerry Hatcher/Monterey Bay Aquarium Research Institute; 10.10.b4: D.L. Divins/National Geophysical Data Center; 10.11.c2: After T. Affolter and J-P Gratier, Journal of Geophysical Research, 2004; 10.11.d1: After F. Diegel, American Association of Petroleum Geologists, Memoir 65; 10.13.a2–10.13.a7: After J. Pindell and L. Kennan, GCSS-CPM Conference, 2001.

CHAPTER 11

11.03.c1: J. Shaw, *Science*, 1999; 11.04.b1: *U.S. Geological Survey Bulletin 2146-D*, 11.04.mtb1: King and H. Beikman/U.S. Geological Survey; 11.04.mtb2: A. Bally, Geologic Society of America DNAG, 1989; 11.08.a2: P. King, Princeton University Press, 1977; 11.09.mtb1: W. Nokleberg and others, U.S. Geological Survey Open-File Report 97-161; 11.10.a1: U.S. Geological Survey Tapestry of Time; 11.10.b1: Geologic Survey of Canada/Digital Data Cornell University; 11.10.mtb1: After E. Moores and others, Geological Society of America Special Paper 338, 1999; 11.12.a6–a11, 11.12.mtb1: After L. Fichter, 1993/ James Madison University.

CHAPTER 12

12.00.a1: Travel time: Kenji Satake/Geological Survey of Japan, Rupture and epicenters: Atul Nayak/Scripps Institution of Oceanography; 12.01.d3: After J. Zucca/Lawrence Livermore National Laboratory; 12.03.a1: Paula Dunbar/ National Oceanographic and Atmospheric Administration; 12.06.a1: U.S. Geological Survey; 12.06.a2: Data from National Earthquake Information Center/U.S. Geological Survey; 12.06.a4: Data from National Earthquake Information Center/U.S. Geological Survey; 12.06.b1: Data from National Earthquake Information Center/U.S. Geological Survey; 12.06.b2: After N. Short, 2006/National Aeronautics and Space Administration; 12.07.c1: After C. Stover, 1993, U.S. Geological Survey Professional Paper 1527; I. Wong/Utah Geological Surveyó Public Information Series 76; 12.08.a1: Paula Dunbar/National Oceanographic and Atmospheric Administration; 12.08.a7: Kathleen M. Haller/ U.S. Geological Survey; 12.11.b1: National Oceanic and Atmospheric Administration; 12.11.b1: After J. Calzia, 2005, EOS Transactions AGU (86)52; 12.11.c1, c2: Y. Klinger and others, *Bulletin of the Seismological Society of America*, 2003; 12.11.mtb1: After Earthscope/U.S. Geological Survey; 12.12.a1: D. Giardani, Global Seismic Hazard Assessment Program; 12.12.a2: U.S. Geological Survey; 12.12.a3: After R. Wesson and others, 1999, U.S. Geological Survey Map I-2679; 12.12.a4: After F. Klein and others, 2000, U.S. Geological Survey Map I-2724; 12.12.b1: G. Plafker eds., U.S. Geological Survey Circular 1045, 1989; 12.12.b2: U.S. Geological Survey; 12.12.c2: U.S. Geological Survey; 12.12.c3: U.S. Geological Survey; 12.13.a1: Kathleen M. Haller/U.S. Geological Survey; 12.15.c1: Incorporated Research Institutions for Seismology; 12.15.c2: Joint Earth Science Education Initiativeó Royal Society of Chemistry; 12.16.a3: Edward Garnero/Arizona State University; 12.16.c1, c3: Edward Garnero/Arizona State University; 12.17.a1: E. Eckel, U.S. Geological Society Professional Paper 546, 1970; 12.17.c1, c2, c3: After G. Plafker, U.S. Geological Society Professional Paper 543, 1969.

CHAPTER 13

13.01.a1: After M. Hackworth/Idaho State University; 13.04.a1: Earth Observatory/National Aeronautics and Space Administration; 13.04.a3: Goddard Space Flight Center/National Aeronautics and Space Administration; 13.04.b1: National Severe Storms Laboratory—National Oceanic and Atmospheric Administration; 13.04.c2: After N. Short, 2006, National Aeronautics and Space Administration; 13.05.a1: After United Nations Atlas of the Oceans; 13.05.b1: International Arctic Science Committee; 13.05.c3, c4: National Weather Center—National Oceanic and Atmospheric Administration; 13.07.b1: National Center for Atmospheric Research; 13.08.b1, c1: U.S. Geological Survey; 13.08.d1: After P. Reich—World Soil Resources/

United States Department of Agriculture; 13.10.a1–13.10.a4 National Academy of Sciences; 13.10.a6–13.10.a8 National Academy of Sciences; 13.11.a4: Carbon Dioxide Information Analysis Center Oak Ridge National Laboratory; 13.11.a5: National Research Council, 2006; 13.11.a6: National Research Council, 2006; 13.12.c3, c4: After C. Duncan, 2003, Geology, 31, 75–78; 13.13.b1: After W. Hargrove and R. Luxmoore, 1998 Oak Ridge National Laboratory; 13.14.a1: After National Hurricane Center—National Oceanic and Atmospheric Administration, 2005; 13.14.b1, b2, c1: After W. Gray and P. Klotzbach, 2004/Colorado State University.

CHAPTER 14

14.09.mtb1: After S. Dutch/University of Wisconsin, Green Bay; 14.14.b2: After PACE21 Network, European Science Foundation; 14.14.c1: After J. Feth, U.S. Geological Survey Professional Paper 424B, 1961; 14.14.mtb1: After L. Topinka, Cascade Volcano Observatory/U.S. Geological Survey; 14.15.b3: University of British Columbia; 14.16.c3: Personal correspondence with Emi Ito, University of Minnesota, 2009; 14.18.d1: U.S. Geological Survey Fact Sheet 076-00.

CHAPTER 15

15.09.mtb1: After G. Kiersch, *Civil Engineering*, 1964; 15.11.b1: U.S. Geological Survey Open-File Report 97-0289; 15.11.b2: E. Brabb and others, U.S. Geological Survey Map 2329, 1999; 15.13.a3, 15.13.b1–15.13.c2: D. Varnes and W. Savage, *U.S. Geological Survey Bulletin 2130*, 1996.

CHAPTER 16

16.03.b1: After J. Mount, 1995, *California Rivers and Streams: The Conflict Between Fluvial Process and Land Use*, University of California Press; 16.04.a2: C. Bailey/William and Mary College, 1998; 16.10.a1: U.S. Geological Survey Tapestry of Time; 16.10.a2: U.S. Geological Survey; 16.10.a3–a4: Martin Jakobsson/Stockholm Geo Visualisation Laboratory; 16.13.a1: K. Wahl, U.S. Geological Survey Circular 1120B; 16.13.a2: C. Parett, U.S. Geological Survey

Circular 1120-A; 16.13.a5: U.S. Geological Survey; 16.13.b2: Data from R. Maddox and others, *Monthly Weather Review*, 1977; 16.14.b1: Data from National Water Information System, U.S. Geological Survey; 16.14.b2: Data from National Water Information System, U.S. Geological Survey; 16.15.a11: After D. Topping, 2003, U.S. Geological Survey Professional Paper 1677.

CHAPTER 17

17.01.a2: After U.S. Geological Survey; 17.02.a4: U.S. Geological Survey; 17.11.a1: High Plains Regional Ground Water Study/U.S. Geological Survey; 17.11.a2: Texas Water Development Board; 17.11.b1: U.S. Geological Survey; 17.11.b2: Texas Water Development Board; 17.11.c1, 17.11.c2: U.S. Geological Survey.

CHAPTER 18

18.00.a2: R. Pollastro, *U.S. Geological Survey Bulletin 2202-H;* 18.00.a3: Various sources; 18.00.mtb1: National Mining Association; 18.02.c5: W. Perry, *U.S. Geological Survey Bulletin 2146-D*, 1997; 18.04.a1: Energy Information Administration, DOE; 18.04.b1: U.S. Geological Survey Fact Sheet 021-01, 2001; 18.04.c2: Energy Information Administration, DOE; 18.04.d2: Energy Information Administration, DOE; 18.05.d1: T. Moore, U.S. Geological Survey Open-File Report 98-34, 1999; 18.10.b2: U.S. Geological Survey Mineral Resources; 18.14.a1: P. King and H. Beikman/U.S. Geological Survey; 18.14.a2: Wyoming Geological Survey; 18.14.b1, 18.14.b2, 18.14.b3: Wyoming Geological Survey.

CHAPTER 19

19.01.a2: NASA and E. Karkoschka/University of Arizona; 19.02.b1: R. Evans/NASA/HST Comet Science Team; 19.03.c1: NASA/Johns Hopkins University; 19.05.a3: NASA; 19.07.a2: Erich Karkoschka/University of Arizona/NASA; 19.07.c2: Modified from A. Feild/NASA/ESA; 19.08.b1: NASA/JPL; 19.08.mtb1: P.R. Christensen/NASA/JPL/Arizona State University.

INDEX

fjords, 412
Flagstaff, Arizona elevation, 301
flash floods, 21, 382, 490, 493
flood basalts
    causes, 146–47, 311
    mass extinctions from, 255, 257
    oceanic plateaus from, 276
floodplains
    agriculture in, 497
    assessing histories, 260
    formation of, 477, 490
    of low-gradient rivers, 482, 483
    sandstone formation in, 186
    as sedimentary environments, 171
    siltstone formation in, 188
    terrace formation, 488–89
floods. *See also* river channels; rivers
    Big Thompson River
        (Colorado, 1976), 493
    Colorado River history, 497
    destruction from, 487, 491, 492–93
    flash floods, 21, 382, 490, 493
    investigation activity, 498–99
    judging danger of, 260
    Lake Missoula (Montana), 425
    levee failures, 483
    measuring, 494–95
    Mississippi River (1993), 492
    normal flow versus, 490–91
    from overtopped reservoir, 455
    overview of common causes, 491
    runoff variations and, 487
    from sea-level changes, 388, 412
    sedimentary layers formed by, 183
    from tectonic activity, 388
    from volcanic melting of ice, 149, 491
Florida, 285, 392, 393
flow bands, 155
flow cells, 367
flows, 453, 456–57
flowstone, 513
flow within Earth, 359
fluid magma, 119
fluid pressures in pore spaces, 206
fluorine, 95, 148
fluorite, 97, 549
fluvial processes, 170
foci of earthquakes, 328
fold and thrust belts, 225, 306
folds
    basic features, 212–13
    with cleavage, 225
    erosion of, 227
    extreme examples, 215
    with faults, 224–25
    in landslides, 462
    mountain building from, 303
    observing in landscapes, 202, 203
    petroleum in, 530
    in salt deposits, 288
foliation, 214, 219
footprints, fossilized, 245
footwalls, 210
Foraminifera, 269
forecasting earthquakes, 351

force, defined, 204
forearc basins, 304, 305
foreland basins, 304, 306, 307, 323
foreshocks, 351
forest soils, 447
formations defined, 182
formations in caves, 513
Fort Tejon earthquake
        (California, 1857), 352
fossil fuels. *See* coal; natural gas; oil
fossils
    continental drift support from, 42
    data obtained from, 7
    dating rock layers from,
        37, 248–49, 258
    in drill cores, 269
    as environmental evidence, 194, 197
    formation of, 244–46
    in Front Range layers, 199
    in limestone, 180, 190
    similarity between continents, 41
    in tar pits, 529
    in terranes, 315
Four Corners region (U.S.), 130
fracture resistance of Earth's layers, 9
fractures. *See also* faults; joints
    crosscutting, 237
    effects on slope stability, 451
    with faults, 226
    groundwater in, 506, 507
    impact on weathering, 175, 439, 445
    in minerals, 78, 82
    observing in landscapes, 26
    petroleum in, 529
    as response to stresses, 206, 207
    surface processes causing, 174, 439
    types, 208–9
fracture zones, 52–53
framework silicates, 87
Fran, Hurricane (1996), 408
Frances, Hurricane (2004), 392
Fraser River Valley (British
        Columbia), 168–69
freeze-thaw cycles, 451
freshwater, 17, 502, 504–5.
        See also water
fringing reefs, 284
Front Range foothills, 198–99
frost heaving, 451
frost wedging, 174, 439
fuel cells, 543
Fuji, Mount, 51, 151
full moon, 567
fumes, water pollution from, 519
funnel clouds, 373

G

Ga (giga-annum), 36
gabbro, 92, 109, 122
gaining rivers and streams, 511
Galápagos Islands, 67, 68, 277
galaxies, 559
galena, 91, 549
Galilei, Galileo, 568

Galileo spacecraft, 561
gamma decay, 112
Ganymede, 568, 569
Garnero, Ed, 359
garnets
    basic features, 89
    human use, 100
    in metamorphic rocks, 217, 229
gas centrifuges, 539
gases. *See also* natural gas
    gathering data about, 34
    as guide to volcanic eruptions, 162
    hazards, 148
    in magma, 118, 141
gas fields, 530
gas giants, 19, 559, 568–73
gas hydrate, 534, 535
gasoline contamination, 40
Gaspra, 561
gas pressure, 118
gas stations, 518
gauging stations, 494
gem-quality minerals, 101
gentle slopes, 32, 177
geochemical sampling, 520
geochemistry, 550
geographic information systems, 515
geologic cross sections, 537
geologic diagrams, 33
geologic events, determining
        sequences, 29
geologic features, observing
        in landscapes, 27
geologic histories. *See also* ages;
        geologic timescale
    of Colorado, 199
    of Grand Canyon, 262–63
    investigation activity, 264–65
    reasons to explore, 260–61
    reconstruction, 28–29, 133,
        200–201, 258–59
    regional differences due to, 6
    Upheaval Dome investigation, 48–49
    western United States, 322–23
geologic maps
    basic features, 30, 31
    to guide field work, 228, 550
geologic timescale
    development of, 250–51
    fauna in, 246–47
    stages of, 37, 250
geologists
    acceptance of continental drift, 43
    data gathering approaches,
        34–35, 38–39
    determination of site evolution, 28
    diagrams used by, 33
    major tasks, 44–45
    Mediterranean Sea studies, 25
    Sierra Nevada studies, 133
    time and rate estimates by, 36–37
geomagnetic polarity timescale, 270
geophones, 269, 354
geophysical surveys, 354, 537
geophysical techniques, 551

geophysicists, 355
geothermal energy, 542, 555
geothermal gradient, 113
geysers, 510, 542
glacial erratics, 426
glacial outwash, 423
glacial periods, 428, 431
Glacier National Park (Montana), 2, 421
glaciers
    Antarctic melting, 432–33
    Baltic Sea formation by, 282
    basic features, 416–17
    breccia formation from, 184
    Cape Cod formation by, 407
    climate data from, 384, 385
    continental drift support from, 43
    effects on rivers, 487
    erosion, 2, 420–21
    evidence for, 426–27
    flooding hazards from, 149
    formation of, 418, 430–31
    during ice ages, 397
    salt, 289
    sea level effects, 414, 415
    sediment formation and transport,
        12, 16, 171, 416, 422–24
    snow accumulation in, 16
    as setting of Earth's water, 502
glasses, 106, 107, 110
glass making, 101
Glen Canyon Dam (Arizona), 541
global climate change
    causing glaciation, 430–31
    global warming causes, 386–87
    global warming evidence, 384–85
    investigation activity, 434–35
    potential sea-level rise, 432–33
    since last ice age, 7
Global Positioning System
    as guide to volcanic eruptions, 163
    measuring plate movements with, 65
    measuring uplift with, 297
    studying landslides with, 462
global warming. *See* global
        climate change
Glomar Challenger, 25
gneiss domes, 231
gneisses, 214, 216, 217
gneissic foliation, 214, 216
Gobi Desert (Mongolia), 381
gold, 544, 546–47
Gondwana
    Africa and South America in, 69
    development of concept, 42
    formation and division, 290, 291, 318
    role in Appalachian evolution, 321
grabens, 211, 222
graded beds, 183, 194, 287
graded streams, 475
grade of ore, 544
gradients
    of river systems, 472, 479, 482–83
    of slopes, 32
Grand Canyon (Arizona), 240, 262–63
Grand Junction (Colorado), 497

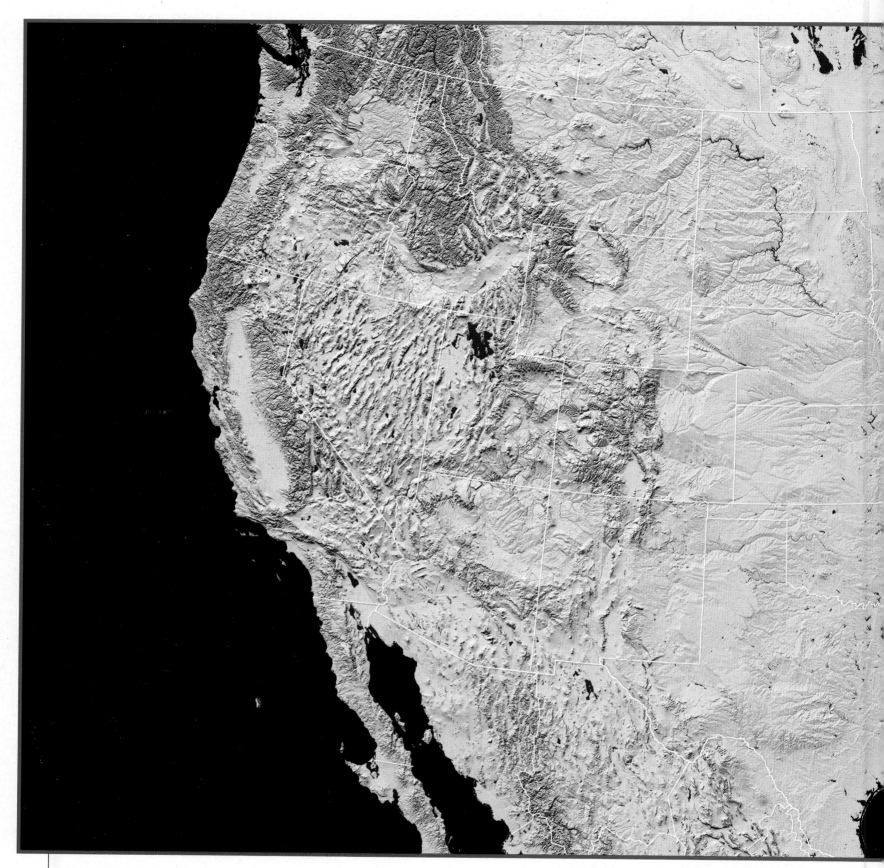

Shaded-Relief Map of the Conterminous United States and Adjacent Areas

This map simulates shading on the land surface as if lighted from the upper left. It includes the southern part of Canada and the northern part of Mexico.

Source: Paul Morin

Geologic Time Scale

(ages in millions of years before present)

0	Latest Neogene
	Neogene
	Paleogene
65	Cretaceous
	Jurassic
	Triassic
251	Permian
	Pennsylvanian
	Mississippian
	Devonian
	Silurian
	Ordovician
	Cambrian
542	Late Proterozoic
	Middle Proterozoic
	Early Proterozoic
	Late Archean
	Middle Archean
4000	Early Archean
	Glacial Ice
	Age Unknown

Cenozoic

Mesozoic

Cenozoic and Mesozoic

Paleozoic

Paleozoic and Precambrian

Paleozoic and Mesozoic

Precambrian

Sedimentary Rocks and Deposits

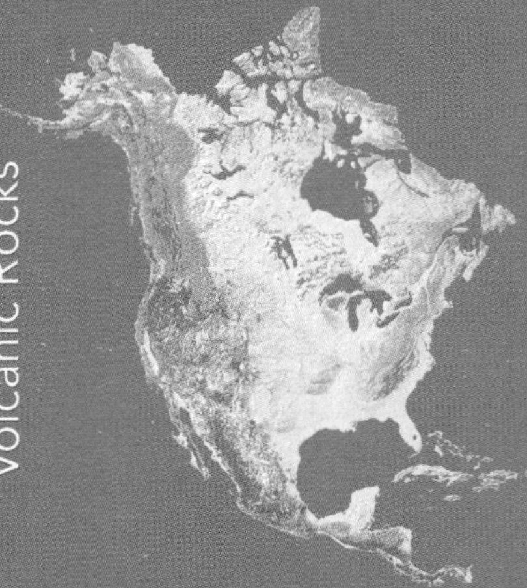

Volcanic Rocks

Plutonic Rocks

Metamorphic Rocks

Spacing of Boxes is Not Proportional
to Duration of Time Period